Dictionary of Plant Toxins

Dictionary of Plant Toxins

EDITORS

Jeffrey B. Harborne FRS

University of Reading, Reading, UK

AND

Herbert Baxter

Consultant, Ilford, Essex, UK

ASSOCIATE EDITOR

Gerard P. Moss

Queen Mary and Westfield College, London, UK

JOHN WILEY & SONS

Chichester · New York · Brisbane · Toronto · Singapore

National 01243 779777
International (+44) 1243 779777
e-mail (for orders and customer service enquiries): cs-books@wiley.co.uk
Visit our Home Page on http://www.wiley.co.uk
or http://www.wiley.com

Reprinted February 1997, June 1997, January 1998

Other Wiley Editorial Offices

John Wiley & Sons, Inc., 605 Third Avenue,
New York, NY 10158-0012, USA

Jacaranda Wiley Ltd, 33 Park Road, Milton,
Queensland 4064, Australia

John Wiley & Sons (Canada) Ltd, 22 Worcester Road,
Rexdale, Ontario M9W 1L1, Canada

John Wiley & Sons (Asia) Pte Ltd, 2 Clementi Loop #02-01,
Jin Xing Distripark, Singapore 0512

Library of Congress Cataloging-in-Publication Data

Dictionary of plant toxins / editor, J. B. Harborne; executive editor,
H. Baxter.
p. cm.
Includes bibliographical references and index.
Contents: Phytotoxins.
ISBN 0-471-95107-2 (v. 1: alk. paper)
1. Plant toxins–Dictionaries. I. Harborne, J. B. (Jeffrey B.)
II. Baxter, Herbert, 1928–
RA1250.D53 1996
615.9'52'03–dc20 96-11164 CIP

British Library Cataloguing in Publication Data

A catalogue record for this book is available from the British Library

ISBN 0 471 95107 2

Typeset in 10/12pt Times by Page Bros (Norwich) Ltd
Printed and bound in Great Britain by Bookcraft, Midsomer-Norton, Somerset
This book is printed on acid-free paper responsibly manufactured from sustainable forestation, for which at least two trees are planted for each one used for paper production.

Contents

List of Contributors

Herbert Baxter *Consultant, Ilford, Essex, UK*

Jeffrey B. Harborne FRS *University of Reading, UK*

Gerard P. Moss *Queen Mary and Westfield College, London, UK*

Introduction

It has been recognised since remote antiquity that plants can contain substances that are harmful or poisonous to the human race. This is reflected in the common names of such dangerous plants. Thus the fruits of deadly nightshade, *Atropa belladonna*, and of black nightshade, *Solanum nigrum,* are recognisably more poisonous that those of the woody nightshade, *Solanum dulcamara.* Again, the water dropwort, *Oenanthe crocata*, is also known as cowbane (i.e. not good for cows) and the poisonous roots are referred to as dead men's fingers or five finger death.

However, it is wrong to think that plant toxins, the subject of this dictionary, are dangerous only to ourselves, the animals of the farmyard and our domestic pets. On this view, relatively few plants are really poisonous and those that are will contain a deadly alkaloid such as atropine, coniine or solanine. This anthropocentric viewpoint of the plant kingdom completely neglects the fact that plants which may be relatively harmless to mammals may be highly toxic to birds, fish, molluscs and insects. In marine environments, algae may protect themselves from molluscan and fish predation by producing distasteful chemicals. On land, plant insecticides such as nicotine, the pyrethrins and the rotenoids may be well known but there are a very considerable number of plant toxins, mostly unfamiliar, which are inimical to insect life. Why otherwise would so many phytophagous insects confine themselves to feeding on only a handful of plant species?

The purpose of this dictionary is to list those plant toxins which are known or which are likely to have adverse effects on any and every form of animal life. For the sake of completeness, compounds principally known for their antifungal or antibacterial properties are included since these substances generally have considerable biological activity and some may be dietarily ingested and produce harmful effects in animals. Any listing of plant toxins is, however, even today handicapped by our considerable ignorance on the toxic principles of many poisonous plants. While the toxins of temperate floras are reasonably well studied, less is known of the toxic properties of the rich tropical rain forest floras. In this dictionary, the term 'plant' is used to include all vascular plants (angiosperms, gymnosperms, pteridophytes, bryophytes) and algae. For the sake of completeness, toxins of blue-green algae are included in this volume, although many scientists regard these organisms as being allied (as cyanobacteria) with the bacterial kingdoms.

A striking feature of most plants which are poisonous to human beings is that these plants are also known for their curative properties. Thus, they often have a long history of use as herbal medicines. This is hardly surprising since the active principle of such a herb may provide the cure of a particular disease at one dose and can equally cause the death of the individual at another. The classic example is the cardenolide digitoxin, from leaves of the foxglove, *Digitalis purpurea*, which is an extremely valuable drug for treating the heart condition angina pectoris. However, poisoning is quite common from an overdose of the drug while two or three leaves of the plant can make up a lethal amount. This link between toxicity and medicinal use is summarised in the title of an article by the late Norman Bisset (1991) 'One man's poison, another man's medicine?' and in the title of the book by Blackwell (1990) which is called *Poisonous and Medicinal Plants*. Because of this recognised link, the medicinal properties of the plant toxins are included where appropriate in this dictionary.

For each entry in the main body of this work, common names with synonyms, chemical class and subclass are given. The chemical structure, with indicated stereochemistry, is combined with Chemical Abstracts service registry number, molecular weight and chemical formula. The plant sources (with family names) are followed by brief notes on biological activity and a reference is given, usually to the isolation from one of the mentioned sources. Some further comments follow about plant sources and the major classes of plant toxins.

Some of the more familiar plants poisonous to humans are listed in Table 1. A more comprehensive listing can be found in Frohne and Pfander (1984). Details of the toxins responsible for the poisonous effects can be found in the main body of this dictionary (see plant names index) but even with well-known plants, it is possible that not all the active constituents have been fully identified. This listing largely refers to European or northern temperate plants but in other countries, other plant species may be more important.

The plants in Table 1 produce their harmful effects when leaves, bulbs, fruits or seeds or the extracts thereof are taken internally. There are also other plant species, which produce their effects when handled. These are allergenic plants, which cause skin rashes and blisters. Such effects vary according to the sensitivity of the individual and can be considerably enhanced by sunlight (so-called photosensitisation). Allergenic dermatitis is not normally fatal, but individuals can take much time to recover and allergenic constituents must be considered as toxins in the broad sense of the word. A fuller account of allergenic plants can be found in Mitchell and Rook (1979).

The number of plant species toxic to farm animals is very much greater than those that may be harmful to humans. Cattle and sheep are exposed to a wide range of native plant species in pasturelands around the world, but especially in North America, southern Africa and Australia. There is only space here for a small representative sample of such plants (Table 2). Some are considerably more important to farmers than others. It has been estimated, for example, that some 50% of cattle deaths due to grazing on poisonous plants are caused by species of *Senecio*. There is no sharp division between plants poisonous to humans and domestic animals. It is mainly a matter of exposure and it may be seen that yew and water dropwort appear in both tables.

It is noticeable that native grazing animals (deer, rabbits, etc.) avoid eating these dangerous plants and rarely suffer poisoning. Those that do may be able to detoxify the poison successfully. Kangaroos in Western Australia can safely graze on species of *Gastrolobium*, which contain highly poisonous fluoracetate (see under monofluoracetic acid). Even domestic cattle will generally avoid some of the poisonous plants (e.g. *Senecio* and *Asclepias* spp.) because of the bitter taste of the toxins present. They may, however, be 'forced' to eat this plant material either through overgrazing or by the accidental admixture of poisonous plant leaf with the hay fodder. Much is known of the toxins responsible for livestock poisoning and the great majority of active principles have been included in the dictionary. A useful recent reference to plants poisonous to farm animals is the book of James *et al.* (1992).

The toxicity of a plant chemical is always relative, dependent on the dose taken in a given time period, the age and state of health of the animal, the mechanism of absorption and mode of excretion. The steroidal alkaloid solanine, for example, is present in all domestic potatoes but the amount present is so low that it is rarely a dietary hazard. It is only when enormously large amounts of solanine accumulate in tubers that have been exposed above the soil surface and become 'greened' that death from solanine poisoning is a reality. In such cases, victims have no time to adapt to dealing with the toxin and, unless they are sick, they die from respiratory failure. Whether death occurs on intake of a toxin depends therefore on whether the animal has time to become accustomed to small amounts of the poison in the diet, i.e. whether it has been able to develop a detoxification mechanism. Some human diets contain poisonous plants in which the toxic principles have been removed by leaching and processing. This is true of cassava, *Manihot esculenta*, which contains potentially lethal amounts of cyanogenic glycoside (see under linamarin). Toxic effects may alternatively be avoided by geophagy, i.e. eating the plant material mixed with clay, which binds to the toxin. The Ahmara Indians in Peru, for example, practise geophagy when consuming wild potato tubers, which contain otherwise lethal amounts of steroidal alkaloids (Johns, 1990).

Toxins often have the role of feeding repellents, since plants usually advertise their presence by a warning signal of a visual or olfactory nature. Thus animals may be made aware of the presence of the toxins even before they start feeding. Mustard oils, for example, which occur in crucifers in bound form and are toxic to most insects, have a pungent acrid smell and are probably emitted continuously in trace amounts from the living plant. Other immediate warnings of danger may be provided through visual means by toxins deposited on the surface of leaves and other or-

Table 1. Same representative plants poisonous to humans

Latin name	Common name*	Plant part(s) that are poisonous†
Aconitum napellus	Monkshood	Leaf, tuber
Arum maculatum	Lord and ladies	Fruit
Atropa belladonna	Deadly nightshade	Fruit
Cicuta virosa	Cowbane	Leaf
Conium maculatum	Hemlock	Leaf, seed
Convallaria majalis	Lily of the valley	Fruit
Cotoneaster horizontalis	Cotoneaster	Leaf, seed
Cytisus scoparius	Broom	Fruit
Datura stramonium	Thorn apple	Fruit
Euonymus europaeus	Spindle	Fruit
Ilex aquifolium	Holly	Fruit
Laburnum anagyroides	Laburnum	Seed
Lonicera periclymenum	Honeysuckle	Fruit
Oenanthe crocata	Water dropwort	Root
Ricinus communis	Castor bean	Whole plant
Solanum dulcamara	Bittersweet	Fruit
Solanum nigrum	Black nightshade	Fruit
Sorbus aucuparia	Rowan	Fruit
Taxus baccata	Yew	Needle, fruit
Veratrum album	White hellebore	Root

*Many of these plants have several common names.
†The concentrations of toxin in fruits usually varies considerably with the degree of ripeness. Often birds can eat these fruits with impunity.

Table 2. Representative plant species poisonous to livestock

Latin name	Common name	Poisonous plant part(s)†
*Asclepias syriaca**	Milkweed	Leaf
*Astragalus lentiginosus**	Locoweed	Leaf
*Brassica campestris**	Cabbage	Aerial parts
Colchicum autumnale	Autumn crocus	Bulb, leaf
Dichapetalum cymosum	Gifblaar	Leaf
*Echium plantagineum**	Paterson's curse	Leaf
Galega officinalis	Goat's rue	Leaf
*Gossypium barbadense**	Cotton	Seed oil
Hymenoxys odorata	Bitterweed	Leaf
Hypericum perforatum	St. John's wort	Leaf
Lantana camara	Lantana	Aerial parts
*Lupinus albus**	Lupin	Leaf
Oenanthe crocata	Water dropwort	Root
Parthenium hysterophorus	Parthenium	Aerial parts
Pteridium aquilinum	Bracken	Frond
*Quercus robur**	Oak	Acorn, leaf
*Rhododendron ponticum**	Rhododendron	Aerial parts
Senecio jacobaea	Ragwort	Leaf
*Swainsona canescens**	Swainsona	Leaf
Taxus baccata	Yew	Leaf, fruit

*Most species in these genera are also poisonous.
†Cattle and sheep are most commonly affected, but toxic symptoms may also be observed in horses, pigs, goats and domestic fowl.

gans. Potentially toxic secondary compounds may occur in the surface waxes. Alternatively, glandular hairs on the leaf may secrete a toxic quinone, as in *Primula obconica* (see under primin). Again, chemical defence is often 'advertised' in woody plants when they exude resins from bark and fruit.

In the case of hydrogen cyanide, intact cyanophoric plants release no HCN, since the substrates and enzymes for HCN production are located in different organelles. It is only when the leaf is damaged by herbivores that the substrate and enzyme come together to produce the poison, which has a clear warning in its 'odour of bitter almonds'. In the case of alkaloids and saponins, the warning signal is only received after the animal has started feeding, in the form of a bitter taste. Most alkaloids and saponins are known to be bitter. Many other plant compounds are bitter, especially the triterpenoid cucurbitacins of the cucumber family, which clearly provide the basis of repellency to herbivores in these plants. The latexes which flow within plants such as chicory, dandelion, and other composites also have an obvious role in herbivore deterrence, since the latex itself may interfere with feeding and may additionally contain bitter toxins among its constituents.

Toxins may be found among almost every class and subclass of secondary plant constituent, although some subclasses (e.g. the pyrrolizidine alkaloids) may have many more toxic members than others. Plant toxins may be conveniently divided into two groups: these containing nitrogen; and those without.

Of the various nitrogen-based plant toxins the simplest in structure are the non-protein amino acids. These are widely present in plants and may be directly toxic inasmuch as they are anti-metabolites of one or other of the 20 protein amino acids. In the simplest case (see azetidine 2-carboxylic acid) they may be mistakenly incorporated into protein synthesis; the organism produces unnatural enzymic protein which cannot function properly and death of the organism ensues. The toxic effect of other non-protein amino acids may sometimes be more complex. 3,4-Dihydroxyphenylalanine or L-DOPA, which is harmful to insects, interferes with the activity of tyrosinase, an enzyme essential to the hardening and darkening of the insect cuticle.

There are about 300 known structures of these plant amino acids (Rosenthal, 1982). While they are found in a number of unrelated families, they are particularly characteristic of legumes and occur mainly in the seeds. Other examples of toxic non-protein amino acids which can be found in this dictionary are canavanine, β-cyanoalanine and hypoglycin.

Another structurally simple class of nitrogenous toxins are the cyanogenic glycosides (e.g. linamarin, prunasin). They are toxic only when broken down enzymically with the release of hydrogen cyanide. The primary site of action of HCN is on the cytochrome system; terminal respiration is inhibited, oxygen starvation occurs at the cellular level and rapid death ensues. An up-to-date review of the cyanogenic glycosides is that of Seigler and Brinker (1993). Like HCN, nitrite is toxic to a wide range of organisms and some plants, notably species of *Astragalus*, accumulate glucosides of organic nitro compounds such as miserotoxin which are toxic due to the release of nitrite. The toxicity of miserotoxin mainly affects cattle, although the human nervous system is not completely immune to nitrite poisoning. Incidentally, the legume genus *Astragalus*, which contains many species with miserotoxin or related nitro compounds (Stermitz *et al.*, 1972), is heterogeneous in its toxic components. Other species accumulate selenium amino acids such as selenocystathionine, which also poisons livestock.

Glucosinolates (mustard oil glycosides) are closely related biosynthetically to cyanogenic glycosides and they can also be toxic to animals when they occur in sufficient amount in plants, as in wild species of *Brassica* (see under sinigrin). Toxic symptoms include severe gastroenteritis, salivation, diarrhoea and irritation of the mouth. Toxicity is actually due to the release of isothiocyanates (mustard oils), which are highly vesicant in their action. A further hazard of these substances is due to the fact that during their release from bound forms, the isothiocyanates produced can undergo rearrangement in part to the corresponding thiocyanates. The latter substances are harmful because they are goitrogenic, and produce hyperthyroidism in mammals. The toxicity of glucosinolates to insects has been demonstrated by Erickson and Feeny (1974) who found that caterpillars of the black swallowtail butterfly (*Papilio polyxenes*) were killed by being fed on celery leaves which had been infiltrated with sinigrin (at a concentration of 0.1% per fresh weight of leaf). A review of the biological effects of plant glucosinolates is that of Chew (1988).

The most familiar class of plant toxins are the alkaloids. These substances have been used since time immemorial for poisoning purposes, an extract of hemlock leaves being used by the ancient Greeks to put the philosopher Socrates to death. The physiological effects of alkaloids on the central nervous system in humans have been widely studied and alkaloids are utilised in modern medicine for a variety of purposes. There are at least 10 000 alkaloids of known structure. These bases occur widely, albeit sporadically, in the angiosperms, being present in about 20% of higher plant families. The term alkaloid covers an enormous range of chemical structures, from the simple monocyclic piperidine, coniine of hemlock *Conium maculatum*, to the hexacyclic alkaloid solanine of *Solanum tuberosum.* Not all alkaloids are highly toxic; few are as dangerous as, say, atropine, the principal toxin of deadly nightshade, *Atropa belladonna*. Nevertheless, most alkaloids are liable to show some toxic effects if ingested in any quantity over an extended period of time.

One highly toxic group of alkaloids are the pyrrolizidines. They are well known as the toxic principles of plants poisonous to cattle, e.g. species of the genus *Senecio.* They are also dangerous to humans. The identification of echimidine and intermedine in leaves of a reputedly harmless plant, comfrey (*Symphytum officinale*), should be noted since comfrey has not only been used medicinally but also has been recommended by 'natural food' promoters as an item of human diet. The alkaloid content of the leaves of both *S. officinale* and the hybrid *S.* × *uplandicum* (known as Russian comfrey) can be as much as 0.15% dry weight; furthermore the crude leaf extracts have been shown to produce chronic hepatotoxicity in rats. Clearly, to avoid liver damage, one should avoid eating these plants in any quantity on a regular basis. A complete listing of all pyrrolizidine-containing plants can be found in the review of Hartmann and Witte (1955).

Another toxic group of plant alkaloids harmful to cattle are diterpenoid based, exemplified by aconitine from *Aconitum.* They represent a significant cause of cattle poisoning in the rangelands of the western United States. They occur in species of larkspur (*Delphinium*) which are grazed by the cattle. Some of these alkaloidal structures are significantly more toxic than others (Manners *et al.*, 1995). The toxicology of these alkaloids has been reviewed by Benn and Jacyno (1983).

While the general toxicity of plant alkaloids in mammals, and especially in humans and farm animals, is widely recognised, their teratogenic effects have only recently been recorded. Adult female cattle and sheep may imbibe alkaloids in their diet in insufficient amount to cause their death, but as a result of feeding on alkaloid-containing plants, congenital defects may occur in their offspring. Among alkaloids implicated in this way are the pyrrolizidine group, the nicotine group, those of *Lupinus* and also the simple piperidine derivative, coniine of hemlock. The malformed offspring usually suffer skeletal damage and defects of the digits or of the palate. Such livestock have very limited survival rates.

One final group of toxic alkaloids should be mentioned; a series of polyhydroxy alkaloids with a structural resemblance to sugars has been encountered in plants (Fellows *et al.*, 1986). They are simple molecules in which the ether oxygen atom of a monosaccharide is replaced by nitrogen. Two typical structures are deoxymannojirimycin, which was found in the seeds of the legume *Lonchocarpus sericeus*, and castanospermine, which occurs in the seeds of *Castanospermum australe.* These alkaloids have been termed 'sugar-shaped weapons of plants' because as sugar analogues they are able to inhibit the enzymes of animal carbohydrate metabolism. Deoxymannojirimycin thus inhibits α-mannosidase and castanospermine α-glucosidase. Toxic effects in animals ingesting them may be due to their ability to arrest carbohydrate breakdown. Indeed, swainsonine, a third member of the group which occurs in the leaves of *Swainsona*, a legume pasture plant, is toxic to grazing cattle, causing neurological symptoms. This is due to the accumulation of mannose-based oligosaccharides, which the animal's α-mannosidase is prevented from breaking down.

While plant proteins are not usually thought of as being toxic, there are a few which are highly dangerous to animals. One is abrin, the main protein of the seed of *Abrus precatorious* (Leguminosae), the lethal dose of which in humans is as little as half a milligram. Since the seeds, which are attractively coloured red and black, are employed by African natives for making necklaces, fatalities due to abrin poisoning do occasionally occur. Like most proteins, abrin can be denatured by heating and the toxic effects disappear when the temperature is raised above 65°C. A second well-known protein toxin is ricin, the protein of the castor bean, *Ricinus communis.* It is a protoplasmic poison, the lethal dose, as measured in mice, being 0.001 μg ricin nitrogen/g body weight.

A number of legume seeds, e.g. soybean, *Glycine max*, contain proteins which are trypsin inhibitors. While these are not toxic as such, they presumably have a protective role against animal feeding, since they reduce the nutritional value of the protein in seeds containing them. Another class of proteins present in legume seeds are the phytohaemagglutinins, so-called because of their ability to coagulate the erythrocytes of human blood. These glycoproteins are toxic to bruchid beetles and can have adverse effects when legume seeds are included in human diets (see D'Mello *et al.*, 1991).

Many non-nitrogen-containing secondary metabolites are also known to be toxic to animals. There are many poisonous compounds which are terpenoids or even fairly simply hydrocarbons. For example, many of the plant extracts used as arrow poisons by natives in Africa contain cardiac glycosides, such as ouabain, as active ingredients.

These steroidal substances are heart poisons. Again, the so-called 'five-finger death' caused by the consumption by humans or cattle of the oddly shaped roots of water dropwort, *Oenanthe crocata*, is caused by the presence of polyacetylene hydrocarbons, such as oenanthetoxin.

One of the simplest of all non-nitrogenous toxins is monofluoroacetic acid, CH_2FCO_2H, which occurs in the South African plant, *Dichapetalum cymosum.* It is poisonous because it stops respiration, through inhibition of the Krebs (tricarboxylic acid) cycle; the fatal dose in man is 2–5 mg/kg body weight. Fluoracetic acid is taken into the cycle instead of acetic acid, and is metabolised to fluorocitric acid. It is the enzyme aconitase which refuses this substrate as a substitute for citric acid and thus causes the respiratory cycle to stop. Another toxic organic acid in plants such as rhubarb (in the leaves) is oxalic acid $(CO_2H)_2$. Oxalate is only really toxic when associated with the sodium or potassium ion to give a soluble salt; the calcium salt, by contrast, is insoluble and may pass through the animal body without being absorbed. In spite of its simple structure, its mode of action is poorly understood, although it is conceivable that it interferes with terminal respiration by inhibiting the key enzyme succinic dehydrogenase. The fatal dose of oxalic acid is quite high; only plants which contain 10% or more as dry weight are likely to be harmful to mammals (see Keeler *et al.*, 1978).

Among the terpenoids, two particularly toxic groups are the cardiac glycosides (or cardenolides) and the saponins. The cardenolides are mainly found in plants of Asclepiadaceae (in the milkweeds) and of Apocynaceae (e.g. oleander) but there are other notable occurrences, e.g. in *Digitalis* (Scrophulariaceae). Their effects on animals have been reviewed by Malcolm (1991). By contrast, the saponins have a very wide occurrence in over 100 plant families (Hostettmann and Marston, 1995). They characteristically cause haemolysis of blood erythrocytes and are more toxic to cold-blooded rather than to warm-blooded animals.

There are other terpenoid structures in addition; for example, the toxic principles of *Rhododendron* leaves and flowers are diterpenes. Also, the sesquiterpene lactones, compounds widely distributed in the Compositae, include some substances which are either toxic in insects or repellent in having allergenic skin effects in animals (Mitchell and Rook, 1979). A few sesquiterpene lactones are poisonous to farm animals (e.g. hymenovin in *Hymenoxys odorata*), while quite a number are cytotoxic, having antitumour activity (Rodriguez *et al.*, 1976). The sesquiterpene lactones are often bitter-tasting as are another group of terpenoid toxins, the monoterpene lactones or iridoids. Iridoids occur in plants both in the free state (e.g. nepetalactone in the volatile oil of catmint) and in glycosidic form (e.g. aucubin) from which the free toxin is liberated after enzymic hydrolysis.

Some non-nitrogenous toxins in plants are also notable for causing photosensitisation in farm animals. The quinone hypericin of *Hypericum perforatum*, for example, is a photodynamic compound which is absorbed by the animal and enters peripheral circulation. When exposed to sunlight, the animal as a result becomes susceptible to sunburn and other damage; serious necrosis of the skin can occur, with subsequent infection and starvation. Among other photodynamic compounds present in plants are furanocoumarins, such as psoralen, which are responsible for photosensitisation in sheep which have fed on spring parsley *Cymopterus watsonii* (see Keeler, 1975). Through their photodynamic properties, furanocoumarins are also toxic to most insects. Other plant molecules shown to cause toxicity in animals through photosensitisation are the polyacetylenes and thiophenes (Towers, 1980). Some examples of these compounds (e.g. α-terthienyl) can be found in this dictionary.

General books on plant toxins include Keeler and Tu (1983, 1991) and D'Mello *et al.* (1991). The poisonous effects of plants on humans are reviewed in Cooper and Johnson (1989), Lewis and Elvin-Lewis (1977), Frohne and Pfander (1984) and Lampe and McCann (1985). The poisonous effects of plants on farm animals are documented in Keeler (1975), Keeler *et al.* (1978), James *et al.* (1992). CAB International have produced three volumes of a bibliography on plant poisoning in animals (Hails and Crane, 1983; Hails, 1986, 1994), while a listing of plants poisonous to insects can be found in Grainge and Ahmed (1988). Finally, it should be mentioned that an interactive identification system for recognising poisonous plants in Britain and Ireland has been published by HMSO (1995) in compact disk format.

J. B. Harborne
31 October 1995

References

Benn, M. H. and Jacyno, J. M. (1983) Toxicology and pharmacology of diterpenoid alkaloids, in: *Alkaloids, Chemical and Biological Perspectives*, Vol. 1 (Pelletier, S. W., ed.), John Wiley & Sons, New York, pp. 153–210.

Bisset, N. G. (1991) One man's poison, another man's medicine? *J. Ethnopharmacol.*, **32**, 71–81.

Blackwell, W. H. (1990) *Poisonous and Medicinal Plants*, Prentice-Hall, Englewood Cliffs, New Jersey, 329 pp.

Chew, F. S. (1988) Biological effects of glucosinolates, in *Biologically Active Natural Products* (Cutler, H. G., ed.), American Chemical Society, Washington DC, pp. 155–181.
Cooper, M. R. and Johnson, A. W. (1984) *Poisonous Plants in Britain and their Effects on Animals and Man*, HMSO, London.
D'Mello, J. P. F., Duffus, C. M. and Duffus, J. H. (1991) *Toxic Substances in Crop Plants*, Royal Society of Chemistry, Cambridge, 339 pp.
Erickson, J. M. and Feeny, P. (1974) *Ecology*, **55**, 103–111.
Fellows, L. E., Evans, S. V., Nash, R. J. and Bell, E. A. (1986) Polyhydroxy plant alkaloids as glycosidase inhibitors and their possible ecological role, in *Natural Resistance of Plants to Pests* (Green, M. B. and Hedin, P. A., eds), American Chemical Society, Washington, pp. 72–78.
Frohne, D. and Pfander, H. J. (1984) A *Colour Atlas of Poisonous Plants*, translated from the German by N. G. Bisset, Wolfe Publishing Co., London, 291 pp.
Grainge, M. and Ahmed, S. (1988) Handbook of Plants with Pest-control Properties, John Wiley & Sons, New York, 470 pp.
Hails, M. R. and Crane, T. D. (1983) *Plant Poisoning in Animals, 1960–1979*, CAB International, Wallingford, Oxford, 153 pp.
Hails, M. R. (1986) *Plant Poisoning in Animals, No. 2, 1980–1982*, CAB International, Wallingford, Oxon, 69 pp.
Hails, M. R. (1994) *Plant Poisoning in Animals, No. 3, 1981–1992,* CAB International, Wallingford, Oxon, 281 pp.
Hartmann, T. and Witte, L. (1995). Chemistry, biology and chemoecology of the pyrrolizidine alkaloids, in *Alkaloids, Chemical and Biological Perspectives*, Vol. 9 (Pelletier, S. W., ed.), Pergamon Press, Oxford, pp. 155–234.
Hostettmann, K. and Marston, A. (1955) *Saponins*, Cambridge University Press, Cambridge, 548 pp.
Hostettmann, K., Marston, A., Maillard, M. and Hamburger, M. (1995) *Phytochemistry of Plants used in Traditional Medicine*, Clarendon Press, Oxford, 408 pp.
James, L. F., Keeler, R. F., Bailey, E. M., Cheeke, P. R. and Hegarty, M. P. (1992) *Poisonous plants*, Iowa State University Press, Ames, 661 pp.
Johns, T. (1990) *With Bitter Herbs They Shall Eat It*, University of Arizona Press, Tucson.
Keeler, R. F. (1975) Toxins and teratogens of higher plants, *Lloydia*, **38**, 56–86.
Keeler, R. F., van Kampen, K. R. and James, L. F. (eds) (1978) *Effects of Poisonous Plants on Livestock*, Academic Press, New York, 600 pp.
Keeler, R. F. and Tu, A. T. (1983) Plant and fungal toxins, Vol. 1 in *Handbook of Natural Toxins* (Tu, A. T., ed.) Marcel Dekker, New York.
Keeler, R. F. and Tu, A. T. (1991) Toxicology of plant and fungal compounds, Vol. 6 in *Handbook of Natural Toxins* (Tu, A. T., ed.) Marcel Dekker, New York.
Lampe, K. F. and McCann, M. A. (1985) *AMA Handbook of Poisonous and Injurious Plants*, American Medical Association, Chicago.
Lewis, W. H. and Elvin-Lewis, M. P. F. (1977) *Medical Botany: Plants Affecting Man's Health*, John Wiley & Sons, New York, 515 pp.
Malcolm, S. B. (1991) Cardenolide-mediated interactions between plants and herbivores, in *Herbivores, Their Interactions with Secondary Plant Metabolites*, Vol. 1 (Rosenthal, G. A. and Berenbaum, M. R., eds), Academic Press, San Diego, pp. 251–286.
Manners, G. D., Panter, K. E. and Pelletier, S. W. (1995) Structure–activity relationships of norditerpenoid alkaloids in toxic larkspur species, *J. Nat. Prod.*, **58**, 863–869.
Mead, R. J., Oliver, A. J., King, D. R. and Hubach, P. H. (1985) *Oikos*, **44**, 55–60.
Mitchell, J. and Rook, A. (1979) *Botanical Dermatology*, Greenglass, Vancouver, 787 pp.
Rodriguez, E., Towers, G. H. N. and Mitchell, J. C. (1976) Biological activities of sesquiterpene lactones, *Phytochemistry*, **15**, 1573–1580.
Rosenthal, G. A. (1982) *Plant Nonprotein Amino and Imino Acids*, Academic Press, New York, 273 pp.
Seigler, D. S. and Brinker, A. (1993) Characterisation of cyanogenic glycosides from plants, in *Methods in Plant Biochemistry, Vol. 8. Alkaloids and Sulphur Compounds* (Waterman, P. G., ed.), Academic Press, London, pp. 51–132.
Stermitz, F. R., Lowry, W. T., Norris, F. A., Buckeridge, F. A. and Williams, M. C. (1972) Aliphatic nitro compounds from *Astragalus* species. *Phytochemistry*, **11**, 1117–1124.
Towers, G. H. N. (1980) Photosensitizers from plants and their photodynamic action. *Prog. Phytochem.*, **6**, 183–202.

Abrin 1

Abrus agglutinin

Protein
(Lectin) MW 63 000 - 67 000

Seeds of jequirity, *Abrus precatorius* (Leguminosae).

Abrins, the mixture of related proteins in the seed, are extremely toxic; the extract of one seed can cause fatal poisoning. LD_{50} in mice is 2 mg/kg body-weight. They are potent haemagglutinins. Abrins are more toxic to tumour cells than to normal cells, and they have been used experimentally in cancer research. The leaf of the plant is free of these toxic proteins and is used as a source of the sweetening agent glycyrrhizin. Dermatitis can occur in those wearing necklaces made from the the seeds of jequirity.

Oliver-Bever, B. E. P., Medicinal Plants in Tropical West Africa, 1986, 230, CUP.

Absinthin 2

Absinthiin; Absynthin

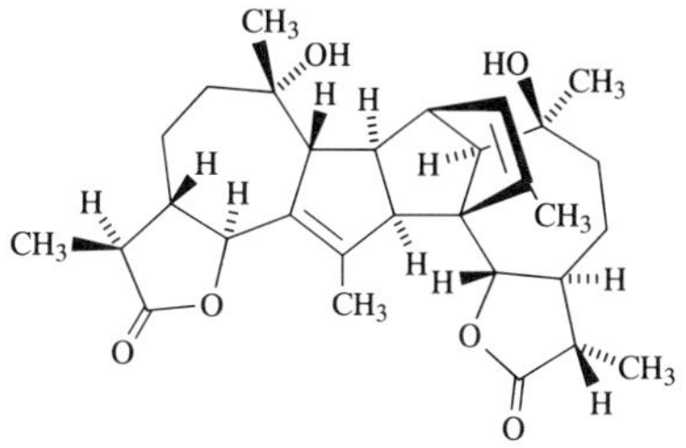

Sesquiterpenoid lactone
(Dimeric guaianolide)

$C_{30}H_{40}O_6$ [1362-42-1] MW 496.64

Whole plant of wormwood, *Artemisia absinthinum* and of *A. sieversiana* (Compositae).

Ingestion may cause nervousness, convulsions and even death. Used as an anthelmintic, a bitter tonic and for flavouring alcoholic beverages such as vermouth.

Beauhaire, J., *Tetrahedron Lett.,* 1980, **21**, 3191.

Abyssinone VI 3

2′,4,4′-Trihydroxy-3,5-diprenylchalcone

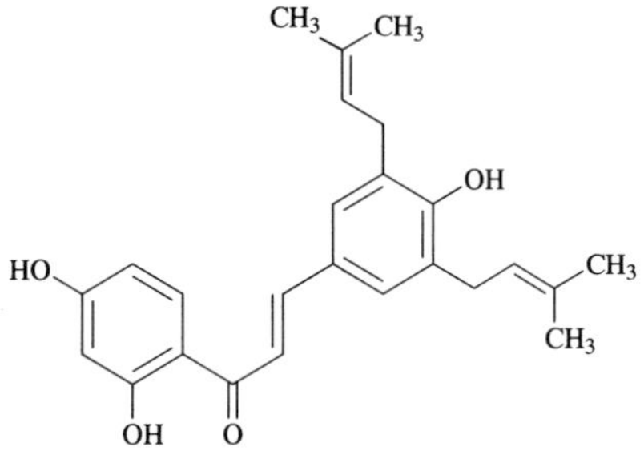

Chalcone

$C_{25}H_{28}O_4$ [77263-12-8] MW 392.49

Roots of East African medicinal plant *Erythrina abyssinica* (Leguminosae).

Strong inhibitor of rabbit platelet aggregation. Antipeptic activity.

Kamat, V. S., *Heterocycles,* 1981, **15**, 1163.

Acacetin 4

Apigenin 4′-methyl ether; Buddleoflavonol; Linarigenin

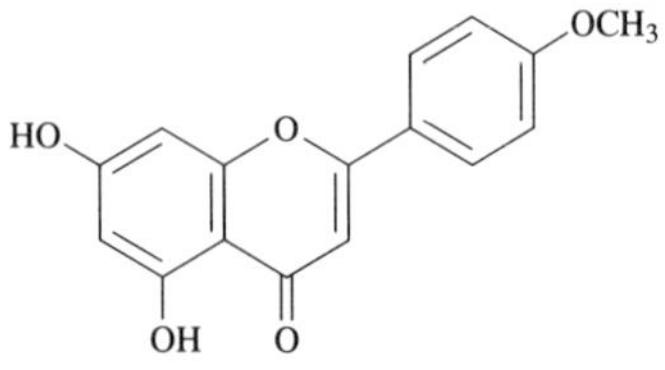

Flavone

$C_{16}H_{12}O_5$ [480-44-4] MW 284.27

Leaf bud exudates of birch trees, *Betula* spp. (Betulaceae). Also on the leaf surface of *Eupatorium tinifolium* (Compositae), in the farinose exudate of ferns such as *Notholaena* spp. (Polypodiaceae). In glycosidic form is relatively widespread in the leaves and flowers of higher plants, e.g. in petals of toadflax, *Linaria vulgaris* (Scrophulariaceae).

When occurring in birch bud exudates, it has contact allergenic activity. It is an inhibitor of lens aldose reductase, iodothyronine deiodinase and of histamine release from rat peritoneal mast cells.

Shibata, S., *Yakugaku Zasshi,* 1960, **80**, 620.

Acacipetalin 5

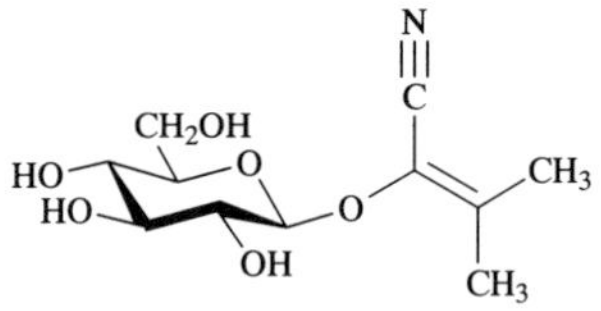

Cyanogenic glycoside

$C_{11}H_{17}NO_6$ [644-68-8] MW 259.26

Acacia sieberiana and *A. hebeclada* (Leguminosae). It may be present in the plant as the isomer proacacipetalin (q.v.).

Bound toxin, releasing poisonous cyanide on hydrolysis.

Ettlinger, M. G., *J. Chem. Soc., Chem. Commun.,* 1977, 952.

Acalyphin 6

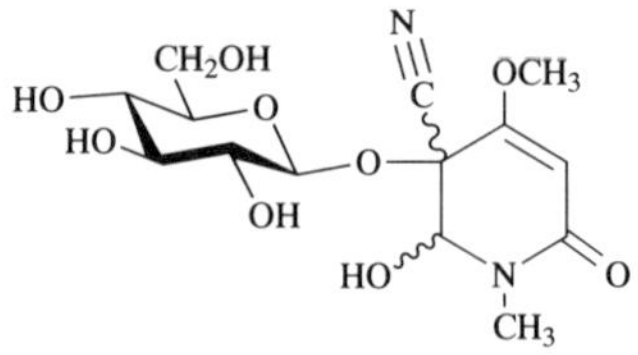

Cyanogenic glycoside

$C_{14}H_{20}N_2O_9$ [81861-72-5] MW 360.32

All parts of *Acalypha indica* (Euphorbiaceae).

Toxic to all life. Causes death to livestock grazing on this plant. Aerial parts and roots have been used in folk medicine.

Nahrstedt, A., *Phytochemistry,* 1982, **21**, 101.

Acamelin 7

6-Methoxy-2-methyl-4,7-benzofurandione;
2-Methyl-6-methoxy-4,7-benzofurandione

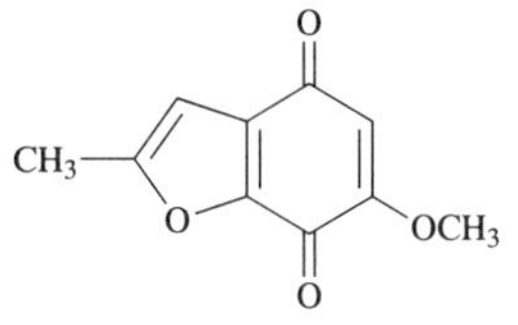

Benzoquinone

$C_{10}H_8O_4$ [74161-27-6] MW 192.17

Heartwood of Australian blackwood, *Acacia melanoxylon* (Leguminosae).

Allergen, causing contact dermatitis and bronchial asthma (together with 2,6-dimethoxybenzoquinone) in workers handling the wood of this tree.

Schmalle, H. W., *Tetrahedron Lett.,* 1980, **21**, 149.

(−)-Acanthocarpan 8

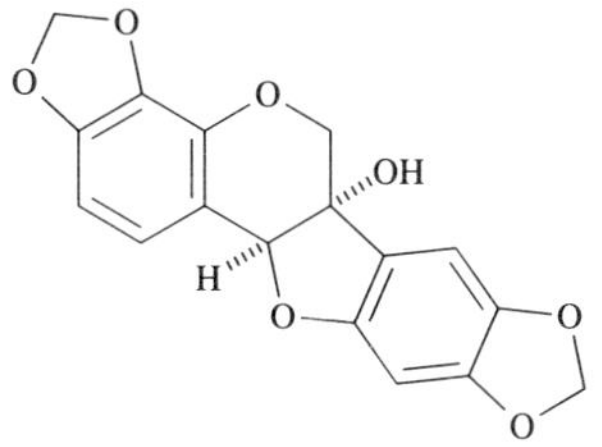

Isoflavonoid
(Pterocarpan)

$C_{17}H_{12}O_7$ [70285-12-0] MW 328.28

In fungally affected leaves of *Caragana acanthophylla* and of *Tephrosia bidwillii* (Leguminosae).

Antifungal properties (phytoalexin).

Ingham, J. L., *Phytochemistry,* 1982, **21**, 2969.

Acanthoidine 9

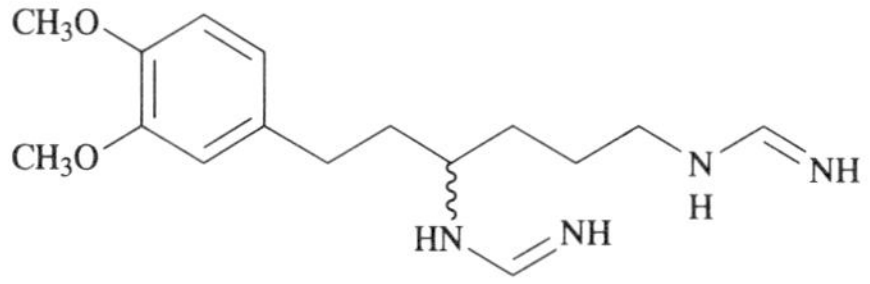

Miscellaneous alkaloid

$C_{16}H_{26}N_4O_2$ [6793-30-2] MW 306.41

Dried stalks of *Carduus acanthoides* and other *Carduus* spp. (Compositae).

Hypertensive activity in humans.

Frydman, B., *Tetrahedron,* 1962, **18**, 1063.

Acerosin 10

5,7,3′-Trihydroxy-6,8,4′-trimethoxyflavone

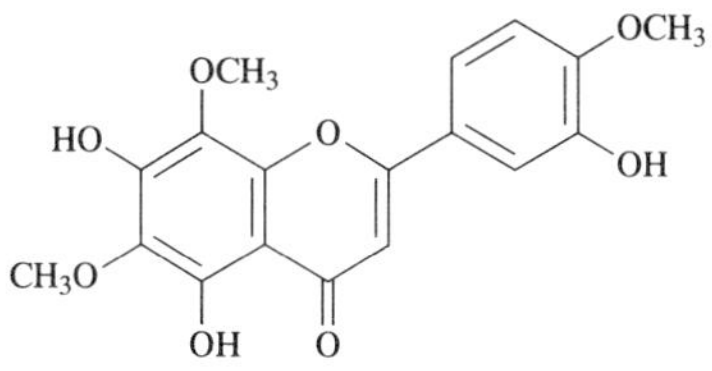

Flavone

$C_{18}H_{16}O_8$ [15835-74-2] MW 360.32

Iva acerosa (Compositae), aerial parts of *Helianthus strumosus* (Compositae), in *Gardenia* spp. (Rubiaceae), in seeds of *Vitex negundo* (Verbenaceae) and in fruit peel of *Citrus reticulata* (Rutaceae).

Causes disruption of later stages of spermatogenesis in dogs. Has oestrogenic properties.

Bhargava, S. K., *Plant. Med. Phytother.,* 1984, **18**, 74.

Acetic acid 11

Ethanoic acid; Methane carboxylic acid; Vinegar acid

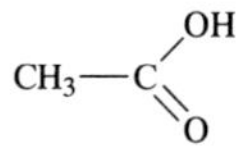

Organic acid

$C_2H_4O_2$ [64-19-7] MW 60.05

Wide occurrence in plant volatiles, both free (in trace amounts) and in ester form. The amyl ester, for example, is one of the flavour principles of banana, *Musa sapientum* (Muraceae).

Is a strong acid; when ingested, causes a collapse of the circulatory system. Causes severe burns. It has a useful bacteriocidal activity at a concentration above 5%. Also used as an acidulant and flavouring agent in the food industry.

Martindale, The Extra Pharmacopaeia, 30th ed., 1993, 1329, The Pharmaceutical Press.

Acetovanillone 12

Apocynin; Apocynine; 4′-Hydroxy-3′-methoxyacetophenone

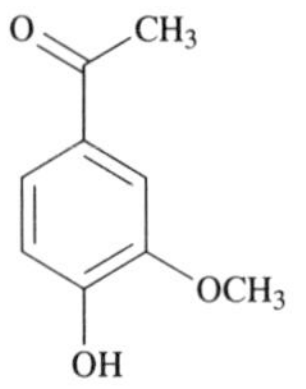

Phenolic ketone

$C_9H_{10}O_3$ [498-02-2] MW 166.18

Sapwood of *Photinia davidiana* and other Rosaceae trees, both in the free state and bound as the glucoside [531-28-2]. Occurs free in rhizomes of *Apocynum cannabinum* (Apocynaceae), rhizomes of *Iris* spp. (Iridaceae), bulbs of *Buphane distiche* (Amaryllidaceae) and *Echinocereus engelmannii* (Cactaceae).

Antifungal activity.

Ivanova, S. Z., *Khim. Prir. Soed.,* 1976, **12**, 107.

1′-Acetoxychavicol acetate 13

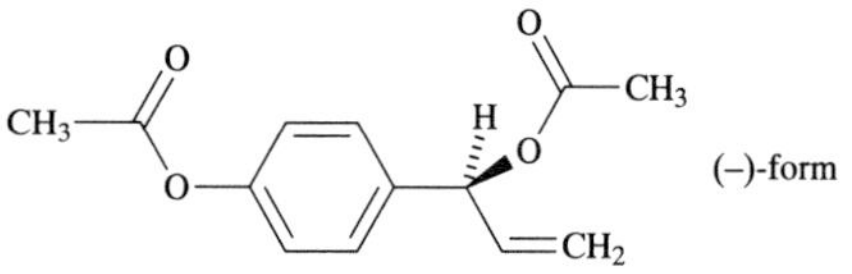

(–)-form

Phenylpropanoid

$C_{13}H_{14}O_4$ [53890-21-4] MW 234.25
(±)-form [108147-21-3]
(–)-form [52946-22-2]

In the essential oil of the rhizomes and leaves of *Alpinia galanga* (Zingiberaceae).

Antitumour activity against sarcoma 180 ascites in mice, and antifungal activity. The rhizomes of the plant are used to treat fungal skin infections, dysentery and problems of indigestion.

Mitsui, S., *Chem. Pharm. Bull.,* 1976, **24**, 2377.

1′-Acetoxyeugenol acetate 14

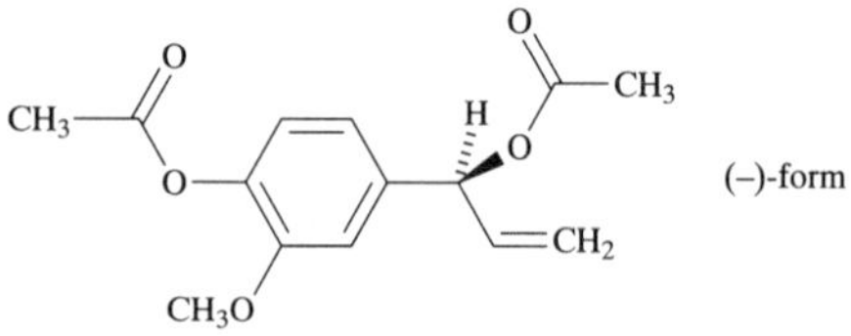

(–)-form

Phenylpropanoid

$C_{14}H_{16}O_5$ [53890-24-7] MW 264.28
(±)-form [108093-58-2]
(–)-form [52946-23-3]

In rhizomes of Siamese ginger, *Alpinia galanga* (Zingiberaceae).

Pungent odour. It shows antitumour activity against sarcoma 180 ascites in mice.

Mitsui, S., *Chem. Pharm. Bull.,* 1976, **24**, 2377.

1-Acetoxy-2-hydroxyheneicosa-12,15-dien-4-one 15

1-Acetoxy-2-hydroxy-4-oxoheneicosa-12,15-diene

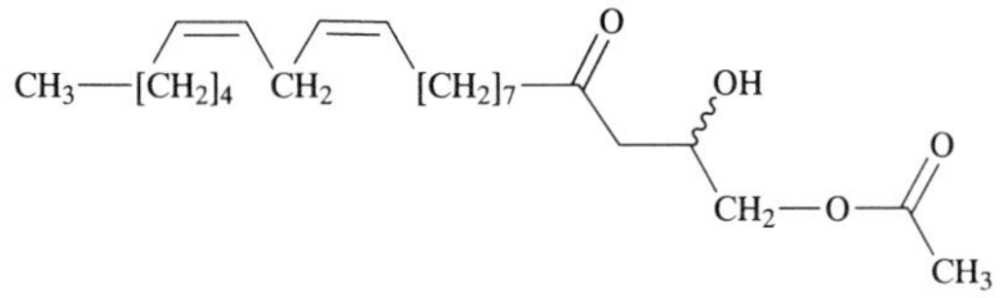

Aliphatic ketone

$C_{23}H_{40}O_4$ [86535-60-6] MW 380.57

Peel of unripe fruit of avocado, *Persea americana* (Lauraceae).

Fungitoxic against *Colletotrichum gloeosporioides*.

Prusky, D., *Physiol. Plant Pathol.,* 1983, **22**, 189.

Acetylcaranine 16

Belamarine; Bellamarine

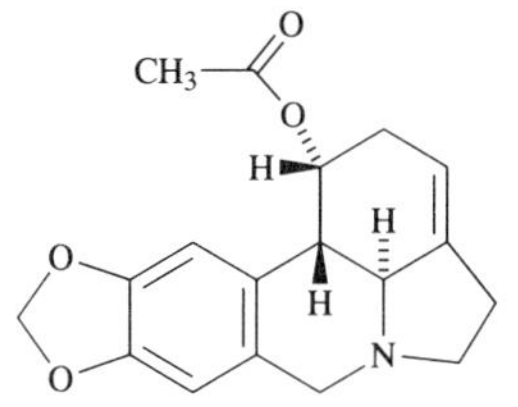

Amaryllidaceae alkaloid
(Galanthan)

$C_{18}H_{19}NO_4$ [477-12-3] MW 313.35

In bulbs of *Amaryllis belladonna* hybrid, *Crinum macrantherum* and *Ammocharis coranica* (Amaryllidaceae).

A uterine stimulant. Also has *in vitro* antineoplastic activity to P388 leukaemia.

Mason, L. H., *J. Am. Chem. Soc.,* 1955, **77**, 1253.

O-Acetylcypholophine 17

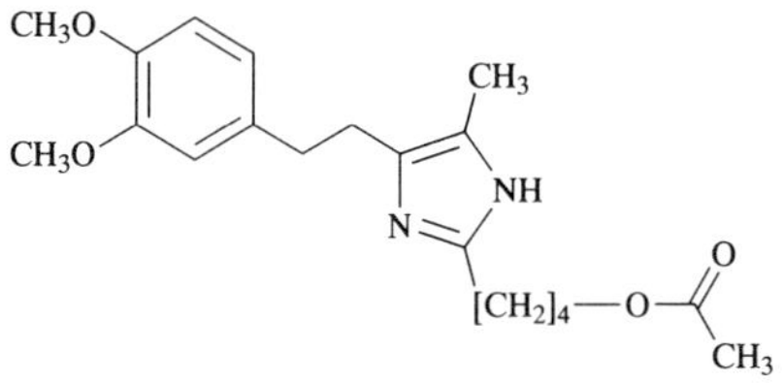

Imidazole alkaloid

$C_{20}H_{28}N_2O_4$ [26482-11-1] MW 360.45

Leaves and stems of *Cypholophus friesianus* (Urticaceae).

Cytotoxic activity.

Hart, N. K., *Aust. J. Chem.,* 1971, **24**, 857.

3-Acetylnerbowdine 18

3-Acetylhaemanthine; 3-Acetylhemanthine; 3-*O*-Acetylnerbowdine

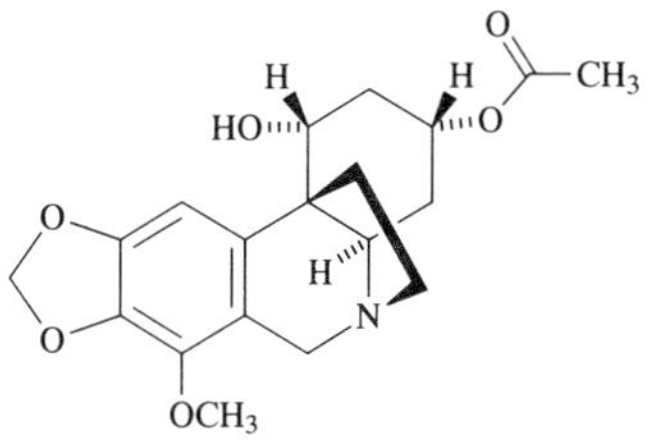

Amaryllidaceae alkaloid
(Crinan)

$C_{19}H_{23}NO_6$ [100196-22-3] MW 361.39

Bulbs of *Boophane distica* and *Nerine crispa* (Amaryllidaceae).

Highly toxic. A constituent of arrow poisons which are used in South Africa for hunting game. The bulb extract shows muscle relaxant, atropine-like properties.

Hauth, H., *Helv. Chem. Acta,* 1963, **46**, 810.

Achillin 19

Santolin

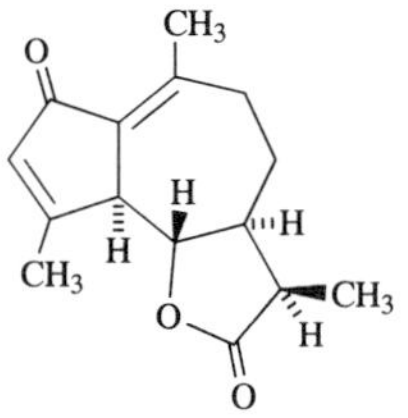

Sesquiterpenoid lactone
(Guaianolide)

$C_{15}H_{18}O_3$ [5956-04-7] MW 246.31

Yarrow, *Achillea millefolium*, other *Achillea* spp. and several *Artemisia* spp. (Compositae).

Probably responsible for allergenic activity of yarrow.

White, E. H., *J. Am. Chem. Soc.,* 1967, **89**, 5511.

Aconifine 20

10-Hydroxyaconitine; Nagarine

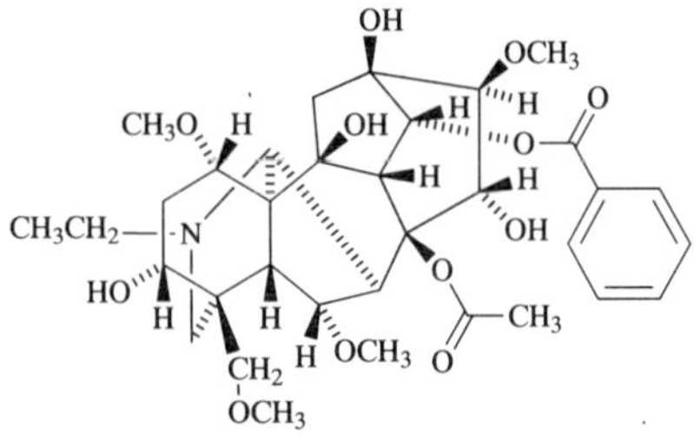

Diterpenoid alkaloid
(Aconitane)

$C_{34}H_{47}NO_{12}$ [41849-35-8] MW 661.75

The roots of *Aconitum karakolicum* and *A. nagarum* (Ranunculaceae).

In most mammals similar toxic effects to aconitine (q.v.) but toxicity in mice is only about half of that of aconitine.

Zhu, Y., *Heterocycles,* 1982, **17**, 607.

Aconitine 21

Acetylbenzoylaconine

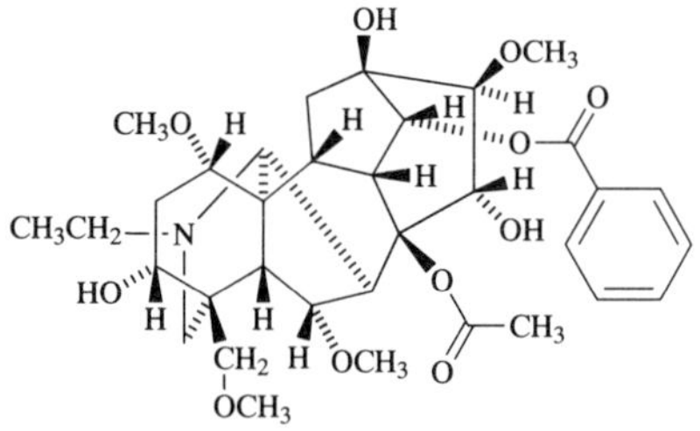

Diterpenoid alkaloid
(Aconitane)

$C_{34}H_{47}NO_{11}$ [302-27-2] MW 645.75

Tuberous roots and other parts of monkshood, *Aconitum napellus* (Ranunculaceae). Widely present in other *Aconitum* spp., e.g. in *A. chasmanthum* from India.

Very potent and quick acting poison, causing slowing of the heart rate and lowering of the blood pressure. Absorption through the skin can be fatal and florists handling monkshood can suffer poisoning. Lethal dose is 3-6 mg, which is present in only a few grams of plant material.

Birnbaum, K. B., *Tetrahedron Lett.,* 1971, 867.

Acrifoline 22

Alkaloid L27; Alkaloid L29

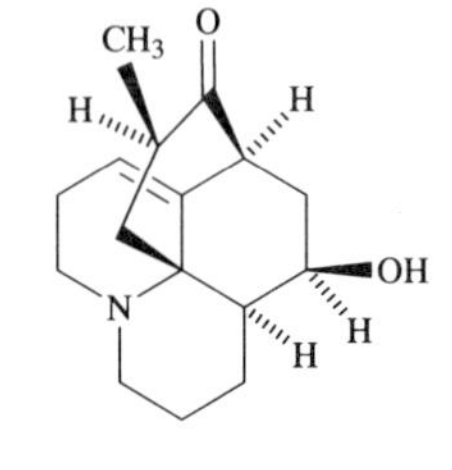

Lycopodium alkaloid
(Lycopodane)

$C_{16}H_{23}NO_2$ [664-24-4] MW 261.36

Whole plant of *Lycopodium solago* and *L. annotinum* var. *acrifolium* (Lycopodiaceae).

Toxic.

French, W. N., *Can. J. Chem.,* 1961, **39**, 2100.

Acronycidine 23

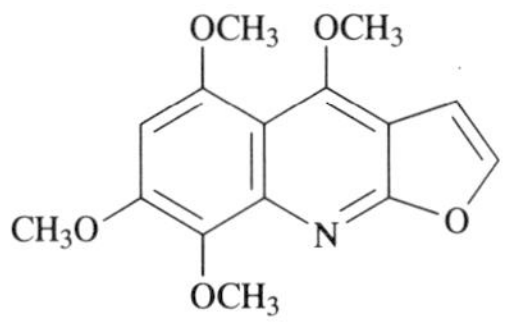
OCH3 OCH3
CH3O
N
O
OCH3

Furoquinoline alkaloid

$C_{15}H_{15}NO_5$ [521-43-7] MW 289.29

Leaves of *Acronychia baueri* and bark of *Melicope fareana* (Rutaceae).

Central nervous system depressant.

Lahey, F.N., *Aust. J. Sci. Res., Ser. A,* 1950, **3**, 155.

Acronycine 24

Acronine; Compound 42339; NSC 403169

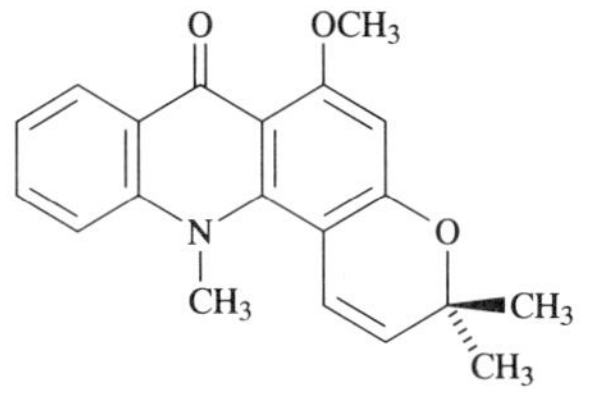
O
OCH3
N
CH3
O
CH3
CH3

Acridone alkaloid

$C_{20}H_{19}NO_3$ [7008-42-6] MW 321.38

Acronychia baueri, A. haplophylla and *Melicope leptococca* (Rutaceae).

Carcinogenic. Potent antitumour agent; possesses a broad-spectrum activity against experimental neoplasts.

Tan, P., *Cancer Res.,* 1973, **33**, 2320.

Acroptilin 25

Chlorohyssopifolin C

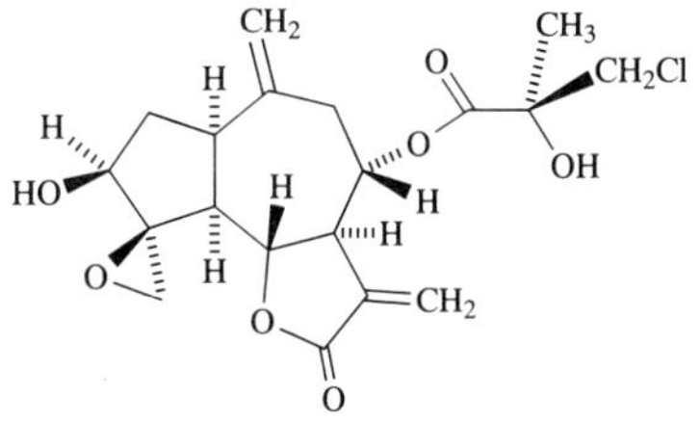
CH2
H
H
HO
O
H
H
O
CH3
CH2Cl
OH
O
H
H
CH2
O
O

Sesquiterpenoid lactone
(Guaianolide)

$C_{19}H_{23}ClO_7$ [41787-75-1] MW 398.84

Acroptilon repens, Centaurea hyssopifolia, C. hyrcanica and *C. linifolia* (Compositae).

Cytotoxic and antitumour activity. Shows strong activity against the pathogenic protozoa *Entamoeba histolytica* and *Trichomonas vaginalis.*

Stevens, K. L., *Cryst. Struct. Commun.,* 1982, **11**, 949.

Acrovestone 26

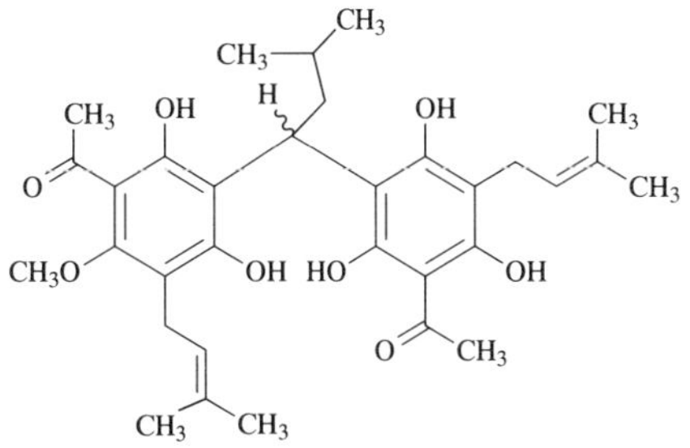
CH3
CH3
H
CH3
OH
OH
CH3
O
CH3
CH3O
OH
HO
OH
O
CH3
CH3
CH3

Phenolic ketone

$C_{32}H_{42}O_8$ [24177-16-0] MW 554.68

Stem and root bark of *Acronychia pedunculata, A. vestita* and *A. laurifolia* (Rutaceae).

Cytotoxic in the human KB tissue culture assay. *Acronychia pedunculata* is used for treating diarrhoea, asthma, ulcers and rheumatism.

Wu, T.S., *J. Nat. Prod.,* 1989, **52**, 1284.

Actinidine 27

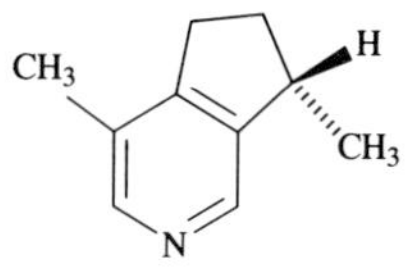

Monoterpenoid alkaloid

$C_{10}H_{13}N$ [524-03-8] MW 147.22

In leaves and galls of *Actinidia polygama* and *A. arguta* (Actinidiaceae), in leaves of *Tecoma radicans* (Bignoniaceae) and in the roots of *Valeriana officinalis* (Valerianaceae).

Causes a remarkable excitation in cats and other Felidae, similar to that of catmint. Toxic to some animals, since the substance occurs in the defensive secretions of ants and rare beetles.

Sakan, T., *Bull. Chem. Soc. Japan,* 1960, **33**, 712.

Acutumidine 28

N-Noracutumine

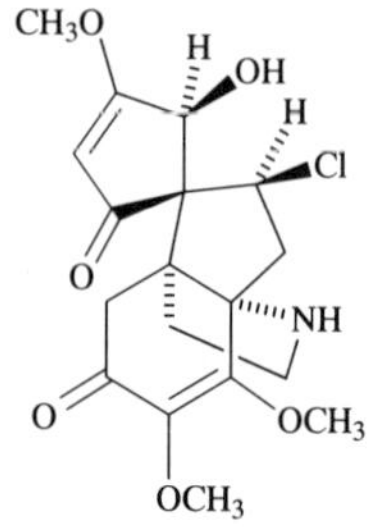

Benzylisoquinoline alkaloid

$C_{18}H_{22}ClNO_6$ [18145-26-1] MW 383.83

Leaves of *Menispermum dauricum* and *Sinomenium acutum* (Menispermaceae).

Stimulates the uterus in experimental animals.

Tomita, M., *Chem. Pharm. Bull.,* 1971, **19**, 770.

Adiantifoline 29

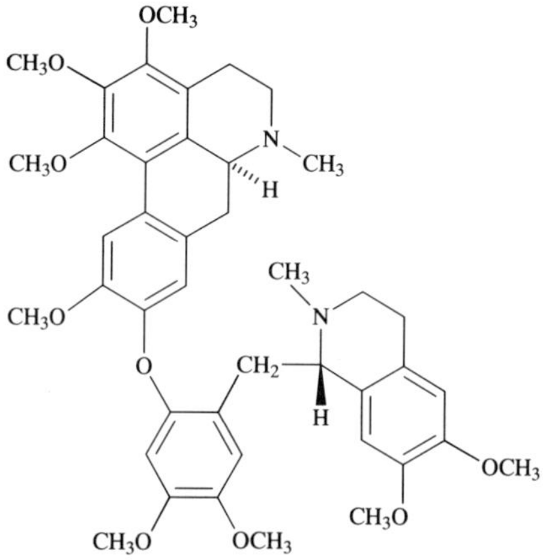

Bisbenzylisoquinoline alkaloid
(Aporphine and benzylisoquinoline)

$C_{42}H_{50}N_2O_9$ [20823-96-5] MW 726.88

Roots and aerial parts of *Thalictrum minus* var. *adiantifolium* (Ranunculaceae).

Toxic to mammals. Intravenous administration to rabbits produces a brief hypotensive response.

Doskotch, R. W., *Tetrahedron Lett.,* 1968, 4999.

Adlumine 30

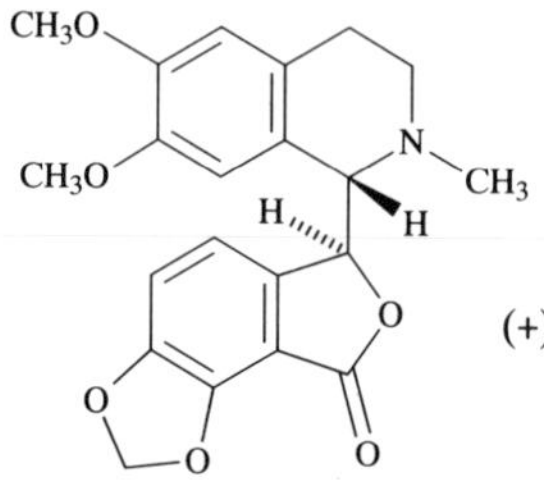

(+)-form

Phthalideisoquinoline alkaloid

$C_{21}H_{21}NO_6$ [524-46-9] MW 383.41
(−)-form [21414-43-7]
(±)-form [38184-69-9]

(+)-Form occurs in *Adlumia fungosa* (Papaveraceae); (−)-form occurs in *Corydalis scouleri, C. sempervirens* and *C. ophiocarpa*. The (±)-form occurs in *Corydalis rosea* (Papaveraceae).

Has activity as a convulsant, cardiac depressor and uterine and intestinal stimulant.

Safe, S., *Can. J. Chem.,* 1964, **42**, 160.

Adonitoxin 31

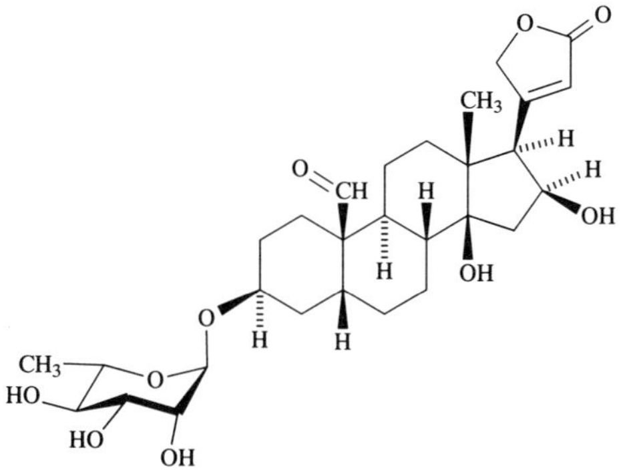

Cardenolide

$C_{29}H_{42}O_{10}$ [17651-61-5] MW 550.65

Leaves of yellow pheasant's-eye, *Adonis vernalis* (Ranunculaceae).

Toxic to mammals and other vertebrates. Poorly absorbed in mammals; human poisoning from this plant is unlikely. In the cat LD_{50} is 0.191 mg/kg body-weight .

Pitra, J., *Collect. Czech. Chem. Commun.,* 1961, **26**, 1551.

Adynerin 32

Adynerigenin 3-diginoside

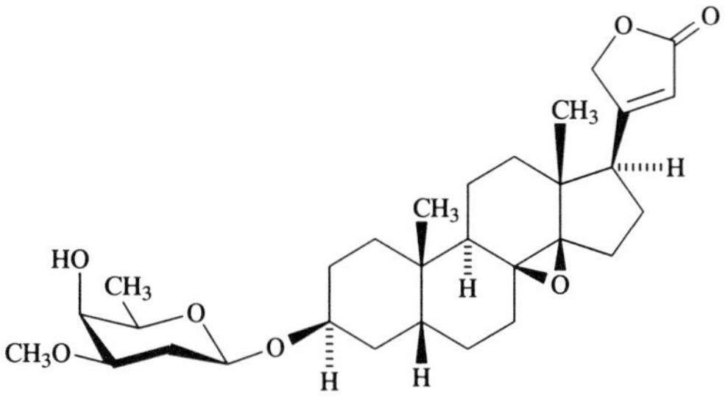

Cardenolide

$C_{30}H_{44}O_7$ [35109-93-4] MW 516.67

Leaves of oleander, *Nerium oleander* (Apocynaceae).

Toxic to mammals and other vertebrates. However, the bitter taste of the cardenolide is a deterrent to excessive consumption of the plant. Cases of poisoning through drinking infusions of oleander as an abortifacient or for committing suicide have been described.

Janiak, P. St., *Helv. Chim. Acta.,* 1963, **46**, 374.

Aescin 33

Escin; Aescusan; Reparil

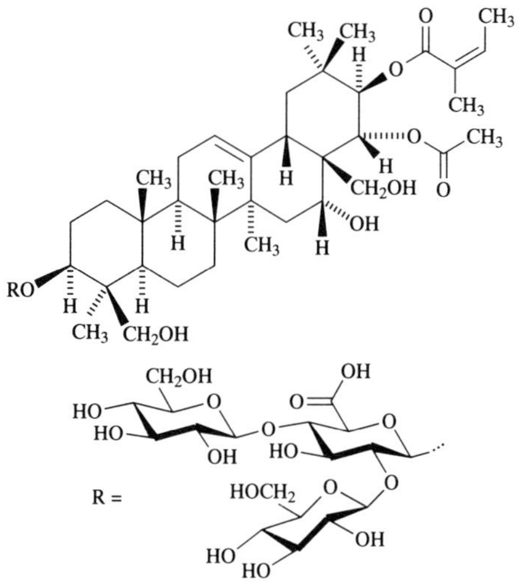

Triterpenoid saponin
(Oleanane)

$C_{55}H_{86}O_{24}$ MW 1131.29

Aescin [6805-41-0] is a mixture of saponins, and occurs in thc fruits (conkcrs) of horsc-chcstnut, *Aesculus hippocastanum* (Hippocastanaceae). The mixture is also called β-aescin [11072-93-8]. The most abundant component in the mixture is shown above [123748-68-5]. Other components are the (*Z*)-isomer and related glycosides.

Strong haemolytic activity. Shows anti-inflammatory, anti-exudative and cancerostatic activities.

Hoppe, W., *Angew. Chem., Int. Ed. Eng.,* 1968, **7**, 547.

Aethusin 34

Ethusin

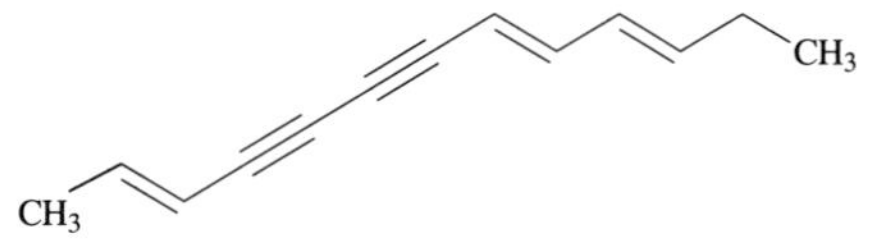

Polyacetylene

$C_{13}H_{14}$ [463-34-3] MW 170.25

In fool's parsley, *Aethusa cynapium* and in roots and/or rhizomes of *Peucedanum carvifola*, *P. austriacum*, *P. rablense* and *P. verticillare* (Umbelliferae).

Toxic principle of fool's parsley, which can cause poisoning in man and livestock. Coniine-like alkaloids have also been reported to occur in this plant.

Bohlmann, F., *Chem. Ber.*, 1960, **93**, 981.

Affinin 35

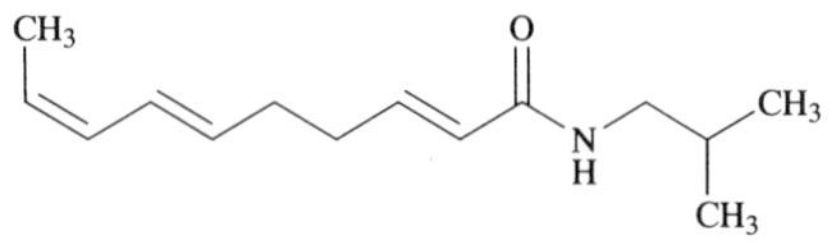

Aliphatic amide

$C_{14}H_{23}NO$ [25394-57-4] MW 221.34

Heliopsis longipes roots (Compositae).

Very potent insecticidal properties.

Crombie, L., *J. Chem. Soc.*, 1963, 4970.

Affinine 36

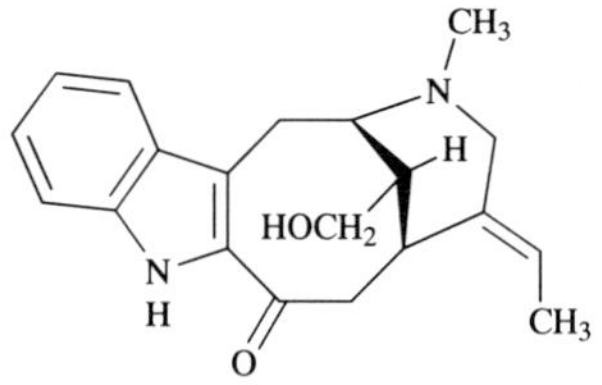

Indole alkaloid
(Vobasan)

$C_{20}H_{24}N_2O_2$ [2134-82-9] MW 324.42

Peschiera affinis, *P. laeta* and *Tabernaemontana psychotrifolia* (Apocynaceae).

Produces gross behavoural changes in the mouse. Causes slight nervous system depression, delayed intention tremors, ataxia, hypothermia and bradypnoea.

Weisbach, J. A., *J. Pharm. Sci.*, 1963, **52**, 350.

Affinisine 37

Dehydroxymethylvoachalotinol

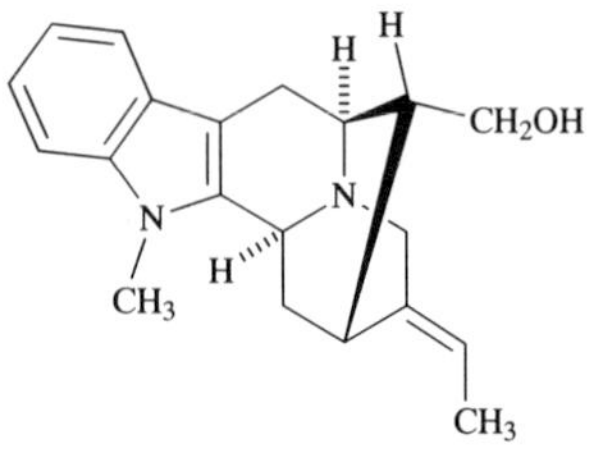

Indole alkaloid
(Sarpagan)

$C_{20}H_{24}N_2O$ [2912-11-0] MW 308.42

Alstonia macrophylla, *Peschiera affinis* and *Tabernaemontana fuchsiaefolia* (Apocynaceae).

Produces central nervous system depression, lachrymation and tremors in the mouse; it has moderate analgesic properties in the rat.

Banerji, A., *Phytochemistry*, 1972, **11**, 2605.

Agavoside A 38

Hecogenin 3-galactoside

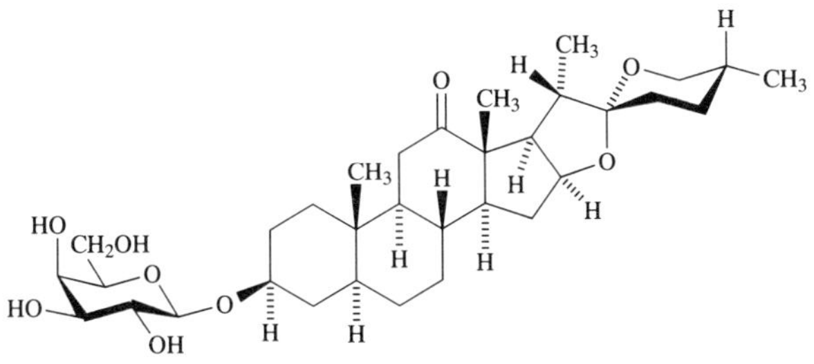

Steroid saponin
(Spirostan)

$C_{33}H_{52}O_9$ [56857-65-9] MW 592.77

Agave americana (Agaraceae).

Active against epidermoid carcinoma of the nasopharynx in tissue culture. Toxic; one of several toxic glycosides present in this plant.

Kintya, P. K., *Tezisy Dokl. Vses. Simp. Bioorg. Khim.*, 1975, 20.

Agrimoniin 39

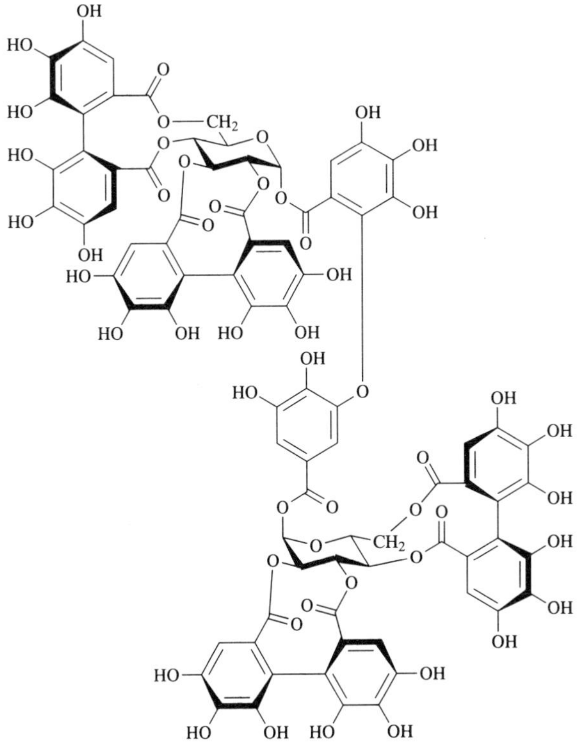

Ellagitannin

$C_{82}H_{54}O_{52}$ [82203-01-8] MW 1871.30

Roots of *Agrimonia pilosa* and *A. japonica* and in *Potentilla kleiniana* (Rosaceae).

Antitumour activity against Sarcoma 180 in mice; it is a moderate inhibitor of induced lipid peroxidation in rat liver mitochondria.

Okuda, T., *Chem. Pharm. Bull.*, 1984, **32**, 2165.

Agroclavine 40

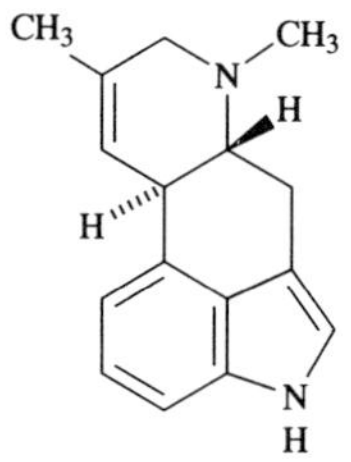

Indole alkaloid
(Ergoline)

$C_{16}H_{18}N_2$ [548-42-5] MW 238.33

Seeds of *Cuscuta*, *Ipomoea* and *Rivea* spp. (Convolvulaceae).

Highly active *in vivo* on the uteri of the rabbit and guinea-pig. Toxic in mammals. Central nervous system excitator acting by stimulation of sympathetic centres.

Chao, J. M., *Phytochemistry,* 1973, **12**, 2435.

Ailanthinone 41

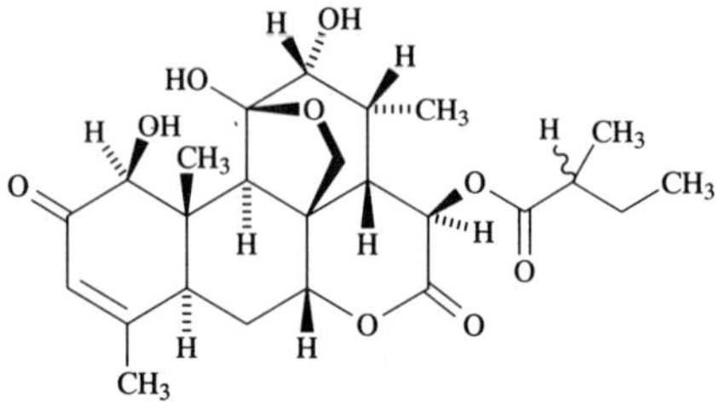

Nortriterpenoid
(Quassane)

$C_{25}H_{34}O_9$ [53683-73-1] MW 478.54

Pierreodendron kerstingii and *Ailanthus altissima* (Simaroubaceae).

Antileukaemic activity *in vivo*.

Kupchan, S. M., *J. Org. Chem.,* 1975, **40**, 654.

Ajaconine 42

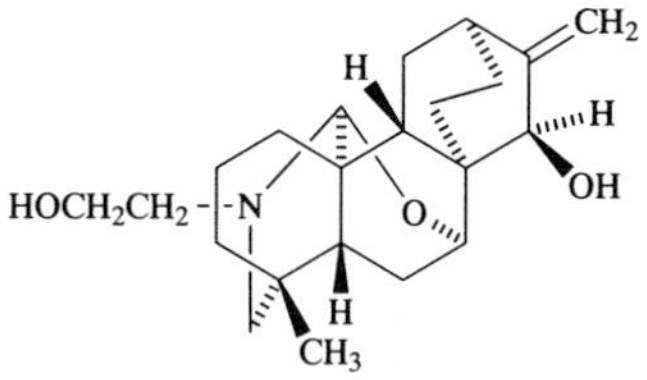

Diterpenoid alkaloid
(Ajaconane)

$C_{22}H_{33}NO_3$ [545-61-9] MW 359.51

Seeds of *Delphinium ajacis*, roots of *Delphinium tatsienense* and whole plant of *Delphinium virescens* (Ranunculaceae).

Poisonous.

Dvornik, D., *Tetrahedron,* 1961, **14**, 54.

Ajmalicine 43

Raubasine; Tetrahydroserpentine; Vincaine; Vinceine; δ-Yohimbine

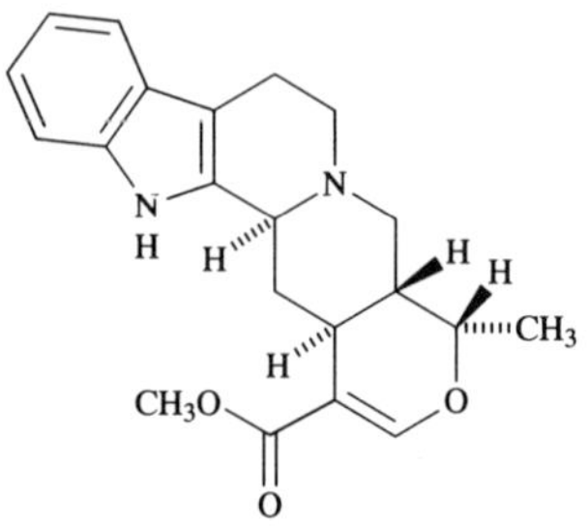

Indole alkaloid
(Oxayohimban)

$C_{21}H_{24}N_2O_3$ [483-04-5] MW 352.43

Bark of *Corynanthe johimbe* (Rubiaceae), roots of many *Rauwolfia* spp. (e.g. *R. serpentina*) and in *Vinca rosea* (Apocynaceae).

Tranquilliser, antihypertensive; used to improve cerebral blood circulation. Toxic.

Wenkert, E., *J. Am. Chem. Soc.,* 1961, **83**, 5037.

Ajmaline 44

Raugalline; Rauwolfine

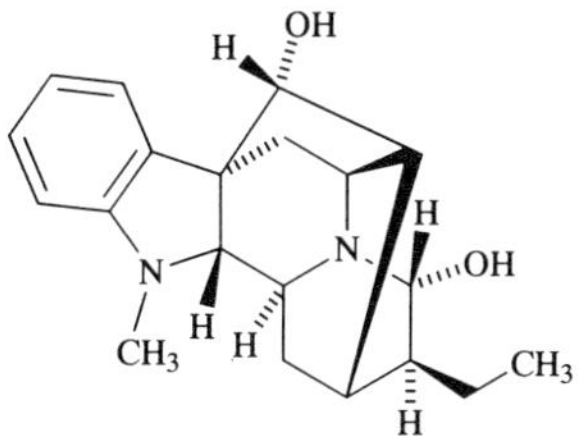

Indole alkaloid
(Ajmalan)

$C_{20}H_{26}N_2O_2$ [4360-12-7] MW 326.44

Melodinus balansae, *Tonduzia longifolia* and roots of *Rauwolfia serpentina* (Apocynaceae).

Coronary dilating and antirrhythmic activities. Toxic.

Schlittler, E., *Arzneim.-Forsch.,* 1974, **24**, 873.

Akagerine 45

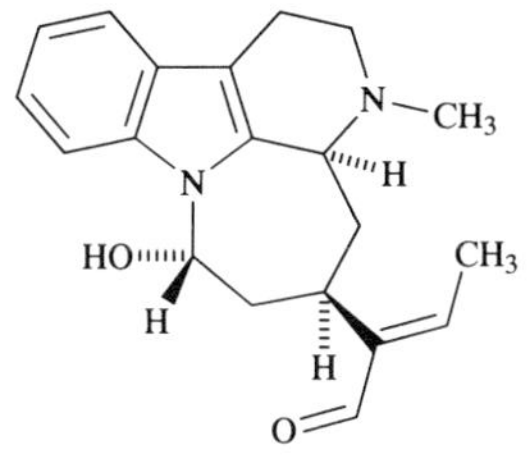

Indole alkaloid
(β-Carboline)

$C_{20}H_{24}N_2O_2$ [56519-07-4] MW 324.42

Root of *Strychnos usambarensis* (Loganiaceae).

100% Toxicity at 10 μg/ml to two cancer cell lines.

Angenot, L., *Tetrahedron Lett.,* 1975, 1357.

Akuammicine 46

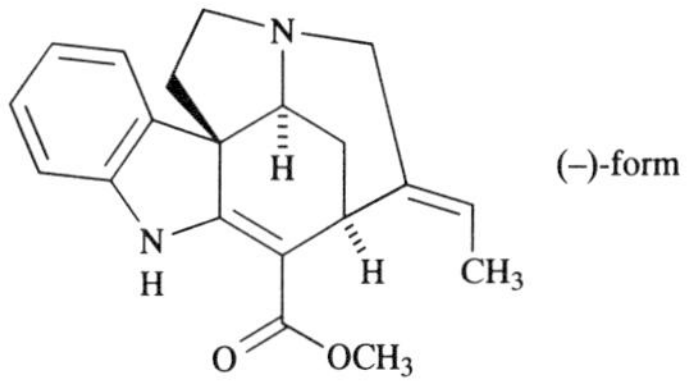

(–)-form

Indole alkaloid
(Curan)

$C_{20}H_{22}N_2O_2$ [639-43-0] MW 322.41
(±)-form [7344-80-1]

Seeds and aerial parts of *Picralima klaineana*, roots of *Vinca rosea* (Apocynaceae).

Gonadotropic (follicular stimulation) activity.

Aghoramurthy, K., *Tetrahedron,* 1957, **1**, 172.

Akuammidine 47

Rhazine

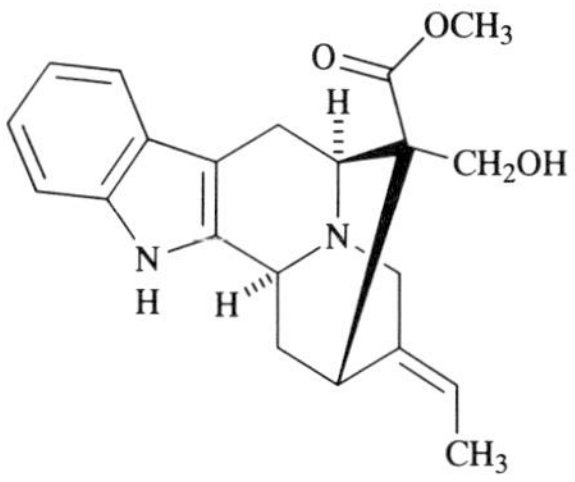

Indole alkaloid
(Sarpagan)

$C_{21}H_{24}N_2O_3$ [639-36-1] MW 352.43

Major alkaloid of *Picralima klaineana* and *Rhazya stricta* (Apocynaceae); also occurs in *Aspidosperma*, *Melodius*, *Vallesia*, *Vinca* and *Voacanga* spp. (all Apocynaceae).

Hypotensive. Is a skeletal muscle relaxant, and has local anaesthetic action about three times as potent as cocaine.

Silvers, S., *Tetrahedron Lett.,* 1962, 339.

Akuammine 48

Vincamajoridine

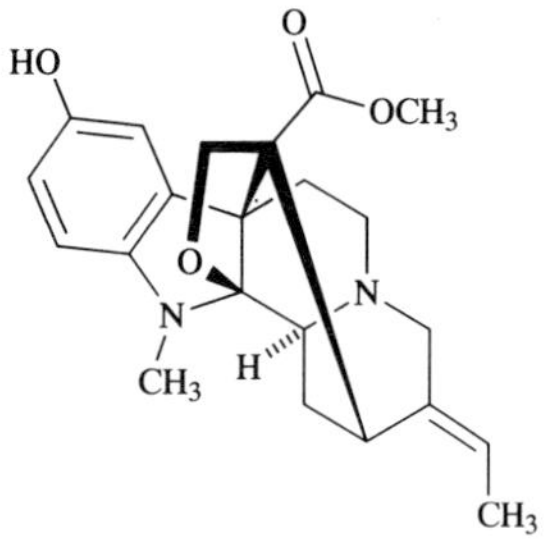

Indole alkaloid
(Akuammilan)

$C_{22}H_{26}N_2O_4$ [3512-87-6] MW 382.46

Seeds of *Picralima klaineana* (Apocynaceae).

Augments the hypertensive effects of adrenaline. Has local anaesthetic action almost equal to that of cocaine. In Ghana the seeds are used to control malaria in the place of quinine.

Olivier, L., *Bull Soc. Chim. France,* 1965, 868.

Alantolactone 49

Alantic anhydride; Helenine

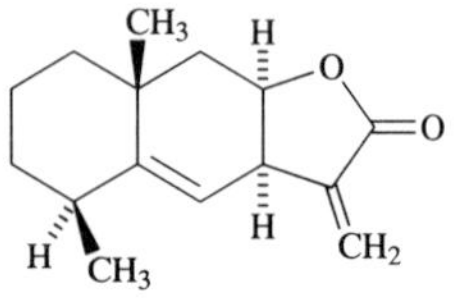

Sesquiterpenoid lactone
(Eudesmanolide)

$C_{15}H_{20}O_2$ [546-43-0] MW 232.32

Oil of elecampane, *Inula helenium*, and in *Inula grandis* and *I. magnifica* (Compositae).

Causes allergic contact dermatitis in humans. Elecampane is widely used in the former Soviet republic for the self-treatment of skin ailments. *In vitro* shows toxic effects on cultures of lymphocytes.

Dupuis, G., *Chem. Biol. Interact.,* 1976, **15**, 205.

Alatolide 50

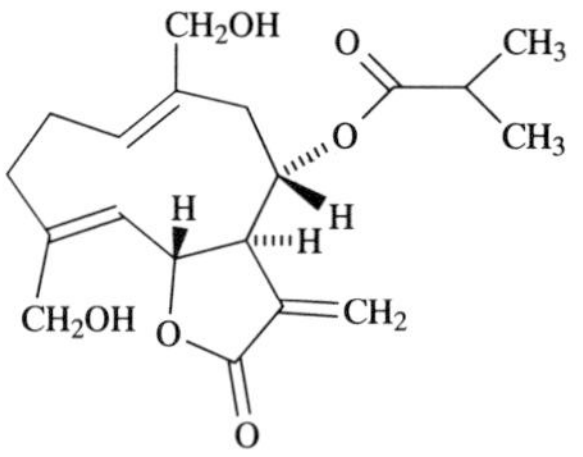

Sesquiterpenoid lactone
(Germacranolide)

$C_{19}H_{26}O_6$ [41929-10-6] MW 350.41

Jurinea alata (Compositae).

Cytotoxic and antitumour activities.

Drozdz, B., *Collect. Czech. Chem. Commun.,* 1973, **38**, 727.

Albanol A 51

Mulberrofuran G

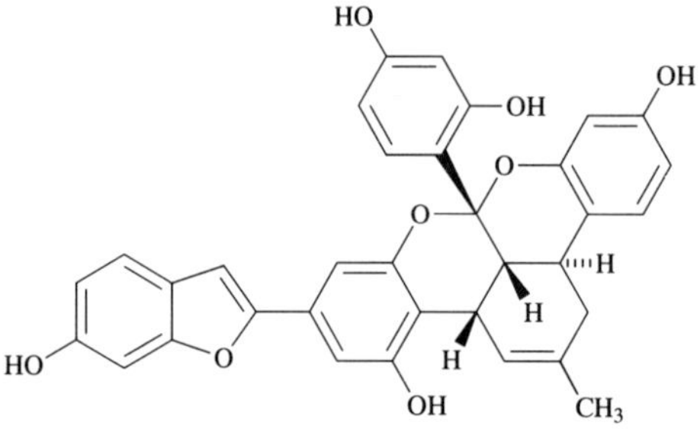

Benzofuran

$C_{34}H_{26}O_8$ [87085-00-5] MW 562.58

Root bark of *Morus lhou* and *M. alba* (Moraceae).

Hypotensive activity.

Rao, A. V. R., *Tetrahedron Lett.,* 1983, **24**, 3013.

Albaspidin AA 52

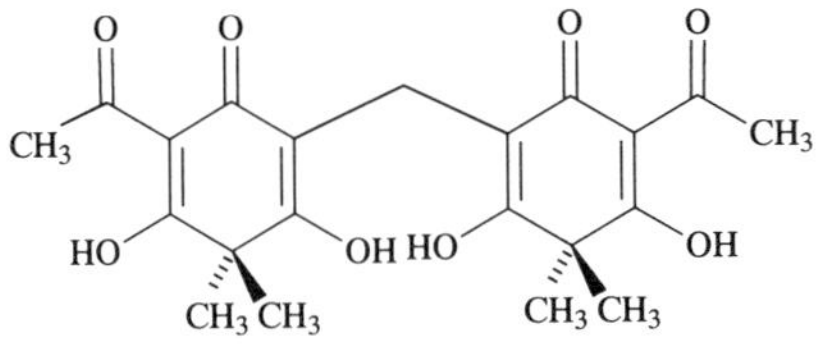

Phenolic ketone

$C_{21}H_{24}O_8$ [3570-40-9] MW 404.42

One of six related albaspidins in the male fern, *Dryopteris filix-mas* and other *Dryopteris* spp. (Polypodiaceae).

Male fern extract is used to treat tapeworms, especially in veterinary practice. Overdose can lead to fatalities. Symptoms of poisoning include cramps, intestinal irritation and visual disturbances. The active principles also include filixic acids (q.v.).

Penttilä, A., *Acta Chem. Scand.*, 1964, **18**, 344.

Alchorneine 53

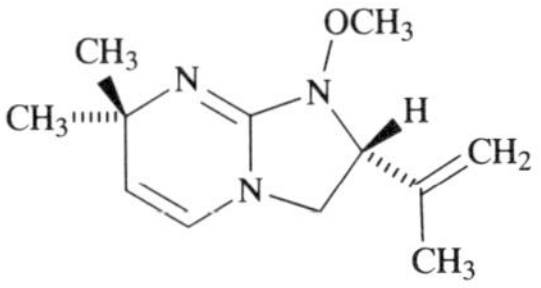

Imidazole alkaloid

$C_{12}H_{19}N_3O$ [28340-21-8] MW 221.30

Alchornea floribunda and *A. hirtella* (Euphorbiaceae).

Spasmolytic agent in dogs. A ganglioplegic parasympathomimetic and a strong vagolytic agent. It is also an inhibitor of intestinal peristalsis.

Khuong-Huu, F., *Tetrahedron*, 1972, **28**, 5207.

Alchornine 54

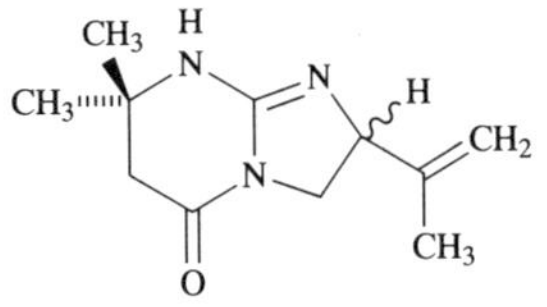

Imidazole alkaloid

$C_{11}H_{17}N_3O$ [25819-91-4] MW 207.28

Leaves and bark of *Alchornea jaranensis* (Euphorbiaceae).

Similar properties to alchorneine (q.v.).

Hart, N. K., *Aust. J. Chem.*, 1970, **23**, 1679.

Alizarin 55

C.I. Mordant Red 11; 1,2-Dihydroxyanthraquinone; Lizarinic acid; Madder

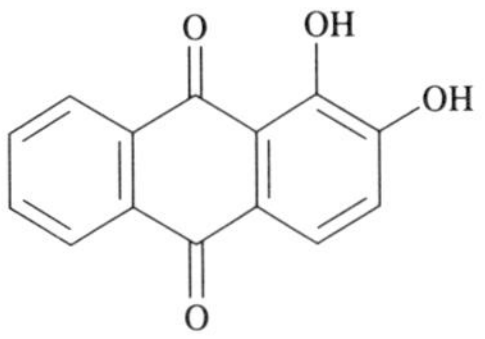

Anthraquinone

$C_{14}H_8O_4$ [72-48-0] MW 240.22

Roots of the madder plant, *Rubia tinctorum*, in *Galium* spp., *Asperula odorata* and in the heartwood of *Morinda citrifolia* (all Rubiaceae). Also in *Rheum palmatum* (Polygonaceae) and as a glycoside in *Libertia coerulescens* (Iridaceae).

Orange-red pigment, the principle of madder, one of the most ancient of dyestuffs. Has been removed from certain brand name cosmetics because it is a sensitiser. Also shows antileukaemic activity.

Murti, V. V. S., *Indian J. Chem.*, 1972, **10**, 246.

Alkannin 56

Alhanin; Alkanna red; Anchusa acid; Anchusaic acid; Anchusin; Arnebin IV

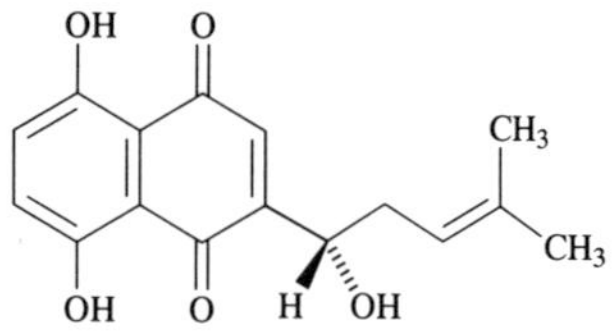

Naphthoquinone

$C_{16}H_{16}O_5$ [517-88-4] MW 288.30

Roots of *Alkanna tinctoria*, *Arnebia nobilis* and *Macrotomia cephalotes* and a leaf surface constituent of *Plagiobothrys arizonicus* (all Boraginaceae).

Red dye, used for cosmetics and food. Astringent properties. At low concentrations has immunomodulatory effects; at higher concentrations has suppressive action in granulocyte and lymphocyte test systems. Used in the treatment of Ulcus cruris.

Brockmann, H., *Justus Liebigs Ann. Chem.*, 1936, **521**, 1.

Alkannin *β,β*-dimethylacrylate 57

Arnebin I; *β,β*-Dimethylacrylalkannin

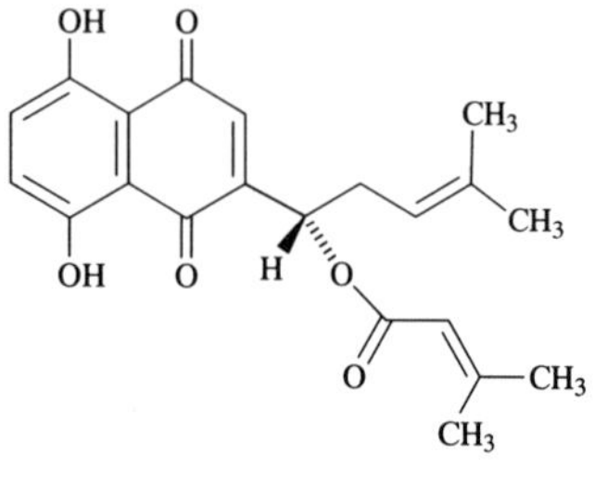

Naphthoquinone

$C_{21}H_{22}O_6$ [5162-01-6] MW 370.40

Roots of *Arnebia nobilis* (Boranginaceae).

Anticancer activity against Walker carcinosarcoma in rats. Also shows wound-healing properties.

Shukla, Y. N., *Phytochemistry*, 1971, **10**, 1909.

Allamandin 58

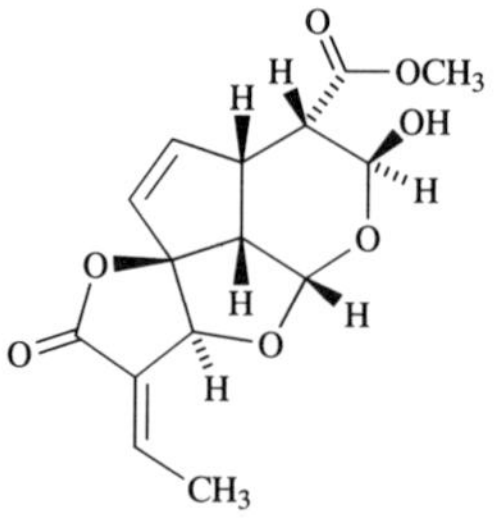

Monoterpenoid
(Iridoid)

$C_{15}H_{16}O_7$ [51820-82-7] MW 308.29

Allamanda cathartica (Apocynaceae).

Antileukaemic and tumour-inhibiting properties. All parts of *Allamanda cathartica* may cause irritant dermatitis and allamandin may be the active principle.

Kupchan, S. M., *J. Org. Chem.*, 1974, **39**, 2477.

Allicin 59

S-Oxodiallyl disulfide

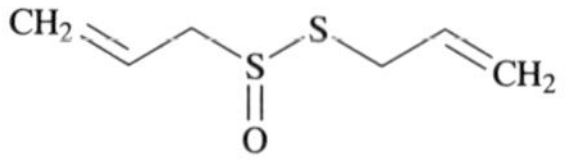

Disulfide

$C_6H_{10}OS_2$ [539-86-6] MW 162.28

Bulbs of onion, *Allium cepa* and garlic, *Allium sativum* (Alliaceae).

Major odour principle of garlic. Has antidiabetic, antihypertensive and antithrombotic activities. Moderately toxic and irritant.

Cavallito, C. J., *J. Am. Chem. Soc.*, 1944, **66**, 1952.

Alliin 60

S-Allyl-L-cysteine *S*-oxide; β-Allylsulfenylalanine

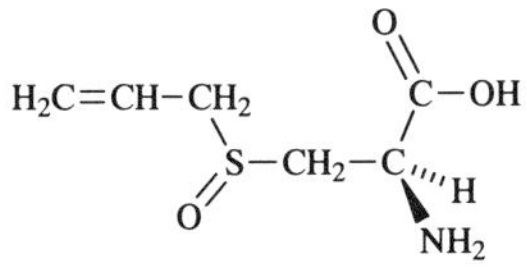

Amino Acid

$C_6H_{11}NO_3$ [556-27-4] MW 177.22

Bulbs of onion, *Allium cepa* and garlic, *Allium sativa* (Alliaceae).

Platelet aggregation inhibitor and thrombotic.

Virtanen, A. I., *Angew. Chem., Int. Ed. Engl.*, 1962, **1**, 299.

Allocryptopine 61

α-Fagarine; Fagarine I; β-Homochelidonine; γ-Homochelidonine; Thalictrimine

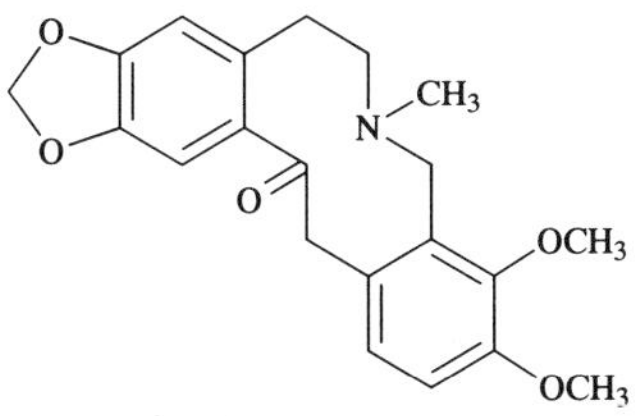

Benzylisoquinoline alkaloid (Protoberberine)

$C_{21}H_{23}NO_5$ [485-91-6] MW 369.42

Occurs in two isomorphic forms, α-allocryptopine [485-91-6] and β-allocryptopine [24240-04-8]. Found in *Bocconia, Chelidonium, Corydalis, Dicentra, Escholtzia, Glaucium* and *Sanguinaria* spp. (Papaveraceae) and *Zanthoxylum* spp. (Rutaceae).

Oxytocic agent. Hypotensive agent, cardiac inhibitor, respiratory stimulant and muscle relaxant.

Deulofeu, V., *J. Org. Chem.*, 1947, **12**, 217.

Alloimperatorin 62

8-Hydroxy-5-prenylpsoralen; 9-Hydroxy-4-prenylpsoralen; Prangenidine

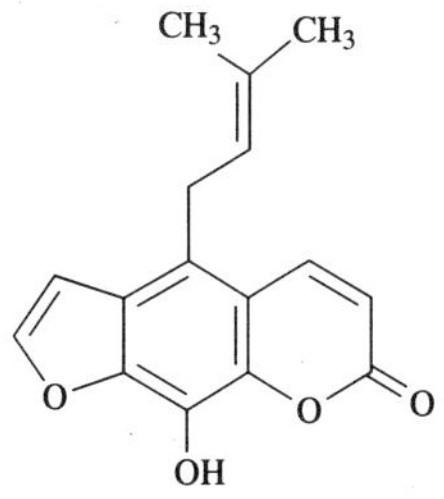

Furocoumarin

$C_{16}H_{14}O_4$ [642-05-7] MW 270.28

Roots of *Prangos pabularia*, seeds of *Heracleum nepalense*, fruits of *Ammi majus* and of *Selinum monnieri* and in aerial parts of *Smyrniopsis armena* (all Umbelliferae). Also in *Aegle marmelos, Poncirus trifoliata* and *Thamnosma montana* (Rutaceae) and in *Zea mays* (Gramineae).

Piscicidal and antispasmodic activity.

Späth, E., *Ber. Dtsch. Chem. Ges.*, 1933, **66**, 1137.

Alnusiin 63

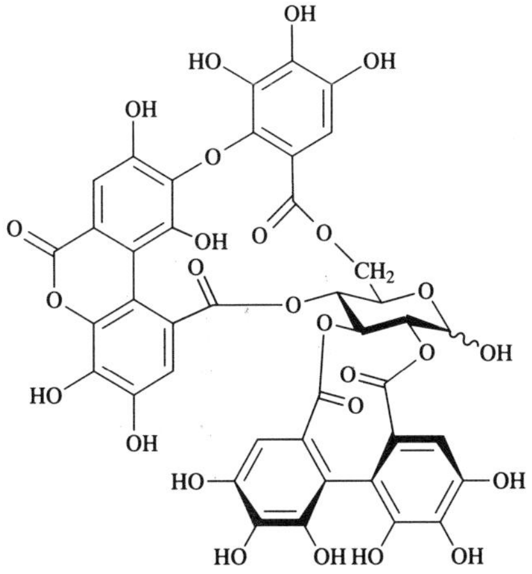

Ellagitannin

$C_{41}H_{26}O_{26}$ [78836-97-9] MW 934.64

Fruit of *Alnus sieboldiana* (Betulaceae).

Inhibits induced lipid peroxidation in the mitochondria of fat cells in rats and the microsomes of rat liver cells. Also shows antitumour activity against Sarcoma 180 in mice.

Yoshida, T., *Heterocycles,* 1989, **29**, 861.

Aloe-emodin 64

Iso-emodin; Rhabarberone; Rottlerin

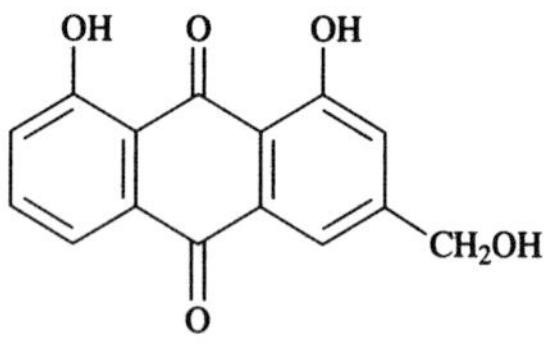

Anthraquinone

$C_{15}H_{10}O_5$	[481-72-1]	MW 270.24

Leaves of *Cassia senna* (Leguminosae), some *Rheum* spp. (Polygonaceae), tubers of *Asphodelus microcarpus*, inflorescences of *Xanthorrhoea australis*, *Aloe* spp. (Liliaceae), leaves of *Oroxylum indicum* (Bignoniaceae), wood of teak, *Tectona grandis* (Verbenaceae).

Cathartic and antileukaemic activities. It is genotoxic in hamster fibroblast mutagenicity assay. Toxic.

Mitter, P. C., *J. Indian Chem. Soc.,* 1932, **9**, 375.

Aloperine 65

Allopterin

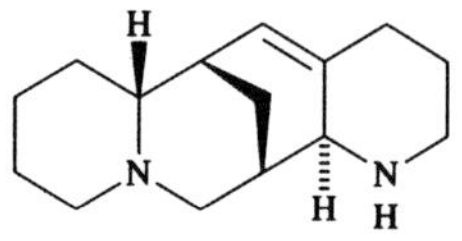

Quinolizidine alkaloid
(Sparteine)

$C_{15}H_{24}N_2$	[56293-29-9]	MW 232.37

Seeds and aerial parts of *Sophora alopecuroides* (Leguminosae).

Toxic, causing paralysis of the central nervous system and of respiratory organs.

Tolkachev, O. N., *Chem. Nat. Compd.,* 1975, **11**, 29.

Alstonine 66

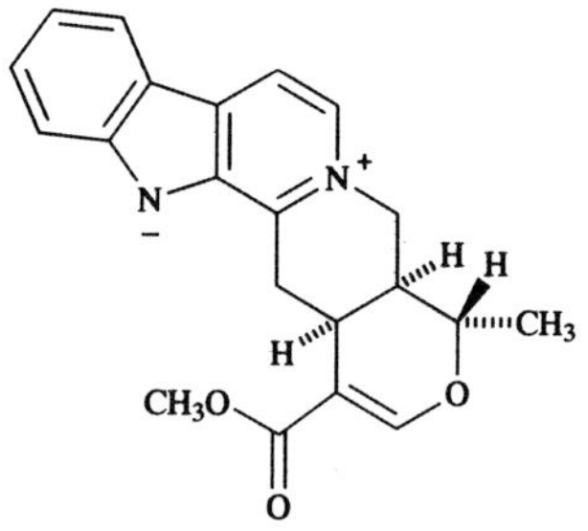

Indole alkaloid
(Oxayohimban)

$C_{21}H_{20}N_2O_3$	[47485-83-6]	MW 348.40

Alstonia constricta, *Rauwolfia hirsuta*, *R. obscura*, *R. vomitoria* and *Vinca rosea* (Apocynaceae).

Neoplasm inihibiting activity.

Elderfield, R. C., *J. Org. Chem.,* 1951, **16**, 506.

Amabiline 67

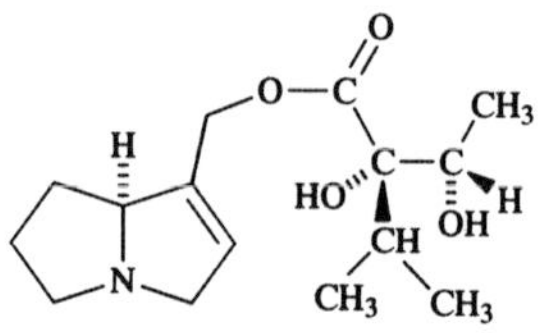

Pyrrolizidine alkaloid

$C_{15}H_{25}NO_4$	[17958-43-9]	MW 283.37

Cynoglossum australe (Boraginaceae).

Hepatotoxic.

Culvenor, C. C. J., *Aust. J. Chem.,* 1967, **20**, 2499.

Amaralin 68

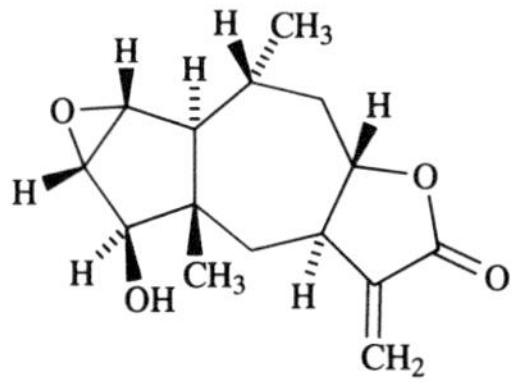

Sesquiterpenoid lactone
(Pseudoguaianolide)

$C_{15}H_{20}O_4$ [6831-10-3] MW 264.33

The sneezeweed, *Helenium amarum* (Compositae).

Strong analgesic action when used subcutaneously.

Lucas, R. A., *J. Org. Chem.,* 1964, **29**, 1549.

Amarogentin 69

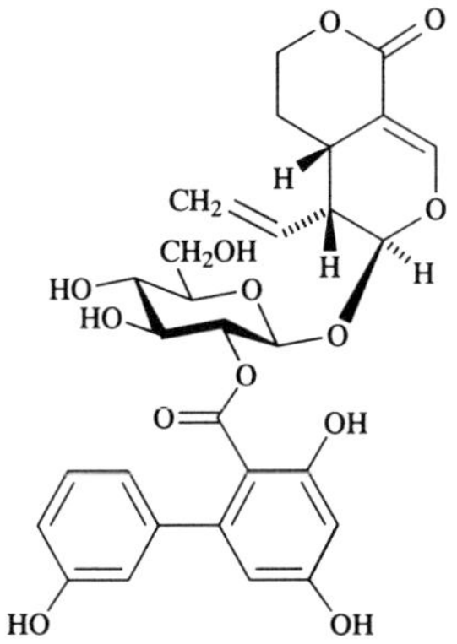

Monoterpenoid
(Seco-iridoid glycoside)

$C_{29}H_{30}O_{13}$ [21018-84-8] MW 586.55

Roots of *Gentiana* spp., e.g. *G. lutea* and roots of Indian gentian, *Swertia* spp. (Gentianaceae).

Very bitter tasting. Bitter principle of gentian root, used as a bitter tonic.

Inouye, H., *Tetrahedron,* 1971, **27**, 1951.

Amaryllisine 70

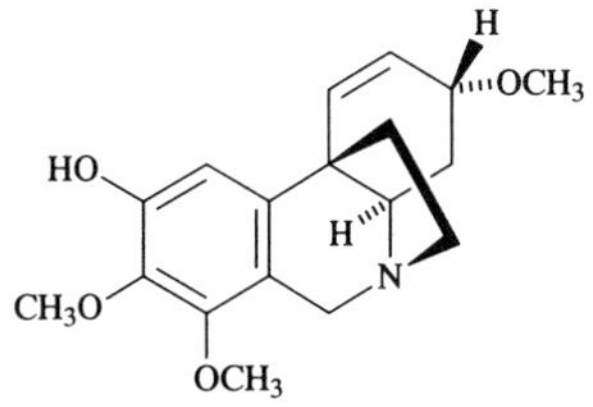

Amaryllidaceae alkaloid
(Crinan)

$C_{18}H_{23}NO_4$ [6874-70-0] MW 317.39

Brunsvigia rosea (Amaryllidaceae).

Toxic to humans. Shows pharmacodynamic activity but too toxic for use as a medicine.

Burlingame, A. L., *J. Am. Chem. Soc.,* 1964, **86**, 4976.

Ambelline 71

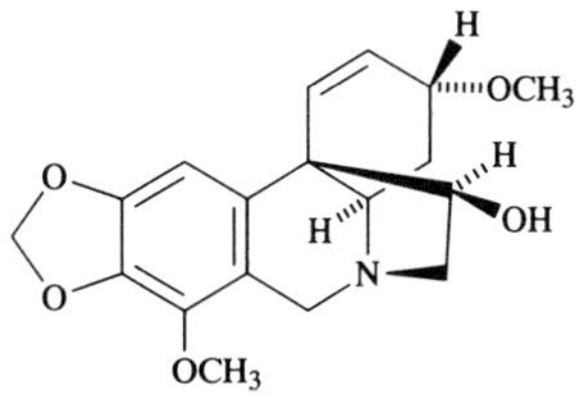

Amaryllidaceae alkaloid
(Crinan)

$C_{18}H_{21}NO_5$ [3660-62-6] MW 331.37

Amaryllis belladonna, *Boophane fischeri*, *Crinum laurentii* and *Hippeastrum* spp. (Amaryllidaceae).

Analgesic activity similar to morphine but too toxic to be used as a medicine; LD_{50} 5 mg/kg body-weight administered subcutaneously to mice.

Naegeli, P., *J. Org. Chem.,* 1963, **28**, 206.

Ambrosin 72

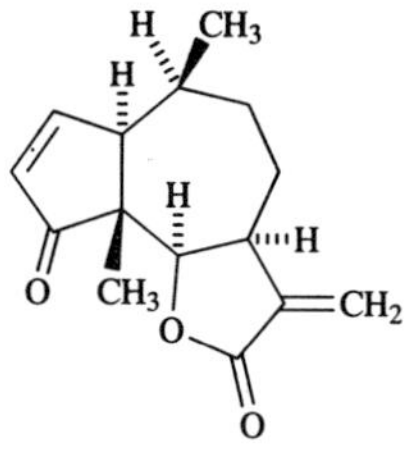

Sesquiterpenoid lactone
(Pseudoguaianolide)

$C_{15}H_{18}O_3$ [509-93-3] MW 246.31

Common ragweed, *Ambrosia artemisiifolia*, other *Ambrosia* spp. and in some *Hymenoclea*, *Iva* and *Parthenium* spp. (Compositae).

Cytotoxic and antitumour activities. Also active against the human parasite *Schistosoma haematobium*. Probably responsible, with other sesquiterpene lactones present, for ragweed dermatitis.

Torrance, S. J., *J. Pharm. Sci.,* 1975, **64**, 887.

Amijitrienol 73

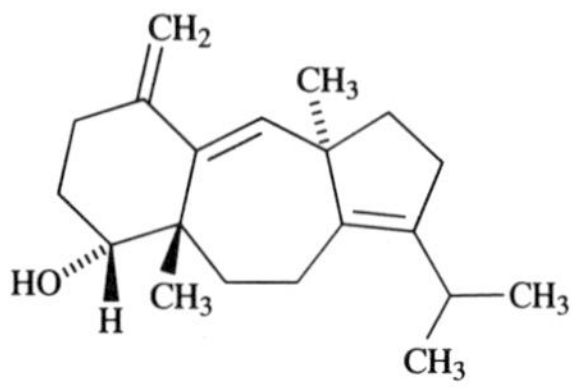

Diterpenoid

$C_{20}H_{30}O$ [87745-20-8] MW 286.46

Brown alga, *Dictyota linearis*.

Antimicrobial activity.

Ochi, M., *Bull. Chem. Soc. Jpn,* 1986, **59**, 661.

γ-Aminobutyric acid 74

4-Aminobutanoic acid; 4-Aminobutyric acid; GABA; Piperidic acid; Piperidinic acid

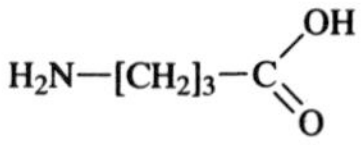

$H_2N—[CH_2]_3—C(=O)OH$

Amino acid

$C_4H_9NO_2$ [56-12-2] MW 103.12

Widespread in seeds, including *Pisum*, *Vicia* and *Phaseolus* spp. (Leguminosae).

Neurotoxic to birds. Inhibitory transmitter at the neuromuscular junction in the central nervous system. Has been used to treat cerebral disorders, including coma, and is antihypertensive.

Synge, R. L. M., *Biochem. J.,* 1951, **48**, 429.

L-α-Amino-γ-oxalylaminobutyric acid 75

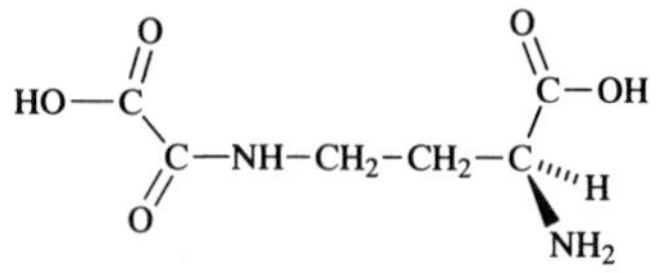

Amino acid

$C_6H_{10}N_2O_5$ [5302-43-2] MW 190.16

Seeds of *Acacia* spp. and of the everlasting pea, *Lathyrus latifolius* (Leguminosae).

Causes neurolathyrism, a neurotoxic syndrome, in humans and domestic animals. This condition may be permanent and cause death.

Bell, E. A., *Phytochemistry,* 1966, **5**, 1211.

L-α-Amino-β-oxalylaminopropionic acid 76

Dencichin

Amino acid

$C_5H_8N_2O_5$ [7554-90-7] MW 176.13

In all parts, including seeds, of *Lathyrus* spp., especially the chickling vetch, *L. sativus* (Leguminosae). Also in *Crotalaria* spp. (Leguminosae) and in *Panax notoginseng* (Araliaceae).

Causes neurolathyrism, a neurotoxic syndrome, in humans and domestic animals. This can become permanent and lead to death. Neurolathyrism is common in times of famine in India, since there is a greater reliance on *L. sativus* as a food plant and the susceptibility of individuals is thus increased.

Bell, E. A., *Phytochemistry,* 1966, **5**, 1211.

Ammodendrine 77

N-Acetyltetrahydroanabasine

(–)-form

Piperidine alkaloid

$C_{12}H_{20}N_2O$ [27542-15-0] MW 208.30
(+)-form [494-15-5]
(±)-form [20824-32-2]

Ammodendron conollyi, several *Sophora* spp., e.g. *S. franchetiana*, *Genista hystrix*, *Thermopsis lupinoides* and *Lupinus formosus* (all Leguminosae).

Toxic, especially to insects. Causes teratogenicity in cows which have been feeding on *Lupinus formosus.*

Tashkhodzhaev, B., *Chem. Nat. Compd.,* 1982, **36**, 631.

Ammothamnine 78

Matrine *N*-oxide; Oxymatrine; Pachycarpidine

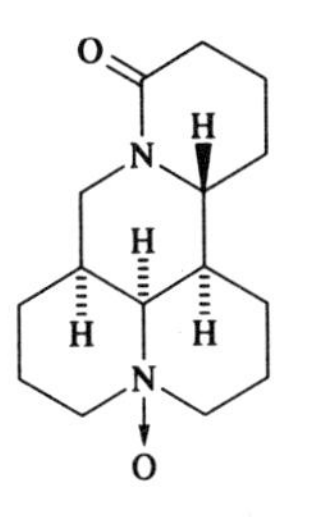

Quinolizidine alkaloid
(Matridine)

$C_{15}H_{24}N_2O_2$ [16837-52-8] MW 264.37

Ammothamnus lehmannii, *A. songorica* and *Sophora flavescens* (Leguminosae).

Toxic.

Ibragimov, B. T., *Kristallografiya.,* 1979, **24**, 45.

Amurensine 79

Xanthopetaline

Benzylisoquinoline alkaloid
(Isopavine type)

$C_{19}H_{19}NO_4$ [10481-92-2] MW 325.37

Papaver alpinum, *P. nudicaule* var. *amurense* and several other *Papaver* spp. (Papaveraceae).

Ameliorates pain. Also used as an expectorant and as a tranquilliser.

Šantavý, F., *Collect. Czech. Chem. Commun.,* 1966, **31**, 4286.

Amurine 80

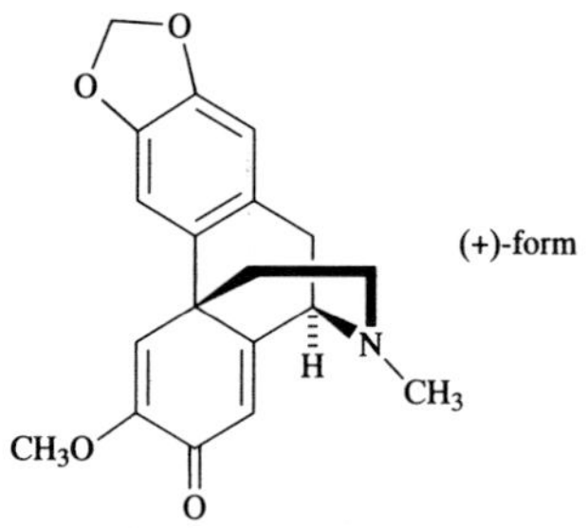

(+)-form

Benzylisoquinoline alkaloid
(Morphinan)

$C_{19}H_{19}NO_4$ [4984-99-0] MW 325.36
(−)-form [77449-68-4]

Papaver auranticum and *P. nudicaule* var. *amurense* (Papaveraceae).

Uses as for amurensine (q.v.).

Döpke, W., *Tetrahedron,* 1968, **24**, 4459.

Amygdalin 81

Amygdaloside; Mandelonitrile β-gentiobioside

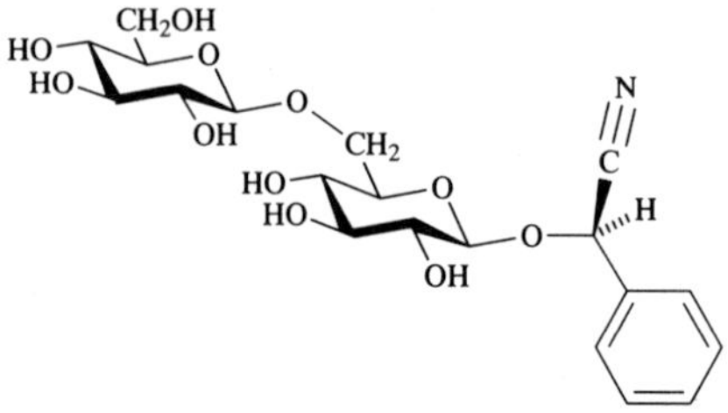

Cyanogenic glycoside

$C_{20}H_{27}NO_{11}$ [29883-15-6] MW 457.43

In bitter almonds, seeds of *Prunus amygdalus* and in the seeds of many other members of the Rosaceae. Also in subterranean parts of *Gerbera jamesonii* cultivars (Compositae).

Is responsible for the bitterness and toxicity of the seed of bitter almond (100 μmol/g) and of apricots (20-80 μmol/g). Minimum lethal dose of hydrogen cyanide in humans is 0.5-3.5 mg/kg body-weight. Has been used, interchangeably with mandelonitrile β-glucuronide, as the drug laetrile for treating cancer, but is ineffective.

Hawarth, W. N., *J. Chem. Soc.,* 1923, 3120.

(−)-Anabasine 82

Nicotimine; Nicotinimine

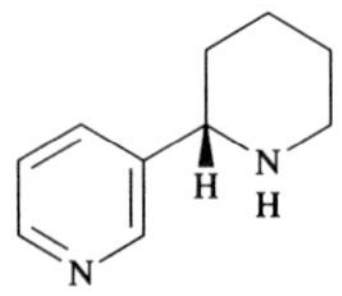

Pyridine alkaloid

$C_{10}H_{14}N_2$ [494-52-0] MW 162.23

Anabasis aphylla (Chenopodiaceae), *Nicotiana acuminata* and other *Nicotiana* spp., *Duboisia myoporoides* (Solanaceae), *Zinnia elegans*, *Zollikoferia eliquiensis* (Compositae) and *Alangium* spp. (Alangiaceae).

Acute and subacute toxicity. It has muscle stimulating action. A major use is to discourage the smoking of tobacco.

Orechoff, A., *Ber. Dtsch. Chem. Ges.,* 1932, **65**, 232.

Anacardic acid 83

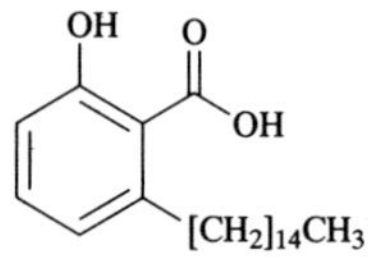

Phenolic acid

$C_{22}H_{36}O_3$ [16611-84-0] MW 348.53

Shell liquid of the cashew nut, *Anacardium occidentale* (Anacardiaceae).

One of four closely related 6-alkylated 2-hydroxybenzoic acids in the shell of the cashew nut, responsible for the acute dermatitis caused in humans handling these nuts. Is a prostaglandin synthase inhibitor and has antitumour activity.

Tyman, J. H. P., *Chem. Soc. Rev.,* 1979, **8**, 499.

Highly toxic and teratogenic. It is responsible for the 'crooked calf' disease caused by the ingestion of leguminous plants which contain this alkaloid. It is a cardiotonic agent, inducing tachycardia.

Keeler, R. F., *J. Toxicol. Environ. Health*, 1976, **1**, 887.

Anacrotine 84

Crotalaburnine

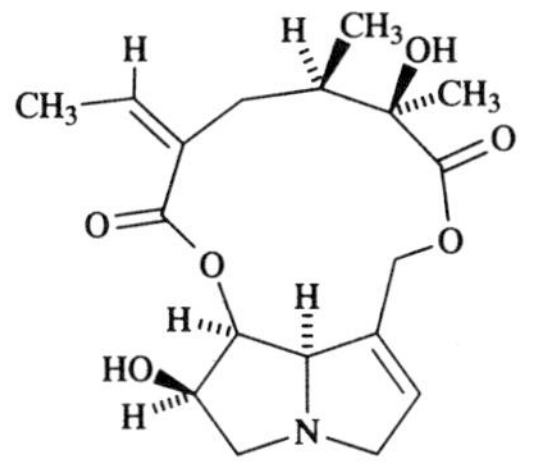

Pyrrolizidine alkaloid
(Senecionan)

$C_{18}H_{25}NO_6$ [5096-49-1] MW 351.40

Crotalaria laburnifolia, *C. anagyroides* and *C. incana* (Leguminosae).

Hepatotoxin and pneumotoxin.

Culvenor, C. C. J., *Chem. Biol. Interact.,* 1976, **12**, 299.

Anagyrine 85

Alkaloid III; Monolupine; Rhombinine

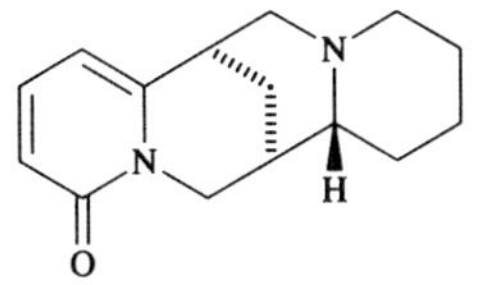

Quinolizidine alkaloid
(Sparteine)

$C_{15}H_{20}N_2O$ [486-89-5] MW 244.34

Seeds of *Anagyris foetida*, *Ulex europaeus* and *Thermopsis chinensis*, whole plant of *Cytisus*, *Genista*, *Lupinus*, *Sophora* and *Ammodendron* spp. (all Leguminosae).

Anatoxin a 86

Very fast death factor

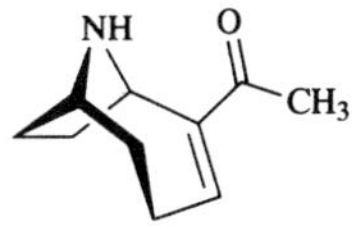

Homotropane alkaloid

$C_{10}H_{15}NO$ [64285-06-9] MW 165.24

Produced by strains of *Anabaena* spp. and *Oscillatoria* spp., a bloom of the latter being responsible for a recent outbreak of poisoning of dogs in Scotland.

A potent post-synaptic depolarising neuromuscular blocking agent. Death can follow ingestion within minutes to a few hours after a sequence of clinical symptoms which includes muscle fasciculations, reduced movement, cyanosis and convulsions.

Codd, G. A., *Nature (London),* 1992, **359**, 110.

Anatoxin a(s) 87

Salivation factor

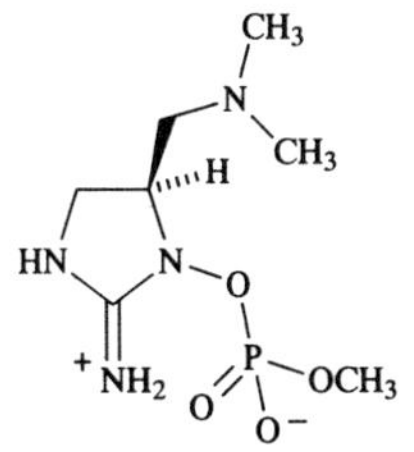

Cylic guanidinium phosphate ester

$C_7H_{17}N_4O_4P$ [103170-78-1] MW 252.21

Produced by strains of the freshwater species *Anabaena flos-aquae*.

A potent cholinesterase inhibitor responsible for a rapid neurotoxicosis in a wide range of animal species. Intraperitoneal LD_{50} in mice *ca* 20 μg/kg body-weight. Symptoms include hypersalivation and death is frequently due to respiratory arrest.

Falconer, I. R., Algal Toxins in Seafoods and Drinking Water, 1993, 187, Academic Press.

Androcymbine 88

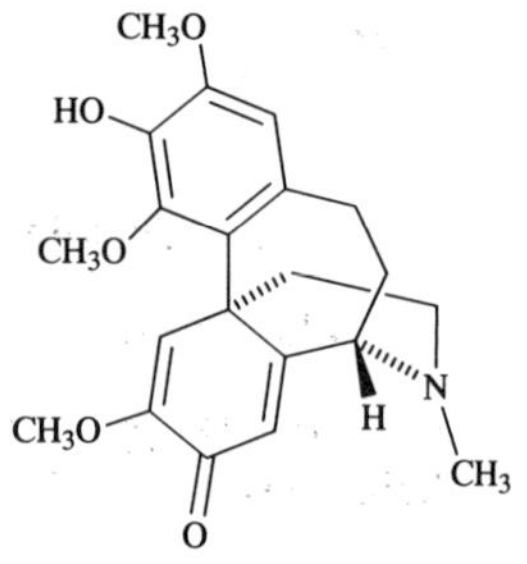

Phenethylisoquinoline alkaloid
(Homomorphinan)

$C_{21}H_{25}NO_5$ [2115-98-2] MW 371.43

Leaves of *Androcymbium melanthoides* var. *stricta* (Liliaceae).

Highly toxic. Is also an irritant, carcinogen and a teratogen.

Battersby, A. R., *J. Chem. Soc., Perkin Trans. 1.*, 1972, 1736.

Anethole 89

Anise camphor; *p*-Propenylanisole

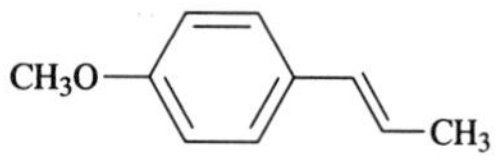

Phenylpropanoid

$C_{10}H_{12}O$ [104-46-1] MW 148.20

In essential oils of *Pimpinella anisum* and fennel, *Foeniculum vulgare* (both Umbelliferae), of *Clausenia anisata* and *Pelea christophersenii* (both Rutaceae), of *Backhousia anisata* (Myrtaceae) and of *Magnolia salicifolia* (Magnoliaceae). Also in the roots of *Artemisia porrecta* and *Aster tartaricus* (Compositae), in *Juniperus rigida* (Cupressaceae) and *Illicium anisatum* (Illiciaceae).

Sweet taste. Stimulates hepatic regeneration in rats and shows spasmolytic activity. Used as a flavouring agent. Moderately toxic.

Naves, Y. R., *Helv. Chim. Acta,* 1960, **43**, 230.

Angelicin 90

Isopsoralen

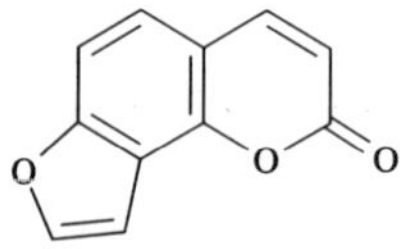

Furanocoumarin

$C_{11}H_6O_3$ [523-50-2] MW 186.17

Roots of *Angelica archangelica*, *Heracleum* spp. and *Selinum vaginatum* (Umbelliferae). Also in seeds of *Psoralea corylifolia* (Leguminosae) and *Castanopsis indica* (Fagaceae) and in *Ficus nitida* (Moraceae).

Slight photosensitising activity. Also shown to have spasmolytic activity. Is a weak photocarcinogen.

Steck, W., *Can. J. Chem.,* 1969, **47**, 2425.

Angularine 91

13,19-Didehydrorosmarinine

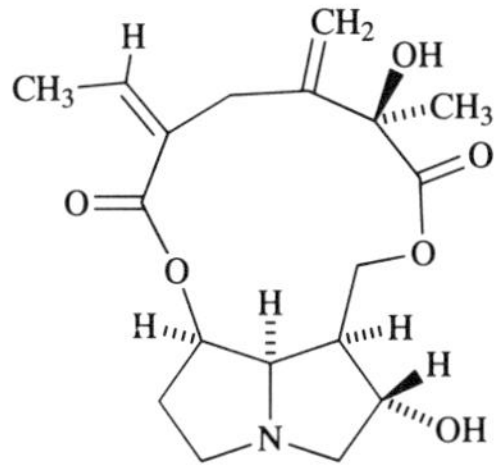

Pyrrolizidine alkaloid
(Senecionan)

$C_{18}H_{25}NO_6$ [1354-37-6] MW 351.40

Senecio angulatus (Compositae).

Hepatotoxic alkaloid causing necrosis of the liver in cattle feeding on this *Senecio* species.

Porter, L. A., *J. Org. Chem.,* 1962, **27**, 4132.

Angustibalin 92

Helenalin acetate

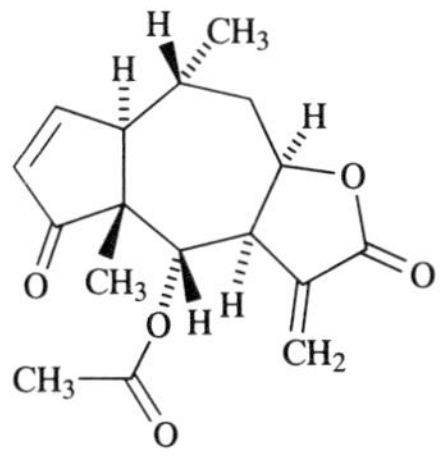

Sesquiterpenoid lactone
(Pseudoguaianolide)

$C_{17}H_{20}O_5$ [10180-86-6] MW 304.34

Balduina angustifolia (Compositae).

Cytotoxic and antitumour activities.

Lee, K. H., *J. Pharm. Sci.,* 1972, **61**, 626.

Angustine 93

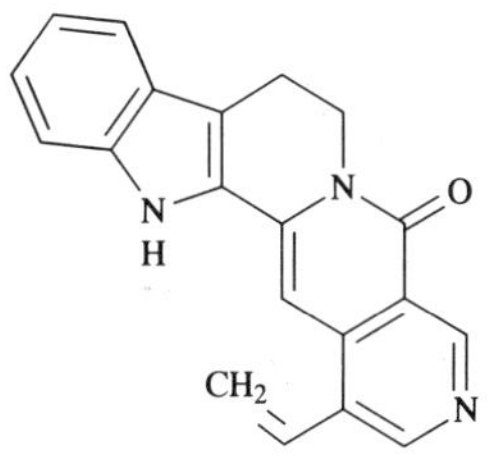

Indole alkaloid
(Azayohimbane)

$C_{20}H_{15}N_3O$ [40041-96-1] MW 313.36

Leaves of *Strychnos angustifolia* (Loganiaceae).

Toxic.

Phillipson, J. D., *Phytochemistry,* 1974, **13**, 973.

Anhydroglycinol 94

3,9-Dihydroxypterocarp-6a-ene

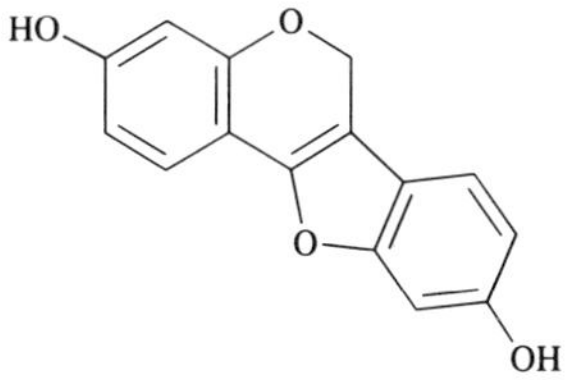

Isoflavonoid
(Pterocarpan)

$C_{15}H_{10}O_4$ [67685-22-7] MW 254.24

In fungally-infected leaves of the winged pea, *Tetragonolobus maritimus* (Leguminosae).

Antifungal properties (phytoalexin).

Ingham, J. L., *Phytochemistry,* 1978, **17**, 535.

Anisatin 95

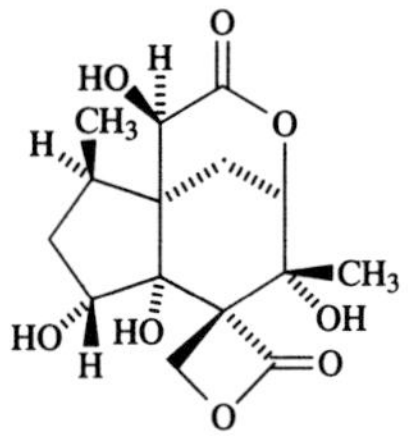

Sesquiterpenoid lactone

$C_{15}H_{20}O_8$ [5230-87-5] MW 328.32

Seeds of Japanese star anise, *Illicium anisatum* (Illiciaceae).

Toxic to humans, causing convulsions.

Yamada, K., *Tetrahedron Lett.,* 1965, 4797.

Anisodamine 96

Alkaloid V; 6β-Hydroxyhyoscyamine

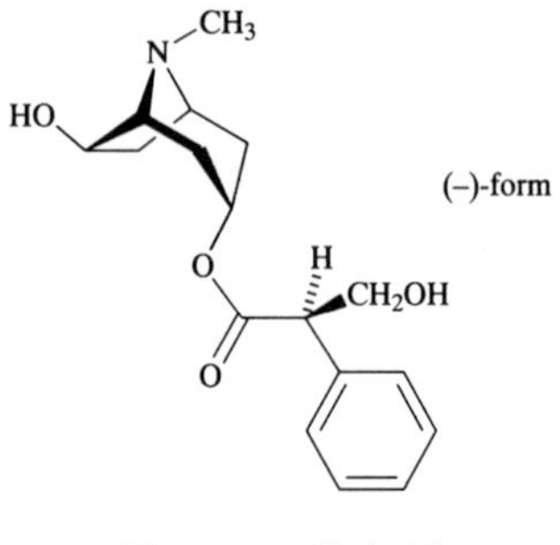

(−)-form

Tropane alkaloid

$C_{17}H_{23}NO_4$ [17659-49-3] (−)-form [55869-99-3] MW 305.37

Datura spp. (Solanaceae).

Reduces acute myocardial infarction in rabbits, prevents haemorrhagic shock in cats and endotoxin-induced shock in dogs. Used as a biochemical tool in studying the mode of action of anaesthetics, because of its effects on liposome formation.

Romeike, A., *Naturwissenschaften,* 1962, **49**, 281.

Anonaine 97

Annonaine

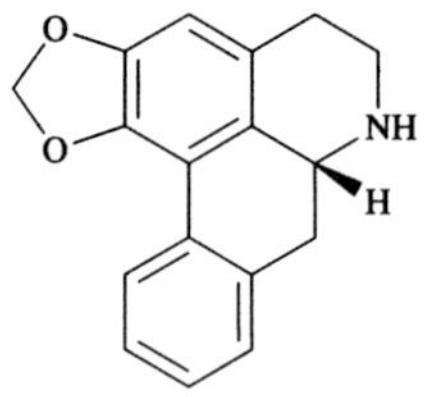

Benzylisoquinoline alkaloid (Aporphine)

$C_{17}H_{15}NO_2$ [1862-41-5] MW 265.31

Annona reticulata (Annonaceae) and *Nelumbo nucifera* (Nelumbonaceae).

Insecticidal.

Marion, L., *Can. J. Research,* 1950, **28B**, 21.

Anopterine 98

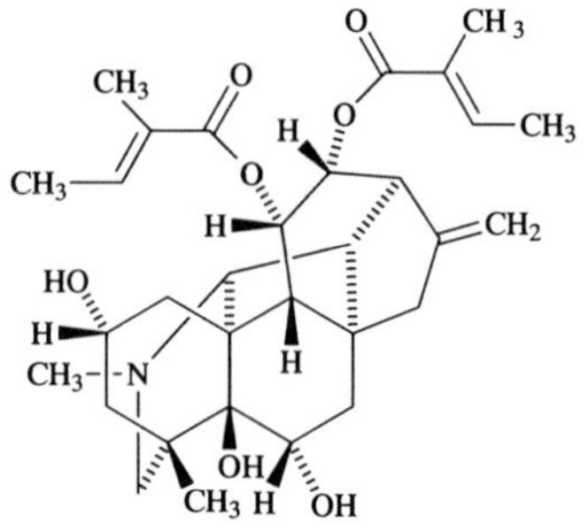

Diterpenoid alkaloid (Cycloveatchane)

$C_{31}H_{43}NO_7$ [38826-62-9] MW 541.68

Leaves and bark of *Anopterus macleayanus*, bark of *Anopterus glandulosus* (Saxifragaceae).

Antitumour activity in mouse leukaemia assays with P-388 and KB-systems.

Hart, N. K., *Aust. J. Chem.,* 1976, **29**, 1295.

Anthemis glycoside A 99

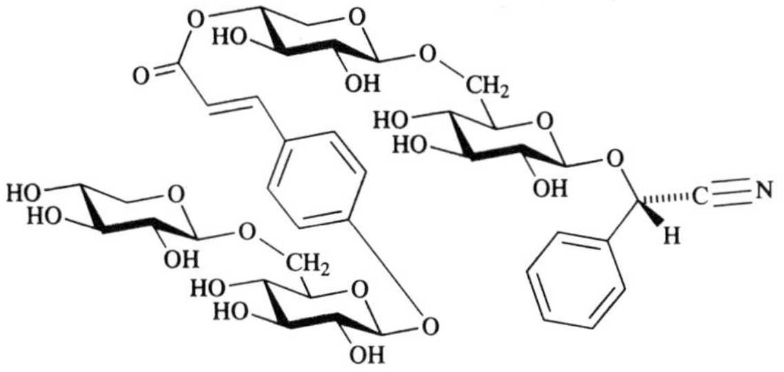

Cyanogenic glycoside

$C_{39}H_{49}NO_{21}$ [89354-48-3] MW 867.81

Cyanogenic achenes of *Anthemis altissima* and probably present in other *Anthemis* spp. (Compositae).

Toxic, releasing hydrogen cyanide on hydrolysis.

Nahrstedt, A., *Planta Med.,* 1983, **49**, 143.

Anthemis glycoside B 100

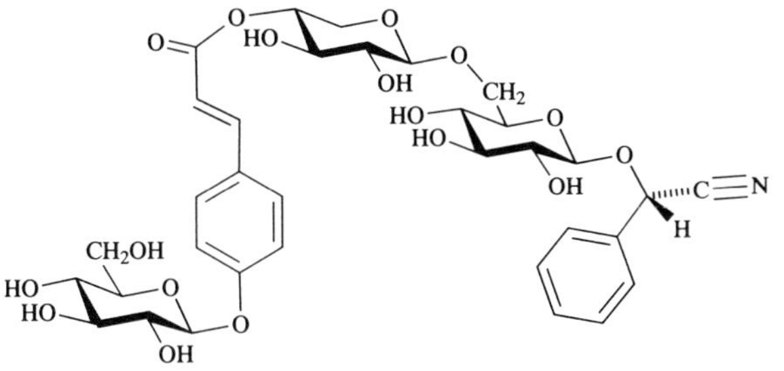

Cyanogenic glycoside

$C_{34}H_{41}NO_{17}$ [89354-49-4] MW 735.70

Cyanogenic achenes of *Anthemis altissima* and probably also present in other *Anthemis* spp. (Compositae).

Toxic, releasing hydrogen cyanide on hydrolysis.

Nahrstedt, A., *Planta Med.,* 1983, **49**, 143.

Anthranoyllycoctonine 101

Inuline

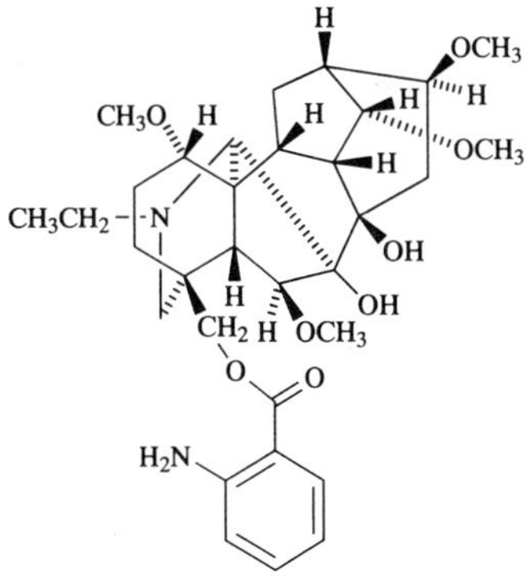

Diterpenoid alkaloid
(Aconitane)

$C_{32}H_{46}N_2O_8$ [22413-78-1] MW 586.72

Roots of *Delphinium barbeyi* and other *Delphinium* spp. (Ranunculaceae), and in *Inula royleana* (Compositae).

Toxic in animals, causing partial loss of motor control and respiratory paralysis, culminating in death preceded by violent convulsions. In mice, it has some potency as a neuromuscular blocking agent.

Pelletier, S. W., *J. Am. Chem. Soc.,* 1981, **103**, 6536.

Anthrogallol 102

Anthrogallic acid

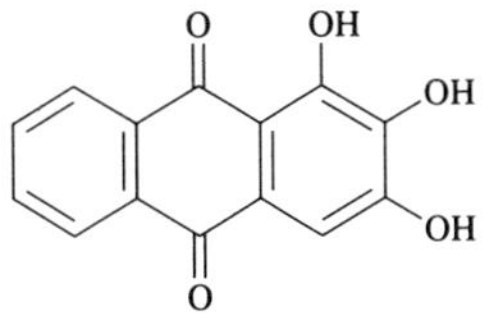

Anthraquinone

$C_{14}H_8O_5$ [602-64-2] MW 256.22

Roots of *Relbunium hypocarpium*, in *Coprosma lucida*, *Rubia tinctorum* and *Hymenodictyon excelsum* (all Rubiaceae).

Immunosuppressive activity *in vitro*; at higher dosage, it shows inhibitory and cytotoxic activity against macrophages, and T- and B-lymphocytes. Toxic properties.

Brew, E. J. C., *J. Chem. Soc. (C),* 1971, 2001.

α-Antiarin 103

Antiarigenin 3-antiaroside

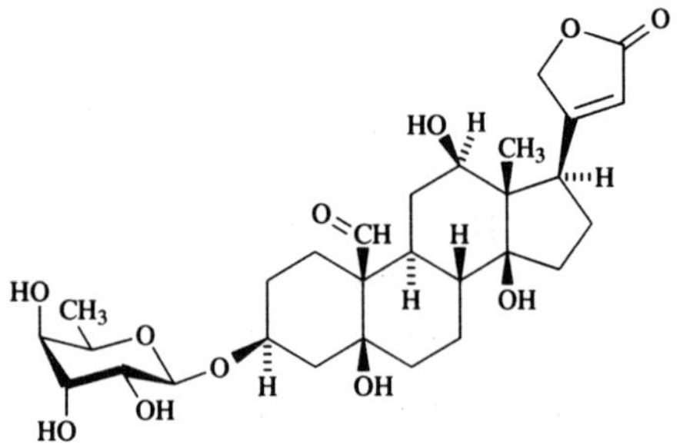

Cardenolide

$C_{29}H_{42}O_{11}$ [23605-05-2] MW 566.65

Latex of upas tree, *Antiaris toxicaria* and of other *Antiaris* spp. (Moraceae).

Very toxic in vertebrates. LD_{50} in cats is 0.116 mg/kg bodyweight. The latex of the upas tree has been used as an arrow poison, and for the purpose of public execution.

Juslen, C., *Helv. Chim. Acta,* 1963, **46**, 117.

Antioside 104

Antiogenin 3-rhamnoside

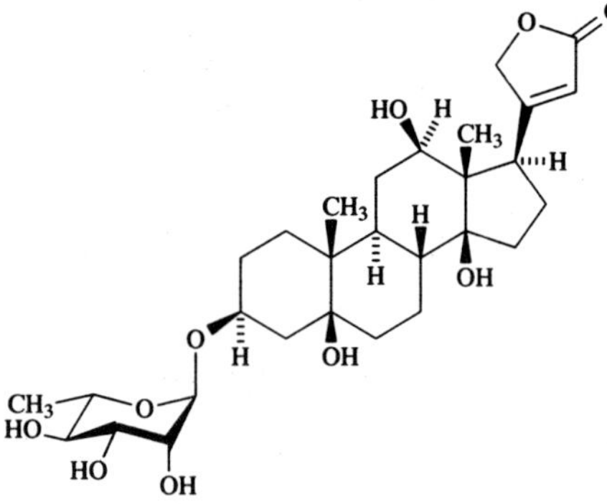

Cardenolide

$C_{29}H_{44}O_{10}$ [3981-16-6] MW 552.66

Latex of upas tree, *Antiaris toxicaria*, and of *A. welwitschii* and *A. decipiens* (Moraceae).

Very toxic to vertebrates. The latex of the upas tree has been used as an arrow poison.

Juslen, C., *Helv. Chim. Acta,* 1962, **45**, 2285.

Antofine 105

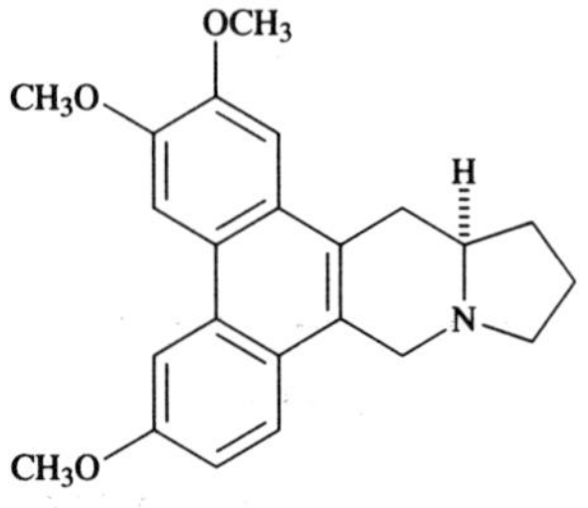

Phenanthroidolizidine alkaloid

$C_{23}H_{25}NO_3$ [32671-82-2] MW 363.46

Antitoxicum funebre, *Vincetoxicum officinale* (Asclepiadaceae) and *Ficus septica* (Moraceae).

Antifungal and antibacterial properties.

Govindachari, T. R., *J. Chem. Soc., Perkin Trans. 1,* 1974, 1161.

Apigenin 106

5,7,4′-Trihydroxyflavone

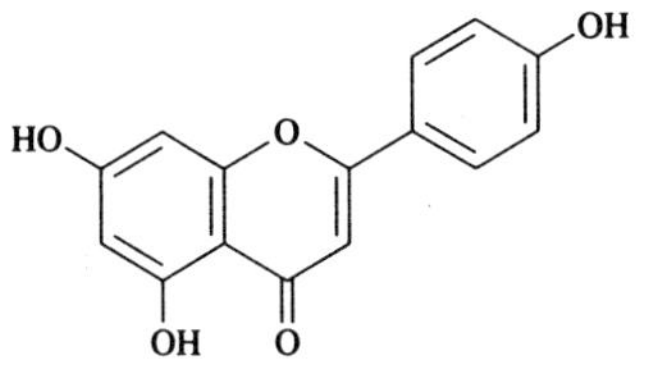

Flavone

$C_{15}H_{10}O_5$ [520-36-5] MW 270.24

On the leaf surface of certain plants, e.g. members of the Labiatae, and in the farinose exudates of fern fronds. Widely present in plants generally in glycosidic form.

Anti-inflammatory, diuretic and hypotensive activities. It inhibits many enzymes. Also, it promotes smooth muscle relaxation.

Harborne, J. B., The Flavonoids: Advances in Research since 1980, 1988, 191, Chapman & Hall.

Apigenin 7,4′-dimethyl ether 107

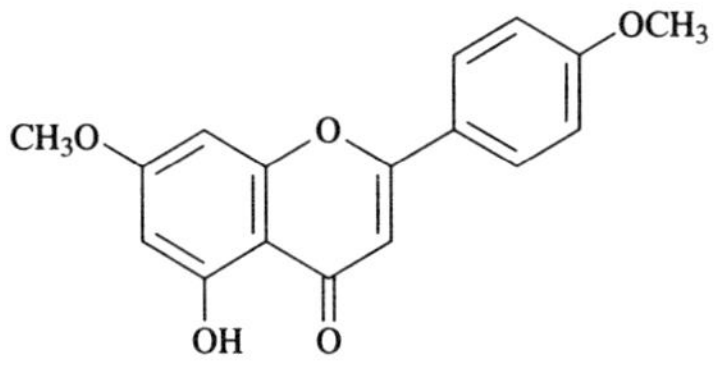

Flavone

$C_{17}H_{14}O_5$ [5128-44-9] MW 298.30

Leaf resin of *Cistus* spp. (Cistaceae), roots of *Rhus undulata* (Anacardiaceae), as an external flavonoid on *Thymus piperella* (Labiatae), *Striga asiatica* (Scrophulariaceae) and *Andrographis paniculata* (Acanthaceae). Many other plant sources are also known.

Anti-inflammatory and antigastric ulcer activities.

Harborne, J. B., The Flavonoids: Advances in Research since 1986, 1994, 263, Chapman & Hall.

Apiole 108

Apiol; Apioline; Parsley camphor

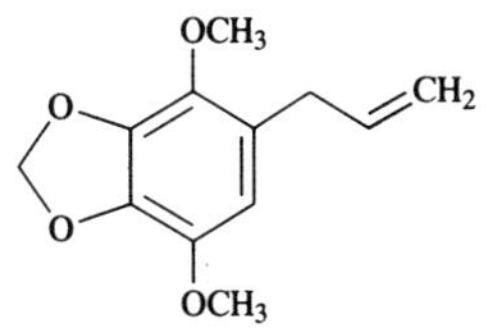

Phenylpropanoid

$C_{12}H_{14}O_4$ [523-80-8] MW 222.24

Essential oils of the seeds of parsley, *Petroselinum crispum*, the roots of *Crithmum maritimum* (both Umbelliferae), camphor wood, *Cinnamomum camphora* (Lauraceae) and the leaves of *Piper angustifolium* (Piperaceae).

In high doses, it may cause short-lived intoxication. Insecticidal and spasmolytic activities.

Sethi, M. L., *Phytochemistry,* 1976, **15**, 1773.

Aplysiatoxin 109

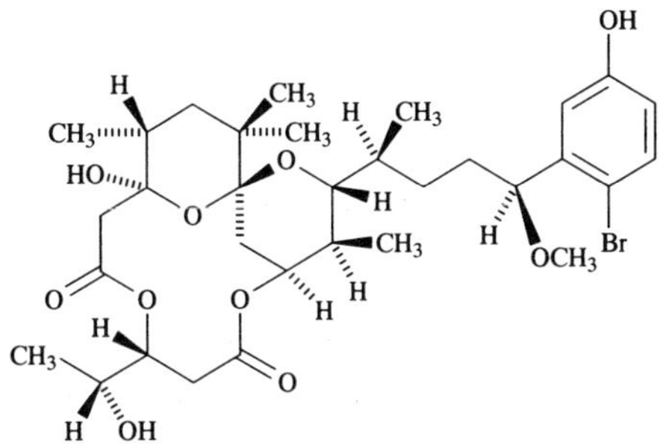

Phenolic bislactone

$C_{32}H_{47}BrO_{10}$ [52659-57-1] MW 671.62

First isolated from the sea hare, *Stylocheilus longicauda* (Aplysiomorpha), but subsequently shown to be produced by cyanobacteria on which this species feeds. Produced by the tropical and subtropical marine species *Lyngbia majuscula.*

Associated with the severe contact dermatitis known as swimmers' itch characterised by an erythematous, pustular folliculitis. Tumour promoting activity has been demonstrated in experimental animals. LD_{100} mg/kg body-weight in mice (intraperitoneal).

Kato, Y., *J. Am. Chem. Soc.,* 1974, **96**, 2245.

Aplysin 110

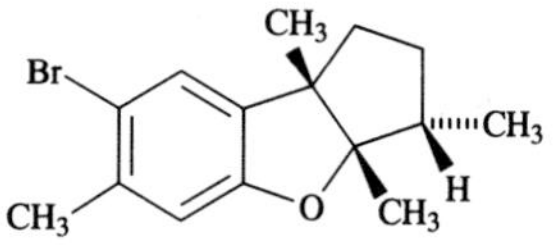

Sesquiterpenoid

$C_{15}H_{19}BrO$ [6790-63-2] MW 295.22

Red alga, *Laurencia pacifica*.

Toxic to marine molluscs.

Cameron, A. F., *J. Chem. Soc., Chem. Commun.*, 1967, 271.

Apoatropine 111

Atropamine; Atropyltropeine

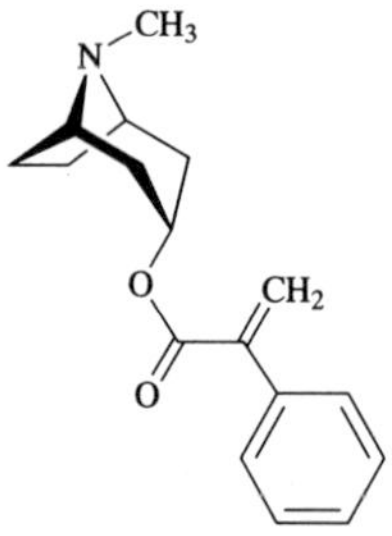

Tropane alkaloid

$C_{17}H_{21}NO_2$ [500-55-0] MW 271.36

Roots of *Atropa belladonna*, also in *Datura pruinosa*, *Hyoscyamus orientalis* and *Physoclaina alaica* (all Solanaceae).

In relatively small doses, it causes respiratory arrest in animals. Antispasmodic agent.

Kagei, K., *Yakugaku Zasshi*, 1980, **100**, 216.

Apoglaziovine 112

N-Methylsparsiflorine

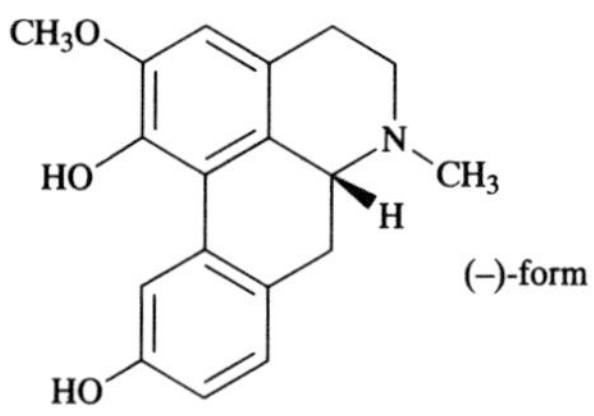

(–)-form

Benzylisoquinoline alkaloid
(Aporphine)

$C_{18}H_{19}NO_3$ [2128-77-0] MW 297.35
(+)-form [18058-59-8]
(±)-form [56261-23-5]

The (−)- and (±)-forms are from the leaves of *Ocotea glaziovii* and the (+)-form from *O. variabilis* (Lauraceae).

Hypotensive agent and tranquilliser.

Cava, M. P., *Tetrahedron Lett.*, 1972, 4647.

Apovincamine 113

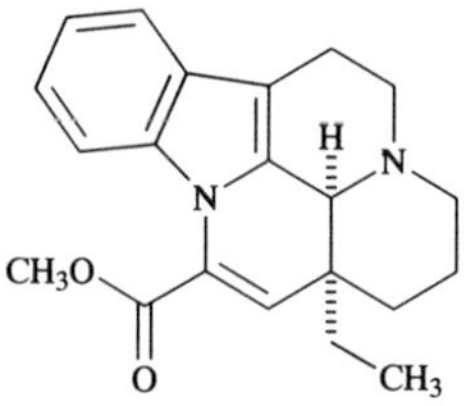

Indole alkaloid
(Eburnamenine)

$C_{21}H_{24}N_2O_2$ [4880-92-6] MW 336.43

Tabernaemontana riedelii and *T. rigida* (Apocynaceae).

Shows purgative and febrifuge activity, as well as central nervous system action.

Cava, M. P., *J. Org. Chem.*, 1968, **33**, 1055.

(−)-Apparicine 114

Gomezina; Pericalline; Tabernoschizine

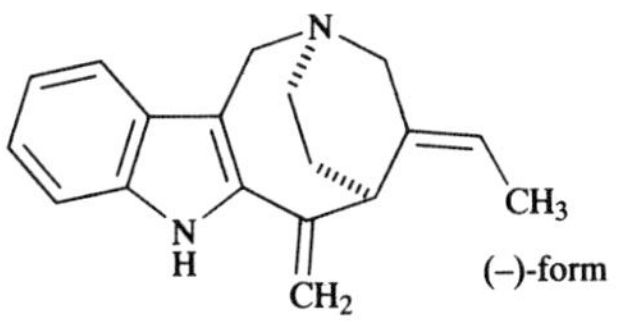

N H CH₂ CH₃ (−)-form

Indole alkaloid
(Vallesamine type)

$C_{18}H_{20}N_2$ [2122-36-3] MW 264.37
(+)-form [3463-93-2]

In *Aspidosperma dasycarpon*, *Tabernaemontana pachysiphon* and *T. divaricata* (Apocynaceae).

It shows pronounced analeptic properties, and is active against polio virus.

Gilbert, B., *Tetrahedron,* 1965, **21**, 1141.

Arborine 115

Glycosine

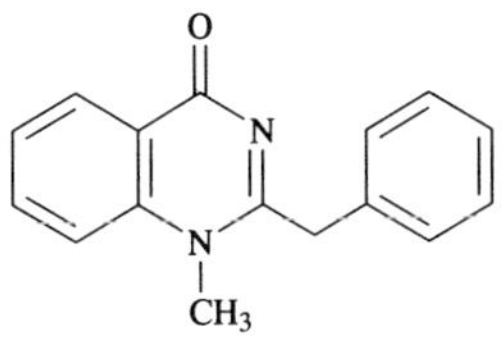

Quinazoline alkaloid

$C_{16}H_{14}N_2O$ [6873-15-0] MW 250.30

Leaves of *Glycosmis arborea,* of rue, *Ruta graveolens* and in the fruits of *Zanthoxylum budrunga* (Rutaceae).

Inhibits peripheral action of acetylcholine and acts as a central hypotensive agent. It shows an inhibitory action against pituitrin on rat uterus. An extract of *Glycosmis arborea* is used in folk medicine for treating fever, anaemia and jaundice.

Brossi, A., The Alkaloids, 1986, **29**, 99, Academic Press.

Arborinine 116

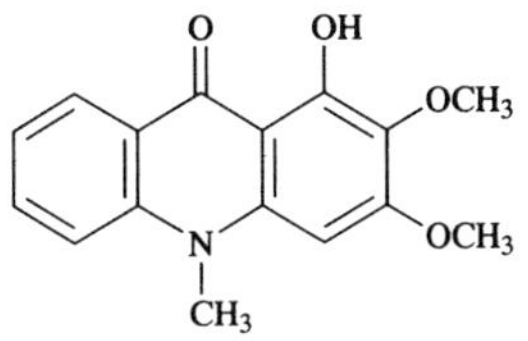

Acridone alkaloid

$C_{16}H_{15}NO_4$ [5489-57-6] MW 285.30

In *Glycosmis arborea*, *Fagara lepieurii*, *Euodia xanthoxyloides*, *Ruta graveolens* and *Teclea natalensis* (Rutaceae).

Spasmolytic activity.

Chakravati, D., *J. Chem. Soc.,* 1953, 3337.

Arbutin 117

Arbutoside; Ericolin; Hydroquinone glucoside

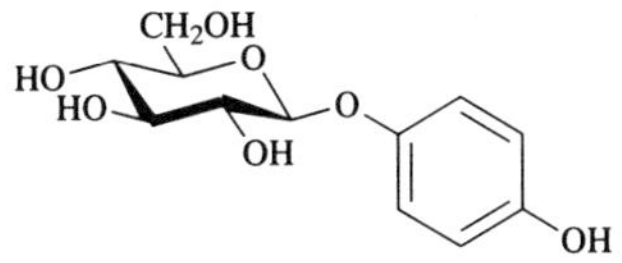

Phenol

$C_{12}H_{16}O_7$ [497-76-7] MW 272.25

Leaves of *Arctostaphylos uva-ursi*, *Vaccinium vitis-idaea* (Ericeae), *Bergenia crassifolia* (Saxifragaceae), *Pyrus communis* (Rosaceae) and *Origanum majorana* (Labiatae).

Diuretic, antitussive and urinary anti-infective activities. It is easily hydrolysed; the gallotanninsin crude plant extracts prevent hydrolysis and thus crude extracts are more effective than the pure compound. It inhibits insulin degradation.

Lutterbach, R., *Helv. Chim. Acta,* 1992, **75**, 2009.

Arcelin 118

Exists as a heterotetramer
M_r 140 000±10 000

Glycoprotein MW 32 000-36 000

Seed of common bean, *Phaseolus vulgaris* (Leguminosae).

Toxic to the bean weevil, *Zabrotes subfasciatus*. The toxicity is due to the stability of arcelin to enzymic hydrolysis and hence the bean weevil dies from starvation.

Minney, B. H. P., *J. Insect Physiol.,* 1990, **36**, 757.

Archangelicin 119

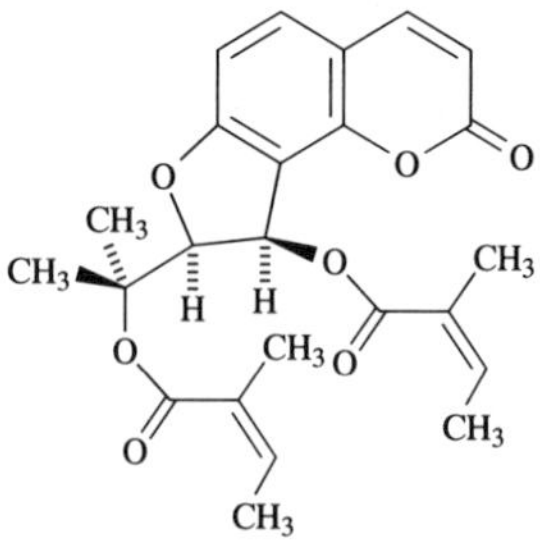

Furocoumarin

$C_{24}H_{26}O_7$ [2607-56-9] MW 426.47

Roots of *Angelica archangelica, A. keiskei, A. longeradiata* and *Cnidium japonicum* (Umbelliferae).

Spasmolytic activity.

Neilsen, B. E., *Acta Chem. Scand.,* 1964, **18**, 932.

(−)-Arctigenin 120

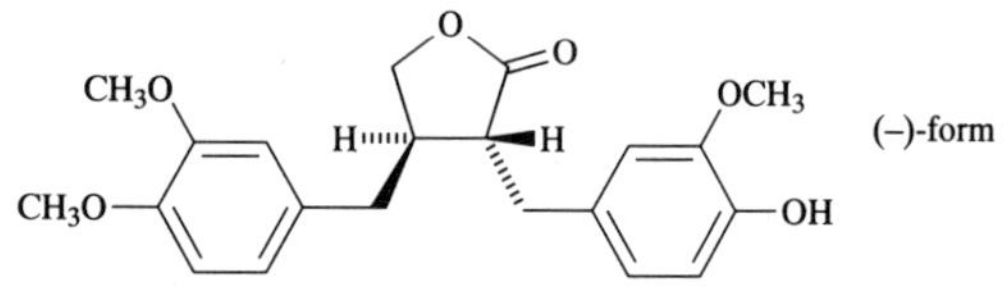

(-)-form

Lignan

$C_{21}H_{24}O_6$ [7770-78-7] MW 372.42
(+)-form [84413-77-4]

(−)-Form occurs in *Trachelospermum asiaticum* (Apocynaceae) and in *Ipomoea carica* (Convolvulaceae). The (+)-form is present in roots of *Wikstroemeria indica* (Thymeleaceae) and, as a glycoside, in fruits of *Arctium lappa, Lappa minor* and *L. tomentosa* (Compositae).

Inhibitor of cyclic adenosine monophosphate phosphodiesterase. It has cytostatic activity in lymphome cell systems.

Bruno, M., *Phytochemistry,* 1991, **30**, 4165.

Arctolide 121

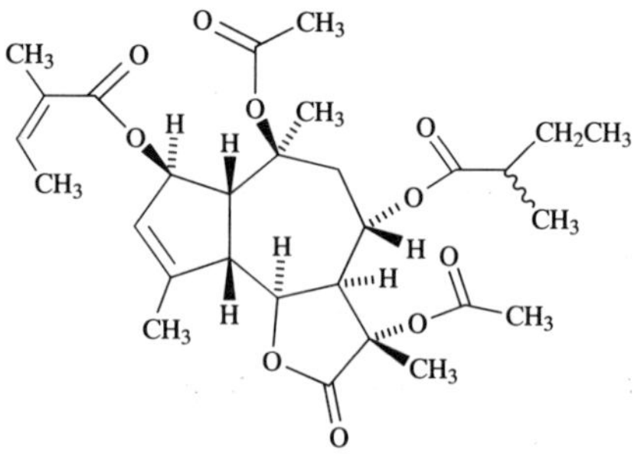

Sesquiterpenoid lactone
(Guaianolide)

$C_{17}H_{20}O_6$ [64390-62-1] MW 320.34

African daisy, *Arctotis grandis* (Compositae).

Cytotoxic and antitumour activities.

Samek, Z., *Collect. Czech. Chem. Comm.,* 1977, **42**, 2217.

Ardisianone 122

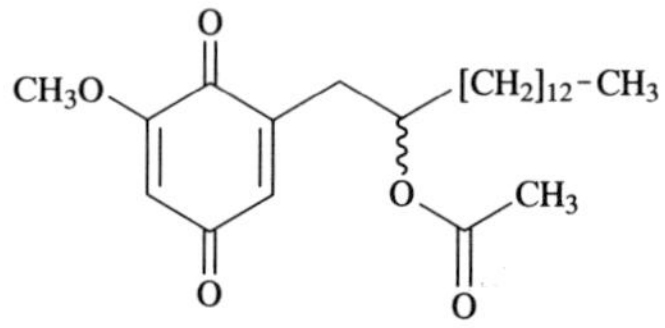

Benzoquinone

$C_{24}H_{38}O_5$ [66398-68-3] MW 406.56

Roots of *Ardisia cornudentata* and *A. quinquegona* (Myrsinaceae).

Inhibits the binding of leukotrienes in various receptor assays.

Chen, Y. P., *Bull. Chem. Soc. Jpn,* 1978, **51**, 943.

Arecoline 123

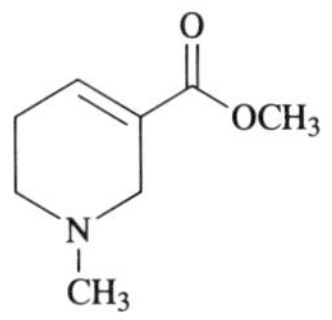

Piperidine alkaloid

$C_8H_{13}NO_2$ [63-75-2] MW 155.20

Major alkaloid of the betel nut, the seed of *Areca catechu* (Palmae).

Markedly toxic; in spite of this, the seeds are used medicinally. It shows a parasympathetic stimulant action and some anthelmintic activity. It is a teniacide for animals, a cathartic in horses and a ruminateric in cattle.

Mannich, C., *Ber. Dtsch. Chem. Ges. B,* 1942, **75**, 1480.

L-Arginine 124

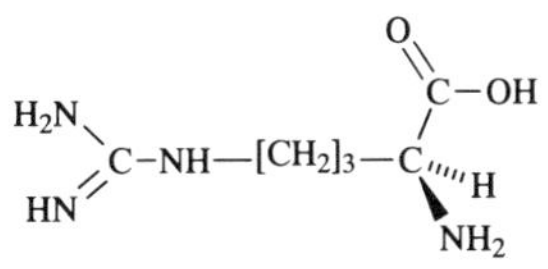

Amino acid

$C_6H_{14}N_4O_2$ [74-79-3] MW 172.20

As a protein amino acid, it is found in all plants. It may be isolated from leaves of alfalfa (lucerne), *Medicago sativa* (Leguminosae), asparagus, *Asparagus officinale* (Liliaceae) and fenugreek, *Trigonella foenum-graecum* (Leguminosae).

It has been used to treat hyperammonaemia. Arginine stimulates the release of growth hormone from the pituitary gland in man.

Greenstein, J. P., Chemistry of the Amino Acids, Part 3, 1961, 1841, Wiley.

Aristolindiquinone 125

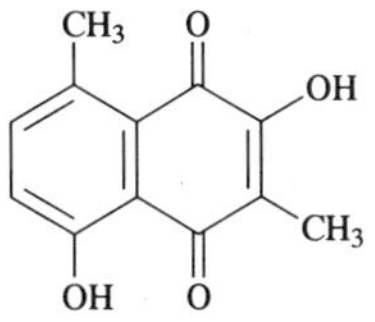

Naphthoquinone

$C_{12}H_{10}O_4$ [86533-36-0] MW 218.21

Isolated from the roots of *Aristolochia indica* (Aristolochiaceae).

Antifertility activity.

Che, C., *Tetrahedron Lett.,* 1983, **24**, 1333.

Aristolochic acid 126

Aristolochic acid-I; Aristolochic acid A

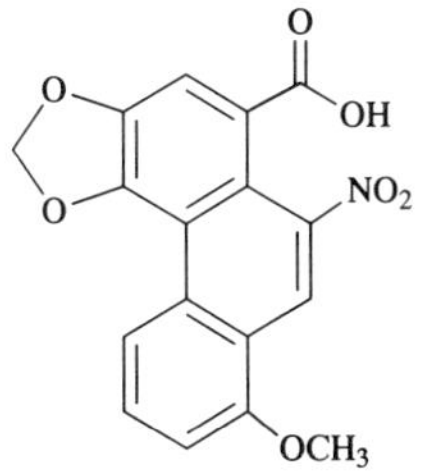

Aromatic nitro compound

$C_{17}H_{11}NO_7$ [313-67-7] MW 341.28

Birthwort, *Aristolochia clematis*, Indian birthwort, *A. indica*, long birthwort, *A. longa* and wild ginger, *Asarum canadense* (Aristolochiaceae). It co-occurs with several closely related structures, which vary in the number and position of hydroxy or methoxy substituents.

Toxic to birds, since the butterfly *Battus archidamus* sequesters and stores it from the food plants of the *Aristolochia* species. It exhibits anti-inflammatory and antifertility activity in animals and is an *in vitro* tumour growth inhibitor. It is also carcinogenic in animals.

Mix, D. B., *J. Nat. Prod.,* 1982, **45**, 657.

Armepavine 127

Evoeuropine

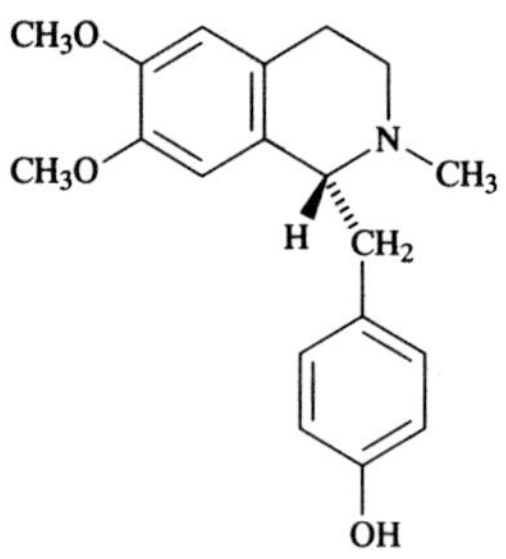

Benzylisoquinoline alkaloid

$C_{19}H_{23}NO_3$ [524-20-9] MW 313.40

In *Papaver caucasicum* and *P. persicum* (Papaveraceae), *Euonymus europaea* (Celastraceae), *Rhamnus frangula* (Rhamnaceae) and *Nelumbo nucifera* (Nymphaeaceae).

Convulsive and irritant agent; produces cardiac irregularity.

Bishay, D. W., *Phytochemistry,* 1973, **12**, 693.

Arnebinol 128

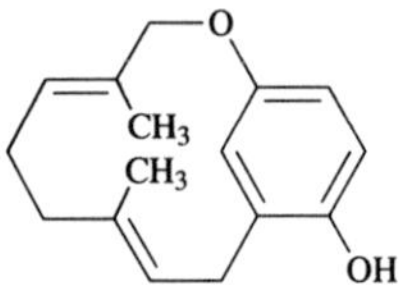

Phenol

$C_{16}H_{20}O_2$ [87064-17-3] MW 244.33

Found in the roots of *Arnebia euchroma* (Boraginaceae).

Inhibits prostaglandin biosynthesis.

Xin-Sheng, Y., *Tetrahedron Lett.,* 1983, **24**, 2407.

Arnebinone 129

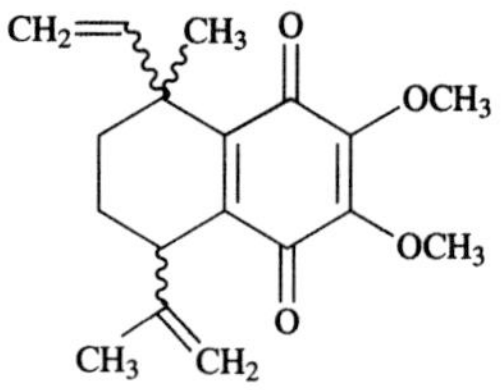

Benzoquinone

$C_{18}H_{22}O_4$ [87255-09-2] MW 302.37

Found in the roots of *Arnebia euchroma* (Boraginaceae).

Inhibits prostaglandin biosynthesis.

Xin-Sheng, Y., *Tetrahedron Lett.,* 1983, **24**, 3247.

Arnicolide A 130

Dihydrohelenalin acetate

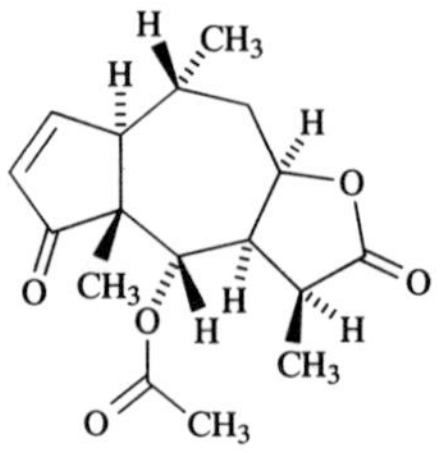

Sesquiterpenoid lactone
(Pseudoguaianolide)

$C_{17}H_{22}O_5$ [36505-53-0] MW 306.36

In *Arnica montana* and *A. longifolia* (Compositae).

Has cytotoxic and antitumour activities.

Willuhn, G., *Planta Med.,* 1979, **37**, 325.

Aromaticin 131

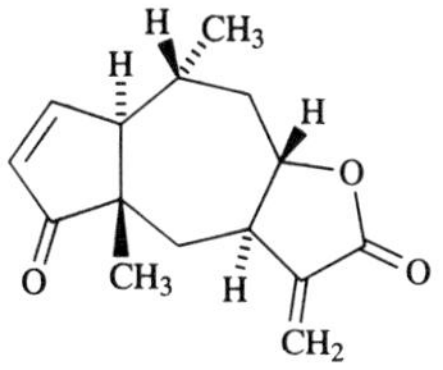

Sesquiterpenoid lactone
(Pseudoguaianolide)

$C_{15}H_{18}O_3$ [5945-42-6] MW 246.31

In *Helenium aromaticum* and *H. amarum* (Compositae).

Cytotoxic and antitumour activities.

Romo, J., *Tetrahedron,* 1964, **20**, 79.

Aromoline 132

Thalicrine

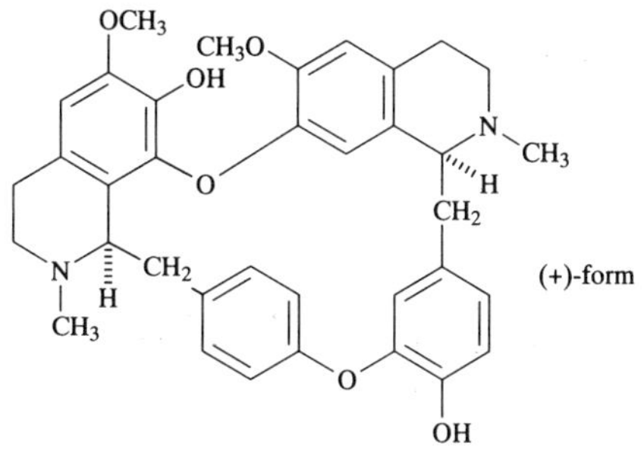

(+)-form

Bisbenzylisoquinoline alkaloid

$C_{36}H_{38}N_2O_6$ [519-53-9] MW 594.71
(−)-form [66288-77-5]

Bark of *Daphnandra aromatica*, *D. tennipes* and *Doryphora aromatica* (Monimiaceae); leaves of *Triclisia patens* (Menispermaceae); roots of *Thalictrum lucidum* (Ranunculaceae) and *Berberis orthobotrys* (Berberidaceae). The (−)-isomer is macolidine [66288-77-5].

Antiprotozoal activity.

Bick, I. R. C., *J. Chem. Soc.,* 1960, 4928.

Arteglasin A 133

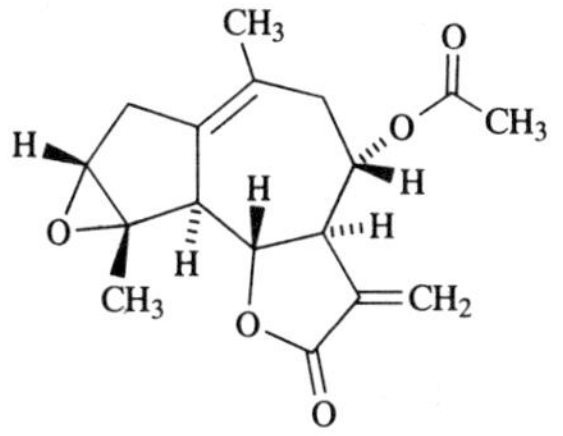

Sesquiterpenoid lactone
(Guaianolide)

$C_{17}H_{20}O_5$ [33204-39-6] MW 304.34

Found in *Artemisia douglasiana* and the garden chrysanthemum, *Dendranthema indicum* (Compositae).

It causes allergic contact dermatitis in humans. Has cytotoxic and antitumour activities.

Lee, K. H., *Phytochemistry,* 1971, **10**, 405.

Arvenososide A 134

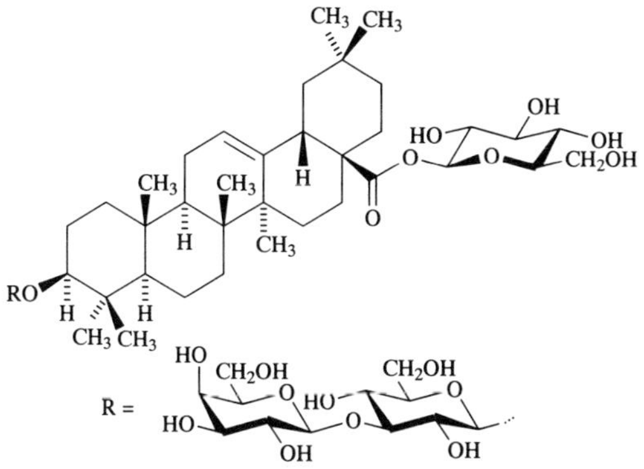

Triterpenoid saponin
(Oleanane)

$C_{48}H_{78}O_{18}$ [110219-89-1] MW 943.14

In the aerial parts of *Calendula arvensis* (Compositae).

Anti-inflammatory activity.

Chemli, R., *Phytochemistry,* 1987, **26**, 1785.

β-Asarone 135

Asarone; *cis*-Asarone; Asarin; Asarum camphor; Asarabacca camphor

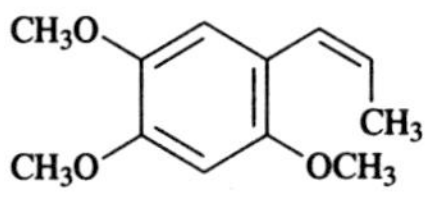

Phenylpropenoid

$C_{12}H_{16}O_3$ [5273-86-9] MW 208.26
α-form [2883-98-9]

In the essential oil of the roots of *Acorus calamus* (Araceae), and of *Asarum europaeum* (Aristolochiaceae), and in *Piper angustifolium* (Piperaceae). It occurs both in the *cis*-form (β-asarone) and in the *trans*-form (α-asarone).

β-Asarone is carcinogenic in animals and calamus oil has been banned from use in the USA. Drugs containing only the α-form are preferred for use in pharmacy. β-Asarone has spasmolytic activity. It also shows anti-algal activity, is an insect chemosterilant and a strong insect attractant.

Patra, A., *J. Nat. Prod.,* 1981, **44**, 668.

Ascaridole 136

Ascaridol; Ascapurin; Ascarisin

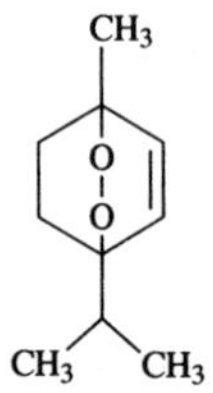

Monoterpenoid
(Menthane)

$C_{10}H_{16}O_2$ [512-85-6] MW 168.24

Principal constituent of the oil of chenopodium, the volatile oil from fresh flowering and fruiting plants of American wormseed, *Chenopodium ambrosioides* var. *anthelminticum* (Chenopodiaceae).

Toxic to mammals. Possesses anthelmintic activity.

Bernhard, R. A., *Phytochemistry,* 1971, **10**, 177.

Asclepin 137

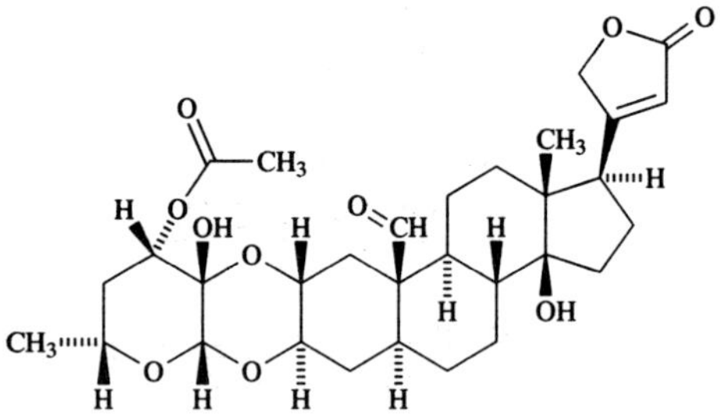

Cardenolide

$C_{31}H_{42}O_{10}$ [36573-63-4] MW 574.67

In the latex of *Asclepias* spp., such as *Asclepias curassavica* (Asclepiadaceae).

Very toxic to vertebrates, LD_{50} intravenously in cats is 0.236 mg/kg body-weight.

Patnaik, G. K., *Arzneim.-Forsch.,* 1978, **28**, 1130.

Asebotoxin II 138

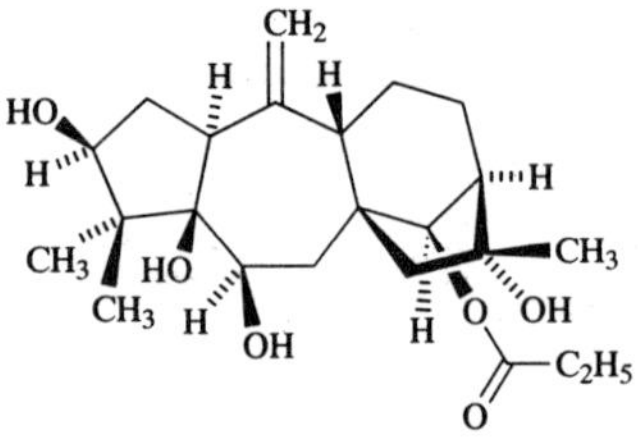

Diterpenoid
(Grayanotoxane)

$C_{23}H_{36}O_6$ [23984-18-1] MW 408.54

Occurs, together with other related asebotoxins, in flowers of *Pieris japonica* (Ericaceae).

Highly toxic.

Hikino, H., *Chem. Pharm. Bull.,* 1970, **18**, 1071.

L-Aspartic acid 139

(*S*)-Aminosuccinic acid; L-Asparagic acid; L-Asparaginic acid

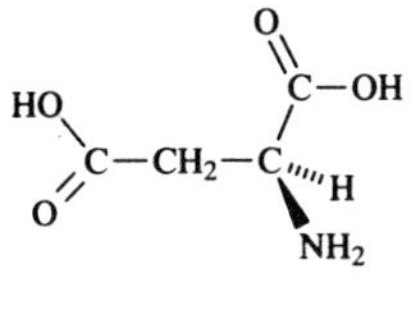

Amino acid

$C_4H_7NO_4$ [56-84-8] MW 133.11

As a protein amino acid, it occurs in all green plants. The free amino acid can be isolated from plants such as coffee, *Coffea* spp. (Rubiaceae), liquorice, *Glycyrrhiza glabra* (Leguminosae) and sugar-cane, *Saccharum officinarum* (Gramineae).

Dietary amino acid. However, if taken in large doses, produces neuro-excitatory symptoms. It is used in the preparation of the sweetening agent aspartame (1-methyl-*N*-L-aspartyl-L-phenylalanine).

Greenstein, J. P., Chemistry of the Amino Acids, Part 3, 1961, 1856, Wiley.

Aspecioside 140

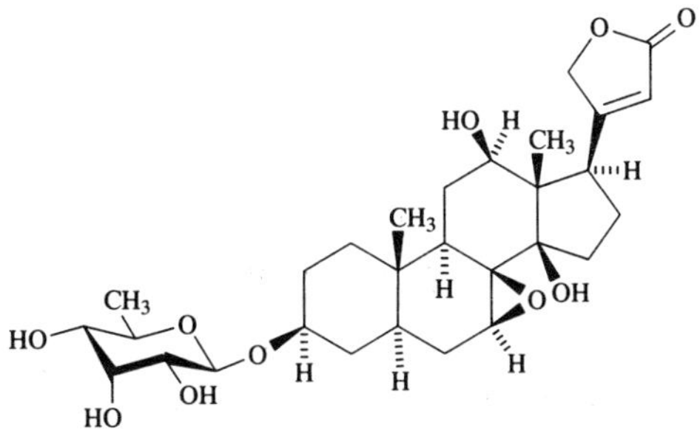

Cardenolide

$C_{29}H_{42}O_{10}$ [101915-75-7] MW 550.65

Latex of *Asclepias speciosa* and *A. syriaca* (Asclepiadaceae).

Toxic to vertebrates. Responsible, with other cardiac glycosides present, for livestock poisoning that occurs in North America after ingestion of *Asclepias* leaves.

Cheung, H. T. A., *J. Chem. Soc., Perkin Trans. 1,* 1986, 61.

Asperuloside 141

Asperulin; Rubichloric acid

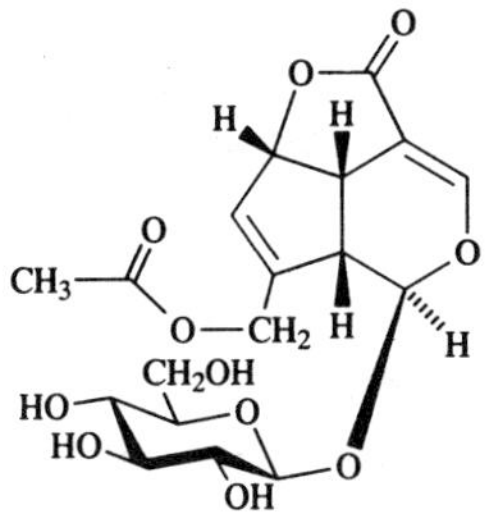

Monoterpenoid
(Iridoid glucoside)

$C_{18}H_{22}O_{11}$ [14259-45-1] MW 414.37

Leaves of *Asperula odorata* and *Galium aparine* (Rubiaceae), of many *Escallonia* spp. (Saxifragaceae) and of *Daphniphyllum macropodum* (Daphniphyllaceae).

Laxative properties.

Swiatek, L., *Herba Pol.,* 1972, **18**, 168.

Aspidospermatine 142

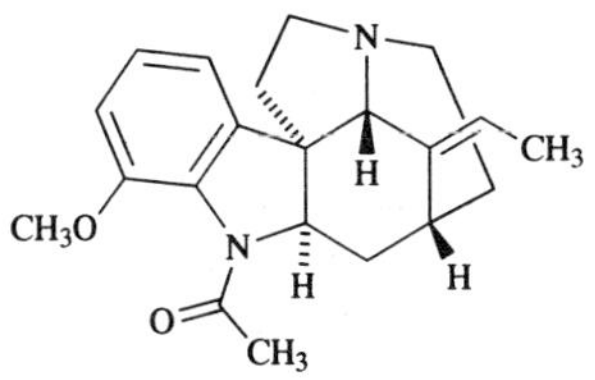

Indole alkaloid
(Condyfolan)

$C_{21}H_{26}N_2O_2$ [5794-14-9] MW 338.45

Aspidosperma quebrachoblanco (Apocynaceae).

Used against dyspnoea in Argentina, South and Central America under the names "quebracho blanco" and "quebracho colorada".

Biemann, K., *J. Am. Chem. Soc.,* 1963, **85**, 631.

Aspidospermine 143

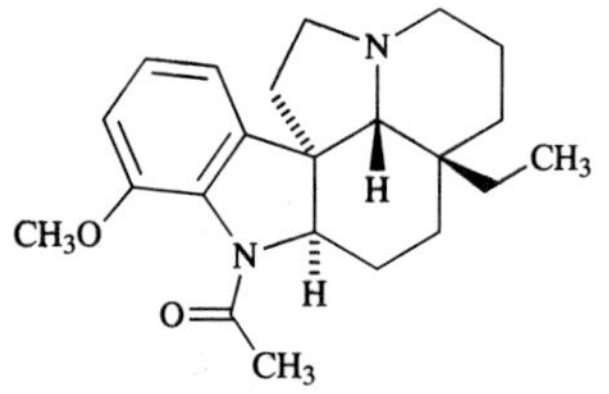

Indole alkaloid
(Aspidospermidine)

$C_{22}H_{30}N_2O_2$ [466-49-9] MW 354.49

Common in *Aspidosperma* and *Vallesia* spp. (Apocynaceae), e.g. in *Aspidosperma quebrachoblanco* and in *Vallesia dichotoma*.

Diuretic and respiratory stimulant activities.

Craven, B. M., *Experientia,* 1968, **24**, 770.

Astragaloside III 144

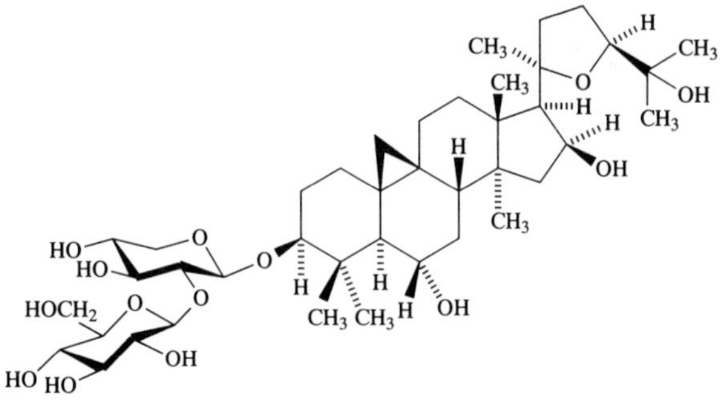

Triterpenoid saponin
(Cycloartane)

$C_{41}H_{68}O_{14}$ [84687-42-3] MW 784.98

In the roots of *Astragalus membranaceus* (Leguminosae).

Inhibits the formation of lipid peroxidase induced by intraperitoneal administration of adriamycin (15 mg/kg body-weight) in rats.

Kitagawa, I., *Chem. Pharm. Bull.,* 1983, **31**, 709.

Astrasieversianin XVI 145

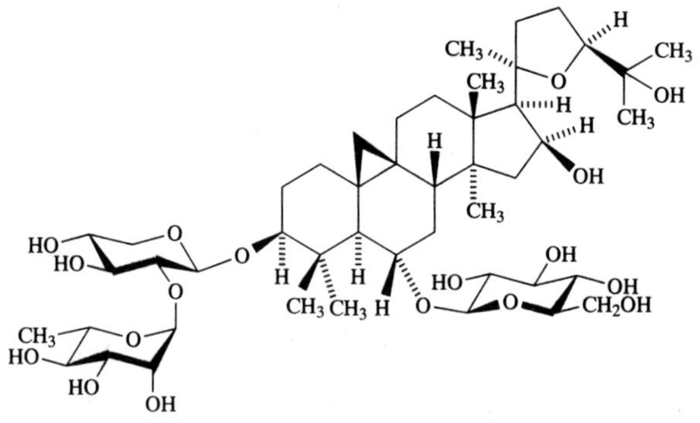

Triterpenoid saponin
(Cycloartane)

$C_{47}H_{78}O_{18}$ [101843-82-7] MW 931.13

Found in *Astragalus sieversianus* (Leguminosae).

Antihypertensive activity.

Gan, L. X., *Phytochemistry,* 1986, **25**, 2389.

Astringin 146

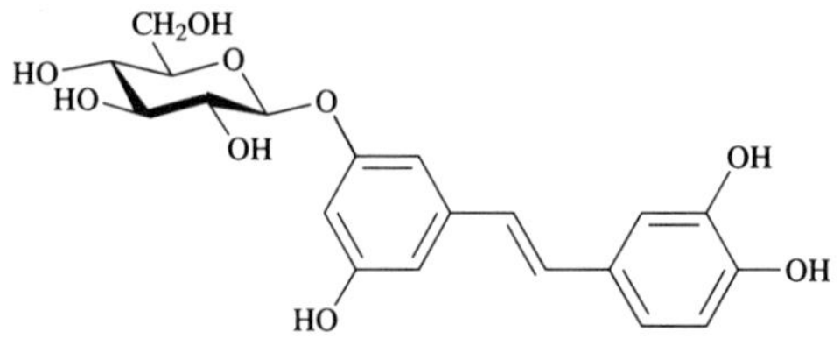

Stilbenoid

$C_{20}H_{22}O_9$ [29884-49-9] MW 406.39

Bark of Sitka spruce, *Picea sitchensis* (Pinaceae), wood of *Angophora cordifolia* and *Eucalyptus* spp. (Myrtaceae).

Antifungal activity, toxic to *Paeolus schweinitzii.*

Pearson, T. W., *Wood Sci.,* 1977, **10**, 93.

Atanine 147

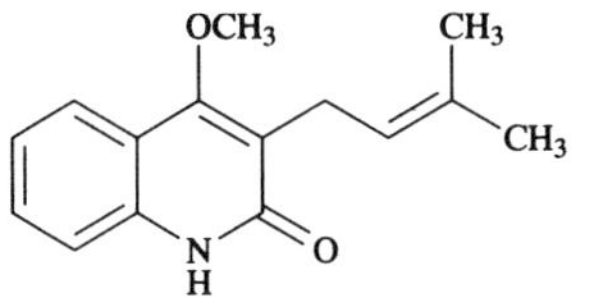

Quinoline alkaloid

$C_{15}H_{17}NO_2$ [7282-19-1] MW 243.31

Fruits of *Evodia rutaecarpa*, leaves of *E. hupehensis* and of *Ravenia spectabilis* and heartwood of *Fagara zanthoxyloides* (Rutaceae).

Anthelmintic activity, effective against the human parasite *Schistosoma mansoni* and the soil nematode *Caenorhabditis elegans*.

Eshiett, I. T., *J. Chem. Soc., C,* 1968, 481.

Athamantin 148

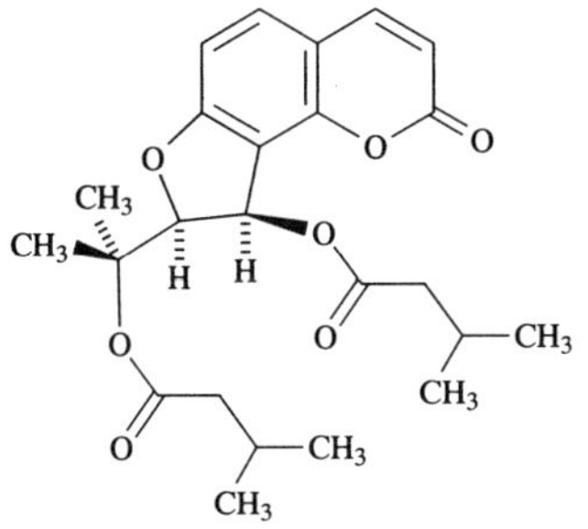

Furocoumarin

$C_{24}H_{30}O_7$ [1892-56-4] MW 430.50

Roots and seeds of *Athamanta oreoselinum*. It also occurs in *Ammi visnaga*, *Angelica sylvestres*, *Athamanta cretensis*, *Libanotis transcaucasica*, *Peucedanum* spp. and *Seseli libanotis* (Umbelliferae).

Spasmolytic activity.

Halpern, O., *Helv. Chim. Acta,* 1957, **40**, 758.

Athyriol 149

3-Methoxy-1,6,7-trihydroxyxanthone;
1,6,7-Trihydroxy-3-methoxyxanthone

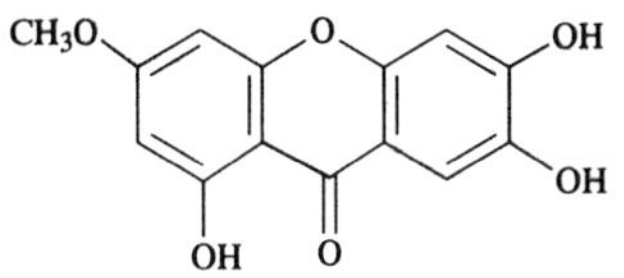

Xanthone

$C_{14}H_{10}O_6$ [28283-84-3] MW 274.23

In the leaves of the fern *Athyrium mesosorum* (Polypodiaceae).

Inhibits xanthine oxidase and therefore of possible use in the treatment of gout.

Ueno, A., *J. Pharm. Soc. Jpn,* 1962, **82**, 1482.

Atractyloside 150

Atractylin; Potassium atractylate

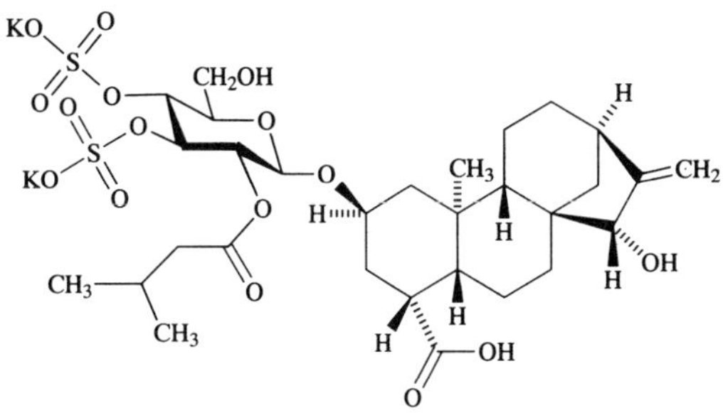

Diterpenoid
(Kaurane)

$C_{30}H_{44}K_2O_{16}S_2$ [17754-44-8] MW 803.00

In aerial parts of *Atractylis gummifera* (Compositae).

Highly toxic in mammals, with strychnine-like activity. LD_{50} intramuscularly in rats is 431 mg/kg body-weight. It is used experimentally because it is a specific inhibitor of ADP transport at the mitochondrial membrane.

Santi, R., Atractyloside: Chemistry Biochemistry and Toxicology, 1978, 33, Piccin Editore, Padua, Italy.

Atropine 151

Tropine tropate

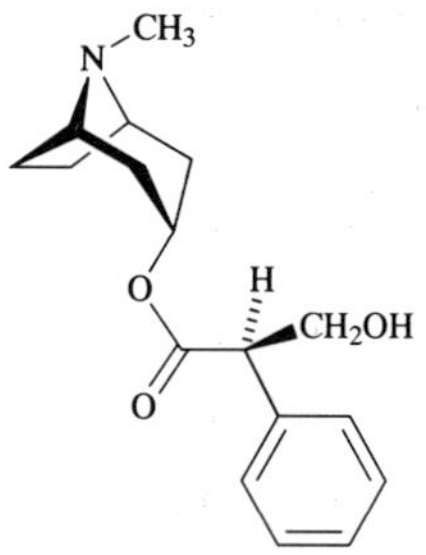

Tropane alkaloid

$C_{17}H_{23}NO_3$ [51-55-8] MW 289.37

Roots, leaves and seeds of deadly nightshade, *Atropa belladonna*, of thornapple, *Datura stramonium* and other Solanaceae. Also known in the optically active form, hyoscyamine (q.v.).

Highly toxic. The lethal dose in humans is 100 mg. It has anticholinergic activity and causes blurred vision, suppressed salivation, vasodilation and delirium. It is used in anaesthesia. In the past, has been used for poisoning. The berries of deadly nightshade are poisonous to children, less so to adults, but other parts of the plant are extremely poisonous. Cats and dogs are very susceptible to the poison, but rabbits are not affected.

Seeger, R., *Dtsch. Apoth. Ztg,* 1986, **126**, 1930.

Aucubin 152

Aucuboside; Rhinanthin; Rhinantin

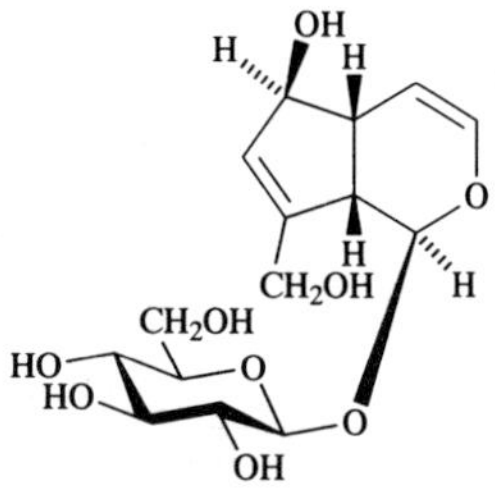

Monoterpenoid
(Iridoid glucoside)

$C_{15}H_{22}O_9$ [479-98-1] MW 346.33

In *Aucuba japonica* (Cornaceae), *Plantago lanceolata* (Plantaginaceae), *Rhinanthus* spp. (Scrophulariaceae) and many other sources.

The aglycone, aucubigenin [64274-28-8], is toxic to mammals and birds; it has antitumour activity. Aucubin itself has laxative and diuretic properties.

Birch, A. J., *J. Chem. Soc.,* 1961, 5194.

Aurantio-obtusin 6-*β*-D-glucoside 153

Glucoaurantioobtusin

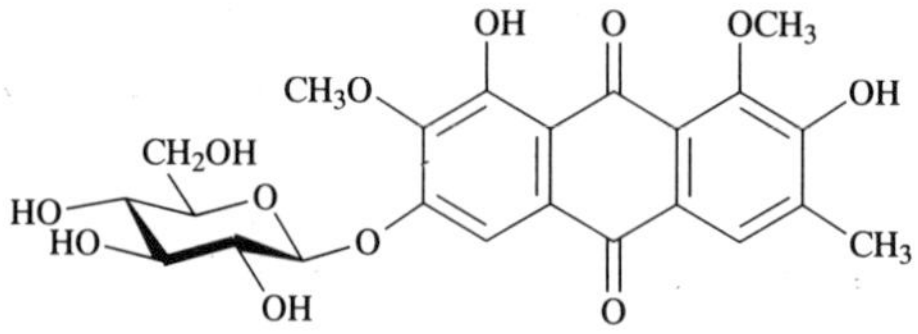

Anthraquinone

$C_{23}H_{24}O_{12}$ [129025-96-3] MW 492.44

Seeds of *Cassia obtusifolia* (Leguminosae).

Strong inhibitor of rat platelet aggregation, but the corresponding aglycone is only moderately active.

Takido, M., *Chem. Abstr.,* 1965, **62**, 5326

Auriculine 154

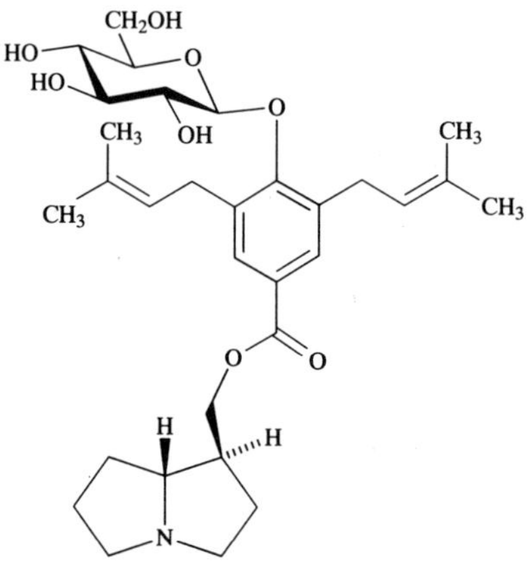

Pyrrolizidine alkaloid

$C_{31}H_{45}NO_8$ [22595-00-2] MW 559.70

In *Liparis auriculata* and *L. loeselii* (Orchidaceae).

Slightly hepatotoxic.

Lindström, B., *Acta Chem. Scand.,* 1971, **25**, 895.

Auriculoside 155

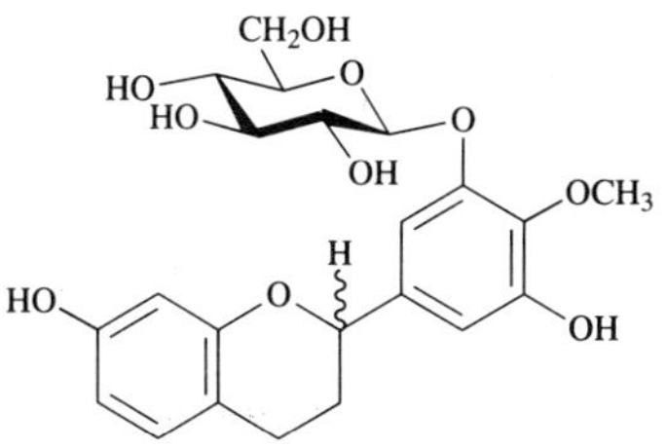

Flavan glycoside

$C_{22}H_{26}O_{10}$ [75871-96-4] MW 450.45

Acacia auriculiformis (Leguminosae).

Minor central nervous system depressant with some antistress/anti-anxiety activity.

Sahai, R., *Phytochemistry,* 1980, **19**, 1560.

Austrobailignan 1 156

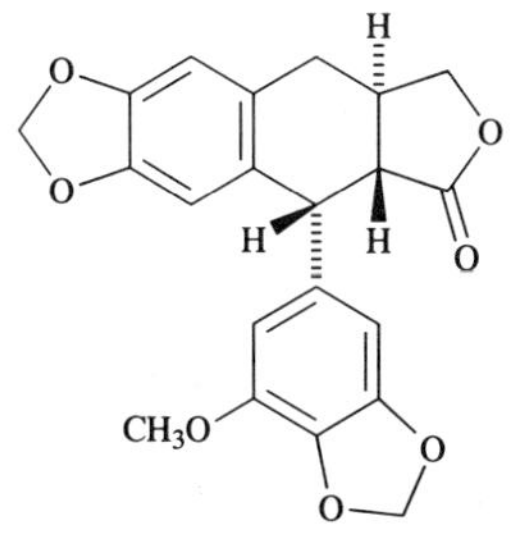

Lignan

$C_{21}H_{18}O_7$ [55955-07-2] MW 382.37

Isolated from *Austrobaileya scandens* (Austrobaileyaceae) and from twigs of *Amyris pinnata* (Rutaceae).

Antitumour activity.

Murphy, S. T., *Aust. J. Chem.,* 1975, **28**, 81.

Autumnolide 157

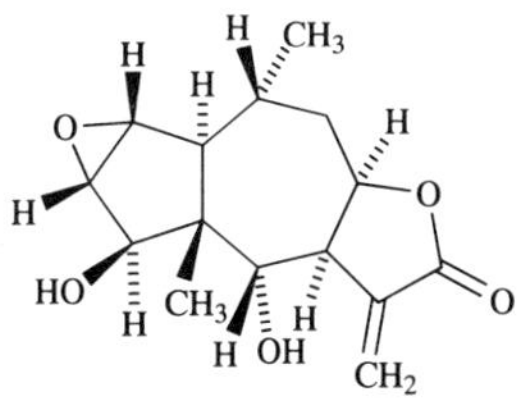

Sesquiterpenoid lactone
(Pseudoguaianolide)

$C_{15}H_{20}O_5$ [20505-32-2] MW 280.32

Occurs in sneezeweed, *Helenium autumnale* (Compositae).

Cytotoxic and antitumour activities.

Herz, W., *J. Org. Chem.,* 1969, **34**, 2915.

Avadharidine 158

Avadkharidine; Awadcharidine

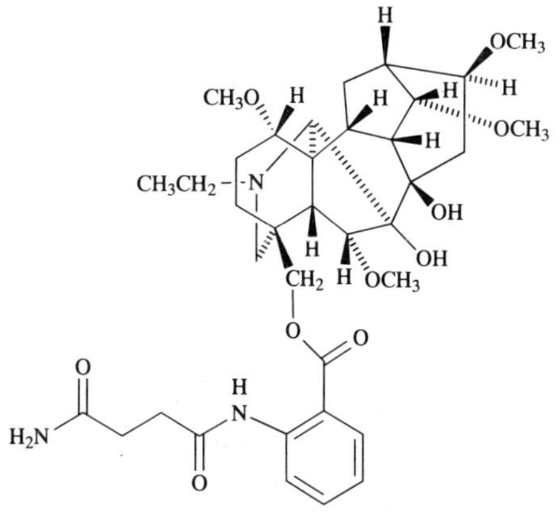

Diterpenoid alkaloid
(Aconitane)

$C_{36}H_{51}N_3O_{10}$ [509-16-0] MW 685.82

Found in the roots of *Aconitum orientale*, *A. finetianum* and *Delphinium cashmirianum* (Ranunculaceae).

Toxic, with curare-like properties. It has a negligible effect on the heart rate of anaesthetised rabbits in intravenous doses up to 5 mg/kg body-weight.

Shamma, M., *J. Nat. Prod.,* 1979, **42**, 615.

Avenacin A-1 159

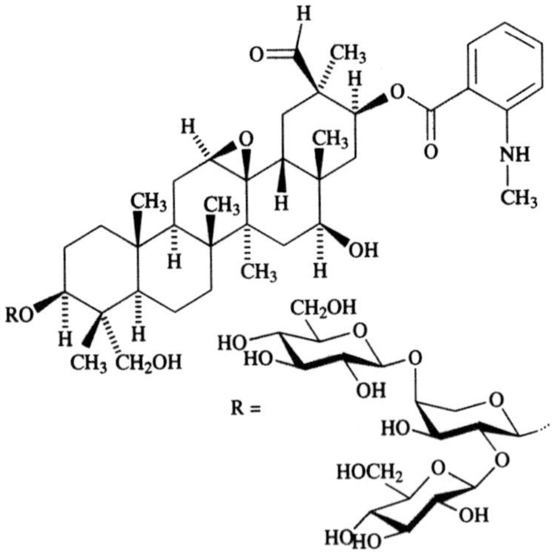

Triterpenoid saponin
(Oleanane)

$C_{55}H_{83}NO_{21}$ [90547-90-3] MW 1094.26

In roots of oats, *Avena sativa* (Gramineae).

Haemolytic activity.

Crombie, W. M. L., *Phytochemistry,* 1986, **25**, 2069.

Avenacin B-2 160

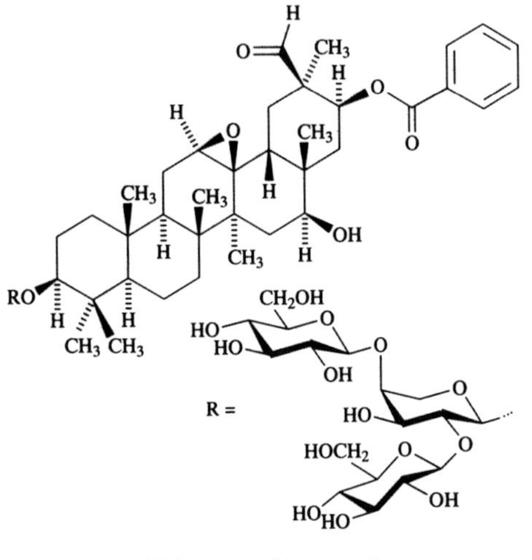

Triterpenoid saponin
(Oleanane)

$C_{54}H_{80}O_{20}$ [90547-93-6] MW 1049.22

In the roots of oats, *Avena sativa* (Gramineae).

Haemolytic activity.

Crombie, W. M. L., *Phytochemistry,* 1986, **25**, 2069.

Avenalumin I 161

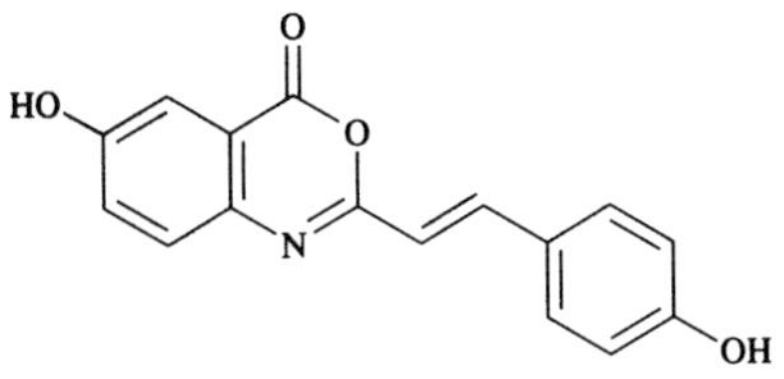

Benzoxazin-4-one

$C_{16}H_{11}NO_4$ [78164-38-2] MW 281.27

Primary oat leaves, *Avena sativa* (Gramineae) infected with *Puccinia coronata*.

Antifungal agent (phytoalexin).

Mayama, S., *Tetrahedron Lett.,* 1981, **22**, 2103.

Avrainvilleol 162

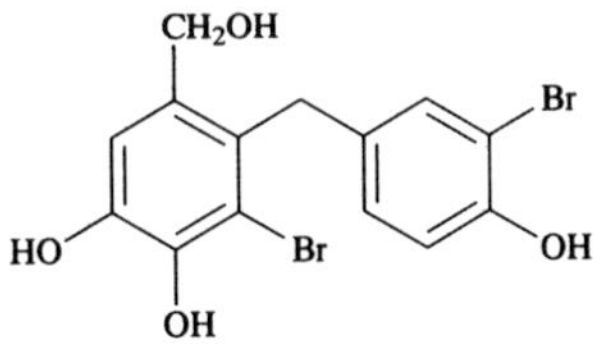

Phenol

$C_{14}H_{12}Br_2O_4$ [87402-66-2] MW 404.05

Green alga, *Avrainvillea nigricans*.

Fish poison.

Sun, H. H., *Phytochemistry,* 1983, **22**, 743.

Ayapin 163

Aiapin

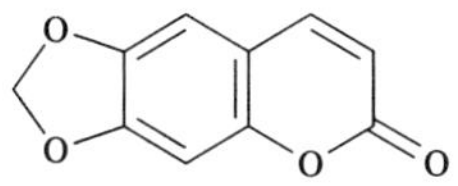

Coumarin

$C_{10}H_6O_4$ [494-56-4] MW 190.16

Leaves and stems of sunflower, *Helianthus annuus* (Compositae) infected with *Helminthosporium carbonum*.

Antifungal agent (phytoalexin).

Dieterman, L. J., *Arch. Biochem. Biophys.*, 1964, **106**, 275.

Azadirachtin 164

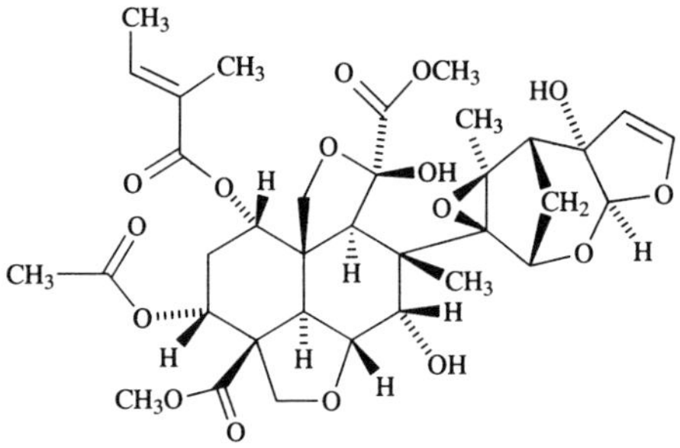

Nortriterpenoid
(Limonane)

$C_{35}H_{44}O_{16}$ [11141-17-6] MW 720.72

Seeds of neem tree, *Azadirachta indica* (Meliaceae). Several closely related structures (e.g. Azadirachtin H) occur in minor amounts together with the major toxin.

Potent insect antifeedant activity, e.g. against *Spodoptera* and *Locusta* species. Licensed as a natural insecticide in USA and other countries. Extracts of neem seed are used in folk medicine in India.

Ley, S. V., *J. Chem. Soc., Chem. Commun.*, 1992, 1304.

L-Azetidine-2-carboxylic acid 165

L-2-Azetidinecarboxylic acid

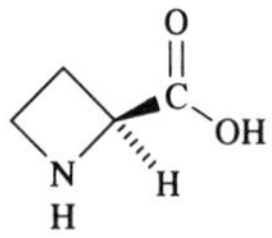

Amino acid

$C_4H_7NO_2$ [2133-34-8] MW 101.11

Rhizome and fresh foliage of many liliaceous plants, e.g. Solomon's seal, *Polygonatum multiflorum*, lily-of-the-valley, *Convallaria majalis* and squill, *Drimia maritima* (Liliaceae). Also recorded in lesser amounts in sugar beet, *Beta vulgaris* (Chenopodiaceae) and *Delonix regia* (Leguminosae).

Larvicide and also causes aberrations in the development of chick embryos. These effects are due to the competitive inhibition of proline uptake and incorporation, with particular reference to collagen synthesis.

Cromwell, N. H., *Chem. Rev.*, 1979, **79**, 331.

B

Baccharinoid B21 166

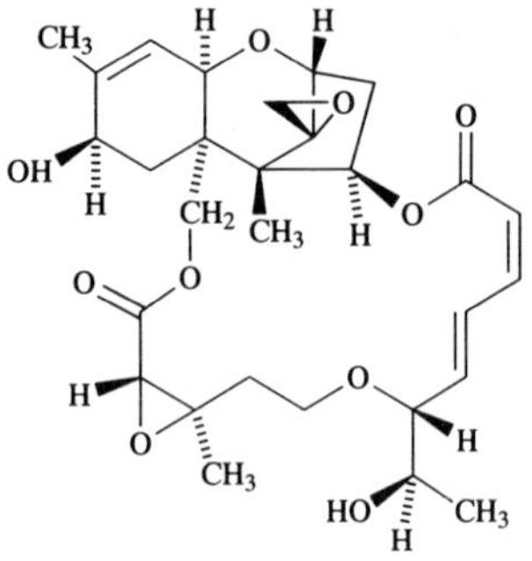

Sesquiterpenoid
(Trichothecane)

$C_{29}H_{38}O_{10}$ [105563-54-0] MW 546.61

One of a number of closely related toxins in roots and aerial parts of *Baccharis megapotamica* (Compositae).

Cause of livestock poisoning following grazing on this plant.

Jarvis, B. B., *J. Org. Chem.*, 1987, **52**, 45.

Baicalein 167

5,6,7-Trihydroxyflavone; Noroxylin

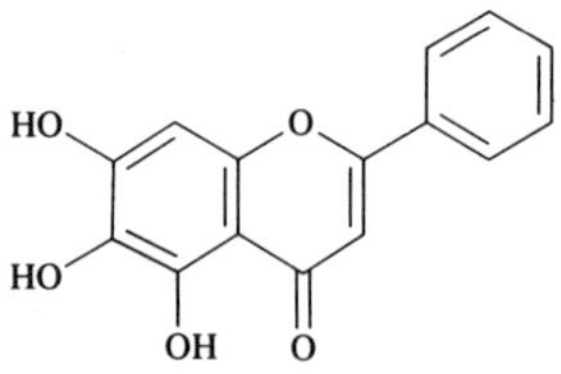

Flavone

$C_{15}H_{10}O_5$ [491-67-8] MW 270.24

Roots and leaves of many *Scutellaria* spp. (Labiatae), often as the 7-glucuronide, baicalin (q.v.). The 7-rhamnoside occurs in *Scutellaria galericulata* and the 6-glucoside and 6-glucuronide in leaves of *Oroxylum indicum* (Bignaniaceae).

Inhibits glyoxylase-I and platelet lipoxygenase. It shows antipromotion activity in carcinogenesis, and anti-inflammatory, anti-allergic and diuretic activities. High concentrations of baicalein prolong the clotting time of fibrinogen by thrombin.

Schönberg, A., *J. Am. Chem. Soc.*, 1955, **77**, 5390.

Baicalin 168

Baicalein 7-glucuronide; Baicalein 7-*O*-glucuronide; Baicaloside

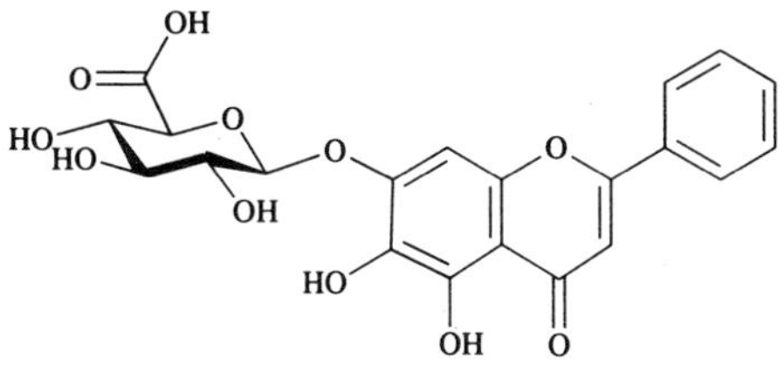

Flavone

$C_{21}H_{18}O_{11}$ [21967-41-9] MW 446.37

Found in the roots of *Scutellaria baicalensis* (Labiatae).

Diuretic and anti-allergic activities. In high concentrations (0.1-1mM), it prolongs the clotting time of fibrinogen by thrombin.

Shibata, K., *Acta Phytochim.*, 1923, **1**, 105.

Baileyin 169

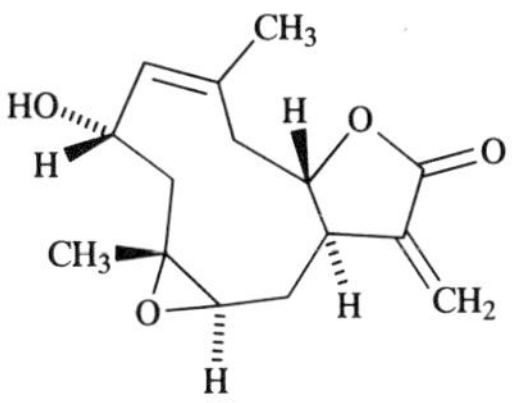

Sesquiterpenoid lactone (Germacranolide)

$C_{15}H_{20}O_4$ [27875-37-2] MW 264.32

Occurs in *Baileya multiradiata* and *B. pleniradiata* (Compositae).

Cytotoxic and antitumour activities.

Waddell, T. G., *Phytochemistry,* 1969, **8**, 2371.

Bakkenolide A 170

Fukinanolide

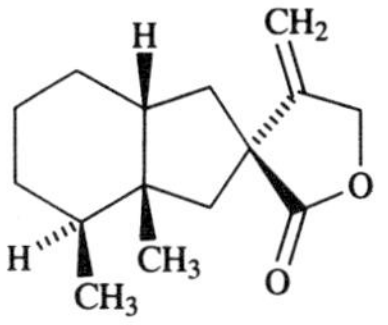

Sesquiterpenoid lactone (Bakkenolide)

$C_{15}H_{22}O_2$ [19906-72-0] MW 234.34

Senecio pyramidatus, *Cacalia hastata*, *Ligularia calthaefolia*, *Homogyne alpina* and many *Petasites* spp. (Compositae).

An insect antifeedant. Possesses cytotoxic and antitumour activities.

Abe, N., *Tetrahedron Lett.,* 1968, 369.

Baliospermin 171

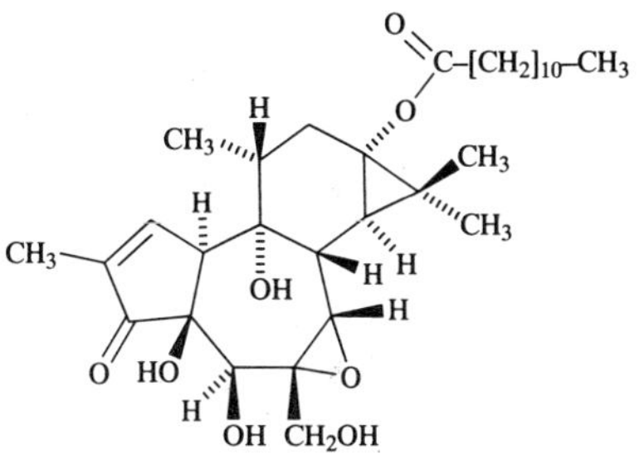

Diterpenoid (Tigliane)

$C_{32}H_{50}O_8$ [66583-56-0] MW 562.74

Roots of *Baliospermum montanum* (Euphorbiaceae).

Extracts of *Baliospermum* plants are drastic purgatives. Baliospermin has cytotoxic activity.

Ogura, M., *Planta Med.,* 1978, **33**, 128.

Barbaloin 172

Aloin A

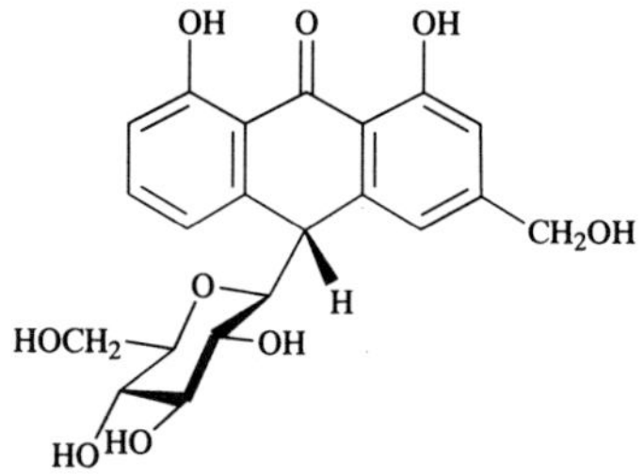

Anthraquinone
(Anthrone)

$C_{21}H_{21}O_9$ [1415-73-2] MW 417.39

Leaves of *Aloe vera, A. ferox* and *A. perryi* (Liliaceae).

Used commercially as a purgative.

Thomson, R. H., Naturally Occurring Quinones, 2nd edn, 1971, 400, Academic.

Barbinine 173

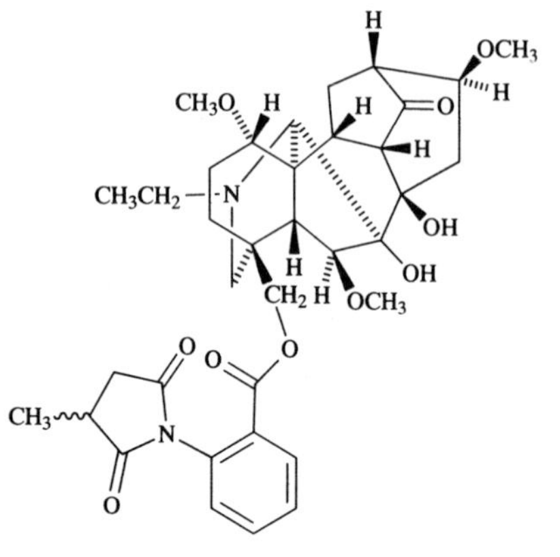

Diterpenoid alkaloid
(Aconitane)

$C_{36}H_{46}N_2O_{10}$ [123497-99-4] MW 666.77

Delphinium barbeyi (Ranunculaceae).

One of several poisonous alkaloids in larkspurs responsible for cattle deaths in North America. The LD_{50} on intravenous injection in mice is 57 mg/kg body-weight.

Pelletier, S. W., *Phytochemistry,* 1989, **28**, 1521.

Bayogenin 3-glucoside 174

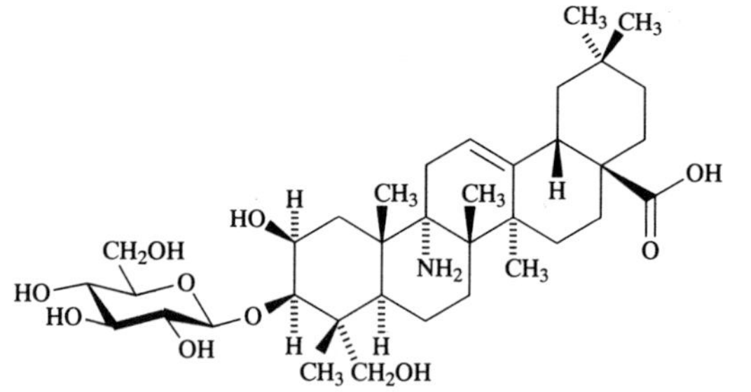

Triterpenoid saponin
(Oleanane)

$C_{36}H_{58}O_{10}$ [104513-86-2] MW 650.85

Roots of *Dolichos kilimandscharicus* (Leguminosae).

Antifungal activity at a concentration of 2.5 μg/ml and molluscicidal at a concentration of 7.5 μg/l.

Marston, A., *Phytochemistry,* 1988, **27**, 1325.

(+)-Bebeerine 175

Aristolochine; Curine; Chondodendrine; Chondrodendrine; Pelosine

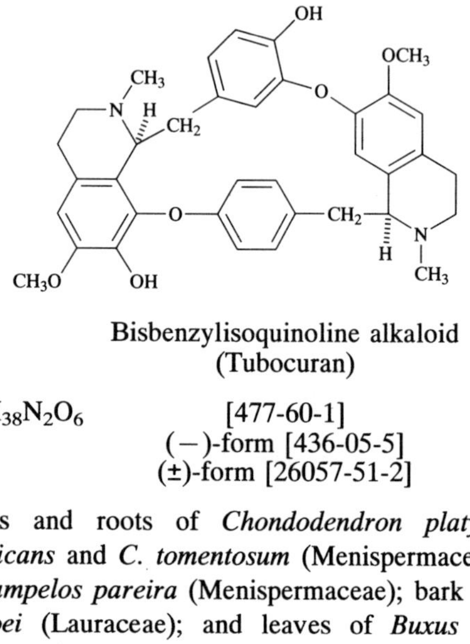

(+)-form

Bisbenzylisoquinoline alkaloid
(Tubocuran)

$C_{36}H_{38}N_2O_6$ [477-60-1] MW 594.71
(−)-form [436-05-5]
(±)-form [26057-51-2]

Stems and roots of *Chondodendron platyphyllum, C. candicans* and *C. tomentosum* (Menispermaceae); stems of *Cissampelos pareira* (Menispermaceae); bark of *Nectandra rodioei* (Lauraceae); and leaves of *Buxus sempervirens* (Buxaceae). This alkaloid occurs in both the (+)- and (−)-forms. The (±)-form is called hayatine.

It has cardiac action, and is an antimalarial drug.

Cava, M. P., *Phytochemistry,* 1969, **8**, 2341.

Bellidifolin 176

3-Methoxy-1,5,8-trihydroxyxanthone; 1,5,8-Trihydroxy-3-methoxyxanthone; Bellidifolium

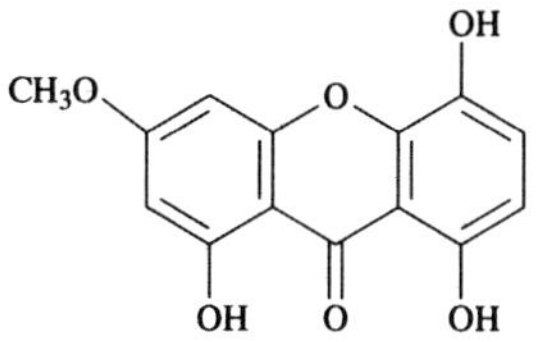

Xanthone

$C_{14}H_{10}O_6$ [2798-25-6] MW 274.23

Aerial parts of *Gentiana lactea* and *Swertia chirata* (Gentianaceae).

Tuberculostatic agent, and also an inhibitor of the enzyme monoamine oxidase A. It is antihepatotoxic and has a strong mutagenic activity in the *Salmonella typhimurium* test.

Markham, K. R., *Tetrahedron,* 1964, **20**, 991.

Benzoyltropein 177

Tropine benzoate

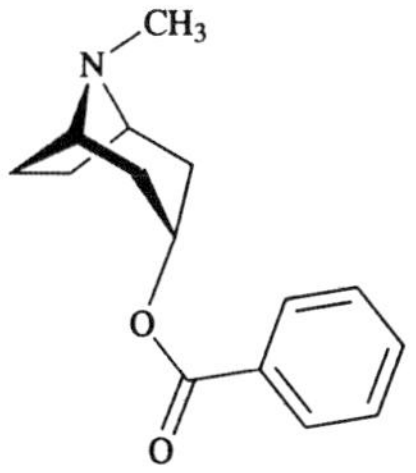

Tropane alkaloid

$C_{15}H_{19}NO_2$ [19145-60-9] MW 245.32

Found in *Erythroxylum coca* (Erythroxylaceae), *Crossostylis ebertii* and *Bruguiera sexangula* (Rhizophoraceae).

On hydrolysis, it gives rise to tropine (q.v.), which is poisonous.

Beyerman, H. C., *Rec. Trav. Chim.,* 1956, **75**, 1445.

Berbamine 178

Berbenine

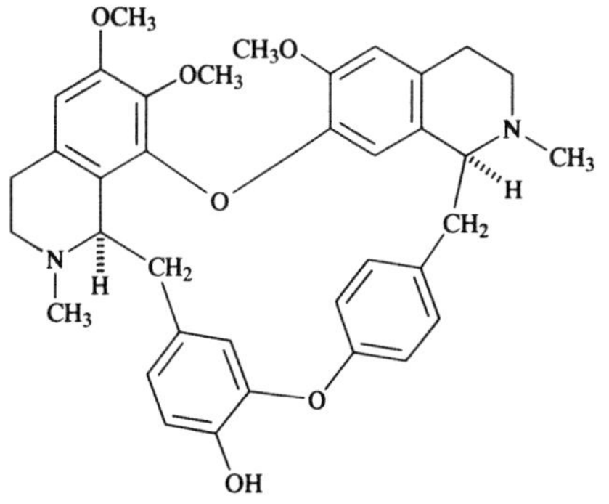

Bisbenzylisoquinoline alkaloid (Berbaman)

$C_{37}H_{40}N_2O_6$ [478-61-5] MW 608.73

Atherosperma moschatum (Monimiaceae), *Berberis thunbergii, B. vulgaris* and *Mahonia aquifolium* (Berberidaceae), *Pycnarrhena manillensis* and *Stephania sasakii* (Menispermaceae).

A strong curarising agent and highly toxic. Has tumour inhibiting activity. In animal experiments it exhibits spasmolytic and vasodilatating action.

Akasu, M., *Phytochemistry,* 1976, **15**, 471.

Berberastine 179

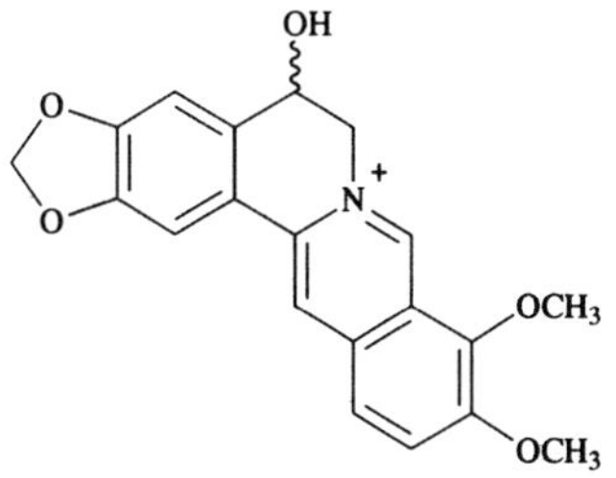

Benzylisoquinoline alkaloid (Berbine)

$[C_{20}H_{18}NO_5]^+$ [2435-73-6] MW 352.37

Rhizomes of *Coptis japonica*, roots of *Hydrastis canadensis*, stems of *Zanthorrhiza simplicissima* (all Ranunculaceae) and roots of *Berberis laurina* (Berberidaceae).

Toxicity, activity and uses as for berberine (q.v.).

Ikuta, A., *J. Nat. Prod.*, 1984, **47**, 189.

Berberine 180

Umbellatine; Berbericine; Natural Yellow 18

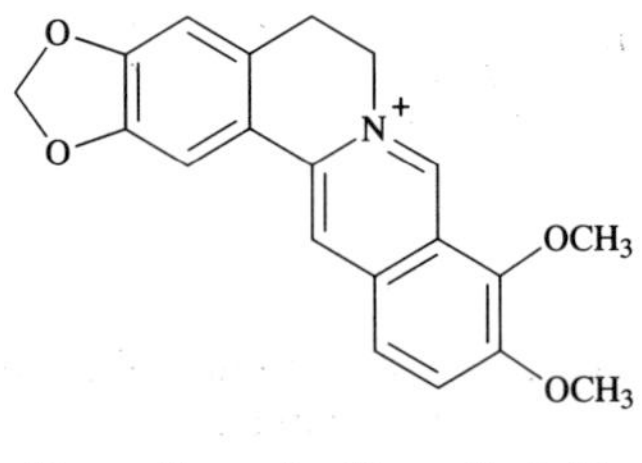

Benzylisoquinoline alkaloid
(Berbine)

$[C_{20}H_{18}NO_4]^+$ [2086-83-1] MW 336.37

Yellow pigment of the common barberry, *Berberis vulgaris* (Berberidaceae). Found in many plant families including Annonaceae (*Coelocline* spp.), Berberidaceae (*Berberis, Mahonia, Nandina* spp.), Menispermaceae (*Archangelisia* spp.), Papaveraceae (*Argemone, Chelidonium, Corydalis* spp.), Rutaceae (*Evodia, Toddalia, Zanthoxylum* spp.) and Ranunculaceae (*Coptis, Thalictrum* spp.).

Moderately toxic. LD_{50} in humans is 27.5 mg/kg bodyweight. It causes cardiac damage, dyspnoea and hypotension. It is a useful drug, since it has bitter stomachic, antimalarial, antipyretic, anthelmintic and cytotoxic activities.

Brossi, A., The Alkaloids, 1986, **28**, 95, Academic Press.

Bergamottin 181

Bergaptol geranyl ether; 5-Geranyloxypsoralen; Bergaptin

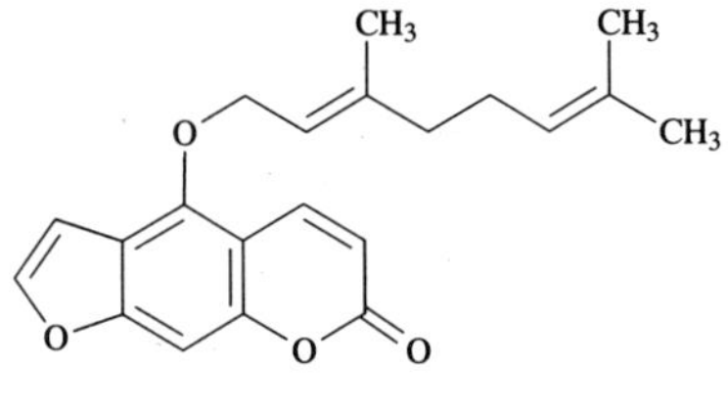

Furocoumarin

$C_{21}H_{22}O_4$ [7380-40-7] MW 338.40

Oil of bergamot, *Citrus bergamia*; also lemon oil and oils of other *Citrus* spp. (Rutaceae) and *Daucus carota* (Umbelliferae).

Ovicidal to *Leptinotarsa decemlineata*.

Späth, E., *Ber. Dtsch. Chem. Ges.*, 1937, **70**, 2272.

Bergapten 182

Bergaptene; Heraclin; Majudin; 5-Methoxypsoralin

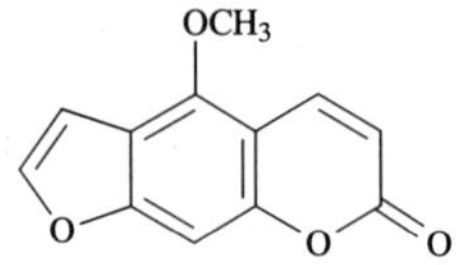

Furocoumarin

$C_{12}H_8O_4$ [484-20-8] MW 216.19

Widespread in the Rutaceae, e.g. in oil of bergamot, *Citrus bergamia* and in rue, *Ruta graveolens*. Common in the Umbelliferae; in *Heracleum, Ligusticum, Angelica, Ammi, Seseli, Levisticum, Pimpinella* and *Petroselinum* spp.. It has been found in aerial parts of the tomato, *Lycopersicon esculentum* (Solanaceae).

Toxic to fish, toads and snails. It has been used as an ingredient of suntan preparations, because of its UV absorbtive properties, but was recently shown to have lethal and clastogenic effects on mammalian cells in culture. Used also in the treatment of leukoderma and psoriasis, but is less effective than psoralen.

Karrer, W., Konstitution und Vorkommen der Organischen Pflanzenstoffe, 2nd edn, **Part 1**, 1981, 293, Birkhäuser.

Betagarin 183

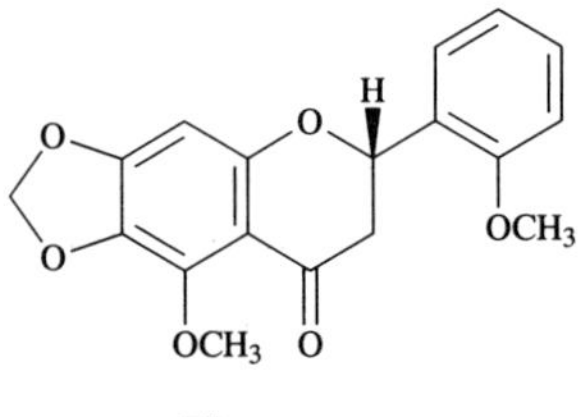

Flavanone

$C_{18}H_{16}O_6$ [60132-69-6] MW 328.32

Leaves of sugar beet, *Beta vulgaris* (Chenopodiaceae) infected with *Cercospora beticola*.

Antifungal agent (phytoalexin).

Geigert, J., *Tetrahedron,* 1973, **29**, 2703.

Betavulgarin 184

Isoflavone

$C_{17}H_{12}O_6$ [51068-94-1] MW 312.28

Leaves of sugar beet, *Beta vulgaris* (Chenopodiaceae) infected with *Cercospora beticola*.

Antifungal agent (phytoalexin).

Geigert, J., *Tetrahedron,* 1973, **29**, 2703.

(−)-Betonicine 185

4-Hydroxyproline betaine; Achillein

Pyrrolidine alkaloid

$C_{17}H_{13}NO_3$ [515-25-3] MW 159.19

In *Betonica officinalis*, *Marrabium vulgare* and *Stachys sylvatica* (Labiatae) and in *Achillea moschata* and *A. millefolium* (Compositae).

Anti-inflammatory activity.

Patchett, A. A., *J. Am. Chem. Soc.,* 1957, **79**, 185.

Betulin 186

Betulinol; Betulol; Trochol

Triterpenoid
(Lupane)

$C_{30}H_{50}O_2$ [473-98-3] MW 442.73

In the outer bark of birch, *Betula* spp. (Betulaceae), often in high concentrations; the outer cortical layer of *B. platyphylla* contains 35% betulin. It occurs elsewhere in angiosperms, mainly in trees and shrubs.

Antitumour activity. It is active against the Walker carcinoma 256 tumour system.

Hayek, E. W. H., *Phytochemistry,* 1989, **28**, 2229.

(+)-Bicuculline 187

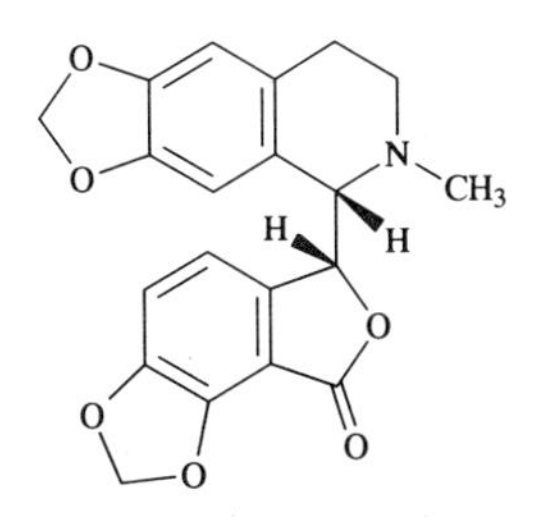

Phthalideisoquinoline alkaloid

$C_{20}H_{17}NO_6$ [485-49-4] MW 367.36

Whole plant of *Adlumia fungosa*, *Corydalis thalictrifolia* and *C. incisa* (Papaveraceae).

Effective γ-aminobutyric acid antagonist and is widely used in neurological research. Toxic.

Edwards, O. E., *Can. J. Chem.,* 1961, **39**, 1801.

Bikhaconitine 188

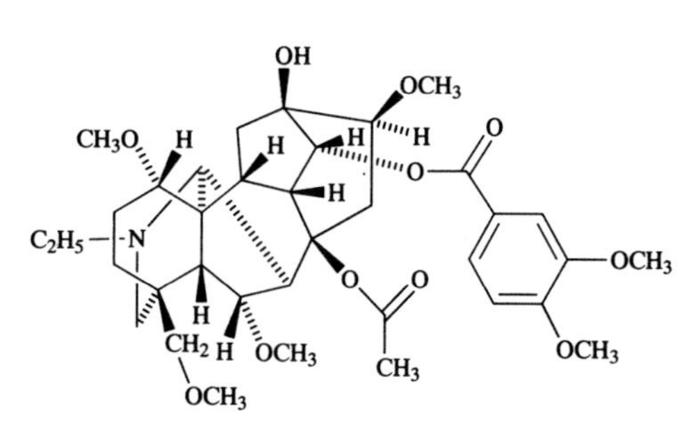

Diterpenoid alkaloid
(Aconitane)

$C_{36}H_{51}NO_{11}$ [6078-26-8] MW 673.80

Roots of *Aconitum spicatum*, *A. ferox* and *A. violaceum* (Ranunculaceae).

Highly toxic, producing respiratory paralysis and a direct action upon the heart, terminating in ventricular fibrillation. The plants are used extensively as a poison in India ('bikh' means "poison").

Klasek, A., *Lloydia,* 1972, **35**, 55.

Bilobol 189

(15:1)-Cardol

Phenol

$C_{21}H_{34}O_2$ [22910-32-6] MW 318.50

Sarcotesta of *Ginkgo biloba* fruit (Ginkgoaceae), in *Anacardium occidentale* and *Schinus terebinthifolius* (Anacardiaceae).

Antitumour activity against Sarcoma 180 ascites in mice.

Cirigottis, K. A., *Aust. J. Chem.,* 1974, **27**, 345.

Biochanin A 190

Pratensol

Isoflavone

$C_{16}H_{12}O_5$ [491-80-5] MW 284.27

In chickpea, *Cicer arietinum*, clover, *Trifolium* spp., *Baptisia* spp. and *Dalbergia* spp. (Leguminosae). Also in fruit of *Cotoneaster pannosa* (Rosaceae) and trunkwood of *Virola cadudifolia* (Myristicaceae).

Has oestrogenic activity and also hypolipidaemic properties.

Ingham, J. L., *Fortschr. Chem. Org. Naturst.,* 1983, **43**, 52.

Bipindoside 191

Bipindogenin 3-digitaloside

Cardenolide

$C_{30}H_{46}O_{10}$ [6246-79-3] MW 566.69

Strophanthus sarmentosus var. *senegambiae* and *S. thollonis* (Apocynaceae).

Toxic to vertebrates.

Fechtig, B., *Helv. Chim. Acta,* 1960, **43**, 1570.

(+)-α-Bisabolol 192

Bisabolol

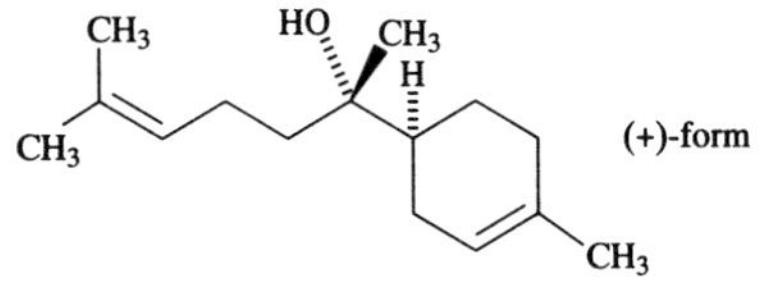

(+)-form

Sesquiterpenoid
(Bisabolane)

$C_{15}H_{26}O$ [515-69-5] MW 222.37
(−)-α-form [23089-26-1]

A commonly occurring constituent of plant essential oils. The (−)-form occurs in chamomile oil from chamomile, *Matricaria chamomilla* (Compositae), whereas the (+)-form is found in buds of *Populus balsamifera* (Salicaceae).

Less toxic than other similar sesquiterpenoids, e.g. guaiazulene. It has useful anti-inflammatory activity.

Jakovlev, V., *Arzneim.-Forsch.*, 1969, **19**, 615.

2,4-Bis(prenyl)phenol 193

2,4-Diprenylphenol

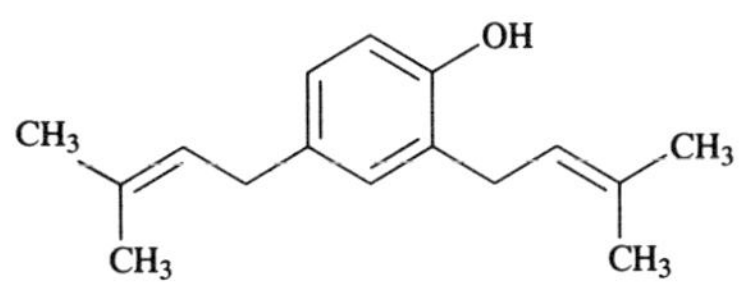

Phenol

$C_{16}H_{22}O$ [55824-31-2] MW 230.35

The marine alga *Encyothalia cliftonii* (Sporochnales).

Feeding deterrent to the herbivorous sea urchin *Tripneustes esculentus*.

Roussis, V., *Phytochemistry,* 1993, **34**, 107.

Bocconine 194

Chelirubine

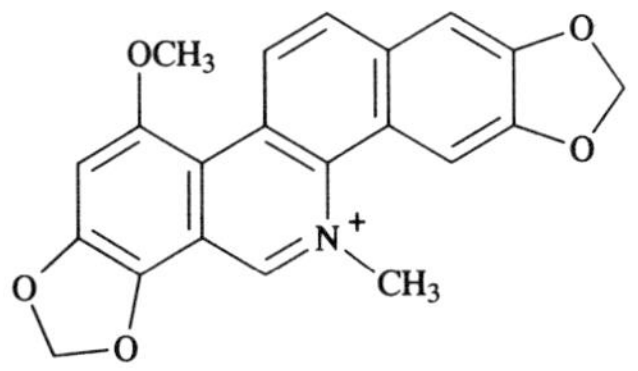

Benzylisoquinoline alkaloid
(Chelidonine)

$[C_{21}H_{16}NO_5]^+$ [18203-11-7] MW 362.36

In *Bocconia cordata* (Papaveraceae) and is widely distributed in plants of the Papaveraceae and Fumariaceae.

Toxic to nematodes. Used medicinally as a local anaesthetic.

Slavík, J., *Collect. Czech. Chem. Comm.,* 1968, **33**, 1619.

Borneol 195

endo-Borneol; Bornyl alcohol; Camphol

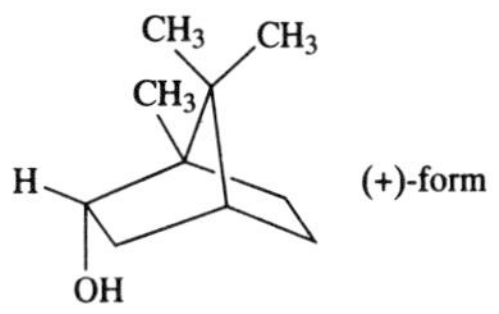

(+)-form

Monoterpenoid

$C_{10}H_{18}O$ [507-70-0] MW 154.25
(+)-form [464-43-7]
(−)-form [464-45-9]

(+)-Form is the main constituent of the volatile oil of Borneo camphor, distilled from *Dryobalanops aromatica* (Dipterocarpaceae). It is also present in oil of spike, from leaves and flowering tips of *Lavandula spica* (Labiatae) and in the volatile oil of nutmeg, *Myristica fragrans* (Myristicaceae). The (−)-form occurs in Ngai camphor, from *Blumea balsamifera* (Compositae).

Toxic to mammals, affecting the central nervous system. Used in perfumery.

Karrer, W., Konstitution und Vorkommen der Organischen Pflanzenstoffe, 2nd edn, 1981, 94, Birkhäuser.

Borrecapine 196

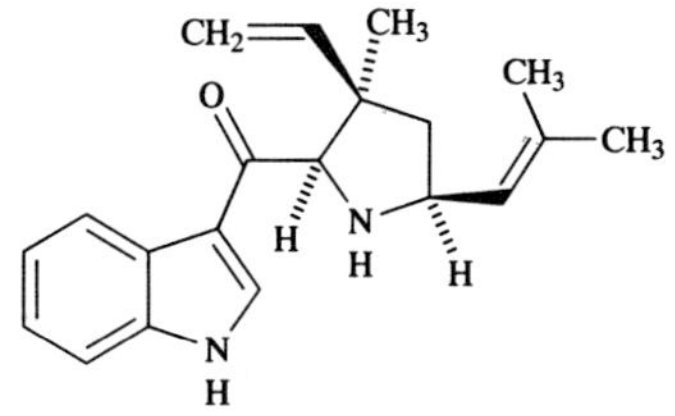

Indole alkaloid

$C_{20}H_{24}N_2O$ [66408-14-8] MW 308.42

In *Borreria capitata* (Rubiaceae).

Emetic, expectorant and astringent properties. It is used as a substitute for ipecacuanha.

Jössang, A., *Tetrahedron Lett.,* 1977, 4317.

Boschnialactone 197

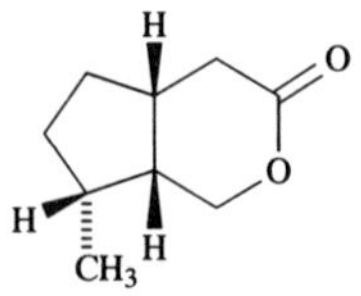

Monoterpenoid
(Iridoid)

$C_9H_{14}O_2$ [17957-87-8] MW 154.21

In *Boschniakia rossica* (Orobanchaceae).

Excitatory activity towards cats and other Felidae, similar to nepetalactone from catnip (q.v.).

Sakan, T., *Tetrahedron,* 1967, **23**, 4635.

Bovoside A 198

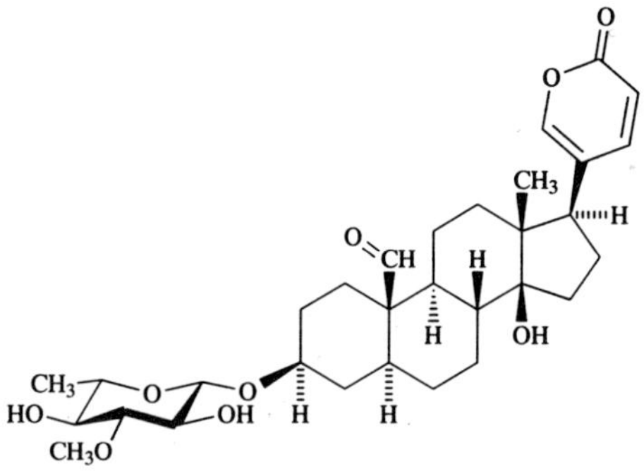

Bufadienolide

$C_{31}H_{44}O_9$ [11028-14-1] MW 560.68

Bulbs of sea onion, *Bowiea volubilis* and of *B. kilimandscharica* (Hyacinthaceae). It co-occurs with several closely related cardenolides.

Poisonous, with an LD_{50} in cats of 0.12 mg/kg body-weight. Has caused death in South Africa due to native people taking an overdose of the plant as a herbal remedy. Symptoms of poisoning are vomiting, salivation, palpitations and cramps, followed by death. It has been used in lower doses to procure abortions.

Katz, A., *Helv. Chim. Acta,* 1953, **36**, 1417.

Brassilexin 199

Brassilexine

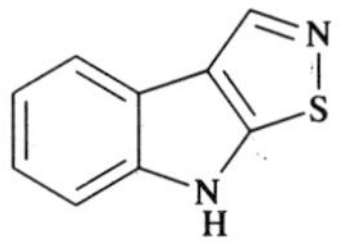

Indole

$C_9H_6N_2S$ [119752-76-0] MW 174.23

Fungally infected leaves of *Brassica juncea* (Cruciferae).

Antifungal agent (phytoalexin).

Devys, M., *Tetrahedron Lett.,* 1988, **29**, 6447.

Brassinin 200

Brassinine

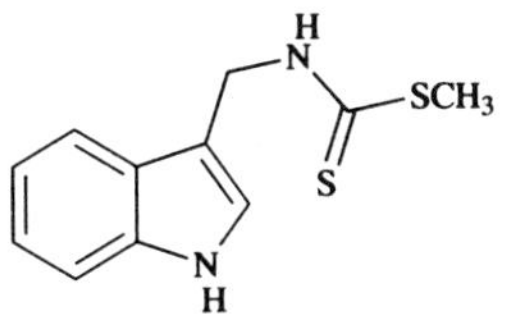

Indole

$C_{11}H_{12}N_2S_2$ [105748-59-2] MW 236.36

One of four indole derivatives produced in fungally infected leaves of the radish, *Raphanus sativus* (Cruciferae).

Antifungal agent (phytoalexin).

Takasugi, M., *J. Chem. Soc., Chem. Commun.*, 1986, 1077.

Brevetoxin A 201

T46 Toxin; Toxin GB-1

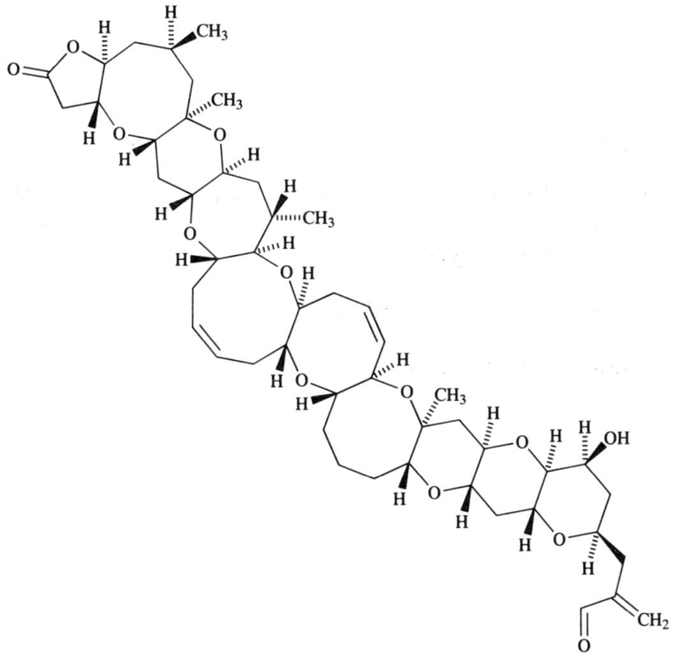

Polyether

$C_{49}H_{70}O_{13}$ [85087-27-0] MW 867.08

Florida red tide alga, *Gymnodium breve* along with brevetoxin B (q.v.) and several other closely similar toxins.

Toxic to fish at a concentration of 3 μg/ml. The oral LD_{50} in mice is 0.3 mg/kg body-weight.

Shimizu, Y., *J. Am. Chem. Soc.*, 1986, **108**, 514.

Brevetoxin B 202

Toxin GB-2; T34 Toxin

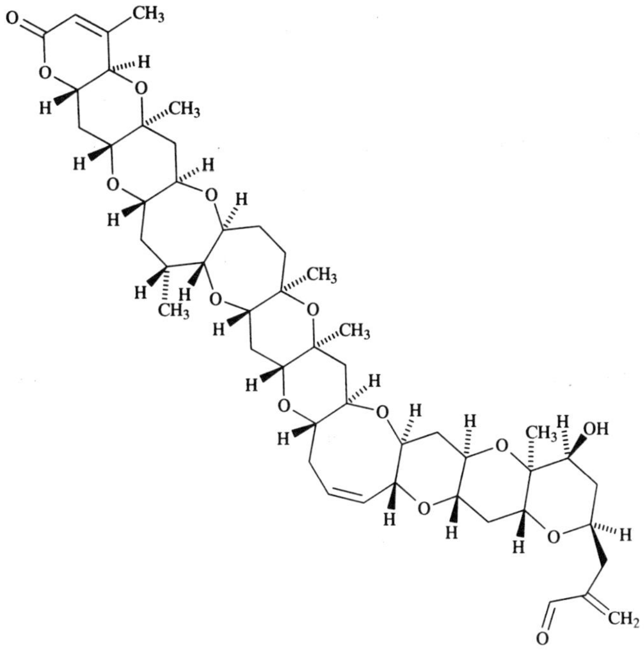

Polyether

$C_{50}H_{70}O_{14}$ [79580-28-2] MW 895.10

The Florida red tide alga, *Gymnodinium breve*.

Fish poison.

Lin, Y. Y., *J. Am. Chem. Soc.*, 1981, **103**, 6773.

Brevicolline 203

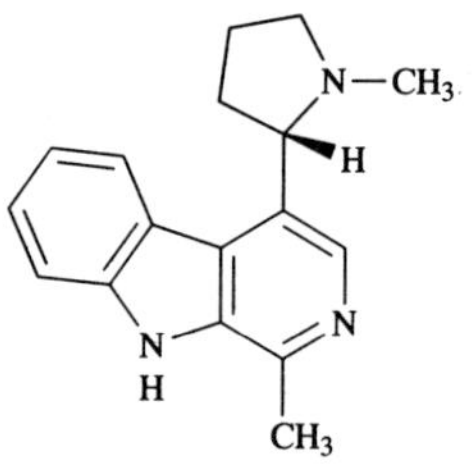

Indole alkaloid
(β-Carboline)

$C_{17}H_{19}N_3$ [20069-02-7] MW 265.36

In the seeds of sedge, *Carex brevicollis* (Cyperaceae).

Significant effect on uterine contractibility in animals.

Vember, P. A., *Khim. Prir. Soedin.*, 1967, **3**, 249.

Brevilin A 204

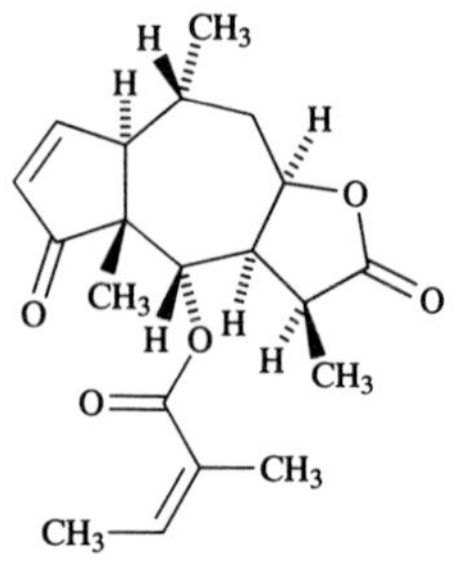

Sesquiterpenoid lactone
(Pseudoguaianolide)

$C_{20}H_{26}O_5$ [16503-32-5] MW 346.42

Centipeda minima and *Helenium alternifolium* (Compositae).

Antiprotozoal activity.

Herz, W., *J. Org. Chem.*, 1968, **33**, 2780.

3-Bromo-4,5-dihydroxybenzyl alcohol 205

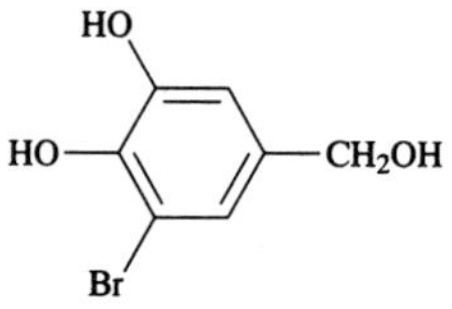

Phenol

$C_7H_7BrO_3$ [52897-61-7] MW 219.03

Green alga, *Arrainvillea nigricans.*

Antitumour activity with an ID_{50} of 8.9 μg/ml. Also antimicrobial.

Glombitza, K. W., *Planta Med.*, 1974, **25**, 105.

Broussonin A 206

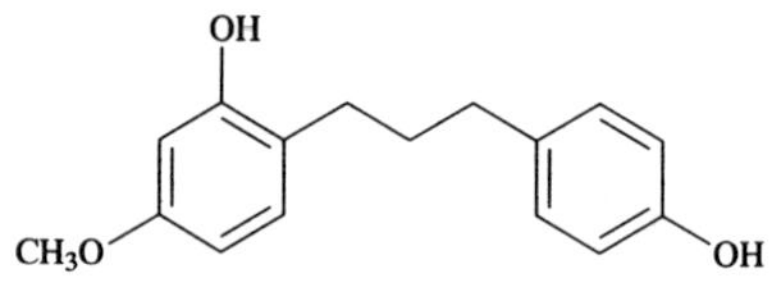

Diphenylpropanoid

$C_{16}H_{18}O_3$ [73731-87-0] MW 258.32

Occurs, together with several related structures, in shoots of paper mulberry, *Broussonetia papyrifera* (Moraceae) infected with *Fusarium solani.*

Antifungal agent (phytoalexin).

Takasugi, M., *Chem. Lett.*, 1980, 339.

Bruceantin 207

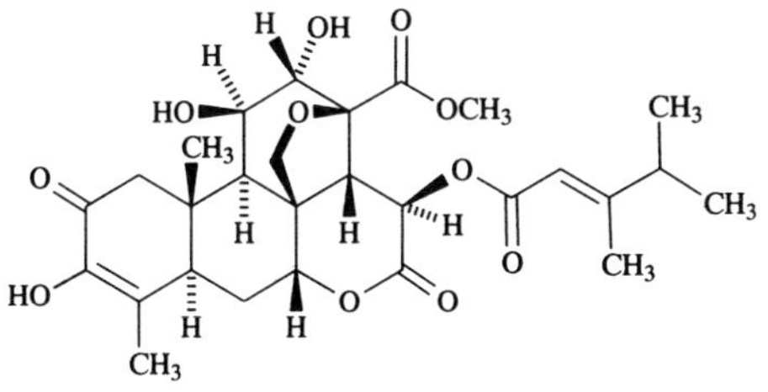

Nortriterpenoid
(Quassane)

$C_{28}H_{36}O_{11}$ [41451-75-6] MW 548.59

In the tree *Brucea antidysenterica* (Simaroubaceae).

Antileukaemic activity *in vitro*; amoebicidal *in vitro*. Has undergone clinical trials as an anticancer agent, but found to be too toxic.

Kutney, S. M., *J. Org. Chem.*, 1975, **40**, 648.

Bruceine B 208

Brucein B

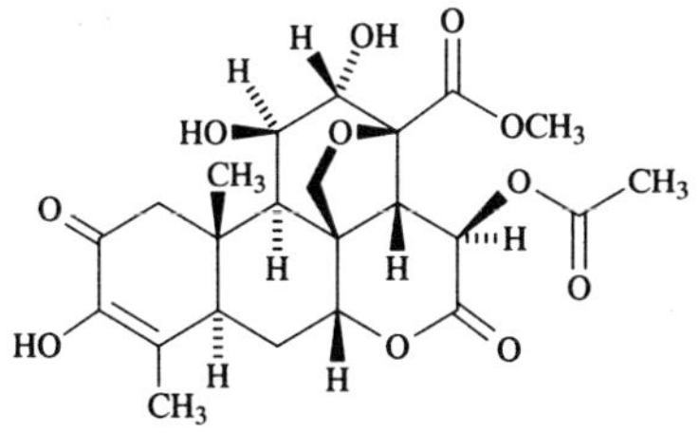

Nortriterpenoid
(Quassane)

$C_{23}H_{28}O_{11}$ [25514-29-8] MW 480.47

Seeds of *Brucea amarissima* (Simaroubaceae).

Significant insecticidal toxicity.

Polonsky, J., *Experientia*, 1967, **23**, 424.

Bruceoside A 209

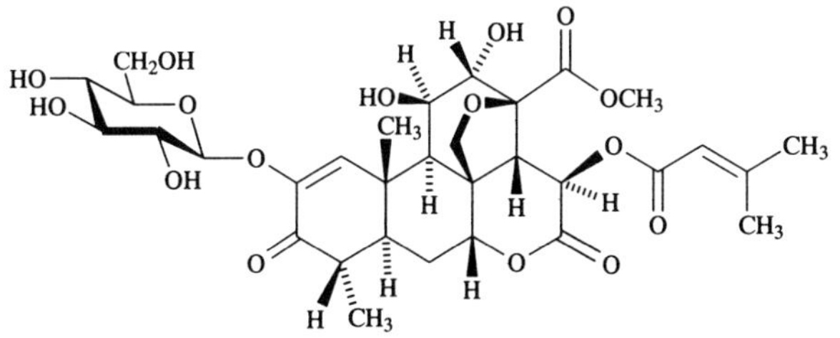

Nortriterpenoid
(Quassane)

$C_{32}H_{42}O_{16}$ [63306-30-9] MW 682.68

Seeds of *Brucea javanica* (Simaroubaceae).

Antileukaemic activity.

Li, X., *Chem. Abstr.*, 1981, **95**, 280

Brucine 210

10,11-Dimethoxystrychnine

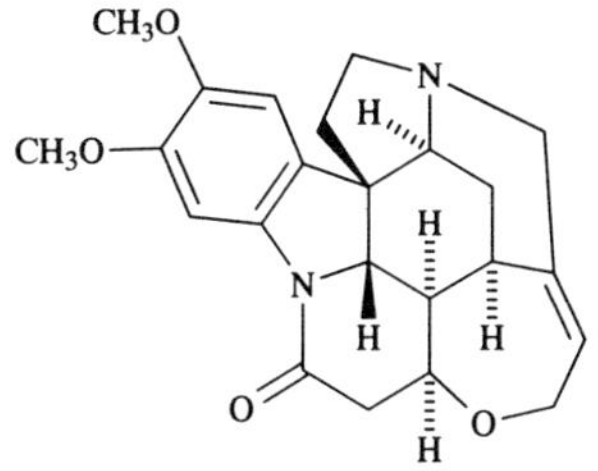

Indole alkaloid
(Strychnidine)

$C_{23}H_{26}N_2O_4$ [357-57-3] MW 394.47

In bark, wood and seed of several *Strychnos* spp. such as *S. ignatii*, *S. nux-vomica* and *S. aculeata* (Loganiaceae).

Central nervous system stimulant, resembling strychnine (q.v.), but less toxic. The lethal dose of brucine in humans is about 200 mg.

Robinson, R., *Prog. Org. Chem.*, 1952, **1**, 1.

Bryophyllin A 211

Bufadienolide

$C_{26}H_{32}O_8$ [105608-32-0] MW 472.54

Bryophyllum pinnatum (Crassulaceae).

Cytotoxic activity, with an ED_{50} of 14 ng/ml against KB cells.

Yamagishi, T., *Chem. Pharm. Bull.,* 1988, **36**, 1615.

Bryotoxin A 212

Triterpenoid
(Bufadienolide)

$C_{32}H_{42}O_{12}$ [101329-50-4] MW 618.68

Flowers of *Bryophyllum tubiflorum* (Crassulaceae). It co-occurs in the flowers with two closely related bufadienolides, bryotoxins B and C, [105608-31-9] and [105608-32-0], respectively.

The flowers cause cattle poisoning in Queensland and New South Wales. Bryotoxin A is the most toxic constituent, the fatal dose being 10 mg/kg body-weight. Poisoned cattle may die suddenly without symptoms or may be cyanotic, dyspnoeic, weak and ataxic before collapse and rapid death.

Capon, R. J., *J. Chem. Res., Synop.,* 1985, 333.

Buddledin A 213

Sesquiterpenoid
(Caryophyllane)

$C_{17}H_{24}O_3$ [62346-20-7] MW 276.38

In root bark of *Buddleja davidii* (Buddlejaceae) together with the related sesquiterpenes buddledin B and buddledin C.

Toxic to fish.

Yoshida, T., *Chem. Pharm. Bull.,* 1978, **26**, 2535.

Budlein A 214

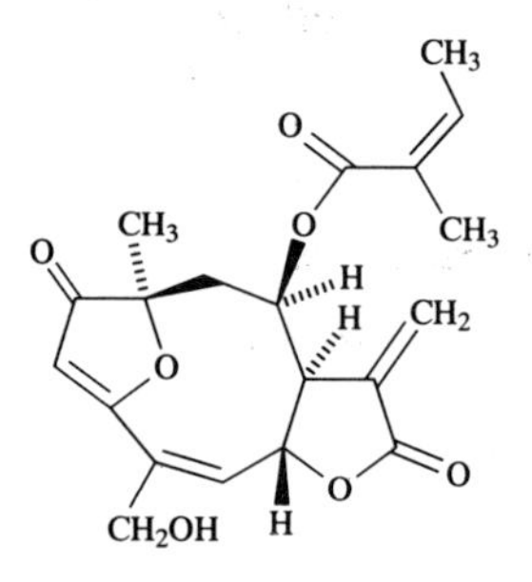

Sesquiterpenoid lactone
(Germacranolide)

$C_{20}H_{22}O_7$ [59481-48-0] MW 374.39

In *Viguiera buddleiaeformis* and *V. angustifolia* (Compositae).

Cytotoxic and antitumour activities.

Romo de Vivar, A., *Phytochemistry,* 1976, **15**, 525.

Bufotenine 215

5-Hydroxy-*N,N*-dimethyltryptamine; *N,N*-Dimethylseratonin; Mappine

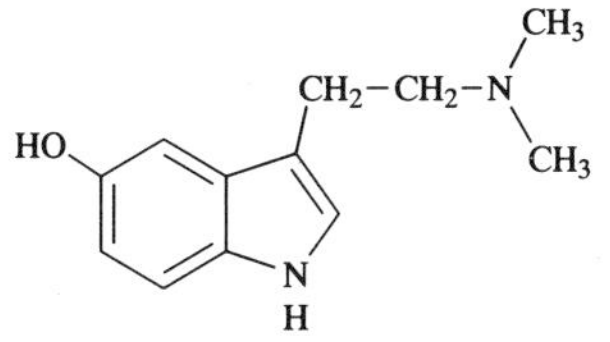

Indole alkaloid

$C_{12}H_{16}N_2O$ [487-93-4] MW 204.27

Flowers of the reed, *Arundo donax* (Gramineae) and seeds and leaves of *Piptadenia peregrina* and *P. macrocarpa* (Leguminosae). Also occurs in the poisonous secretion of toads and in some mushrooms.

Hallucinogenic, causing other mental effects including anxiety and perceptual disturbances. Autonomic effects, such as an increase in blood pressure and dilatation of the pupil are also produced.

Schultes, R. E., *Planta Med.,* 1976, **29**, 330.

Bulbocapnine 216

N-Methyl-launobine

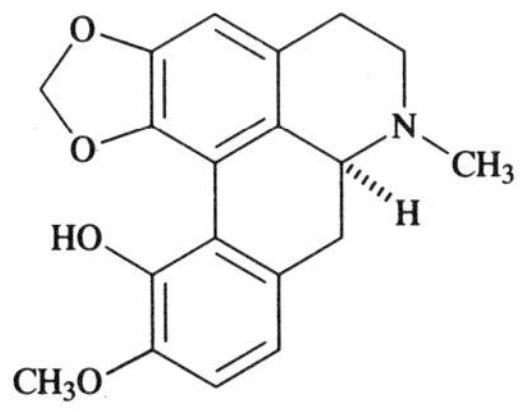

Benzylisoquinoline alkaloid (Aporphine)

$C_{19}H_{19}NO_4$ [298-45-3] MW 325.36

In most *Corydalis* spp., e.g. *C. bulbosa*, in *Fumaria officinalis*, *Glaucum flavum* and *G. pulchrum* (Papaveraceae).

Toxic. Causes cataleptic and sedative effects, and also potentiates some hypnotic drugs such as pentobarbital.

Manske, R. H. F., The Alkaloids, 1954, **4**, 119, Academic Press.

Burseran 217

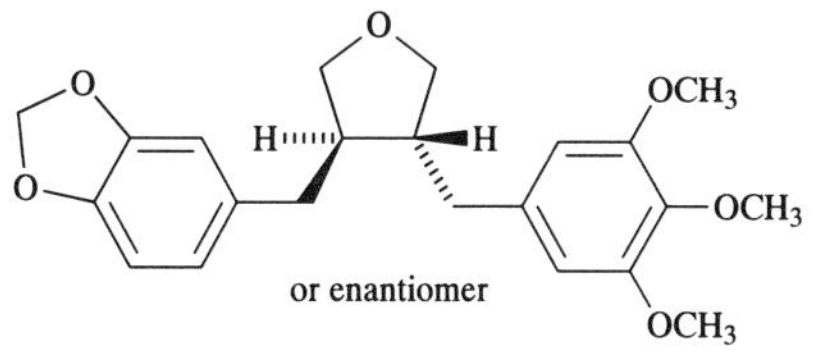

or enantiomer

Lignan

$C_{22}H_{26}O_6$ [23284-23-3] MW 386.44

Stems and leaves of *Bursera microphylla* (Burseraceae).

Antitumour activity.

Cole, J. R., *J. Pharm. Sci.,* 1969, **58**, 175.

Butrin 218

Butin 7,3′-diglucoside

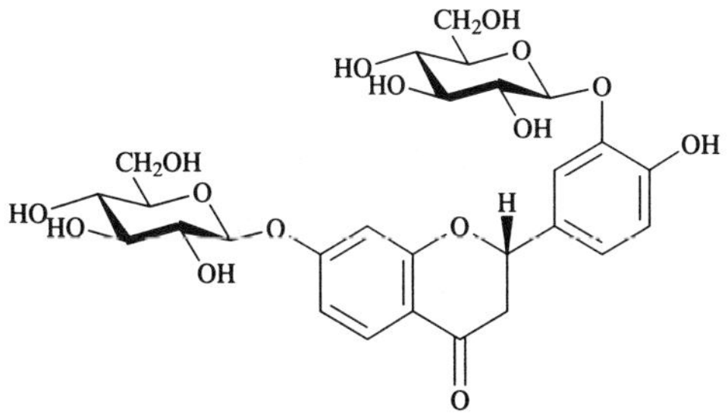

Flavanone *O*-glucoside

$C_{27}H_{32}O_{15}$ [492-13-7] MW 596.54

Flowers of *Butea monosperma* (Leguminosae).

Antihepatotoxic activity. In India, flowers of *Butea monosperma* are used directly in the treatment of hepatic disorders and of viral hepatitis.

Lal, J. B., *J. Indian Chem. Soc.,* 1935, **12**, 262.

3-Butylidene-7-hydroxyphthalide 219

Senkyunolide B

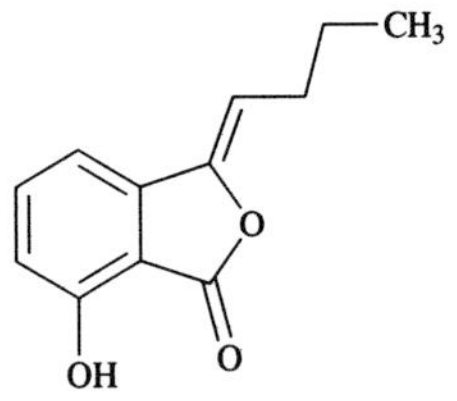

Phthalide

$C_{12}H_{12}O_3$ [93236-67-0] MW 204.23

Rhizomes of *Ligusticum wallichii* (Umbelliferae).

Increases coronary blood flow in dog heart. The crude drug has haemodynamic and analgesic effects.

Pushan, W., *Phytochemistry,* 1984, **23**, 2033.

1-*tert*-Butyl-3-methylbenzene 220

m-tert-Butyltoluene

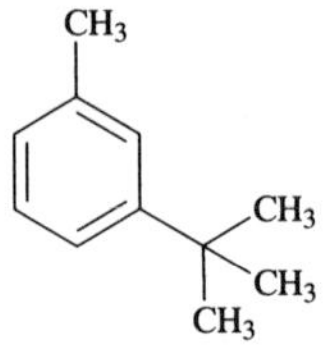

Aromatic hydrocarbon

$C_{11}H_{16}$ [1075-38-3] MW 148.25

In pine resin oil, *Pinus* spp. (Pinaceae).

Moderately toxic by inhalation.

Schlatter, M. J., *J. Am. Chem. Soc.,* 1953, **75**, 367.

1-*tert*-Butyl-4-methylbenzene 221

p-tert-Butyltoluene

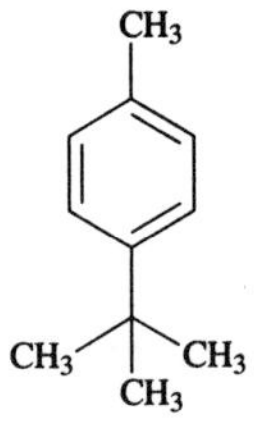

Aromatic hydrocarbon

$C_{11}H_{16}$ [98-51-1] MW 148.25

In essential oil of *Nepeta leucophylla* (Labiatae).

Highly toxic by inhalation.

Olah, G. A., *J. Am. Chem. Soc.,* 1964, **86**, 1060.

Butyrylmallotochromene 222

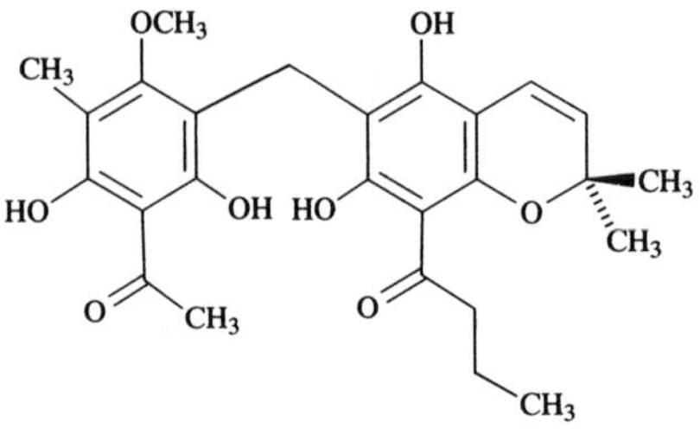

Chromene

$C_{26}H_{30}O_8$ [116979-51-2] MW 470.52

Pericarp of *Mallotus japonicus* (Euphorbiaceae).

Moderately cytotoxic against KB-cell lines. In Japan the bark of *Mallotus* is used medicinally for treatment of ulcers and cancer.

Fujita, A., *J. Nat. Prod.,* 1988, **51**, 708.

Buxamine E 223

Buxamine

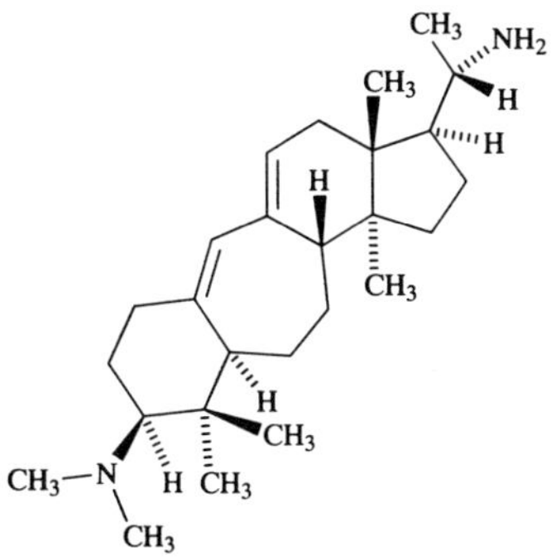

Steroid alkaloid
(Buxane)

$C_{26}H_{44}N_2$ [14317-17-0] MW 384.65

Leaves and seeds of *Buxus* spp., notably *B. sempervirens*. Several *N*-methyl derivatives (buxamines A,B,C and G) co-occur in these plants.

Strongly purgative.

Khuong-Huu, F., *Tetrahedron,* 1966, **22**, 3321.

C

(−)-Caaverine 224

1-Hydroxy-2-methoxynoraporphine

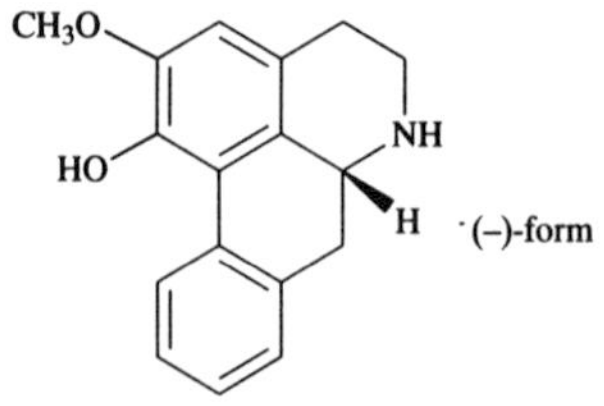

CH_3O HO NH H ·(−)-form

Benzylisoquinoline alkaloid
(Aporphine)

$C_{17}H_{17}NO_2$ [6899-64-5] MW 267.33
(+)-form [54383-30-1]

In *Symplocos celastrinea* (Symplocaceae), *Liriodendron tulipifera* (Magnoliaceae), *Isolona pilosa* (Annonaceae) and *Ocotea glaziovii* (Lauraceae).

Extracts of the bark of these trees, which grow in Brazil and North Paraguay, are toxic to farm animals.

Tschesche, R., *Tetrahedron,* 1964, **20**, 1435.

Cadaverine 225

Animal coniine; Pentamethylenediamine; Pentane-1,5-diamine

$NH_2—[CH_2]_5—NH_2$

Aliphatic amine

$C_5H_{14}N_2$ [462-94-2] MW 102.18

Seedlings of a number of leguminous plants, including pea, *Pisum sativum*, subterranean clover, *Trifolium subterraneum*, Indian pea, *Lathyrus sativus*, soybean, *Glycine max*. Also present in the stonecrop, *Sedum acre* (Crassulaceae). Emitted in the odour of human corpses. Also present in the anal sac secretion of the red fox.

Toxic; it is also a skin irritant and sensitiser.

Smith, T. A., *Prog. Phytochem.,* 1977, **4**, 27.

Cafestol 226

Cafesterol; Coffeol

Diterpenoid
(Kaurane)

$C_{20}H_{28}O_3$ [469-83-0] MW 316.44

In coffee bean oil, from *Coffea arabica* and other *Coffea* spp. (Rubiaceae).

Anti-inflammatory activity.

Kaufmann, H. P., *Chem. Ber.,* 1963, **96**, 2489.

Caffeine 227

Coffeine; Guaranine; 3-Methyltheobromine; Thein; Theine; 1,3,7-Trimethylxanthine

Purine alkaloid

$C_8H_{10}N_4O_2$ [58-08-2] MW 194.19

In the coffee bean, *Coffea* spp. (Rubiaceae), tea leaves, *Camellia sinensis* (Theaceae), maté, *Ilex paraguayensis* (Aquifoliaceae), guarana, *Paullinia cupana* (Sapindaceae) and cola, *Cola acuminata* (Sterculiaceae).

Toxic to some insects. Central nervous system stimulant, causing wakefulness and increased mental activity, diuretic, cardiac and respiratory stimulant. It is used medicinally to enhance analgesic activity.

Arnaud, M. J., *Prog. Drug Res.,* 1987, **31**, 273.

Cajanol 228

Isoflavonoid
(Isoflavanone)

$C_{17}H_{16}O_6$ [61020-70-0] MW 316.31

In fungally-infected leaves of pigeon pea, *Cajanus cajan* (Leguminosae).

Antifungal properties (phytoalexin).

Ingham, J. L., *Z. Naturforsch.,* 1979, **34C**, 159.

Calactin 229

Cardenolide

$C_{29}H_{40}O_9$ [20304-47-6] MW 532.63

Latex of *Calotropis procera* and of *Asclepias curassavica* (Asclepidaceae).

Toxic to vertebrates. LD_{50} intravenously in cats is 0.12 mg/kg body-weight.

Al-Said, M. S., *Phytochemistry,* 1988, **27**, 3245.

Calanolide A 230

Pyranocoumarin

$C_{22}H_{26}O_5$ [142632-32-4] MW 370.45

Calophyllum lanigerum (Guttiferae).

Antiviral activity.

Kashman, Y., *J. Med. Chem.,* 1992, **35**, 2735.

Calaxin 231

Sesquiterpenoid lactone
(Germacranolide)

$C_{19}H_{20}O_6$ [30412-86-3] MW 344.36

In *Helianthus ciliaris* and *Calea axillaris* (Compositae).

Cytotoxic and antitumour activities.

Ortega, A., *Rev. Latinoam. Quim.,* 1970, **1**, 81.

Calebassine 232

C-Calebassine; *C*-Curanine II; *C*-Strychnotoxine; *C*-Toxiferine II

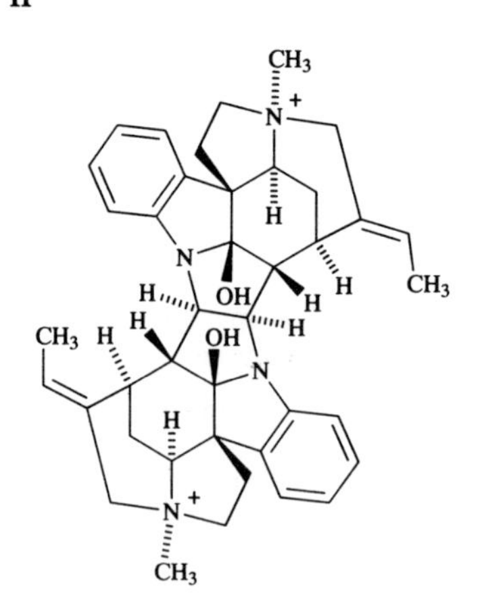

Bisindole alkaloid

$[C_{40}H_{48}N_4O_2]^{2+}$ [7257-29-6] MW 616.85

In *Strychnos* spp., including *S. triervis*, *S. mittschelichii* and *S. divaricans* (Loganiaceae).

Highly toxic. Potent neuromuscular blocking agent. A component of calabash curare, it is used as an arrow poison in South America.

Hesse, M., *Helv. Chim. Acta,* 1961, **44**, 2211.

Callicarpone 233

Diterpenoid
(Abietane)

$C_{20}H_{28}O_4$ [5938-11-4] MW 332.44

In leaves of *Callicarpa candicans* (Verbenaceae).

Toxic to fish.

Kawazu, K., *Agric. Biol. Chem.,* 1967, **31**, 498.

Calligonine 234

Elaeagnine; Tetrahydroharman

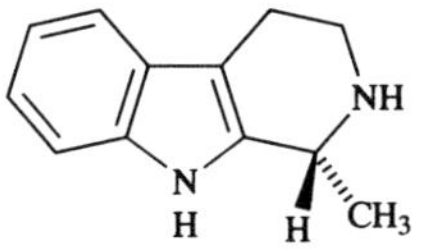

Indole alkaloid
(β-Carboline)

$C_{12}H_{14}N_2$ [2254-36-6] MW 186.26
($\pm$)-form [525-40-6]

Major alkaloid in root of *Calligonum minimum* (Polygonaceae), from *Elaeagnus augustifolia* (Elaeagnaceae), *Petalostylis labicheoides* (Leguminosae) and *Banisteriopsis argentea* (Malphigiaceae).

Causes substantial and lasting depression of blood pressure, comparable with reserpine (q.v.).

Paris, P. R., *Bull. Soc. Chim. Fr.,* 1957, 780.

Calophyllin B 235

Guanandin; 6-(3,3-Dimethylallyl)-1,5-dihydroxyxanthone

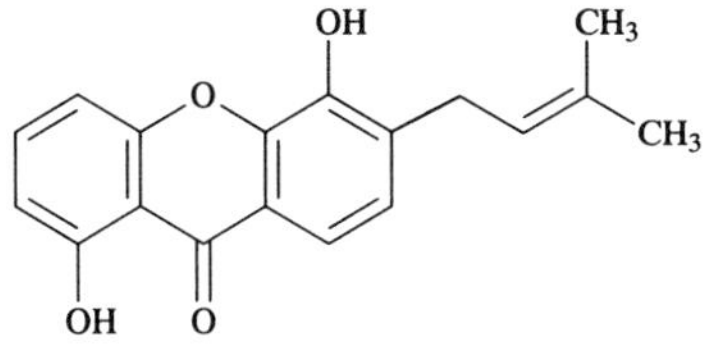

Xanthone

$C_{18}H_{16}O_4$ [17623-60-8] MW 296.32

In *Calophyllum inophyllum* and *C. bracteatum* (Guttiferae).

Anti-inflammatory activity.

Govindachari, T. R., *Indian J. Chem.,* 1968, **6**, 57.

Calophyllolide 236

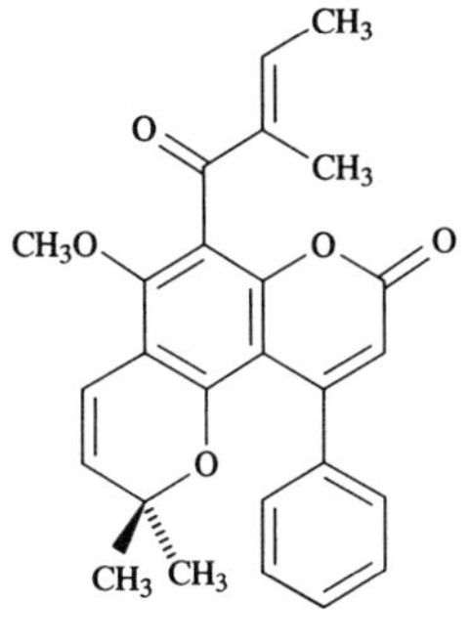

Neoflavone

$C_{26}H_{24}O_5$ [548-27-6] MW 416.47

In nuts of *Calophyllum inophyllum* (Guttiferae).

Anti-inflammatory activity. Its reported anticoagulant activity has been disputed. Toxic.

Polonsky, J., *Bull. Chem. Soc. Fr.,* 1957, 1079.

Calotropin 237

Pekilocerin A

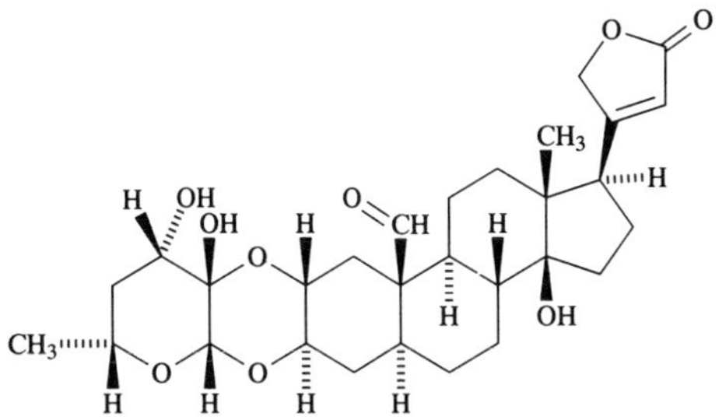

Cardenolide

$C_{29}H_{40}O_9$ [1986-70-5] MW 532.63

In latex of *Calotropis procera* (Asclepiadaceae).

Very toxic to vertebrates. LD_{50} intraperitoneally in male Swiss Webster mice 9.8 mg/kg body-weight; LD_{50} intravenously in cats is 0.11 mg/kg body-weight. In Africa used as an arrow poison.

Cheung, H. T. A., *J. Chem. Soc., Perkin Trans. 1,* 1983, 2827.

Calycanthine 238

(+)-form

Alkaloid

$C_{22}H_{26}N_4$ [595-05-1] MW 346.48

Seeds of *Calycanthus glaucus*, *C. floridus*, *C. occidentalis* and *C. praecox* (Calycanthaceae) and in the dried stems and leaves of *Palicourea alpina* (Rubiaceae).

Highly toxic; can cause violent convulsions, paralysis and cardiac depression. It shows uterine stimulant activity. Due to the presence of calycanthine, fruits of *Calycanthus floridus* cause sheep and cattle intoxication in Tennessee, USA.

Woodward, R. B., *Proc. Chem. Soc., London,* 1960, 76.

Calystegin B_2 239

Calystegine B_2; $1\alpha,2\beta,3\alpha,4\beta$-Tetrahydroxy-*nor*-tropane

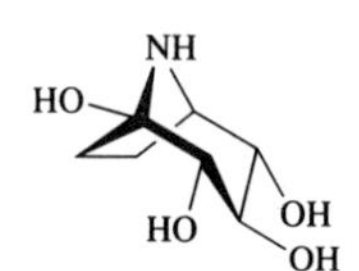

Tropane alkaloid

$C_7H_{13}NO_4$ [127414-85-1] MW 175.18

Calystegia sepium, *Convolvulus arvensis* and the weir vine, *Ipomoea* sp. (Convolvulaceae), tubers and leaves of the potato, *Solanum tuberosum*, leaves of *S. dimidiatum* and *S. kurebense*, fruits of aubergine, *S. melongena* (Solanaceae).

Toxic principle, biological activity being due to its inhibitory effects on specific glycosidases *in vivo*. Probably responsible for the neurological disorder 'crazy cow syndrome' produced in cattle that have fed on *Solanum dimidiatum* leaves in Texas. Also responsible, together with swainsonine (q.v.), for the sheep and cattle poisoning in Queensland due to feeding on weir vine.

Molyneux, R. J., *Arch. Biochem. Biophys.,* 1990, **304**, 81.

Camalexin 240

3-(2-Thiazolyl)-1*H*-indole

Indole

$C_{11}H_8N_2S$ [135531-86-1] MW 200.28

Infected leaves of *Camalium sativa* (Cruciferae) together with the 6-methoxy derivative [135531-87-2].

Antifungal agent (phytoalexin).

Browne, L. M., *Tetrahedron,* 1991, **47**, 3909.

Camellidin I 241

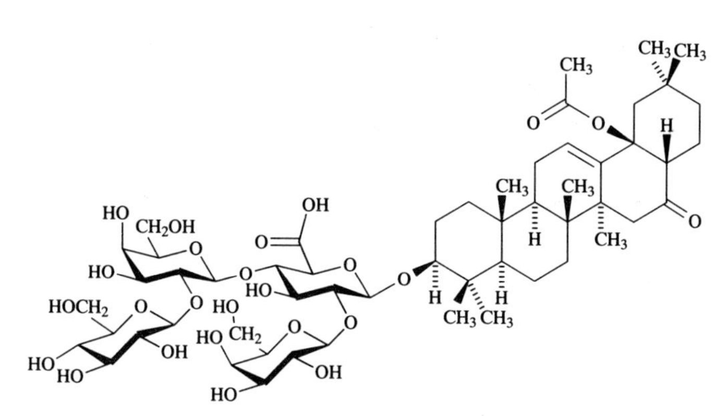

Triterpenoid saponin
(Oleanane)

$C_{55}H_{86}O_{25}$ [96827-22-4] MW 1147.27

Leaves of *Camellia japonica* (Theaceae).

Antifungal activity, interfering with spore germination at a dose of 30-100 p.p.m.

Nishino, C., *J. Chem. Soc., Chem. Commun.*, 1986, 720.

Camellidin II 242

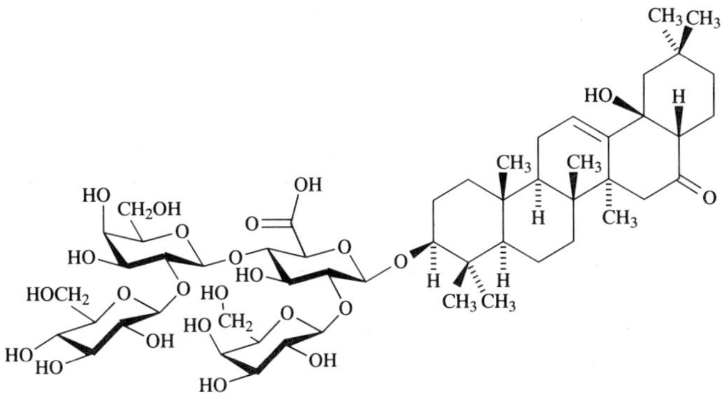

Triterpenoid saponin
(Oleanane)

$C_{53}H_{84}O_{24}$ [96827-23-5] MW 1105.24

Leaves of *Camellia japonica* (Theaceae).

Antifungal and antifeedant activities.

Numata, A., *Chem. Pharm. Bull.*, 1987, **35**, 3948.

Camphor 243

Bornan-2-one; Camphan-2-one

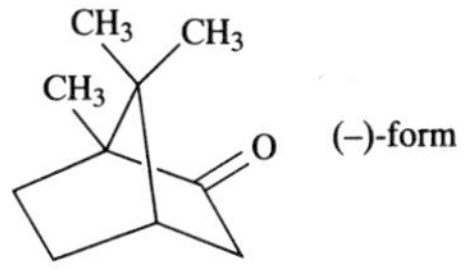

(–)-form

Monoterpenoid
(Bornane)

$C_{10}H_{16}O$ [76-22-2] MW 152.24
(+)-form [464-49-3]
(–)-form [464-48-2]

The (+)-form occurs in Japanese oil of camphor from the tree *Cinnamomum camphora* (Lauraceae). The (–)-form, matricaria camphor, occurs in feverfew, *Tanacetum parthenium* (Compositae), in *Artemisia* spp. (Compositae) and in *Lavendula* spp. (Labiatae).

Irritant; it affects the central nervous system and is moderately toxic to humans. It is used commercially as a moth repellent. Other uses are as a rubifacient and mild analgesic, and as a topical anti-pruritic.

Freudenberg, K., *Justus Liebigs Ann. Chem.*, 1954, **587**, 213.

Camptothecin 244

Camptothecine

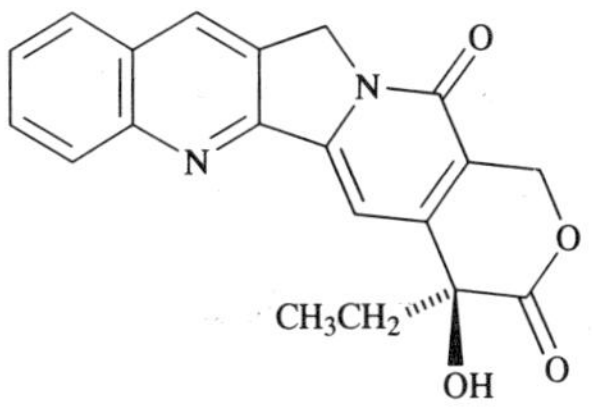

Quinoline alkaloid

$C_{20}H_{16}N_2O_4$ [7689-03-4] MW 348.36

Fruit, stemwood, bark and leaf of *Camptotheca acuminata* (Nyssaceae) and in *Mappia foetida* (Icacinaceae).

High toxicity in both farm animals and humans. In China it causes poisoning in goats who feed on the leaves of *Camptotheca acuminata*. It has notable antitumour and antileukaemia activities and is widely employed in China, despite severe side effects, to treat various forms of cancer.

Brossi, A., The Alkaloids, 1986, **27**, 1, Academic Press.

Canaliculatol 245

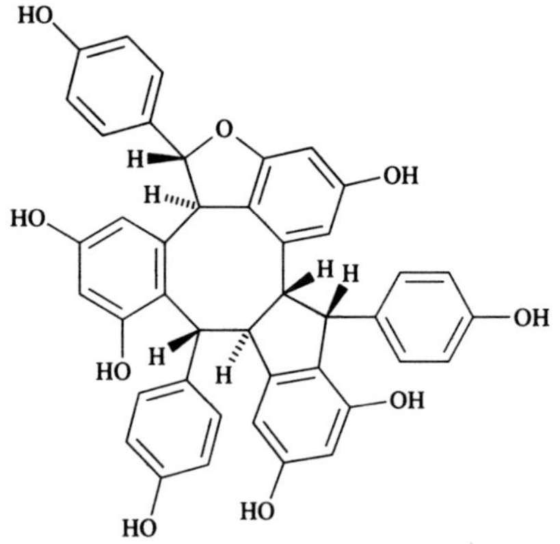

Stilbene trimer

$C_{42}H_{32}O_9$ [114488-83-4] MW 680.71

Bark of *Stemonoporus canaliculatus* (Dipterocarpaceae).

Antifungal activity against *Cladosporium cladosporioides*.

Bokel, M., *Phytochemistry,* 1988, **27**, 377.

L-Canaline 246

(*S*)-2-Amino-4-(aminoxy)butyric acid; *O*-Aminohomoserine

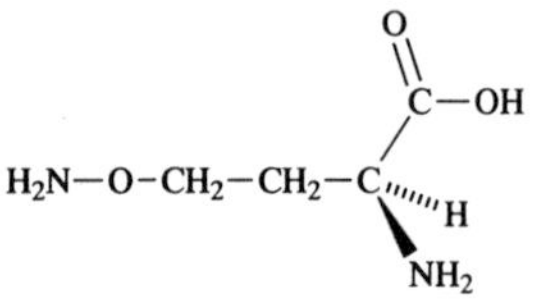

Amino acid

$C_4H_{10}N_2O_3$ [496-93-5] MW 134.14

Widespread occurrence in the seeds of Leguminosae, particularly those containing canavanine (q.v.) (from which it is easily formed), including the jackbean, *Canavalia ensiformis*.

As a non-protein amino acid, it has antimetabolic activity in a variety of systems, although its mode of action is unknown. It is potentially toxic, especially to insects, where it affects the central nervous system.

Greenstein, J. P., Chemistry of the Amino Acids, Part 3, 1961, 2622, Wiley.

L-Canavanine 247

2-Amino-4-(guanidinoxy)butyric acid

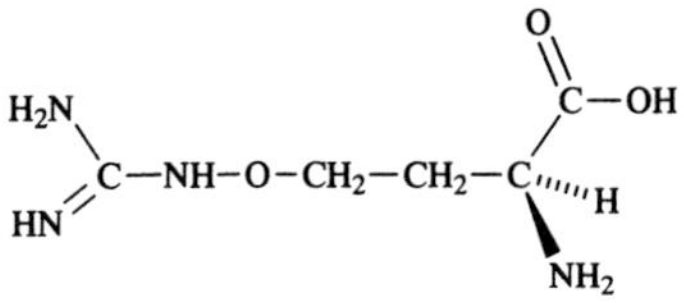

Amino acid

$C_5H_{12}N_4O_3$ [543-38-4] MW 176.18

Widespread in the seeds of the Leguminosae, including the jackbean, *Canavalia ensiformis*, the first known source.

It is cytotoxic in assorted cultures of human and animal cells and inhibits placental alkaline phosphatase in humans. These effects appear to be caused by canavanine acting as an antimetabolite and thereby blocking arginine uptake.

Rosenthal, G. A., *Q. Rev. Biol.,* 1977, **52**, 155.

Candicine 248

Maltotoxin

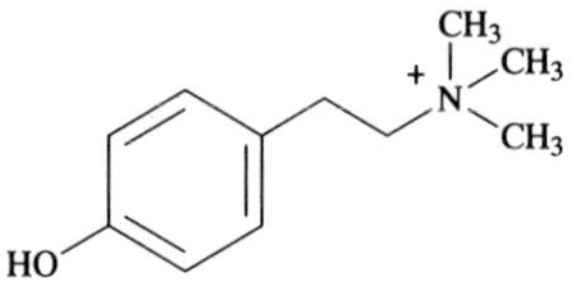

Alkaloid

$[C_{11}H_{18}NO]^+$ [6656-13-9] MW 180.27

Young roots of *Trichocereus candidans* and *T. lampochlorus* (Cactaceae), *Fagara* spp. (Rutaceae), *Hordeum vulgare* (Gramineae), *Magnolia grandiflora* (Magnoliaceae), *Phellodendron amurense* (Rutaceae) and *Lysichitum camtschatcense* (Araceae).

It has a nicotine-like action on the central nervous system. In animals, it provokes hypertension, while large doses have a curare-like action. The LD_{50} in rats is 50 mg/kg bodyweight.

Smith, T. A., *Phytochemistry,* 1977, **16**, 9.

Candimine 249

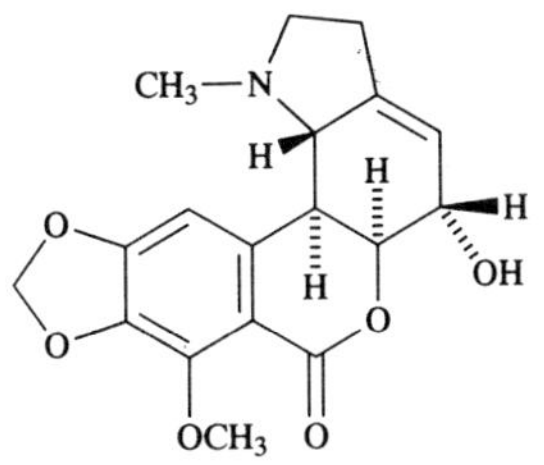

Amaryllidaceae alkaloid
(Lycorenan)

$C_{18}H_{19}NO_6$ [24585-19-1] MW 345.35

Bulbs of *Hippeastrum candidum* (Amaryllidaceae).

Toxic, causing death to animals by respiratory paralysis.

Döpke, W., *Arch. Pharm. (Weinheim),* 1962, **295**, 920.

Candletoxin A 250

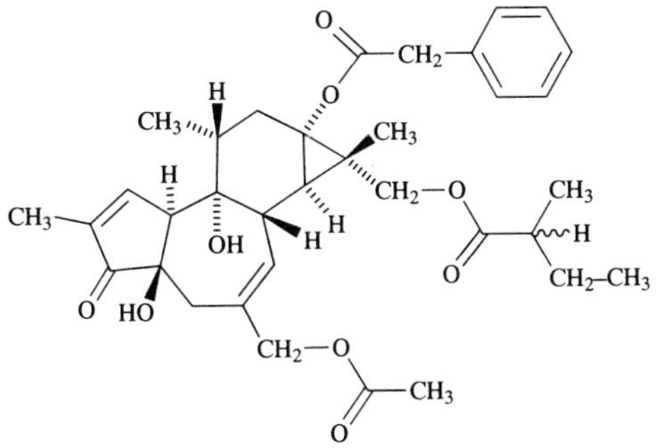

Diterpenoid
(Tigliane)

$C_{35}H_{44}O_9$ [64854-99-5] MW 608.73

Latex of *Euphorbia poisonii* (Euphorbiaceae). Co-occurs with candletoxin B [64854-99-4], which has a similar structure but lacks the acetyl group at C-20.

Toxic principle of this plant.

Schmidt, R. J., *Experientia,* 1977, **33**, 1197.

Canin 251

Chrysartemin A

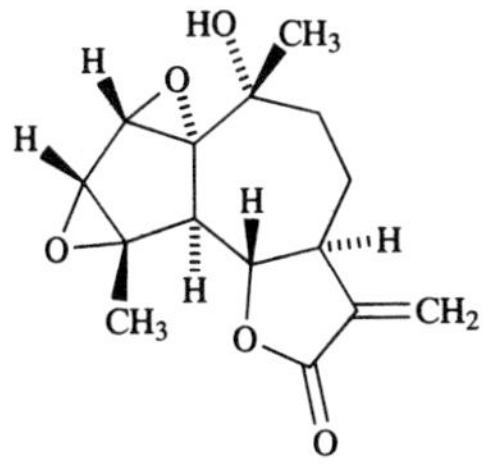

Sesquiterpenoid lactone
(Guaianolide)

$C_{15}H_{18}O_5$ [24959-84-0] MW 278.30

In *Artemisia cana* and other *Artemisia* spp., in feverfew, *Tanacetum parthenium* and in *Handelia trichophylla* (Compositae).

Toxic to insects. Cytotoxic and antitumour activities.

Lee, K. H., *Phytochemistry,* 1969, **8**, 1515.

Cannabichromene 252

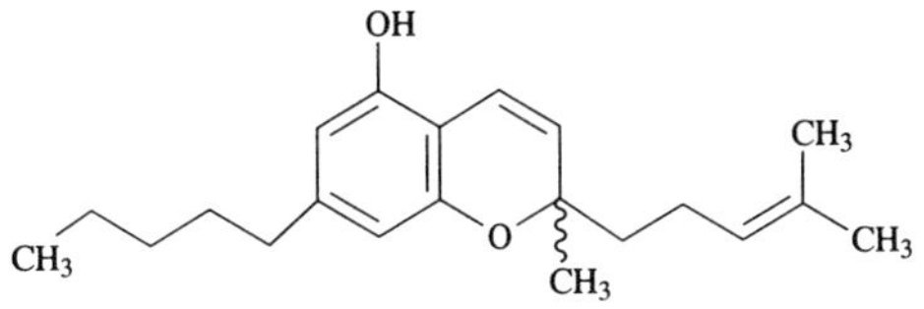

Chromene

$C_{21}H_{30}O_2$ [20675-51-8] MW 314.47

In the cannabis plant, *Cannabis sativa* (Cannabaceae).

An active constituent of hashish. Has anti-inflammatory activity. Also protects erythrocytes from hypotonic lysis.

Claussen, F., *Tetrahedron,* 1966, **22**, 1477.

Canthin-6-one 253

Canthinone

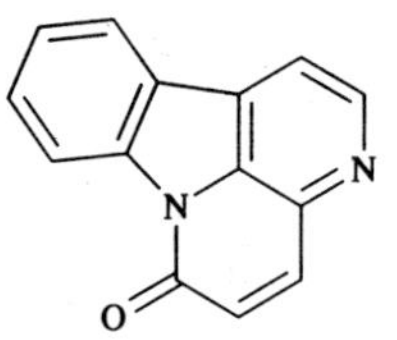

Indole alkaloid

$C_{14}H_8N_2O$ [479-43-6] MW 220.23

Wood and leaves of *Pentaceras australis* and *Zanthoxylum suberosum* (Rutaceae) and in *Picrasma crenata* (Simaroubaceae).

It is cytotoxic to guinea-pig keratinocytes.

Rosenkranz, H. J., *Justus Liebigs Ann. Chem.*, 1966, **691**, 159.

Capaurine 254

Capaurimine 10-methyl ether

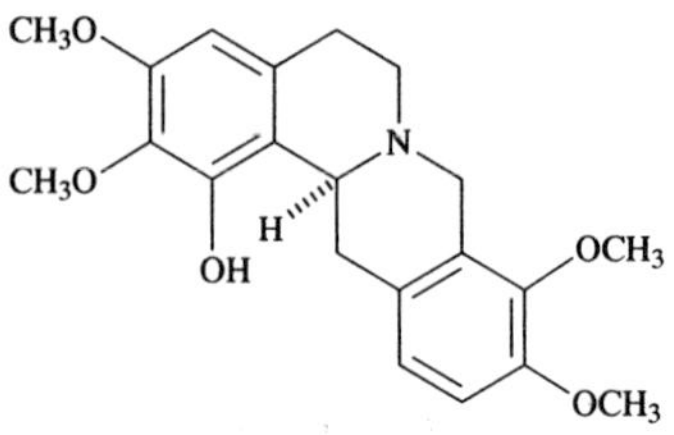

Benzylisoquinoline alkaloid
(Berberine)

$C_{21}H_{25}NO_5$ [478-14-8] MW 371.43

In *Corydalis aurea*, *C. micrantha*, *C. montana* and *C. pallida* (Fumariaceae).

Uterine stimulant activity.

Shimanouchi, H., *Acta Crystallogr., Ser. B.*, 1969, **25**, 1310.

Capsidiol 255

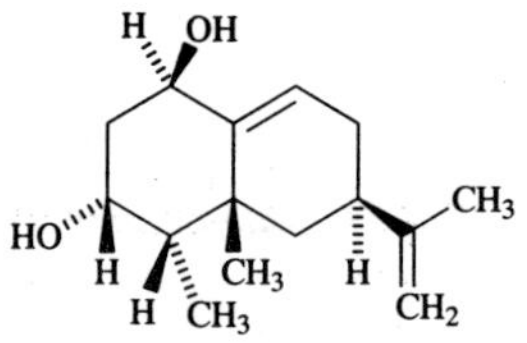

Sesquiterpenoid
(Eremophilane)

$C_{15}H_{24}O_2$ [37208-05-2] MW 236.35

Fruit of the pepper, *Capsicum frutescens* (Solanaceae) infected with *Monilinia fruticola* and leaves of the tobacco, *Nicotiana tabacum* infected with tobacco mosaic virus.

Antifungal agent (phytoalexin).

Birnbaum, G. I., *Can. J. Chem.*, 1974, **52**, 993.

Caracurine V 256

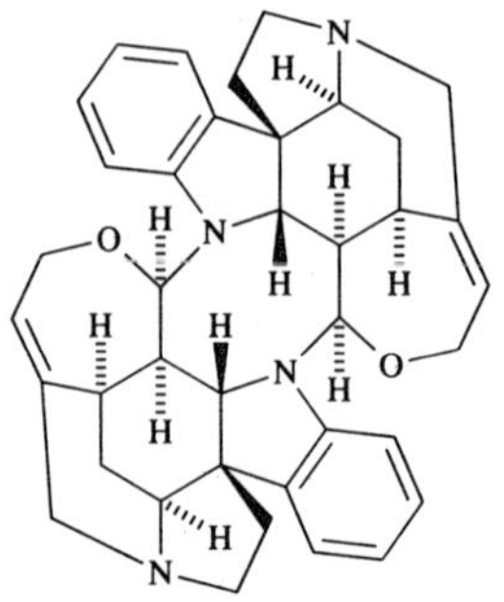

Bisindole alkaloid

$C_{38}H_{40}N_4O_2$ [630-87-5] MW 584.76

Stem bark of *Strychnos dolichothyrsa* (Loganiaceae).

Toxic. A neuromuscular blocking agent.

Verpoorte, R., *J. Pharm. Sci.*, 1978, **67**, 171.

Caranine 257

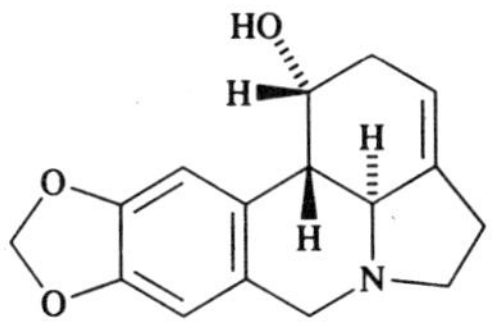

Amaryllidaceae alkaloid
(Galanthan)

$C_{16}H_{17}NO_3$ [477-12-3] MW 271.32

Bulbs of *Amaryllis belladonna*, *Ammocharis coranica* and *Clivia defixum* (Amaryllidaceae).

Toxic to animals, killing them by respiratory paralysis. The analgesic action and duration of effect of this alkaloid approach those of morphine (q.v.) and codeine (q.v.).

Warnhoff, E. W., *J. Am. Chem. Soc.*, 1957, **79**, 2192.

Carboxyatractyloside 258

Gummiferin

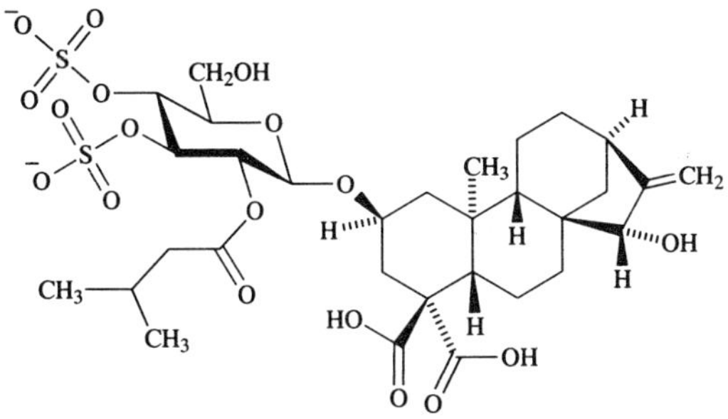

Diterpenoid
(Kaurane)

$[C_{31}H_{46}O_{18}S_2]^{2-}$ [33286-30-5] MW 770.83

Seeds and cotyledons of cocklebur, *Xanthium strumarium* and aerial parts of *Iphiona aucheri* (both Compositae).

Highly toxic. Causes a problem, mainly in swine, who eat cocklebur at the dicotyledon stage. The lethal dose in pigs is 0.75-2% body-weight and in calves 1% body-weight. The true leaves have harmless levels of toxin present. Carboxyatractyloside readily undergoes decarboxylation to atractyloside (q.v.). These two compounds co-occur in *Iphiona aucheri*, which is eaten by camels with fatal consequences.

Brookes, K. B., *S. Afr. J. Chem.*, 1983, **36**, 65.

(15:1)-Cardanol 259

Ginkgol; Anacardol; 3-(8-Pentadecenyl)phenol

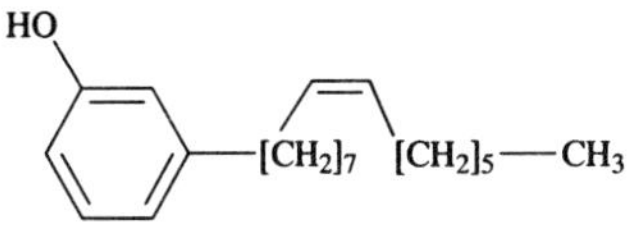

Phenol

$C_{21}H_{34}O$ [501-28-8] MW 302.50

Fruit of the Brazilian pink pepper tree, *Schinus terebinthifolius* (Anacardiaceae); fruit of *Ginkgo biloba* (Ginkgoaceae) and in the lipid oil of the cashew nut, *Anacardium occidentale* (Anacardiaceae).

It is responsible for the reported toxic effects of pink pepper. It is a skin irritant and inhibits cyclo-oxygenase and 5-lipoxygenase. It has antitumour activity against Sarcoma 180 ascites.

Tyman, J. H. P., *Chem. Soc. Rev.*, 1979, **8**, 499.

Cardiospermin 260

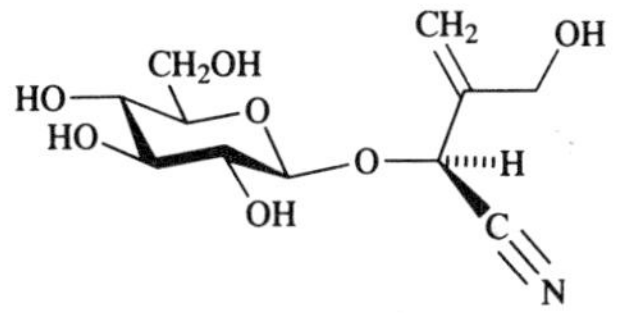

Cyanogenic glycoside

$C_{11}H_{17}NO_7$ [54525-10-9] MW 275.26

Leaves of *Cardiospermum grandiflorum*, *C. hirsutum* and *Heterodendron oleaefolium* (Sapindaceae). Is accompanied by cardiospermin sulfate in leaves and stems of *Cardiospermum grandiflorum*. It occurs as the *p*-hydroxybenzoate and *p*-hydroxycinnamate (esterified through the aglycone hydroxyl) in *Sorbaria arborea* (Rosaceae).

Toxic due to the release of cyanide.

Seigler, D. S., *Phytochemistry,* 1974, **13**, 2330.

Car-3-ene 261

3-Carene; Δ^3-Carene; Isodiprene

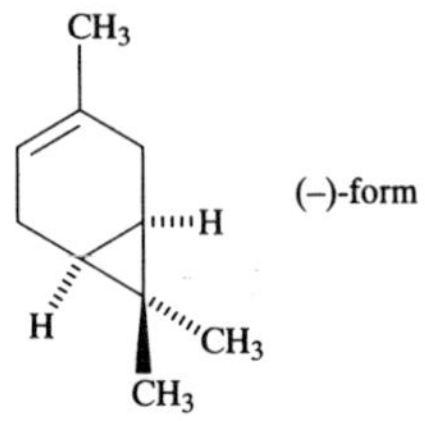

(–)-form

Monoterpenoid
(Carane)

$C_{10}H_{16}$ [13466-78-9] MW 136.24
(+)-form [498-15-7]
(–)-form [20296-50-8]

Wide occurrence in many essential oils. In gymnosperms, it is found in turpentine oils from *Pinus*, *Picea* and *Abies* spp. (Pinaceae) and especially from *Pinus sylvestris* and *P. longifolia*.

Irritant.

Karrer, W., Konstitution und Vorkommen der organischen Pflanzenstoffe, 2nd edn, 1981, 31, Birkhäuser.

(±)-Carnegine 262

Pectenine; *N*-Methylsalsolidine

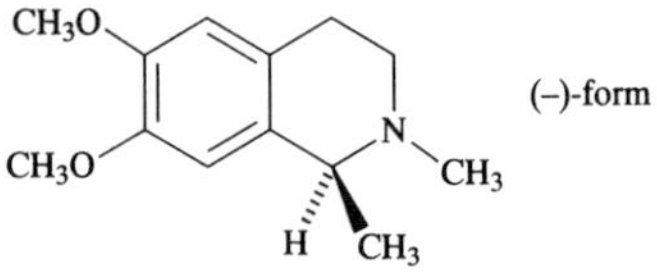

(–)-form

Isoquinoline alkaloid

$C_{13}H_{19}NO_2$ [490-53-9] MW 221.30

In *Carnegiea gigantea* and *Cereus pectenaboriginum* (Cactaceae) and *Haloxylon salicornicum* and *H. articulatum* (Chenopodiaceae).

Provokes convulsions in warm-blooded animals. Toxic.

Battersby, A. R., *J. Chem. Soc.,* 1960, 1214.

Carolinianine 263

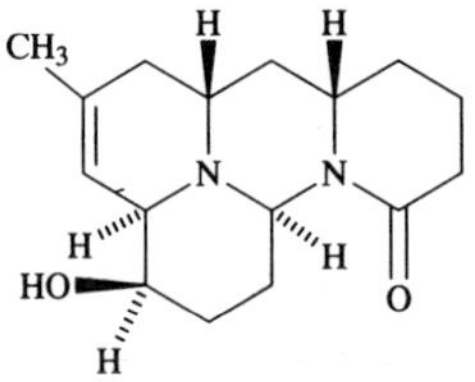

Lycopodium alkaloid

$C_{16}H_{24}N_2O_2$ [36101-39-0] MW 276.38

In *Lycopodium carolinianum* var. *affine* (Lycopodiaceae).

Moderately toxic.

Miller, N., *Bull. Soc. Chim. Belg.,* 1971, **80**, 629.

Carpacin 264

Isosafrole methyl ether

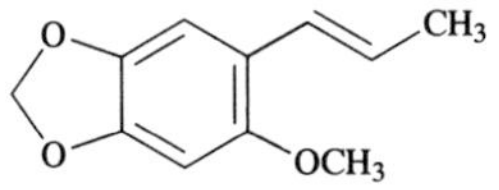

Phenylpropanoid

$C_{11}H_{12}O_3$ [23953-63-1] MW 192.21

In the bark of the carpano tree, *Cinnamomum* sp. (Lauraceae) and in *Justicia prostrata* (Acanthaceae).

Toxic to insects. It has weak sedative action. *Justicia* plants are used as an antidepressant in Asiatic folk medicine.

Mohondas, J., *Aust. J. Chem.,* 1969, **22**, 1803.

Carpaine 265

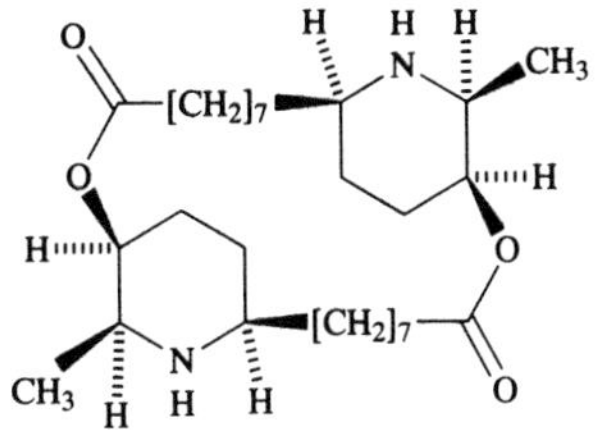

Bispiperidine alkaloid

$C_{27}H_{50}N_2O_4$ [3463-92-1] MW 466.71

Seeds and leaves of the pawpaw, *Carica papaya* and of *Vasconcellea hastata* (Caricaceae).

Cardiotonic agent and a potent amoebicide.

Govindachari, T. R., *J. Chem. Soc.*, 1955, 1563.

Cascaroside A 266

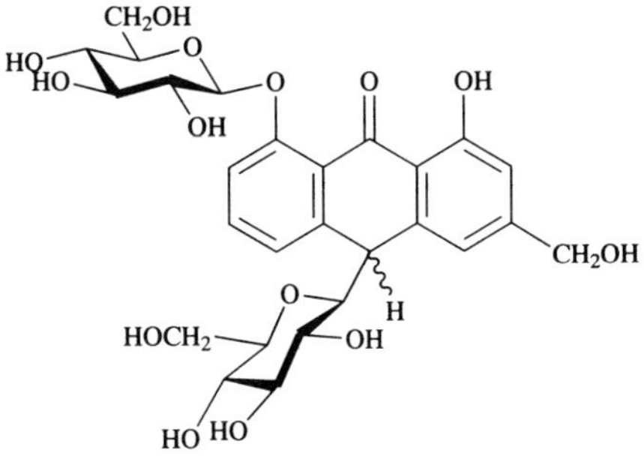

Anthrone

$C_{27}H_{32}O_{14}$ [53823-08-8] MW 580.54

Dried bark of *Rhamnus purshiana* (Rhamnaceae).

Cathartic principle of *Rhamnus* bark.

Wagner, H., *Z. Naturforsch.*, 1976, **31B**, 267.

Cassaidine 267

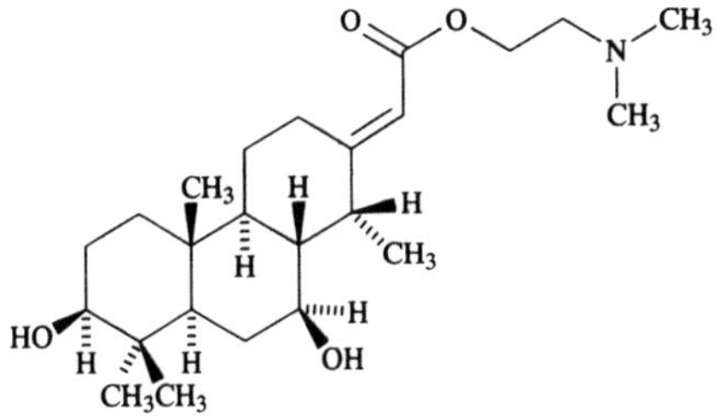

Diterpenoid alkaloid
(Cassane)

$C_{24}H_{41}NO_4$ [26296-41-3] MW 407.59

In the bark of *Erythrophleum guineense* (Leguminosae), together with cassaine (q.v.).

Toxic, with a digitalis-like effect on the heart and a very strong local anaesthetic action.

Ruzicka, L., *Helv. Chim. Acta*, 1940, **23**, 753.

Cassaine 268

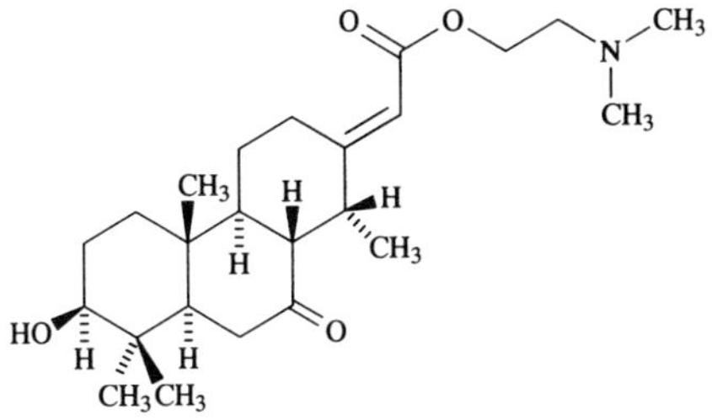

Diterpenoid alkaloid
(Cassane)

$C_{24}H_{39}NO_4$ [468-76-8] MW 405.58

In the bark of *Erythrophleum guineense* (Leguminosae), together with cassaidine (q.v.). Also found in the leaves of *E. chlorostachys*.

Convulsant action. Cardiotonic and cardiotoxic with digitalis-like activity. It has only moderate anaesthetic activity, (c.f. cassaidine). It is dangerous to livestock browsing on leaves of *Erythrophleum chlorostachys*.

Turner, R. B., *J. Am. Chem. Soc.*, 1966, **88**, 1766.

Cassyfiline 269

Cassythine

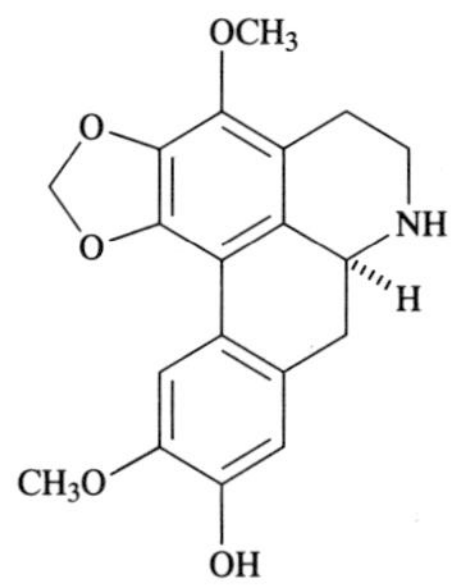

Benzylisoquinoline alkaloid
(Aporphine)

$C_{19}H_{19}NO_5$ [4030-51-7] MW 341.36

Stems of *Cassytha filiformis* (Lauraceae).

Tetanic action in animals.

Johns, S. R., *Aust. J. Chem.,* 1966, **19**, 297.

(+)-Cassythicine 270

N-Methylactinodaphnine

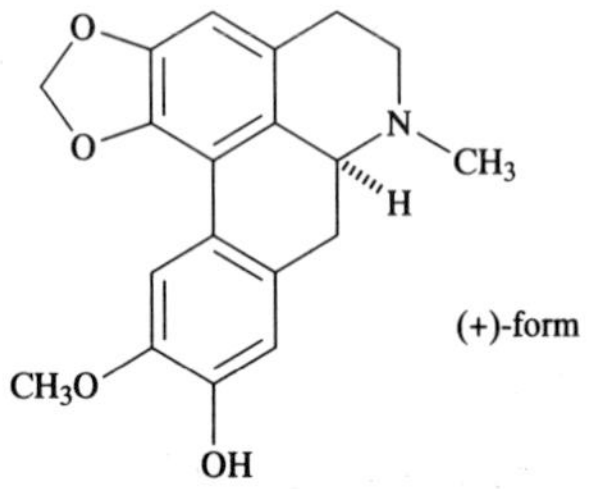

(+)-form

Benzylisoquinoline alkaloid
(Aporphine)

$C_{19}H_{19}NO_4$ [5890-28-8] MW 325.36

In *Cassytha malantha* and *Cassia glabella* (Lauraceae). The (−)-form is found in *Annona glabra* (Annonaceae).

Cytotoxic activity.

Cava, M. P., *J. Org. Chem.,* 1968, **33**, 2443.

Castalagin 271

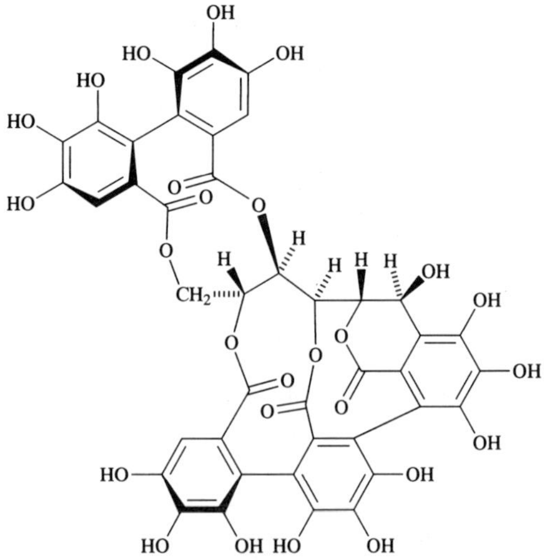

Ellagitannin

$C_{41}H_{26}O_{26}$ [24312-00-3] MW 934.64

Major leaf tannin of *Thilea glaucocarpa* (Combretaceae). Also obtained from leaves of *Quercus sessiliflora*, *Q. stenophylla* and *Castanea sativa* (Fagaceae). Co-occurs with the 1β-hydroxy isomer, vescalagin [36001-47-5].

Responsible for cattle poisoning in Brazil, following ingestion of *Thilea* or *Quercus* leaves. Vescalagin, the 1β-hydroxy isomer, appears to be equally toxic.

Mayer, W., *Justus Liebigs Ann. Chem.,* 1967, **707**, 182.

Castanospermine 272

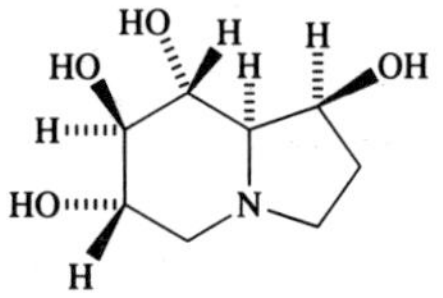

Indolizine alkaloid

$C_8H_{15}NO_4$ [79831-76-8] MW 189.21

Seeds of *Castanospermum australe* (Leguminosae) which grows in Australia, and pods of *Alexa leiopetala* from Brazil.

Unripe seeds of *Castanospermum* cause severe gastrointestinal irritation and sometimes death when eaten by horses and cattle. Australian Aborigines use the seeds as food after soaking in water and roasting. Castanospermine is toxic to insects and a potent inhibitor of α- and β-glucosidases. It also reduces the ability of the human immunodeficiency virus (HIV) to infect cultured cells and has potential for treating AIDS.

Nash, R. J., *Phytochemistry,* 1988, **27**, 1403.

Castelanone 273

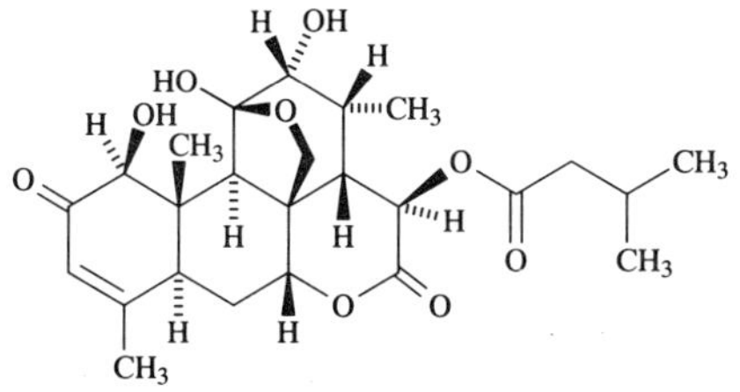

Nortriterpenoid
(Quassane)

$C_{25}H_{34}O_9$ [70424-54-3] MW 478.54

Root bark of *Castela tweediei* (Simaroubaceae).

In vitro antiviral activity against the oncogenic Rous sarcoma virus.

Polonsky, J., *C. R. Acad. Sci., Ser. C,* 1979, **288**, 269.

Casuarictin 274

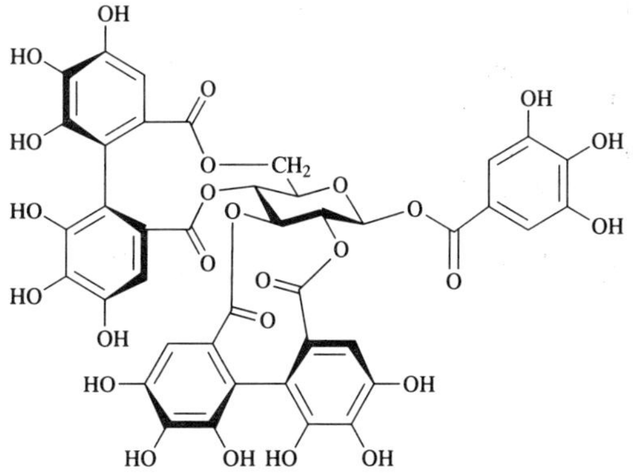

Ellagitannin

$C_{41}H_{28}O_{26}$ [79786-00-8] MW 936.66

In *Casuarina stricta* (Casuarinaceae), *Stachyurus praecox* (Stachyuraceae), *Psidium guajava, Syzygium aromaticum* and *Eucalyptus viminalis* (Myrtaceae), *Quercus* spp. (Fagaceae) and *Rubus* spp. (Rosaceae).

One of several oak tannins responsible for poisoning in cattle who eat fresh sprouts or green unripe acorns of *Quercus* species. There is evidence of biodegradation of the tannin *in vitro* to simple phenols, which enter the blood and are the true toxins. Casuarictin has *in vitro* antihepatotoxic activity, due to enzyme inhibitory action on glutamine-pyruvic transaminase.

Okuda, T., *J. Chem. Soc., Perkin Trans. 1,* 1983, 1765.

Catalpol 275

Catalpinoside

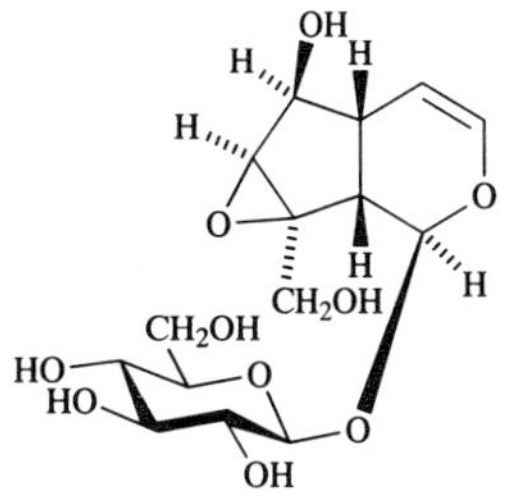

Monoterpenoid glycoside
(Iridane)

$C_{15}H_{22}O_{10}$ [2415-24-9] MW 362.33

In *Catalpa* spp. (Bignoniaceae), *Veronica* spp. (Serophulariaceae), *Plantago* spp. (Plantaginaceae) and *Buddleja* spp. (Buddlejaceae). Often co-occurs with aucubin (q.v.).

Very bitter taste. Is sequestered and stored by some butterflies, e.g. *Euphydryas* spp. to protect from bird predation. Birds tasting the butterflies avoid the toxicity by vomiting. Catalpol also has diuretic and laxative activities.

El-Naggar, L. J., *J. Nat. Prod.*, 1980, **43**, 649.

Catalposide 276

Catalpin

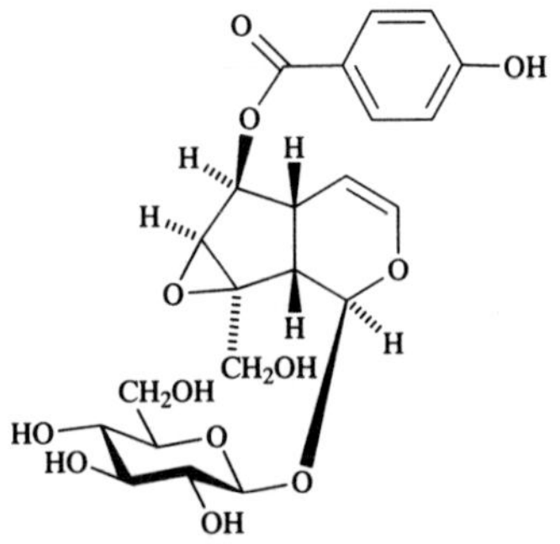

Monoterpenoid glycoside
(Iridane)

$C_{22}H_{26}O_{12}$ [6736-85-2] MW 482.44

In *Catalpa speciosa* (Bignoniaceae) and *Veronica persica* (Serophulariaceae).

Toxic to birds and antifeedant to insects. Diuretic and laxative properties in man.

Bobbitt, J. M., *J. Org. Chem.*, 1966, **31**, 500.

(+)-Catechin 277

Catechinic acid; Catechol; Catechuic acid; (+)-Cyanidanol; (+)-Cyanidan-3-ol

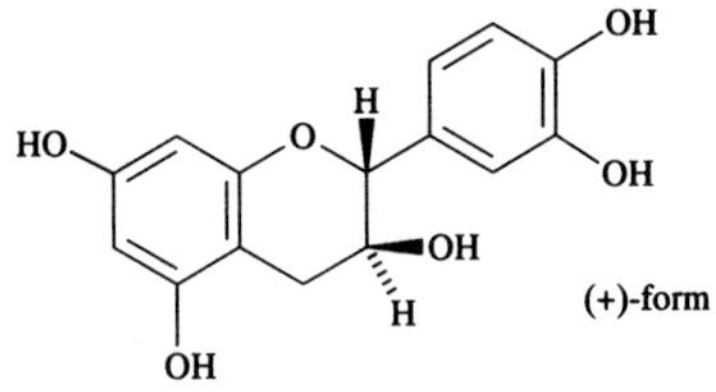

(+)-form

Flavan-3-ol

$C_{15}H_{14}O_6$ [154-23-4] MW 290.27

Widespread occurrence in Nature in woody plants, in bark, heartwood and leaf. Present, for example, in willow catkin, *Salix caprea* (Salicaceae).

Biologically very active. It is used as a haemostatic drug and in the treatment of various liver diseases, especially acute hepatitis. However, prolonged treatment with (+)-catechin can induce several adverse reactions, most of them immunomediated, such as haemolysis, acute renal failure and skin rashes.

Karrer, W., Konstitution und Vorkommen der organischen Pflanzenstoffe, 1st edn, 1958, 692, Birkhäuser.

Catechol 278

Pyrocatechol; Pyrocatechin; 1,2-Dihydroxybenzene; 1,2-Benzenediol

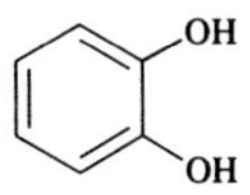

Phenol

$C_6H_6O_2$ [120-80-9] MW 110.11

Uncommon in plants. In leaves of *Populus* spp. (Salicaceae), in grapefruit, *Citrus paradisi* (Rutaceae), in avocado fruit, *Persea americana* (Lauraceae) and as the monoglucoside in the leaves of *Gaultheria* spp. (Ericaceae).

Convulsive agent and may cause eczematous dermatitis. It has been used as a topical antiseptic.

Karrer, W., Konstitution und Vorkommen der organischen Pflanzenstoffe, 1st edn, 1958, 78, Birkhäuser.

Catharanthine 279

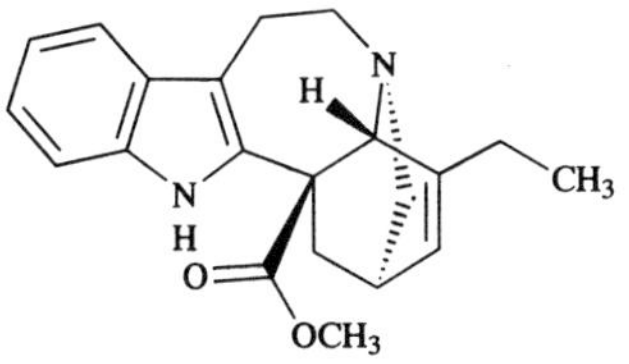

Indole alkaloid
(Ibogamine)

$C_{21}H_{24}N_2O_2$ [2468-21-5] MW 336.43

In periwinkle, *Vinca rosea* (Apocynaceae).

Hypoglycaemic activity.

Neuss, N., *Tetrahedron Lett.*, 1961, 206.

D-Cathine 280

Katine; Norpseudoephedrine; Pseudonorephedrine; ψ-Norephedrine; Nor-ψ-ephedrine

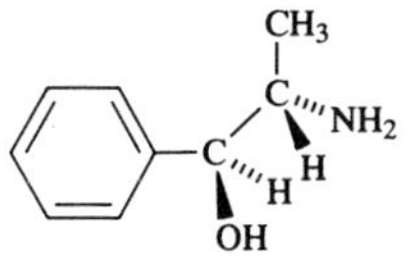

Aromatic amine

$C_9H_{13}NO$ [37577-07-4] MW 151.21

In the kat (or khat) plant, *Catha edulis*, in *Maytenus krukorii* (Celastraceae) and in the mother liquors of *Ephedra* spp. (Ephedraceae) after removal of ephedrine.

Central nervous system stimulant, mild euphoriant and anorexic. Khat is used as a stimulant drink in Arab countries.

Smith, T. A., *Phytochemistry,* 1977, **16**, 9.

D-Cathinone 281

(*S*)-2-Amino-1-oxo-1-phenylpropane; α-Benzoylethylamine

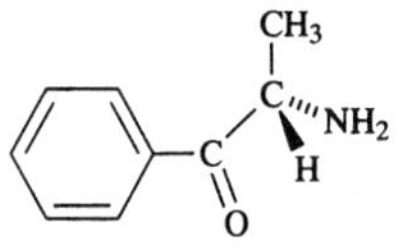

Aromatic amine

$C_9H_{11}NO$ [71031-15-7] MW 149.19

In the kat (or khat) plant, *Catha edulis*, and in *Maytenus krukorii* (Celastraceae).

It is a central nervous system stimulant, mild euphoriant and anorexic.

Kirchner, G., *Justus Liebigs Ann. Chem.*, 1959, **625**, 104.

Caulerpenyne 282

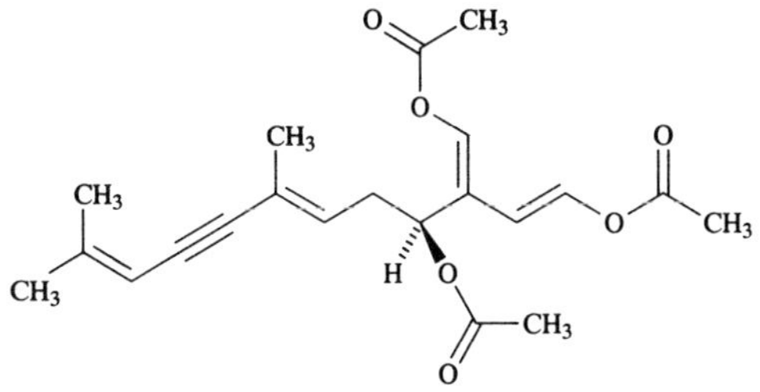

Acyclic sesquiterpenoid

$C_{21}H_{26}O_6$ [70000-22-5] MW 374.43

Caulerpa taxifolia, green alga.

Toxic to marine ciliates.

Amico, V., *Tetrahedron Lett.*, 1978, 3593.

Caulophylline 283

Methylcytisine

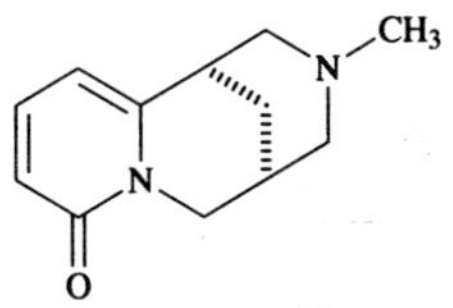

Quinolizidine alkaloid
(Cytisine)

$C_{12}H_{16}N_2O$ [486-86-2] MW 204.27

In broom, *Cytisus laburnum* and *Spartium junceum* and in many related plants of the Leguminosae; also leaves and fruits of *Caulophyllum thalictroides* (Berberidaceae).

Toxic. It is repellent to snails. Children exhibit symptoms of poisoning after eating *Caulophyllum* fruits.

Manske, R. H. F., The Alkaloids, 1953, **3**, 119, Academic Press.

Celapanine 284

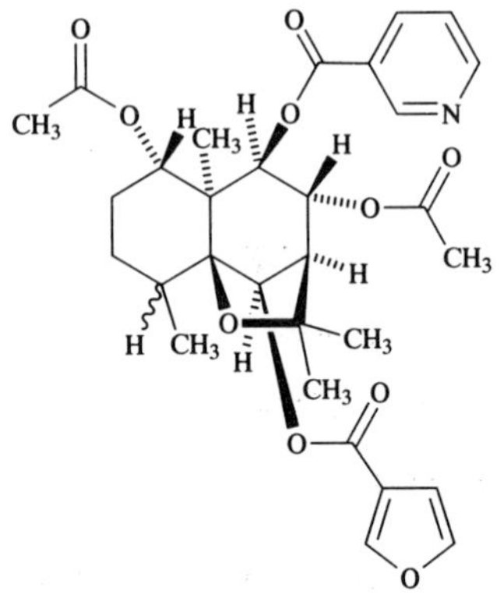

Sesquiterpenoid alkaloid
(Eudesmane)

$C_{30}H_{35}NO_{10}$ [52658-32-9] MW 569.61

One of the five structurally closely related alkaloids in *Celastrus paniculata* (Celastraceae).

The oil from the seeds is a powerful stimulant, also used for relieving rheumatic pains and for paralysis. In the sub-Himalayan region, the juice from the leaves is used as an antidote for opium poisoning.

Wagner, H., *Tetrahedron,* 1975, **31**, 1949.

α-Cembrenediol 285

2,7,11-Cembratrien-4,6-diol; 4,8,13-Duvatriene-1,3-diol

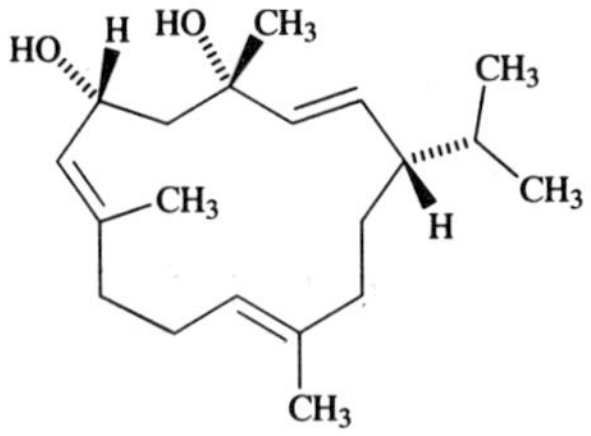

Diterpenoid
(Cembrane)

$C_{20}H_{34}O_2$ [57605-80-8] MW 306.49
β-form [57605-81-9]

Present as a mixture with an isomer (β-cembrenediol) on the leaf surface of *Nicotiana sylvestris* and *N. tabacum* (Solanaceae).

Antifungal activity, with an ED_{50} of 20 μg/ml towards *Peronospora hyoscyami* (syn. *P. tabacina*).

Cruickshank, I. A. M., *Phytopathol. Z.,* 1977, **90**, 243.

Centdarol 286

3-Himachalene-2,7-diol

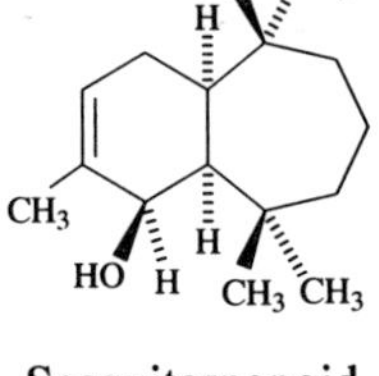

Sesquiterpenoid
(Himachalane)

$C_{15}H_{26}O_2$ [57308-24-4] MW 238.37

Wood of *Cedrus deodara* (Pinaceae).

Spasmolytic activity.

Kulshreshtha, D. K., *Phytochemistry,* 1975, **14**, 2237.

Cephalocerone 287

4,5-Methylenedioxy-6-hydroxyaurone

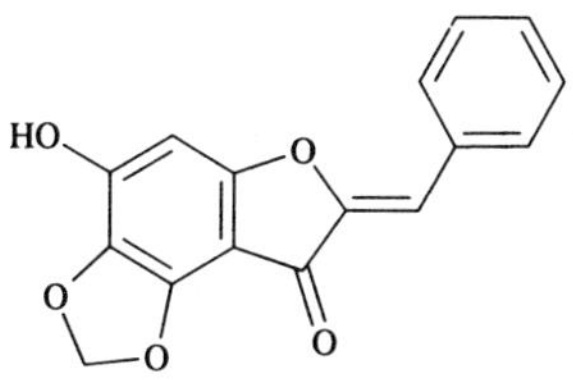

Flavonoid
(Aurone)

$C_{16}H_{10}O_5$ [135383-79-8] MW 282.25

Chitin-induced in cell cultures of the cactus, *Cephalocereus senilis* (Cactaceae).

Antifungal agent (phytoalexin), toxic to the cactus rot, *Erwinia* species.

Paré, P. W., *Phytochemistry,* 1991, **30**, 1133.

Cephalomannine 288

Taxol B

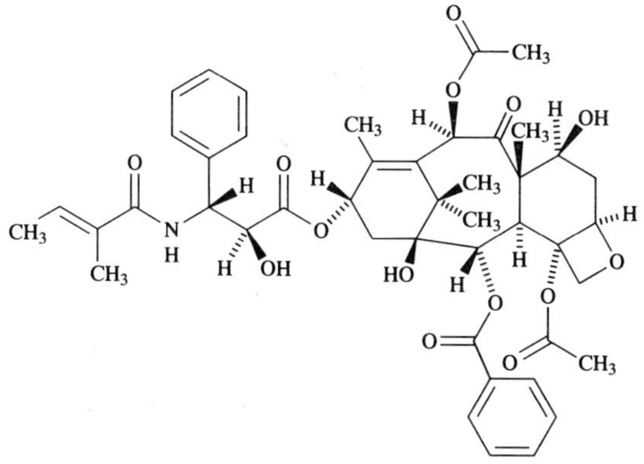

Diterpenoid alkaloid
(Taxane)

$C_{45}H_{53}NO_{14}$ [71610-00-9] MW 831.91

In all parts of the yew tree, *Taxus baccata* (Taxaceae).

It is cytotoxic in all culture experiments. Has antileukaemic and antitumour activities.

Miller, R. W., *J. Org. Chem.,* 1981, **46**, 1469.

Cephalotaxine 289

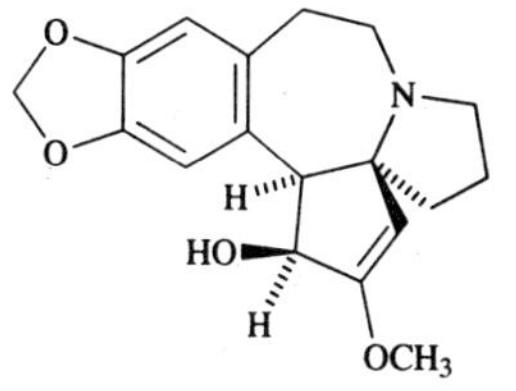

Cephalotaxine alkaloid

$C_{18}H_{21}NO_4$ [24316-19-6] MW 315.37

In *Cephalotaxus harringtonia*, *C. fortunei* and *C. wilsoniana* (Cephalotaxaceae).

Antileukaemic activity.

Arora, S. K., *J. Org. Chem.,* 1976, **41**, 551.

Cernuine 290

Alkaloid L32

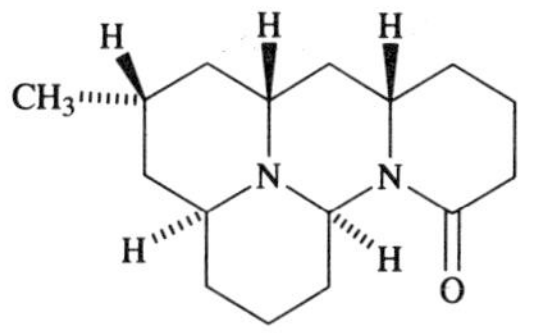

Lycopodium alkaloid

$C_{16}H_{26}N_2O$ [6880-84-8] MW 262.40

In *Lycopodium cernuum* and *L. carolinianum* (Lycopodiaceae).

Slightly toxic.

Ayer, W. A., *Tetrahedron Lett.,* 1964, 2201.

α-Chaconine 291

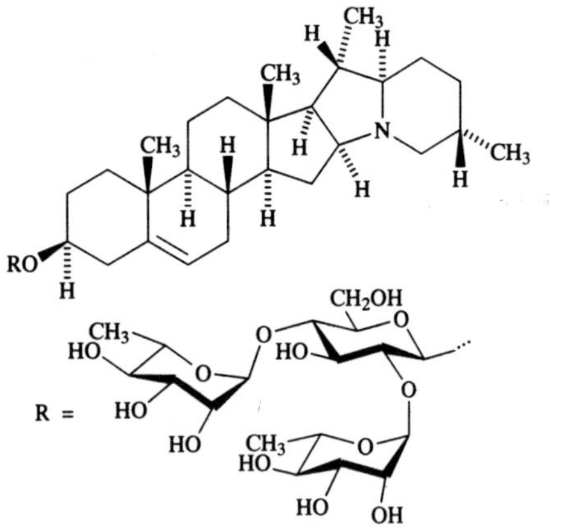

Steroidal alkaloid
(Solanidane)

$C_{45}H_{73}NO_{14}$ [20562-03-2] MW 852.07

In tubers and other parts of the wild potato, *Solanum chacoense* and of the domestic potato, *S. tuberosum* (Solanaceae) and also in black nightshade, *S. nigrum*. It is also found in *Notholiron hyacinthinum* and *Veratrum stenophyllum* (Liliaceae).

One of the toxic constituents of the potato tuber, along with solanine (q.v.). *Solanum chacoense* extracts are resistant to Colorado beetle feeding.

Keeler, R. F., Alkaloids: Chemical and Biological Perspectives, 1984, **4**, 389, Wiley.

Chaksine 292

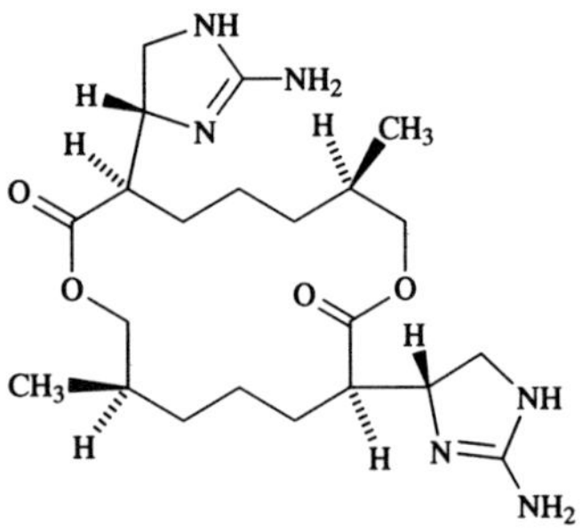

Bismonoterpenoid alkaloid
(Imidazole)

$C_{22}H_{38}N_6O_4$ [486-53-3] MW 450.59

In seeds, leaves and roots of *Cassia absus* (Leguminosae), together with isochaksine.

Causes respiratory paralysis in mice, vasoconstriction in rats and contraction of the ileum in guinea-pigs.

Voelter, W., *Angew. Chem., Int. Ed. Engl.,* 1985, **24**, 959.

Chalepensin 293

3-(α,α-Dimethylallyl)psoralen; Xylotenin

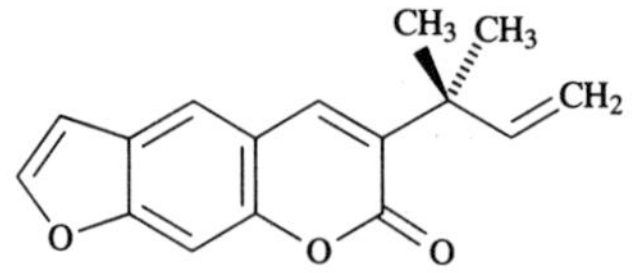

Furocoumarin

$C_{16}H_{14}O_3$ [13164-03-9] MW 254.29

In root, stem and leaf of the rue plant, *Ruta graveolens* (Rutaceae). Also present in *Psilopeganum sinense* and *Boenninghausenia* spp. (all Rutaceae).

Produces antifertility activity in rats at nontoxic doses.

Delle Monache, F., *Gazz. Chim. Ital.,* 1976, **106**, 681.

Chamaecynone 294

Chamecynone

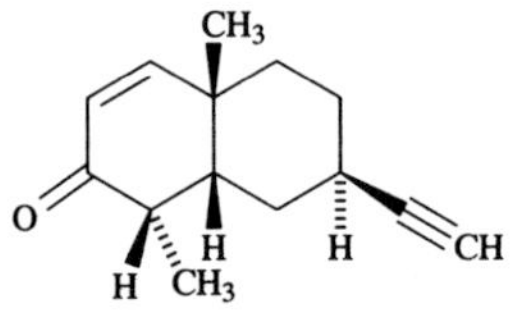

Norsesquiterpenoid
(Eudesmane)

$C_{14}H_{18}O$ [10208-54-5] MW 202.30

Wood of *Chamaecyparis formosensis* (Cupressaceae).

Toxic to termites.

Nozoe, T., *Tetrahedron Lett.,* 1966, 3663.

Chamazulene 295

Dimethulene; 7-Ethyl-1,4-dimethylazulene; Camazulene; Lindazulene

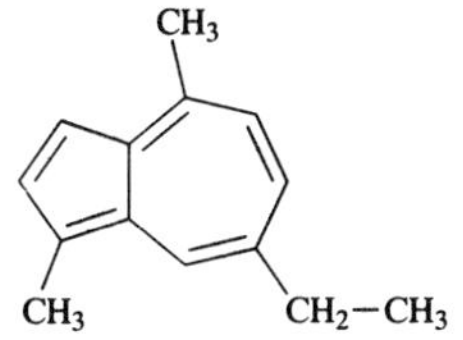

Sesquiterpenoid
(Azulene)

$C_{14}H_{16}$ [529-05-5] MW 184.28

Blue oil produced during distillation of chamomile, *Matricaria chamomilla*, wormwood, *Artemisia absinthum* and yarrow, *Achillea millefolium* (Compositae) from the sesquiterpenoid lactones such as matricin, achillin and artabsin (q.v.) present in these plants.

Toxic. Anti-inflammatory and antipyretic activities.

Meisels, A., *J. Am. Chem. Soc.*, 1953, **75**, 3865.

Chamissonin diacetate 296

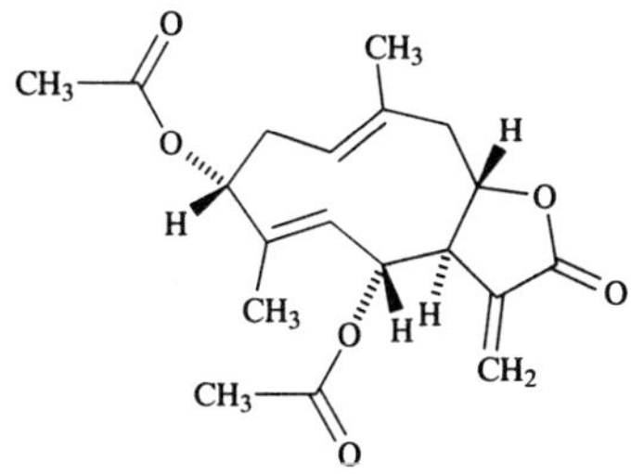

Sesquiterpenoid lactone
(Germacranolide)

$C_{19}H_{24}O_6$ [24112-95-6] MW 348.40

In *Ambrosia acanthicarpa* (Compositae).

Cytotoxic and antitumour activities.

L'Homme, M. F., *Tetrahedron Lett.*, 1969, 3161.

Chanoclavine-I 297

Chanoclavine; Secaclavine

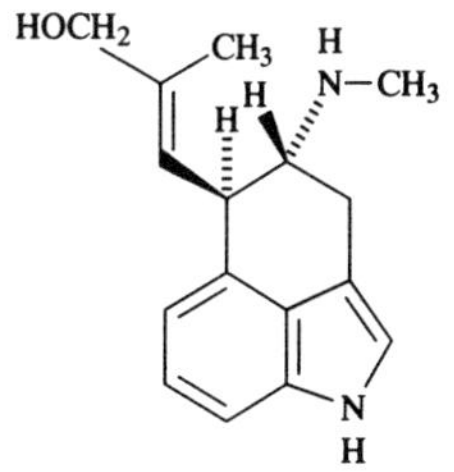

Indole alkaloid
(Ergot)

$C_{16}H_{20}N_2O$ [2390-90-0] MW 256.35

In *Rivea corymbosa*, *Ipomoea argyrophylla*, *I. violacea* and *I. tricolor* (Convolvulaceae). These form the drug 'ololiuqui' used by Central American Indians. The C-10 epimer, chanoclavine-II [1466-08-6], is also present in the same sources. Both isomers also occur in the ergot fungus, *Claviceps purpurea*.

Hallucinogenic. It is the major active principle of ololiuqui.

Hofmann, A., *Helv. Chim. Acta,* 1957, **40**, 1358.

Chaparrin 298

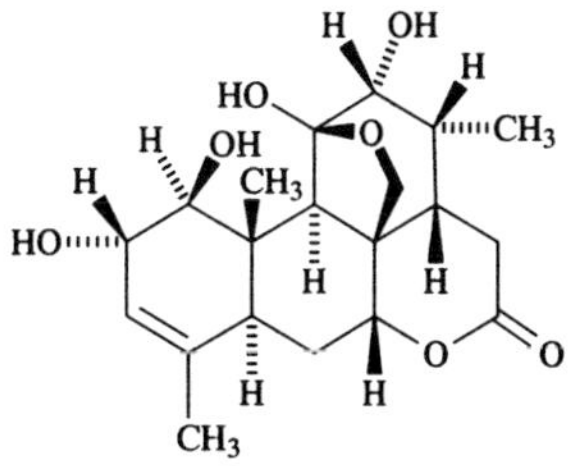

Nortriterpenoid
(Quassane)

$C_{20}H_{28}O_7$ [4616-50-6] MW 380.44

In aerial parts of chaparro amargosa, *Castela nicholsoni* (Simaroubaceae).

'Chaparro amargosa' is used as a medicine for amoebic dysentery in Mexico.

Davidson, T. A., *Can. J. Chem.,* 1965, **43**, 2996.

Chaparrinone 299

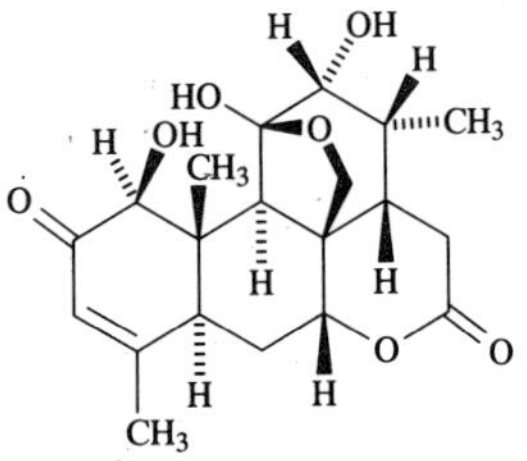

Nortriterpenoid
(Quassane)

$C_{20}H_{26}O_7$ [22611-34-3] MW 378.42

In seeds of *Quassia undulata* and of *Ailanthus altissima* (Simaroubaceae).

In vitro antiviral activity against the oncogenic Rous sarcoma virus. Tigloyloxy-6α-chaparrinone [69423-70-7] has *in vivo* antileukaemic activity.

Polonsky, J., *Bull. Soc. Chim. Fr.*, 1965, 2793.

Chaulmoogric acid 300

Hydnocarpylacetic acid

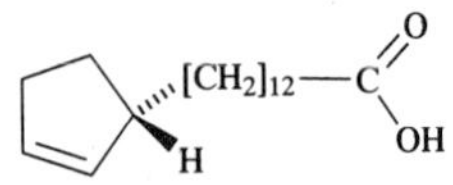

Fatty acid

$C_{18}H_{32}O_2$ [29106-32-9] MW 280.45

As the glyceride ester in chaulmoogra oil (27%) extracted from seeds of *Hydnocarpus wightiana* (Flacourtiaceae). Also present in the seed oils of *Carpotroche brasiliensis*, *Lindackeria dentata* and *Buchnerodendron speciosum* (Flacourtiaceae).

The ethyl ester has been used in the treatment of leprosy.

Mislow, K., *J. Am. Chem. Soc.*, 1955, **77**, 3807.

Chimaphylin 301

Chimaphilin

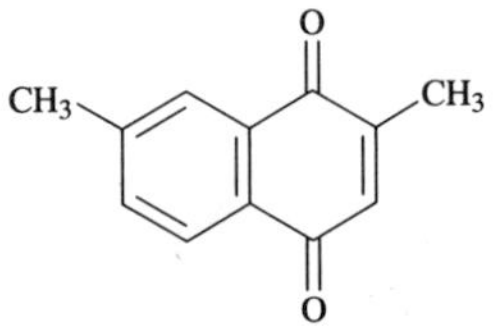

Naphthoquinone

$C_{12}H_{10}O_2$ [482-70-2] MW 186.21

In *Chimaphila corymbosa*, *Pyrola incarnata* and other members of the Pyrolaceae.

Moderately active as a phagocytose inhibitor of human granulocytes; at low dosage it stimulates phagocytose. Has been used as a urinary antiseptic.

Thomson, R. H., Naturally Occurring Quinones, 2nd edn., 1971, 199, Academic Press.

Chinensin I 302

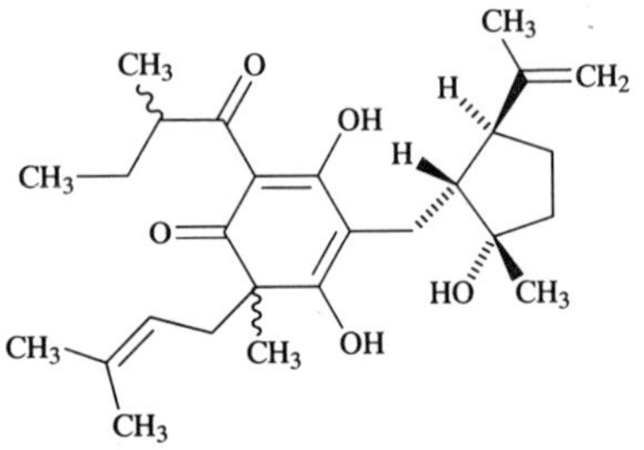

Phloroglucinol derivative

$C_{27}H_{40}O_5$ [110383-37-4] MW 444.61

Flowers of *Hypericum chinense* (Guttiferae).

Antimicrobial properties.

Nagai, M., *Chem. Lett.*, 1987, 1337.

Chinensin II 303

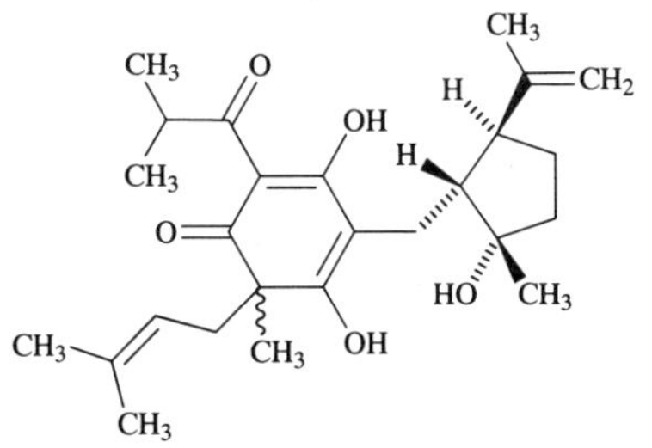

Phloroglucinol derivative

$C_{26}H_{38}O_5$ [110383-38-5] MW 430.58

Flowers of *Hypericum chinense* (Guttiferae).

Antimicrobial properties.

Nagai, M., *Chem. Lett.,* 1987, 1337.

Chlorohyssopifolin A 304

Centaurepensin; Hyrcanin

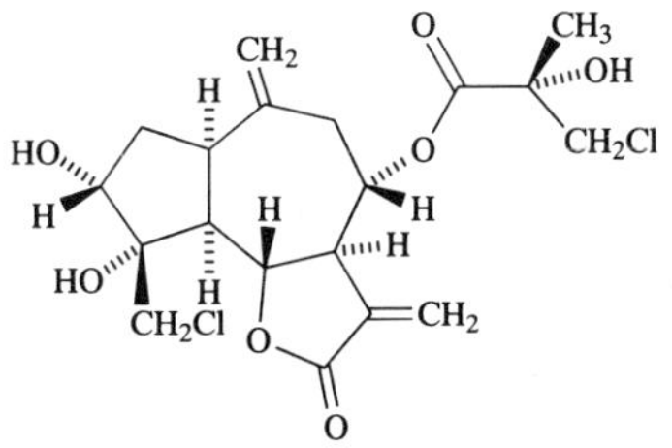

Sesquiterpenoid lactone
(Guaianolide)

$C_{19}H_{24}Cl_2O_7$ [37006-36-3] MW 435.30

In *Centaurea hyssopifolia* and Russian knapweed, *Acroptilon repens* (Compositae).

One of several sesquiterpenoid lactones which could be responsible for the toxicity of Russian knapweed to livestock. Has cytotoxic and antitumour activities.

Gonzalez, A. G., *J. Chem. Soc., Perkin Trans. 1,* 1976, 1663.

Chlororepdiolide 305

Cebellin E

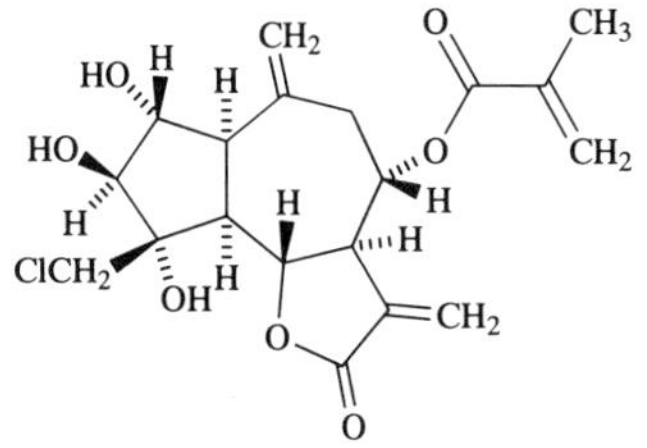

Sesquiterpenoid lactone
(Guaianolide)

$C_{19}H_{23}ClO_7$ [106566-98-7] MW 398.84

In aerial parts of Russian knapweed, *Centaurea repens* (Compositae).

One of several sesquiterpene lactones in Russian knapweed which contribute to its toxicity to livestock. It has cytotoxic activity.

Stevens, K. L., *J. Nat. Prod.,* 1986, **49**, 833.

Chrysanthemic acid 306

Chrysanthemumic acid; Chrysanthemum monocarboxylic acid

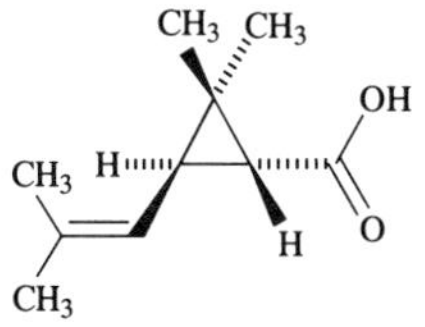

Monoterpenoid

$C_{10}H_{16}O_2$ [10453-89-1] MW 168.24

Esters of chrysanthemic acid with pyrethrolone and cinerolone (pyrethrin I and cinerin I, respectively, q.v.) occur in the cultivated pyrethrum plant, *Tanacetum cinerariifolium* (Compositae).

Irritates eyes and mucosa. The esters or pyrethrins have strong insecticidal properties and are used for killing pest insects.

Arlt, D., *Angew. Chem., Int. Ed. Engl.,* 1981, **20**, 703.

Chrysarobin 307

1,8-Dihydroxy-3-methylanthrone; Chrysophanol anthrone; Chrysophanic acid 9-anthrone

Anthrone

$C_{15}H_{12}O_3$ [491-58-7] MW 240.26

Ferreirea spectabilis and *Andira araroba* (Leguminosae); found in *Cassia* spp. (Leguminosae) and *Rumex* spp. (Polygonaceae).

Irritant and allergen. Goa powder derived from the timber of *Andira araroba* is an irritant to the respiratory tract of wood workers.

King, F. E., *J. Chem. Soc.,* 1952, 4580.

Chrysazin 308

Danthron; Dantron; 1,8-Dihydroxyanthraquinone

Anthraquinone

$C_{14}H_8O_4$ [117-10-2] MW 240.22

In roots of *Rheum palmatum* (Polygonaceae) and in leaves and stems of *Xyris semifuscata* (Xyridaceae).

At high dosage, there is immunosuppressive activity in macrophages and lymphocyte cell test systems. It shows cathartic activity and is a purgative for use in veterinary practice.

Martindale, The Extra Pharmacopoeia, 30th edn, 1993, 881, The Pharmaceutical Press.

Chrysin 309

5,7-Dihydroxyflavone; Chrysinic acid

Flavone

$C_{15}H_{10}O_4$ [480-40-0] MW 254.25

In leaf buds of *Populus* spp. (Salicaceae), in heartwood of *Pinus* spp. (Pinaceae) and in the leaves of *Escallonia* spp. (Saxifragaceae). Also occurs in glycoside combination, e.g. the 7-glucuronide [35775-49-6] is present in the leaves of *Scutellaria galericulata* (Labiatae).

Anti-inflammatory activity. It inhibits some enzyme activities (e.g. lens aldose reductase) but induces others (e.g. oestrogen synthetase).

Hauteville, M., *Bull. Soc. Chim. Fr.,* 1973, 1784.

Chrysophanol 310

Chrysophanic acid; 3-Methylchrysazin; 1,8-Dihydroxy-3-methylanthraquinone; Rheic acid

Anthraquinone

$C_{15}H_{10}O_4$ [481-74-3] MW 254.24

Widespread in nature. In *Rumex* and *Rheum* spp. (Polygonaceae), in senna leaves, *Cassia senna* (Leguminosae), in fruits and heartwood of *C. siamea* (Leguminosae), in *Rhamnus purshiana* bark (Rhamnaceae) and in teak wood, *Tectona grandis* (Verbenaceae).

Toxic to termites in teak wood. It is used as a natural red dye.

Thomson, R. H., Naturally Occurring Quinones, Recent Advances, 3rd edn, 1987, 382, Chapman and Hall.

Chrysophanol 8-glucoside 311

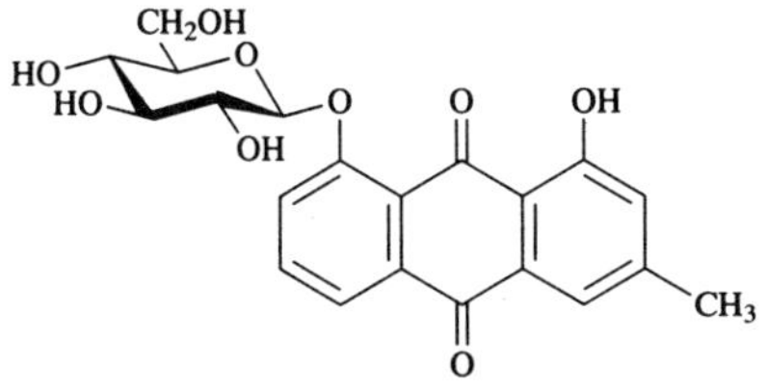

Anthraquinone

$C_{21}H_{20}O_9$ [13241-28-6] MW 416.38

In rhizomes of *Rheum moorcroftianum* (Polygonaceae) and flowers of *Woodfordia fruticosa* (Lythraceae).

Inhibits mobilisation of human spermatozoa.

Thomson, R. H., Naturally Occurring Quinones, Recent Advances, 3rd edn, 1987, 382, Chapman and Hall.

Cibarian 312

1,6-Di(3-nitropropanoyl)glucoside

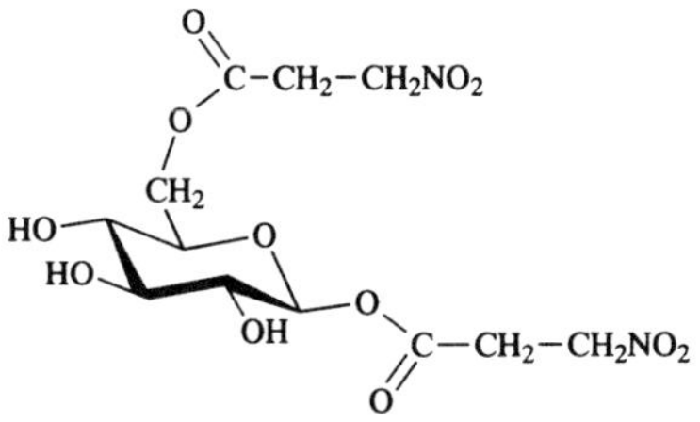

Nitro compound

$C_{12}H_{18}N_2O_{12}$ [39797-90-5] MW 382.28

In aerial parts of *Astragalus cibarius*, *A. canadensis* var. *brevidens* and *A. canadensis* var. *mortonii*, *A. falcatus* and *A. flexuosus*. Also present in *Lotus pedunculatus* and *Coronilla varia* (Leguminosae).

A nitrotoxin, potentially poisonous to livestock feeding on *Astragalus* species (q.v. miserotoxin).

Hutchins, R. F. N., *J. Chem. Ecol.*, 1984, **10**, 81.

Cichoralexin 313

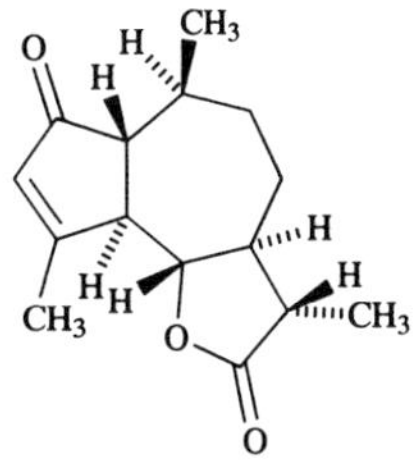

Sesquiterpenoid lactone
(Guaianolide)

$C_{15}H_{20}O_3$ [132296-37-8] MW 248.32

Leaves of chicory, *Cichorium intybus* (Compositae) infected with *Pseudomonas cichorii*.

Antifungal agent (phytoalexin).

Monde, K., *Phytochemistry*, 1990, **29**, 3449.

Cichoriin 314

6,7-Dihydroxycoumarin 7-glucoside; Cichorioside

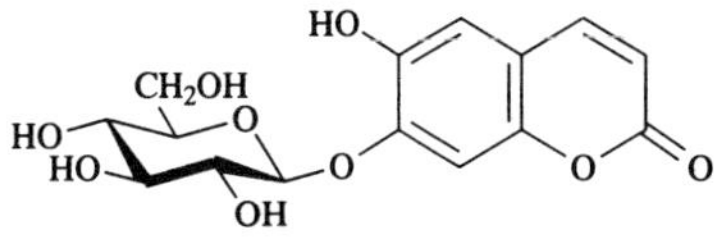

Coumarin

$C_{15}H_{16}O_9$ [531-58-8] MW 340.29

Flowers of *Cichorium intybus* and of *Sonchus* spp. (Compositae); leaves of *Fraxinus* spp. (Oleaceae).

Antifeedant activity against locusts.

Merz, K. W., *Arch. Pharm.*, 1932, **270**, 476.

Cicutoxin 315

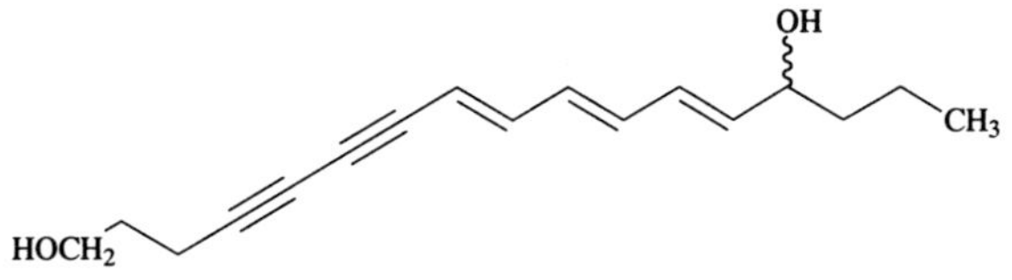

Acetylenic

$C_{17}H_{22}O_2$ [505-75-9] MW 258.36

In cowbane, *Cicuta virosa*, *C. douglasii* and *C. maculata*, where it occurs mainly in the root (Umbelliferae).

The poisonous principle of cowbane, which has caused death in livestock and man. The fatal dose of fresh green material from *Cicuta douglasii*, a North American species, is about 57 g for a sheep, 312 g for a cow and 227 g for a horse. Pigs are relatively resistant to poisoning by cicutoxin. It shows antileukaemic activity.

Konoshima, T., *J. Nat. Prod.*, 1986, **49**, 1117.

Ciguatoxin 316

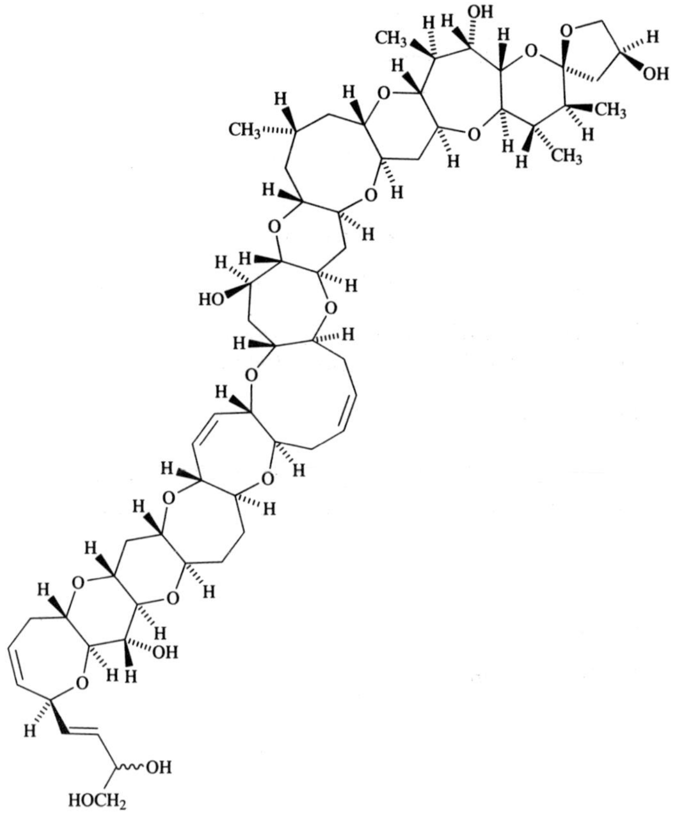

Polyether

$C_{60}H_{86}O_{19}$ [11050-21-8] MW 1111.33

Produced by the dinoflagellate *Gambierdiscus toxicus* and dietarily sequestered by several fish species including *Gymnothorax javanicus* and by shellfish.

Humans eating contaminated fish suffer 'ciguatera' food poisoning, which can be fatal.

Murata, M., *J. Am. Chem. Soc.*, 1990, **112**, 4380.

Cimifugin 317

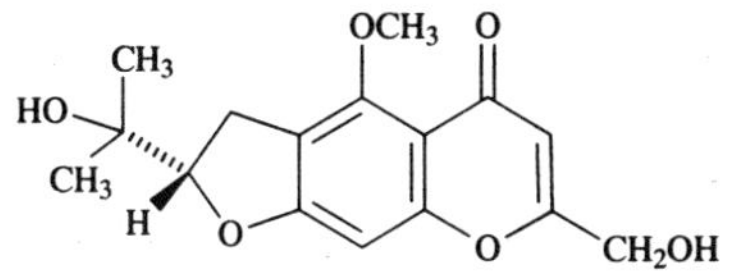

Chromone

$C_{16}H_{18}O_6$ [37921-38-3] MW 306.32

In roots and rhizomes of *Ledebouriella seseloides* (Umbelliferae), in *Cimicifuga simplex* and *Eranthis pinnatifida* (Ranunculaceae) and as a glucoside in *Angelica japonica* (Umbelliferae).

Hypotensive activity in animals. The roots of *Ledebouriella* are used as a diaphoretic, an analgesic and an antipyretic in Chinese medicine. Bitter taste.

Kondo, Y., *Chem. Pharm. Bull.,* 1972, **20**, 1940.

Cinchonidine 318

Cinchocatine; Cinchonan-9-ol; α-Quinidine

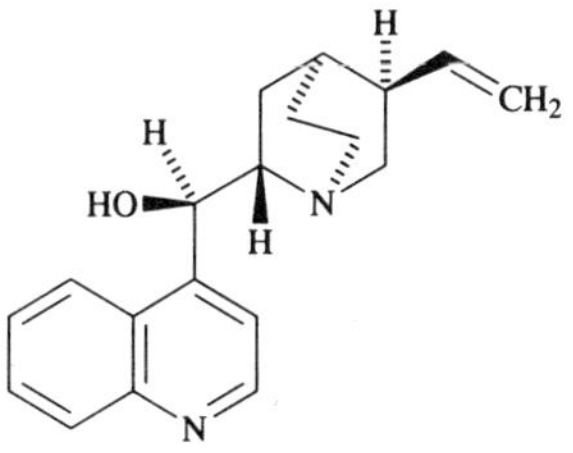

Quinoline alkaloid

$C_{19}H_{22}N_2O$ [485-71-2] MW 294.40

In *Cinchona succirubra*, *C. tucujensis* and *Remijia* spp. (Rubiaceae), in the leaves of olive, *Olea europaea* and of *Ligustrum vulgare* (Oleaceae). It co-occurs in some of these plants with its stereoisomer, cinchonine [118-10-5].

Antimalarial activity. It is used in chemistry as a resolving agent.

Manske, R. H. F., The Alkaloids, 1953, **3**, 1, Academic Press.

1,8-Cineole 319

Eucalyptol; Cajeputol; 1,8-Epoxy-*p*-menthane

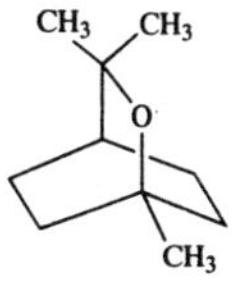

Monoterpenoid
(Menthane)

$C_{10}H_{18}O$ [470-82-6] MW 154.25

Oil of eucalyptus, distilled from fresh leaves of *Eucalyptus globulus* and some other *Eucalyptus* spp. (Myrtaceae), oil of Levant wormseed from flowers of *Artemisia maritima* var. *stechmannia* (Compositae) and in oil of leaves and twigs of *Melaleuca leucadendron* (Myrtaceae). Also widespread as a minor constituent of many other plant essential oils.

Low toxicity in rats (oral LD_{50} 2.48 mg/kg body-weight) but in man, relatively small doses of between 5 and 30 ml have caused severe respiratory depression, coma and death. Is well tolerated by mammals such as the koala bear which feed on eucalypt leaves. Has anthelmintic, expectorant and antiseptic properties. Used as a flavouring in foods.

Karrer, W., Konstitution und Vorkommen der Organischen Pflanzenstoffe, 2nd edn, **Part 1**, 1981, 130, Birkhauser.

Cinerin I 320

Chrysanthemum monocarboxylic acid cinerolone ester

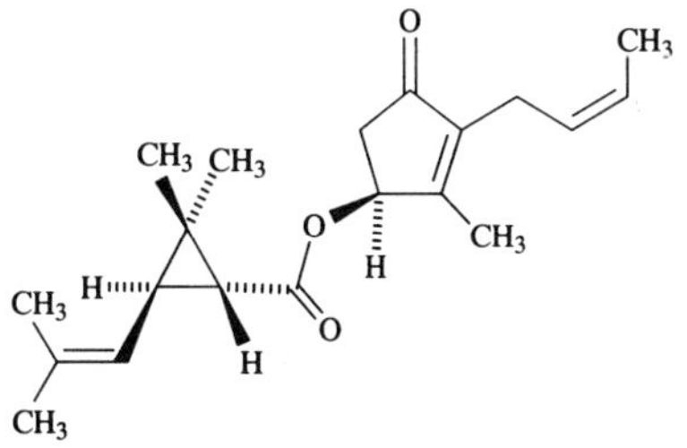

Monoterpenoid

$C_{20}H_{28}O_3$ [97-12-1] MW 316.44

In flowers of the pyrethrum plant, *Tanacetum cinerariifolium* (Compositae).

Toxic to most insects and widely used as an insecticide. It can cause convulsions, diarrhoea, respiratory paralysis and liver and kidney damage in humans.

Crombie, L., *Fortschr. Chem. Org. Naturst.*, 1961, **19**, 120.

Cinerin II 321

Chrysanthemum dicarboxylic acid monomethyl ester cinerolone ester

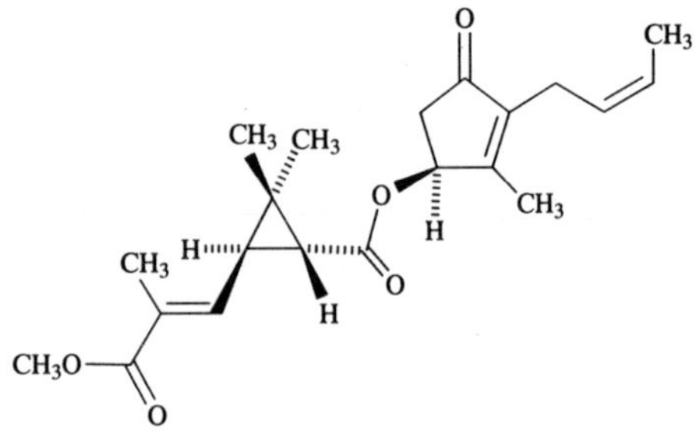

Monoterpenoid

$C_{21}H_{28}O_5$ [121-20-0] MW 360.45

In flowers of the pyrethrum plant, *Tanacetum cinerariifolium* (Compositae).

Toxic to most insects and widely used as an insecticide. It can cause convulsions, diarrhoea, respiratory paralysis and liver and kidney damage in humans.

Bramwell, A. F., *Tetrahedron*, 1969, **25**, 1727.

Cinnamic acid 322

3-Phenylacrylic acid

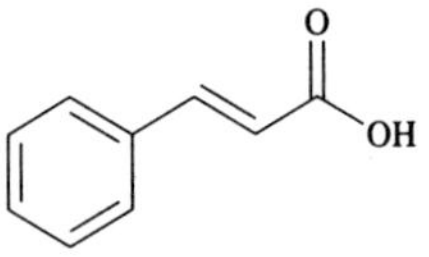

Phenylpropanoid

$C_9H_8O_2$ [621-82-9] MW 148.16

Widespread in nature, in both free and esterified forms. In the essential oils of *Cistus ladaniferus* (Cistaceae), *Cinnamomum* spp. (Lauraceae), *Alpinia* spp. (Zingiberaceae) and *Lilium candidum* (Liliaceae). Also present in various resins, e.g. styrax, *Liquidambar orientalis* (Hamamelidaceae) and in bud exudates, e.g. of *Populus* spp. (Salicaceae).

It can produce contact dermatitis, especially in alkyl ester form. May be responsible for the dermatitis caused by handling propolis from honey bees. It shows spasmolytic activity.

Martindale, The Extra Pharmacopoeia, 30th edn, 1993, 1516, The Pharmaceutical Press.

Cinnamoylcocaine 323

Cinnamylcocaine; Cinnamoylmethylecgonine; Ecgonine cinnamate methyl ester; Cinnamoylecgonine methyl ester

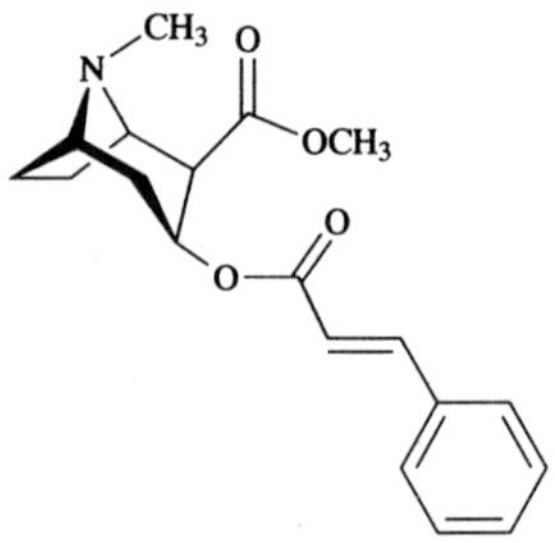

Tropane alkaloid

$C_{19}H_{23}NO_4$ [521-67-5] MW 329.40

In coca leaves, *Erythroxylum coca* and in quantity, in Javanese coca, *E. truxillense* (Erythroxylaceae).

Usually hydrolysed as a crude extract and converted to the well known drug of addiction, cocaine (q.v.).

Moore, J. M., *J. Assoc. Off. Anal. Chem.*, 1973, **56**, 1199.

Cirsilineol 324

Anisomelin; Eupatrin; Fastigenin

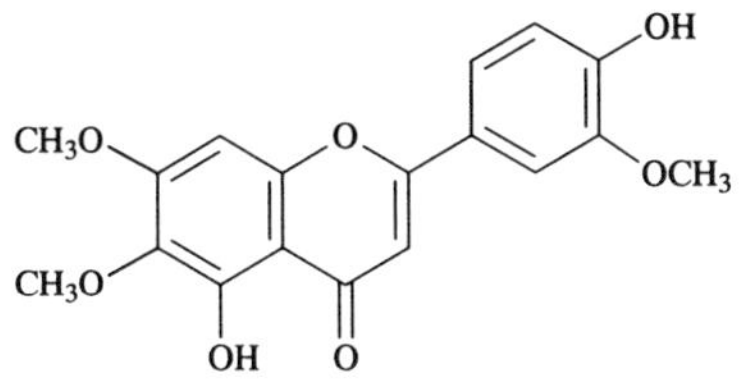

Flavone

$C_{18}H_{16}O_7$ [41365-32-6] MW 344.32

On the leaf surface of many Labiatae, e.g. thyme, *Thymus vulgaris*, *Salvia tomentosa* and several *Sideritis* spp.. Also in the Compositae, on the surface of e.g. *Artemisia hispanica* and in the Chenopodiaceae in the external leaf wax of *Chenopodium ambrosioides*.

Cirsilineol and its 8-methoxy derivative [16520-78-8] (also present in thyme) show spasmolytic activity.

Devi, G., *Indian J. Chem., Sect. B,* 1979, **17**, 84.

Cirsiliol 325

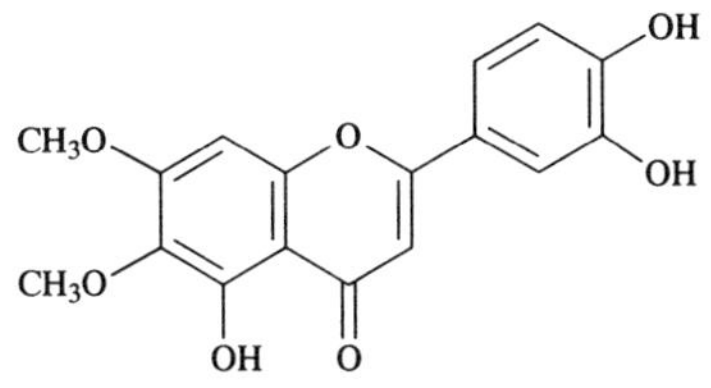

Flavone

$C_{17}H_{14}O_7$ [34334-69-5] MW 330.29

Occurs on the leaf surface of many Compositae, e.g. the thistle, *Cirsium lineare*, and Labiatae, e.g. sage, *Salvia officinalis*.

Potent and relatively selective inhibitor of arachidonate 5-lipoxygenase.

Matsuura, S., *Chem. Pharm. Bull.,* 1973, **21**, 2757.

Citronellal 326

3,7-Dimethyloct-6-enal; Rhodinal

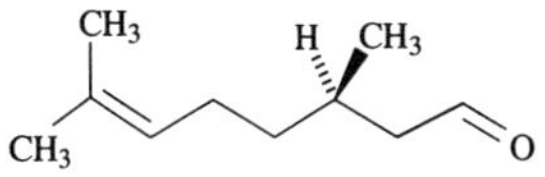

Monoterpenoid

$C_{10}H_{18}O$ [106-23-0] MW 154.25
(*R*)-form [2385-77-5]
(*S*)-form [5949-05-3]

Main constituent of Java and Ceylon oils of citronella, distilled from fresh leaves of *Cymbogon nardus* (Gramineae). It occurs in volatile oils of some *Eucalyptus* spp., e.g. *E. citriodora* (Myrtaceae), in oil of balm, *Melissa officinalis* (Labiatae) and oil of lemon, *Citrus limon* (Rutaceae).

Toxic to insects and used as an insect repellent. Has antiseptic and sedative action.

Karrer, W., Konstitution und Vorkommen der organischen Pflanzenstoffe, 2nd edn., 1981, 103, Birkhäuser.

Cleistanthin A 327

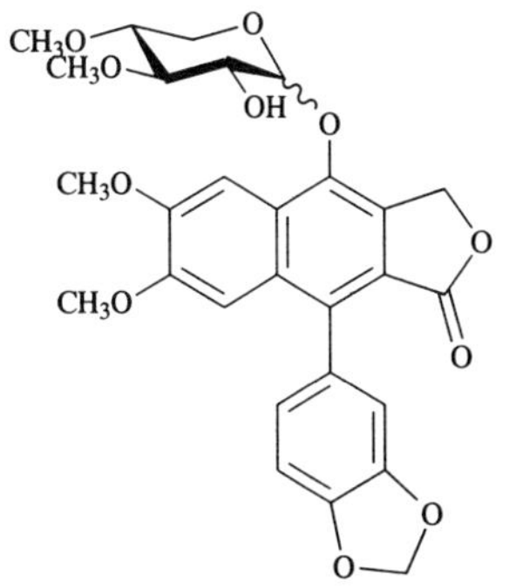

Lignan

$C_{28}H_{28}O_{11}$ [25047-48-7] MW 540.52

In the heartwood of *Cleistanthus collinus* and *C. patulus* (Euphorbiaceae).

Increases neutrophilic granulocyte count and prevents experimentally induced granulocytopenia. Toxic.

Anjaneyulu, A. S. R., *Phytochemistry,* 1975, **14**, 1875.

Cleomiscosin A 328

Cleosandrin

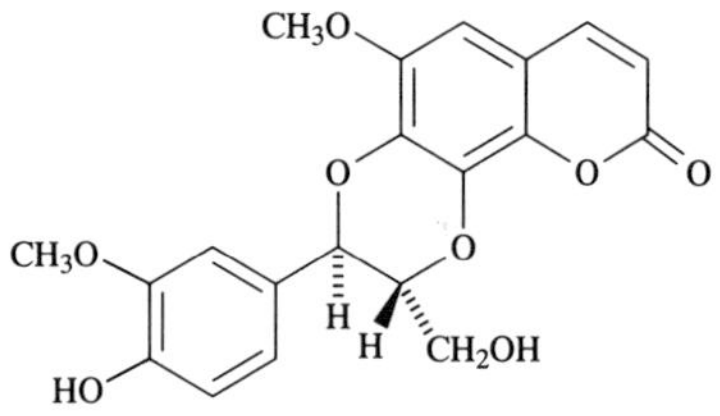

Coumarin

$C_{20}H_{18}O_8$ [76948-72-6] MW 386.36

Stemwood and stembark of *Soulamea soulameoides*, wood of *Simaba multiflora* (Simaroubaceae), bark of *Matayba arborescens* (Sapindaceae) and seed of *Cleome icosandra* (Capparidaceae).

Anticancer and antileukaemic activities. It also shows antihepatotoxic action *in vitro*.

Arisawa, M., *J. Nat. Prod.*, 1984, **47**, 300.

Clerodin 329

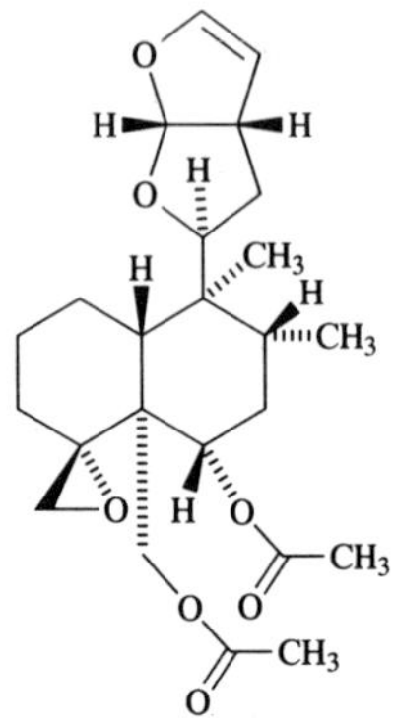

Diterpenoid
(Clerodane)

$C_{24}H_{34}O_7$ [464-71-1] MW 434.53

Leaves and twigs of the Indian bhat tree, *Clerodendrum infortunatum* (Verbenaceae) and aerial parts of *Scutellaria violacea* (Labiatae).

Antifungal activity against *Fusarium oxysporum*. The leaf extract of the Indian bhat tree is used as a vermifuge.

Rogers, D., *J. Chem. Soc., Chem. Commun.*, 1979, 97.

Clivorine 330

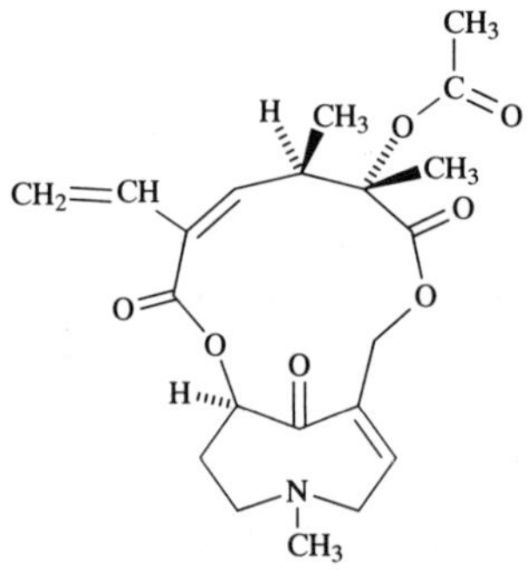

Pyrrolizidine alkaloid
(Senecionan)

$C_{21}H_{27}NO_7$ [33979-15-6] MW 405.45

In *Anchusa officinalis* (Boraginaceae) and in *Ligularia clivorum*, *L. dentata* and *L. elegans* (Compositae).

It is both hepatotoxic and carcinogenic.

Birnbaum, K. B., *Tetrahedron Lett.*, 1971, 3421.

Cnicin 331

Cynisin; Centaurin

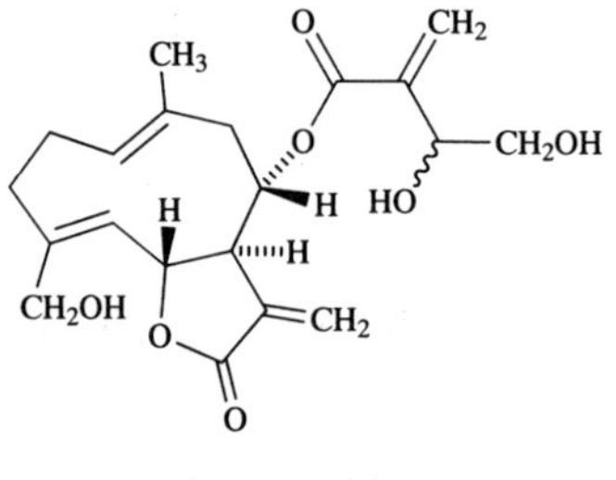

Sesquiterpenoid lactone
(Germacranolide)

$C_{20}H_{26}O_7$ [24394-09-0] MW 378.42

In the blessed thistle, *Cnicus benedictus* and in many *Centaurea* spp. (Compositae).

Bitter taste. It is an insect antifeedant and shows antitumour activity.

Suchý, M., *Collect. Czech. Chem. Commun.*, 1962, **27**, 2398.

Cocaine 332

Benzoylmethylecgonine; Methylbenzoylecgonine; Benzoylecgonine methyl ester

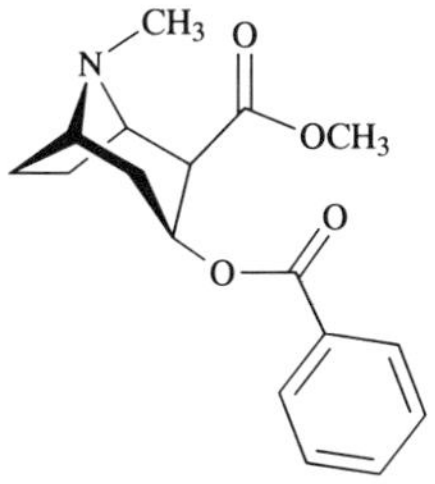

Tropane alkaloid

$C_{17}H_{21}NO_4$ [50-36-2] MW 303.36

In coca leaves, *Erythroxylum coca* and other *Erythroxylum* spp. (Erythroxylaceae).

Central nervous system stimulant and narcotic, subject to widespread abuse. An overdose can prove fatal. It is a local anaesthetic used mainly in ophthalmology, and a mydriatic.

Ashley, R., Cocaine, its history, uses and effects, 1975, Warner Books, N.Y..

Cocculolidine 333

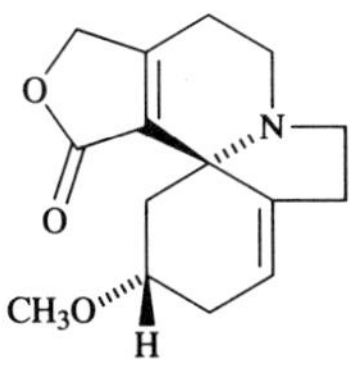

Benzylisoquinoline alkaloid
(Seco-erythrinan)

$C_{15}H_{19}NO_3$ [13497-04-6] MW 261.32

Leaves and rhizomes of *Cocculus trilobus* and fruit of *C. carolinus* (Menispermaceae).

Insecticidal and antifeedant activities.

Wada, K., *Agric. Biol. Chem.*, 1968, **32**, 1187.

Codeine 334

3-*O*-Methylmorphine; Methylmorphine

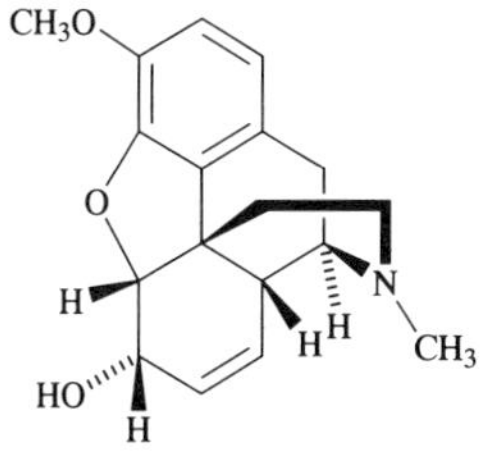

Benzylisoquinoline alkaloid
(Morphinan)

$C_{18}H_{21}NO_3$ [76-57-3] MW 299.37

In opium, the dried latex of *Papaver somniferum* (Papaveraceae), together with morphine and thebaine (q.v.).

Spasmolytic, narcotic, analgesic and antitussive. It is used extensively as a painkiller, since it is not addictive (c.f. morphine). It is also valuable for treating diarrhoea and as a cough suppressant. It has only 25% of the toxicity of morphine.

Bentley, K. W., Chemistry of the Morphine Alkaloids, 1954, 57, Oxford University Press.

Codonopsine 335

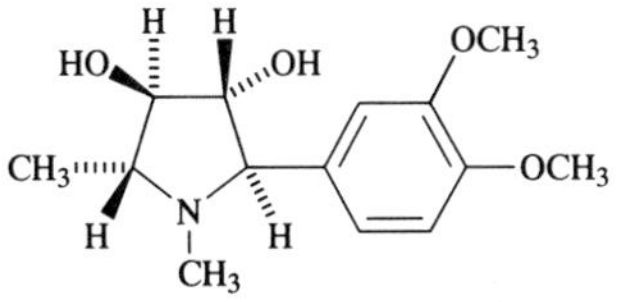

Pyrrolidine alkaloid

$C_{14}H_{21}NO_4$ [26989-20-8] MW 267.33

In *Codonopsis clematidea* (Campanulaceae).

Decreases blood pressure in cats at doses of 20 mg/kg body-weight.

Matkhalikova, S. F., *Chem. Nat. Compd.*, 1969, **5**, 528.

Coelogin 336

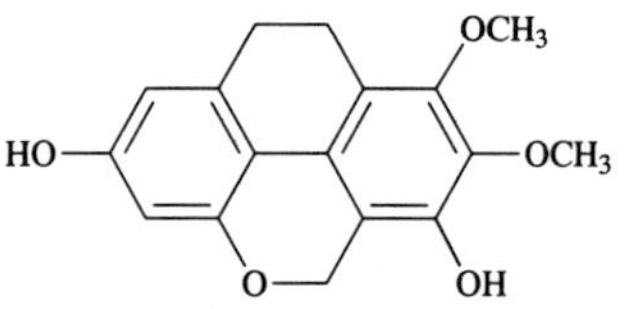

Stilbenoid
(Phenanthrene)

$C_{17}H_{16}O_5$ [82358-31-4] MW 300.31

In aerial parts of *Coelogyne ovalis* and *C. cristata* (Orchidaceae).

Spasmolytic activity.

Majumder, P., *J. Chem. Soc., Perkin Trans. 1,* 1982, 1131.

Colchicine 337

Colchiceine methyl ether

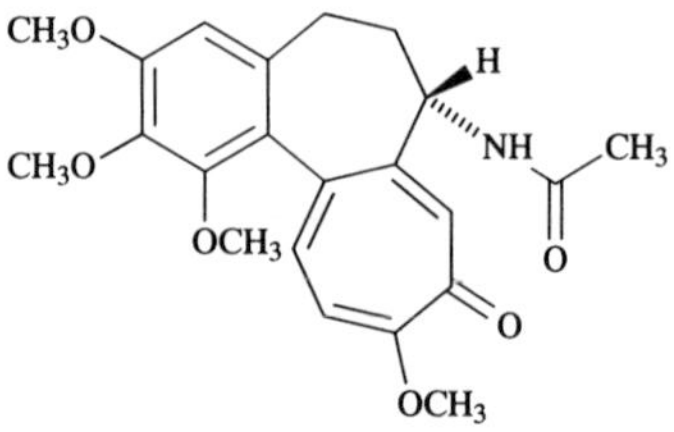

Alkaloid

$C_{22}H_{25}NO_6$ [64-86-8] MW 399.44

In all parts of the plant meadow saffron, *Colchicum autumnale* (Liliaceae). Also present in other *Colchicum* spp. and in the tubers of the glory lily, *Gloriosa superba* (Liliaceae).

Highly toxic, with a lethal dose of about 10 mg in humans. Sixteen fatalities have been recorded in Europe over a period of 30 years. The seeds of *Colchicum* have been used to commit murder. Colchicine is also an irritant, a carcinogen and a teratogen. In spite of its adverse effects, it has been used successfully to relieve the pain of acute gout.

Brossi, A., The Alkaloids, 1984, **23**, 1, Academic Press.

Colpol 338

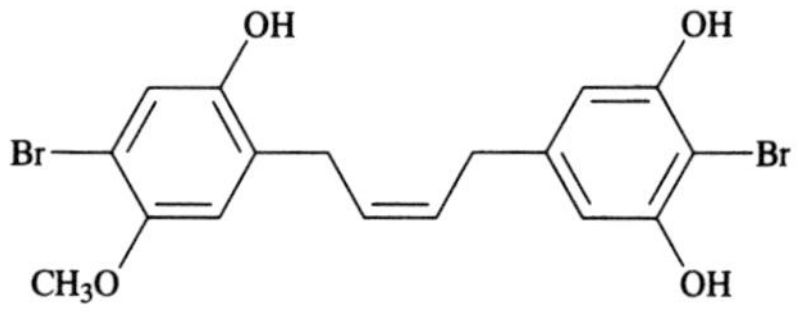

Brominated phenol

$C_{17}H_{16}Br_2O_4$ [151013-34-2] MW 444.12

Brown alga, *Colpomenia sinuosa.*

Cytotoxic activity.

Green, D., *J. Nat. Prod.,* 1993, **56**, 1201.

Columbianetin 339

Dihydro-oroselol; Zosimol

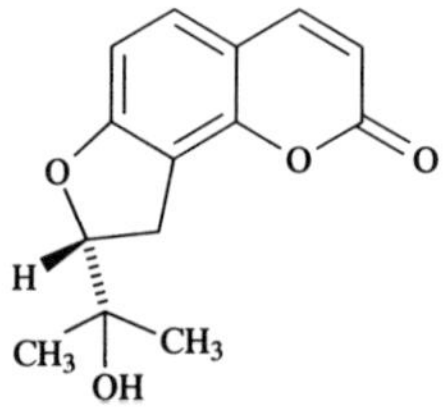

Furocoumarin

$C_{19}H_{14}O_4$ [52842-47-4] MW 246.26
(*S*)-form [3804-70-4]

Stalks of celery, *Apium graveoleus* (Umbelliferae) infected with *Botrytis cinerea.*

Antifungal agent (phytoalexin), active against *Botrytis cinerea* with an EC_{50} of 36 μg/ml. The concentrations in celery tissue (38 μg/g fresh-weight) are sufficient to inhibit growth of celery pathogens *in vivo.*

Nielsen, B. E., *Acta Chem. Scand.,* 1964, **18**, 2111.

Concanavalin A 340

Con A

Protein
(Lectin)
[11028-71-0] MW 51 000

In the seeds of jack bean, *Canavalia ensiformis* (Leguminosae).

Agglutinates many different kinds of cells. It inhibits the growth of tumour cells in experimental animals and restores a normal growth pattern to virus-transformed fibroblasts in tissue culture.

Pusztai, A., Plant Lectins, 1991, 1, CUP.

Condelphine 341

14-Acetylisotalatizidine

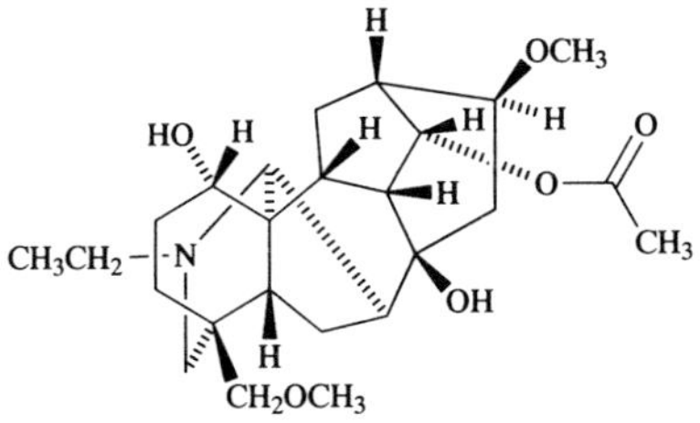

Diterpenoid alkaloid
(Aconitane)

$C_{25}H_{39}NO_6$ [7633-69-4] MW 449.59

In *Delphinium confusum*, *D. denudatum*, *D. nuttallianum* and *Aconitum delphinifolium* (Ranunculaceae).

It is a neuromuscular poison, with useful pharmacological properties. It is both a curare-like neuromuscular blocker and a hypotensive agent. It has been used clinically in the former U.S.S.R. in the treatment of neurological disorders. One of several related alkaloids responsible for the cattle poisoning that occurs in British Columbia following ingestion of *Delphinium nuttallianum*.

Pelletier, S. W., *J. Am. Chem. Soc.*, 1967, **89**, 4146.

Conessine 342

Neriine; Rochessine; Roquessine; Wrightine

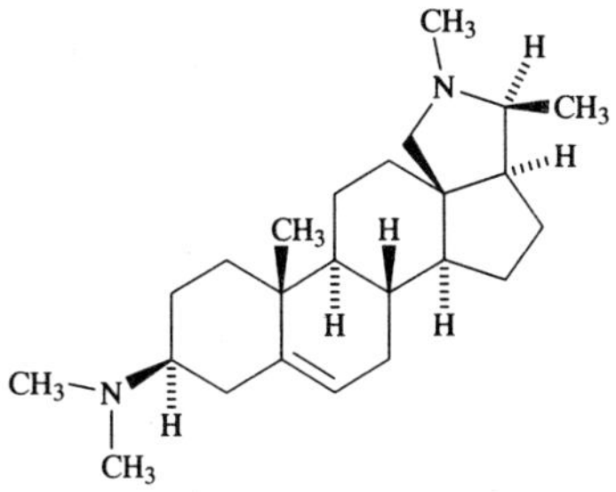

Steroid alkaloid
(Conane)

$C_{24}H_{40}N_2$ [546-06-5] MW 356.60

In *Holarrhena pubescens* (kurchibark), *H. antidysenterica*, *H. febrifuga* and other *Holarrhena* spp. (Apocynaceae).

An anti-amoebic agent useful for the treatment of dysentery. Has local anaesthetic properties. Is narcotic to frogs but not to mammals.

Favre, H., *J. Chem. Soc.*, 1953, 1115.

γ-Coniceine 343

2,3,4,5-Tetrahydro-6-propylpyridine;
2-Propyl-Δ^1-piperideine

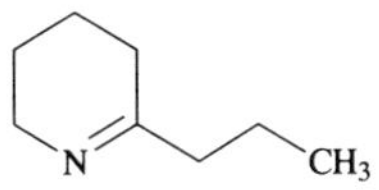

Piperidine alkaloid

$C_8H_{15}N$ [1604-01-9] MW 125.21

Co-occurs with coniine (q.v.) in hemlock, *Conium maculatum* (Umbelliferae). It is also found in *Aloe gililandii*, *A. ballyi* and *A. ruspoliana* (Liliaceae).

Like coniine, it is highly toxic in man and other mammals. It is also teratogenic.

Fairbairn, J. W., *Phytochemistry*, 1961, **1**, 38.

Coniferyl alcohol 344

4-Hydroxyeugenol

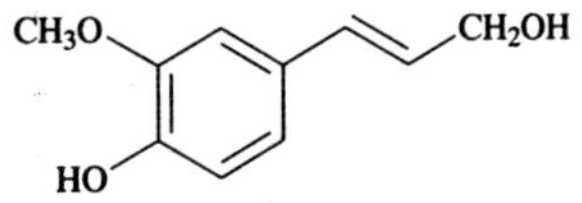

Phenylpropanoid

$C_{10}H_{12}O_3$ [458-35-5] MW 180.20

Leaves of flax, *Linum usitatissimum* (Linaceae) infected with *Melampsora lini*, together with coniferyl aldehyde (q.v.).

Antifungal agent.

Keen, N. T., *Physiol. Plant Pathol.,* 1979, **14**, 265.

Coniine 345

(*S*)-2-Propylpiperidine; Cicutine; Conicine

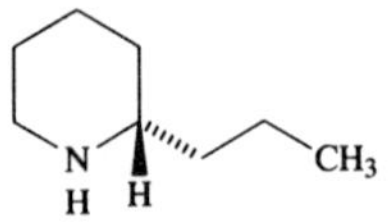

Piperidine alkaloid

$C_8H_{17}N$ [458-88-8] MW 127.23

Throughout the plant of hemlock, *Conium maculatum* (Umbelliferae). Also formed in the pitcher plant, *Sarracenia flava* (Sarraceniaceae) to paralyse insect prey.

Extremely toxic, causing paralysis of motor nerve endings. Fatalities have occurred after consumption of leaves, seeds (alkaloid concentration up to 3.5%) or roots. Teratogenic effects have been recorded after cows and swine have eaten hemlock during pregnancy. Hemlock was used by the Greeks to execute criminals. It has a mousy odour. Hemlock has been used medicinally as a sedative and antispasmodic.

Fairbairn, J. W., *Phytochemistry,* 1961, **1**, 38.

Conocarpan 346

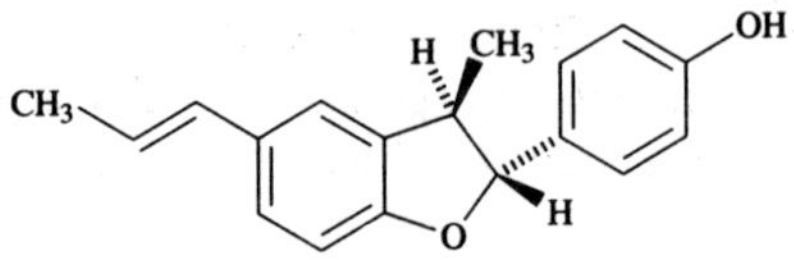

Neolignan

$C_{18}H_{18}O_2$ [56319-02-9] MW 308.14

Wood of *Conocarpus erectus* (Combretaceae), leaves of *Piper decurrens* (Piperaceae) and roots of *Krameria cystisoides* (Krameriaceae).

Insecticidal.

Hayashi, T., *Phytochemistry,* 1975, **14**, 1085.

Conocurvone 347

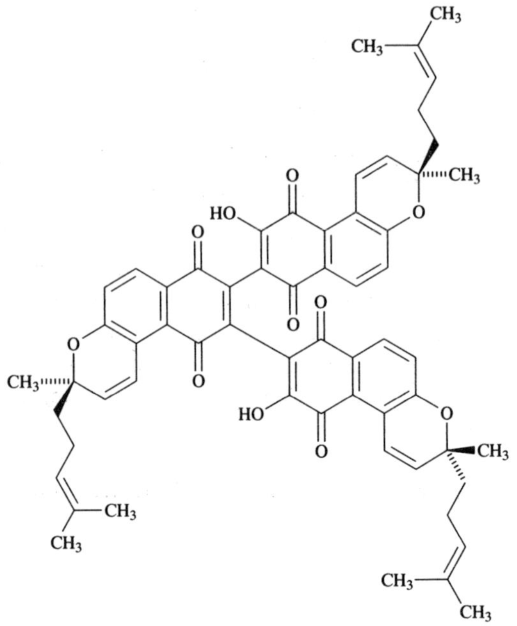

Trimeric naphthoquinone

$C_{60}H_{56}O_{11}$ [149572-31-6] MW 953.10

Conospermum sp. (Proteaceae).

Potent antiviral activity, especially against HIV.

Decosterd, L. A., *J. Am. Chem. Soc.,* 1993, **115**, 6673.

Convallamaroside 348

Convallamarin

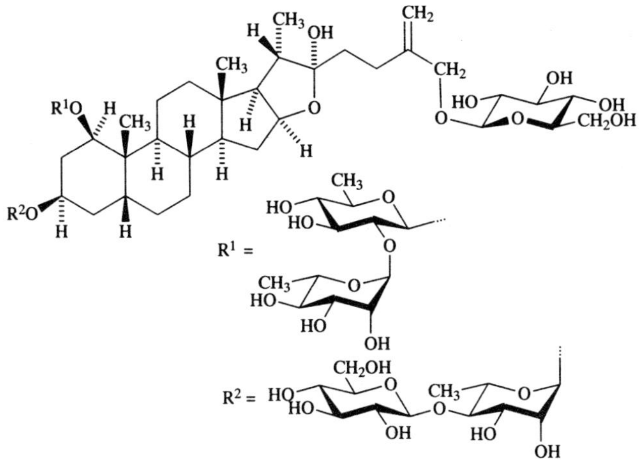

Steroid saponin
(Furostan)

$C_{57}H_{94}O_{27}$ [52591-05-6] MW 1211.36

In roots of lily-of-the-valley, *Convallaria majalis* (Liliaceae) and in *Adonis vernalis* (Ranunculaceae).

Strong haemolytic activity and a local irritant. Lily-of-the-valley is a poisonous plant but the main toxin is the cardenolide, convallatoxin (q.v.).

Tschesche, R., *Chem. Ber.*, 1973, **106**, 3010.

Convallatoxin 349

Strophanthidin 3-rhamnoside; Convallaton; Corglykon; Korglykon; Convallotoxoside

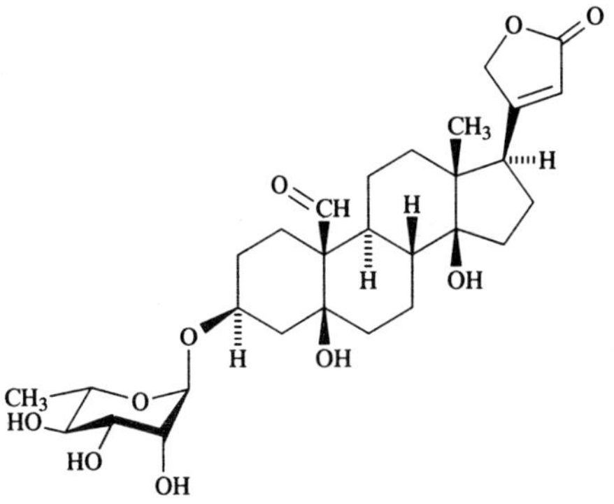

Cardenolide

$C_{29}H_{42}O_{10}$ [508-75-8] MW 550.65

The major cardenolide from the flowers (0.4% dry weight) and leaves (0.13% dry weight) of the lily-of-the-valley, *Convallaria majalis* (Liliaceae). Also present in star-of-Bethlehem, *Ornithogalum umbellatum* (Liliaceae) and in the upas tree, *Antiaris toxicaria* (Moraceae).

Very toxic to vertebrates. The minimum lethal dose intravenously in frogs is 0.3 mg/kg body-weight. Although it is the major toxin of lily-of-the-valley, human poisoning is rare. This is partly because the toxin is poorly absorbed orally and partly because the pulp of the attractive red berries has only traces of cardenolide. Convallatoxin is used as a cardiotonic.

Kubelka, W., *Phytochemistry*, 1974, **13**, 1805.

Convalloside 350

Strophanthidin 3-glucorhamnoside; Bogoroside

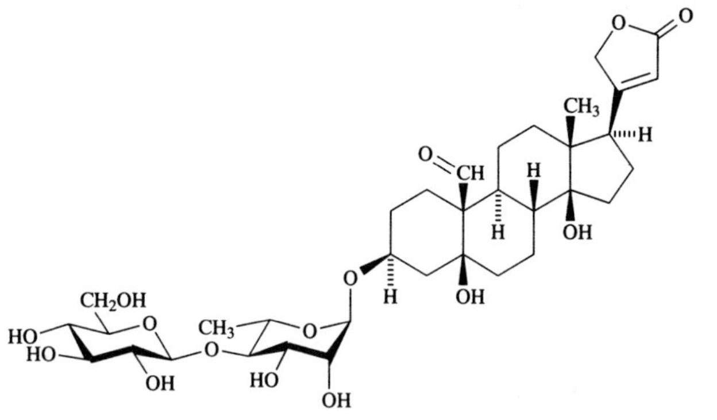

Cardenolide

$C_{35}H_{52}O_{15}$ [13473-51-3] MW 712.79

Mainly in the seeds (0.45% dry weight) of lily-of-the-valley, *Convallaria majalis* (Liliaceae). Also in the latex of the upas tree, *Antiaris toxicaria* (Moraceae).

Replaces convallatoxin (q.v.), the toxic principle of leaves and flowers, as the major toxin of seeds of lily-of-the-valley. Also replaces it as the major leaf cardenolide in East European, as distinct from West European, plants.

Brandt, R., *Helv. Chim. Acta,* 1966, **49**, 2469.

Corchoroside A 351

Strophanthidin 3-(D-boivinoside)

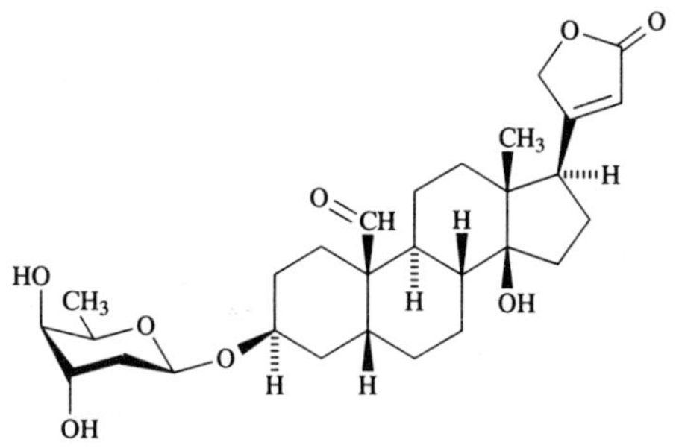

Cardenolide

$C_{29}H_{42}O_9$ [508-76-9] MW 534.65

In seed and leaves of jute, *Corchorus olitorius* (Tiliaceae).

The presence of this and related cardenolides in jute seed has caused poisoning in pigs and cattle. By contrast, poultry appear to be unaffected after feeding on jute seed.

Gonzalez, A. G., *An. Quim.,* 1975, **71**, 97.

Coriamyrtin 352

Coriamyrtione

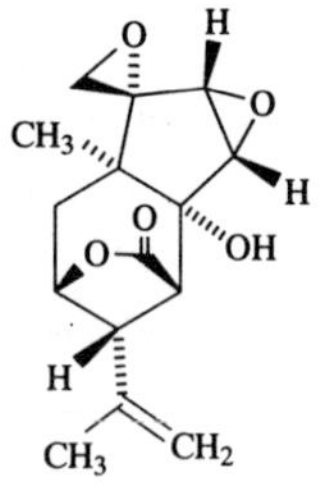

Sesquiterpenoid lactone
(Tutinanolide)

$C_{15}H_{18}O_5$ [2571-81-5] MW 278.30

In the leaves of *Coriaria myrtifolia* and *C. japonica* (Coriariaceae).

Highly toxic to mammals, insects and other animals. It causes extreme excitation to the central nervous system. Structurally, it is closely related to picrotin and picrotoxin (q.v.), which are used as fish poisons. Goats are poisoned by coriamyrtin after grazing on *Coriaria myrtifolia*.

Okuda, T., *Tetrahedron Lett.,* 1965, 4191.

Coriariin A

353

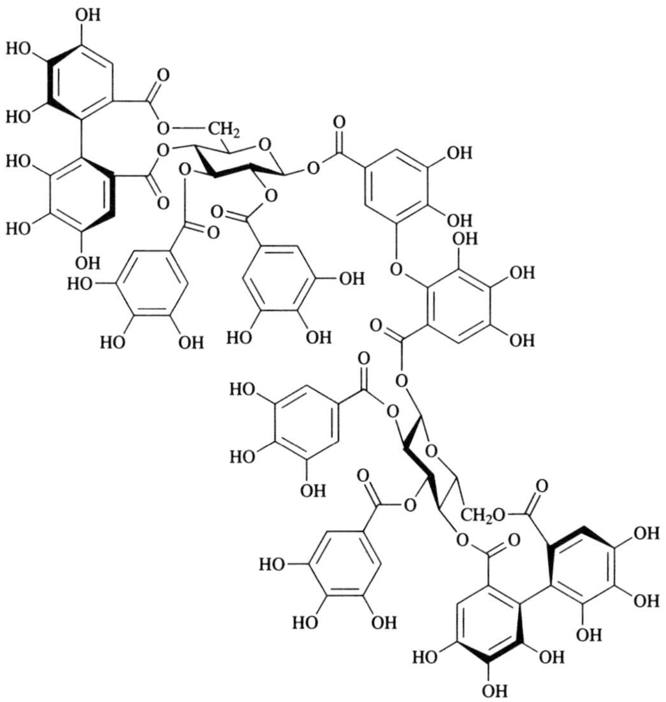

Ellagitannin

$C_{82}H_{58}O_{52}$ [89871-78-3] MW 1875.33

Leaves of *Coriaria japonica* (Coriariaceae).

Antitumour activity.

Hatano, T., *Chem. Pharm. Bull.,* 1986, **34**, 4092.

Corilagin

354

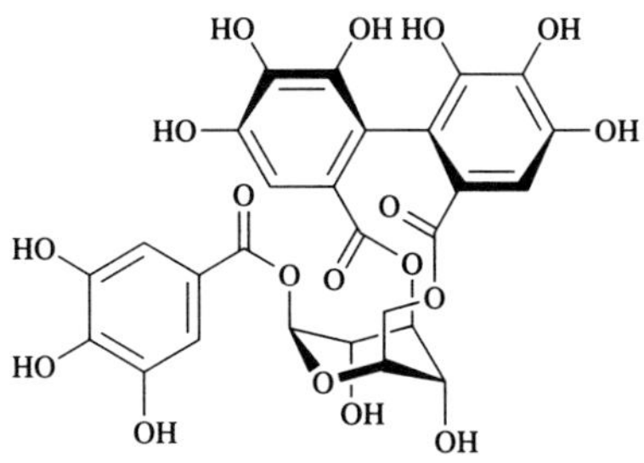

Gallotannin

$C_{27}H_{22}O_{18}$ [23094-69-1] MW 634.40

In *Terminalia chebula* (Combretaceae), *Acer* spp. (Aceraceae), *Poupartia fordii* (Anacardiaceae) and *Sapium japonicum* (Euphorbiaceae). One of several components which make up the commercially available 'tannic acid' prepared from plant galls of *Rhus semialata* (Anacardiaceae).

In vitro antihepatotoxic activity due to inhibitory action on glutamine-pyruvic transaminase and also on induced lipolysis in rat liver microsomes. 'Tannic acid' is used in veterinary practice as an astringent and haemostatic.

Matsuda, H., *Chem. Pharm. Bull.,* 1966, **14**, 877.

Cornudentanone

355

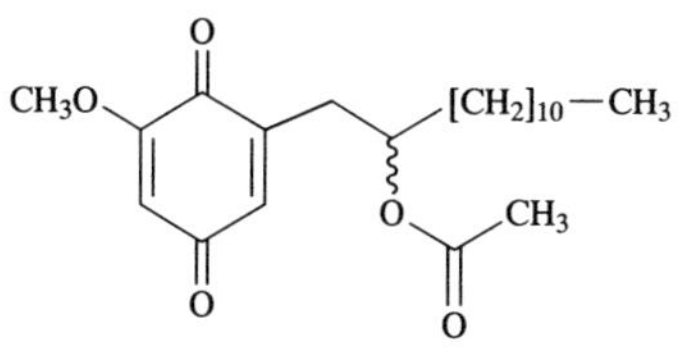

Benzoquinone

$C_{22}H_{34}O_5$ [110976-06-1] MW 378.51

In roots of *Ardisia cornudentata* (Myrsinaceae).

Inhibits binding of leukotrienes in varous receptor assays.

Tian, Z., *Phytochemistry,* 1987, **26**, 2361.

Coronarian

356

2,6-Di(3-nitropropanoyl)-α-glucoside

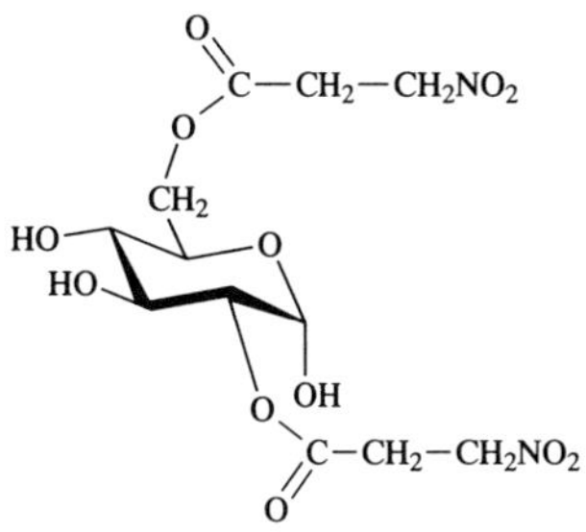

Nitro compound

$C_{12}H_{18}N_2O_{12}$ [63505-68-0] MW 382.28

In *Astragalus cibarius*, *A. falcatus* and *A. flexuosus* (Leguminosae).

Nitro toxin. One of several glucosides, like miserotoxin (q.v.), responsible for livestock poisoning after the consumption of some *Astragalus* species.

Hutchins, R. F. N., *J. Chem. Ecol.,* 1984, **10**, 81.

Coronaridine 357

Carboxymethoxyibogamine

Indole alkaloid
(Ibogamine)

$C_{21}H_{26}N_2O_2$ [467-77-6] MW 338.45

In *Tabernaemontana coronaria* and *Tabernanthe iboga* (Apocynaceae).

Diuretic and oestrogenic activities in animals. Also, it is cytotoxic in *in vitro* systems.

Gorman, M., *J. Am. Chem. Soc.,* 1960, **82**, 1142.

Coronopilin 358

Sesquiterpenoid lactone
(Pseudoguaianolide)

$C_{15}H_{20}O_4$ [2571-81-5] MW 264.32

In many *Ambrosia* spp. such as *A. psilostachia* var. *coronopifolia* (hence its name), in *Hymenoclea salsola*, in *Iva* spp., and in *Parthenium* spp., notably *P. hysterophorus* (Compositae).

It is an insect antifeedant, which affects the growth and development of insects. It is also one of several sesquiterpene lactones present in *Parthenium hysterophorus* which causes allergic contact dermatitis in humans. The allergic effects can be severe and require hospital treatment.

Geissman, T. A., *J. Org. Chem.,* 1964, **29**, 2553.

Costunolide 359

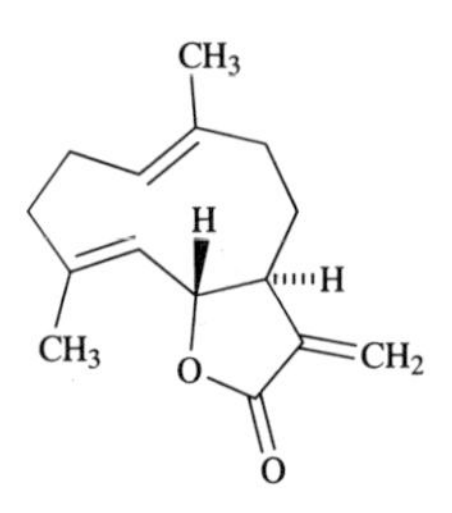

Sesquiterpenoid lactone
(Germacranolide)

$C_{15}H_{20}O_2$ [553-21-9] MW 232.32

In costus root oil, *Saussurea lappa*, but relatively widespread in members of the Compositae. Also recorded in the bay laurel, *Laurus nobilis* (Lauraceae).

Causes allergic contact dermatitis in humans handling plants containing it. It is active against the parasitic trematode *Schistosoma mansonii*, which is the causative agent of schistosomiasis in man. Also has antitumour activity.

Rao, A. S., *Tetrahedron,* 1960, **9**, 275.

α-Cotonefuran 360

Cotonefuran; 2,7-Dihydroxy-3,4,6-trimethoxydibenzofuran

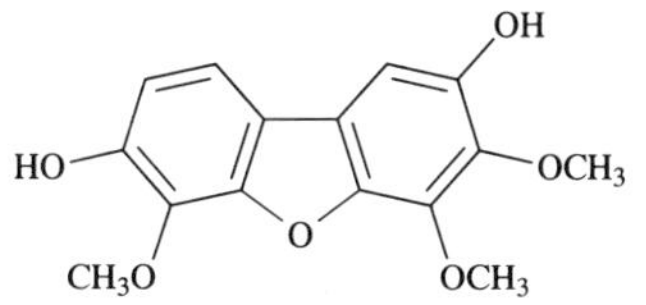

Dibenzofuran

$C_{15}H_{14}O_6$ [93973-22-9] MW 290.27

Sapwood of *Cotoneaster acutifolius* and of *Mespilus germanica* (Rosaceae) infected with the fungus *Nectria cinnabarina*.

Antifungal agent (phytoalexin), with an ED_{50} in the range 14-35 μg/ml.

Kokobun, T., *Phytochemistry,* 1995, **38**, 57.

β-Cotonefuran 361

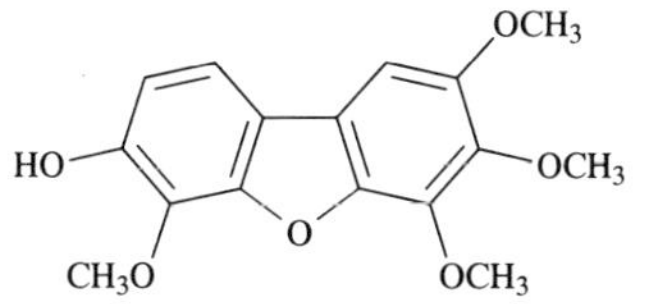

Dibenzofuran

$C_{16}H_{16}O_6$ [161748-46-5] MW 304.30

Sapwood of *Cotoneaster acutifolius* (Rosaceae) infected with *Nectria cinnabarina*.

Antifungal agent (phytoalexin), inhibiting spore germination at a concentration of 16-80 p.p.m.

Kokubun, T., *Phytochemistry,* 1995, **38**, 57.

γ-Cotonefuran 362

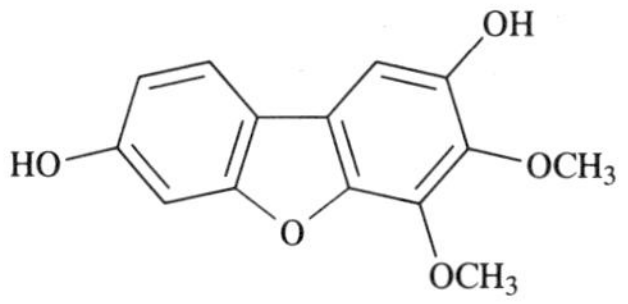

Dibenzofuran

$C_{14}H_{12}O_5$ [161748-47-6] MW 260.25

Sapwood of *Cotoneaster acutifolius* (Rosaceae) infected with *Nectria cinnabarina*.

Antifungal agent (phytoalexin), inhibiting spore germination at a concentration of 29-84 p.p.m.

Kokubun, T., *Phytochemistry,* 1995, **38**, 57.

Cotyledoside 363

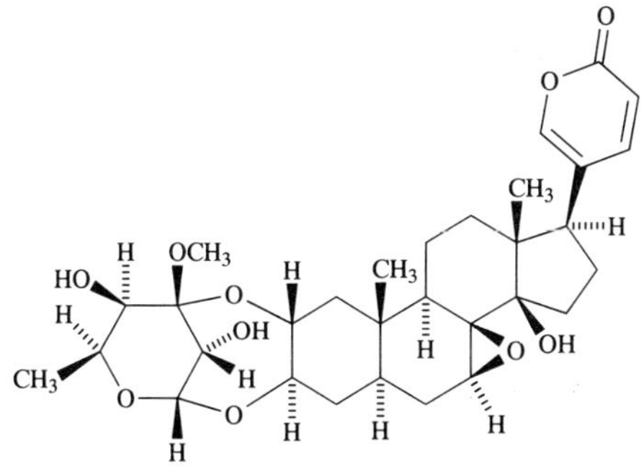

Bufadienolide

$C_{31}H_{42}O_{10}$ [57364-74-6] MW 574.67

Leaves of *Cotyledon orbiculata* and other *Cotyledon* spp. (Crassulaceae).

Toxic, causing livestock poisoning in South Africa. An intravenous dose of 0.05 mg/kg body-weight is sufficient to kill a sheep.

Steyn, P. S., *J. Chem. Soc., Perkin Trans. 1,* 1984, 965.

Coumarin 364

2*H*-1-Benzopyran-2-one; 1,2-Benzopyrone; *cis-o*-Coumarinic acid lactone; Tonka bean camphor; Coumarone

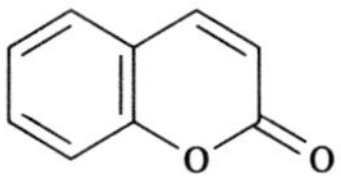

O O

Coumarin

$C_9H_6O_2$ [91-64-5] MW 146.15

One of the most widespread coumarins of the plant kingdom. It occurs in ferns, e.g. members of the Polypodiaceae, and in gymnosperms, such as the Pinaceae. High concentrations may be found in common pasture plants of the Leguminosae and Graminae. It often occurs *in vivo* in bound glucosidic form, with the free coumarin being released during tissue damage. Coumarin is responsible for the smell of newmown hay.

Has a haemorrhagic effect, causing liver damage in rats and dogs. It has piscicidal activity and inhibits larval development in houseflies.

Murray, R. D. H., The Natural Coumarins, 1982, 1, Wiley.

Coumestrol 365

3,9-Dihydroxycoumestan; Cumostrol

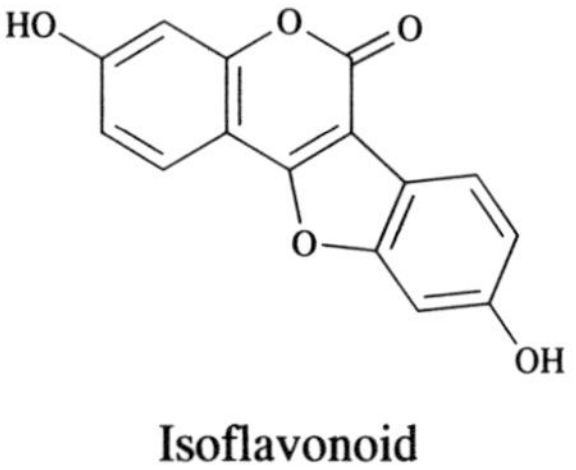

HO O O O OH

Isoflavonoid
(Coumestan)

$C_{15}H_8O_5$ [479-13-0] MW 268.23

In leaves of *Medicago* spp. and clover, *Trifolium* spp. (Leguminosae). Also found in spinach leaf, *Spinacea oleracea* (Chenopodiaceae).

Oestrogenic activity, being 30 times more active than the related isoflavone genistein (q.v.). Coumestrol may adversely affect the fertility of sheep and cattle which feed on clover, but in small amounts, it may be beneficial in increasing lactation in cows, contributing to the so-called "spring flush" in milk yield.

Ingham, J. L., *Fortschr. Chem. Org. Naturst.*, 1983, **43**, 1.

p-Cresol 366

4-Methylphenol; *p*-Cresylic acid; *p*-Hydroxytoluene; Taurylic acid

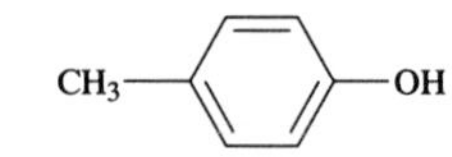

CH_3 OH

Phenol

C_7H_8O [106-44-5] MW 108.14

In leaves of *Morus* spp. (Moraceae), in the wood oil of *Chamaecyparis formosensis* (Cupressaceae) and in the essential oil of *Pimpinella anisum* (Umbelliferae).

Toxic and caustic agent. Cresols are used in veterinary practice as local antiseptics, parasiticides and disinfectants.

Lin, Y. T., *J. Chinese Chem. Soc., Ser II,* 1955, **2**, 91.

Crinamine 367

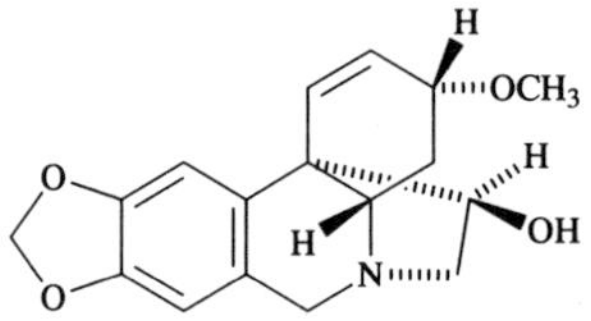

H OCH_3 H O H OH O N

Amaryllidaceae alkaloid
(Crinane)

$C_{17}H_{19}NO_4$ [639-41-8] MW 301.34

In *Crinum asiaticum* and other *Crinum* spp. (Amaryllidaceae).

Toxic, with an oral LD_{50} in dogs of 10 mg/kg body-weight. It is a powerful transient hypotensive agent in dogs. Also it shows respiratory depressant activity.

Wildman, W. C., *J. Am. Chem. Soc.*, 1958, **80**, 6465.

Crinasiadine 368

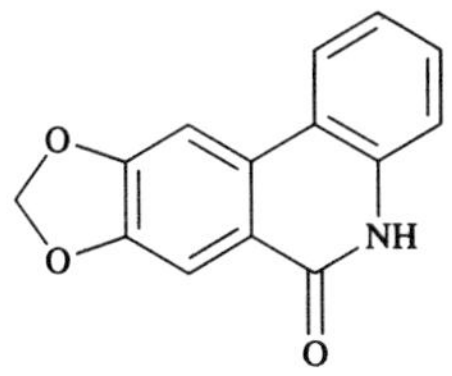

Amaryllidaceae alkaloid
(Narciclasine)

$C_{14}H_9NO_3$ [40141-86-4] MW 239.23

In flowering bulbs of *Crinum asiaticum* (Amaryllidaceae).

Tumour-inhibiting activity.

Ghosal, S., *J. Chem. Res., Synop.,* 1985, 100.

Crinasiatine 369

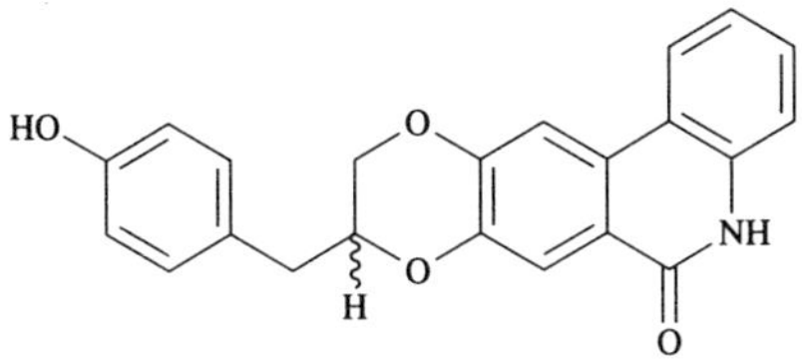

Amaryllidaceae alkaloid

$C_{22}H_{17}NO_4$ [97682-69-4] MW 359.38

In the flowering bulbs of *Crinum asiaticum* (Amaryllidaceae).

Tumour-inhibiting activity.

Ghosal, S., *J. Chem. Res., Synop.,* 1985, 100.

Crotanecine 370

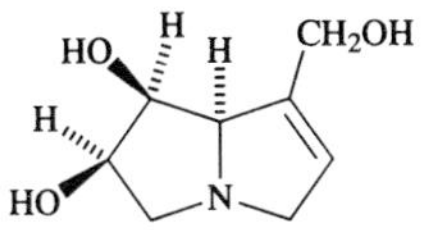

Pyrrolizidine alkaloid

$C_8H_{13}NO_3$ [5096-50-4] MW 171.20

Present, together with its esters (e.g. anacrotine, q.v.) in seeds and other parts of *Crotalaria* spp. (Leguminosae).

Hepatotoxic when present in esterified form. It is not clear whether the free base is toxic or not.

Atal, C. K., *Tetrahedron Lett.,* 1966, 537.

Crotin 371

Protein
(Albumin)
[8001-56-7]

Seed of *Croton tiglium* (Euphorbiaceae).

Poisonous.

Schindler, H., *Arzneimittelforschung,* 1953, **3**, 314.

Cryogenine 372

Vertine

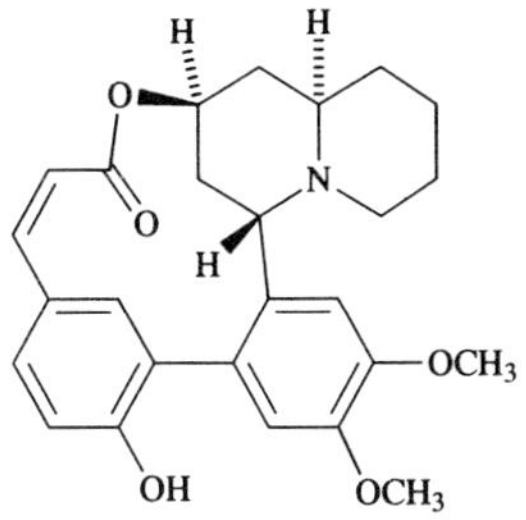

Quinolizine alkaloid

$C_{26}H_{29}NO_5$ [10308-13-1] MW 435.52

In *Decodon verticillatus*, *Heimia salicifolia*, *H. myrtifolia* and *Lagerstroemia fauriei* (Lythraceae).

Anti-inflammatory, sedative and hypotensive activities.

Ferris, J. P., *J. Am. Chem. Soc.,* 1971, **93**, 2942.

Cryptolepine 373

5-Methyl-5*H*-quindoline

Indoloquinoline alkaloid

$C_{16}H_{12}N_2$ [480-26-2] MW 232.28

In *Cryptolepsis triangularis* and *C. sanguinolenta* (Asclepiadaceae).

Hypotensive activity. Unusually for an alkaloid, cryptolepine is deep purple in colour.

Gellért, E., *Helv. Chim. Acta,* 1951, **34**, 642.

Cryptopleurine 374

Phenanthroquinazoline alkaloid

$C_{24}H_{27}NO_3$ [482-22-4] MW 377.48

In *Cryptocarya pleurosperma* (Lauraceae), *Boehmeria cylindrica* (Urticaceae) and *Cissus rheifolia* (Vitidaceae).

It shows potent cytotoxic activity against human nasopharyngeal cells. Very active inhibitor of protein synthesis.

Al-Shamma, A., *Phytochemistry,* 1982, **21**, 485.

Cuauchichicine 375

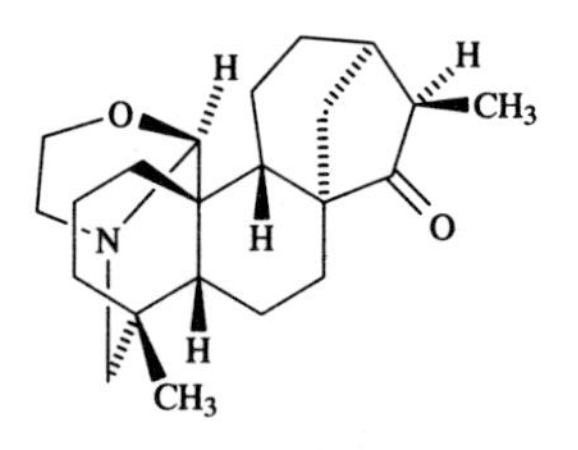

Diterpenoid alkaloid
(Veatchane)

$C_{22}H_{33}NO_2$ [545-60-8] MW 343.51

In the bark of *Garrya laurifolia* and of *G. ovata* var. *lindheimeri* (Garryaceae). It co-occurs with the related diterpenoid alkaloids garryfolina, lindheimerine and ovatine (q.v.).

In mice, it is the most toxic of the veatchine (q.v.)-like alkaloids. Thus, it causes tremors, lowers the blood pressure and increases the heart rate.

Djerassi, C., *J. Am. Chem. Soc.,* 1955, **77**, 4801.

Cubebin 376

Lignan

$C_{20}H_{20}O_6$ [18423-69-3] MW 356.38

In the unripe fruit of the wild pepper, *Piper cubeba* (Piperaceae) and in the roots and shoots of *Aristolochia triangularis* (Aristolochiaceae).

Toxic to insects. Effective blocking agent of the gut microsomal mono-oxygenase of the pest insect, the corn borer (*Ostrinia nubilalis*). Has a possible medical use as a urinary antiseptic.

Batterbee, J. E., *J. Chem. Soc., C,* 1969, 2470.

Cucurbitacin A 377

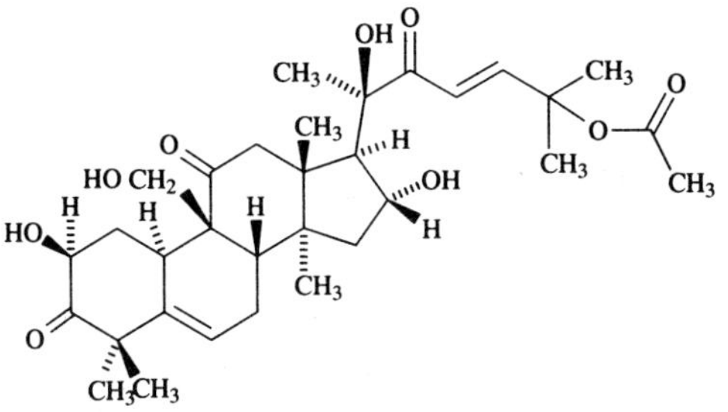

Triterpenoid
(Cucurbitacin)

$C_{32}H_{46}O_9$ [6040-19-3] MW 574.71

In seeds and roots of *Cucumis hookeri*, *C. leptodermis* and *C. myriocarpus* (Cucurbitaceae).

It has an intensely bitter taste and is very toxic to mammals. The LD_{50} intravenously in rabbits is 0.7 mg/kg body-weight.

Lavie, D., *Fortschr. Chem. Org. Naturst.*, 1971, **29**, 307.

Cucurbitacin B 378

Amarin; 1,2-Dihydro-α-elaterin

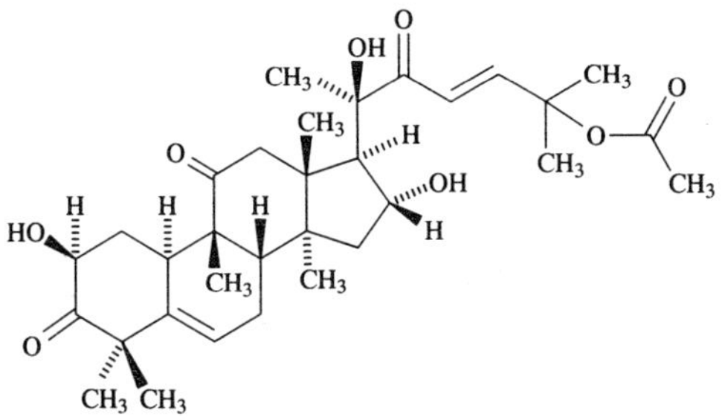

Triterpenoid
(Cucurbitacin)

$C_{32}H_{46}O_8$ [6199-67-3] MW 558.71

Commonly occurring in seeds of many species of Cucurbitaceae, e.g. *Cucumis africanus*, and also found in some *Iberis* spp. (Cruciferae).

Highly toxic to mammals. The LD_{50} when applied intravenously to rabbits is 0.5 mg/kg body-weight, while the oral LD_{50} in mice is 5 mg/kg body-weight. It is a feeding deterrent to most insects, apart from cucumber beetles which use it as a feeding attractant. It also shows cytotoxic and antitumour activities but its significant toxicity prevents any medical uses.

Lavie, D., *Fortschr. Chem. Org. Naturst.*, 1971, **29**, 307.

Cucurbitacin D 379

Elatericin A

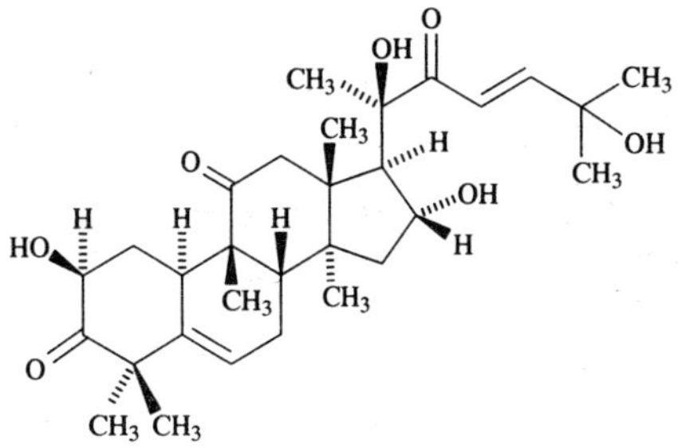

Triterpenoid
(Cucurbitacin)

$C_{30}H_{44}O_7$ [3877-86-9] MW 516.68

In many *Cucumis* spp. and other members of the cucurbit species (Cucurbitaceae), in some *Iberis* spp. (Cruciferae) and in *Crinodendron hookerianum* (Elaeocarpaceae).

Toxic, with a bitter taste. It is a feeding deterrent to most insects (e.g. bees and wasps), although an attractant to cucumber beetles. It exhibits cytotoxic and antitumour activities.

Lavie, D., *J. Org. Chem.*, 1963, **28**, 1790.

Cucurbitacin E 380

α-Elaterin

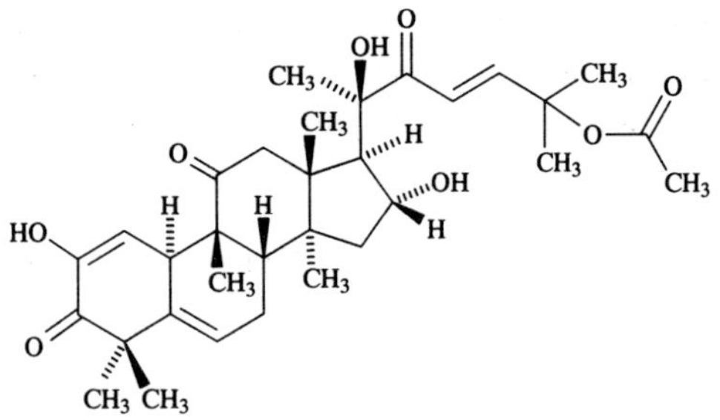

Triterpenoid
(Cucurbitacin)

$C_{32}H_{44}O_8$ [18444-66-1] MW 556.70

In seeds and roots of many species of Cucurbitaceae, e.g. in *Ecballium elaterium*. Occurs as the 2-*O*-β-D-glucoside in *Citrullus colocynthis* (Cucurbitaceae). Also found in many *Iberis* spp. (Cruciferae) and in *Gratiola officinalis* (Scrophulariaceae).

It shows some toxicity to mammals. The oral LD_{50} in mice is 340 mg/kg body-weight. It exhibits cytotoxic and antitumour activities.

Lavie, D., *Fortschr. Chem. Org. Naturst.*, 1971, **29**, 307.

Cucurbitacin I 381

Ibamarin; Elatericin B

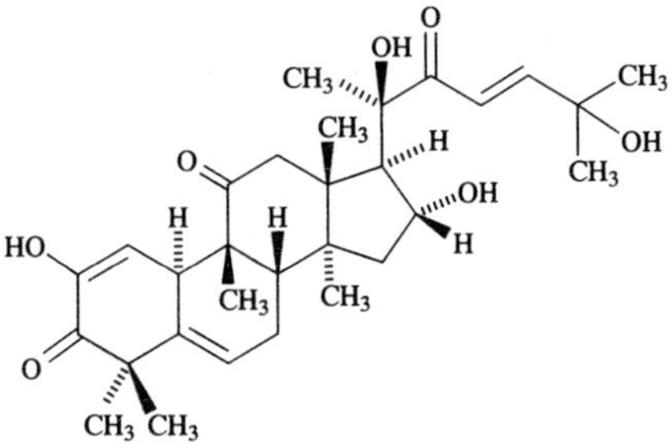

Triterpenoid
(Cucurbitacin)

$C_{30}H_{42}O_7$ [2222-07-3] MW 514.66

In many *Citrullus* spp. (Cucurbitaceae) and in *Iberis amara* (Cruciferae). Also present in the European medicinal plant hedge hyssop, *Gratiola officinalis* (Scrophulariaceae).

Cytotoxic and antitumour activities.

de Kock, W. T., *J. Chem. Soc.*, 1963, 3828.

Cucurbitacin O 382

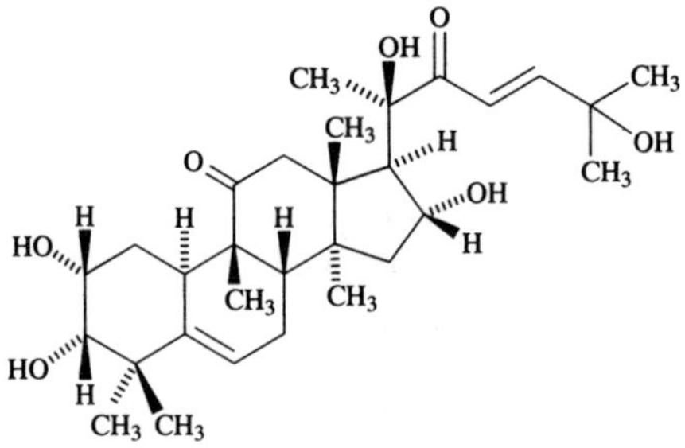

Triterpenoid
(Cucurbitacin)

$C_{30}H_{46}O_7$ [25383-23-7] MW 518.69

In *Brandegea bigelovii* (Cucurbitaceae).

Cytotoxic activity.

Kupchan, S. M., *J. Org. Chem.*, 1970, **35**, 2891.

Cucurbitacin P 383

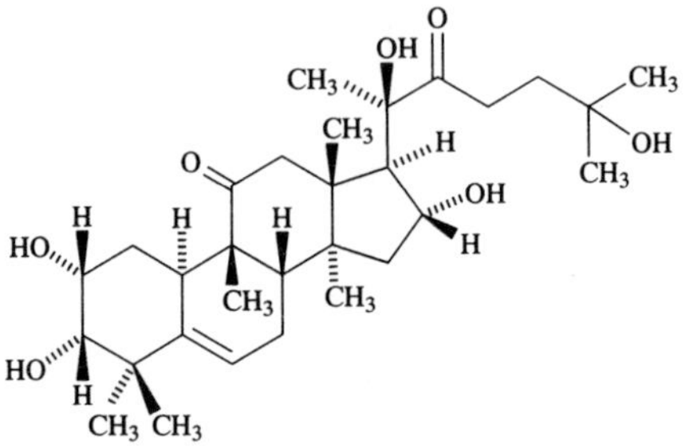

Triterpenoid
(Cucurbitacin)

$C_{30}H_{48}O_7$ [25383-26-0] MW 520.71

In *Brandegea bigelovii* (Cucurbitaceae).

Cytotoxic activity.

Kupchan, S. M., *J. Org. Chem.*, 1970, **35**, 2891.

Cucurbitacin Q 384

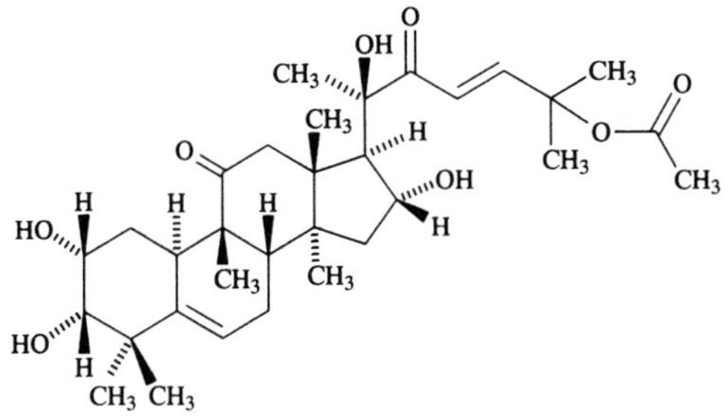

Triterpenoid
(Cucurbitacin)

$C_{32}H_{48}O_8$ [25383-25-9] MW 560.73

In *Brandegea bigelovii* (Cucurbitaceae).

Cytotoxic activity.

Kupchan, S. M., *J. Org. Chem.*, 1970, **35**, 2891.

Cudraisoflavone A 385

Auriculasin

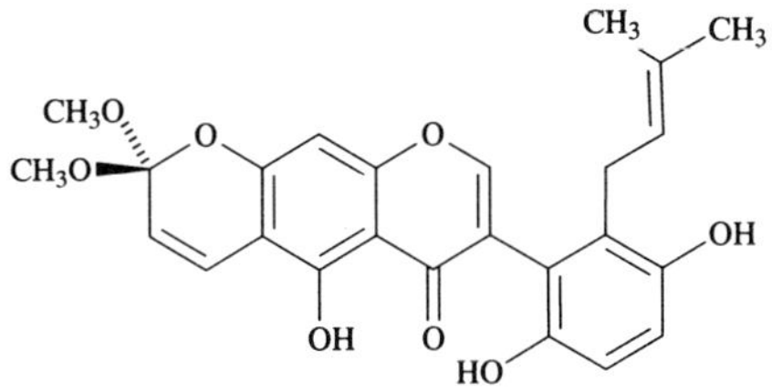

Isoflavone

$C_{25}H_{24}O_6$ [60297-37-2] MW 420.46

Cudrania cochinchinensis (Moraceae).

Cytotoxic activity at a concentration of 1 μg/ml.

Sun, N. J., *Phytochemistry*, 1988, **27**, 951.

Cularicine 386

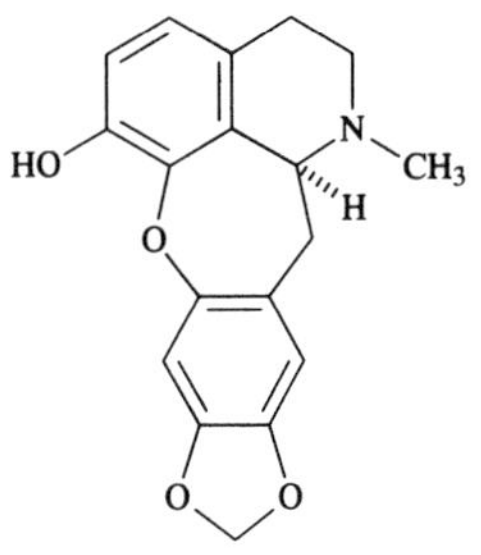

Benzylisoquinoline alkaloid
(Cularine)

$C_{18}H_{17}NO_4$ [2271-08-1] MW 311.34

In *Corydalis clariculata* (Fumariaceae).

Cytotoxic.

Manske, R. H. F., *Can. J. Chem.*, 1965, **43**, 989.

Cularidine 387

O^7-Demethylcularine; Alkaloid F10

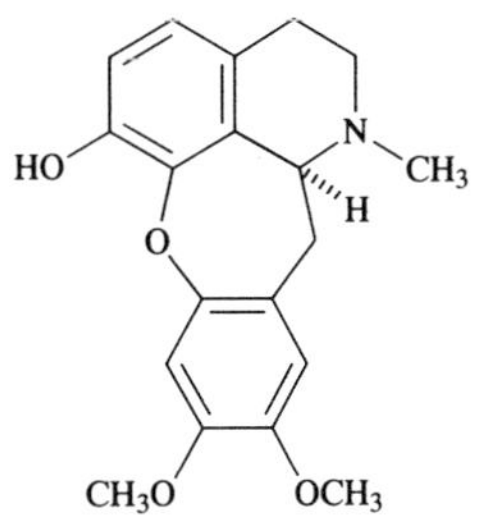

Benzylisoquinoline alkaloid
(Cularine)

$C_{19}H_{21}NO_4$ [5140-50-1] MW 327.38

In *Corydalis claviculata* and *Dicentra cucullaria* (Fumariaceae).

Cytotoxic.

Manske, R. H. F., *Can. J. Chem.*, 1966, **44**, 561.

Cularimine 388

Alkaloid F30

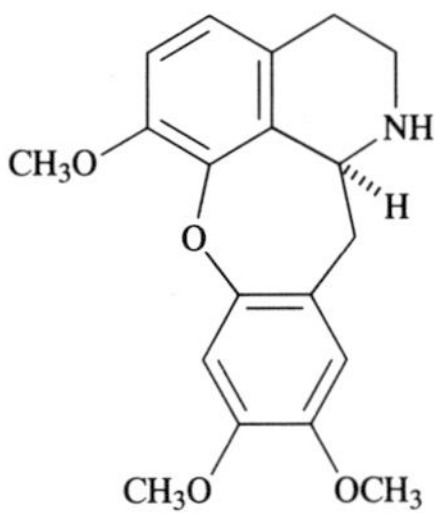

CH3O NH H O CH3O OCH3

Benzylisoquinoline alkaloid
(Cularine)

$C_{19}H_{21}NO_4$ [479-42-5] MW 327.38

In *Dicentra eximia*, *Ceratocapnos palaestinus* and *Corydalis claviculata* (Fumariaceae).

Antineoplastic activity.

Manske, R. H. F., *J. Am. Chem. Soc.,* 1950, **72**, 55.

Cularine 389

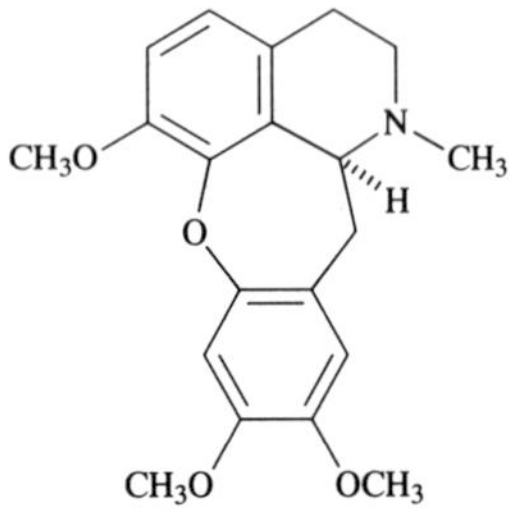

CH3O N CH3 H O CH3O OCH3

Benzylisoquinoline alkaloid
(Cularine)

$C_{20}H_{23}NO_4$ [479-39-0] MW 341.41

In *Berberis valdiviana* (Berberidaceae) and in *Corydalis claviculata* and *Dicentra eximia* (Fumariaceae).

Cytotoxic activity. It stimulates the uterus and increases heart tone and contractility. It is hypotensive in the rabbit and anaesthetic to the rabbit cornea.

Manske, R. H. F., *J. Am. Chem. Soc.,* 1950, **72**, 55.

C-Curarine 390

C-Curarine I

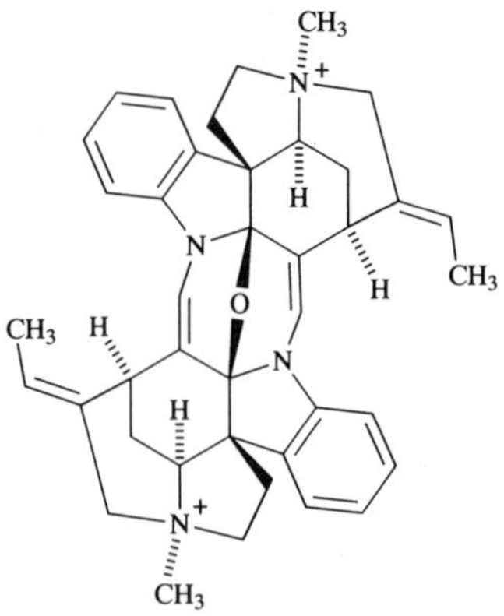

CH3 N+ H N H CH3 CH3 H O N H N+ CH3

Bisindole alkaloid

$[C_{40}H_{44}N_4O]^{2+}$ [7168-64-1] MW 596.82

In bark of *Strychnos froesii*, *S. mittschelichii*, *S. solimoensana* and *S. divaricans* (Loganiaceae).

Extremely toxic, acting as a neuromuscular blocking agent. Curarine is responsible for the paralytic effects of most Amazonian arrow poisons. It is also a component of Calabash curare.

Nagyvary, J., *Tetrahedron,* 1961, **14**, 138.

Curcin 391

Protein
(Lectin)

Seed of *Jatropha curcas* and *J. natalensis* (Euphorbiaceae).

Poisonous, but less toxic than ricin (q.v.).

Morgue, M., *Bull. Soc. Chim. Biol.,* 1961, **43**, 517.

Curcumin 392

Diferuloylmethane; Turmeric yellow; Turmeric colour

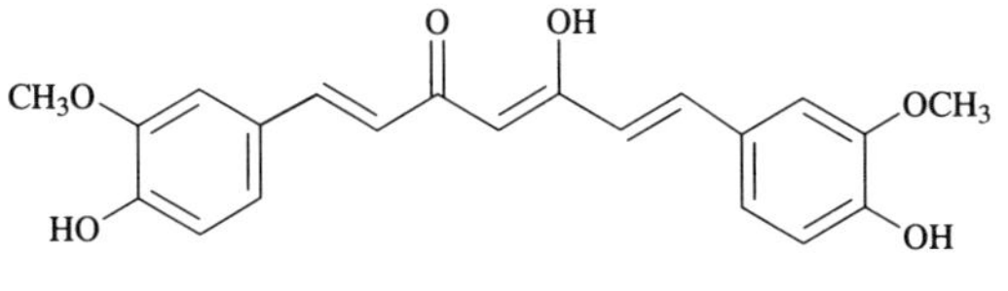

Phenylpropanoid

$C_{21}H_{20}O_6$ [458-37-7] MW 368.39

In the roots of the spice plants *Curcuma longa*, *C. aromatica*, *C. xanthorrhiza* and *C. amada* (Zingiberaceae).

The yellow colouring matter of *Curcuma* roots. It is anti-inflammatory and cytotoxic. It reduces the cholesterol level in the blood and also controls blood sugar.

Pabon, H. J. J., *Rec. Trav. Chim.*, 1964, **83**, 379.

Cusparine 393

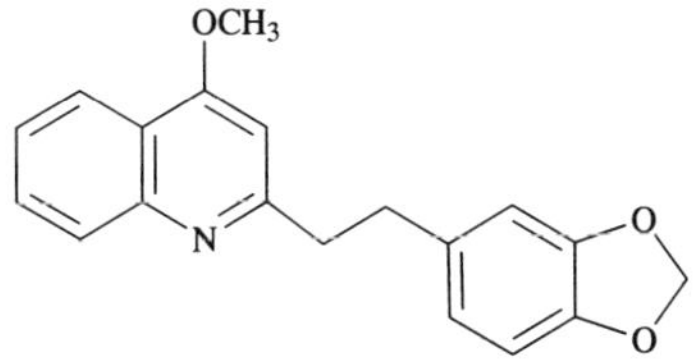

Quinoline alkaloid

$C_{19}H_{17}NO_3$ [529-92-0] MW 307.35

In the so-called angostura bark of *Galipea officinalis* (Rutaceae).

Antispasmodic properties. Angostura bark has antipyretic, antidysenteric and bitter tonic properties. Hence, in folk medicine, angostura serves as a febrifuge, an antidiarrhoeic and as a bitter tonic. However, it is *not* used in the drinks called 'angostura bitters'.

Späth, E., *Ber. Dtsch. Chem. Ges.*, 1924, **57**, 1243.

L-*β*-Cyanoalanine 394

3-Cyano-L-alanine; (*S*)-α-Amino-*β*-cyanopropionic acid; 3-Cyanoalanine; *β*-Cyano-α-alanine

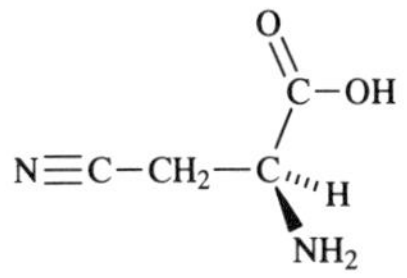

Non-protein amino acid

$C_4H_6N_2O_2$ [6232-19-5] MW 114.10

In the seeds of the common vetch, *Vicia sativa* and other *Vicia* spp. (Leguminosae). Also present in *Vicia sativa* seeds is the γ-glutamyl derivative.

Neurotoxin. Causes convulsions and death when injected subcutaneously into rats at a concentration of 20 mg/kg body-weight. The adverse effects may be due to an inhibition of the action of the vitamin B_6, since pyridoxal hydrochloride delays and alleviates these effects. Poisonous to most forms of life, including insects. Addition of 0.1% to the diet of the bruchid beetle, *Callosobruchus maculatus*, is fatal.

Bell, E. A., *Biochem. J.*, 1965, **97**, 104.

Cyanoviridin RR 395

Cyanoginosin RR; Microcystin RR

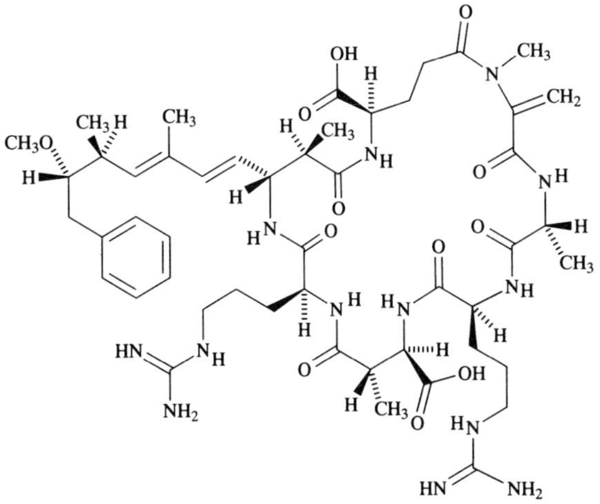

Cyclic peptide

$C_{49}H_{75}N_{13}O_{12}$ [111755-37-4] MW 1038.21

The cyanophyte *Microcystis viridis*.

Toxic.

Tsukuda, T., *Tetrahedron Lett.*, 1989, **30**, 4245.

Cycasin 396

β-D-Glucosyloxyazoxymethane; Methylazoxymethanol β-D-glucoside

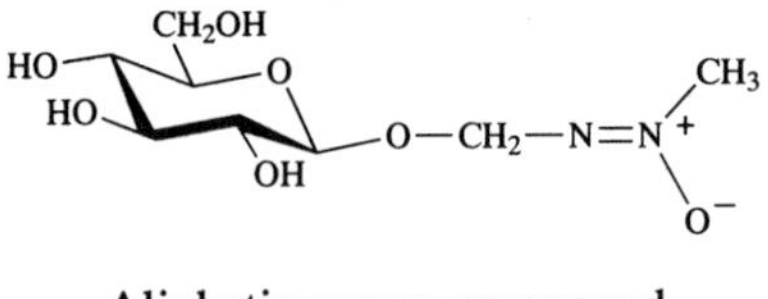

Aliphatic azoxy compound

$C_8H_{16}N_2O_7$ [14901-08-7] MW 252.22

In the seeds of the sago palm, *Cycas circinalis*, of *C. revoluta* and of other *Cycas* spp. (Cycadaceae).

Responsible for the toxicity of cycad palms to livestock. Cycasin is hepatotoxic, carcinogenic and teratogenic in cattle and sheep. Insects seem to be immune to its toxic effects. Thus, the caterpillar of the hairstreak butterfly, *Eumaeus atala*, sequesters cycasin from the cycad food plants and stores it in the adult for defensive purposes.

Whiting, M. G., *Econ. Bot.*, 1963, **17**, 271.

Cyclamin 397

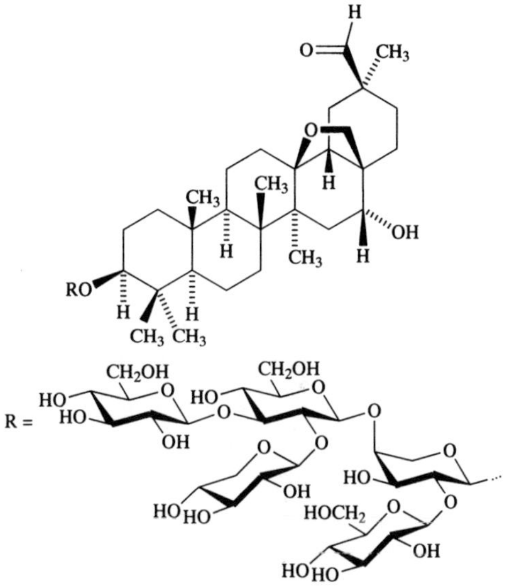

Triterpenoid saponin
(Oleanane)

$C_{58}H_{94}O_{27}$ [23643-76-7] MW 1223.37

Corms of the garden cyclamen, *Cyclamen europaeum* and of the wild *C. purpurascens* (Primulaceae).

Very high haemolytic index and very toxic. As little as 300 mg of ingested corm of *Cyclamen purpurascens* may cause vomiting, diarrhoea and abdominal cramps. Abortions may occur in pregnant women. Larger quantities of fresh corm cause severe intoxication that can end fatally. Cyclamin is extremely toxic to fish, whereas pigs tolerate quantities of poisonous cyclamen tubers without apparent ill-effects.

Tschesche, R., *Justus Liebigs Ann. Chem.*, 1969, **721**, 194.

Cyclobrassinin sulfoxide 398

Cyclobrassinine sulfoxide

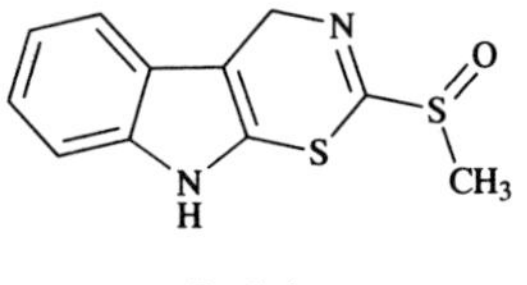

Indole

$C_{11}H_{10}N_2OS$ [128722-96-3] MW 250.35

Fungally infected leaves of *Brassica juncea* (Cruciferae).

Antifungal agent (phytoalexin).

Devys, M., *Phytochemistry,* 1990, **29**, 1087.

Causes allergenic contact dermatitis in those handling these liverworts.

Asakawa, Y., *Bull. Soc. Chim. Fr.,* 1976, 1465.

Cyclobuxine D 399

Cyclobuxine; Alkaloid A

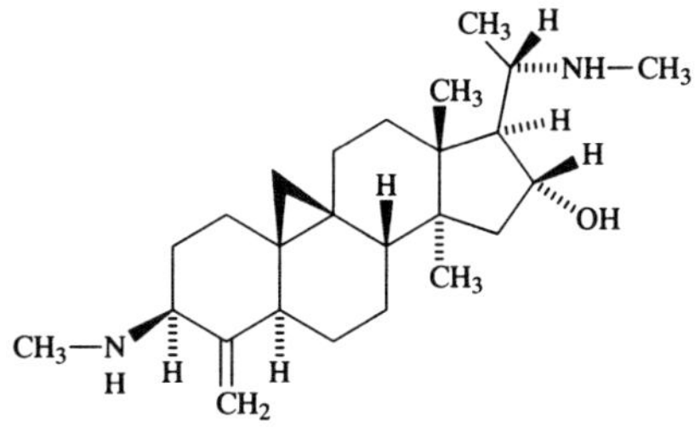

Steroid alkaloid
(Buxane)

$C_{25}H_{42}N_2O$ [2241-90-9] MW 386.62

In leaves and seeds of box, *Buxus sempervirens*, and of *B. harlandi*, *B. hyrcana*, *B. microphylla* and *B. wallichiana* (Buxaceae).

Strong purgative. It is also anti-inflammatory and hypertensive. The alkaloids in the leaves of box are poisonous to farm animals. No cases of human poisoning have yet been reported.

Brown, K. S., *J. Am. Chem. Soc.*, 1964, **86**, 4414.

(+)-*cis*-*β*-Cyclocostunolide 400

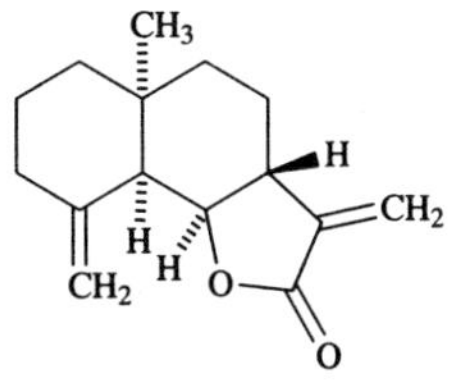

Sesquiterpenoid lactone
(Eudesmanolide)

$C_{15}H_{22}O_2$ [62870-72-8] MW 232.32

The liverworts *Frullania tamarisci*, *F. dilatata* and *F. nisquallensis*.

Cyclopamine 401

11-Deoxyjervine; Alkaloid V

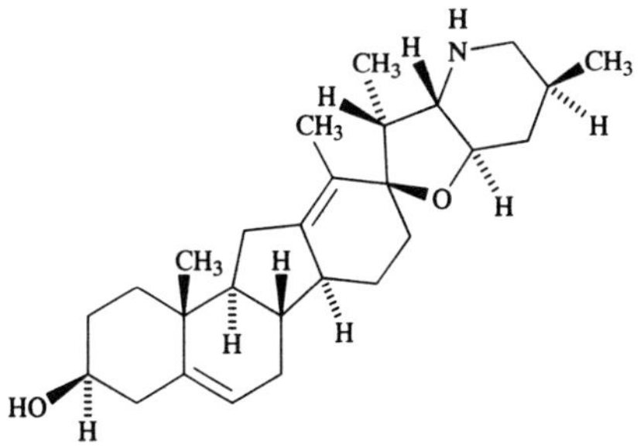

Steroid alkaloid
(Jervanine)

$C_{27}H_{41}NO_2$ [4449-51-8] MW 411.63

In *Veratrum californicum* and white hellebore, *V. album* (Liliaceae).

Severe teratogen, responsible for cyclopic malformation in sheep grazing on *V. californicum*. The alkaloids of white hellebore are highly toxic. The lethal dose in man is 20 mg/kg body-weight, equivalent to 1-2 g dried root.

Keeler, R. F., *Phytochemistry,* 1969, **8**, 223.

Cycloprotobuxine C 402

Alkaloid L

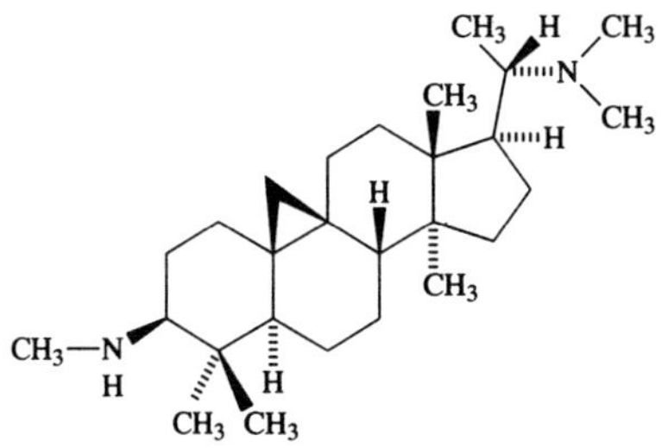

Steroid alkaloid
(Buxane)

$C_{27}H_{48}N_2$ [1936-70-5] MW 400.69

In the common box shrub, *Buxus sempervirens*, also in *B. balearica* and *B. malayana* (Buxaceae). Co-occurs with the closely related cycloprotobuxines A, D and F [2278-38-8, 2255-38-1 and 36151-05-0 respectively].

Seeds and leaves of box are poisonous to livestock, due to the alkaloids present. Cycloprotobuxine C has a purgative action.

Nakano, T., *Tetrahedron Lett.*, 1964, 3679.

Cyclovirobuxine C 403

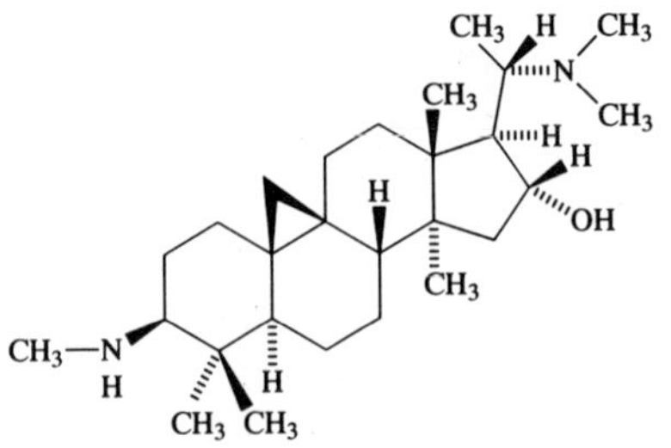

Steroid alkaloid
(Buxane)

$C_{27}H_{48}N_2O$ [30276-35-8] MW 416.69

In *Buxus malayana*, *B. sempervirens*, *B. argentea*, *B. microphylla* and *B. wallichiana* (Buxaceae). It co-occurs with the related cyclovirobuxines D and F [860-79-7 and 100676-13-9 respectively].

Seeds and leaves of *Buxus sempervirens* are poisonous to livestock because of the alkaloids present. All these alkaloids have purgative properties. Cyclovirobuxine C has been proposed as a useful anti-arrhythmic agent.

Khuong-Huu-Laine, F., *Ann. Pharm. Fr.*, 1970, **28**, 211.

Cylindrospermopsin 404

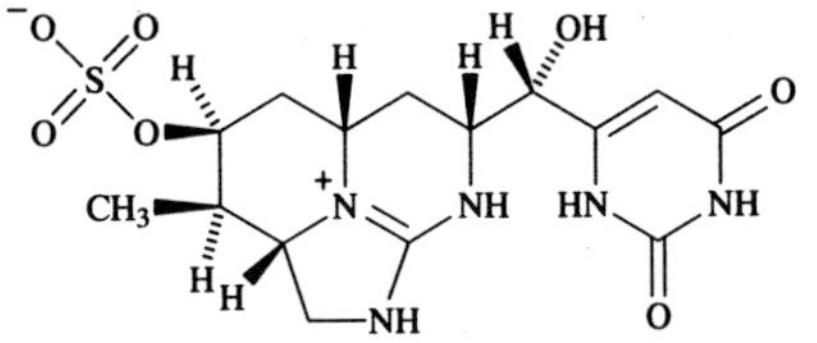

Cyclic guanidinium alkaloid

$C_{15}H_{21}N_5O_7S$ [143545-90-8] MW 415.43

Produced by the tropical species *Cylindrospermopsis raciborskii* isolated from domestic drinking water in Palm Island, North Queensland, Australia.

Responsible for hepatoenteritis in humans. Intraperitoneal LD_{50} in mice 2.1 mg/kg body-weight at 24 h and 0.2 mg/kg at 5-6 days. Principal lesions centrilobular to massive hepatocyte necrosis with damage also of kidneys, adrenal glands, lungs and intestines.

Ohtani, I., *J. Am. Chem. Soc.*, 1992, **114**, 7941.

Cymarin 405

Strophanthidin 3-cymaroside; *k*-Strophanthin; *k*-Strophanthin-α; Cimarin

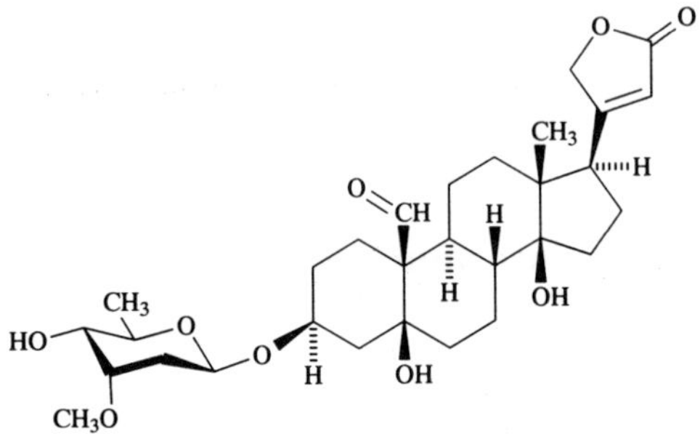

Cardenolide

$C_{30}H_{44}O_9$ [508-77-0] MW 548.67

In leaves of *Strophanthus kombé* (Apocynaceae), yellow pheasant's eye, *Adonis vernalis* (Ranunculaceae) and *Castilloa elastica* (Moraceae).

Very toxic to vertebrates. The LD_{50} when injected intravenously in cats is 0.095 mg/kg body-weight. The presence of cymarin and other cardenolides in *Adonis vernalis* makes it a poisonous plant. Although it is used as a drug plant in Germany, human poisoning is unlikely because cymarin is poorly absorbed following oral ingestion.

Wyss, E., *Helv. Chim. Acta,* 1966, **43**, 664.

p-Cymene 406

Cymene; *p*-Cymol; Dolcymene; *p*-Isopropyltoluene; 1-Isopropyl-4-methylbenzene

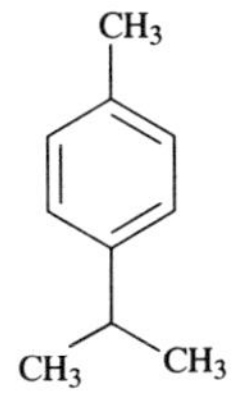

Aromatic hydrocarbon

$C_{10}H_{14}$ [99-87-6] MW 134.22

Present in the essential oils of many plants including American wormseed, *Chenopodium ambrosiodes* (Chenopodiaceae), cumin seed, *Cuminum cyminum*, carum seed, *Carum copticum* (Umbelliferae) and thyme, *Thymus vulgaris* (Labiatae).

Toxic to mammals. The oral LD_{50} in rats is 4.75 g/kg body-weight. *p*-Cymene is used as a local analgesic in rheumatic conditions.

Karrer, W., Konstitution und Vorkommen der organischen Pflanzenstoffe, 2nd edn, 1981, 23, Birkhäuser.

Cynafoside B 407

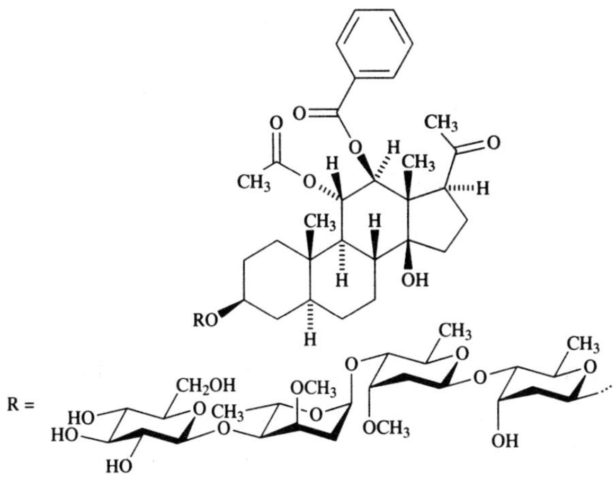

R =

Steroid saponin
(Pregnane)

$C_{56}H_{84}O_{21}$ [100694-58-4] MW 1093.27

Whole plants of *Cynanchum africanum* (Asclepiadaceae).

Major toxin, with six other closely related pregnane glycosides, of the South African coastal plant, *Cynanchum africanum.* In grazing sheep, it causes nervous stimulation, tetanic seizure and paralysis. Lethal dose in guinea pigs is 65 mg/kg body-weight.

Steyn, P. S., *S. Afr. J. Chem.,* 1989, **42**, 29.

Cynaropicrin 408

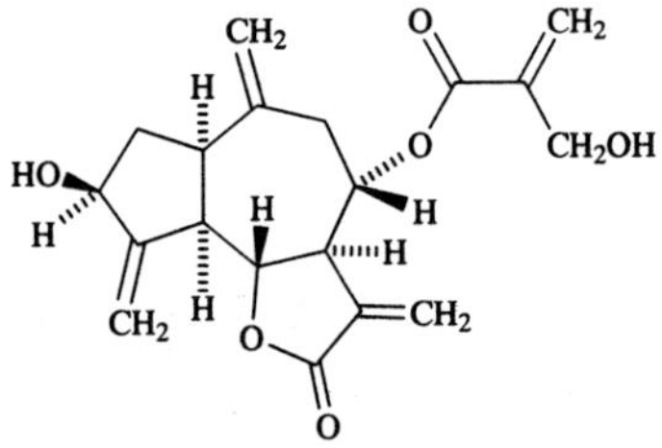

Sesquiterpenoid lactone
(Guaianolide)

$C_{19}H_{22}O_6$ [35730-78-0] MW 346.38

In *Cynara cardunculus*, *C. scolymus*, *Saussurea amara*, *Amberboa muricata* and several *Centaurea* spp. (Compositae).

Cytotoxic and antitumour activities.

Corbella, A., *J. Chem. Soc., Chem. Commun.*, 1972, 386.

Cypripedin 409

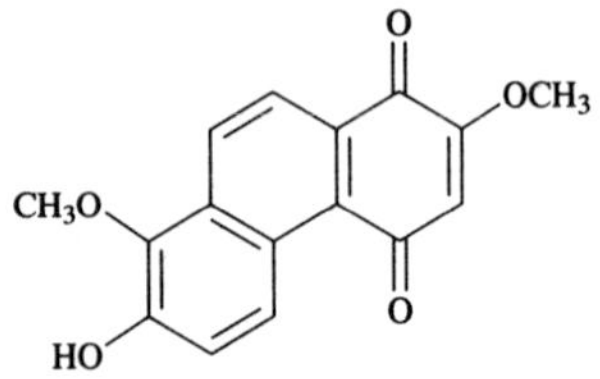

Naphthoquinone

$C_{16}H_{12}O_5$ [8031-72-9] MW 284.27

In the leaves of the ornamental plant *Cypripedium calceolus* (Orchidaceae).

Responsible for contact dermatitis in humans caused by handling leaves of *C. calceolus*.

Schmalle, H., *Naturwissenschaften*, 1979, **66**, 527.

Cytisine 410

Baptitoxine; Citisine; Sophorine; Ulexine

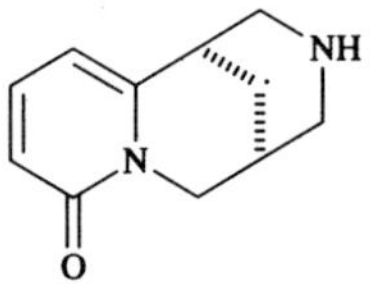

Quinolizidine alkaloid
(Cytisine)

$C_{11}H_{14}N_2O$ [485-35-8] MW 190.25

In the seeds of laburnum, *Laburnum anagyroides*, a popular ornamental tree. It is also widely present in *Anagyris*, *Baptisia*, *Cytisus*, *Genista*, *Sophora* and *Thermopsis* spp. (all Leguminosae).

Highly toxic alkaloid. The LD_{50} intraperitoneally in mice is 18 mg/kg body-weight. It is the cause of poisoning in humans (usually young children) and animals following ingestion of laburnum seeds. Cytisine is teratogenic in rabbits and poultry. It is a respiratory stimulant, with a nicotine-like activity and is hallucinogenic. It is a feeding deterrent to snails.

Govindachari, T. R., *J. Chem. Soc.*, 1957, 3839.

D

Damsin 411

2,3-Dihydroambrosin

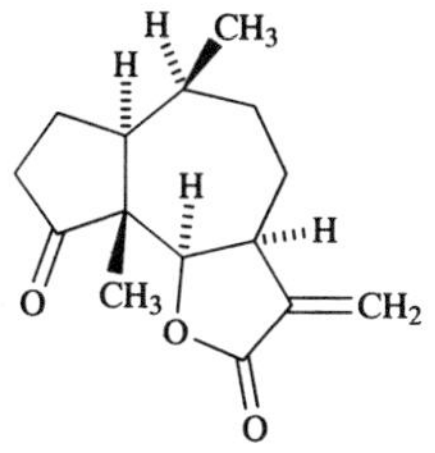

Sesquiterpenoid lactone (Pseudoguaianolide)

$C_{15}H_{20}O_3$ [1216-42-8] MW 248.32

In *Ambrosia maritima* (called 'damsissa' in Egypt) and many other *Ambrosia* spp.. Also present in *Parthenium bipinnatifidum* (Compositae).

Cytotoxic, schistosomicidal and molluscicidal activities.

Suchý, M., *Coll. Czech. Chem. Commun.*, 1963, **28**, 2257.

Daphneticin 412

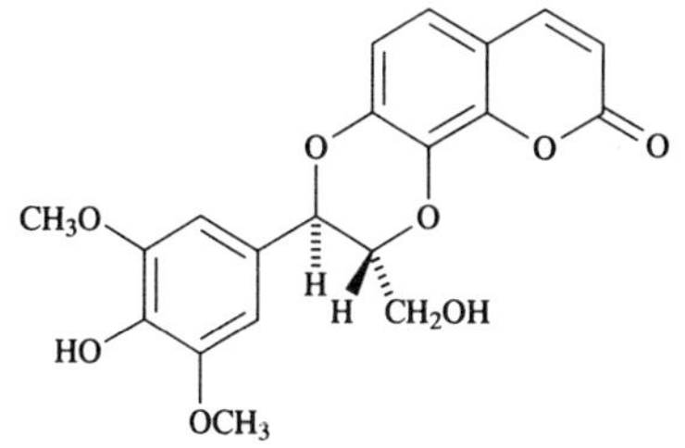

Coumarin

$C_{20}H_{18}O_8$ [83327-22-4] MW 386.36

In roots and stems of *Daphne tangutica* (Thymelaeaceae).

Cytotoxic activity against Walker-256 carcinosarcoma-ascites cells.

Zhuang, L. G., *Phytochemistry*, 1983, **22**, 617.

Daphnetoxin 413

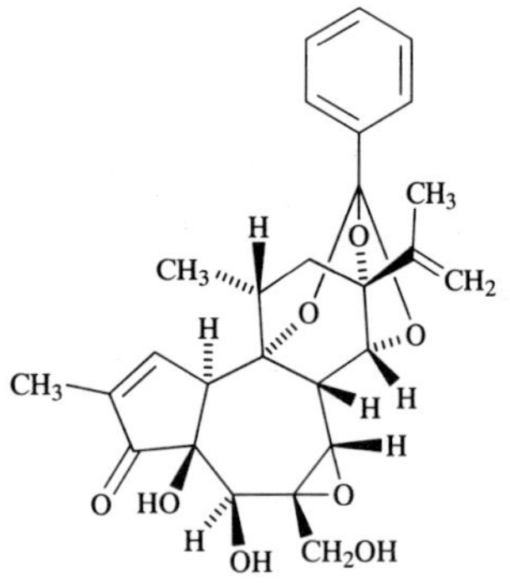

Diterpenoid
(Daphnetoxane)

$C_{27}H_{30}O_8$ [28164-88-7] MW 482.53

In the bark of mezereon, *Daphne mezereum* (Thymelaeaceae).

Highly toxic. Oral LD_{50} is 0.25 mg/kg body-weight. Daphnetoxin, together with mezerein (q.v.), constitute the toxic principles of the mezereon plant. Serious toxic symptoms can follow ingestion of relatively small pieces of bark, leaf, flower or fruit. Children are chiefly at risk from sampling the pleasant-smelling flowers or the shiny red berries.

Stout, G. H., *J. Am. Chem. Soc.,* 1970, **92**, 1070.

Daphnoline 414

Trilobamine

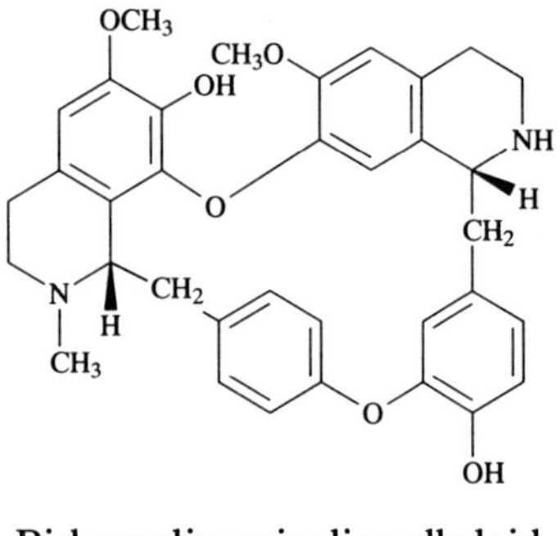

Bisbenzylisoquinoline alkaloid
(Oxyacanthan)

$C_{35}H_{36}O_6$ [479-36-7] MW 580.68

In *Daphnandra micrantha* (Monimiaceae) and in *Cocculus trilobus* (Menispermaceae).

Vasodilator, oedematous agent, central nervous system depressant and respiratory paralytic.

Bick, I. R. C., *Aust. J. Chem.,* 1980, **33**, 225.

Dauricine 415

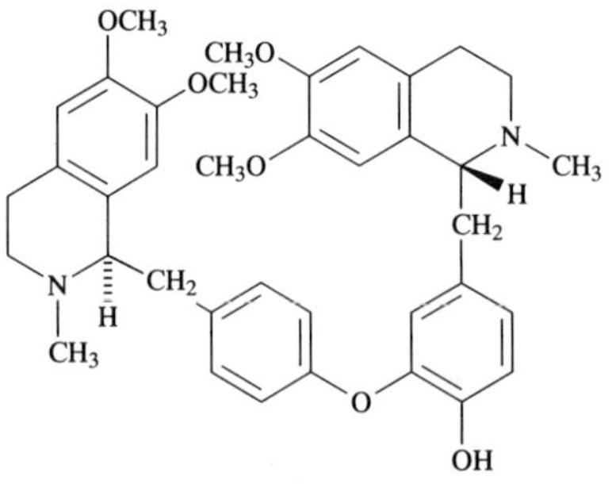

Bisbenzylisoquinoline alkaloid

$C_{38}H_{44}N_2O_6$ [524-17-4] MW 624.78

In *Menispermum dauricum* and *M. canadense* (Menispermaceae), and in *Polyalthia nitidissima* (Annonaceae).

Toxic alkaloid. The LD_{50} when injected intraperitoneally into mice is 6 mg/kg body-weight. It is a weak curarising agent, and also has anaesthetic and anti-inflammatory properties.

Kametani, T., *Tetrahedron Lett.,* 1964, 2771.

14-Deacetylnudicauline 416

Deacetylandersoline

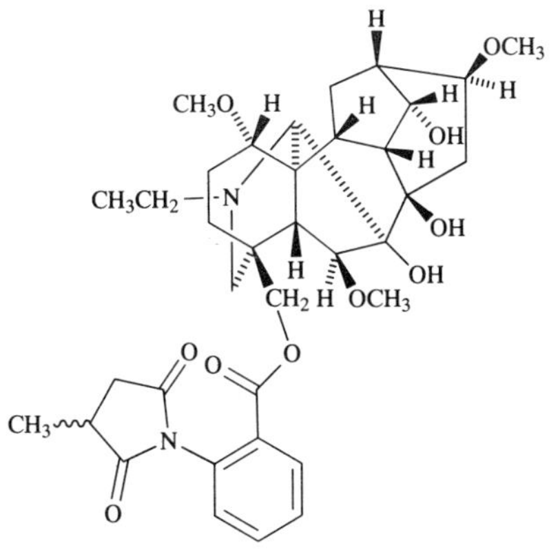

Diterpenoid alkaloid
(Aconitane)

$C_{36}H_{48}N_2O_{10}$ [119347-24-9] MW 668.78

Delphinium andersonii (Ranunculaceae).

One of several poisonous alkaloids in larkspurs causing cattle death in North America. The intravenous LD_{50} in mice is 4 mg/kg body-weight.

Pelletier, S. W., *Heterocycles,* 1988, **27**, 2387.

Debromoaplysiatoxin 417

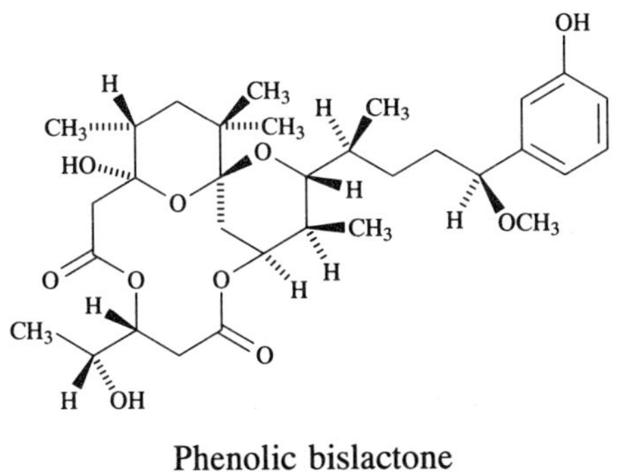

Phenolic bislactone

$C_{32}H_{48}O_{10}$ [52423-28-6] MW 592.73

Isolated from the marine and subtropical species of cyanobacteria, *Lyngbia majuscula* as well as from mixed cultures of *Schizothrix calcicola* and *Oscillatoria nigroviridis.*

Associated with severe contact dermatitis known as swimmers' itch, characterised by erythematous, pustular folliculitis. Shown to have tumour promoting activity in experimental animals. Minimum lethal dose 0.2 mg/kg body-weight in mice.

Moore, R. E., *Pure Appl. Chem.,* 1982, **54**, 1919.

Debromoisocymobarbatol 418

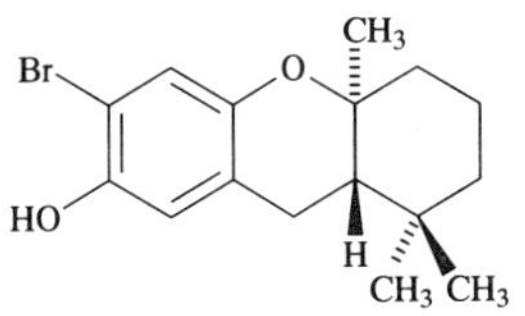

Xanthene

$C_{16}H_{21}BrO_2$ [146428-60-6] MW 325.25

Green alga, *Cymopholia barbata.*

Fish feeding inhibitor.

Park, M., *Phytochemistry,* 1992, **31**, 4115.

Decoside 419

Decogenin 3-oleandroside

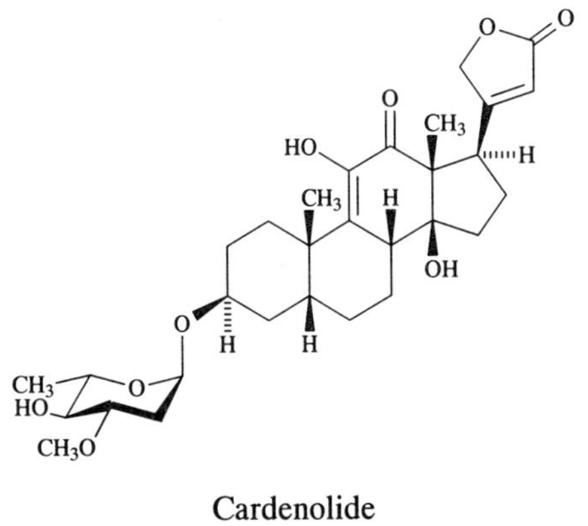

Cardenolide

$C_{30}H_{42}O_9$ [111508-63-5] MW 546.66

In *Strophanthus divaricatus* (Apocynaceae).

Toxic to vertebrates.

Schindler, O., *Helv. Chim. Acta,* 1955, **38**, 140.

Decuroside III 420

Nodakenetin cellobioside

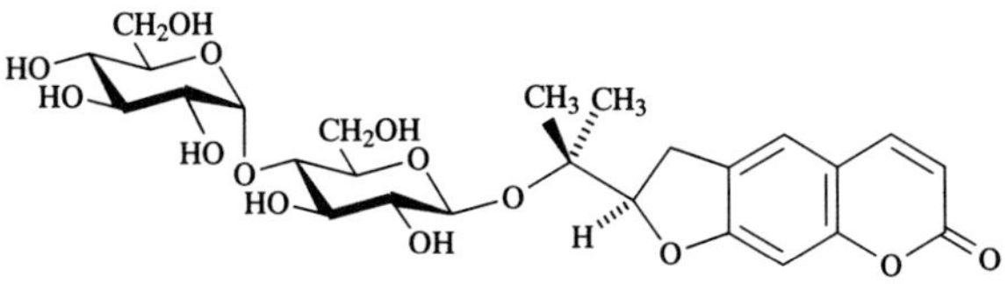

Coumarin glycoside

$C_{26}H_{34}O_{14}$ [96638-81-2] MW 570.55

In roots of *Peucedanum decursivum* (Umbelliferae).

Inhibits human platelet aggregation *in vitro*.

Matano, Y., *Planta Med.,* 1986, 135.

Dehydrocycloguanandin 421

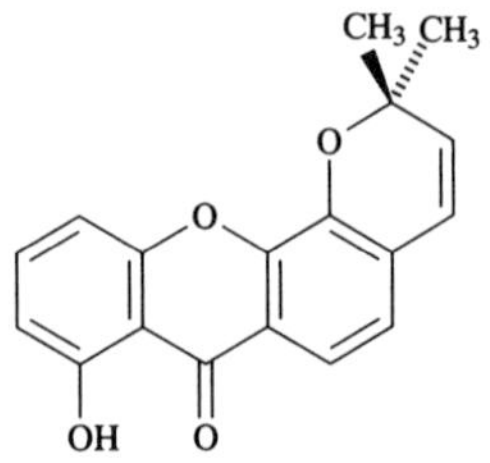

Xanthone

$C_{18}H_{14}O_4$ [17623-63-1] MW 294.31

In *Calophyllum brasiliense, C. inophyllum* and *Mesua ferrea* (Guttiferae).

Anti-inflammatory activity.

Gottlieb, O. R., *Tetrahedron,* 1968, **24**, 1601.

Dehydrojuvabione 422

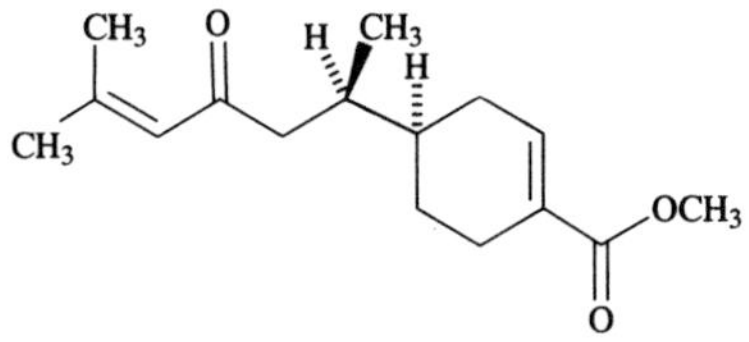

Sesquiterpenoid
(Bisabolane)

$C_{16}H_{24}O_3$ [16060-78-9] MW 264.36

In wood and root of the balsam fir, *Abies balsamea* (Pinaceae).

Juvenile hormone mimic, causing an arrest in embryonic development, with lethal side-effects, in some insects.

Černý, V., *Tetrahedron Lett.,* 1967, 1053.

Dehydromyodesmone 423

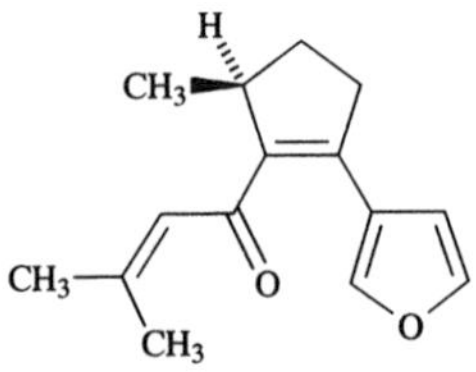

Sesquiterpenoid
(Furanoid ketone)

$C_{15}H_{18}O_2$ [39031-79-3] MW 230.31

In the essential oil of the leaves of *Myoporum deserti*, 'Theodore' variety (Myosporaceae).

Stock poisoning occurs in Australia, following the consumption of this shrub. Dehydromyodesmone is one of several toxins present.

Blackburne, I. D., *Aust. J. Chem.,* 1972, **25**, 1779.

Dehydrongaione 424

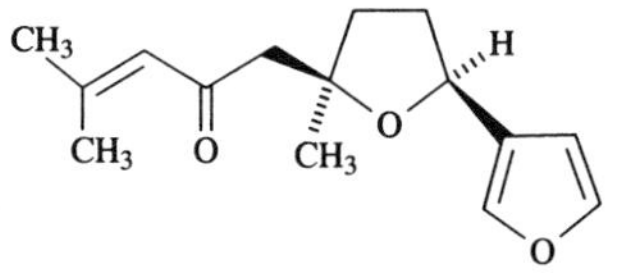

Sesquiterpenoid
(Furanoid ketone)

$C_{15}H_{20}O_3$ [41059-84-1] MW 248.32

In the stock-poisoning Australian shrub, *Myoporum deserti* (Myoporaceae).

One of several toxins responsible for stock poisoning by this shrub. It causes liver damage whem given intraperitoneally to mice.

Pridham, J. B., Terpenoids in Plants, 1967, 147, Academic Press.

Dehydrosafynol 425

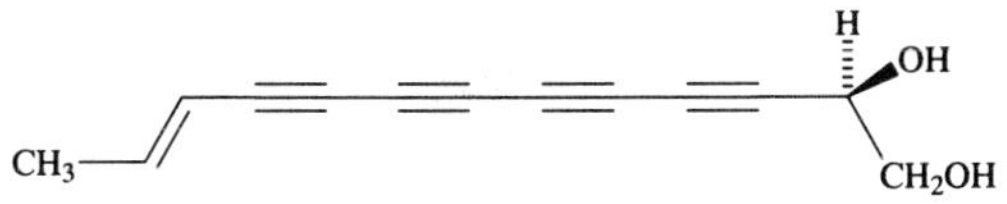

Polyacetylene

$C_{13}H_{10}O_2$ [1540-85-8] MW 198.22

Safflower, *Carthamus tinctorius* (Compositae) infected with *Phytophthora drechsleri*.

Antifungal agent (phytoalexin).

Allen, E. H., *Phytopathology,* 1971, **61**, 1107.

Dehydrotremetone 426

5-Acetyl-2-isopropenylbenzofuran

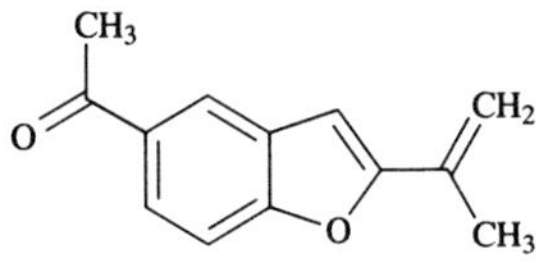

Benzofuran

$C_{13}H_{12}O_2$ [3015-20-1] MW 200.24

In *Eupatorium urticaefolium* and *Haplopappus heterophyllus* (Compositae).

It is responsible, with toxol and tremetone (q.v.), for milk sickness after human consumption of milk from cattle feeding on *Eupatorium urticaefolium*. It passes through the cow without being metabolised and then contaminates the milk. Dehydrotremetone is also toxic to goldfish.

Elix, J. A., *Aust. J. Chem.,* 1971, **24**, 93.

Dehydrozaluzanin C 427

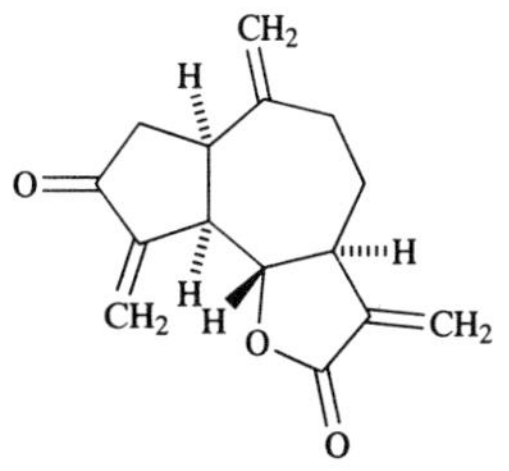

Sesquiterpenoid lactone
(Guaianolide)

$C_{15}H_{16}O_3$ [16836-47-8] MW 244.30

Mannozia maronii (Compositae).

Antiprotozoal activity.

Fournet, A., *Phytother. Res.,* 1993, **7**, 111.

Deidaclin 428

Deidamin

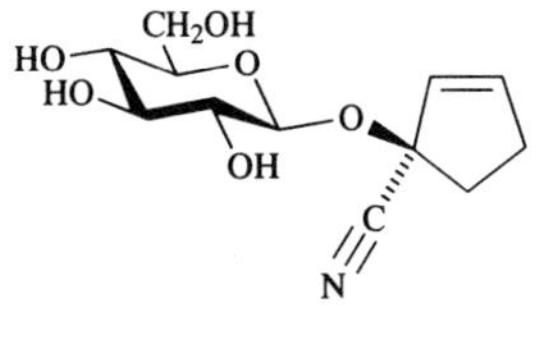

(*R*)-form

Cyanogenic glycoside

$C_{12}H_{17}NO_6$ [88824-26-4] MW 271.27

In leaves of *Deidamia clematoides*; also in *Passiflora coriaceae*, *Tetrapathea tetranda*, *Adenia fruticosa*, *A. spinosa* and *A. globosa* (all Passifloraceae). The (*S*)-form, known as tetraphyllin A [34323-06-3], occurs in *Adenia globosa* and *Tetrapathaea tetrandra*.

Toxic, through the release of cyanide on hydrolysis.

Clapp, R. C., *J. Am. Chem. Soc.*, 1970, **92**, 6378.

Delavaine A 429

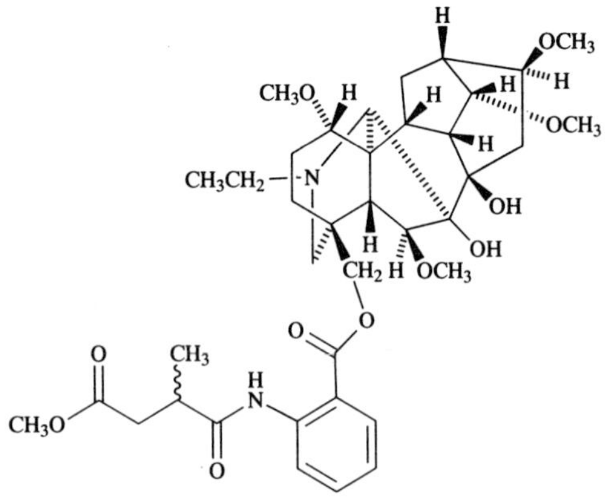

Diterpenoid alkaloid
(Aconitane)

$C_{38}H_{54}N_2O_{11}$ [109291-57-8] MW 714.85

Delphinium delavayi var. *pogonanthum* (Ranunculaceae).

Mammalian toxin. The LD_{50} on intravenous injection in mice is 3.3 mg/kg body-weight. Delavaine B [109314-15-0] is a closely similar alkaloid in *D. delavayi* and is equally toxic.

Pelletier, S. W., *Heterocycles*, 1986, **24**, 1853.

Delcorine 430

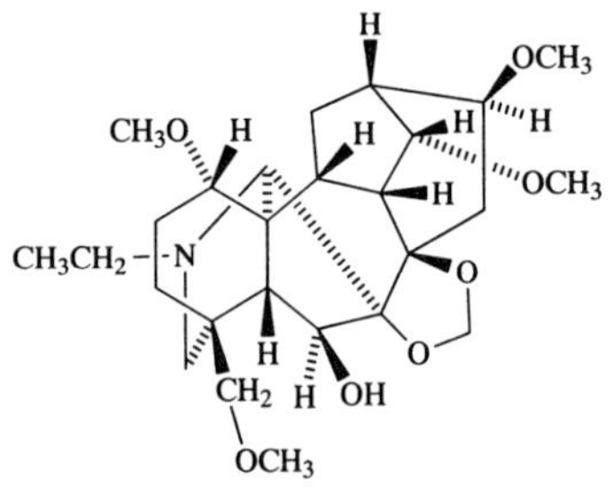

Diterpenoid alkaloid
(Aconitane)

$C_{26}H_{41}NO_7$ [52358-55-1] MW 479.61

In *Delphinium corumbosum* and *D. ternatum* (Ranunculaceae).

Moderately toxic. Intravenous doses of 5-15 mg/kg body-weight in cats and dogs cause a fall in blood-pressure and a temporary respiratory depression. It inhibits the contractions of isolated rat intestine and induces contractions of the guinea-pig uterus; it is a ganglionic blocking agent.

Pelletier, S. W., *Can. J. Chem.*, 1980, **58**, 1875.

Delcosine 431

Iliensine; Ilienzine; Alkaloid C; Delphamine; Lucaconine; Takaobase I; Takawobase I

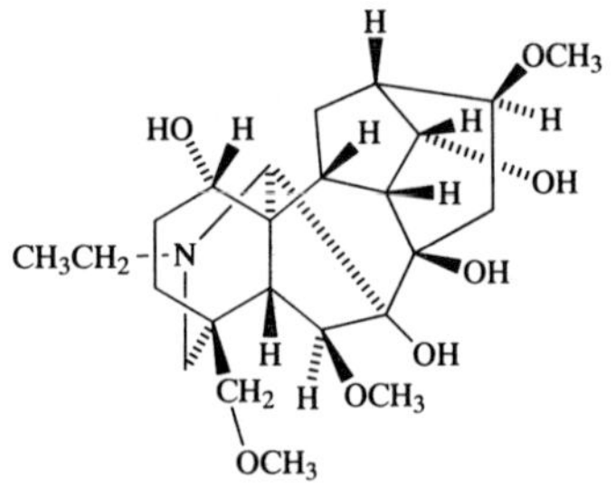

Diterpenoid alkaloid
(Aconitane)

$C_{24}H_{39}NO_7$ [545-56-2] MW 453.58

In *Aconitum ibukiense*, *Delphinium ajacis* and *D. consolida* (Ranunculaceae).

Poisonous to cold-blooded animals, but less toxic to warm-blooded creatures. The acute toxicity is about three times higher than that of lycoctonine (q.v.). In anaesthetised cats, doses of 10 mg/kg body-weight produce a substantial fall in blood-pressure. Also, it is insecticidal.

Skarić, V., *Can. J. Chem.*, 1960, **38**, 2433.

Delphinine

432

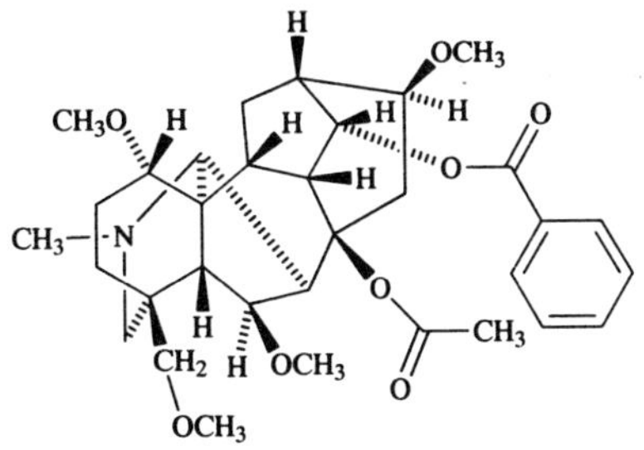

Diterpenoid alkaloid
(Aconitane)

$C_{33}H_{45}NO_9$ [561-07-9] MW 599.72

In seeds of *Delphinium staphisagria* and in roots of *Atragene sibirica* (Ranunculaceae).

The toxicity is slightly less than that of the closely related aconitine (q.v.). The most characteristic symptom of poisoning in farm animals is respiratory depression, which is the primary cause of death.

Birnbaum, K. B., *Tetrahedron Lett.*, 1971, 867.

Delsoline

433

Acomonine; 14-*O*-Methyldelcosine;
18-*O*-Methylgigactonine

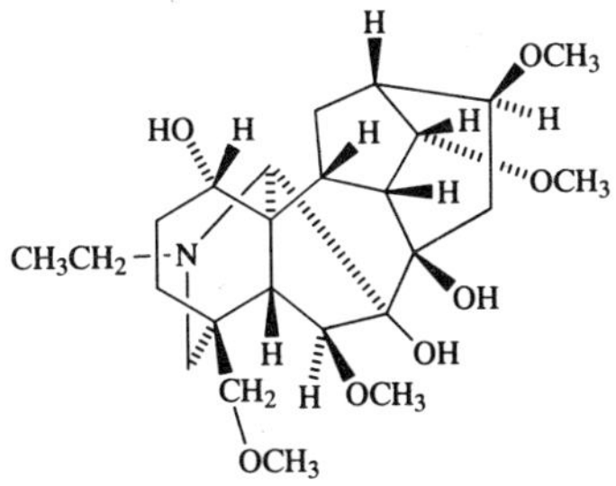

Diterpenoid alkaloid
(Aconitane)

$C_{25}H_{41}NO_7$ [509-18-2] MW 467.60

In *Delphinium consolida* and *Aconitum monticola* (Ranunculaceae).

Parenteral administration to mice causes weakness in the extremities, clonic convulsions and respiratory depression. It also shows insecticidal activity.

Skarić, V., *Can. J. Chem.*, 1960, **38**, 2433.

Deltaline

434

Eldeline; Delphelatine

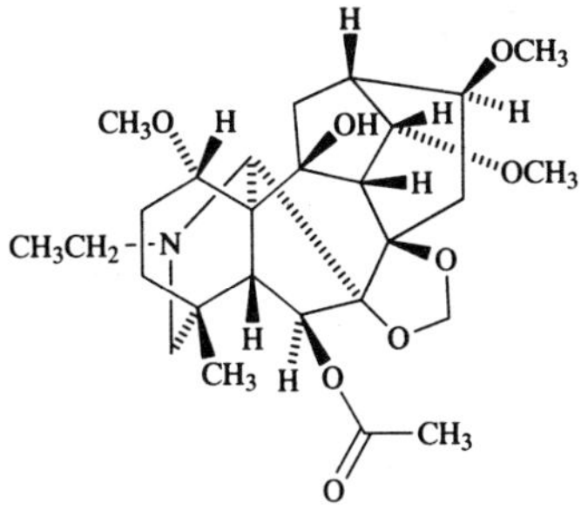

Diterpenoid alkaloid
(Aconitane)

$C_{27}H_{41}NO_8$ [6836-11-9] MW 507.62

In *Delphinium barbeyi*, *D. occidentale*, *D. elatum* and several other *Delphinium* spp. (Ranuncculaceae).

Livestock poisoning following the consumption of *Delphinium barbeyi* is mainly due to the high concentration present of this alkaloid, together with lycoctonine (q.v.). Intravenous administration of 20 mg/kg body-weight to cats lowered the blood-pressure, but respiration was not affected.

Pelletier, S. W., *J. Am. Chem. Soc.*, 1981, **103**, 6536.

Deltonin 435

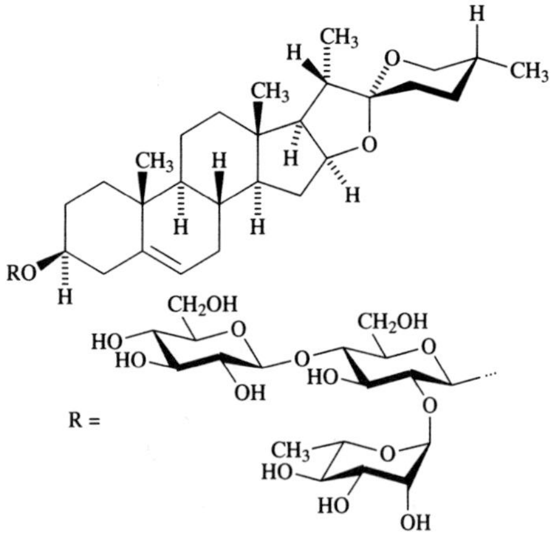

Steroid saponin
(Spirostan)

$C_{45}H_{72}O_{17}$ [55659-75-1] MW 885.06

In roots of *Dioscorea deltoidea* (Dioscoreaceae).

Potent haemolytic toxin to small mammals. Could be used as a rodenticide.

Paseshnichenko, V. A., *Prikl. Biokhim. Mikrobiol.,* 1975, **11**, 94.

Deltoside 436

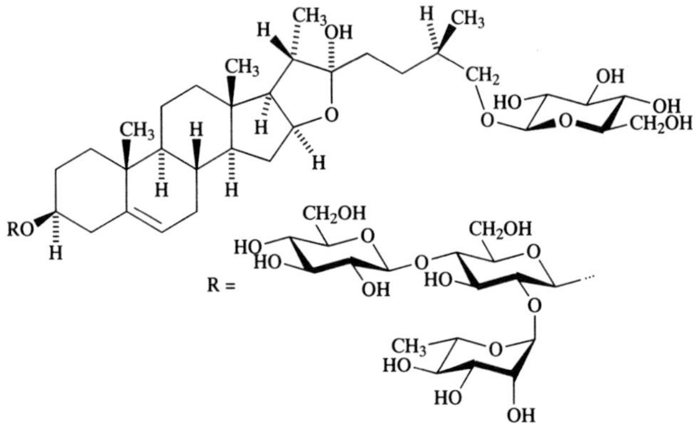

Steroid saponin
(Furostan)

$C_{51}H_{84}O_{23}$ [62751-68-2] MW 1065.21

In roots of *Dioscorea deltoidea* (Dioscoreaceae) and in *Ophiopogon planiscapus* (Liliaceae).

Potent haemolytic toxin to small mammals. Could be used as a rodenticide.

Watanabe, N., *Chem. Pharm. Bull.,* 1983, **31**, 3486.

Demethylbatatasin IV 437

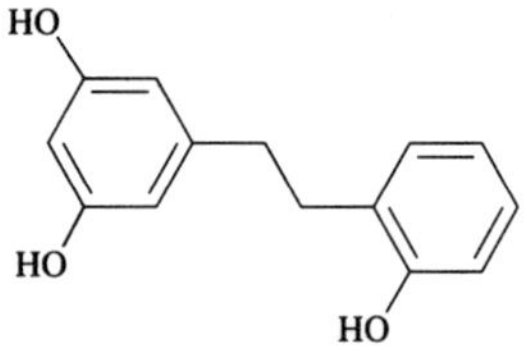

Stilbenoid

$C_{14}H_{14}O_3$ [113276-63-4] MW 230.26

Fungally-infected tubers of *Dioscorea rotundifolia* (Dioscoreaceae).

Antifungal (phytoalexin) and antibacterial activity.

Fagboun, D. E., *Phytochemistry,* 1987, **26**, 3187.

(+)-Demethylcoclaurine 438

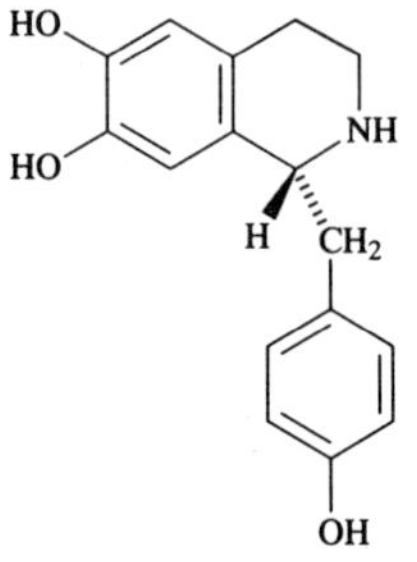

Benzylisoquinoline alkaloid

$C_{16}H_{17}NO_3$ [106032-53-5] MW 271.32
(±)-form [5483-65-2]

In *Aconitum japonicum* (Ranunculaceae). The racemic form is known as higenamine.

Cardiac stimulant.

Kosuge, T., *Yakugaku Zasshi,* 1978, **98**, 1370.

4′-Demethyldeoxypodophyllotoxin 439

Lignan

$C_{21}H_{20}O_7$ [3590-93-0] MW 384.39

In the roots of *Polygala paena*, in the stems, leaves and flowers of *P. macradenia* (Polygalaceae) and in *Hyptis verticillata* (Labiatae).

Antimitotic, antileukaemic and antitumour activities.

Polonsky, J., *Bull. Soc. Chim. Fr.*, 1962, 1722.

4′-Demethylpodophyllotoxin 440

Lignan

$C_{21}H_{20}O_8$ [40505-27-9] MW 400.39

In Indian podophyllin, namely the resin of *Podophyllum hexandrum* (Podophyllaceae) and in *Polygala polygaena* (Polygalaceae).

A compound with antitumour, antimitotic and cathartic activities.

Nadkarni, M. V., *J. Am. Chem. Soc.*, 1953, **75**, 1308.

Demissine 441

β-Demissidine glycoside

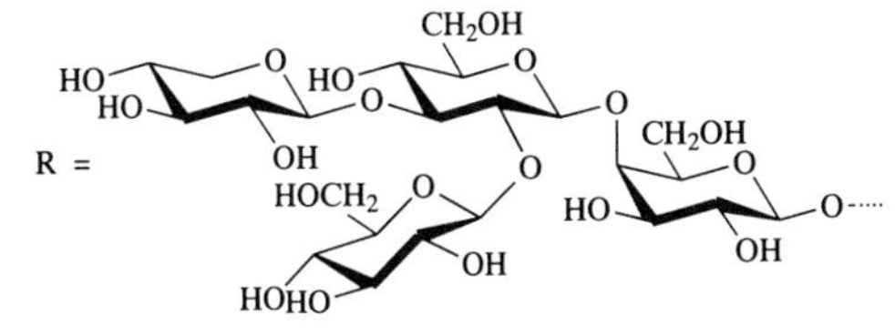

Steroid alkaloid
(Solanidane)

$C_{50}H_{83}NO_{20}$ [6077-69-6] MW 1018.20

In the leaves of *Solanum demissum*, *S. chacoense*, *S. jamesii* and other wild tuber-bearing *Solanum* spp.. Also in *Lycopersicon pimpinellifolium*, a wild tomato species (Solanaceae).

A toxic alkaloid of wild potato species, with haemolytic activity. Is a feeding repellent to the Colorado potato beetle and has been bred into the domestic potato to provide Colorado beetle resistance. On hydrolysis, it gives demissidine [474-08-8], which is a repellent to the cataract beetle.

Schreiber, K., *Z. Naturforsch.*, 1963, **18B**, 471.

Dendrobine 442

Dendroban-1-one

Sesquiterpenoid alkaloid
(Dendrobane)

$C_{16}H_{25}NO_2$ [2115-91-5] MW 263.38

In *Dendrobium nobile* and other *Dendrobium* spp. (Orchidaceae).

A toxic alkaloid, which lowers the blood-pressure and which, in large doses, reduces cardiac activity. On intravenous injection into animals, it produces convulsions.

Onaka, T., *Chem. Pharm. Bull.,* 1966, **12**, 506.

Denudatine 443

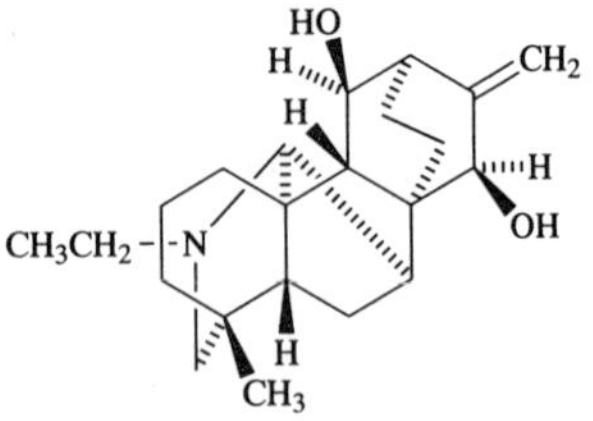

Diterpenoid alkaloid
(Atidane)

$C_{22}H_{33}NO_2$ [26166-37-0] MW 343.51

In the roots of *Delphinium denudatum* and present in other *Delphinium* spp. (Ranunculaceae).

Less toxic than most other *Delphinium* alkaloids. Has no effect on blood-pressure or respiration in anaesthetised dogs. However, it markedly inhibits the contraction of isolated rabbit duodenal strip.

Götz, M., *Tetrahedron Lett.,* 1969, 4369.

11-Deoxocucurbitacin I 444

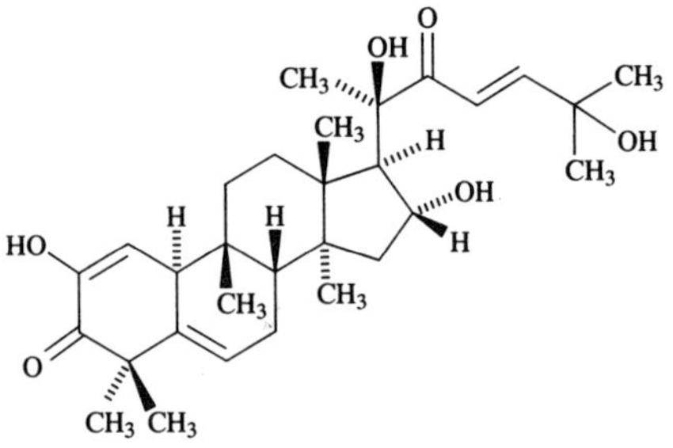

Triterpenoid
(Cucurbitacin)

$C_{30}H_{44}O_6$ [98941-76-5] MW 500.68

In *Desfontainia spinosa* (Desfontainiaceae). Also present in glycosidic form as spinoside A (q.v.).

Cytotoxic and antileukaemic activities.

Amonkar, A. A., *Phytochemistry,* 1985, **24**, 1803.

Deoxyelephantopin 445

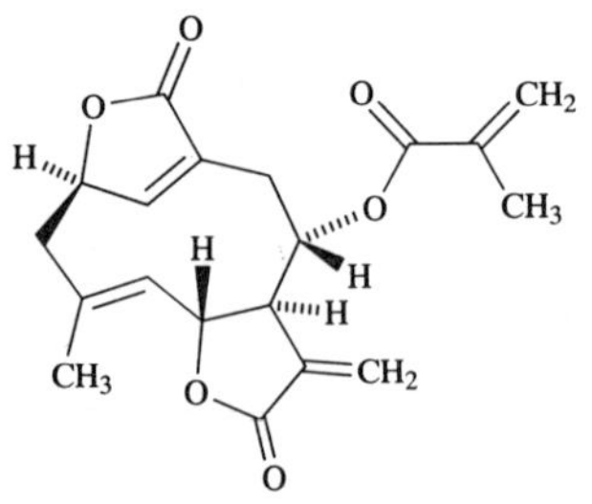

Sesquiterpenoid lactone
(Germacranolide)

$C_{19}H_{20}O_6$ [29307-03-7] MW 344.36

In *Elephantopus carolinianus* and *E. scaber* (Compositae).

Cytotoxic and antitumour properties.

Kurokawa, T., *Tetrahedron Lett.,* 1970, 2863.

6-Deoxyjacareubin 446

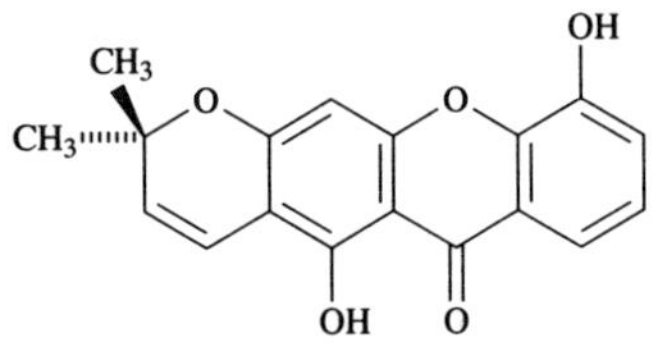

Xanthone

$C_{18}H_{14}O_5$ [16265-56-8] MW 310.31

In *Calophyllum zeylanicum* and in the trunkwood of *Kielmeyera speciosa* (Guttiferae).

Antimicrobial, anti-inflammatory and anti-ulcerogenic properties.

de Oliveira, G. G., *An. Acad. Bras. Cienc.,* 1966, **38**, 421.

8-Deoxylactucin 447

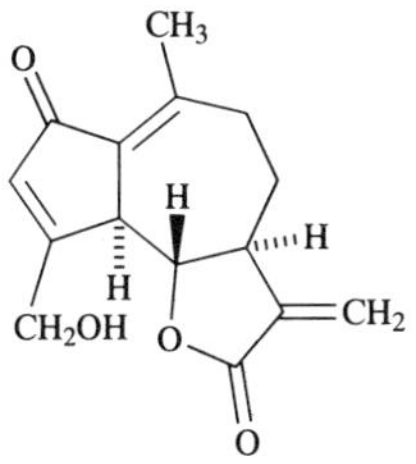

Sesquiterpenoid lactone
(Guaianolide)

$C_{15}H_{16}O_4$ [65725-10-2] MW 260.29

In all parts (except the flowers) of the chicory plant, *Cichorium intybus*; also present in leaves of the prickly lettuce, *Lactuca seriola* (Compositae).

Very bitter taste. Antifeedant to locusts. Possesses both cytotoxic and antitumour activities.

St. Pyrek, J., *Rocz. Chem.*, 1977, **51**, 2165.

Deoxylapachol 448

2-Prenyl-1,4-naphthoquinone

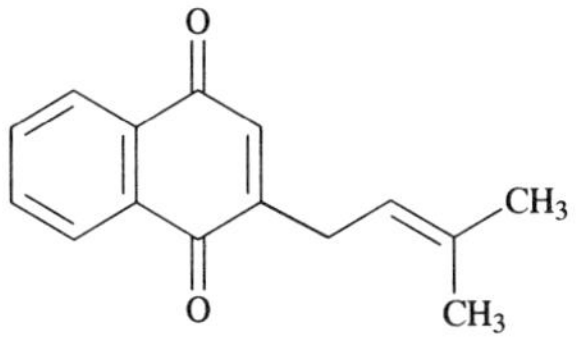

Naphthoquinone

$C_{15}H_{14}O_2$ [3568-90-9] MW 226.27

In teak wood, *Tectona grandis* (Verbenaceae) and in wood of catalpa tree, *Catalpa ovata* (Bignoniaceae).

Major allergen in teak, causing allergic skin reactions in carpenters working with teak. It also shows antitermite activity.

Burnett, A. R., *J. Chem. Soc., C.*, 1967, 2100.

Deoxyloganin 449

7-Deoxyloganin

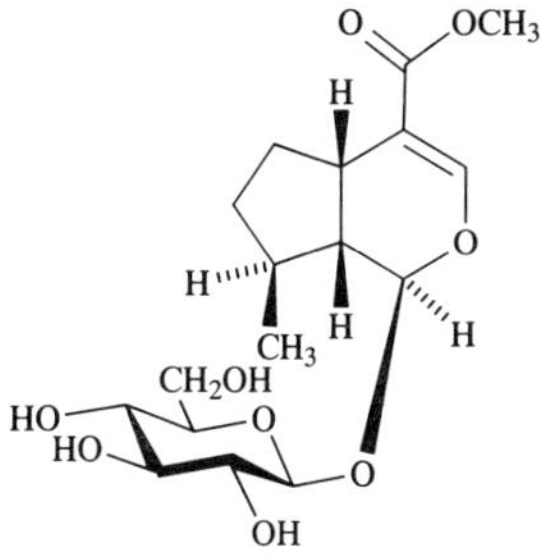

Monoterpenoid
(Iridoid glucoside)

$C_{17}H_{26}O_9$ [26660-57-1] MW 374.39

In many *Strychnos* (Loganiaceae), *Vinca* (Apocynaceae) and *Menyanthes* spp. (Menyanthaceae). First isolated from the fruits of *S. nux-vomica*.

Gives a toxic aglycone on hydrolysis. The glycoside has laxative properties.

Battersby, A. R., *J. Chem. Soc. C*, 1970, 826.

Deoxymannojirimycin 450

DMJ; Antibiotic LU 1; LU 1

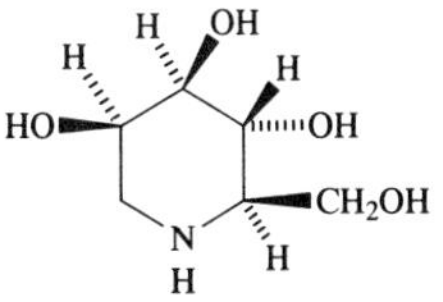

Piperidine alkaloid

$C_6H_{13}NO_4$ [84444-90-6] MW 163.17

In seeds of *Lonchocarpus sericeus* and *L. costaricensis* (Leguminosae).

Is a mannosidase inhibitor, and affects glycoprotein processing.

Fellows, L. E., *J. Chem. Soc. Chem. Commun.*, 1979, 977.

Deoxynojirimycin 451

DNJ; Moranoline; Antibiotic S-GI; S-GI

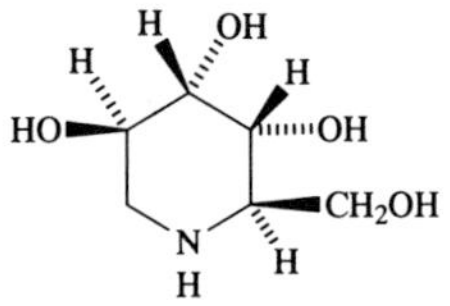

Piperidine alkaloid

$C_6H_{13}NO_4$ [19130-96-2] MW 163.17

In the bark of the mulberry tree, *Morus alba* and of other *Morus* spp. (Moraceae).

An inhibitor of α-glucosidase in the mammalian gut.

Ermert, P., *Helv. Chim. Acta,* 1991, **74**, 2043.

Deoxynupharidine 452

α-Nupharidine

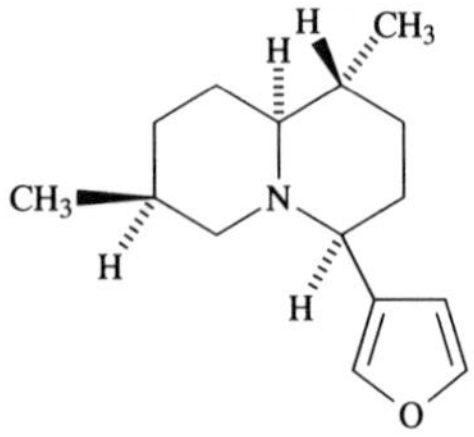

Sesquiterpenoid alkaloid
(Nupharidine)

$C_{15}H_{23}NO$ [1143-54-0] MW 233.35

In rhizomes of the water-lily, *Nuphar japonicum*, of *N. luteum* and of various *Nymphaea* spp. (Nymphaeaceae). Co-occurs with the 7-epimer, 7-epideoxynupharidine [28463-31-2].

Mildly toxic. The rhizomes are used in Eastern Europe and Asia for sedative and narcotic purposes.

Wong, C. F., *J. Org. Chem.,* 1970, **35**, 517.

Deoxypodophyllotoxin 453

Anthricin; Hernandion; Silicicolin; Desoxypodophyllotoxin

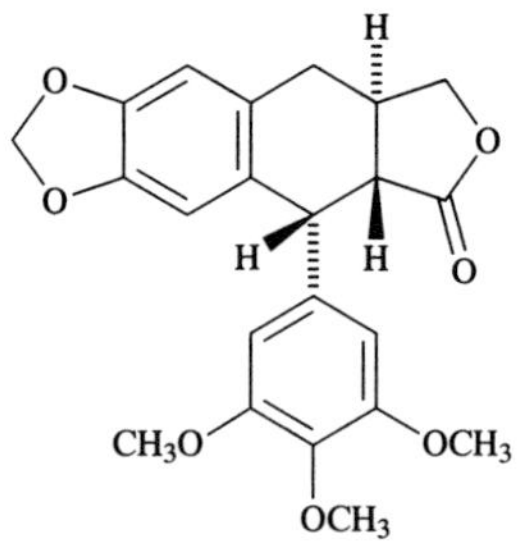

Lignan

$C_{22}H_{22}O_7$ [19186-35-7] MW 398.41

In the roots of *Anthriscus sylvestris* (Umbelliferae), in the seed oil of *Hernandia ovigera* (Hernandiaceae), in the needles and berries of *Juniperus sabina* and in *Libocedrus* spp. (Cupressaceae); also in the rhizomes and roots of *Podophyllum pleianthum* (Podophyllaceae).

Antitumour, antimitotic and antiviral activities.

Kofod, H., *Acta Chem. Scand.,* 1955, **9**, 346.

Desacetoxymatricarin 454

Axillin; Leucodin; Leukodin; Leucomosin

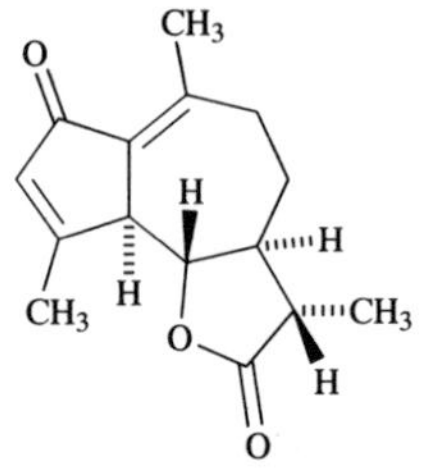

Sesquiterpenoid lactone
(Guaianolide)

$C_{15}H_{18}O_3$ [17946-87-1] MW 246.31

In *Matricaria suffruticosa* and in several *Achillea* and *Artemisia* spp. (Compositae).

Cytotoxic and antitumour properties.

Martinez V, M., *J. Nat. Prod.,* 1988, **51**, 221.

Desacetyleupaserrin 455

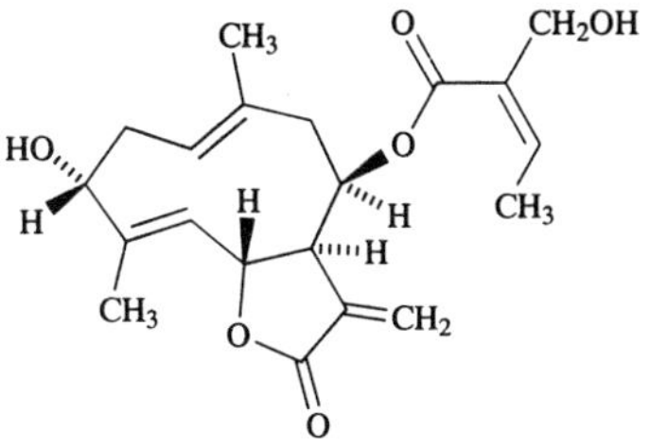

Sesquiterpenoid lactone
(Germacranolide)

$C_{20}H_{26}O_6$ [38456-39-2] MW 362.42

In *Eupatorium semiserratum*, *E. mikanioides* and in several *Helianthus* spp. (Compositae).

Toxic to certain insect larvae, affecting their development. Has cytotoxic and antitumour properties.

Herz, W., *J. Org. Chem.*, 1980, **45**, 489.

Deserpidine 456

Canescine; 11-Desmethoxyreserpine; Harmonyl; Raunormine; Recanescine; Reserpidine

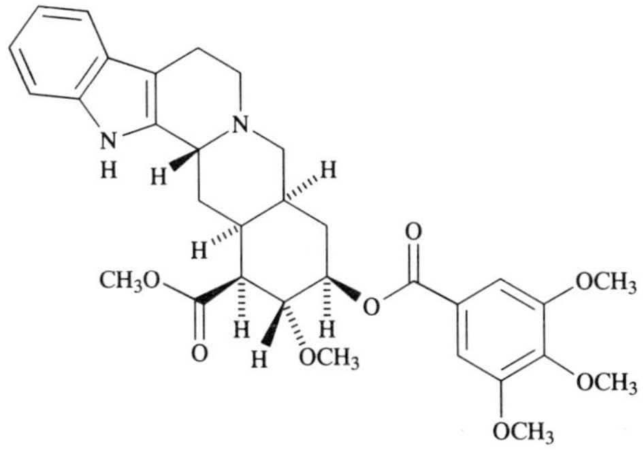

Indole alkaloid
(Yohimbane)

$C_{32}H_{38}N_2O_8$ [131-01-1] MW 578.66

In the roots of *Rauwolfia canescens*, *R. cubana* and *R. littoralis* (Apocynaceae).

Antihypertensive properties. Could be used as a tranquilliser, although it is toxic.

MacPhillamy, H. B., *J. Am. Chem. Soc.*, 1955, **77**, 4335.

Dhurrin 457

p-Hydroxymandelonitrile glucoside

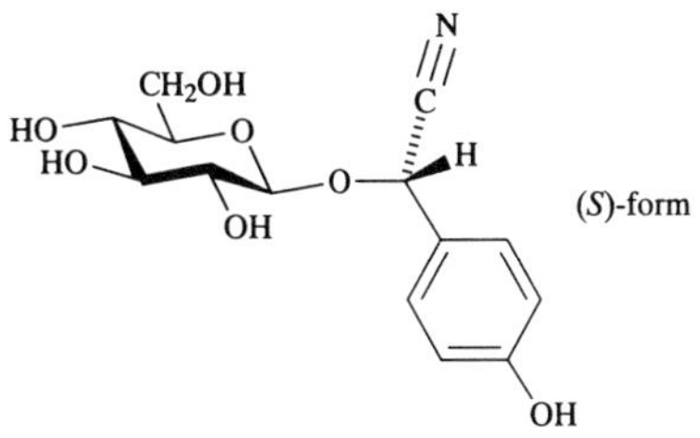

(*S*)-form

Cyanogenic glycoside

$C_{14}H_{17}NO_7$ [499-20-7] MW 311.29

First isolated from cultivated forms of *Sorghum* spp. and later shown to be present in other grasses (Gramineae) where the (*R*)-epimer taxiphyllin (q.v.) is also is isolated. It occurs erratically in dicotyledons, e.g. in *Trochodendron aralioides* (Trochodendraceae), *Borago officinalis* (Boraginaceae), *Platanus* spp. (Platanaceae) and in *Macadamia ternofolia* (Proteaceae).

Releases cyanide on hydrolysis. The high dhurrin content of young *Sorghum* seedlings has led to accidental cattle poisoning. *Sorghum* seeds, which are widely eaten as human food in India and Africa, are non-cyanogenic and hence quite safe dietary components.

Mao, C., *Phytochemistry,* 1965, **4**, 297.

Diallyl disulfide 458

Allyl disulfide; Di-2-propenyl disulfide

$CH_2{=}CH{-}CH_2{-}S{-}S{-}CH_2{-}CH{=}CH_2$

Sulfur compound

$C_6H_{10}S_2$ [2179-57-9] MW 146.28

Principal constituent of the oil of garlic, *Allium sativum* (Alliaceae). Also occurs in many other *Allium* spp. including the onion, *Allium cepa*.

Insecticidal properties.

Schultz, O. E., *Pharmazie,* 1965, **20**, 441.

Diallyl sulfide 459

Allyl sulfide; Thioallyl ether; Oil garlic;
Di-2-propenyl sulfide

$CH_2{=}CH{-}CH_2{-}S{-}CH_2{-}CH{=}CH_2$

Sulfur compound

$C_6H_{10}S$ [592-88-1] MW 114.21

In garlic, *Allium sativum* (Alliaceae).

Garlic odour principle. It is a strong irritant to eyes and skin. Shows anticancer properties.

Bernhard, R. A., *Arch. Biochem. Biophys.*, 1964, **107**, 137.

L-α,γ-Diaminobutyric acid 460

DABA; (*S*)-2,4-Diaminobutyric acid

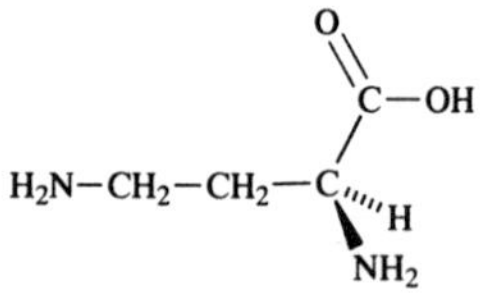

Amino acid

$C_4H_{10}N_2O_2$ [1758-80-1] MW 118.14

In seeds of everlasting pea, *Lathyrus sylvestris* and *L. latifolius* and of *Acacia* spp. (Leguminosae). Also present in Solomon's seal, *Polygonatum multiflorum* (Liliaceae) and some Compositae species.

Poisonous. It inhibits carbamoyltransferase, causing chronic ammonia toxicity in rats. It also impairs liver function and causes tremors, convulsions and weakness of hind limbs.

Ressler, C., *Science,* 1961, **134**, 188.

Dianthalexin 461

Dianthalexine

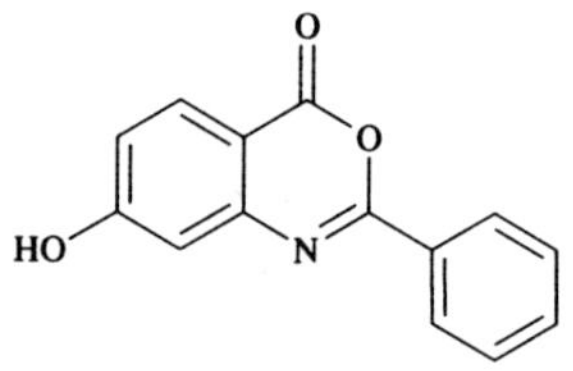

Benzoxazinone

$C_{14}H_9NO_3$ [85915-62-4] MW 239.23

Infected leaves of carnation, *Dianthus caryophyllus* (Caryophyllaceae).

Antifungal agent (phytoalexin).

Bouillant, M. L., *Tetrahedron Lett.,* 1983, **24**, 51.

Dianthramide A 462

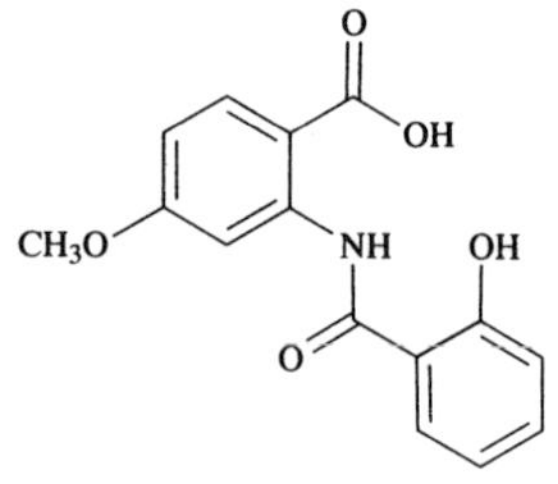

Anthranilic acid

$C_{15}H_{13}NO_5$ [93289-90-8] MW 287.27

Leaves of carnation, *Dianthus caryophyllus* (Caryophyllaceae) infected with *Phytophthora parasitica*.

Antifungal agent (phytoalexin).

Ponchet, M., *Phytochemistry,* 1984, **23**, 1901.

Dicoumarol 463

Dicumarol; Dicumol; Dicoumarin; Dufalone; Melitoxin

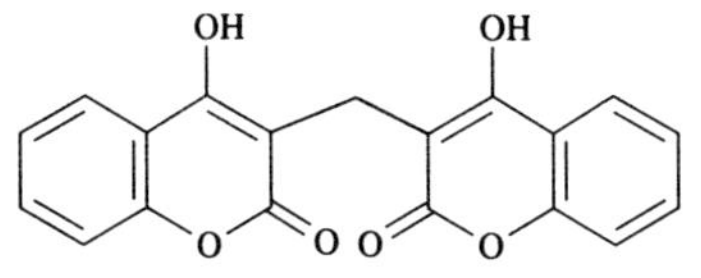

Coumarin

$C_{19}H_{12}O_6$ [66-76-2] MW 336.30

Originates from *o*-coumaric acid via 4-hydroxycoumarin in decomposing hay of both *Anthoxanthum* (Gramineae) and *Melilotus* spp. (Leguminosae).

Haemorrhagic disorders and even death may be caused by cattle eating 'spoiled sweet clover' containing dicoumarol. Also has anticoagulant properties. Related compounds are used as anticoagulants in medicine or to kill rodents.

Griminger, P., *J. Nutr.*, 1987, **117**, 1325.

Dicrotaline 464

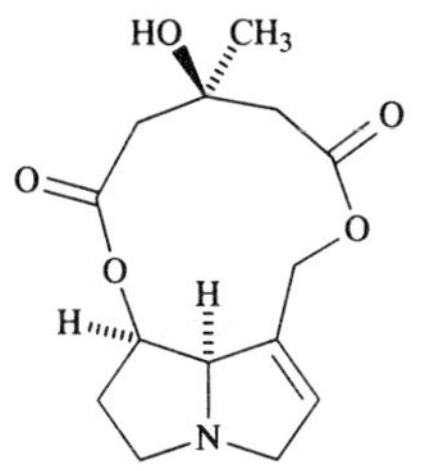

Pyrrolizidine alkaloid

$C_{14}H_{19}NO_5$ [480-87-5] MW 281.31

Crotalaria dura and *C. globifera* (Leguminosae).

Causes chronic respiratory disease in horses and mules which consume these plants in South Africa. Carcinogenic.

Adams, R., *J. Am. Chem. Soc.*, 1953, **75**, 2377.

Dictamnine 465

Dictamine

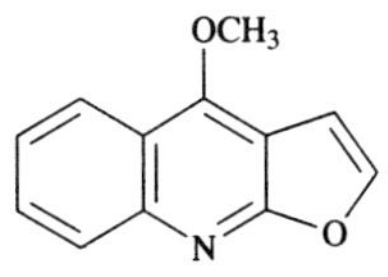

Quinoline alkaloid

$C_{12}H_9NO_2$ [484-29-7] MW 199.21

Occurs widely in plants of the Rutaceae. Present in the roots of *Dictamnus albus*, in *Aegle marmelos* and *Adiscanthus fusciflorus* as well as in various *Haplophyllum*, *Ruta* and *Zanthoxylum* spp.

Strong muscle contractant. Also has DNA-binding effects.

Asahina, Y., *Ber. Dtsch. Chem. Ges.*, 1930, **63**, 2045.

Dictyochromenol 466

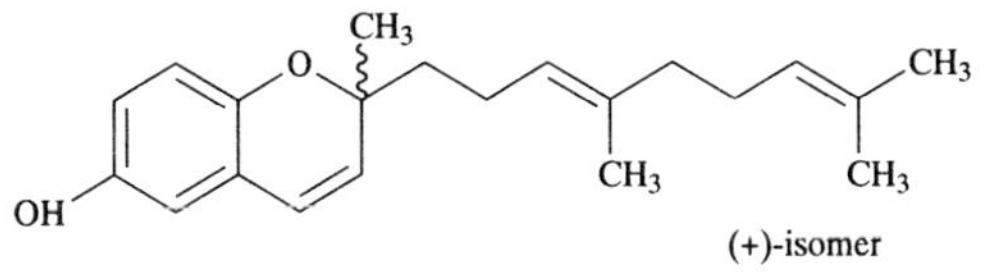

(+)-isomer

Chromene

$C_{21}H_{28}O_2$ [93398-32-4] MW 312.45

Brown alga, *Dictyopteris undulata*.

Poisonous to fish at a concentration of 27 p.p.m.

Dave, M. N., *Heterocycles*, 1984, **22**, 2301.

Diepomuricanin A 467

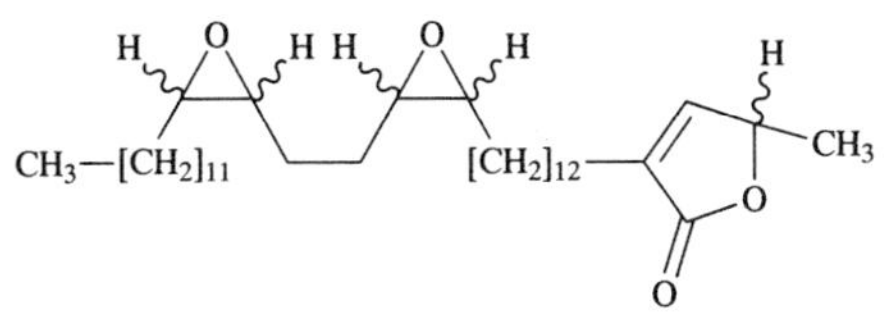

Acetogenin

$C_{35}H_{62}O_4$ [142733-57-1] MW 546.87

Seed of *Annona muricata* (Annonaceae).

In vitro cytotoxicity against KB cells, with an ED_{50} of 0.03 μg/ml.

Roblot, F., *Phytochemistry,* 1993, **34,** 281.

Digiferrugineol 468

1-Hydroxy-2-hydroxymethylanthraquinone; Digiferruginol

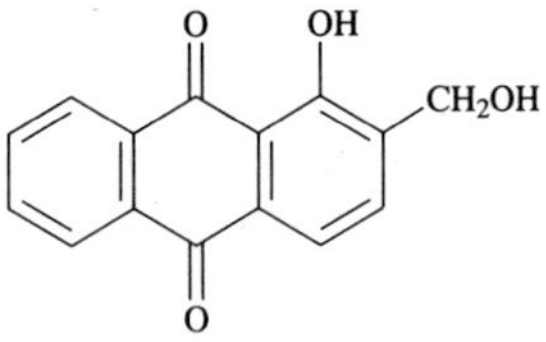

Anthraquinone

$C_{15}H_{10}O_4$ [24094-45-9] MW 254.24

In roots of *Streptocarpus dunnii* (Gesneriaceae), in leaves of *Digitalis ferruginea* (Scrophulariaceae), and in roots of *Morinda parvifolia* (Rubiaceae).

Cytotoxic *in vitro* to the KB-human epidermoid carcinoma of the nasopharynx.

Imre, S., *Z. Naturforsch.,* 1973, **28C**, 471.

Diginatin 469

Diginatigenin 3-tridigitoxoside

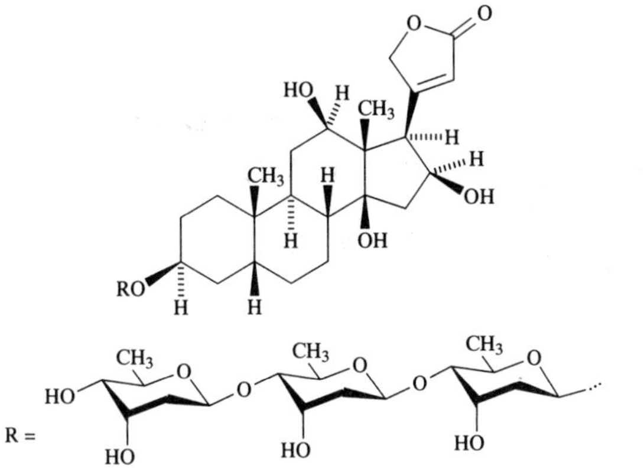

Cardenolide

$C_{41}H_{64}O_{15}$ [52589-12-5] MW 796.95

In *Digitalis lanata* (Scrophulariaceae).

Toxic to vertebrates. It is used as a cardiotonic.

Angliker, E., *Justus Liebigs Ann. Chem.,* 1957, **607**, 131.

Digitalin 470

Gitoxigenin 3-glucosyldigitaloside; Digitalinum verum; Diginorgin

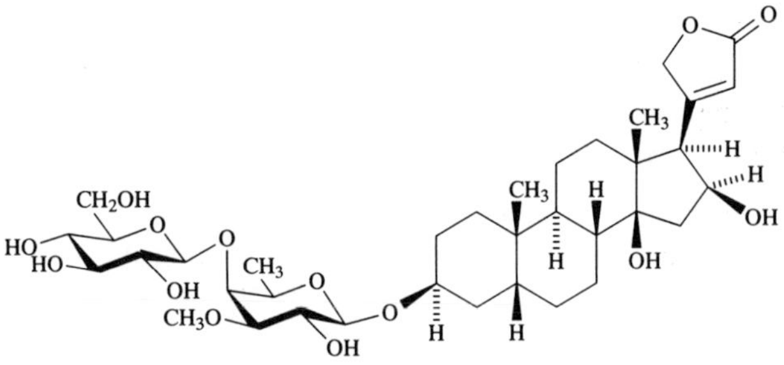

Cardenolide

$C_{36}H_{56}O_{14}$ [752-61-4] MW 712.83

In seeds of the foxglove, *Digitalis purpurea* (Scrophulariaceae) and in roots of *Adenium honghel* (Apocynaceae).

Toxic to vertebrates. It is used as a cardiotonic.

Rittel, W., *Helv. Chim. Acta,* 1952, **35**, 434.

Digitoxin 471

Digitophyllin; Digitoxigenin 3-tridigitoxoside; Cardigin; Carditoxin; Lanatoxin

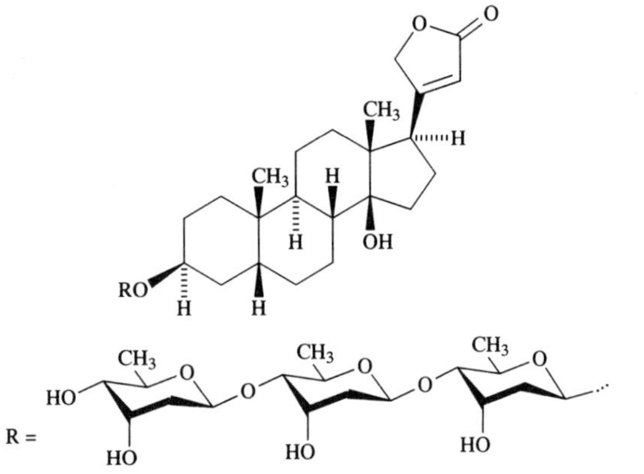

Cardenolide

$C_{41}H_{64}O_{13}$ [71-63-6] MW 764.95

In leaves of the foxglove, *Digitalis purpurea* (Scrophulariaceae).

Major toxic principle of the foxglove. The LD_{50} when administered orally is 60 mg/kg body-weight in guinea-pigs and 0.18 mg/kg body-weight in cats. The LD_{50} when administered intravenously is 0.4 mg/kg body-weight in cats. It is used as a cardiotonic at carefully controlled doses. Human poisoning from ingestion of foxglove leaves is rare, because of the intensive bitter taste and the fact that spontaneous vomiting usually takes place.

Drakenberg, T., *Can. J. Chem.*, 1990, **68**, 272.

Digoxin 472

Digoxigenin 3-tridigitoxoside; Cordioxil; Davoxin; Digacin

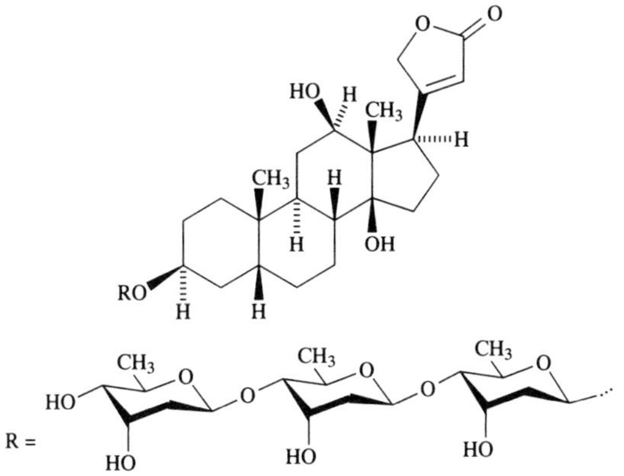

Cardenolide

$C_{41}H_{64}O_{14}$ [20830-75-5] MW 780.95

In *Digitalis lanata* and *D. orientalis* (Scrophulariaceae).

Toxic to vertebrates. Derivatives of digoxin are used as a cardiotonic.

Foss, P. R. B., *Anal. Profiles Drug Subst.*, 1980, **9**, 207.

Dihydrocornin aglycone 473

1-Hydroxydihydrocornin aglycone

α-isomer

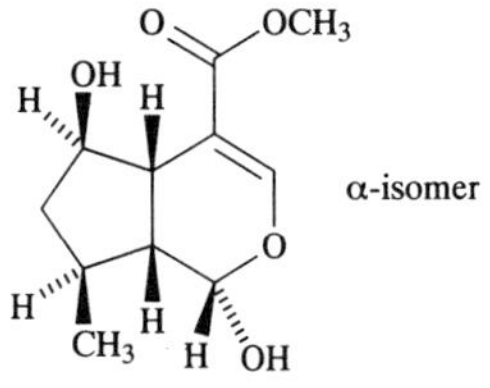

Monoterpenoid
(Iridane)

$C_{11}H_{16}O_5$ [145512-33-0] MW 228.24

Leaves of *Alibertia macrophylla* (Rubiaceae) as an epimeric mixture of α- and β-hydroxy derivatives [β-isomer 138751-01-6].

Antifungal activity against *Aspergillus niger* and three other fungi.

Young, M. C. M., *Phytochemistry,* 1992, **31**, 3433.

Dihydrogriesenin 474

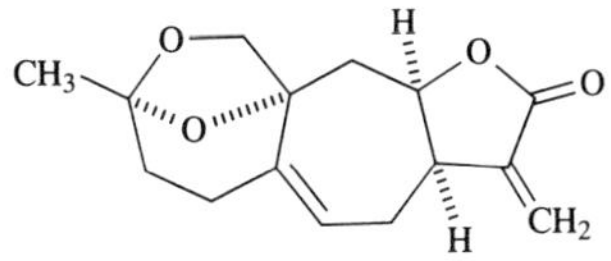

Sesquiterpenoid lactone
(Secoguaianolide)

$C_{15}H_{18}O_4$ [20087-05-2] MW 262.31

In *Geigeria africana, G. aspera* and *G. filifolia* (Compositae).

Toxic to mammals. One of several toxins in *Geigeria* species, which are commonly known as 'vomiting bush', because grazing livestock suffer a vomiting disease after consuming the leaves.

DeKock, W. T., *Tetrahedron,* 1968, **24**, 6037.

Dihydromethysticin 475

Pseudomethysticin; ψ-Methysticin

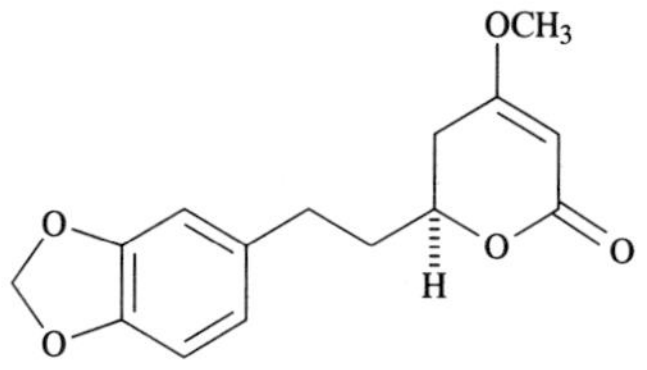

Lactone

$C_{15}H_{16}O_5$ [19902-91-1] MW 276.29

In the rhizomes of *Piper methysticum* (Piperaceae) and in *Aniba gigantifolia* (Lauraceae).

Spasmolytic activity.

Sauer, H., *Planta Med.,* 1967, **15**, 443.

Dihydropiperlongumine 476

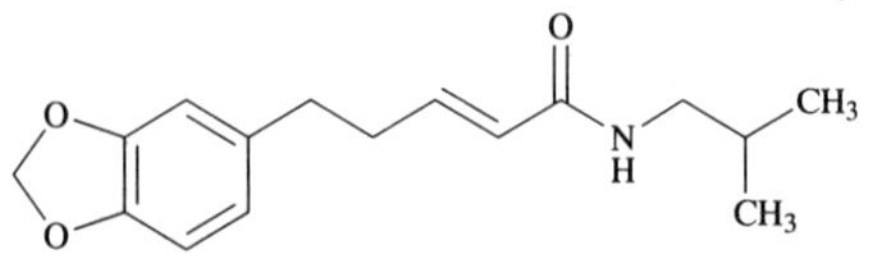

Aliphatic amide

$C_{16}H_{21}NO_3$ [23512-53-0] MW 275.35

Leaves of *Piper tuberculatum* (Piperaceae).

Toxic to mosquito larvae.

Dwuma-Badu, D., *Phytochemistry,* 1976, **15**, 822.

Dihydrosamidin 477

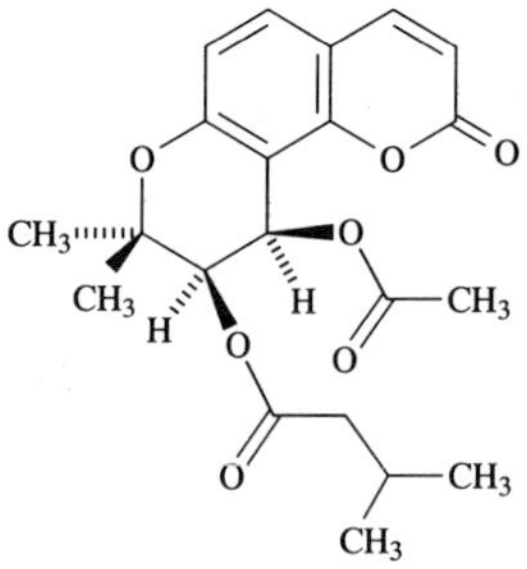

Coumarin

$C_{21}H_{12}O_7$ [6005-18-1] MW 376.32

In the fruits and flowers of *Ammi visnaga* (Umbelliferae).

Vasodilatory agent, used in human medicine.

Murray, R. D. H., *Fortschr. Chem. Org. Naturst.,* 1978, **35**, 199.

Dihydrosanguinarine 478

Alkaloid B

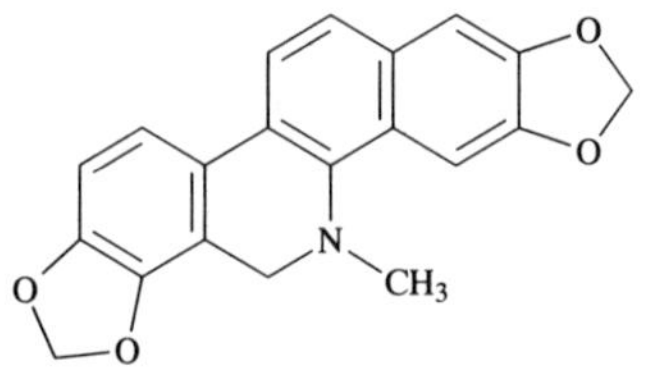

Benzophenanthridine alkaloid

$C_{20}H_{15}NO_4$ [3606-45-9] MW 333.34

In *Fumaria parviflora*, *F. vaillantii*, *Corydalis gigantea* (Fumariaceae), *Eschscholzia californica* (Papaveraceae) and *Pteridophyllum* spp. (Sapindaceae).

Toxic to mammals. Cytotoxic and anti-inflammatory, and has a positive inotropic effect on the heart. It inhibits various enzymes including ATP-ase. Used at high doses over a long period, it causes glaucoma. Dihydrosanguinarine is used in dentifrices because of its antiplaque activity.

Stermitz, F. R., *Phytochemistry,* 1972, **11**, 2644.

Dihydrowyerone 479

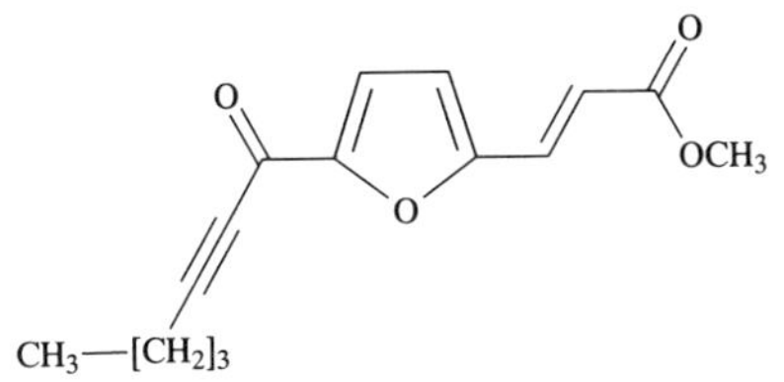

Acetylene

$C_{15}H_{16}O_4$ [20450-54-8] MW 260.29

Broad bean, *Vicia faba* and lentil, *Lens culinaris* (Leguminosae) infected with *Botrytis cinerea*.

Antifungal agent (phytoalexin).

Robeson, D. J., *Phytochemistry,* 1978, **17**, 807.

7,4′-Dihydroxyflavan 480

Demethylbroussin

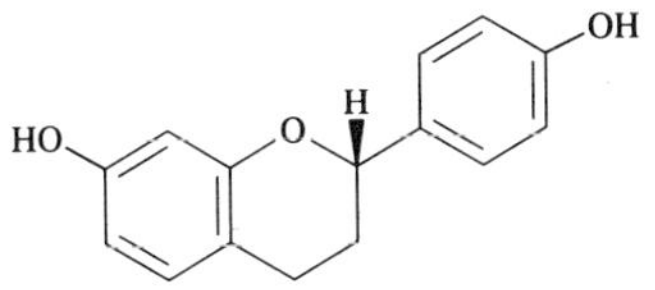

Flavan

$C_{15}H_{14}O_3$ [494-48-4] MW 242.27
(*S*)-form [82925-54-0]

Fungally infected bulbs of the daffodil, *Narcissus pseudonarcissus* (Amaryllidaceae) and constitutively present in in stems of *Bauhinia manca* (Leguminosae).

Fungitoxic to *Botrytis cinerea* with an ED_{50} of 65 μg/ml. Also antibacterial activity.

Coxon, D. T., *Phytochemistry,* 1980, **19**, 889.

2′,6′-Dihydroxy-4′-methoxyacetophenone 481

4-*O*-Methylphloracetophenone

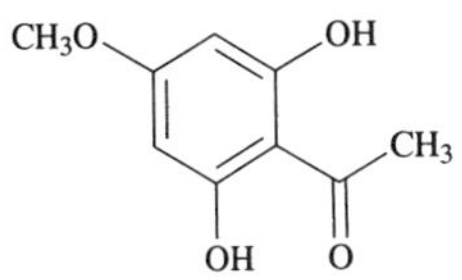

Acetophenone

$C_9H_{10}O_4$ [7507-89-3] MW 182.18

Fungally-infected roots of *Sanguisorba minor* (Rosaceae).

Antifungal agent (phytoalexin).

Kokubun, T., *Phytochemistry,* 1994, **35**, 331.

3′,4′-Dihydroxy-7-methoxyflavan 482

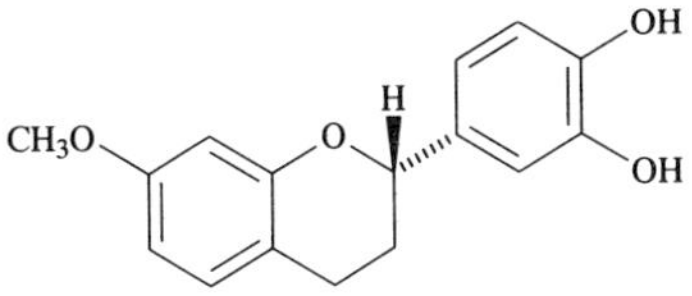

Flavan

$C_{16}H_{16}O_4$ [116384-19-1] MW 272.30

Stem of *Bauhinia manca* (Leguminosae).

Fungitoxic, e.g. to *Saprolegnia asterophora*.

Achenbach, H., *Phytochemistry,* 1988, **27**, 1835.

1,7-Dihydroxy-4-methoxyxanthone 483

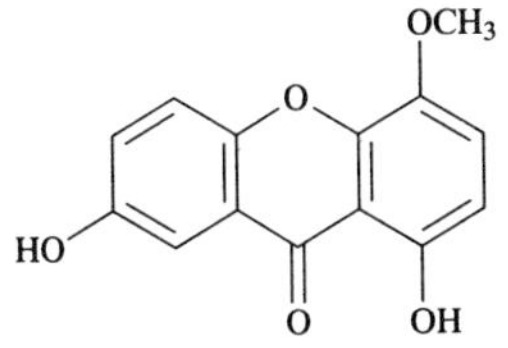

Xanthone

$C_{14}H_{10}O_5$ [87339-76-2] MW 258.23

Roots of *Polygala nyikensis* (Polygalaceae).

Antifungal activity against *Cladosporium cucumerinum*.

Delle Monache, F., *Phytochemistry,* 1983, **22**, 227.

7,4′-Dihydroxy-8-methylflavan 484

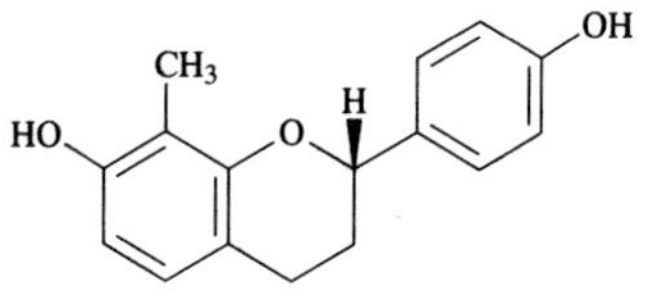

Flavan

$C_{16}H_{16}O_3$ [75412-98-5] MW 256.30

Fungally infected bulbs of the daffodil, *Narcissus pseudonarcissus* (Amaryllidaceae); constitutively present in the resin of the dragon tree, *Dracaena draco* (Agaraceae).

Fungitoxic against *Botrytis cinerea*, with an ED_{50} of 32 μg/ml. Also antibacterial activity.

Coxon, D. T., *Phytochemistry,* 1980, **19**, 889.

Dillapiole 485

Dill apiole; Dillapiol

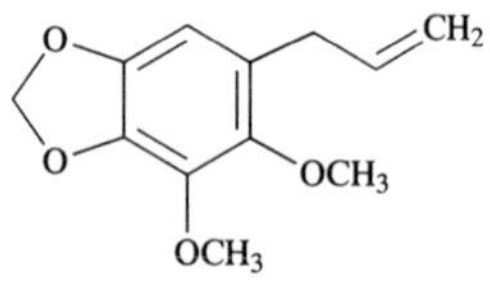

Phenylpropanoid

$C_{12}H_{14}O_4$ [484-31-1] MW 222.24

In the essential oil of dill, *Anethum graveolens*, and in the fruit oil of *Ligusticum scotinum*, *Orthodon formosanus* and *Crithmum maritimum* (Umbelliferae). It also occurs in leaves of *Piper aduncum* and *P. novae-hollandae* (Piperaceae), in *Erigeron* spp. (Compositae) and in leaves of *Laurelia serrata* (Monimiaceae).

Insecticide and molluscicide.

Bernhard, H. O., *Helv. Chim. Acta,* 1978, **61**, 2273.

Dilophic acid 486

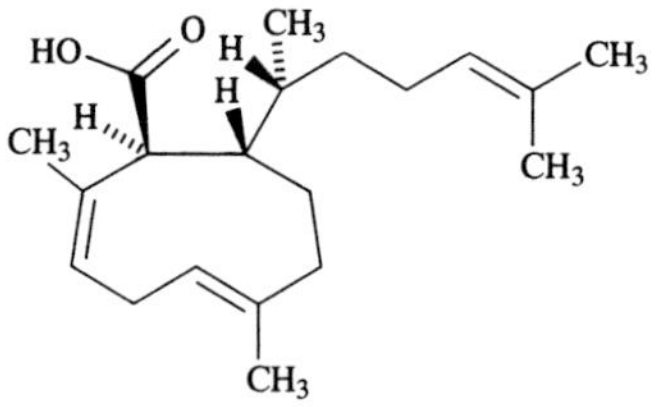

Diterpenoid
(Xenicane)

$C_{20}H_{32}O_2$ [108864-15-9] MW 304.47

Brown alga, *Dilophus guineensis.*

Poisonous to fish at a concentration of 50 μg/ml; antibacterial at a concentration of 100 μg/ml.

Schlenk, D., *Phytochemistry,* 1987, **26**, 1081.

2,6-Dimethoxybenzoquinone 487

2,6-Dimethoxyquinone; 2,6-Dimethoxy-1,4-benzoquinone

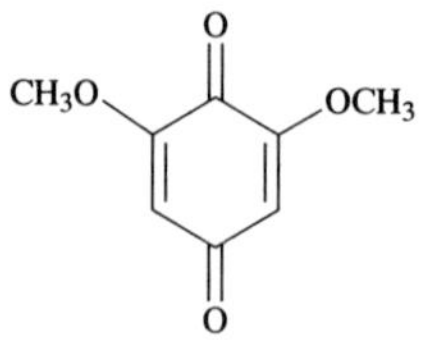

Benzoquinone

$C_8H_8O_4$ [530-55-2] MW 168.15

Fairly widespread as a trace component in woody tissues. Present in the heartwood of *Acacia melanoxylon* (Leguminosae), in wheat grains, *Triticum vulgare* (Gramineae), in the aerial parts of *Adonis vernalis* (Ranunculaceae) and in the roots of *Rauwolfia vomitoria* (Apocynaceae).

Together with acamelin (q.v.) responsible for the contact dermatitis of *Acacia melanoxylon* wood. Shows *in vitro* cytotoxicity in P-388 lymphocytic leukaemia tests.

Thomson, R. H., Naturally Occuring Quinones, 2nd edn, 1971, 106, Academic Press.

3,4-Dimethoxydalbergione 488

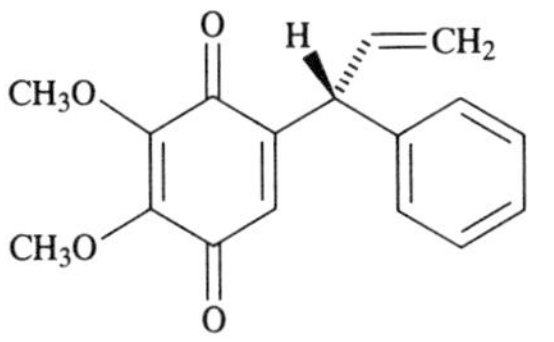

Benzoquinone

$C_{17}H_{16}O_4$ [41043-20-3] MW 284.31
(*R*)-form [3755-64-4]

In the heartwood of *Machaerium scleroxylon*, *M. pedicellatum*, *Prosopis kuntzei* and *Piptadenia macrocarpa* (Leguminosae).

Considered to be the dermatitic agent of *Machaerium* wood.

Eyton, W. B., *Tetrahedron,* 1965, **21**, 2697.

2,6-Dimethoxyphenol 489

Syringol

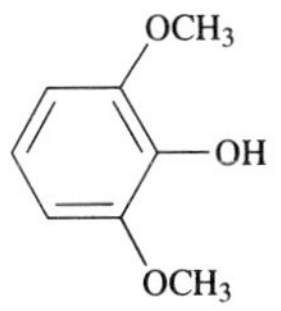

Phenol

$C_8H_{10}O_3$ [91-10-1] MW 154.17

In stems of *Mucuna birdwoodiana* (Leguminosae) and in maple syrup, *Acer saccharum* (Aceraceae).

Toxic to humans. Inhibits prostaglandin synthase and has an inhibitory effect on rabbit platelet aggregation.

Filipic, V. J., *J. Food Sci.,* 1965, **30**, 1008.

Dimethyl disulfide 490

Methyl disulfide

$CH_3—S—S—CH_3$

Sulfur compound

$C_2H_6S_2$ [624-92-0] MW 94.20

A volatile released from the tissue of many *Allium* spp. (Alliaceae) i.e. onion, garlic, etc.

Highly unpleasant odour. Toxic.

Schultz, O. E., *Pharmazie,* 1965, **20**, 441.

N,N-Dimethyltryptamine 491

3-(2-Dimethylaminoethyl)indole; DMT

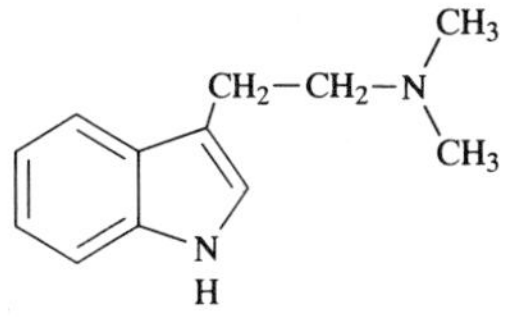

Aromatic amine

$C_{12}H_{16}N_2$ [61-50-7] MW 188.27

In the leaves of *Prestonia amazonica* (Apocynaceae), the flowers of the reed, *Arundo donax* (Gramineae), the seeds and leaves of *Piptadenia peregrina* and in cowhage, *Mucuna pruriens* (Leguminosae).

Psychotomimetic activity, including hallucinations, anxiety and perceptual disorders. It also causes autonomic effects such as hypertension and pupillary dilatation.

Fish, M. S., *J. Am. Chem. Soc.,* 1955, **77**, 5892.

Dinophysistoxin 1 492

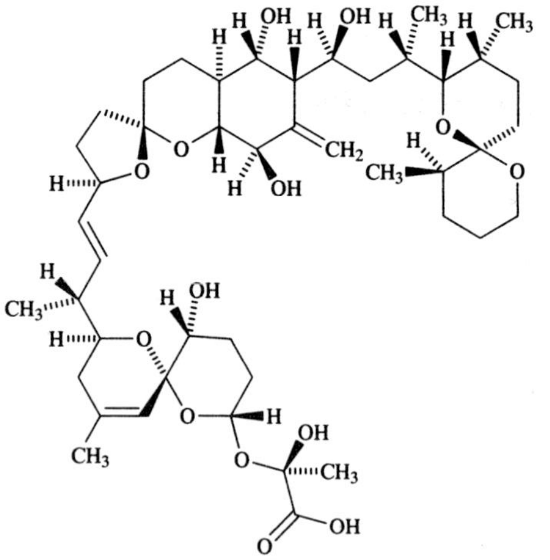

Polyether

$C_{45}H_{70}O_{13}$ [81720-10-7] MW 819.04

Produced by the dinoflagellate alga *Dinophysis fortii* and dietarily sequestered by scallops and mussels.

Component toxin of diarrhoetic shellfish poisoning.

Murata, W., *Nippon Suisan Gakkaishi,* 1982, **48**, 549.

Dinophysistoxin 2 493

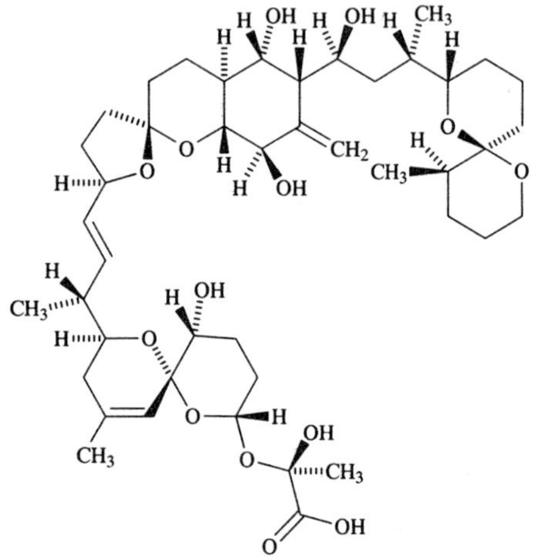

Polyether

$C_{44}H_{68}O_{13}$ [139933-46-3] MW 805.01

Produced by the algae belonging to *Dynophysis* spp. and dietarily sequestered by shellfish.

Component toxin of diarrhoetic shellfish poisoning.

Hu, T., *J. Chem. Soc., Chem. Commun.,* 1992, 39.

Dioncopeltine A 494

Triphyopeltine

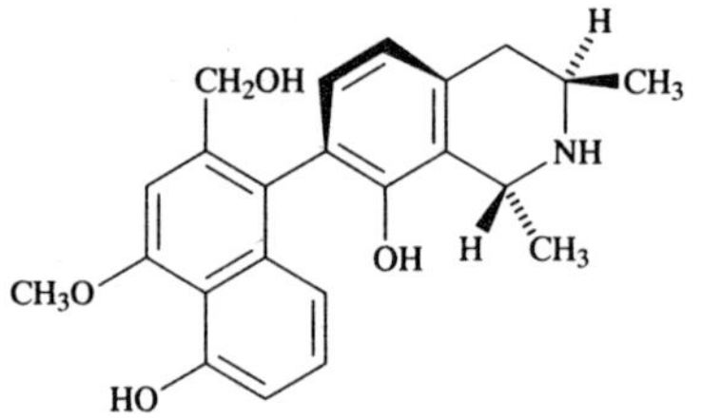

Naphthoisoquinoline alkaloid

$C_{23}H_{25}NO_4$ [60158-81-8] MW 379.46

Dioncophyllum thollonii and *Triphyophyllum peltatum* (Dioncophyllaceae).

Antiprotozoal activity.

Lavault, M., *Planta Med. (Suppl.),* 1980, 17.

Dioscin 495

Collettiside III

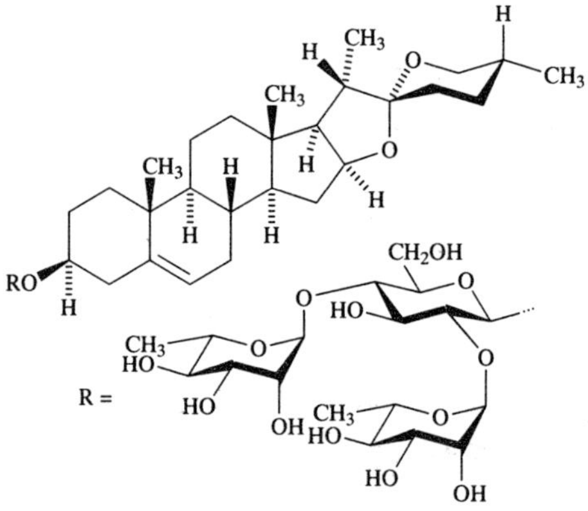

Steroidal saponin
(Spirostan)

$C_{45}H_{72}O_{16}$ [19057-60-4] MW 869.05

Occurs as protodioscin (q.v.) in *Dioscorea* spp. (Dioscoreaceae), *Costus* spp. (Zingiberaceae), *Paris* and *Trillium* spp. (Liliaceae) and *Trigonella* spp. (Leguminosae).

Potent haemolytic toxin to small mammals.

Kawasaki, T., *Chem. Pharm. Bull.,* 1962, **10**, 703.

Dioscorine 496

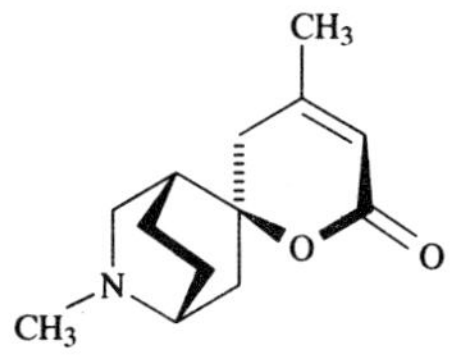

Piperidine alkaloid

$C_{13}H_{19}NO_2$ [3329-91-7] MW 221.30

In the tubers of *Dioscorea hirsuta* and *D. hispida* (Dioscoreaceae).

Highly toxic to humans.

Page, C. B., *J. Chem. Soc.,* 1964, 4811.

Diosgenin 497

Nitogenin; Discorea sapogenin

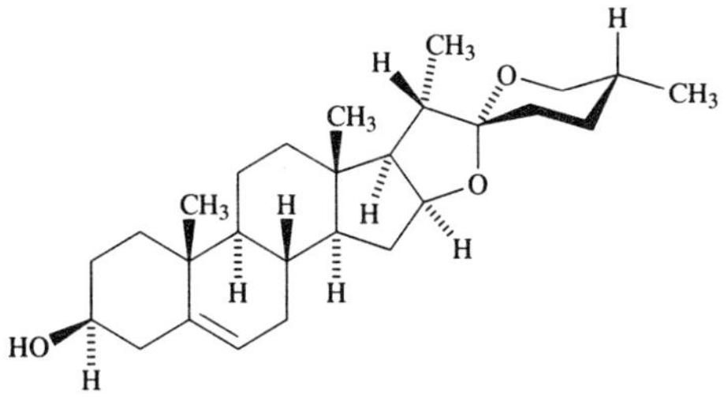

Steroidal sapogenin
(Spirostan)

$C_{27}H_{42}O_3$ [512-04-9] MW 414.63

Obtained by acid hydrolysis of many different saponins, e.g. deltonin, deltoside, dioscin, (q.v.) etc., from *Dioscorea* spp. (Dioscoreaceae), *Costus* spp. (Zingiberaceae), *Panicum* spp. e.g. *P. coloratum* and *P. dichotomiflorum* (Gramineae), as well as other sources.

Implicated in the poisoning of sheep, following the ingestion of the grasses *Panicum coloratum* and *Panicum dichotomiflorum*. The symptoms of poisoning include hepatogenous photosensitisation. The calcium salt of epismilagenin glucuronide has been recovered from the bile of poisoned sheep, presumably formed from the diosgenin-based saponins in the grasses.

Karrer, W., Konsitution und Vorkommen der organischen Pflanzenstoffe, 2nd edn, 1981, 500, Birkhäuser.

Diosphenol 498

Barosma camphor; Buchu camphor; 2-Hydroxypiperitone; 2-Hydroxy-*p*-menth-1-en-3-one; Buccocamphor

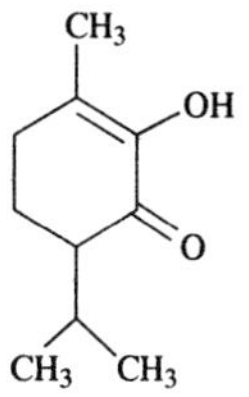

Monoterpenoid
(Menthane)

$C_{10}H_{16}O_2$ [490-03-9] MW 168.24

In the volatile oil of buchu leaves, *Barosma betulina*, *B. serratifolia* and *B. crenulata* (Rutaceae). Also present in the oil of *Cymbopogon densiflorus* (Gramineae) and of *Mentha rotundifolia* (Labiatae).

Diuretic activity.

Ohashi, M., *Bull. Chem. Soc. Jpn,* 1976, **49**, 2292.

Diospyrin 499

Euclin

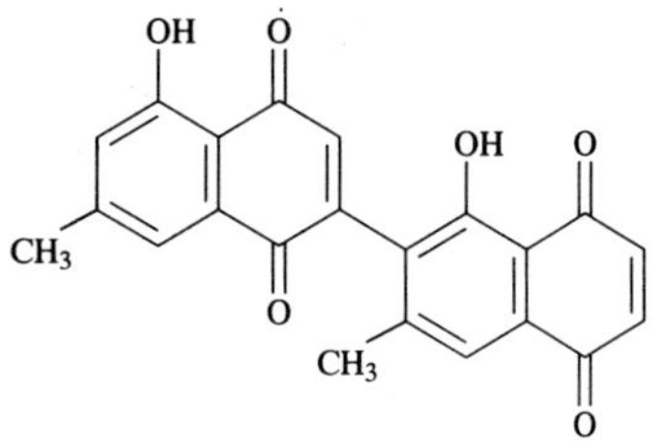

Binaphthoquinone

$C_{22}H_{14}O_6$ [28164-57-0] MW 374.35

In root, bark, wood and leaf of *Diospyros* and *Euclea* spp. (Ebenaceae).

It is immunostimulating at low doses, but cytotoxic at higher doses.

Ganguly, A. K., *Tetrahedron Lett.,* 1966, 3373.

Diosquinone 500

2′,3′-Epoxydiospyrin

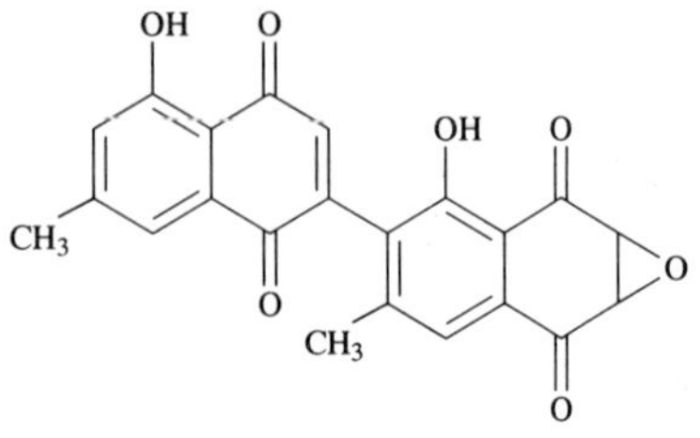

Dinaphthoquinone

$C_{22}H_{14}O_7$ [50886-69-6] MW 390.35

Root of *Diospyros mespiliformis* and *D. tricolor* (Ebenaceae).

Antibacterial properties; especially active against *Staphylococcus aureus* at a concentration of 3 μg/ml. Orange pigment.

Lillie, T. J., *J. Chem. Soc., Perkin Trans. 1,* 1980, 1161.

Dipentene 501

dl-Limonene; Inactive limonene; (±)-Limonene

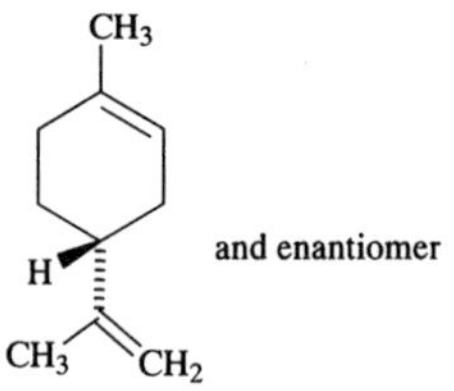

and enantiomer

Monoterpenoid

$C_{10}H_{16}$ [7705-14-8] MW 136.24

Widespread in essential oils of angiosperms. In oil of bergamot, *Citrus aurantium* var. *bergamia* (Rutaceae), oil of cubeb, *Piper cubeba* (Piperaceae), oil of citronella, *Cymbopogon* spp. (Gramineae) and oils of many umbellifer fruits. *See also* limonene.

Skin irritant and sensitiser. It also shows expectorant and sedative properties.

Karrer, W., Konstitution und Vorkommen der organischen Pflanzenstoffe, 2nd edn, 1981, 31, Birhhäuser.

Diphyllin 502

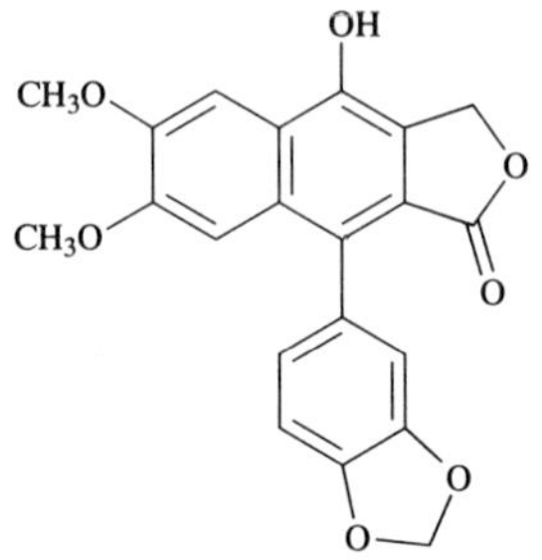

Lignan

$C_{21}H_{16}O_7$ [22055-22-7] MW 380.35

In aerial parts of *Haplophyllum hispanicum* (Rutaceae), in *Diphylleia grayi* (Podophyllaceae) and in *Justicia hayatai* (Acanthaceae). Also occurs in the heartwood of *Cleistanthus collinus* (Euphorbiaceae).

Toxic to fish. Antitumour activity.

Horii, Z., *J. Chem. Soc., Chem. Commun.,* 1968, 653.

Diplophyllin 503

ent-Alloalantolactone

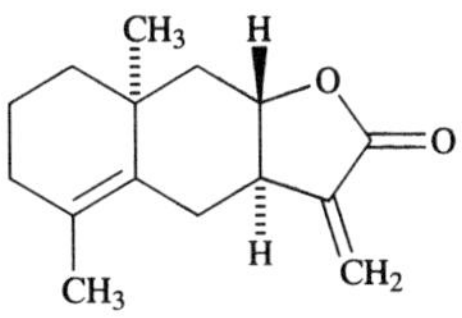

CH_3 H O O H CH_2 CH_3

(–)-form

Sesquiterpenoid lactone
(Eudesmanolide)

$C_{15}H_{20}O_2$ [64340-42-7] MW 232.32
(+)-form[64340-41-6]

The liverworts *Diplophyllum albicans* and *Chiloscyphus polyanthus*.

Toxic to fish, killing them at a concentration of 6.7 p.p.m. within 240 min. Cytotoxic, with an ED_{50} of 2.1 μg/ml in the KB cell test.

Ohta, Y., *Tetrahedron,* 1977, **33**, 617.

Disenecionyl *cis*-khellactone 504

cis-Khellactone disenecionate

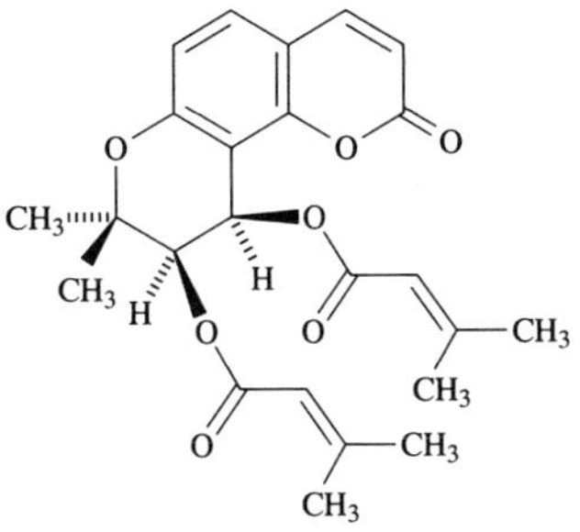

O O O CH_3 O CH_3 H H O O CH_3 CH_3 O CH_3 CH_3

Coumarin

$C_{24}H_{26}O_7$ [54676-88-9] MW 426.47

In roots of *Seseli incanum*, *S. libanotis* and other *Seseli* spp. (Umbelliferae).

Spasmolytic and coronary vasodilatory activities.

Murray, R. D. H., *Fortschr. Chem. Org. Naturst.,* 1978, **35**, 199.

Diterpenoid EF-D 505

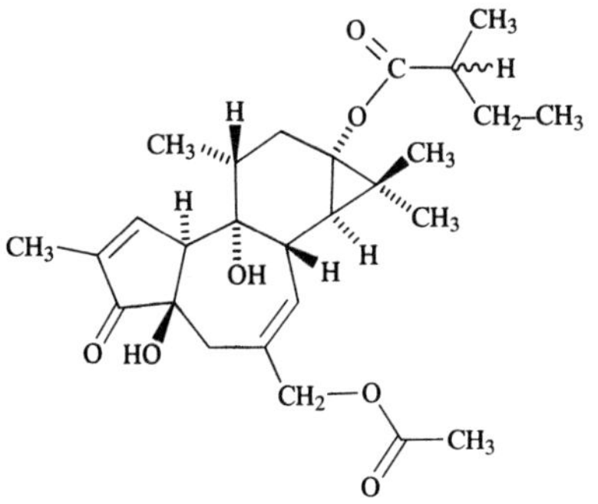

O C CH_3 H CH_2–CH_3 O H CH_3 CH_3 CH_3 H CH_3 OH H H O HO CH_2–O CH_3 O

Diterpenoid
(Tigliane)

$C_{27}H_{38}O_7$ [25090-73-7] MW 474.59

In *Euphorbia fortissima* (Euphorbiaceae).

Skin irritant.

Kinghorn, A. D., *J. Pharm. Pharmacol.,* 1975, **27**, 329.

Diterpenoid SP-II 506

ent-16β,17-Dihydroxykauran-19-oic acid

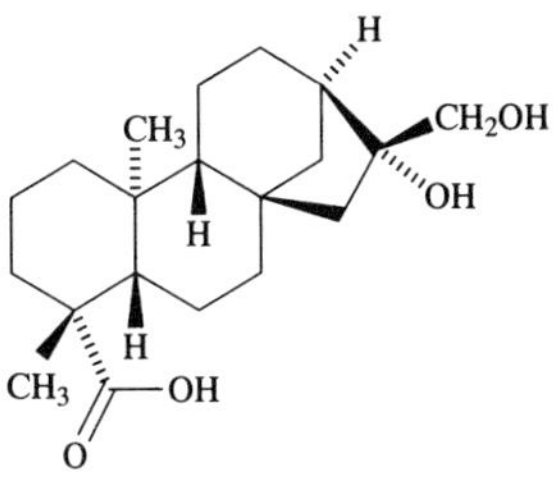

H CH_3 CH_2OH OH H H CH_3 OH O

Diterpenoid
(Kaurane)

$C_{20}H_{32}O_4$ [3301-61-9] MW 336.47

In *Sigesbeckia pubescens* (Compositae).

It is a powerful antihypertensive agent, with anti-inflammatory activity.

Han, K. D., *Chem. Abstr.,* 1976, **85**, 68174

Dithyreanitrile 507

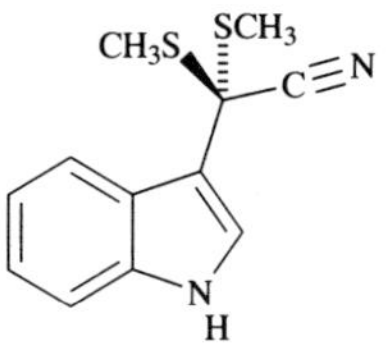

Cyanogen

$C_{13}H_{14}N_2OS_2$ [128717-80-6] MW 278.40

Seeds of *Dithyrea wislizenii* (Cruciferae).

Antifeedant at 1% concentration against the fall armyworm, *Spodoptera frugiperda*, and the European corn borer, *Ostrinia nubilalis* .

Powell, R. G., *Experientia,* 1991, **47**, 304.

Divaricoside 508

Sarmentogenin 3-oleandroside

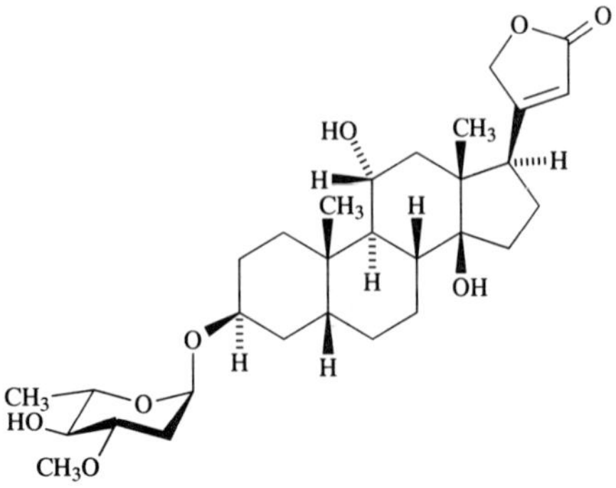

Cardenolide

$C_{30}H_{46}O_8$ [508-84-9] MW 534.69

In seed and leaf of *Strophanthus divaricatus* (Apocynaceae).

Toxic to vertebrates.

Schindler, O., *Helv. Chim. Acta,* 1954, **37**, 667.

Divostroside 509

Sarmentogenin 3-diginoside

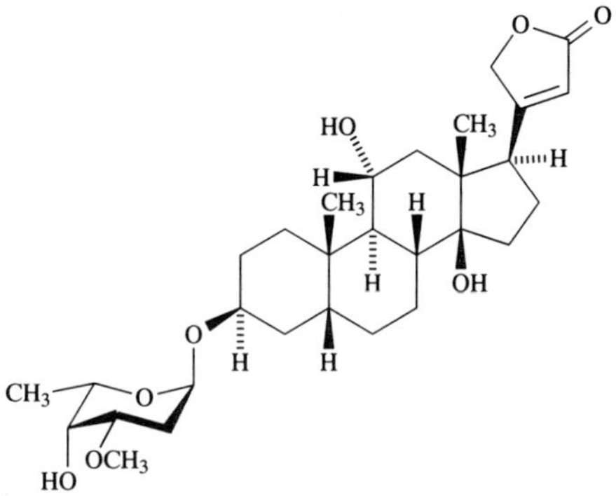

Cardenolide

$C_{30}H_{46}O_8$ [76704-78-4] MW 534.69

In seed and leaf of *Strophanthus divaricatus* (Apocynaceae).

Toxic to vertebrates.

Renkonen, O., *Helv. Chim. Acta,* 1959, **42**, 182.

L-Djenkolic acid 510

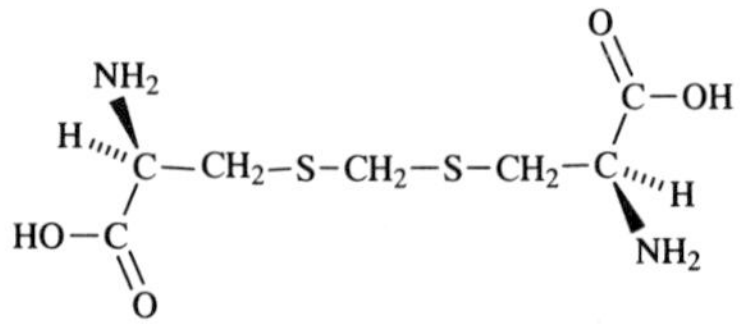

Amino acid

$C_7H_{14}N_2O_4S_2$ [498-59-9] MW 254.33

In the djenkol bean, *Pithecolobium lobatum*, and in *P. bubalinum*, *Albizia lophanta* and in *Acacia* and *Mimosa* spp. (Leguminosae).

Consumption of the djenkol bean, which takes place in Java, can lead to some toxic effects. The acid crystallises out from the urine if sufficient of the bean is eaten. It causes acute kidney malfunction and possible blocking of the urine flow. It has been reported to produce painful swelling of the genitalia in young children.

Gmelin, R., *Z. Naturforsch.,* 1957, **12B**, 687.

DMDP 511

2,5-Di(hydroxymethyl)-3,4-dihydroxypyrrolidine

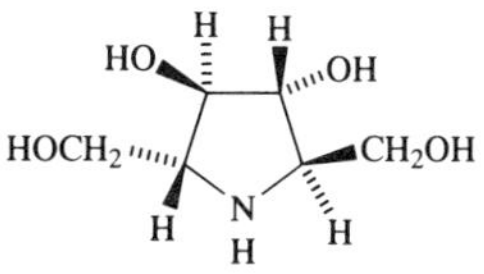

Pyrrolidine alkaloid

$C_6H_{13}NO_4$ [59920-31-9] MW 163.17

In the leaves of *Derris elliptica* and the seeds of *Lonchocarpus sericeus* (Leguminosae).

Toxic to armyworms and antifeedant to locusts. It is a potent inhibitor of viral glycoprotein processing glucosidase I and of insect α- and β-glucosidases.

Evans, S. V., *Phytochemistry,* 1985, **24**, 1953.

Dolichodial 512

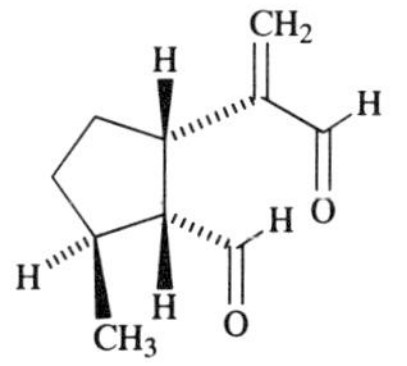

Monoterpenoid
(Iridane)

$C_{10}H_{14}O_2$ [5951-57-5] MW 166.22

In *Teucrium marum* (Labiatae).

Lachrymatory. Insect repellent.

Pagnoni, U. M., *Aust. J. Chem.,* 1976, **29**, 1375.

L-Dopa 513

Dopa; Levodopa; 3,4-Dihydroxy-L-phenylalanine; 3-Hydroxy-L-tyrosine; Laevodopa

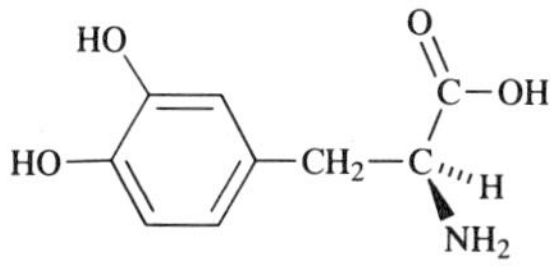

Amino acid

$C_9H_{11}NO_4$ [59-92-7] MW 197.19

In broad, horse or fava beans, *Vicia faba*. Also present in some *Lupinus* and *Mucuna* spp. (Leguminosae). The concentration in some *Mucuna* seed can reach 6 to 9% dry-weight.

It is toxic to beetles and other insects. In humans, it is thought to be responsible for favism, a haemolytic anaemia associated with individuals deficient in glucose 6-phosphate dehydrogenase and who have consumed broad beans. L-Dopa is used to treat Parkinson's disease, a neurological disorder characterised by tremors, rigidity and hypokinesis.

Gomez, R., *Anal. Profiles Drug Subst.,* 1976, **5**, 189.

Dopamine 514

4-(2-Aminoethyl)benzene-1,2-diol;
4-(2-Aminoethyl)pyrocatechol;
3,4-Dihydroxyphenethylamine;
α-(3,4-Dihydroxyphenyl)-β-aminoethane;
3-Hydroxytyramine

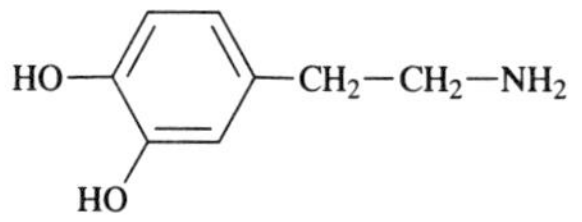

Aromatic amine

$C_8H_{11}NO_2$ [51-61-6] MW 153.18

Present in high concentrations (6-12 μmol/g fresh-weight) in banana peel and in the wild banana, *Musa sapientum* (Musaceae). Occurs in mescal buttons, *Colophora williamsii* (Cactaceae), in broom, *Cytisus scoparius* (Leguminosae) and in *Hermidium alipes* (Nyctaginaceae).

It is a neurotransmitter in the human brain; a direct acting sympathomimetic with effects on both α- and β-adrenergic receptors. It also increases cardiac output. It is used medically to treat shock, but is inactive orally and must be administered by intravenous infusion.

Smith, T. A., *Phytochemistry,* 1977, **16**, 9.

Doronine 515

6′-Chlorodeoxyflorindanine

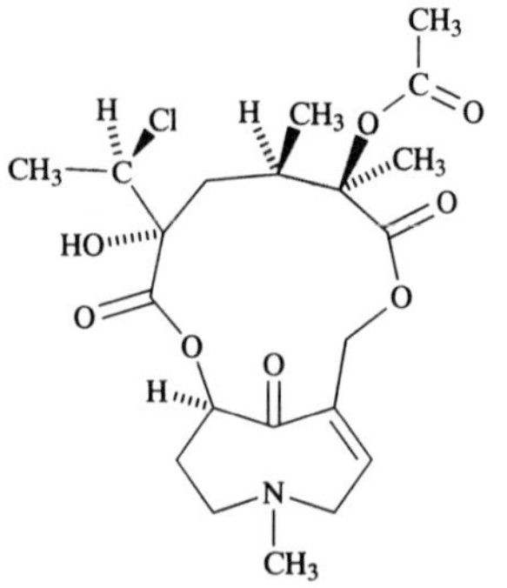

Pyrrolizidine alkaloid
(Secosenecionan)

$C_{21}H_{30}ClNO_8$ [60367-00-2] MW 459.92

In *Doronicum macrophyllum* and *Senecio othonnae* (Compositae).

Toxicity has not yet been established, but it is suspected of being hepatotoxic in mammals.

Alieva, Sh. A., *Chem. Nat. Compd.,* 1976, **12**, 173.

Drummondin A 516

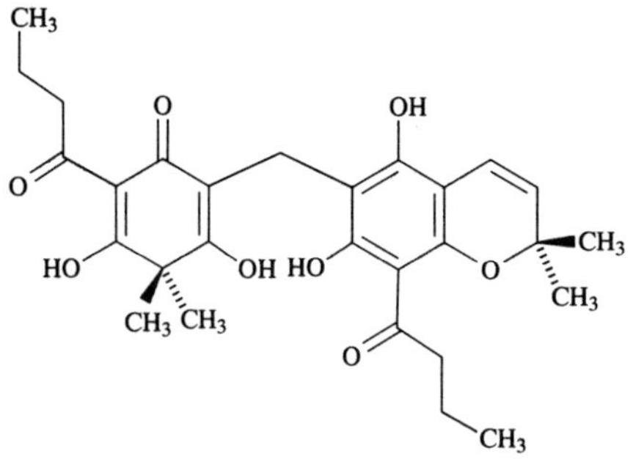

Chromene

$C_{26}H_{30}O_8$ [119171-76-5] MW 470.52

In roots of *Hypericum drummondii* (Guttiferae) together with the closely related drummondins B, C and F [119171-77-6, 119171-78-7 and 122127-73-5] respectively.

Cytotoxic to P-388, KB and human cancer cell line.

Jayasuriya, H., *J. Nat. Prod.,* 1989, **52**, 325.

Dubinidine 517

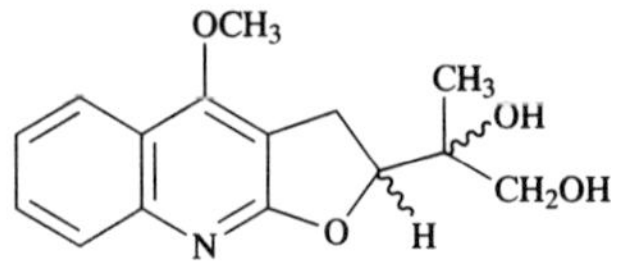

Quinoline alkaloid

$C_{15}H_{17}NO_4$ [22964-77-8] MW 275.30

In leaves of *Haplophyllum dubium* (Rutaceae).

Central nervous system depressant. It has hypothermic activity.

Grundon, M. F., *Tetrahedron Lett.,* 1971, 4727.

E

Eburnamonine 518

Huntericine; Vincamone; Viburnine

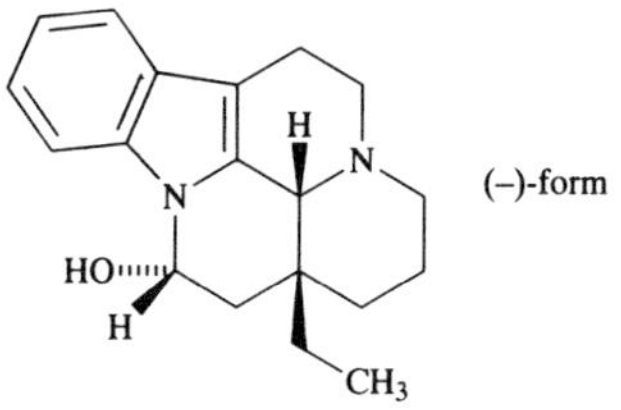

(–)-form

Indole alkaloid
(Eburnamenine)

$C_{19}H_{22}N_2O$ [4880-88-0] MW 294.40
(+)-form [474-00-0]
(±)-form [2580-88-3]

The (+)-form occurs in *Hunteria eburnea*, while the (–)-form is found in the lesser periwinkle, *Vinca minor* (Apocynaceae).

Vasodilator. The (–)-form is used as a drug for stimulating muscle activity.

Trojánek, J., *Chem. Ind. (London),* 1965, 1261.

Ecgonine 519

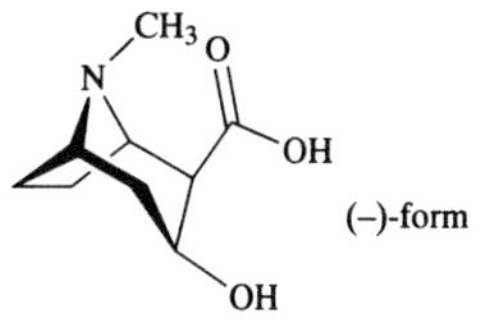

(–)-form

Tropane alkaloid

$C_9H_{15}NO_3$ [481-37-8] MW 185.22

In coca leaves, *Erythroxylum coca* and other *Erythroxylum* spp. (Erythroxylaceae).

Highly toxic by inhalation. It is used as a topical anaesthetic. Has allergenic properties.

Willstätter, R., *Justus Liebigs Ann. Chem.,* 1923, **434**, 111.

Echimidine 520

7-Angelyl-9-echimidinylretronecine

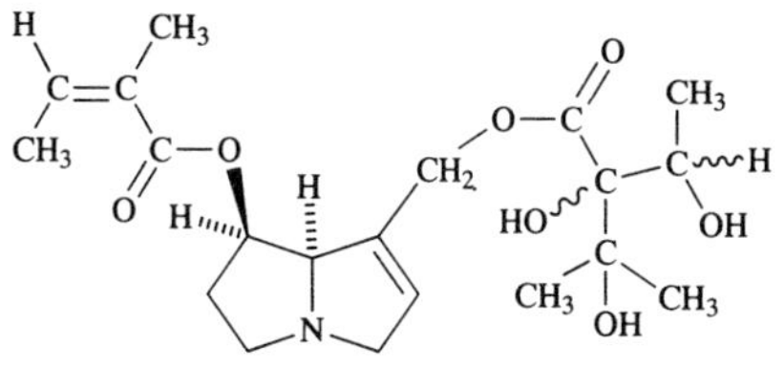

Pyrrolizidine alkaloid

$C_{20}H_{31}NO_7$ [520-68-3] MW 397.47

In the grazing plant Paterson's curse, *Echium plantagineum* and is found contaminating the honey derived from the nectar of this plant. Also present in *E. italicum*, *E. lycopsis*, the common comfrey, *Symphytum officinale*, the tuberous comfrey, *S. tuberosum*, *S. caucasicum* and *S. orientale* (Boraginaceae).

Hepatotoxic in animals. Horses, pigs and calves grazing on Paterson's curse are sensitive to its toxicity, but sheep are less affected.

Furuya, T., *Chem. Pharm. Bull.,* 1968, **16**, 2512.

Echinacoside 521

Phenylpropanoid

$C_{35}H_{46}O_{20}$ [82854-37-3] MW 786.74

In the roots of *Echinacea angustifolia* (Compositae) and in *Cistanche salsa* (Orobanchaceae).

Antihepatotoxic activity.

Becker, H., *Z. Naturforsch.,* 1982, **37C**, 351.

Echinocystic acid 522

Triterpenoid saponin
(Oleanane)

$C_{48}H_{76}O_{19}$ [510-30-5] MW 472.71

As the glycoside scheffleroside [66803-13-2] in *Schefflera capitata* (Araliaceae) and as the glycoside musennin [25480-74-4] in *Albizia anthelmintica* (Leguminosae) .

Haemolytic properties; the glycoside scheffleroside is spermicidal while the glycoside musennin has anthelmintic activity.

Farnsworth, N. R., *Research Frontiers in Fertility Regulation,* 1982, **2**, 1.

Echitamine 523

Ditaine

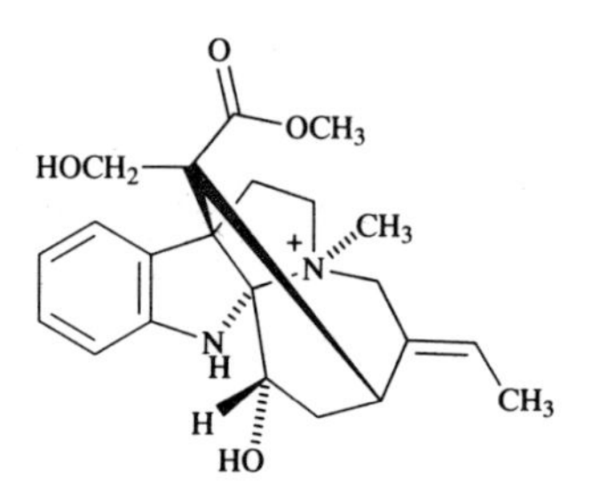

Indole alkaloid

$[C_{22}H_{29}N_2O_4]^+$ [6871-44-9] MW 385.48

In *Alstonia scholaris*, *A. spectabilis* and *A. spatulata* (Apocynaceae).

Highly toxic when taken orally. It is hypotensive. *Alstonia scholaris* bark is used directly as a febrifuge.

Govindachari, T. R., *J. Indian Chem. Soc.,* 1968, **45**, 945.

Echiumine 524

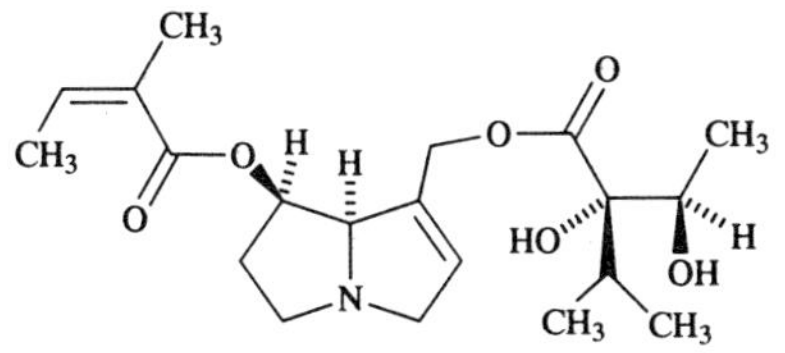

Pyrrolizidine alkaloid

$C_{20}H_{31}NO_6$ [633-16-9] MW 381.47

Echium plantagineum and *Amsinckia intermedia* (Boraginaceae).

Co-occurs with echimidine (q.v.) in this *Echium* species and is jointly responsible for the fatal poisoning of horses which have fed on this plant.

Culvenor, C. C. J., *Aust. J. Chem.*, 1956, **9**, 512.

Elaterinide 525

Colocynthin; Gratiotoxin

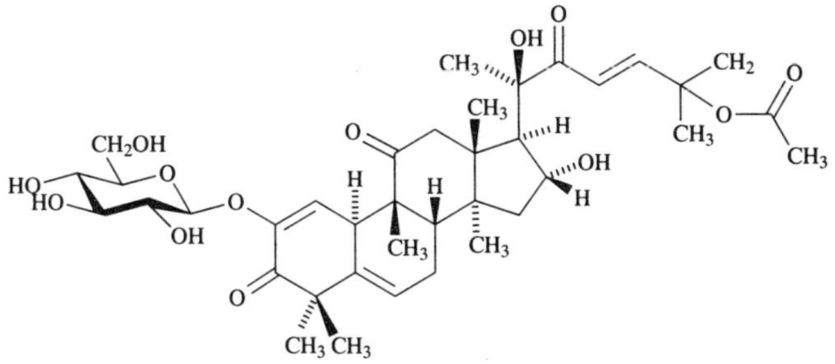

Triterpenoid
(Cucurbitane)

$C_{38}H_{54}O_{13}$ [1398-78-3] MW 718.84

Citrullus lanatus and *C. colocynthis* (Cucurbitaceae).

Toxic.

Ripperger, H., *Tetrahedron,* 1976, **32**, 1567.

Elatine 526

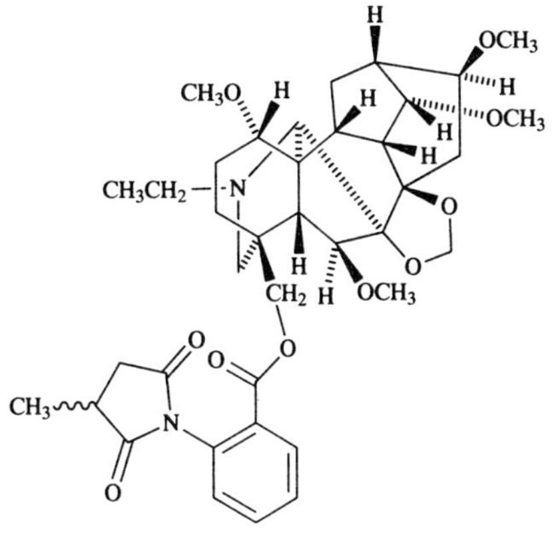

Diterpenoid alkaloid
(Aconitane)

$C_{38}H_{50}N_2O_{10}$ [26000-16-8] MW 694.82

In *Delphinium elatum* (Ranunculaceae).

Potent neuromuscular poison in mammals with classical curariform activity. When compared with tubocurarine (q.v.), it is less toxic and therefore has a wider therapeutic spectrum.

Edwards, O. E., *Can. J. Chem.*, 1982, **60**, 2661.

Eleganin 527

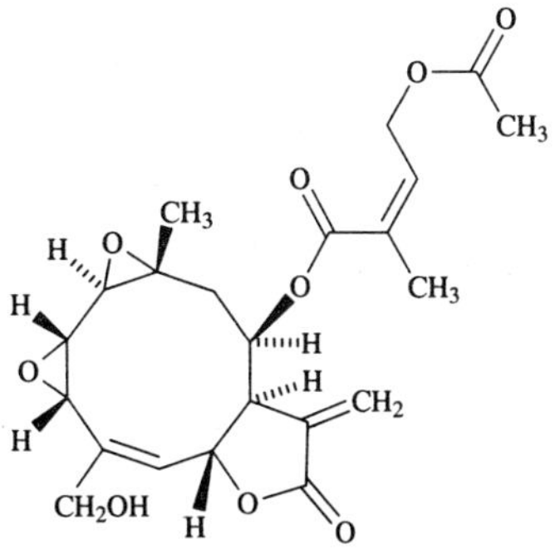

Sesquiterpenoid lactone
(Germacranolide)

$C_{22}H_{26}O_9$ [57498-84-7] MW 434.44

In *Liatris elegans* and *L. scabra* (Compositae).

Cytotoxic and antitumour activities.

Herz, W., *Phytochemistry,* 1975, **14**, 1561.

Elemicin 528

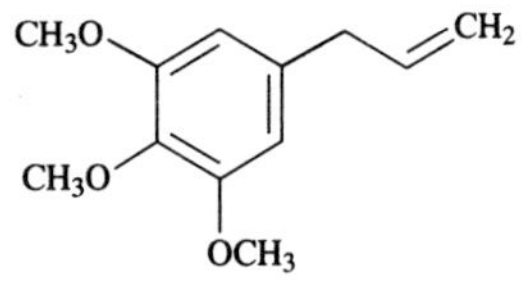

Phenylpropanoid

$C_{12}H_{16}O_3$ [487-11-6] MW 208.26

In the resin of *Canarium commune* (Burseraceae), in the essential oils of *Cinnamomum glanduliferum* wood (Lauraceae), of the grass *Cymbopogon procerus*, of *Boronia pinnata* (Rutaceae), of *Melaleuca bracteata* (Myrtaceae). Also in *Dalbergea spruceata* (Leguminosae), the carrot, *Daucus carota* (Umbelliferae) and nutmeg, *Myristica fragrans* (Myristiceae).

DNA binding activity and also inhibits rabbit platelet aggregation *in vitro*.

Mauthner, F., *Justus Liebigs Ann. Chem.*, 1917, **414**, 250.

Elephantin 529

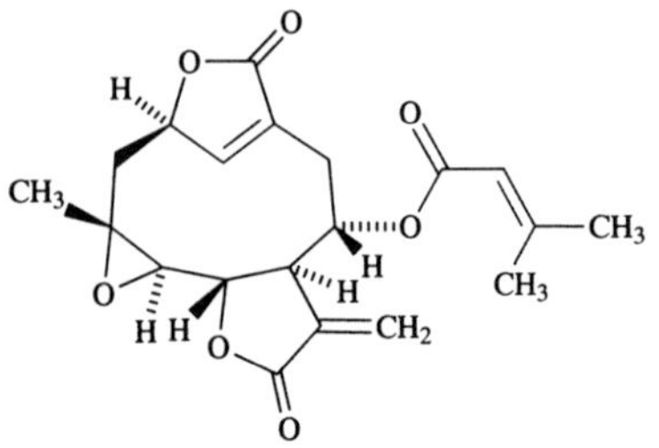

Sesquiterpenoid lactone
(Germacranolide)

$C_{20}H_{22}O_7$ [21899-50-3] MW 374.39

In *Elephantopus elatus* (Compositae).

Cytotoxic and antitumour activities.

Kupchan, S. M., *J. Org. Chem.*, 1969, **34**, 3867.

Elephantopin 530

NSC 100046

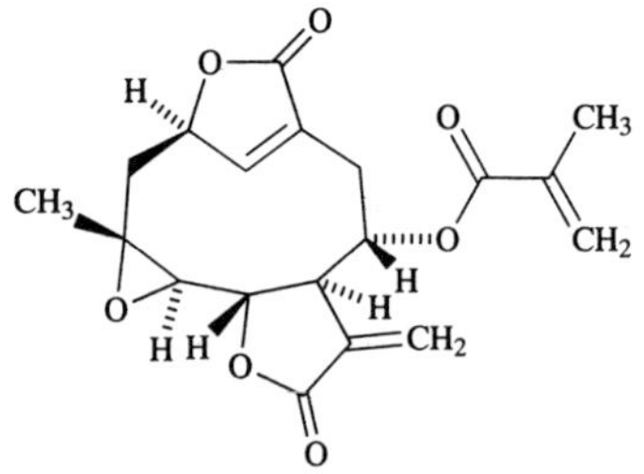

Sesquiterpenoid lactone
(Germacranolide)

$C_{19}H_{20}O_7$ [13017-11-3] MW 360.36

In *Elephantopus elatus* (Compositae).

Cytotoxic and antitumour activities.

Kupchan, S. M., *J. Org. Chem.*, 1969, **34**, 3867.

Ellipticine 531

NSC 71795

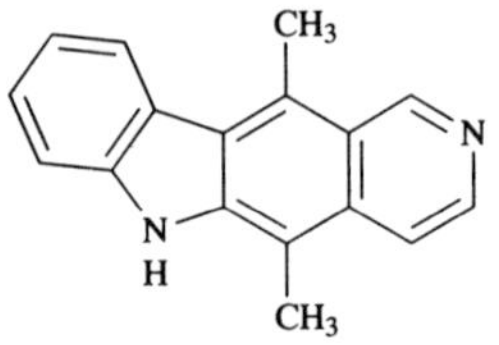

Indole alkaloid
(Pyridocarbazole)

$C_{17}H_{14}N_2$ [519-23-3] MW 246.31

In *Aspidosperma subincanum*, *Bleekeria vitiensis* and *Ochrosia elliptica* (Apocynaceae).

It shows broad antitumour properties, and is also a mutagen. It is also antitrypanosomal *in vitro* against *Trypanosoma cruzi*.

Woodward, R. B., *J. Am. Chem. Soc.*, 1959, **81**, 4434.

Embelin 532

Embelic acid; Embeliaquinone; Oxaloxanthin

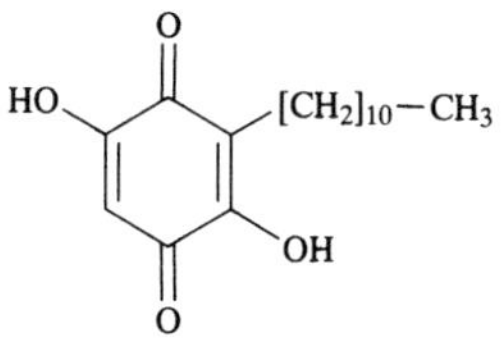

Benzoquinone

$C_{17}H_{26}O_4$ [554-24-3] MW 294.39

In the fruit of *Embelia ribes*, in the root of *Ardisia crenata* and in the berry of *Myrsine africana* (Myrsinaceae). Also in the twig and stem of *Aegiceras corniculatum* (Aegiceraceae).

Potent oral contraceptive, possessing 85% anti-implantation activity in rats. Ammonium salt is used as an anthelmintic; it may be an irritant to mucous membranes causing violent sneezing.

Thompson, R. H., Naturally Occurring Quinones, 3rd edn, 1987, 18, Chapman & Hall.

Emetine 533

Cephaeline methyl ether; Ipecine; NSC 33669

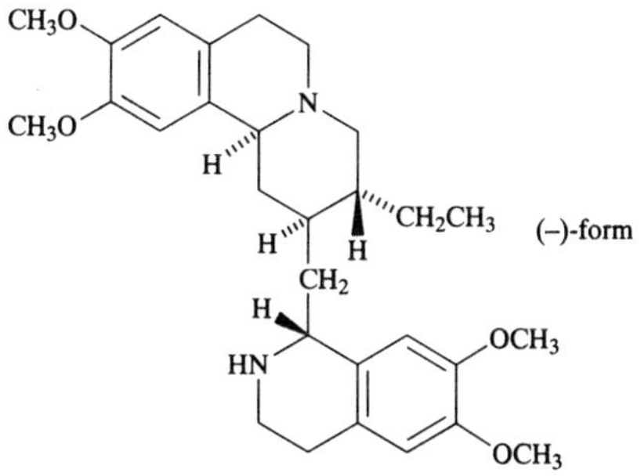

Isoquinoline alkaloid
(Emetane)

$C_{29}H_{40}N_2O_4$ [483-18-1] MW 480.65

In all varieties of ipecacuanha, *Cephaelis ipecacuanha* and in *C. acuminata* (Rubiaceae).

Well known as a plant poison which produces vomiting. The lethal dose is about one gram in humans; it is a cumulative poison over short periods. At the right dose, it is useful as an expectorant and emetic in cases of poisoning and overdosage. It is also used to treat amoebic dysentery, despite the gastrointestinal effects.

Battersby, A. R., *J. Chem. Soc.*, 1959, 1748.

Emodin 534

Archin; Frangula emodin; Frangulic acid; Rheum-emodin

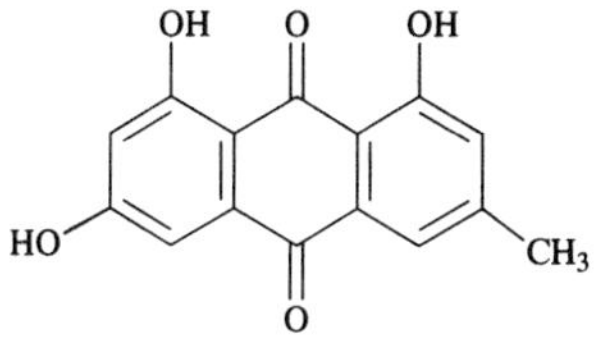

Anthraquinone

$C_{15}H_{10}O_5$ [518-82-1] MW 270.24

In *Rumex* and *Rheum* spp. (Polygonaceae), in the rootbark of *Ventilago calyculata*, in *Rhamnus frangula* (Rhamnaceae), in *Myrsine africana* (Myrsinaceae) and in *Psorospermum glaberrimum* (Guttiferae).

Antileukaemic (P388) and antitumour (Walter Sarcoma) activities. It is also moderately cytotoxic in three human tumour cell lines.

Briggs, L. H., *J. Chem. Soc.*, 1953, 3069.

Emodin 8-glucoside 535

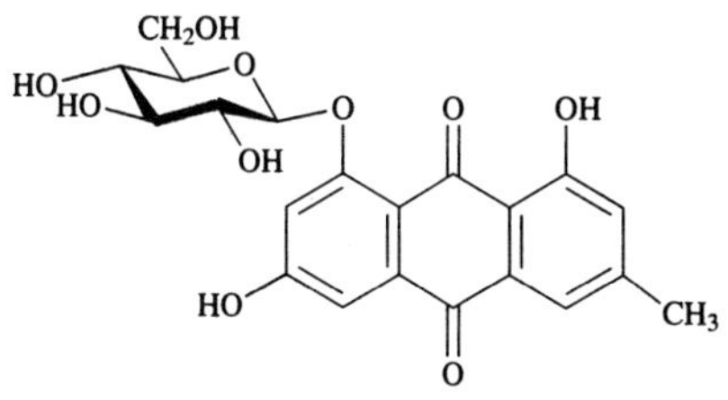

Anthraquinone

$C_{21}H_{20}O_{10}$ [38840-23-2] MW 432.38

In rhizomes of *Rheum moorcroftianum*, in *Polygonum cuspidatum* (Polygonaceae) and in the stembark of *Rhamnus frangula* (Rhamnaceae).

Causes immobilisation of human spermatozoa.

Murakami, T., *Chem. Pharm. Bull.*, 1968, **16**, 2299.

Encelin 536

Anhydrofarinosin

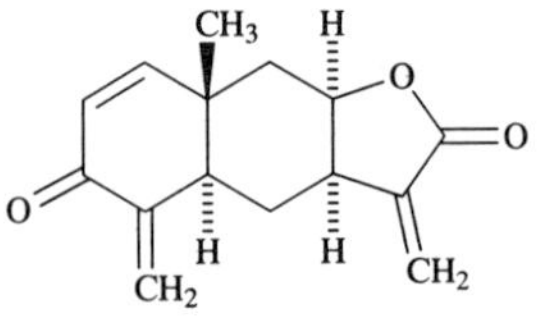

Sesquiterpenoid lactone
(Eudesmanolide)

$C_{15}H_{16}O_3$ [15569-50-3] MW 244.29

In *Encelia farinosa*, *E. virginensis* and *Baltimora recta* (Compositae).

Cytotoxic and antitumour activities.

Geissman, T. A., *J. Org. Chem.*, 1968, **33**, 656.

Enhydrin 537

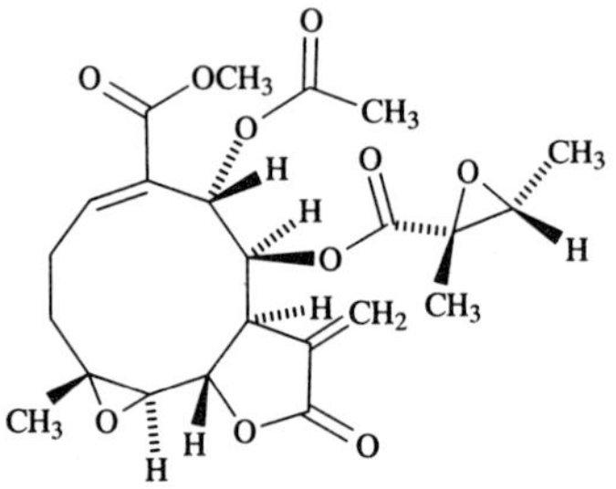

Sesquiterpenoid lactone
(Germacranolide)

$C_{23}H_{28}O_{10}$ [33880-85-2] MW 464.47

In leaves of *Enhydra fluctuans*, in *Melampodium longipilum*, *M. perfoliatum*, *Smallanthus fruticosus* and *S. uvedalius* (Compositae).

Antihypertensive activity.

Joshi, B. S., *Indian J. Chem.*, 1972, **10**, 771.

L-Ephedrine 538

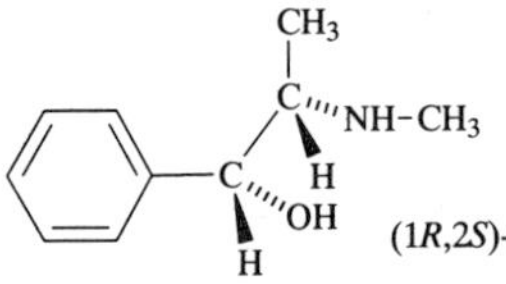

(1*R*,2*S*)-form

Aromatic amine

$C_{10}H_{15}NO$ [299-42-3] MW 165.24

In leaves and stems of Ma-Huang, obtained from *Ephedra* spp.; *E. sinica*, *E. equisetina*, *E. gerardiana* and others (Ephedraceae).

Sympathomimetic activity; it produces peripheral vasoconstriction and raises blood pressure. It is also a central nervous system stimulant. It is used medicinally in treating asthma, rhinitis and sinusitis. The LD_{50} when injected intraperitoneally into mice is 350 mg/kg body-weight.

Martindale, The Extra Pharmacopoeia, 30th edn, 1993, 1244, The Pharmaceutical Press.

Epitulipinolide 539

epi-Tulipinolide

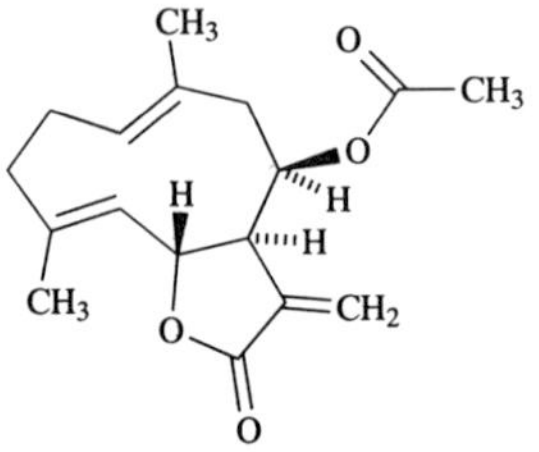

Sesquiterpenoid lactone
(Germacranolide)

$C_{17}H_{22}O_4$ [24164-13-4] MW 290.36

Aerial parts of *Zaluzania pringlei*, *Ambrosia chamissonis* and *A. dumosa* (Compositae) and in root bark of the tulip tree, *Liriodendron tulipifera* (Magnoliaceae).

Cytotoxic and antitumour properties.

Doskotch, R. W., *J. Org. Chem.*, 1970, **35**, 1928.

Epitulipinolide diepoxide 540

epi-Tulipinolide diepoxide

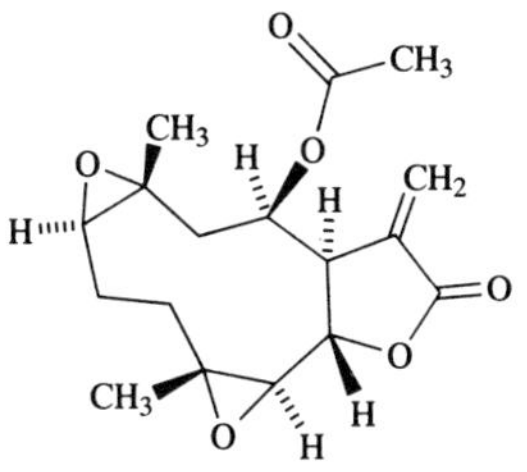

Sesquiterpenoid lactone
(Germacranolide)

$C_{17}H_{22}O_6$ [36815-40-2] MW 322.36

Leaves of the tulip tree, *Liriodendron tulipifera* (Magnoliaceae).

Cytotoxic and antitumour properties. It is an insect antifeedant.

Doskotch, R. W., *Phytochemistry,* 1975, **14**, 769.

Epivoacorine 541

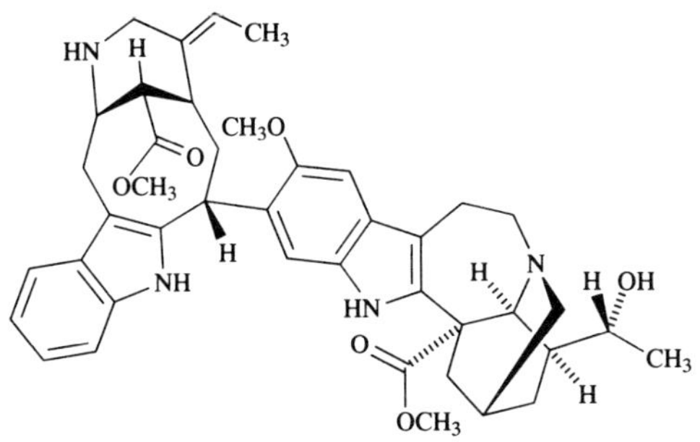

Bisindole alkaloid
(Ibogamine and Vobasan)

$C_{43}H_{52}N_4O_6$ [4835-65-8] MW 720.91

In trunk bark of *Tabernaemontana brachyantha* (Apocynaceae).

Cytotoxic activity.

Patel, M. B., *Phytochemistry,* 1973, **12**, 451.

8-Epixanthatin 542

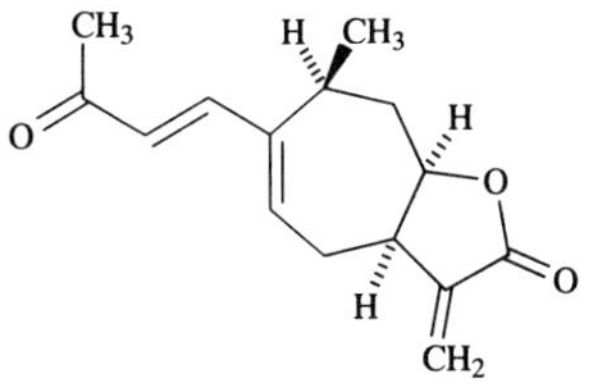

Sesquiterpenoid lactone
(Secoguaianolide)

$C_{15}H_{18}O_3$ [30890-35-8] MW 246.31

Leaves of *Xanthium canadense* (Compositae).

Insect development inhibitor, active against the fruit fly, *Drosophila melanogaster*.

McMillan, C., *Biochem. Syst. Ecol.,* 1975, **3**, 181.

Eremanthine 543

Vanillosmin; Eremanthin

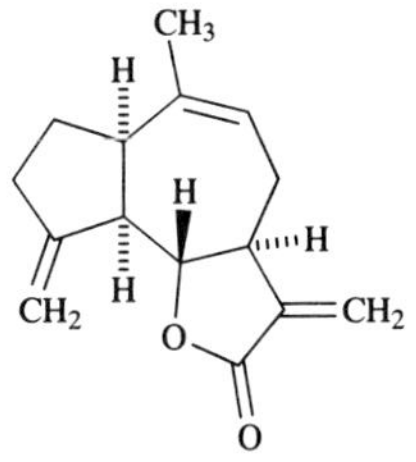

Sesquiterpenoid lactone
(Guaianolide)

$C_{15}H_{18}O_2$ [37936-58-6] MW 230.31

In the heartwood of *Eremanthus elaeagnus* and *E. incanus* and in several members of the *Lychnophora*, *Vanillosmopsis* and *Vernonia* spp. (Compositae).

Schistosomicidal activity.

Vichnewski, W., *Phytochemistry,* 1972, **11**, 2563.

Eremantholide A 544

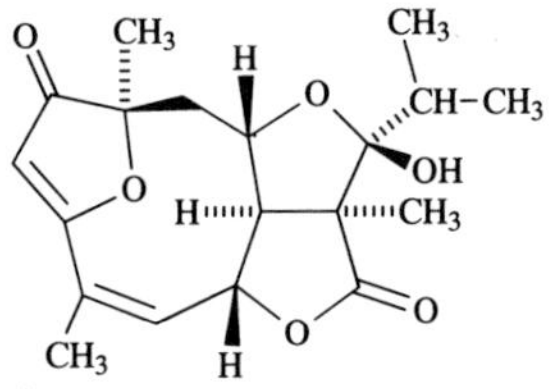

Sesquiterpenoid lactone
(Germacranolide)

$C_{19}H_{24}O_6$ [58030-93-6] MW 348.40

In *Eremanthus bicolor, E. elaeagnus, E. incanus* and *Centratherum punctatum* (Compositae).

Cytotoxic and antitumour activities.

Bohlmann, F., *Phytochemistry,* 1980, **19**, 2663.

Eremofrullanolide 545

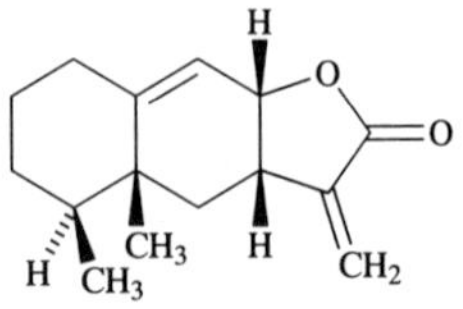

Sesquiterpenoid lactone
(Eudesmanolide)

$C_{15}H_{18}O_2$ [62870-69-3] MW 232.32

The liverworts *Frullania tamarisci, F. dilatata* and *F. nisquallensis*.

Causes allergenic contact dermatitis in those handling these liverworts.

Asakawa, Y., *Bull. Soc. Chim. Fr.,* 1976, 1465.

Ergine 546

Lysergic acid amide; Lysergamide

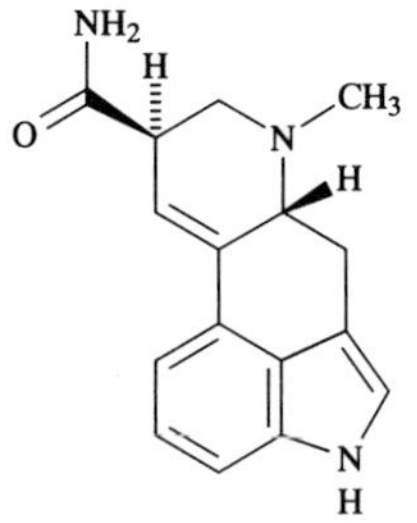

Indole alkaloid
(Ergoline)

$C_{16}H_{17}N_3O$ [478-94-4] MW 267.33

In seeds of *Rivea corymbosa, Ipomoea argyrophylla, I. violacea* and *I. tricolor* (Convolvulaceae). The drug 'ololiuqui' is obtained from these plants and used by Central American Indians. Also present in at least fourteen of the *Argyreia* spp. (also Convolvulaceae).

Hallucinogenic substance. It is the principal active constituent of the native drug ololiuqui.

Chao, J. M., *Phytochemistry,* 1973, **12**, 2435.

Ergometrine 547

Ergobasine; Ergonovine; Ergostetrine; Ergotocine; Lysergic acid propanolamide

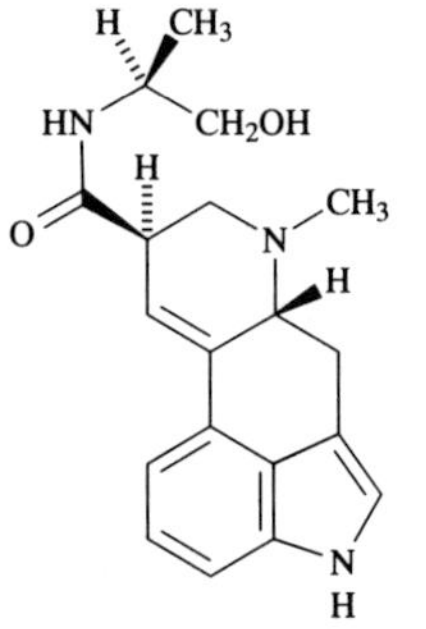

Indole alkaloid
(Ergoline)

$C_{19}H_{23}N_3O_2$ [60-79-7] MW 325.41

In seeds of *Ipomoea argyrophylla* and related Convolvulaceae.

Poisonous, with hallucinogenic effects. It is used clinically to treat post-partum haemorrhage. Also, it inhibits prolactic release, preventing implantation and lactation in women.

Chao, J. M., *Phytochemistry,* 1973, **12**, 2435.

Ergosine 548

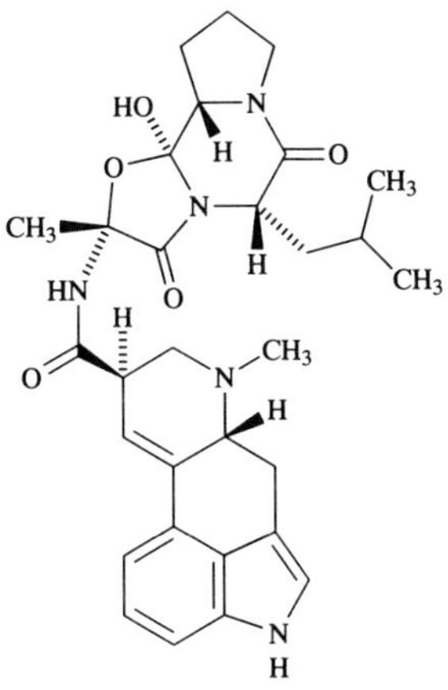

Indole alkaloid
(Ergoline)

$C_{30}H_{37}N_5O_5$ [561-94-4] MW 547.65

In seed of *Ipomoea argyrophylla* and related Convolvulaceae.

Poisonous, causing hallucinations. It is a prolactin release inhibitor, preventing implantation and lactation in women.

Chao, J. M., *Phytochemistry,* 1973, **12**, 2435.

Eriobofuran 549

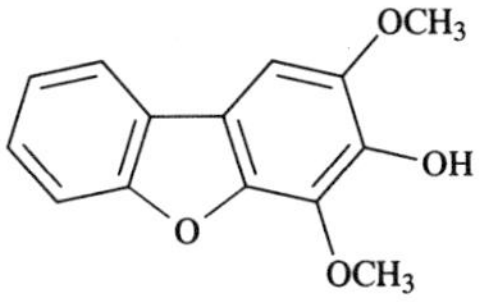

Dibenzofuran

$C_{14}H_{12}O_4$ [97218-06-9] MW 244.25

Leaves of the loquat, *Eriobotrya japonica* (Rosaceae) infected with *Entomosporium eriobotryae*.

Antifungal agent (phytoalexin), inhibiting spore germination and germ tube growth in the pathogenic fungus *Pestalotia funerea*.

Miyakado, M., *Nippon Noyaku Gakkaishi,* 1985, **10**, 101.

Eriocarpin 550

Desglucosyrioside

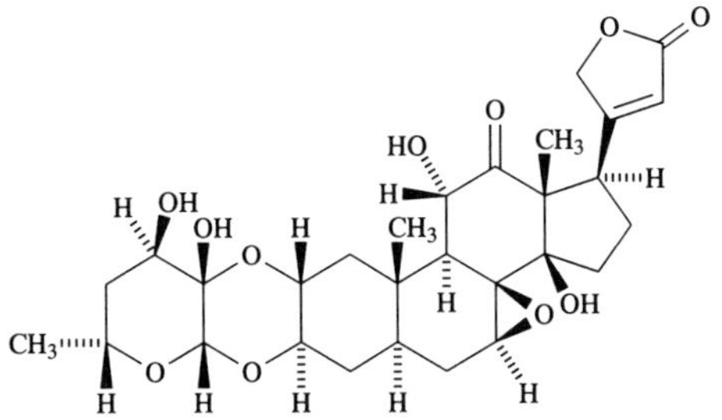

Cardenolide

$C_{29}H_{38}O_{11}$ [66419-09-8] MW 562.61

In the North American milkweed, *Asclepias eriocarpa* and *A. labriformis* (Asclepiadaceae).

Toxic to vertebrates. The LD_{50} when administered intraperitoneally to male Swiss Webster mice is 6.5 mg/kg bodyweight. Livestock poisoning occurs in North America, especially on overgrazed pastures or when cut fodder contains *Asclepias* leaf. It has been estimated that as little as 220 g of dry *Asclepias labriformis* leaf is fatal to a 44 kg sheep.

Cheung, H. T. A., *J. Chem. Soc., Perkin Trans. 1,* 1980, 2169.

Erioflorin acetate 551

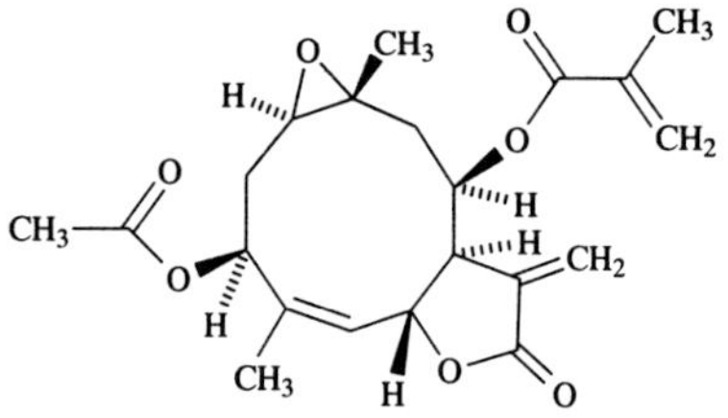

Sesquiterpenoid lactone
(Germacranolide)

$C_{21}H_{26}O_7$ [27542-23-0] MW 390.43

In leaves and stems of *Podanthus ovatifolius* (Compositae).

Cytotoxic and antitumour activities.

Gnecco, S., *Phytochemistry,* 1973, **12**, 2469.

Erioflorin methacrylate 552

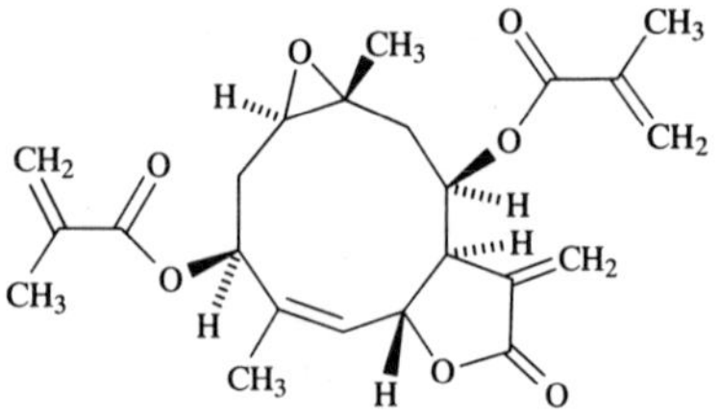

Sesquiterpenoid lactone
(Germacranolide)

$C_{23}H_{28}O_7$ [50186-66-5] MW 416.47

In *Podanthus ovatifolius* (Compositae).

Cytotoxic and antitumour activities.

Gnecco, S., *Phytochemistry,* 1973, **12**, 2469.

Eriolangin 553

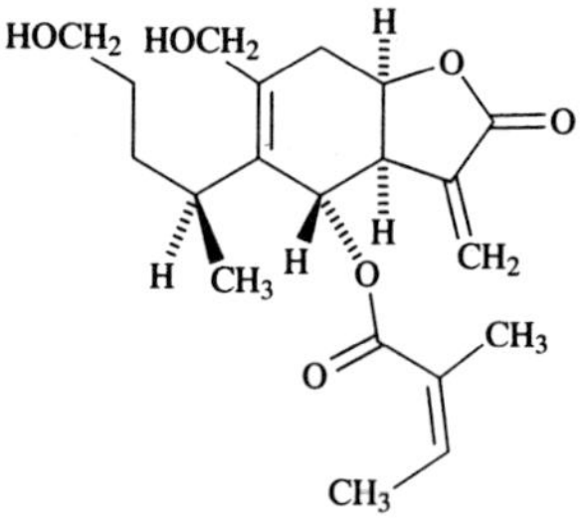

Sesquiterpenoid lactone
(Seco-eudesmanolide)

$C_{20}H_{28}O_6$ [52617-35-3] MW 364.44

In *Eriophyllum lanatum* (Compositae).

Cytotoxic and antitumour activities.

Kupchan, S. M., *J. Chem. Soc., Chem. Commun.,* 1973, 842.

Erucic acid 554

cis-Docosenoic acid

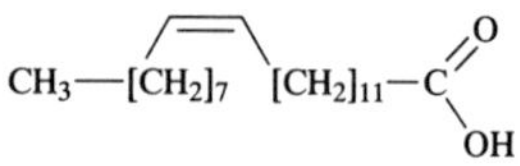

$CH_3—[CH_2]_7—CH=CH—[CH_2]_{11}—COOH$

Fatty acid

$C_{22}H_{42}O_2$ [112-86-7] MW 338.57

Found in bound (triglyceride) form in the seed fats of Cruciferae and Tropaeolaceae, constituting 40-50% of the total fatty acid content of rape, mustard and wallflower seed fats, and up to 80% of the seed fat of *Tropaeolum majus*, the nasturtium.

The edible seed oil from crucifer crops can cause adverse physiological effects when ingested in large amounts by animals such as rats. Modern rape cultivars have been deliberately bred to produce seed oils which are low in erucic acid.

Swain, T., Chemical Plant Taxonomy, 1963, 290, Academic Press.

Erysonine 555

O^3-Demethylerysodine

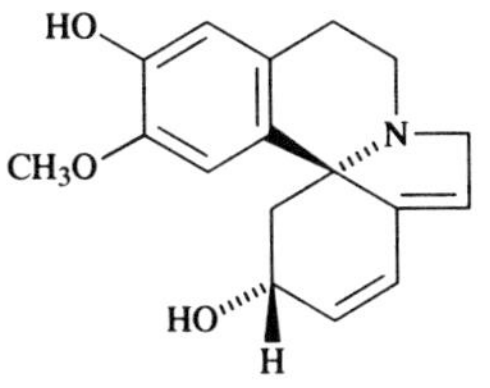

Isoquinoline alkaloid
(Erythrinan)

$C_{17}H_{19}NO_3$ [7290-05-3] MW 285.34

In *Erythrina caribea*, *E. melanacantha* and other *Erythrina* spp. (Leguminosae), principally in the seeds, but also in trunk bark and root bark.

Neuromuscular blocking agent.

Hargreaves, R. T., *J. Nat. Prod.,* 1974, **37**, 569.

Erysotrine 556

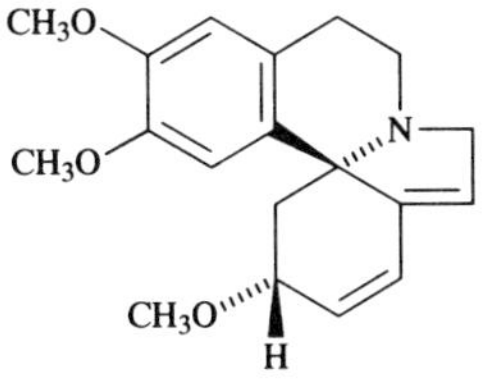

Isoquinoline alkaloid
(Erythrinan)

$C_{19}H_{23}NO_3$ [27740-43-8] MW 313.40

In *Erythrina suberosa* and other *Erythrina* spp. (Leguminosae).

Neuromuscular blocking agent. The leaves and bark of *Erythrina suberosa* have antitumour activity.

Hargreaves, R. T., *J. Nat. Prod.,* 1974, **37**, 569.

Erythratidine 557

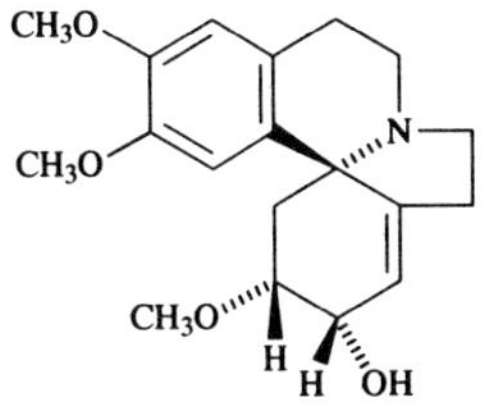

Isoquinoline alkaloid
(Erythrinan)

$C_{19}H_{25}NO_4$ [41431-22-5] MW 231.41

In *Erythrina caribea*, *E. melanacantha* and other *Erythrina* spp. (Leguminosae).

Curare-like neuromuscular blocking agent.

Hargreaves, R. T., *J. Nat. Prod.,* 1974, **37**, 569.

α-Erythroidine 558

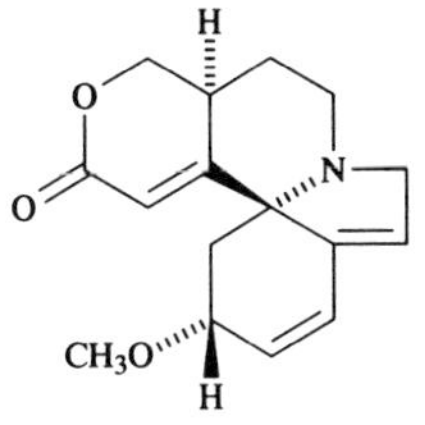

Isoquinoline alkaloid
(Erythrinan)

$C_{16}H_{19}NO_3$ [466-80-8] MW 273.33

In many *Erythrina* spp. (Leguminosae), including flowers of *E. americana*.

Curare-like neuromuscular blocking agent.

Aguilar, M. I., *Phytochemistry,* 1981, **20**, 2061.

β-Erythroidine 559

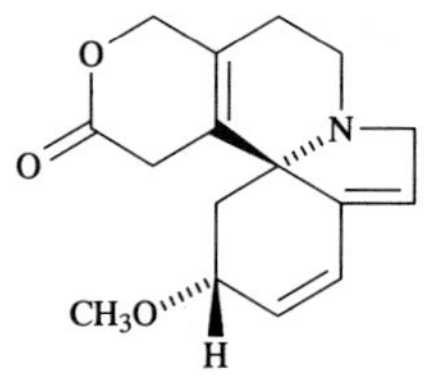

Isoquinoline alkaloid
(Erythrinan)

$C_{16}H_{19}NO_3$ [466-81-9] MW 273.33

In many *Erythrina* spp. (Leguminosae), including flowers of *E. americana*.

Orally active neuromuscular blocking agent, hypnotic, respiratory depressant and hypotensive. It is moderately toxic. The LD_{50} intraperitoneally in mice is 29.5 mg/kg body-weight. Has been used clinically as a curare substitute.

Aguilar, M. I., *Phytochemistry,* 1981, **20**, 2061.

Erythrophleguine 560

6α-Hydroxycassamine

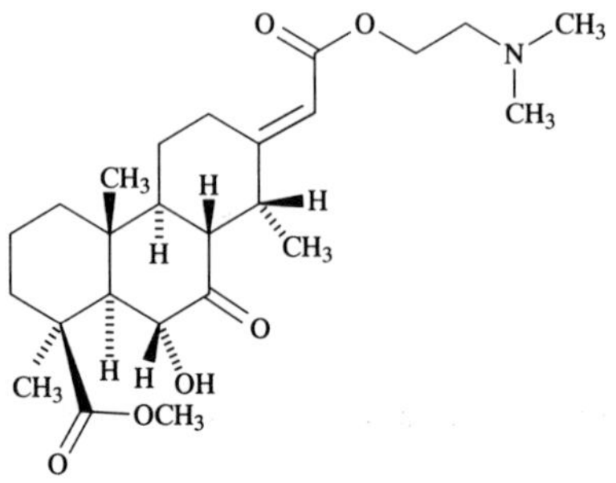

Diterpenoid alkaloid
(Cassane)

$C_{25}H_{39}NO_6$ [4829-28-1] MW 449.59

In the bark of *Erythrophleum guineense* (Leguminosae).

Digitalis-like activity; reduction of the heart frequency, intensification of heart contraction and of diureses.

Thorell, A., *Acta Chem. Scand.,* 1968, **22**, 2835.

Eseramine 561

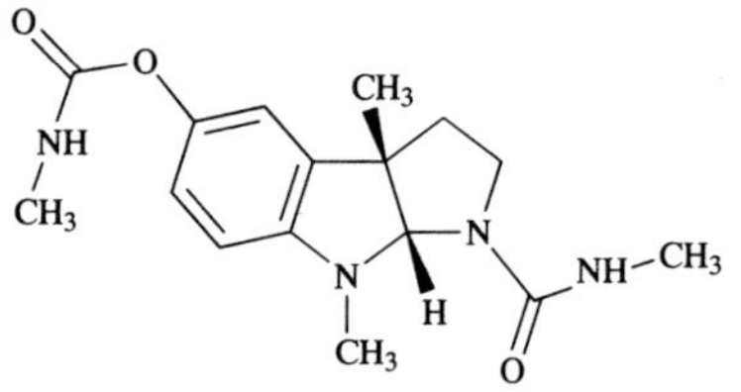

Indole alkaloid
(Physostigmine)

$C_{16}H_{22}N_4O_3$ [6091-57-2] MW 318.38

In the calabar bean, the seeds of *Physostigma venenosum* (Leguminosae).

Anticholinesterase properties. The activity is similar to but weaker than that of physostigmine (q.v.) from the same source.

Robinson, B., *Chem. Ind. (London),* 1964, 459.

Eseridine 562

Eserine aminoxide; Eserine oxide; Geneserine; Physostigmine aminoxide; Physostigmine oxide

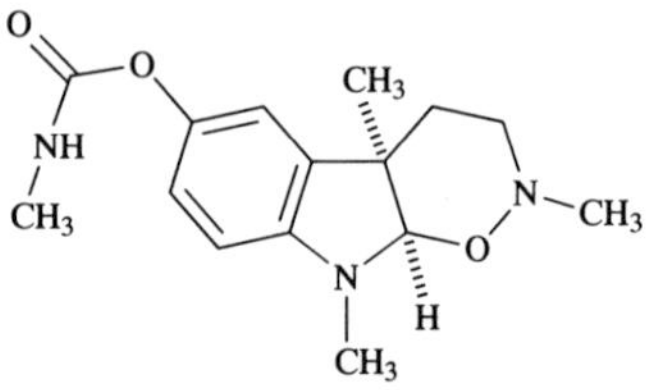

Indole alkaloid
(Secophysostigmine)

$C_{15}H_{21}N_3O_3$ [25573-43-7] MW 291.35

In the calabar bean, the seeds of *Physostigma venenosum* (Leguminosae).

Anticholinesterase properties. Similar in activity to physostigmine (q.v.), the major alkaloid of the calabar bean. Has been used for treating gastrointestinal disorders and chronic dermatoses.

Hootelé, C., *Tetrahedron Lett.,* 1969, 2713.

Espintanol 563

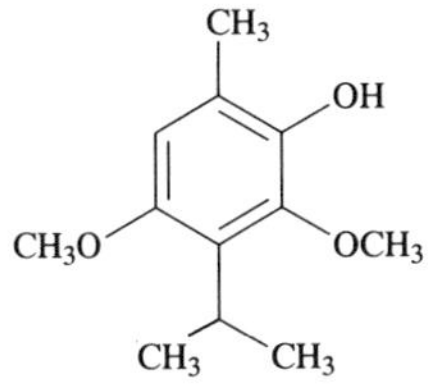

Monoterpenoid
(Menthane)

$C_{12}H_{18}O_3$ [135626-41-4] MW 210.27

Bark of *Oxandra espintana* (Annonaceae).

Antiprotozoal activity.

Hocquemiller, R., *J. Nat. Prod.,* 1991, **54**, 445.

Estragole 564

Esdragole; *p*-Allylanisole; Chavicol methyl ether; Methylchavicol; Isoanethole

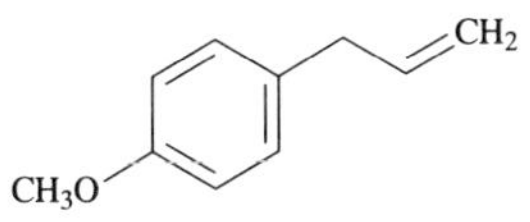

Phenylpropanoid

$C_{10}H_{12}O$ [140-67-0] MW 148.20

Fairly regularly present in plant essential oils. Occurs in the bark of *Persea gratissima* (Lauraceae), the leaf of *Artemisia dracunculus* (Compositae) and leaf of *Agastache rugosa* (Labiatae).

Stimulates liver regeneration. It exhibits hypothermic and DNA binding activities. Used as a food flavour.

Karrer, W., Konstitution und Vorkommen der organischen Pflanzenstoffe, 2nd edn, 1981, 67, Birkhäuser.

(–)-Eudesmin 565

L-Eudesmin; Pinoresinol dimethyl ether

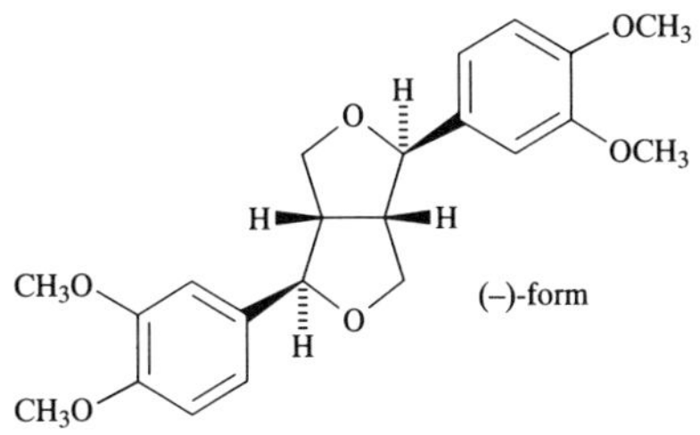

Lignan

$C_{22}H_{26}O_6$ [526-06-7] MW 386.44
(+)-form [29106-36-3]
(±)-form [38759-91-0]

In kino gum, the exudate of *Eucalyptus hemiphloia* (Myrtaceae). The (+)-isomer occurs in the wood of *Araucaria angustifolia* (Araucariaceae) and in the wood of *Humbertia madagascariensis* (Humbertiaceae).

Has antitubercular activity *in vitro*. Also, it has calcium antagonistic activity on taenia coli of guinea-pig.

Vialard, H. M., *C. R. Acad. Sci., Ser. C,* 1968, **266**, 1284.

Eudesobovatol A 566

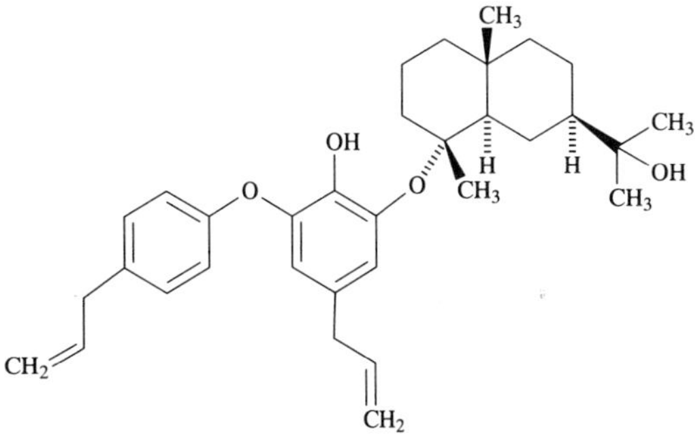

Sesquiterpenoid-Neolignan
(Eudesmane)

$C_{33}H_{44}O_4$ [125916-77-2] MW 504.71

In the bark of *Magnolia obovata* (Magnoliaceae).

Shows neurotrophic activity when tested on neuronal cell cultures of foetal rat cerebral hemisphere.

Fukuyama, Y., *Tetrahedron,* 1992, **48**, 377.

Eugeniin 567

Tellimagrandin II

Ellagitannin

$C_{41}H_{28}O_{26}$ [58970-75-5] MW 936.66

In buds of *Syzygium aromaticum* (Myrtaceae), in *Tellima grandiflora* (Saxifragaceae), *Rosa* spp. (Rosaceae), *Quercus* spp. (Fagaceae) and in leaves of *Coriaria japonica* (Coriariaceae).

Potentially toxic to livestock if leaves containing eugeniin are consumed in quantity. Has antiviral activity against herpes simplex and inhibitory action on adrenaline-induced lipolysis of fat cells of rats.

Nonaka, G. I., *Chem. Pharm. Bull.*, 1980, **28**, 685.

Eupachlorin 568

Sesquiterpenoid lactone
(Guaianolide)

$C_{20}H_{25}ClO_7$ [20071-50-5] MW 412.87

In the round-leaved thoroughwort, *Eupatorium rotundifolium* (Compositae).

Cytotoxic and antitumour properties.

Kupchan, S. M., *J. Org. Chem.*, 1969, **34**, 3876.

Eupachlorin acetate 569

Sesquiterpenoid lactone
(Guaianolide)

$C_{22}H_{27}ClO_8$ [20501-52-4] MW 454.90

The round-leaved thoroughwort, *Eupatorium rotundifolium* (Compositae).

Cytotoxic and antitumour properties.

Kupchan, S. M., *J. Org. Chem.*, 1969, **34**, 3876.

Eupachloroxin 570

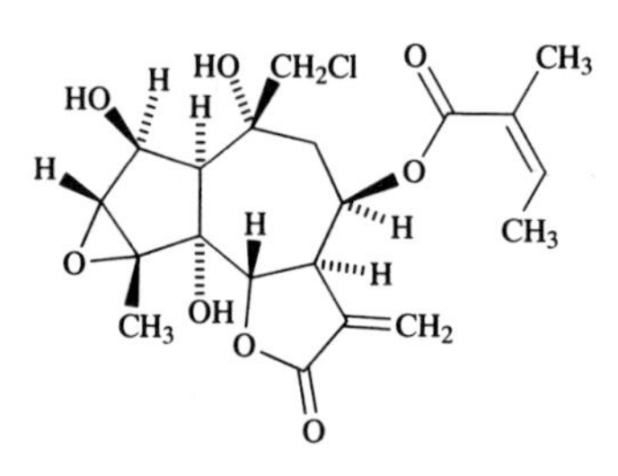

Sesquiterpenoid lactone
(Guaianolide)

$C_{20}H_{25}ClO_8$ [20071-52-7] MW 428.87

The round-leaved thoroughwort, *Eupatorium rotundifolium* (Compositae).

Cytotoxic and antitumour properties.

Kupchan, S. M., *J. Org. Chem.*, 1969, **34**, 3876.

Eupacunin 571

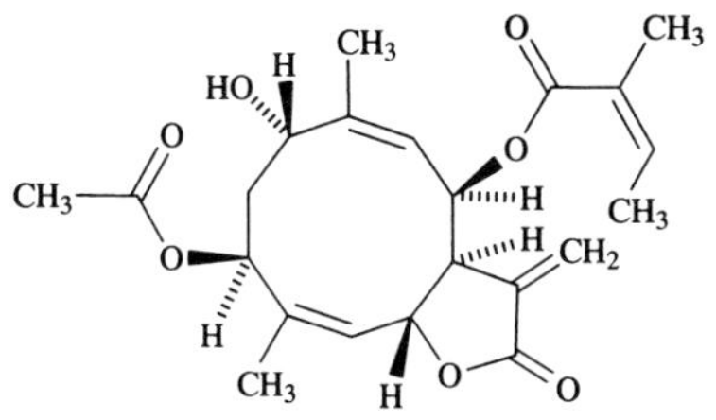

Sesquiterpenoid lactone
(Germacranolide)

$C_{22}H_{28}O_7$ [33854-15-8] MW 404.46

Stems, leaves and flowers of *Eupatorium cuneifolium* and *E. lancifolium* (Compositae).

Cytotoxic and antitumour properties.

Kupchan, S. M., *J. Am. Chem. Soc.,* 1971, **93**, 4914.

Eupacunolin 572

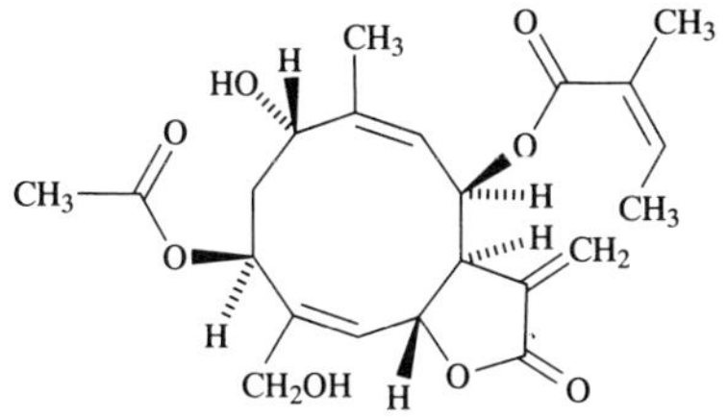

Sesquiterpenoid lactone
(Germacranolide)

$C_{22}H_{28}O_8$ [79491-59-1] MW 420.46

The aerial parts of *Eupatorium cuneifolium* and *E. lancifolium* (Compositae).

Cytotoxic and antitumour properties.

Kupchan, S. M., *J. Org. Chem.,* 1973, **38**, 2189.

Eupacunoxin 573

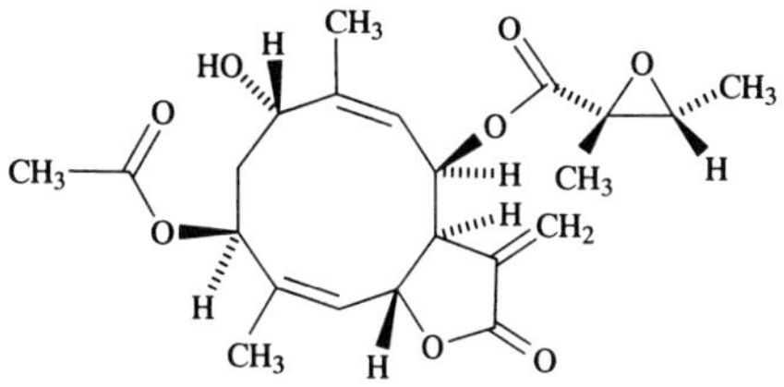

Sesquiterpenoid lactone
(Germacranolide)

$C_{22}H_{28}O_8$ [33853-88-2] MW 420.46

Aerial parts of *Eupatorium cuneifolium* (Compositae).

Cytotoxic and antitumour properties.

Kupchan, S. M., *J. Org. Chem.,* 1973, **38**, 2189.

Eupaformonin 574

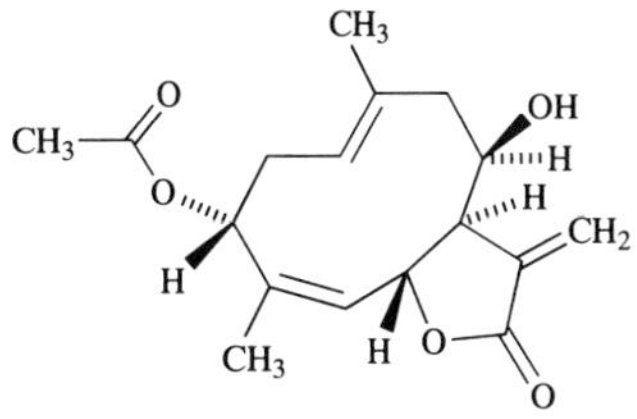

Sesquiterpenoid lactone
(Germacranolide)

$C_{17}H_{22}O_5$ [55520-20-2] MW 306.36

Eupatorium formosanum (Compositae).

Cytotoxic and antitumour properties.

McPhail, A. T., *Tetrahedron Lett.,* 1974, 3203.

Eupaformosanin 575

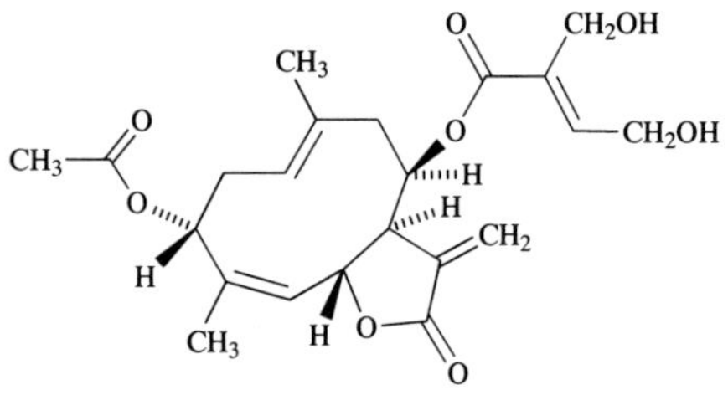

Sesquiterpenoid lactone
(Germacranolide)

$C_{22}H_{28}O_8$ [64439-43-6] MW 420.46

Eupatorium formosanum (Compositae).

Cytotoxic and antitumour properties.

Lee, K. H., *Phytochemistry,* 1977, **16**, 1068.

Eupahyssopin 576

Eupassopin

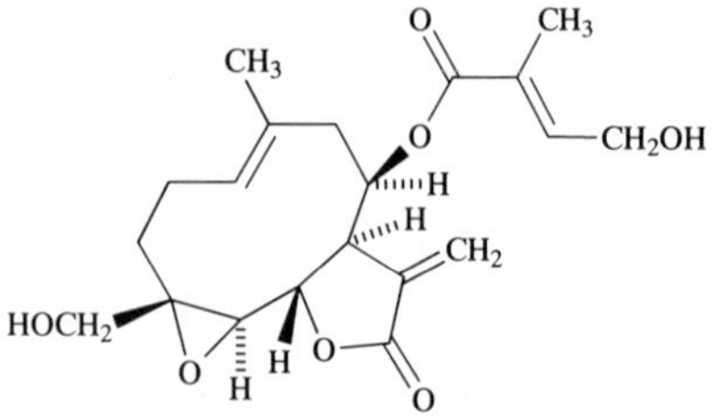

Sesquiterpenoid lactone
(Germacranolide)

$C_{20}H_{26}O_7$ [57718-77-1] MW 378.42

The hyssop-leaved thoroughwort, *Eupatorium hyssopifolium* (Compositae).

Cytotoxic and antitumour properties.

Lee, K. H., *Tetrahedron Lett.,* 1976, 1051.

Euparotin 577

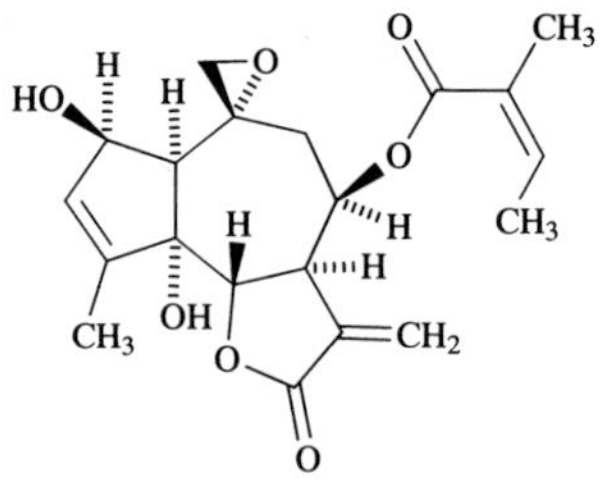

Sesquiterpenoid lactone
(Guaianolide)

$C_{20}H_{24}O_7$ [10191-01-2] MW 376.41

The round-leaved thoroughwort, *Eupatorium rotundifolium* (Compositae).

Cytotoxic and antitumour properties.

Kupchan, S. M., *J. Org. Chem.,* 1969, **34**, 3876.

Euparotin acetate 578

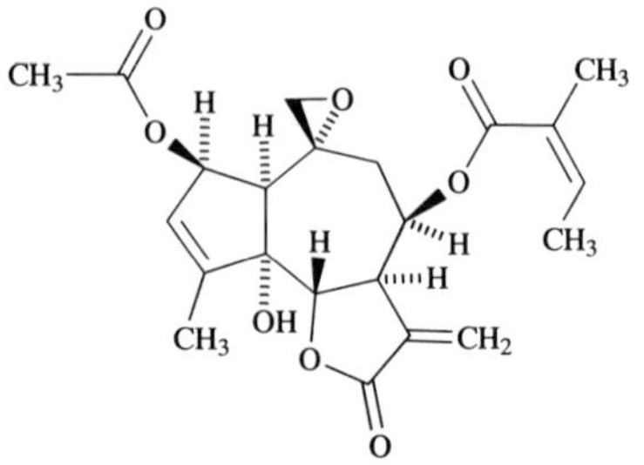

Sesquiterpenoid lactone
(Guaianolide)

$C_{22}H_{26}O_8$ [10215-89-1] MW 418.44

The round-leaved thoroughwort, *Eupatorium rotundifolium* (Compositae).

Cytotoxic and antitumour properties.

Kupchan, S. M., *J. Org. Chem.,* 1969, **34**, 3876.

Eupaserrin 579

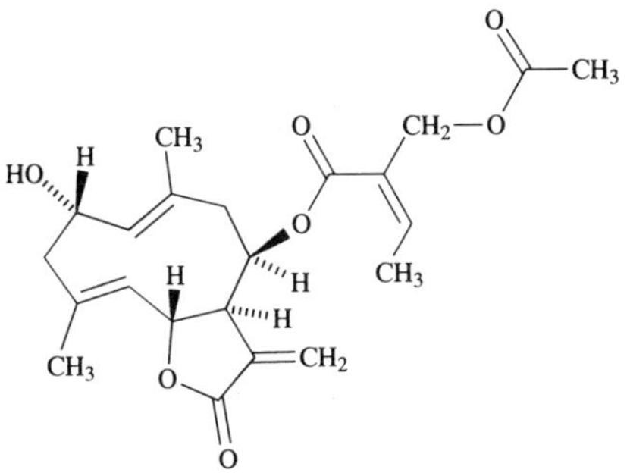

Sesquiterpenoid lactone
(Germacranolide)

$C_{22}H_{28}O_7$ [38456-36-9] MW 404.46

Eupatorium semiserratum and *E. cuneifolium* (Compositae).

Cytotoxic and antitumour properties.

Kupchan, S. M., *J. Org. Chem.*, 1973, **38**, 1260.

Eupatocunin 580

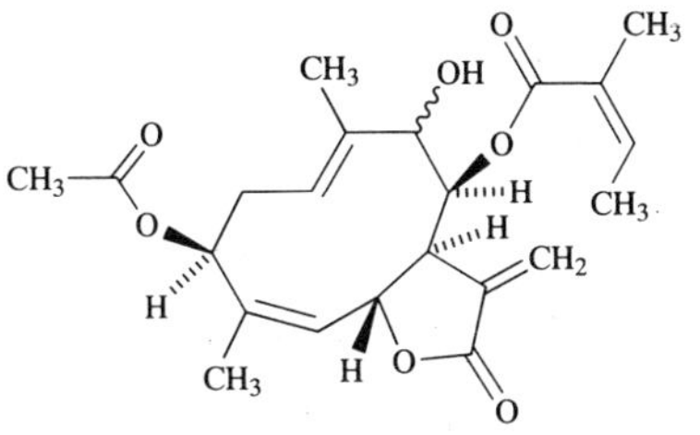

Sesquiterpenoid lactone
(Germacranolide)

$C_{22}H_{28}O_7$ [33853-87-1] MW 404.46

The aerial parts of *Eupatorium cuneifolium* (Compositae).

Cytotoxic and antitumour properties.

Kupchan, S. M., *J. Org. Chem.*, 1973, **38**, 2189.

Eupatocunoxin 581

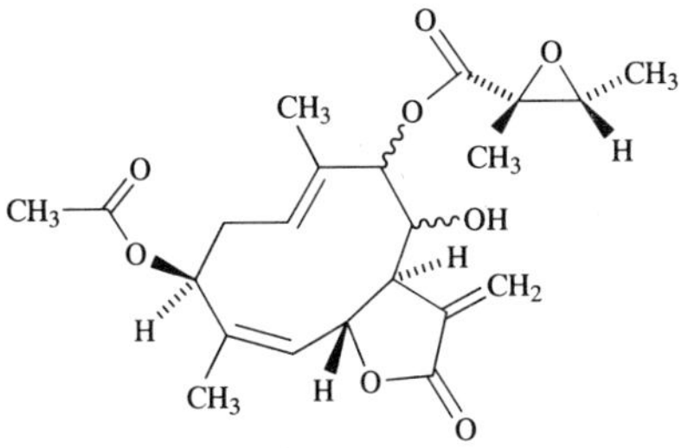

Sesquiterpenoid lactone
(Germacranolide)

$C_{22}H_{28}O_8$ [39204-36-9] MW 420.46

The aerial parts of *Eupatorium cuneifolium* (Compositae).

Cytotoxic and antitumour properties.

Kupchan, S. M., *J. Org. Chem.*, 1973, **38**, 2189.

Eupatolide 582

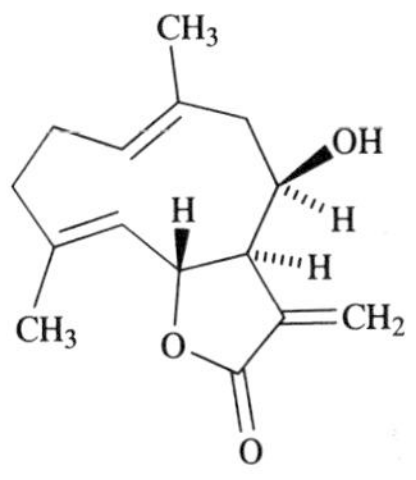

Sesquiterpenoid lactone
(Germacranolide)

$C_{15}H_{20}O_3$ [6750-25-0] MW 248.32

In hemp agrimony, *Eupatorium cannabinum*, in *E. formosanum* and *Helianthus agrophyllus* (Compositae).

Cytotoxic and antitumour activities.

Lee, K. H., *J. Pharm. Sci.*, 1972, **61**, 629.

Eupatoriopicrin 583

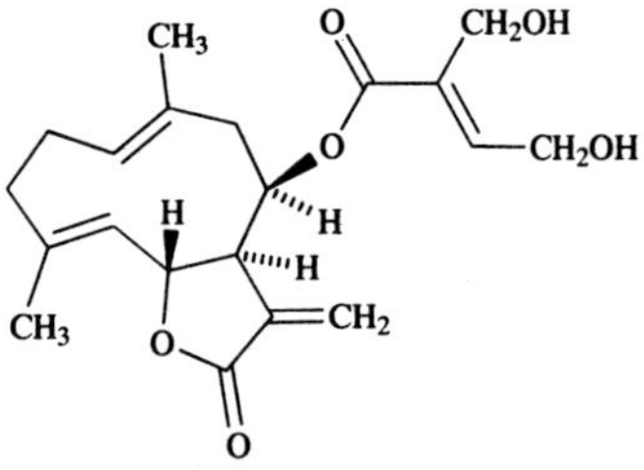

Sesquiterpenoid lactone
(Germacranolide)

$C_{20}H_{26}O_6$ [6856-01-5] MW 362.42

In hemp agrimony, *Eupatorium cannabinum*, in *Chaenactis carphoclinia*, *C. douglassii* and *Eriophyllum stachaedifolium* (Compositae).

Insect antifeedant activity. Cytotoxic and antitumour properties.

Dolejš, L., *Collect. Czech. Chem. Commun.*, 1962, **27**, 2654.

Eupatoroxin 584

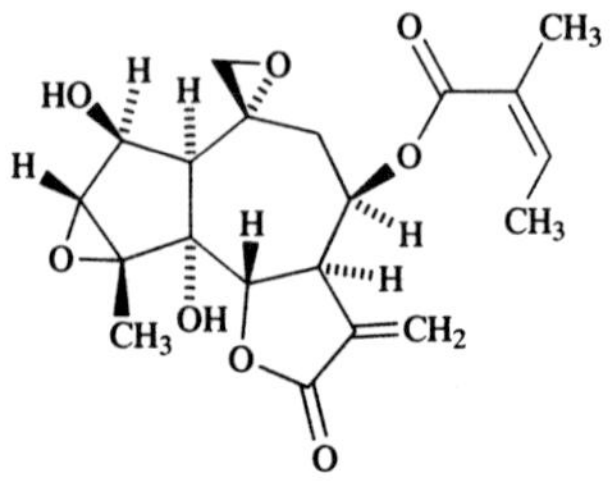

Sesquiterpenoid lactone
(Guaianolide)

$C_{20}H_{24}O_8$ [20071-51-6] MW 392.41

The round-leaved thoroughwort, *Eupatorium rotundifolium* (Compositae).

Cytotoxic and antitumour properties. It co-occurs with the isomeric 10-*epi*-eupatoroxin [20071-54-9], which also displays cytotoxic and antitumour activities.

Kupchan, S. M., *J. Org. Chem.*, 1969, **34**, 3876.

Eupatundin 585

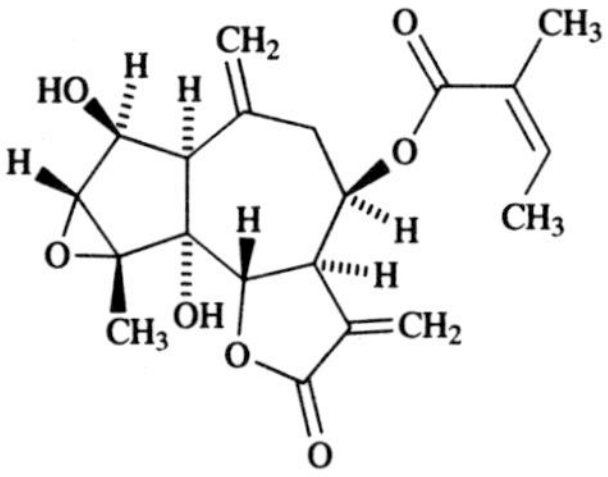

Sesquiterpenoid lactone
(Guaianolide)

$C_{20}H_{24}O_7$ [20071-53-8] MW 376.41

The round-leaved thoroughwort, *Eupatorium rotundifolium* (Compositae).

Cytotoxic and antitumour properties.

Kupchan, S. M., *J. Org. Chem.*, 1969, **34**, 3876.

Europine 586

9-Lasiocarpylheliotridine; Base G

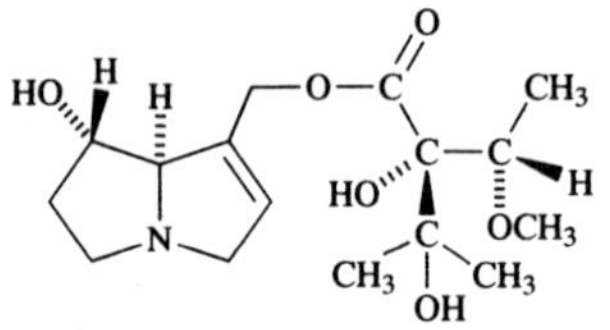

Pyrrolizidine alkaloid

$C_{16}H_{27}NO_6$ [570-19-4] MW 329.39

Heliotropium arbainense, *H. europeum*, *H. maris-mortui*, *H. rotundifolium* and *Trichodesma africana* (Boraginaceae).

Acutely toxic alkaloid. It is hepatotoxic and anticholinergic when fed to rats. Shows antitumour activity.

Bull, L. B., The Pyrrolizidine Alkaloids, 1968, Interscience.

Evomonoside 587

Digitoxigenin 3-α-rhamnoside

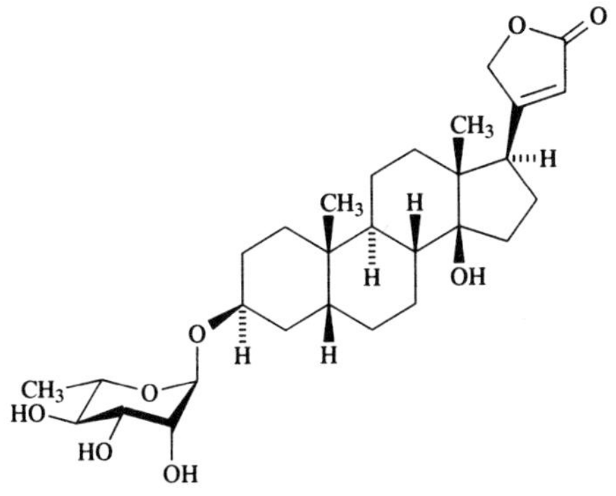

Cardenolide

$C_{29}H_{44}O_8$ [508-93-0] MW 520.66

Seed of *Lepidium apetalum* (Cruciferae) and *Euonymus europaea* (Celastraceae).

Cytotoxic against three human tumour cell lines.

Tamm, Ch., *Helv. Chim. Acta,* 1953, **36**, 1309.

Evoxine 588

Haploperine

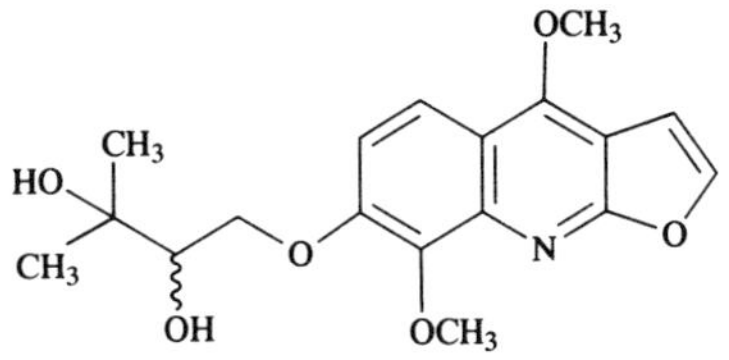

Quinoline alkaloid

$C_{18}H_{21}NO_6$ [522-11-2] MW 347.37

Leaves of *Orixa japonica*, in *Euodia xanthoxyloides*, *Teclea boiviniana*, *Monnieria trifolia* and in various *Haplophyllum* spp. (Rutaceae).

Insect antifeedant. Possesses sedative and spasmolytic effects in animals.

Eastwood, F. W., *Aust. J. Chem.,* 1954, **7**, 87.

F

Fagaramide 589

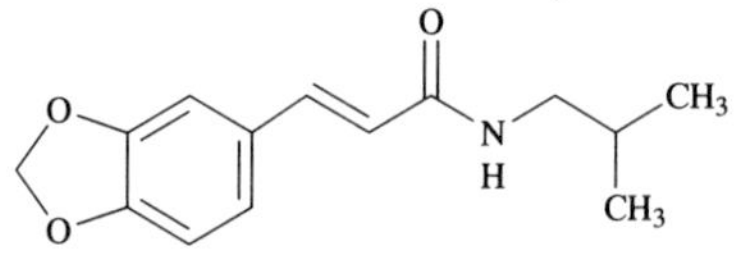

Phenylpropanoid

$C_{14}H_{17}NO_3$ [495-86-3] MW 247.29

In the root bark of *Fagara xanthoxyloides* and of *F. macrophylla* (Rutaceae), in the stem bark of *Anthocleista djalonensis* and *A. vogelli* (Loganiaceae) and in the wood of *Piper novae-hollandiae* (Piperaceae).

Toxic to snails and has an inhibitory effect on insect growth and development. Also possesses anti-inflammatory activity in mammals.

Goodson, J. A., *Biochem. J.*, 1921, **15**, 123.

Fagaronine 590

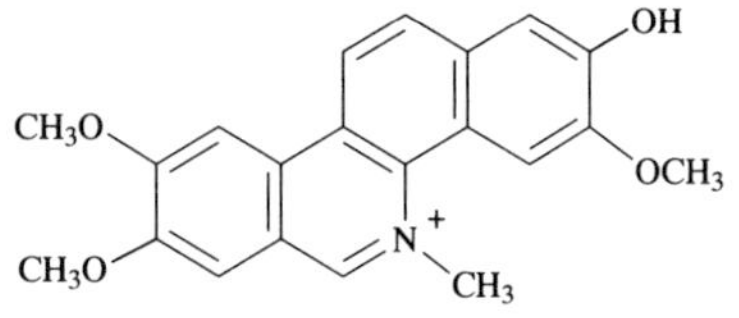

Benzophenanthridine alkaloid

$[C_{21}H_{20}NO_4]^+$ [52259-65-1] MW 350.39

Roots of *Zanthoxylum zanthoxyloides* and of several other species in the Rutaceae.

Antitumour properties. It inhibits reverse transcriptase activity of various RNA oncogenic viruses.

Messmer, W. M., *J. Pharm. Sci.*, 1972, **61**, 1858.

Falaconitine 591

Pyropseudoaconitine

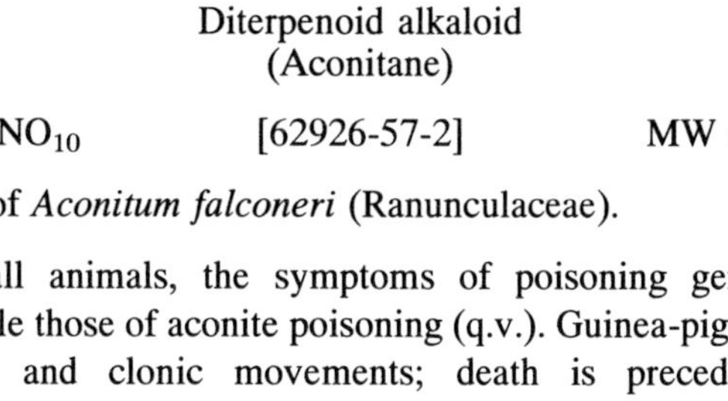

Diterpenoid alkaloid
(Aconitane)

$C_{34}H_{47}NO_{10}$ [62926-57-2] MW 629.75

Roots of *Aconitum falconeri* (Ranunculaceae).

In small animals, the symptoms of poisoning generally resemble those of aconite poisoning (q.v.). Guinea-pigs show paresis and clonic movements; death is preceded by convulsions.

Pelletier, S. W., *J. Chem. Soc., Chem. Commun.*, 1977, 12.

Falcarindiol 592

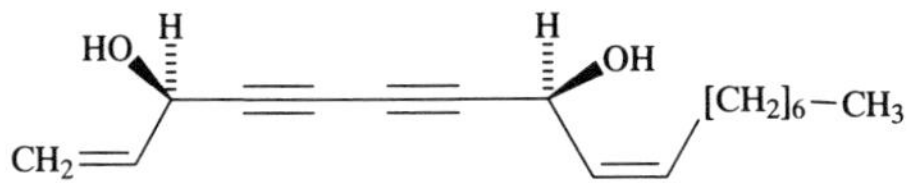

Polyacetylene

$C_{17}H_{24}O_2$ [55297-87-5] MW 260.38

In roots of the common carrot, *Daucus carota* and of *Falcaria vulgaris* (Umbelliferae).

Potentially toxic to mammals. Although the common carrot has been shown to contain both falcarinol (q.v.) and falcarindiol, the levels are so low that toxic effects have never been recorded.

Bohlmann, F., Naturally Occurring Acetylenes, 1973, 195, Academic Press.

Falcarinol 593

Carotatoxin; Panaxynol

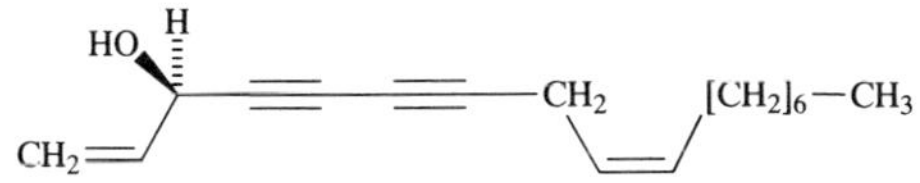

Polyacetylene

$C_{17}H_{24}O$ [21852-80-2] MW 244.38

In roots of the common carrot, *Daucus carota*, of *Falcaria vulgaris* and of water hemlock, *Oenanthe crocata* (Umbelliferae). Also present in the leaves of common ivy, *Hedera helix* and of *Schefflera arboricola* (Araliaceae).

Allergenic and poisonous. Responsible, with oenanthatoxin (q.v.) for the toxicity to mammals of water hemlock. Produces contact dermatitis in humans handling *Hedera* or *Schefflera* plants.

Bohlmann, F., Naturally Occurring Acetylenes, 1973, 194, Academic Press.

Falcarinone 594

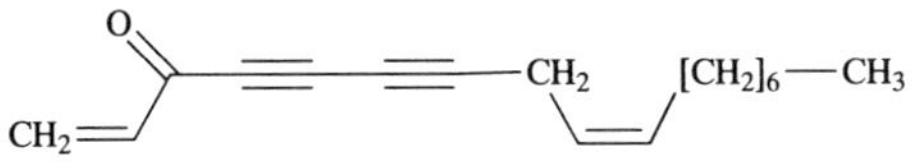

Polyacetylene

$C_{17}H_{22}O$ [4117-11-7] MW 242.36

Roots of *Falcaria vulgaris* and roots and leaves of rough chervil, *Chaerophyllum temulentum* (Umbelliferae). Has been isolated from about eighty other genera in this family.

Toxic. Responsible for livestock poisoning in North America following the consumption of rough chervil.

Bohlmann, F., Naturally Occurring Acetylenes, 1973, 192, Academic Press.

Farinosin 595

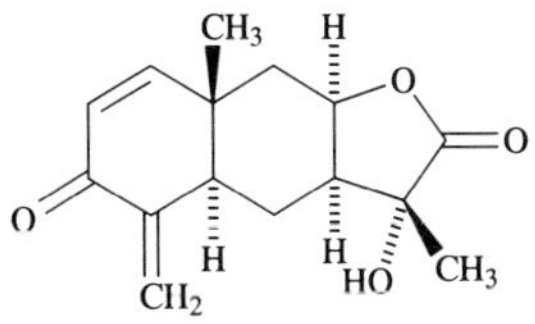

Sesquiterpenoid lactone
(Eudesmanolide)

$C_{15}H_{18}O_4$ [33299-79-5] MW 262.31

Encelia farinosa and *E. virginensis* (Compositae).

Cytotoxic and antitumour properties.

Herz, W., *J. Org. Chem.*, 1968, **33**, 3743.

Farrerol 596

Cyrtopterinetin; 4′-Demethylmatteucinol; 4′,5,7-Trihydroxy-6,8-dimethylflavanone

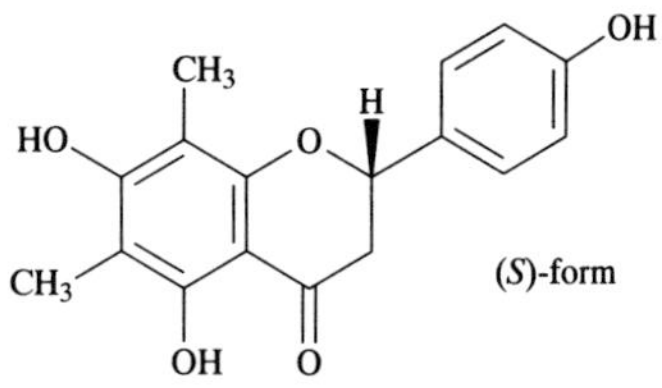

(*S*)-form

Flavanone

$C_{17}H_{16}O_5$ [24211-30-1] MW 300.31

Occurs in combined form as the 5,7-diglucoside in leaves of several *Rhododendron* spp., notably *R. farreria* (Ericaceae) and as the 7-glucoside in fronds of *Cyrtomium* spp. (Pteridophyta).

Expectorant activity; it acts directly on the mucous membrane of the respiratory tract, increasing the secretion of respiratory tract fluid. It is used clinically to treat chronic bronchitis.

Arthur, H. R., *J. Chem. Soc.,* 1955, 3740.

Fastigilin B 597

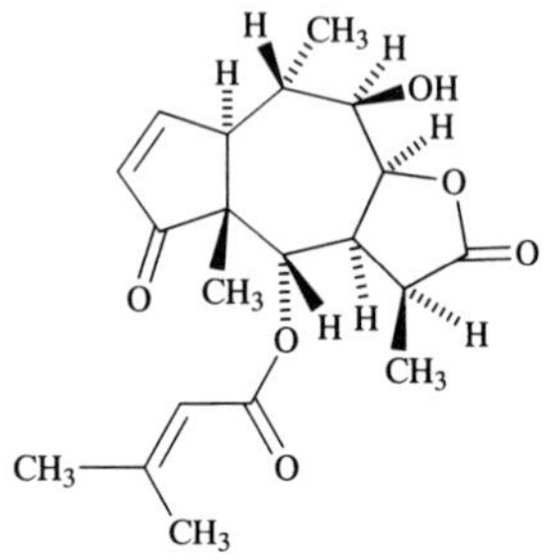

Sesquiterpenoid lactone
(Pseudoguaianolide)

$C_{20}H_{26}O_6$ [6995-11-5] MW 362.42

Baileya multiradiata and *Gaillardia fastigiata* (Compositae).

Cytotoxic and antitumour properties.

Herz, W., *Tetrahedron,* 1966, **22**, 1907.

Fastigilin C 598

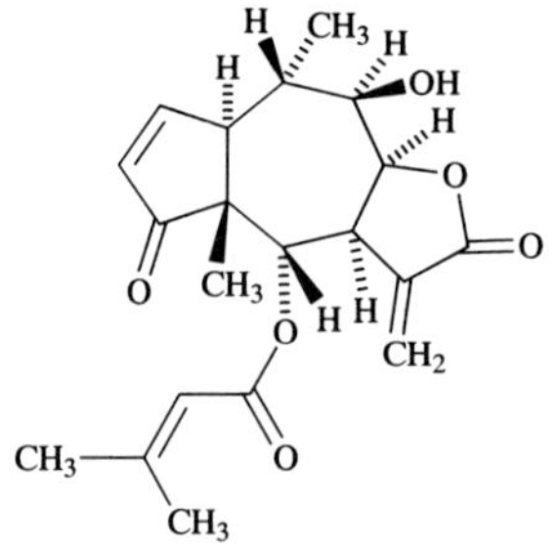

Sesquiterpenoid lactone
(Pseudoguaianolide)

$C_{20}H_{24}O_6$ [6995-12-6] MW 360.41

Baileya multiradiata, *Gaillardia fastigiata* and *Hymenoxys acaulis* (Compositae).

Cytotoxic and antitumour properties.

Herz, W., *Tetrahedron,* 1966, **22**, 1907.

Febrifugine 599

β-Dichroine; *γ*-Dichroine; Dichroine B

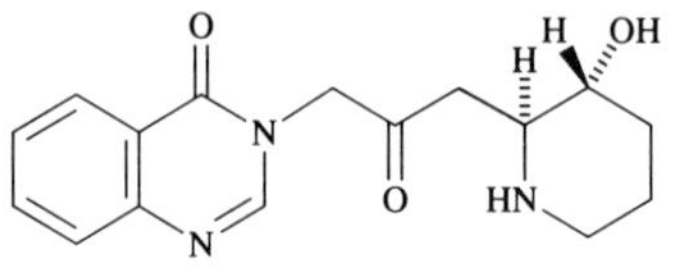

Quinazoline alkaloid

$C_{16}H_{19}N_3O_3$ [24159-07-7] MW 301.35

Roots of *Dichroa febrifuga* and in *Hydrangea* spp. (Saxifragaceae).

A hundred times more active than quinine (q.v.) as an antimalarial drug, but its practical application is severely limited by its toxicity. It also has antipyretic and emetic properties.

Brossi, A., The Alkaloids, 1986, **29**, 99, Academic Press.

Ferprenin 600

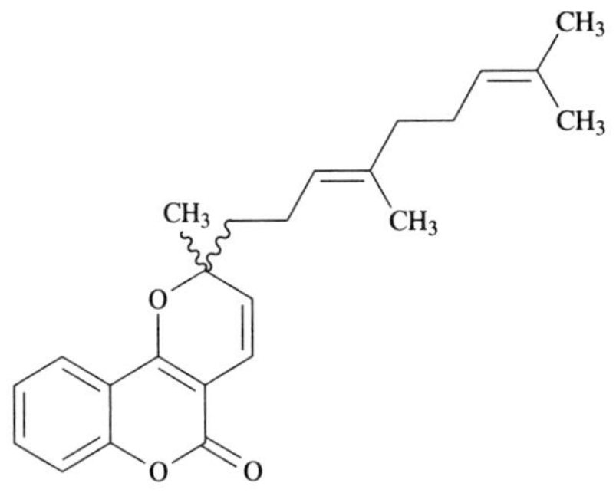

Coumarin

$C_{24}H_{28}O_3$ [114727-96-7] MW 364.48

In latex of *Ferula communis* (Umbelliferae).

Anticoagulant activity. May be responsible, together with ferulinol (q.v.), for the toxicity of *Ferula* in grazing sheep.

Appendino, G., *Phytochemistry,* 1988, **27**, 944.

Ferulenol 601

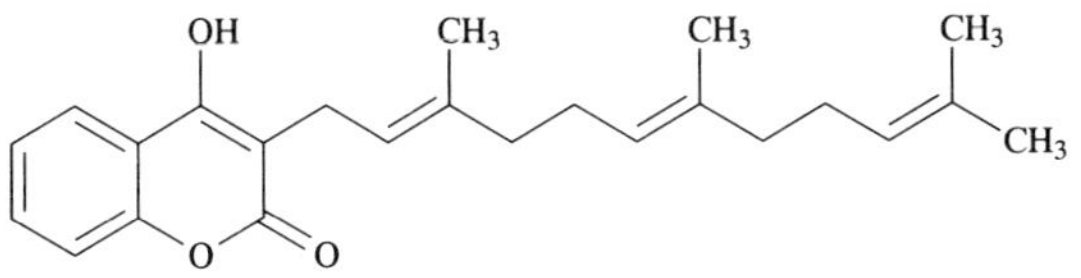

Coumarin

$C_{24}H_{30}O_3$ [6805-34-1] MW 366.50

The latex of the giant fennel, *Ferula communis* (Umbelliferae).

Anticoagulant, affecting blood clotting by increasing prothrombin time. May be responsible for the toxicity of *Ferula* plants to sheep.

Carboni, S., *Tetrahedron Lett.,* 1964, 2783.

Fetidine 602

Foetidine

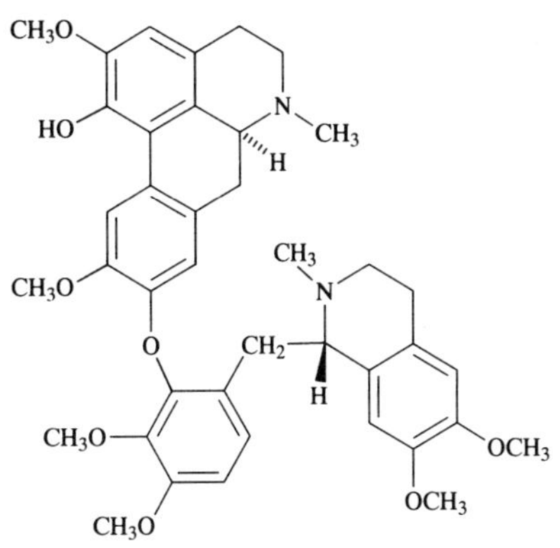

Bisbenzylisoquinoline alkaloid
(Aporphine and benzylisoquinoline)

$C_{40}H_{46}N_2O_8$ [7072-86-8] MW 682.81

Aerial parts of *Thalictrum foetidum* (Ranunculaceae).

Depresses nervous activity in mice. It has hypotensive and anti-inflammatory properties.

Ismailov, Z. F., *Chem. Nat. Compd.,* 1966, **2**, 35.

Ficuseptine 603

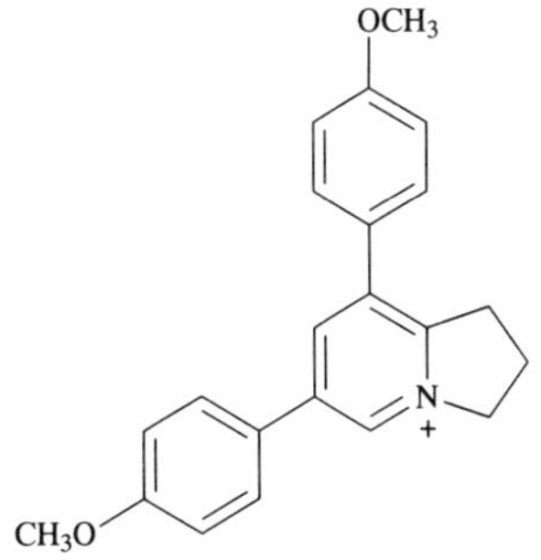

Indolizidine alkaloid

$[C_{22}H_{22}NO_2]^+$ [132923-01-4] MW 332.42

Leaves of *Ficus septica* (Moraceae).

Antibacterial activity.

Baumgartner, B., *Phytochemistry,* 1990, **29**, 3327.

Filixic acid ABA 604

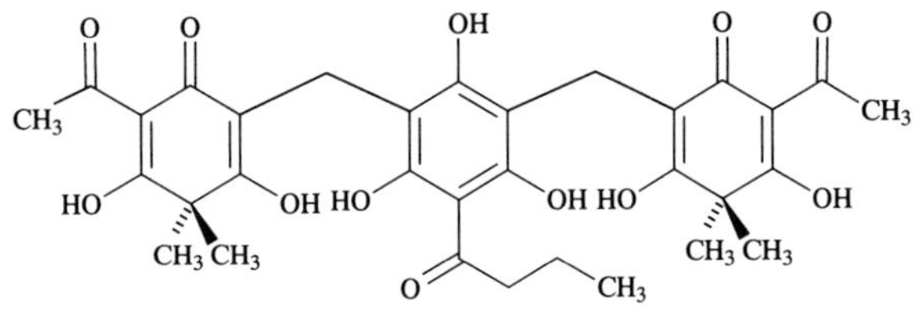

Phenolic ketone

$C_{32}H_{36}O_{12}$ [38226-84-5] MW 612.63

One of six related compounds in the male fern, *Dryopteris filix-mas* and other *Dryopteris* spp. (Polypodiaceae).

Male fern extract is used to treat tapeworms, especially in veterinary practice. Overdoses can lead to fatalities. Symptoms of poisoning include cramps, intestinal irritation and visual disturbances. The active principles also include albaspidins (q.v.).

Hisada, S., *Phytochemistry,* 1972, **11**, 1850.

Florilenalin 605

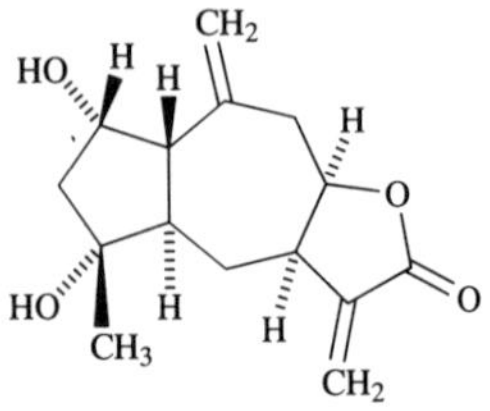

Sesquiterpenoid lactone
(Guaianolide)

$C_{15}H_{20}O_4$ [54964-49-7] MW 264.32

In the sneezeweed, *Helenium autumnale* (Compositae).

Cytotoxic and antitumour properties.

Lee, K. H., *Tetrahedron Lett.,* 1974, 2287.

Flossonol 606

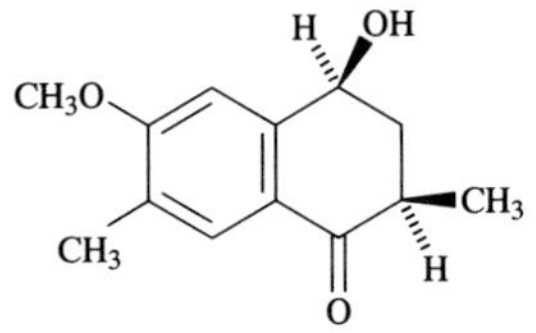

Chroman

$C_{13}H_{16}O_3$ [112936-00-2] MW 220.27

Roots and stems of *Pararistolochia flosavis* (Aristolochiaceae).

Cytotoxic and antileukaemic against PS-cells in culture.

Sun, N. J., *Phytochemistry,* 1987, **26**, 3051.

ω-Fluorooleic acid 607

18-Fluoro-9-octadecenoic acid

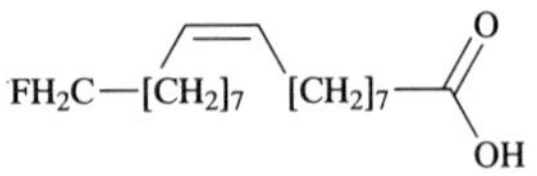

Fatty acid

$C_{18}H_{33}FO_2$ [1478-37-1] MW 300.46

Seed of *Dichapetalum toxicarium* (Dichapetalaceae). Co-occurs with several other monofluorofatty acids, including ω-fluoropalmitic acid (q.v.).

Major toxin of the seed oil, poisonous on skin contact.

Harper, D. B., *Nat. Prod. Rep.,* 1994, **11**, 123.

ω-Fluoropalmitic acid 608

16-Fluoropalmitic acid

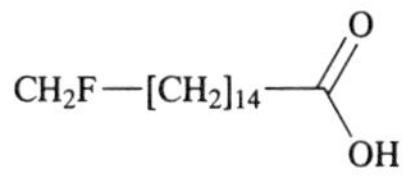

Fatty acid

$C_{16}H_{31}FO_2$ [3109-58-8] MW 274.42

Seed of the West African shrub *Dichapetalum toxicarium* (Dichapetalaceae).

One of the several toxic ω-fluorofatty acids in the seeds of this plant.

Harper, D. B., *Nat. Prod. Rep.,* 1994, **11**, 123.

Formic acid 609

Formylic acid; Methanoic acid

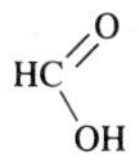

Organic acid

CH_2O_2 [64-18-6] MW 46.03

Present in low concentrations in many fruits, leaves and roots of plants. Higher concentrations are present in stinging nettles, *Urtica dioica* (Urticaceae). Esters of formic acid are commonly present in fruit volatiles.

A strong acid with acute toxicity. Chronic absorption may cause albuminuria and haematuria. Contributes to the painful sting of the stinging nettle.

Martindale, The Extra Pharmacopoeia, 30th edn, 1993, 1371, The Pharmaceutical Press.

Formononetin 610

Daidzein 4′-methyl ether; Biochanin B; Neochanin; Pratol

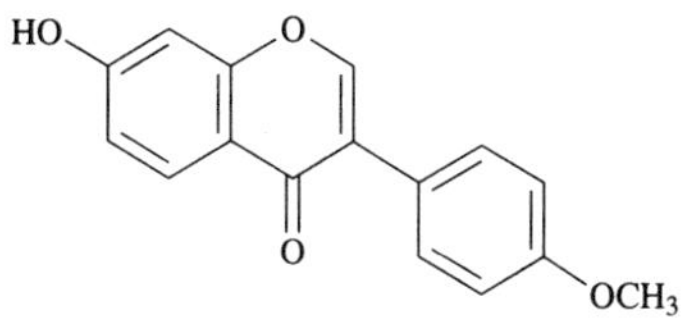

Isoflavone

$C_{16}H_{12}O_4$ [485-72-3] MW 268.27

Present in leaves of all *Trifolium* spp., with relatively high concentrations in *T. subterraneum*. Occurs widely in other Leguminosae, e.g. in the chickpea, *Cicer arietinum* and in liquorice, *Glycyrrhiza glabra*.

Responsible for "clover disease", an infertility problem in Australian ewes which have fed on *Trifolium subterraneum*. It is a pro-oestrogen, being converted *in vivo* to the more active isoflavan, equol [94105-90-5]. It was estimated in 1976 that one million ewes failed to lamb in Australia because of clover disease.

Shutt, D. A., *Endeavour,* 1976, **35**, 110.

Forskolin 611

Coleonol; Colforsin

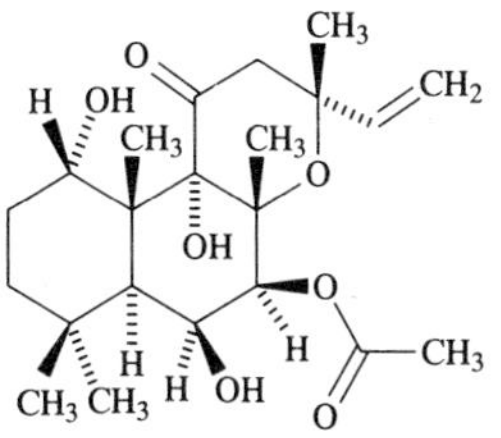

Diterpenoid
(Labdane)

$C_{22}H_{34}O_7$ [66575-29-9] MW 410.51

In leaves and roots of *Coleus forskohlii* (Labiatae).

Cardiovascular activity. Used in the treatment of glaucoma.

Colombo, M. I., *Tetrahedron,* 1992, **48**, 963.

Frangulin A 612

Emodin-3-rhamnoside; Rhamnoxanthin; Franguloside

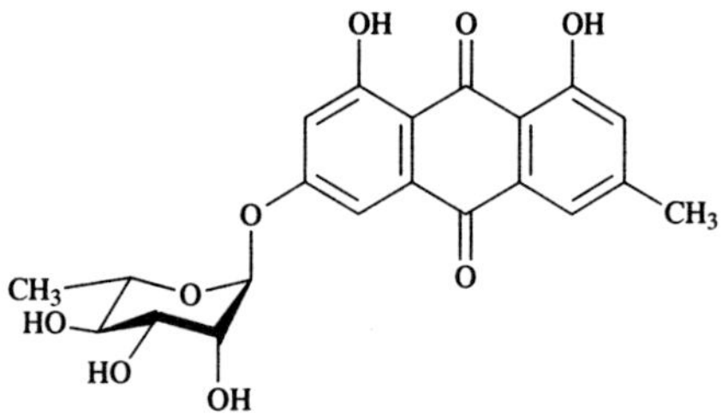

Anthraquinone glycoside

$C_{21}H_{20}O_9$ [521-62-0] MW 416.38

Seed, bark and rootbark of *Rhamnus cathartica* and *R. frangula* (Rhamnaceae).

Cathartic activity.

Hörhammer, L., *Naturwissenschaften,* 1964, **51**, 310.

Frangulin B 613

Emodin-3-apioside

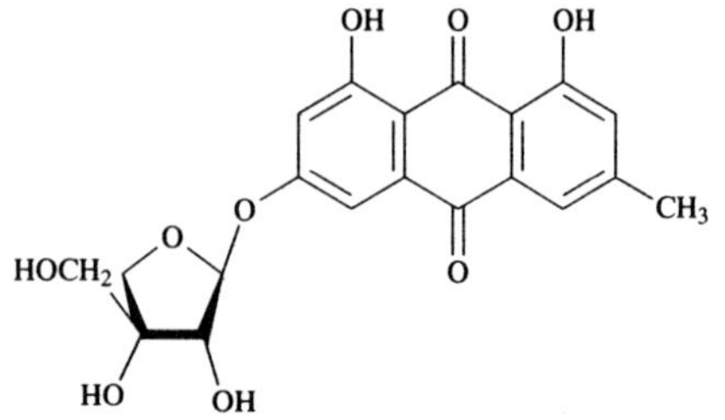

Anthraquinone glycoside

$C_{20}H_{18}O_9$ [14101-04-3] MW 402.36

The seed, bark and rootbark of *Rhamnus cathartica* and *R. frangula* (Rhamnaceae).

Cathartic activity.

Hörhammer, L., *Naturwissenschaften,* 1964, **51**, 310.

(−)-Frullanolide 614

Tournefortiolide

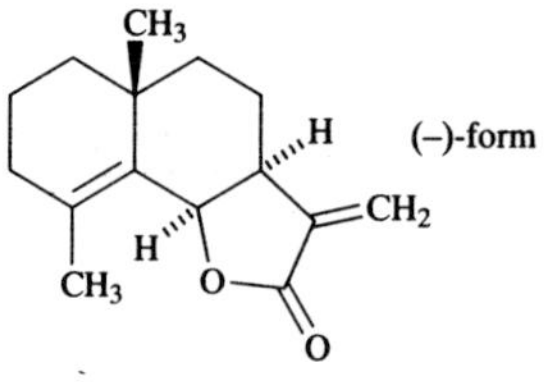

(−)-form

Sesquiterpenoid lactone
(Eudesmanolide)

$C_{15}H_{20}O_2$ [27579-97-1]
(+)-form [40776-40-7] MW 232.32

The (−)-form occurs in the liverwort, *Frullania tamarisci* and the (+)-form in *F. dilatata* (Hepaticae).

Causes allergic contact dermatitis in humans. Both forms are equally active. Since *Frullania* is epiphytic on many tree barks, it notably causes dermatitis in forest workers in France and Canada.

Asakawa, Y., *Bull. Soc. Chim. Fr.,* 1976, 1465.

Fulvine 615

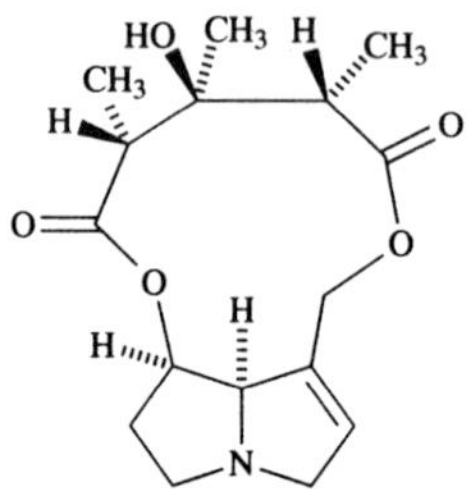

Pyrrolizidine akaloid
(Norcrotalanan)

$C_{16}H_{23}NO_5$ [6029-87-4] MW 309.36

Crotalaria crispa, *C. fulva*, *C. madurensis* and *C. paniculata* (Leguminosae).

Hepatotoxic and pneumotoxic.

Culvenor, C. C. J., *Aust. J. Chem.,* 1963, **16**, 239.

Funtumine 616

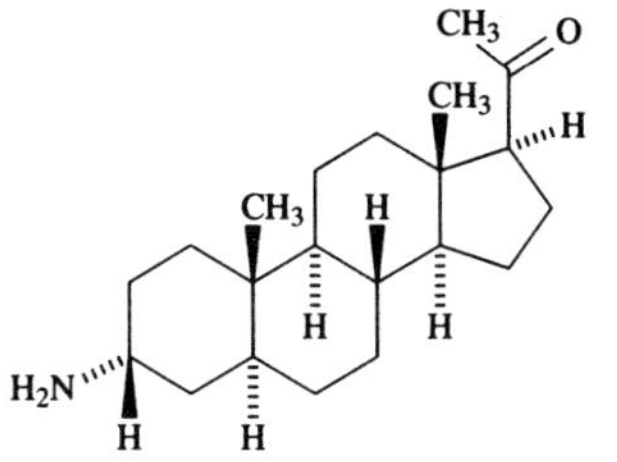

Steroid alkaloid
(Buxane)

$C_{21}H_{35}NO$ [474-45-3] MW 317.52

In the trees *Funtumia elastica* and *F. latifolia* (Apocynaceae), as well as in *Holarrhena febrifuga* and *H. congolensis*.

Hypocholesterolaemic activity and respiratory stimulant.

Janot, M. M., *C. R. Acad. Sci.,* 1958, **246**, 3076.

G

Gabunamine 617

N-Dimethylconoduramine

Bisindole alkaloid
(Ibogamine and Vobasan)

$C_{42}H_{50}N_4O_5$ [66086-99-5] MW 690.88

The stem bark of *Tabernaemontana johnstonii* (Apocynaceae).

Antitumour activity. It is cytotoxic to P-388 lymphocytic leukaemia cells *in vitro*.

Kingston, D. G. I., *J. Pharm. Sci.,* 1978, **67**, 249.

Gabunine 618

N-Demethylconodurine

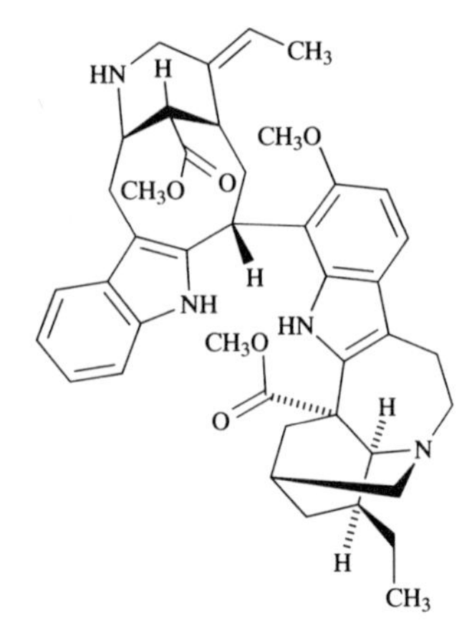

Bisindole alkaloid
(Ibogamine and Vobasan)

$C_{42}H_{50}N_4O_5$ [1357-30-8] MW 690.88

Tabernaemontana johnstonii, *T. holstii* and *Gabunia odoratissima* (Apocynaceae).

Antitumour properties. It is cytotoxic to P-388 lymphocytic leukaemia cells *in vitro*.

Cava, M. P., *Tetrahedron Lett.,* 1965, 931.

Gaillardin 619

Haillardin

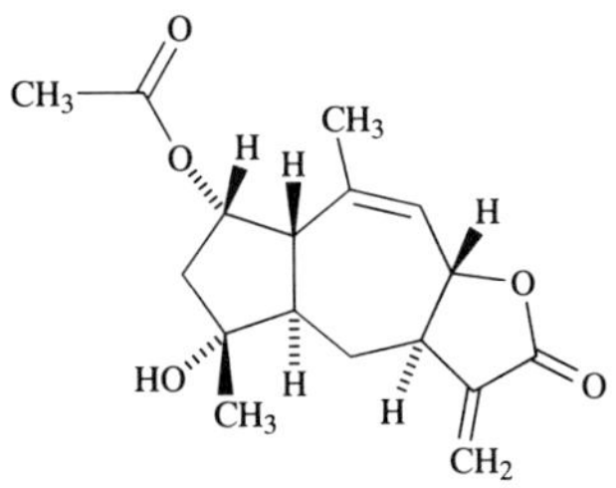

Sesquiterpenoid lactone
(Guaianolide)

$C_{17}H_{22}O_5$ [14682-46-3] MW 306.36

Gaillardia pulchella and several *Inula* spp. (Compositae), including *I. britannica.*

Cytotoxic and antitumour activities.

Dullforce, T. A., *Tetrahedron Lett.,* 1969, 693.

Galangin 620

3,5,7-Trihydroxyflavone; 5,7-Dihydroxyflavonol; Norizalpinin

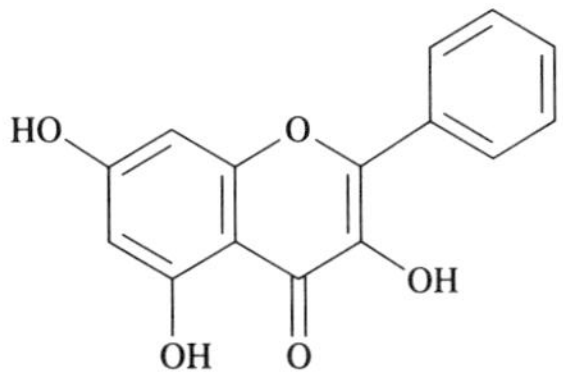

Flavonol

$C_{15}H_{10}O_5$ [548-83-4] MW 270.24

In bud excretions of Salicaceae and Betulaceae, on the leaves of *Escallonia* spp. (Saxifragaceae), in the farinose exudate of fern fronds and on the leaves of Labiatae.

A potent inhibitor of bull seminal cyclo-oxygenase activity.

Chavan, J. J., *J. Chem. Soc.,* 1933, 368.

Galanthamine 621

Galantamine; Lycorimine; Lycoremine; Galanthine

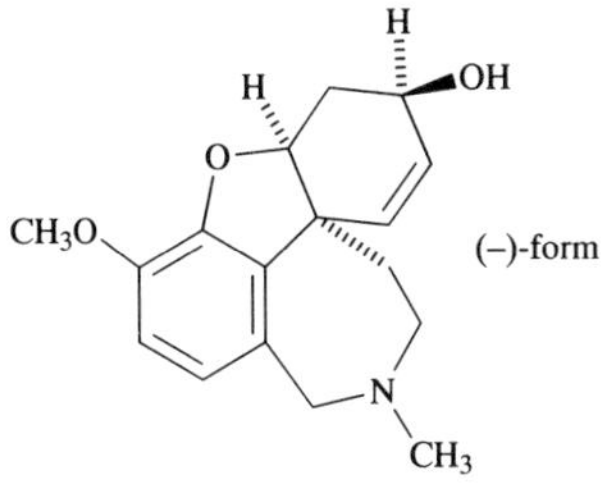

(–)-form

Amaryllidaceae alkaloid
(Galanthamine)

$C_{17}H_{21}NO_3$ [357-70-0] MW 287.36

Occurs in *Crinum, Galanthus, Hippeastrum, Hymenocallis, Leucojum, Lycoris, Narcissus, Pancratium* and *Ungernia* spp. (Amaryllidaceae).

Insecticidal. Inhibits cholinesterase activity reversibly. It shows strong analgesic activity comparable to that of morphine (q.v.) and has been used in Russia in the treatment of nervous diseases. The LD_{50}, when administered subcutaneously to mice, is 11 mg/kg body-weight.

Barton, D. H. R., *J. Chem. Soc.,* 1962, 806.

Galegine 622

N-(3,3-Dimethylallyl)guanidine; Isoamyleneguanidine; (3-Methylbut-2-enyl)guanidine

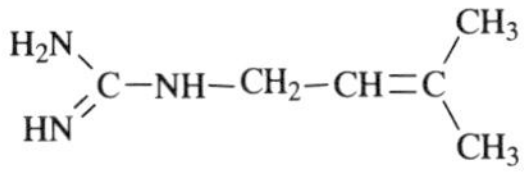

$H_2N(HN=)C{-}NH{-}CH_2{-}CH{=}C(CH_3)_2$

Guanidine

$C_6H_{13}N_3$ [543-83-9] MW 127.19

The leaves, flowers and seeds of goat's rue or French lilac, *Galega officinalis* (Leguminosae) and in *Verbesina encelioides* (Compositae).

Toxic, affecting mitochondrial function. Responsible for poisoning in sheep and goats following the consumption of goat's rue.

Eichholzer, J. V., *Phytochemistry,* 1982, **21**, 97.

Galipine 623

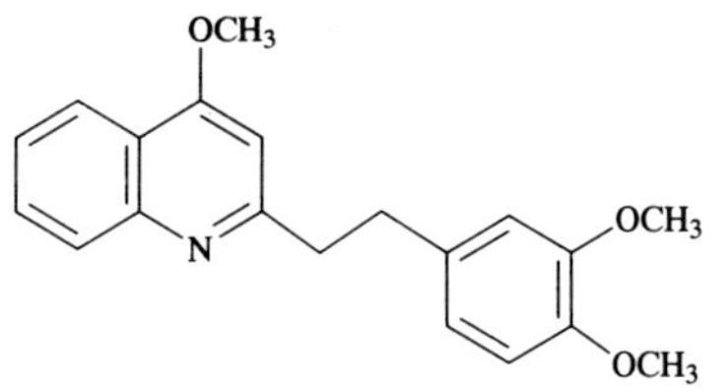

Quinoline alkaloid

$C_{20}H_{21}NO_3$ [525-68-8] MW 323.39

The bark of angostura, *Cusparia febrifuga* (Rutaceae).

Antispasmodic activity.

Späth, E., *Ber. Dtsch. Chem. Ges.*, 1924, **57**, 1687.

Gallic acid 624

3,4,5-Trihydroxybenzoic acid

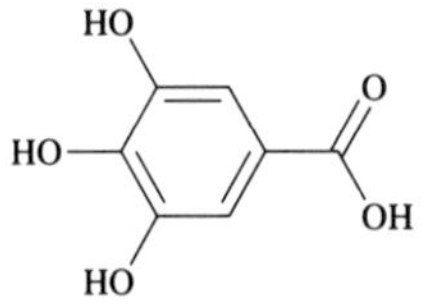

Phenolic acid

$C_7H_6O_5$ [149-91-7] MW 170.12

Relatively widespread in Nature. Present e.g. in *Allanblackia floribunda* (Clusiaceae), *Bridelia micrantha* (Euphorbiaceae), *Dillenia indica* (Dilleniaceae), *Psidium guajava* (Myrtaceae), *Tamarix nilotica* (Tamaricaceae) and *Vitis vinifera* (Vitaceae). This is the parent compound of the gallotannins, such as chebulinic acid (q.v.).

Potentially toxic, if taken in quantity. It shows anti-inflammatory, antitumour, anti-anaphylactic, antimutagenic, choleretic and bronchodilatory activities. Once used in medicine as an astringent and styptic, and in veterinary practice as an intestinal astringent.

Pryce, R. J., *Phytochemistry,* 1972, **11**, 1911.

Gardenoside 625

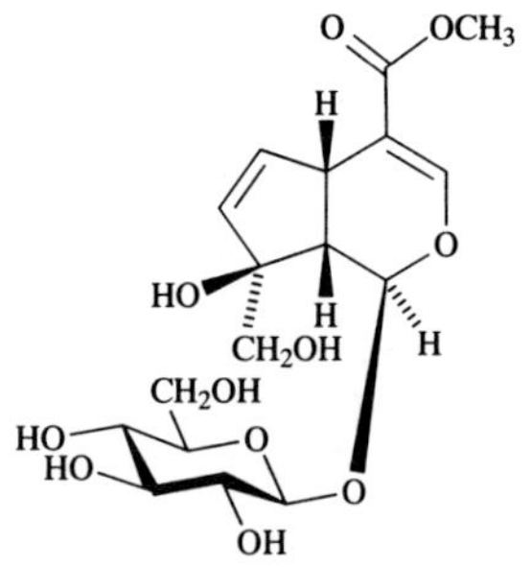

Monoterpenoid
(Iridoid glucoside)

$C_{17}H_{24}O_{11}$ [24512-62-7] MW 404.37

Gardenia jasminoides var. *grandiflora*, *G. jasminoides* var. *radicans* and *Randia canthioides* (Rubiaceae).

Mild laxative.

Bailleul, F., *Phytochemistry,* 1977, **16**, 723.

Garryine 626

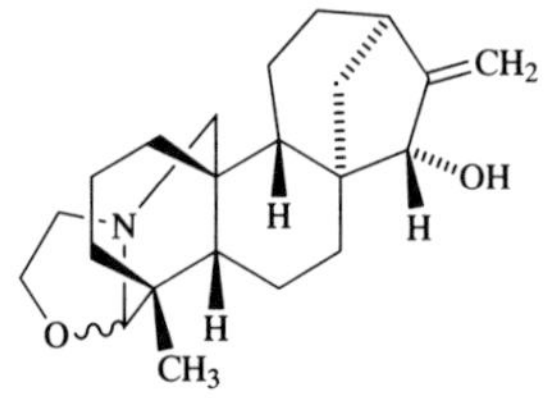

Diterpenoid alkaloid
(Garryine)

$C_{22}H_{33}NO_2$ [561-51-3] MW 343.51

Bark of *Garrya veatchii* (Garryaceae).

The toxicity is relatively high compared with other diterpenoid alkaloids. The poisonous effects in mice include gasping, convulsions and respiratory failure. In an anaesthetised cat, it decreases the blood pressure and the heart rate.

Pelletier, S. W., *Experientia,* 1964, **20**, 1.

Geigerin 627

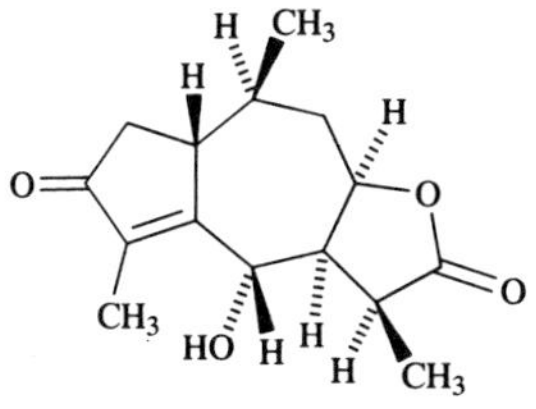

Sesquiterpenoid lactone
(Guaianolide)

$C_{15}H_{20}O_4$ [436-45-3] MW 264.32

Geigera africana and *G. aspera* (Compositae).

Very toxic to mammals. Eating *Geigera* plants has caused mass poisoning among sheep in Southern Africa.

Barton, D. H. R., *Proc. Chem. Soc.,* 1960, 279.

Gelsemicine 628

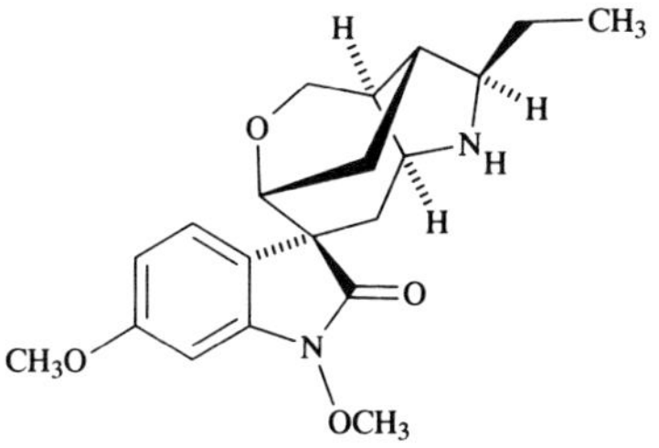

Indole alkaloid
(Gelsedine)

$C_{20}H_{26}N_2O_4$ [6887-28-1] MW 358.44

Roots and rhizomes of the Carolina or yellow jessamine, *Gelsemium sempervirens* and in *Mostuea brunosis* (Loganiaceae).

Highly toxic. Small doses stimulate respiration, while large doses cause respiratory paralysis. It is a central nervous system stimulant. Most toxic of the alkaloids present in *Gelsemium.*

Przybylska, M., *Can. J. Chem.,* 1961, **39**, 2124.

Gelsemine 629

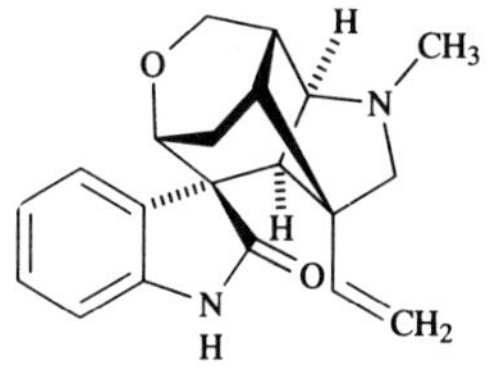

Indole alkaloid
(Gelsemine)

$C_{20}H_{22}N_2O_2$ [509-15-9] MW 322.41

Roots and rhizomes of the Carolina or yellow jessamine, *Gelsemium sempervirens* (Loganiaceae).

Highly toxic. It has caused human poisoning. Like the co-occurring alkaloid gelsemicine (q.v.), it is a central nervous system stimulant.

Teuber, H. J., *Chem. Ber.,* 1960, **93**, 3100.

Gemin A 630

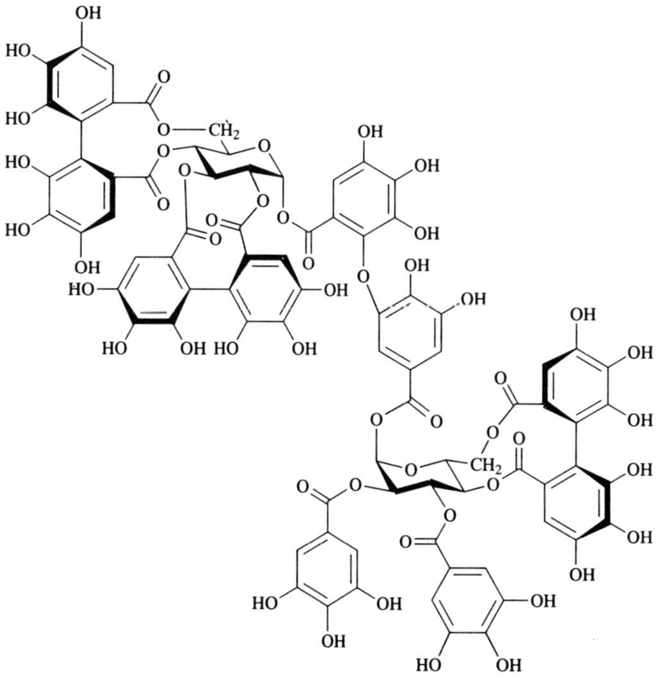

Ellagitannin

$C_{82}H_{56}O_{52}$ [82220-61-9] MW 1873.31

In the leaf of *Geum japonicum* (Rosaceae).

Antitumour properties in mice against Sarcoma 180. It inhibits induced lipolysis in rat liver mitochondria and adrenalin-induced lipolysis in fat cells of rat.

Yoshida, T., *J. Chem. Soc., Perkin Trans. 1,* 1985, 315.

Genipin 631

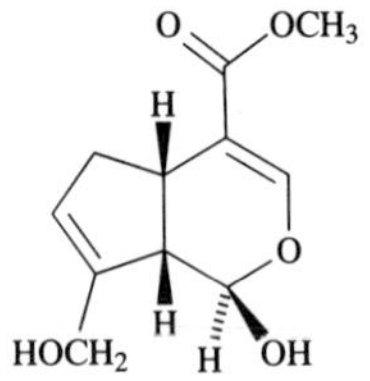

Monoterpenoid
(Iridane)

$C_{11}H_{14}O_5$ [6902-77-8] MW 226.23

Genipa americana (Rubiaceae).

Increases bile flow in mammals.

Büyük, G., *Tetrahedron Lett.,* 1978, 3803.

Geniposide 632

Genipin 1-glucoside

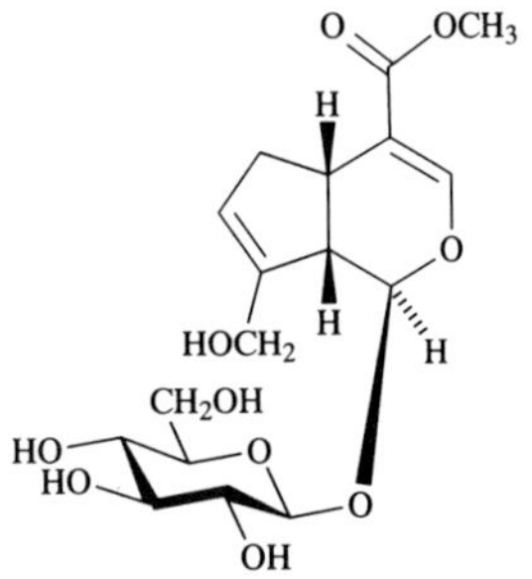

Monoterpenoid
(Iridoid glucoside)

$C_{17}H_{24}O_{10}$ [24512-68-3] MW 388.37

Gardenia spp. (Rubiaceae) and in *Cornus* spp. (Cornaceae).

Laxative properties.

Inouye, H., *Tetrahedron Lett.,* 1969, 2347.

Genistein 633

Prunitol; Sophoricol; Genisteol; Differenol A; 4′,5,7-Trihydroxyisoflavone

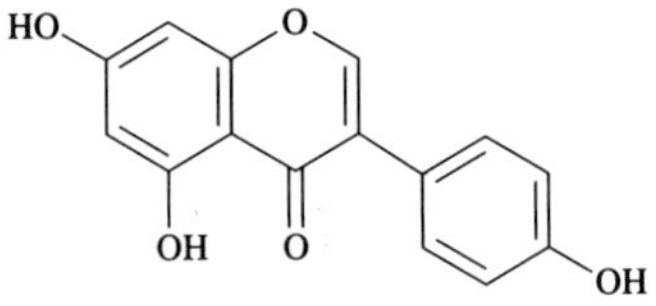

Isoflavone

$C_{15}H_{10}O_5$ [446-72-0] MW 270.24

Relatively widely distributed in plants of the Leguminosae, notably in the seed of soya bean, *Glycine max* and in leaves of many fodder legume genera, *Lupinus* and *Trifolium* spp.. Also present in the wood of cherry trees, *Prunus* spp. (Rosaceae).

Pro-oestrogen, being converted to oestrogenic isoflavan *in vivo.* In ewes feeding on clover, genistein is further metabolised to inactive products and does not cause fertility problems (c.f. formononetin).

Curnow, D. H., *Aust. J. Exp. Biol. Med. Sci.,* 1955, **33**, 243.

Gentianadine 634

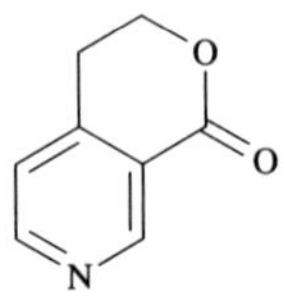

Monoterpenoid alkaloid

$C_8H_7NO_2$ [6790-32-5] MW 149.15

Aerial parts of *Gentiana turkestanorum*, *G. olgae* and *G. olivieri* (Gentianaceae).

Mildly toxic. It shows hypothermic, hypotensive, anti-inflammatory and muscular-relaxant properties.

Samatov, A., *Chem. Nat. Compd.,* 1967, **3**, 150.

Gentianaine 635

Gentiocrucine

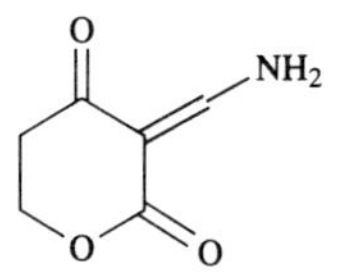

Monoterpenoid alkaloid

$C_6H_7NO_3$ [58213-76-6] MW 141.13

Enicostemma hyssopifolium, *Gentiana olgae*, *G. kaufmanniana*, *G. olivieri*, *G. cruciata* and *G. turkestanorum* (Gentianaceae).

Anti-inflammatory activity.

Ghosal, S., *Tetrahedron Lett.,* 1974, 403.

Gentianamine 636

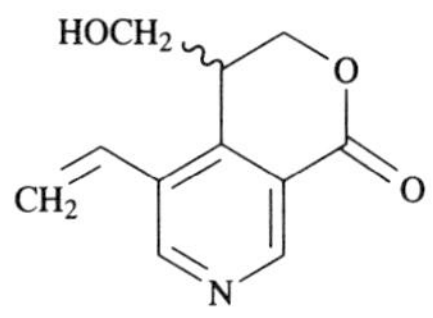

Monoterpenoid alkaloid

$C_{11}H_{11}NO_3$ [22952-54-1] MW 205.21

Gentiana olivieri and *G. turkestanorum* (Gentianaceae).

Anti-inflammatory activity.

Manske, R. H. F., The Alkaloids, 1977, **16**, 431, Academic Press.

Gentianine 637

Erythricine

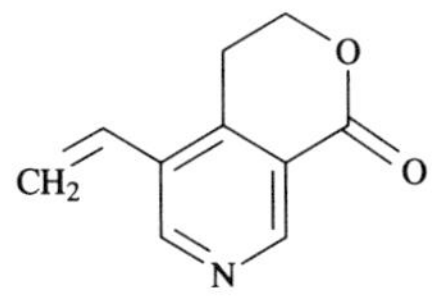

Monoterpenoid alkaloid

$C_{10}H_9NO_2$ [439-89-4] MW 175.19

Gentiana kirilowi, other *Gentiana* spp. and in *Swertia* spp. (Gentianaceae).

Low toxicity. At low doses, it stimulates the central nervous system, but at higher doses it has a paralysing effect. It also exerts hypotensive action and the tonic effects of the gentian liquor are probably due to the presence of this alkaloid.

Govindachari, T. R., *J. Chem. Soc.,* 1957, 2725.

Gentiopicrin 638

Gentiopicroside

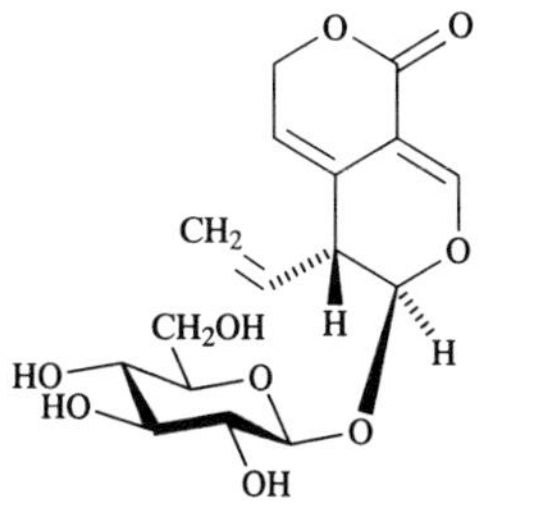

Monoterpenoid
(Seco-iridoid glucoside)

$C_{16}H_{20}O_9$ [20831-76-9] MW 356.33

In roots of the yellow gentian, *Gentiana lutea* and other plants of the Gentianaceae.

Antimalarial activity.

Das, S., *Phytochemistry,* 1984, **23**, 908.

Gentisein 639

1,3,7-Trihydroxyxanthone

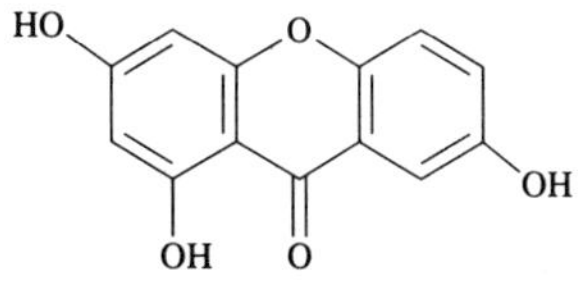

Xanthone

$C_{13}H_8O_5$ [529-49-7] MW 244.20

In *Haploclathra paniculata* and *Hypericum degenii* (Guttiferae) and in *Gentiana lutea* (Gentianaceae).

Tuberculostatic activity.

Atkinson, J. E., *Tetrahedron,* 1969, **25**, 1507.

Geraniin 640

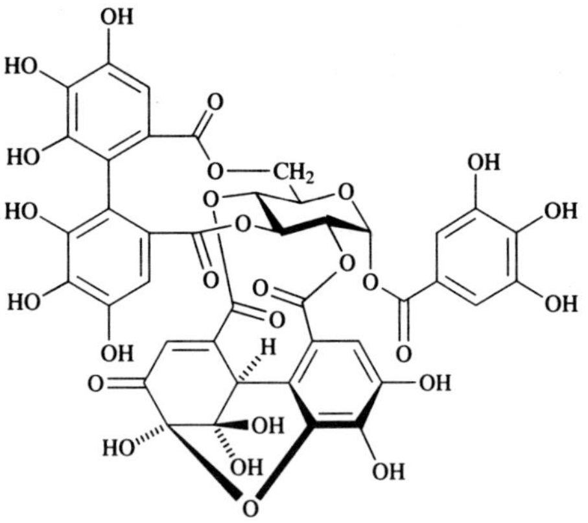

Ellagitannin

$C_{41}H_{28}O_{27}$ [60976-49-0] MW 952.66

Geranium spp. (Geraniaceae), *Erythroxylum coca* (Erythroxylaceae), *Acer* spp. (Aceraceae) and leaves of *Coriaria japonica* (Coriariaceae).

Astringent taste. Inhibits lipid peroxidation in rat liver microsomes and the adrenalin-induced lipolysis in fat cells of rats. On the other hand, it enhances adrenocorticotrophic hormone-induced lipolysis in fat cells.

Okuda, T., *Tetrahedron Lett.*, 1980, **21**, 2561.

Geranylacetone 641

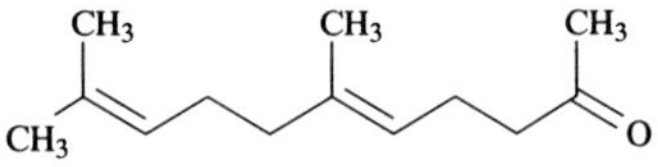

Norsesquiterpenoid

$C_{13}H_{22}O$ [3796-70-1] MW 194.32

In the brown alga, *Cystophora moniliformis*.

Feeding deterrent.

van Altena, I. A., *Aust. J. Chem.*, 1988, **41**, 49.

Geranylhydroquinone 642

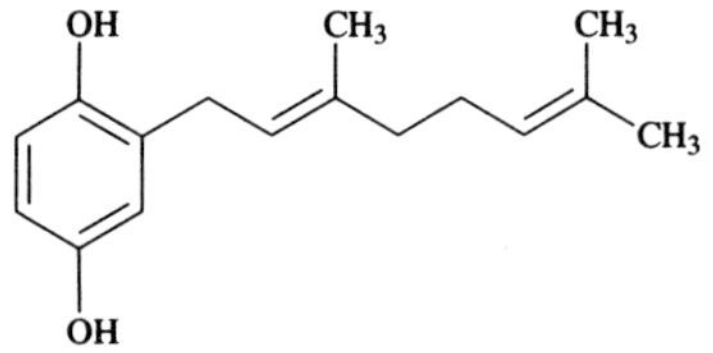

Phenol

$C_{16}H_{20}O_2$ [61977-06-8] MW 244.33

In trichomes of *Phacelia ixodes* (Hydrophyllaceae) leaves, and in the wood of *Cordia alliodora* (Boraginaceae).

Elicits allergic skin reactions and is a potent skin irritant.

Reynolds, G. W., *Phytochemistry*, 1979, **18**, 1567.

Germacrone 643

1(10),4,7(11)-Germacratrien-8-one; Germacrol

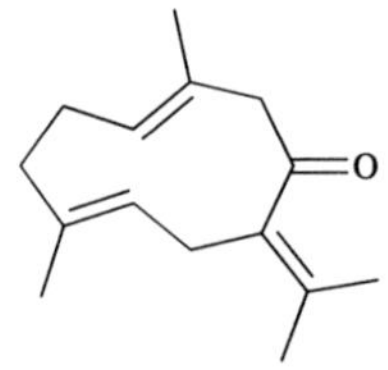

Sesquiterpenoid
(Germacrane)

$C_{15}H_{22}O$ [6902-91-6] MW 218.34

In the leaf oil of Labrador tea, *Ledum groenlandicum*, of *Rhododendron adamsii* (Ericaceae) and of *Geranium macrorrhizum* (Geraniaceae).

Antifeedant to the snowshoe hare, protecting the Labrador tea plant from grazing.

Ognjanov, I., *Collect. Czech. Chem. Commun.*, 1958, **23**, 2033.

Germine 644

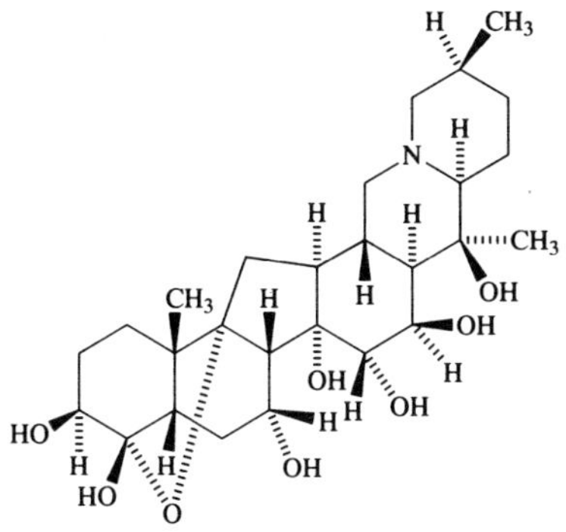

Steroidal alkaloid

$C_{27}H_{43}NO_8$ [508-65-6] MW 509.64

Zigadenus venenosus, *Veratrum* spp. e.g. *V. viride* (Liliaceae).

Toxic. Potent hypotensive agent with characteristic action on the heart, causing irregularity and prolongation of the beat.

Kupchan, S. M., *J. Am. Chem. Soc.,* 1959, **81**, 1913.

Gigantine 645

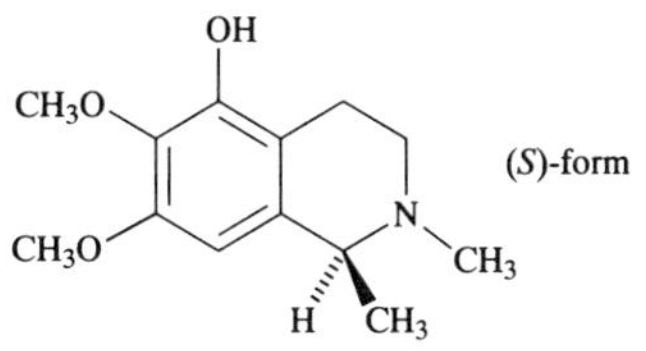

Isoquinoline alkaloid

$C_{13}H_{19}NO_3$ [32829-58-6] MW 237.30

In the giant or saguaro cactus, *Carnegeia gigantea* (Cactaceae).

Toxic to fruit flies. Thought to be hallucinogenic in animals, but unconfirmed in humans.

Kapadin, G. J., *J. Chem. Soc., Chem. Commun.,* 1970, 856.

Gingerenone A 646

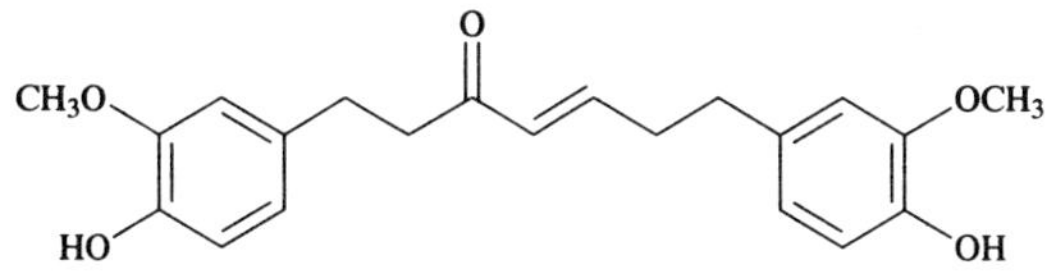

Phenolic ketone

$C_{21}H_{24}O_5$ [128700-97-0] MW 356.42

Rhizomes of ginger, *Zingiber officinale* (Zingiberaceae).

Antifungal activity against the rice pathogen *Pyricularia oryzae*.

Endo, K., *Phytochemistry,* 1990, **29**, 797.

[6]-Gingerol 647

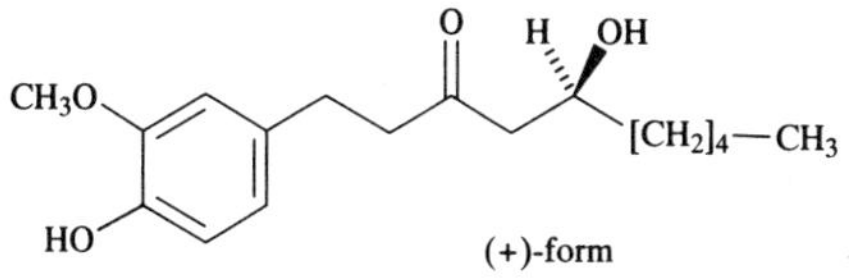

Phenol

$C_{17}H_{26}O_4$ [23513-14-6] MW 294.39

In the rhizomes of ginger, *Zingiber officinale* (Zingiberaceae).

The major pungent principle of ginger. It inhibits cyclooxygenase activity. It has anti-emetic and antiseratogenic properties.

Connell, D. W., *Aust. J. Chem.,* 1969, **22**, 1033.

Ginkgoic acid 648

Ginkgolic acid; Romanicardic acid

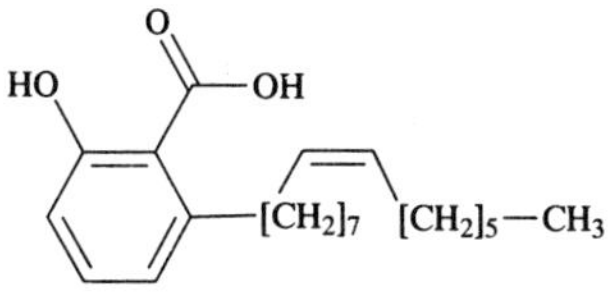

Phenolic acid

$C_{22}H_{34}O_3$ [22910-60-7] MW 346.51

In the fruit of the maidenhair tree, *Ginkgo biloba* (Ginkgoaceae) and in the shell liquid of the cashew nut, *Anacardium occidentale* (Anacardiaceae).

Antitumour and molluscicidal properties.

Tyman, J. H. P., *Chem. Soc. Rev.*, 1979, **8**, 499.

Ginkgolide A 649

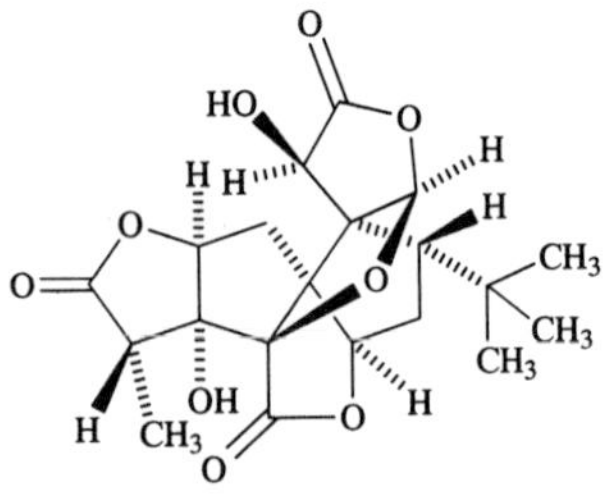

Diterpenoid

$C_{20}H_{24}O_9$ [15291-75-5] MW 408.41

In rootbark and leaves of *Ginkgo biloba*, the maidenhair tree (Ginkgoaceae).

Bitter substance, which is an insect antifeedant. It is one of the active principles of *Ginkgo* leaf preparations, which are used to treat allergic inflammation, asthma and memory-loss.

Sakabe, N., *J. Chem. Soc., Chem. Commun.*, 1967, 259.

Gitorin 650

Gitoxigenin 3-glucoside

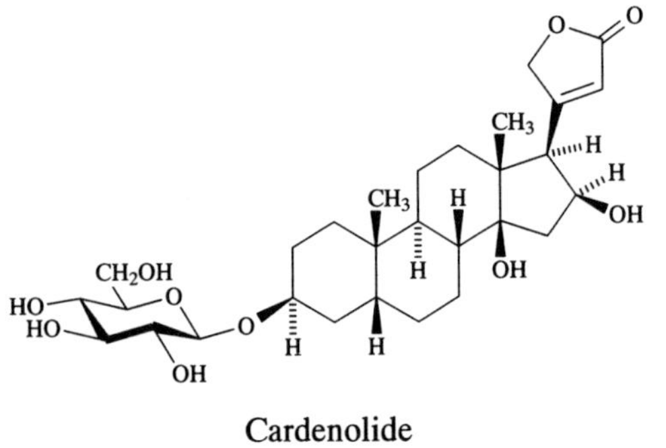

Cardenolide

$C_{29}H_{44}O_{10}$ [32077-87-5] MW 552.66

Digitalis lanata, *D. purpurea*, *D. davisiana* and *D. thapsi* (Scrophulariaceae).

Toxic. LD_{50} is 0.44 mg/kg body-weight in the cat.

Sasakawa, Y., *Chem. Pharm. Bull.*, 1959, **7**, 265.

Gitoxin 651

Anhydrogitalin; Bigitalin; Gitoxoside; Gitoxigenin 3-*O*-tridigitoxoside

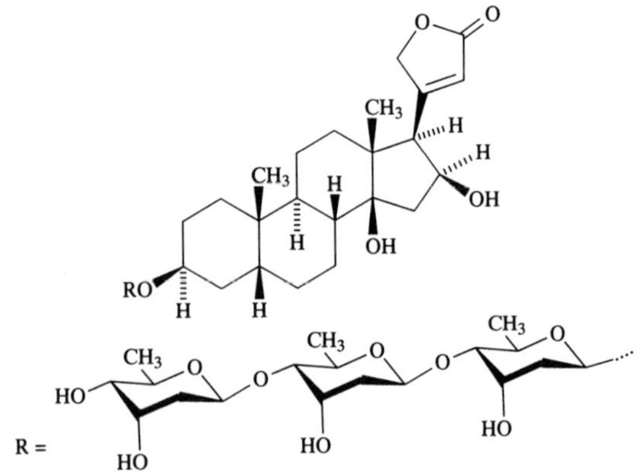

Cardenolide

$C_{41}H_{64}O_{14}$ [4562-36-1] MW 780.95

In foxglove leaf, *Digitalis purpurea* and in *D. lanata* (Scrophulariaceae).

Relatively toxic to vertebrates. It is used medicinally as a cardiotonic.

Faruya, T., *Chem. Pharm. Bull.*, 1970, **18**, 1080.

Glabranin 652

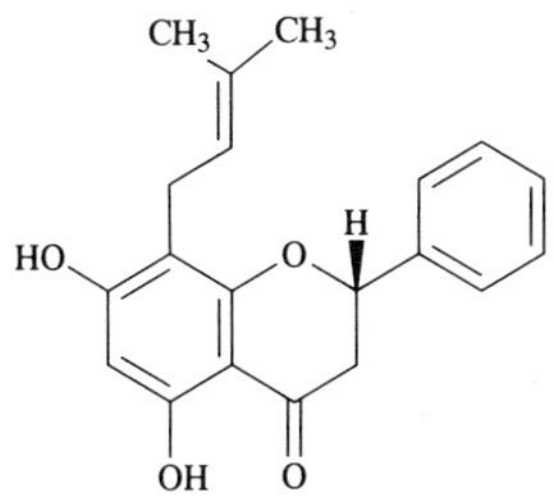

Flavanone

$C_{20}H_{20}O_4$ [41983-91-9] MW 324.38

American liquorice plant, *Glycyrrhiza lepidota* and *Piscidia erythrina* (Leguminosae).

Broad spectrum antimicrobial activities.

Kattaev, N. Sh., *Chem. Nat. Compd.,* 1972, **8**, 790.

Glaucarubinone 653

α-Kirondrin

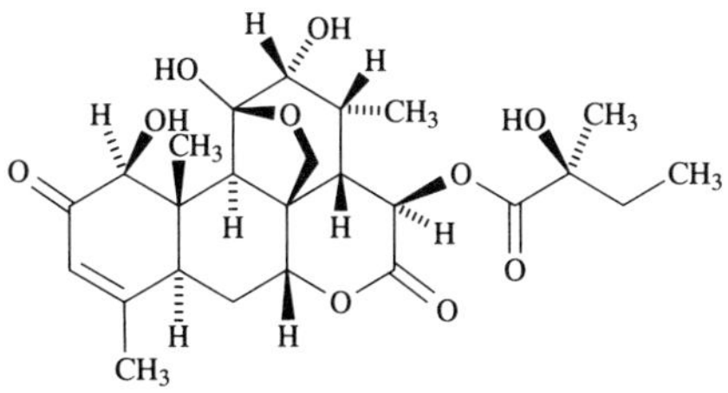

Nortriterpenoid
(Quassinoid)

$C_{25}H_{34}O_{10}$ [1259-86-5] MW 494.54

Seeds of *Quassia undulata* and of *Q. simarouba* and in bitter fruits of *Perriera madagascariensis* (Simaroubaceae).

Bitter taste. Insecticidal. Antimalarial (effective at 0.006 mg/ml) and amoebicidal properties.

Gaudemer, A., *Phytochemistry,* 1965, **4**, 149.

Glaucarubolone 654

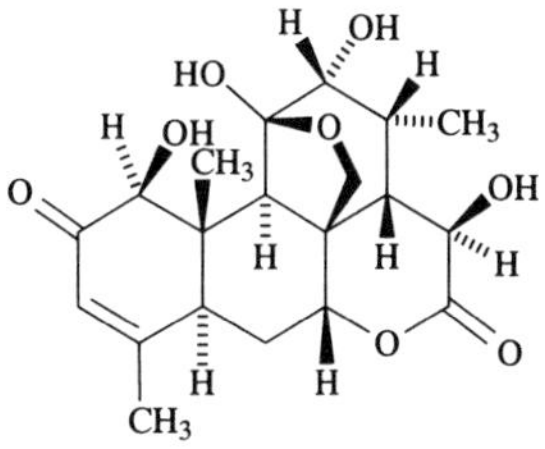

Nortriterpenoid
(Quassinoid)

$C_{20}H_{26}O_8$ [1990-01-8] MW 394.42

Seeds of *Quassia simarouba* and *Q. undulata*, in fruits of *Perriera madagascariensis* and in wood of *Castela nicholsoni* (Simaroubaceae).

Bitter taste. Amoebicidal properties. Antiviral activity *in vitro* against the oncogenic Rous sarcoma virus.

Polonsky, J., *Phytochemistry,* 1965, **4**, 149.

Glaucine 655

Boldine dimethyl ether; Tetramethoxyaporphine

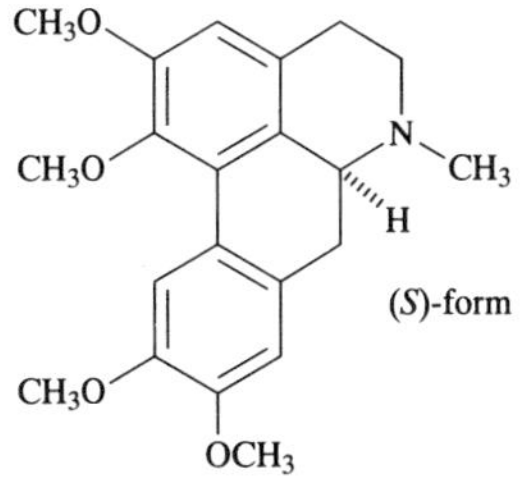

(*S*)-form

Benzylisoquinoline alkaloid
(Aporphine)

$C_{21}H_{25}NO_4$ [475-81-1] MW 355.43

In at least eleven higher plant families, including *Glaucium flavum* (Papaveraceae), *Dicentra eximia* (Fumariaceae) and *Beilschmiedia podagrica* (Lauraceae).

Toxic alkaloid, with an LD_{50} on intravenous administration of 4.8 mg/kg body-weight. It inhibits respiration, lowers blood pressure and blood glucose levels in animals.

Corrodi, H., *Helv. Chim. Acta,* 1956, **39**, 889.

Glaucolide A 656

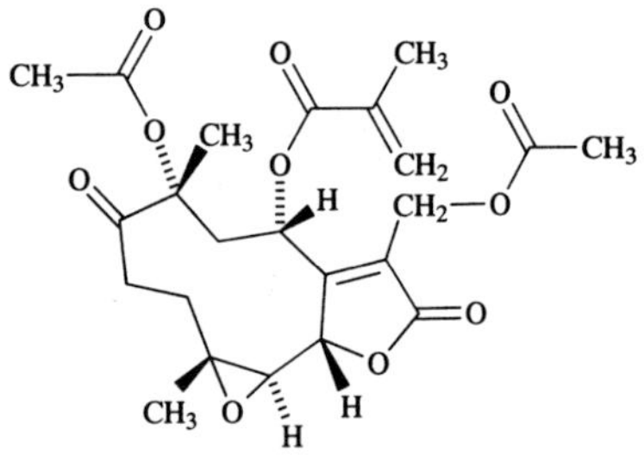

Sesquiterpenoid lactone
(Germacranolide)

$C_{23}H_{28}O_{10}$ [11091-29-5] MW 464.47

In ironweed, *Vernonia* spp., such as *V. glauca* (Compositae).

Feeding deterrent to deer, rabbits and insects. It inhibits the normal development of lepidopteran larvae.

Padolina, W. G., *Tetrahedron,* 1974, **30**, 1161.

Glucocapparin 657

Methylglucosinolate

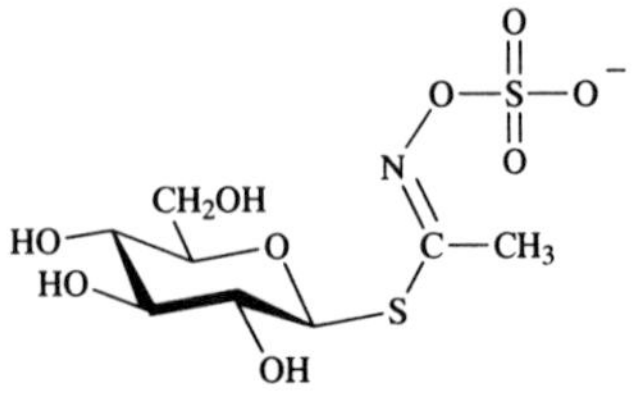

Glucosinolate

$[C_8H_{14}NO_9S_2]^-$ [479-77-8] MW 332.34

Common in the cabbage family (Cruciferae) e.g. in horseradish, *Armoracia lapathifolia* and in cauliflower, *Brassica oleracea* var. *botrytis*. Also present in plants of the caper family (Capparaceae).

The hydrolysis product, methyl isothiocyanate, released when the plant containing it is damaged, is lachrimatory and is partly responsible for the characteristic 'hot' flavour of horseradish.

Kjær, A., *Phytochemistry,* 1981, **20**, 2379.

Glucocheirolin 658

3-(Methylsulfonyl)propylglucosinolate

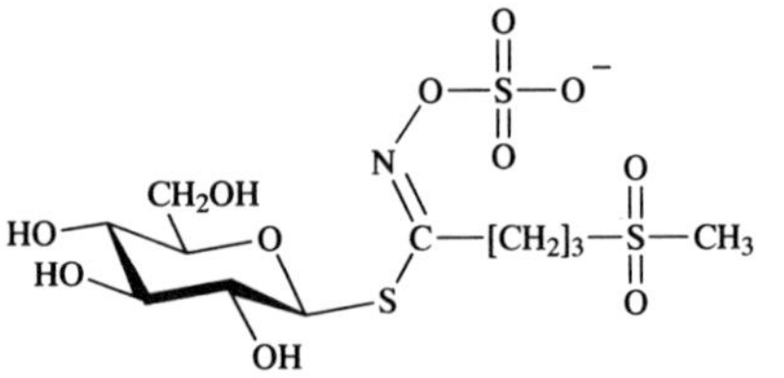

Glucosinolate

$[C_{11}H_{20}NO_{11}S_3]^-$ [15592-36-6] MW 438.48

In horseradish, *Armoracia lapathifolia*, cauliflower, *Brassica oleracea* var. *botrytis*, swede, *B. napus* var. *napobrassica* and turnip, *B. campestris* var. *rapifera* (Cruciferae).

Bound toxin. One of its hydrolysis products, 3-(methylsulfonyl)propylisothiocyanate, is goitrogenic and cytotoxic in animals.

Kjær, A., *Acta Chem. Scand.,* 1959, **13**, 851.

Glucoerysolin 659

4-(Methylsulfonyl)butylglucosinolate

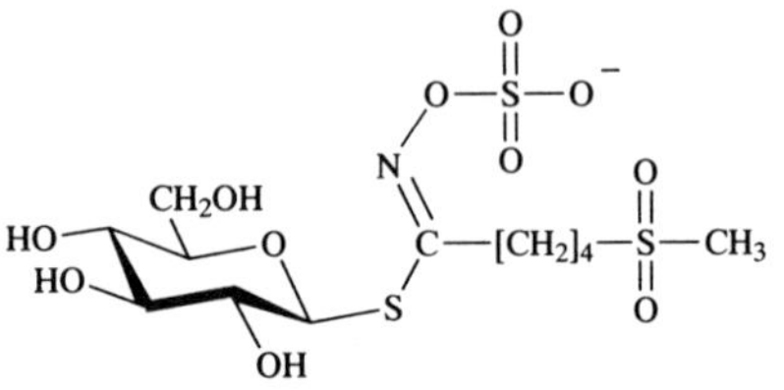

Glucosinolate

$[C_{12}H_{22}NO_{11}S_3]^-$ [22149-26-4] MW 452.50

In swede, *Brassica napus* var. *napobrassica* and in turnip, *B. campestris* var. *rapifera* (Cruciferae).

Bound toxin. The hydrolysis product, erysoline, is cytotoxic in animals.

Gmelin, R., *Acta Chem. Scand.,* 1968, **22**, 2875.

Glucofrangulin A 660

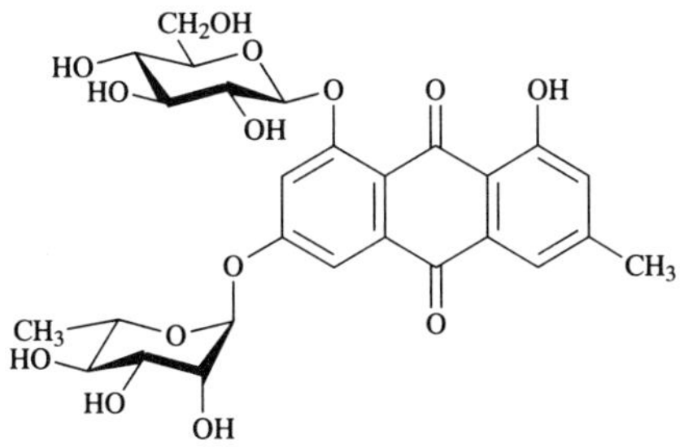

Anthraquinone

$C_{27}H_{30}O_{14}$ [21133-53-9] MW 578.53

Bark and seed of alder buckthorn, *Rhamnus frangula* (Rhamnaceae).

Active principle of the laxative activity of bark or seed extracts. Dose should be carefully controlled, since animal experiments indicate that drastic purging can have fatal consequences. Hence children eating buckthorn berries are at risk.

Thomson, R. H., Naturally Occurring Quinones, 2nd edn, 1971, 421, Academic Press.

Glucolepidiin 661

Ethylglucosinolate

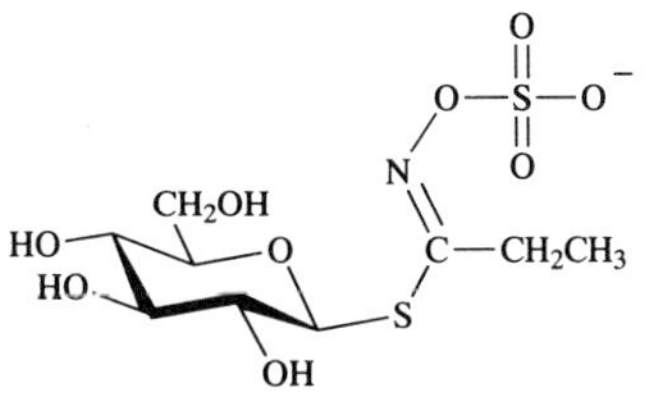

Glucosinolate

$[C_9H_{16}NO_9S_2]^-$ [101144-39-2] MW 346.36

In garden cress, *Lepidium sativum* and horseradish, *Armoracia lapathifolia* (Cruciferae).

Bound toxin. The hydrolysis product, ethyl isothiocyanate, is extremely pungent and garlic-like.

Kjær, A., *Acta Chem. Scand.*, 1954, **8**, 699.

Gluconasturtiin 662

2-Phenylethylglucosinolate

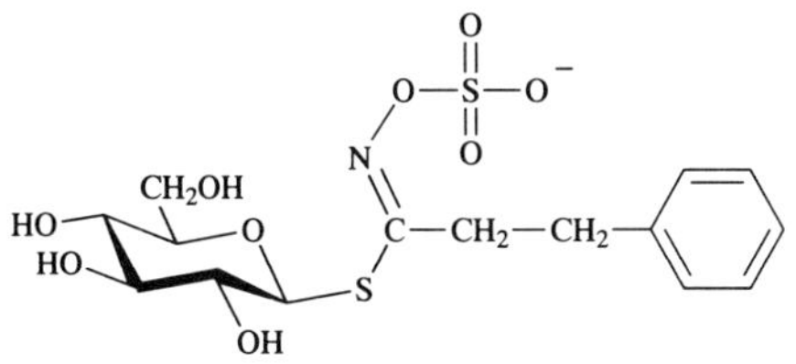

Glucosinolate

$[C_{15}H_{20}NO_9S_2]^-$ [499-30-9] MW 422.46

Widespread in the Cruciferae, including watercress, *Nasturtium officinale*, garden cress, *Lepidium sativum* and black mustard, *Brassica nigra*.

Bound toxin. The hydrolysis product, 2-phenylethylisothiocyanate, produces a tingling sensation on the tongue and is cytotoxic in animals.

Dörnemann, D., *Can. J. Biochem.*, 1974, **52**, 916.

p-Glucosyloxymandelonitrile 663

p-Glucosyloxybenzaldehyde cyanohydrin

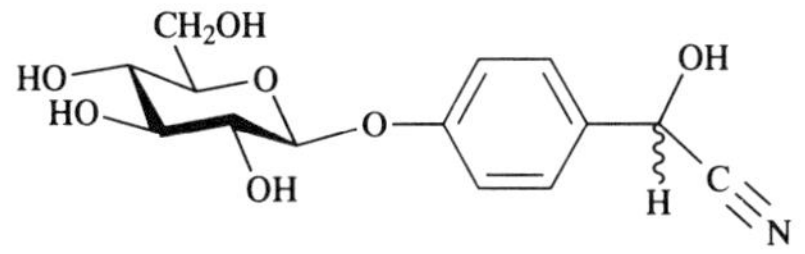

Cyanogenic glycoside

$C_{14}H_{17}NO_7$ [22660-95-3] MW 311.29

Goodia latifolia (Leguminosae) and in *Nandina domestica* (Berberidaceae), as such and as the 4′-caffeic acid ester, nandinin [91919-94-7].

Bound toxin. It can release hydrogen cyanide without the intervention of a β-glucosidase.

Abrol, Y. P., *Phytochemistry,* 1966, **5**, 1021.

Glucosyl taraxinate 664

Taraxinic acid glucosyl ester

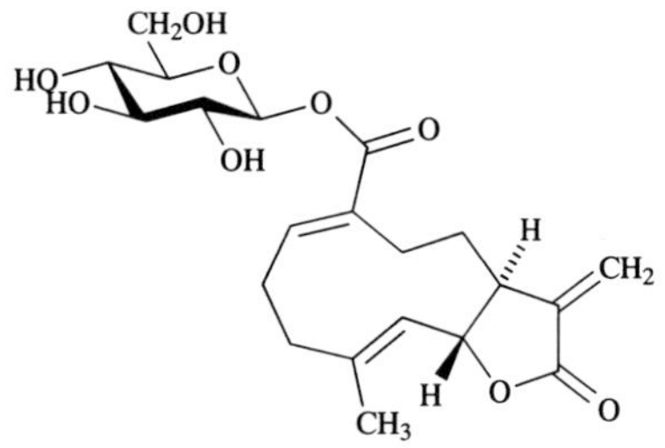

Sesquiterpenoid lactone
(Germacranolide)

$C_{21}H_{28}O_9$ [75911-14-7] MW 424.45

Leaf and root of dandelion, *Taraxacum officinale* (Compositae).

The contact sensitising agent of dandelion and one of the major bitter principles of the leaf.

Hansel, R., *Phytochemistry,* 1980, **19**, 857.

Glucotropaeolin 665

Benzylglucosinolate; Phenylmethylglucosinolate

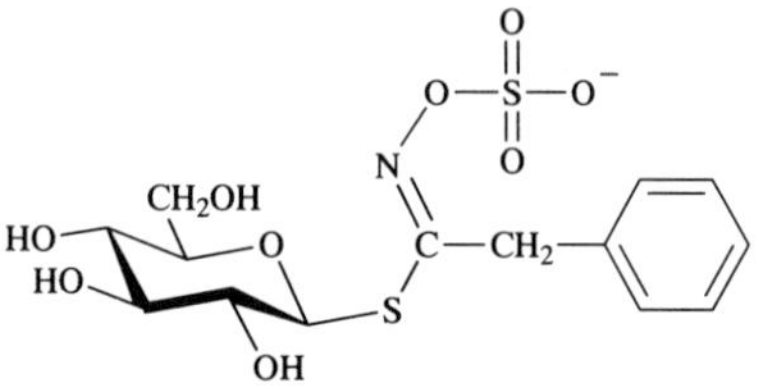

Glucosinolate

$[C_{14}H_{18}NO_9S_2]^-$ [499-26-3] MW 408.43

In nasturtium, *Tropaeolum majus* (Tropaeolaceae), horseradish, *Armoracia lapathifolia* and other Cruciferae; in dry latex of papaya, *Carica papaya* (Caricaceae).

Bound toxin. Benzyl isothiocyanate [622-78-6] and benzyl thiocyanate [3012-37-1] are hydrolysis products; the latter is probably goitrogenic if large amounts are ingested.

Ettlinger, M. G., *J. Am. Chem. Soc.,* 1957, **79**, 1764.

L-γ-Glutamyl-L-hypoglycin 666

Hypoglycin B

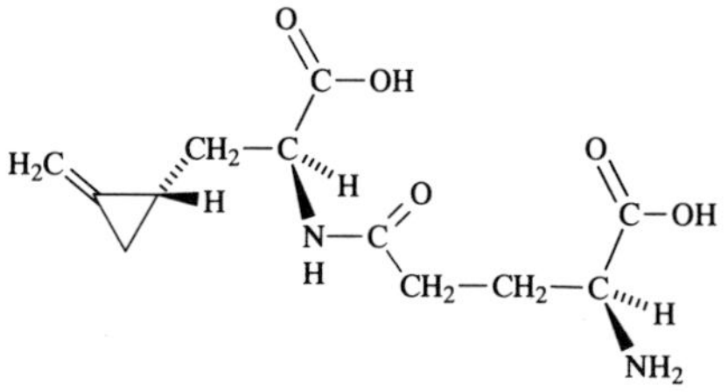

Amino acid

$C_{12}H_{18}N_2O_5$ [502-37-4] MW 270.29

In the arillus and seed of the ackee, *Blighia sapida* (Sapindaceae).

Very toxic. It possesses similar properties, including causing hypoglycaemia, to those of hypoglycin (q.v.). Teratogenic.

Hassall, C. H., *J. Chem. Soc.,* 1960, 4112.

Glutaric acid 667

Pentanedioic acid; 1,3-Propanedicarboxylic acid

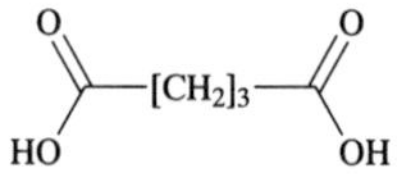

Organic acid

$C_5H_8O_4$ [110-94-1] MW 132.12

In green sugar beet, *Beta vulgaris* (Chenopodiaceae), and fruit of *Prunus cerasus* (Rosaceae).

Moderately toxic.

Polyakov, V. V., *Chem. Nat. Compd.,* 1985, **21**, 795.

Gluten 668

Protein mixture

In wheat flour, *Triticum aestivum*, and to a lesser extent, in barley, maize, oats, rye and other cereal seeds (Gramineae).

Highly toxic to individuals suffering from coeliac disease. Gliadin fractions α, β and γ are responsible.

Pomeranz, Y., Wheat, Chemistry and Technology, 2nd edn, 1971, 227, American Association of Cereal Chemists.

Glutinosone 669

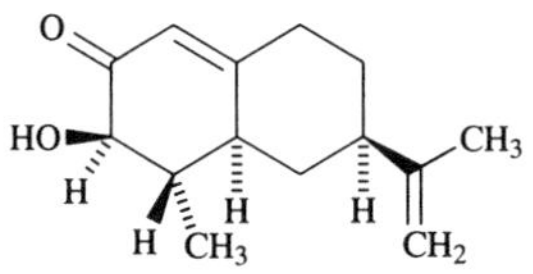

Sesquiterpenoid
(Noreudesmane)

$C_{14}H_{20}O_2$ [55051-94-0] MW 220.31

Leaves of *Nicotiana glutinosa* (Solanaceae) infected with tobacco mosaic virus.

Antifungal agent (phytoalexin).

Burden, R. S., *Phytochemistry,* 1975, **14**, 221.

(−)-Glyceollin I 670

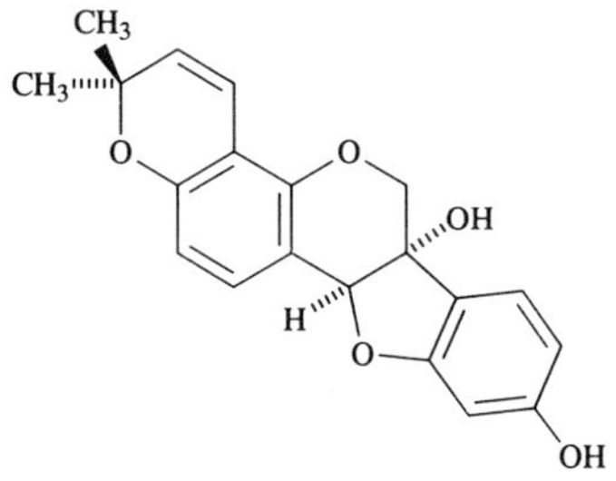

Isoflavonoid
(Pterocarpan)

$C_{20}H_{18}O_5$ [57103-57-8] MW 338.36

Fungally-infected leaves of soya bean, *Glycine max* (Leguminosae).

Antifungal agent (phytoalexin). Toxic to the root-knot nematode *Meloidogyne incognita*. Produces lysis of human blood cells.

Burden, R. S., *Phytochemistry,* 1975, **14**, 1389.

(−)-Glyceollin II 671

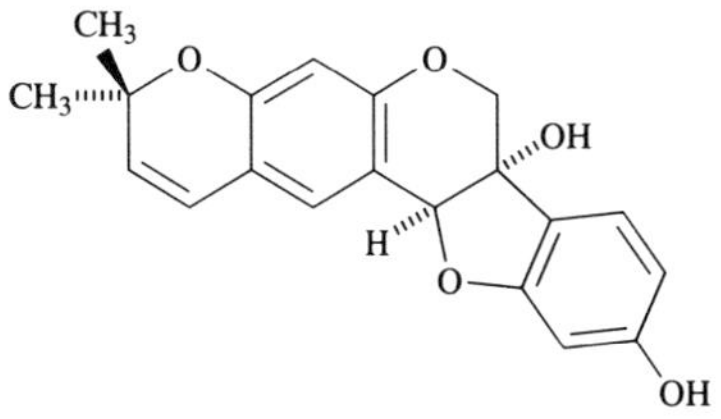

Isoflavonoid
(Pterocarpan)

$C_{20}H_{18}O_5$ [67314-98-1] MW 338.36

Fungally-infected leaves of soya bean, *Glycine max* (Leguminosae).

Antifungal agent (phytoalexin).

Ingham, J. L., *Phytochemistry,* 1981, **20**, 795.

Glyoxylic acid 672

Aldehydoformic acid; Formylformic acid; Glyoxalic acid; Oxoacetic acid; Oxoethanoic acid

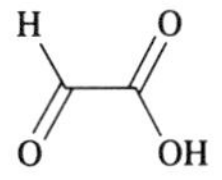

Organic acid

$C_2H_2O_3$ [298-12-4] MW 74.04

In unripe fruit and young green leaves and can be isolated from very young sugar beets, *Beta vulgaris* (Chenopodiaceae); also in potato, *Solanum tuberosum* (Solanaceae).

Irritant and corrosive to the skin.

Towers, G. H. N., *J. Am. Chem. Soc.,* 1954, **76**, 2392.

Gnidicin 673

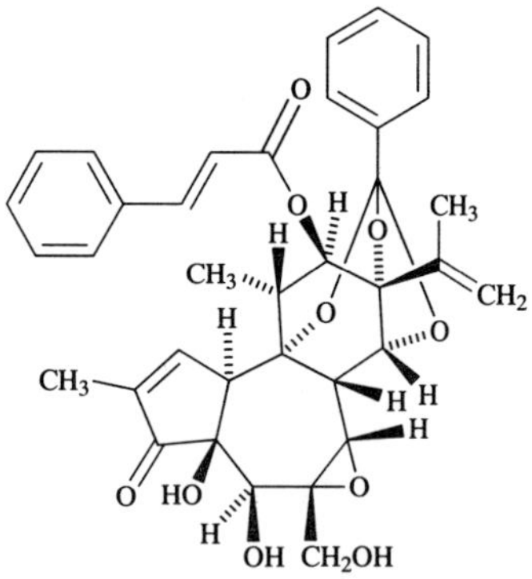

Diterpenoid
(Daphnetoxane)

$C_{36}H_{36}O_{10}$ [55319-39-6] MW 628.68

Gnidia lamprantha (Thymelaeaceae).

Antitumour properties.

Kupchan, S. M., *J. Am. Chem. Soc.,* 1975, **97**, 672.

Gnididin 674

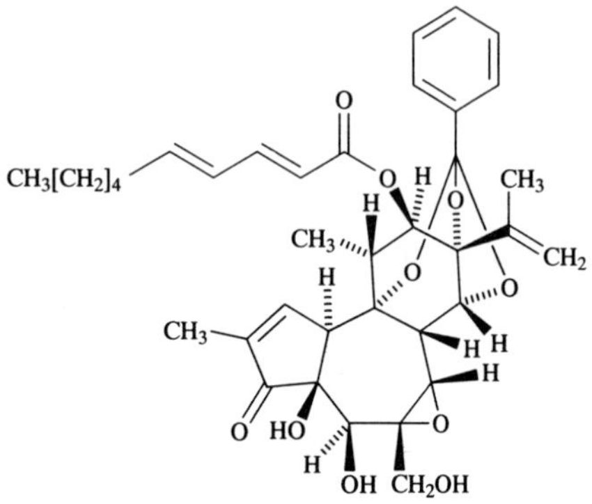

Diterpenoid
(Daphnetoxane)

$C_{37}H_{44}O_{10}$ [55306-11-1] MW 648.75

Gnidia lamprantha (Thymelaeaceae).

Antitumour properties.

Kupchan, S. M., *J. Am. Chem. Soc.,* 1975, **97**, 672.

Gnidilatin 675

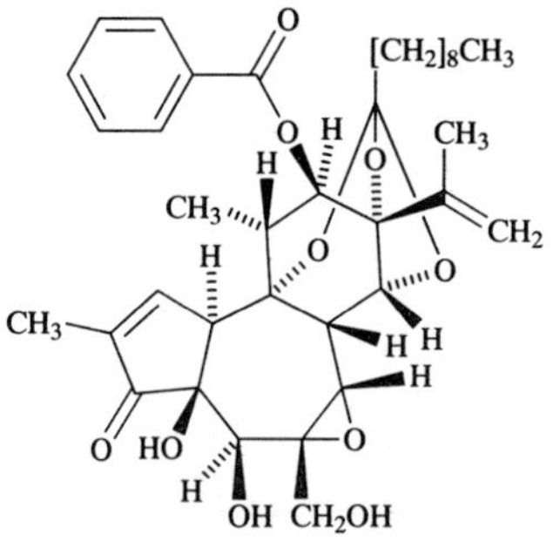

Diterpenoid
(Daphnetoxane)

$C_{37}H_{48}O_{10}$ [60195-69-9] MW 652.78

Gnidia spp. (Thymelaeaceae).

Antileukaemic activity.

Kupchan, S. M., *J. Org. Chem.,* 1976, **41**, 3850.

Gniditrin 676

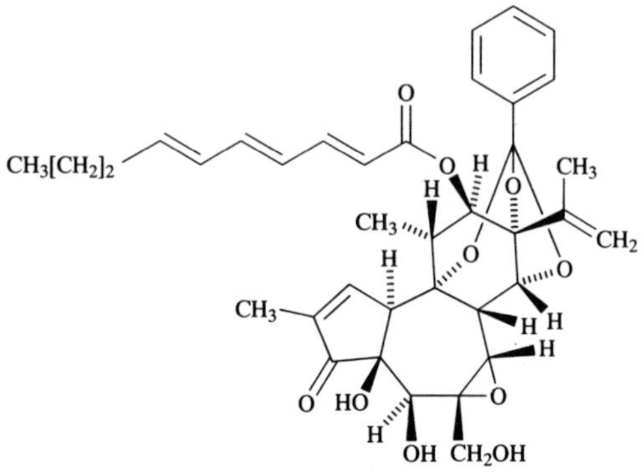

Diterpenoid
(Daphnetoxane)

$C_{37}H_{42}O_{10}$ [55306-10-0] MW 646.73

Gnidia lamprantha (Thymelaeaceae).

Antileukaemic properties.

Kupchan, S. M., *J. Am. Chem. Soc.,* 1975, **97**, 672.

Gofruside 677

Corotoxigenin 3-allomethyloside

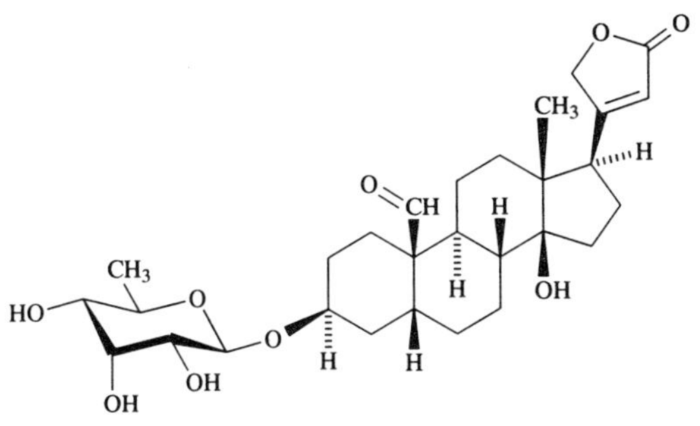

Cardenolide

$C_{29}H_{42}O_9$ [26931-65-7] MW 534.65

Seeds of *Gomphocarpus fruticosus* (Asclepiadaceae).

Toxic to mammals. LD_{50} 0.19 mg/kg body-weight in the cat.

Hunger, A., *Helv. Chim. Acta.,* 1952, **35**, 1073.

Goniodomin A 678

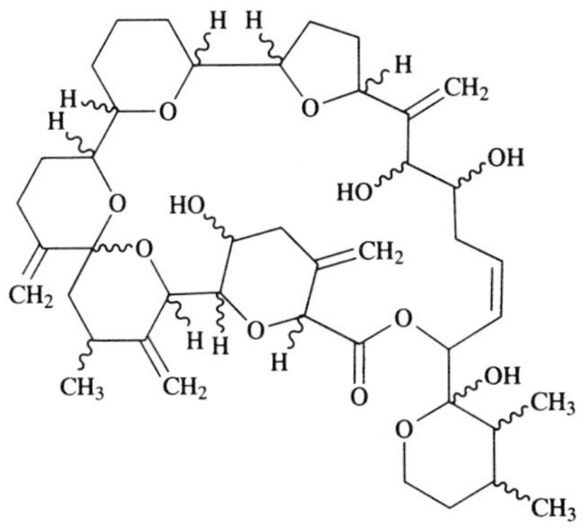

Polyether

$C_{43}H_{60}O_{12}$ [112923-40-7] MW 768.94

Dinoflagellate *Goniodoma pseudogoniaular.*

Inhibits the growth of sea urchin eggs at a concentration of 0.05 μg/ml. Also possesses antifungal activity.

Murakami, M., *Tetrahedron Lett.,* 1988, **29**, 1149.

Goniothalenol 679

Altholactone

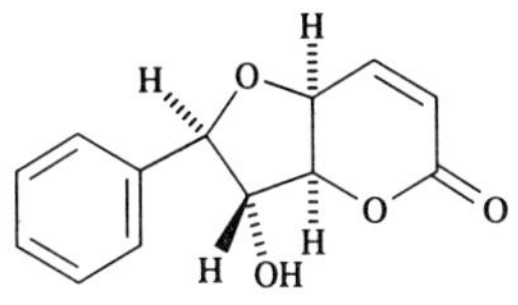

Lactone

$C_{13}H_{12}O_4$ [65408-91-5] MW 232.24

In stembark of *Goniothalamus giganteus* and in a *Polyalthia* sp. (Annonaceae).

Toxic to brine shrimp. Cytotoxic in the P-388 test.

Loder, J. W., *Heterocycles,* 1977, **7**, 113.

Gonyautoxin I 680

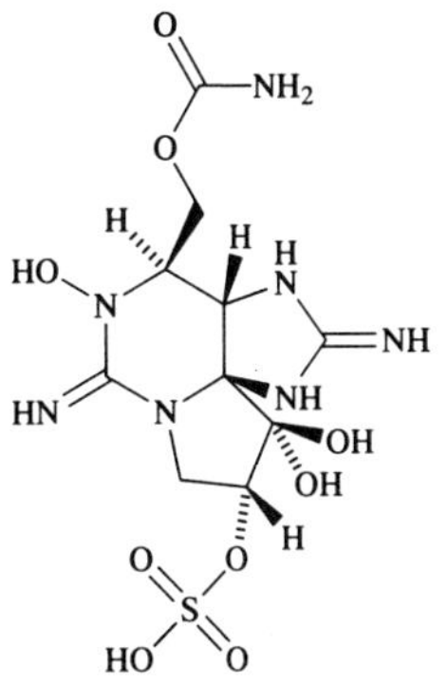

Purine alkaloid

$C_{10}H_{17}N_7O_9S$ [60748-39-2] MW 411.35

One of six very similar toxins produced by algal plankton, *Gonyaulax* spp. and *Protogonyaulax* spp., which are dietarily sequestered by shellfish.

Neurotoxin, causal agent with saxitoxin (q.v.) of shellfish poisoning. Less toxic than saxitoxin because of the presence of the sulfate group.

Hall, S., *Tetrahedron Lett.,* 1984, **25**, 3537.

Gossypol 681

Thespesin

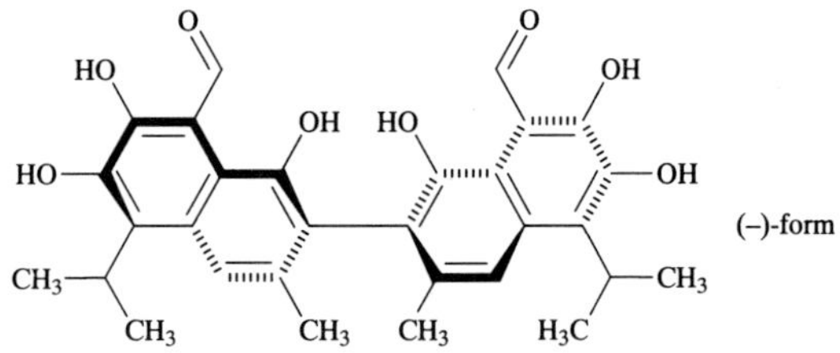

Bis-sesquiterpenoid
(Cadinane)

$C_{30}H_{30}O_8$ [303-45-7] MW 518.56
(+)-form [20300-26-9]
(−)-form [90141-22-3]
(±)-form [40112-23-0]

The racemic form occurs in the seeds of cotton, *Gossypium* spp. and in *Montezuma speciosissima* (Malvaceae). The (+)-isomer occurs in good yield in *Thespesia populnea* (Malvaceae).

Toxic to nonruminant mammals, to birds, insect larvae, nematodes and other animals. Livestock fed cotton seed meal may suffer poisoning if the gossypol content is high. It causes loss of body weight, diarrhoea, cardiac irregularity, haemorrhage and oedema. The (±)- and (−)-forms act as a male contraceptive in man, since they block sperm formation.

King, T. J., *Tetrahedron Lett.,* 1968, 261.

Goyazensolide 682

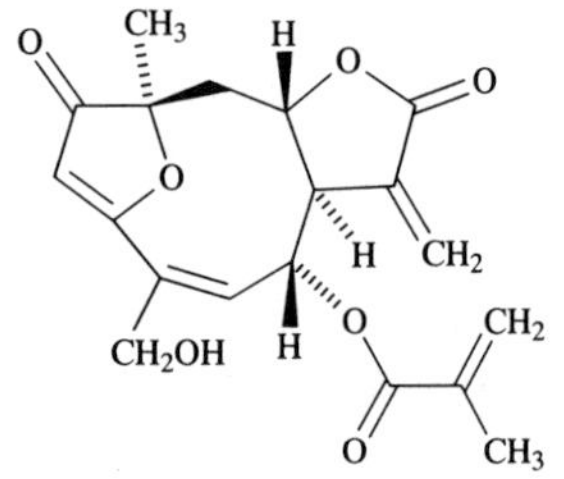

Sesquiterpenoid lactone
(Germacranolide)

$C_{19}H_{20}O_7$ [60066-35-5] MW 360.36

Eremanthus goyazensis, *Lychnophora passerina*, *Vanillosmopsis brasiliensis* and *V. pohlii* (Compositae).

Schistosomicide.

Vichnewski, W., *Phytochemistry,* 1976, **15**, 191.

Gracillin 683

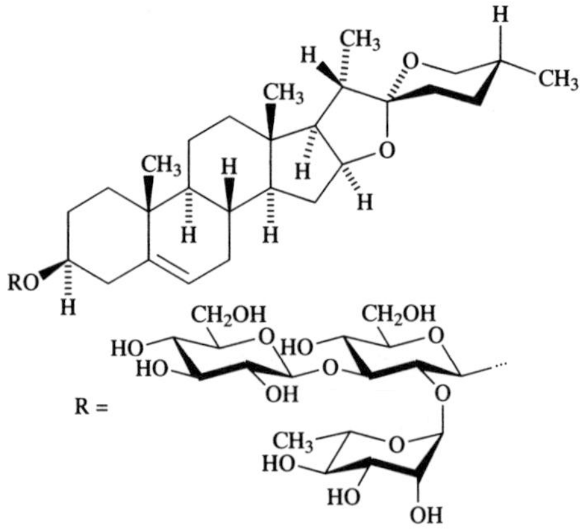

Steroid saponin
(Spirostan)

$C_{45}H_{72}O_{17}$ [19083-00-2] MW 885.06

Costus speciosus (Zingiberaceae).

Strong haemolytic properties.

Tsukamoto, T., *Chem. Pharm. Bull.,* 1956, **4**, 104.

Gradolide 684

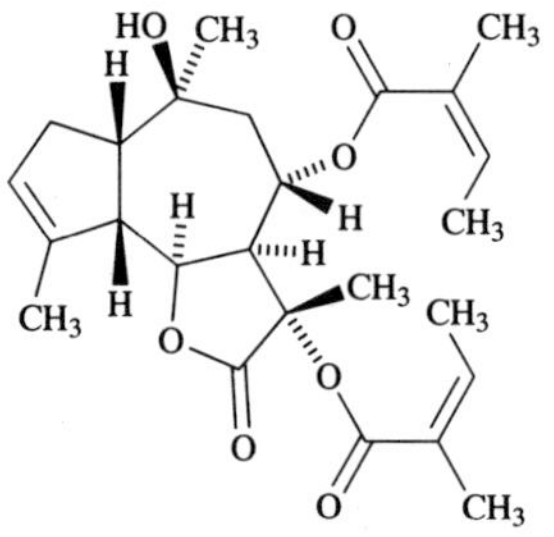

Sesquiterpenoid lactone
(Guaianolide)

$C_{25}H_{34}O_7$ [68852-48-2] MW 446.54

In fruit of *Laserpitium siler* (Umbelliferae).

Insect antifeedant.

Holub, M., *Collect. Czech. Chem. Commun.*, 1978, **43**, 2471.

Gramine 685

3-(Dimethylaminomethyl)indole; Donaxine; Doranine

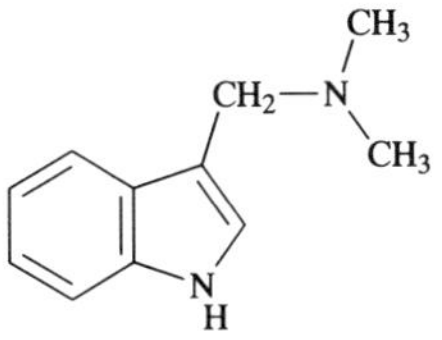

Aromatic amine

$C_{11}H_{14}N_2$ [87-52-5] MW 174.25

In leaves of some varieties of barley, *Hordeum vulgare*, of the reed, *Arundo donax*, of reed canary grass, *Phalaris arundinacea* and other grasses (Gramineae); also in the silver maple, *Acer saccharinum* and *A. rubrum* (Aceraceae) and in some *Lupinus* spp. (Leguminosae).

Partly responsible for the toxic condition in sheep, known as "Phalaris staggers", caused by grazing on *Phalaris* pastures. Also an insect feeding inhibitor.

Leete, E., *Phytochemistry,* 1975, **14**, 471.

Graminiliatrin 686

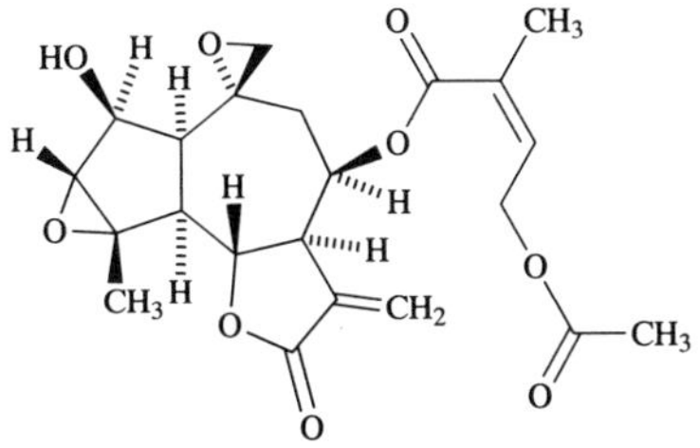

Sesquiterpenoid lactone
(Guaianolide)

$C_{22}H_{26}O_9$ [53142-34-0] MW 434.44

In the blazing star, *Liatris graminifolia* (Compositae).

Cytotoxic and antitumour properties.

Herz, W., *J. Org. Chem.,* 1975, **40**, 199.

Grayanin 687

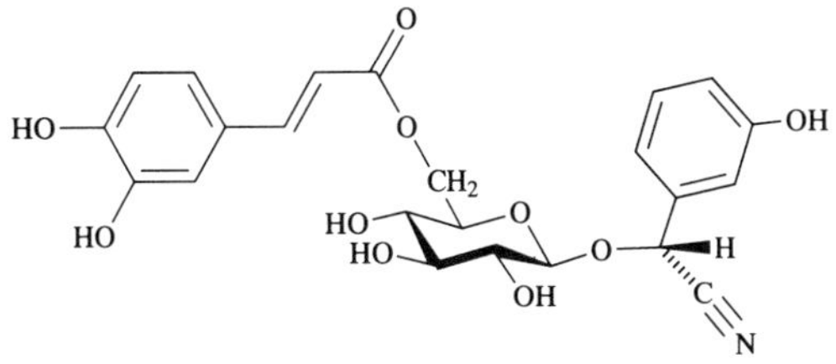

Cyanogenic glycoside

$C_{23}H_{23}NO_9$ [110978-97-7] MW 457.44

Bark of *Prunus grayana* (Rosaceae).

Bound toxin, releasing poisonous cyanide on hydrolysis.

Shimomura, H., *Phytochemistry,* 1987, **26**, 2363.

Grayanotoxin I 688

Acetylandromedol; Andromedotoxin; Asebotoxin; Rhodotoxin

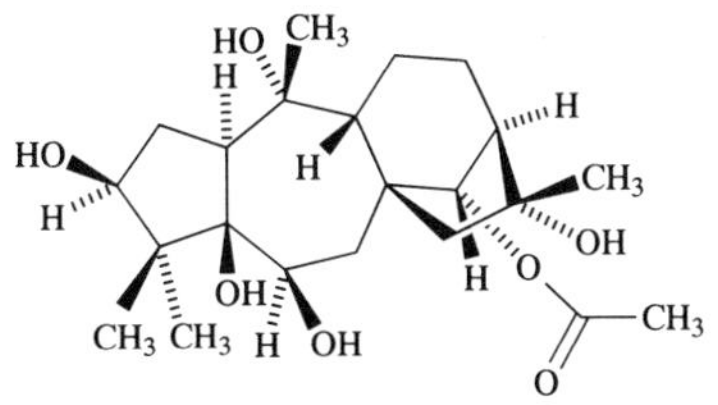

Diterpenoid
(Grayanotoxin)

$C_{22}H_{36}O_7$ [4720-09-6] MW 412.52

Some thirty grayanotoxins and related diterpenes have been found in the leaves of *Kalmia, Leucothoe* and *Rhododendron* spp., e.g. *R. ponticum*, and other Ericaceae.

Toxic. The LD_{50} in mice intraperitoneally is 1.31 mg/kg body-weight. Sheep and goats eating leaves of *Rhododendron ponticum* have suffered poisoning from grayanotoxin. The toxin is also found in the honey of bees which collect nectar from *Rhododendron* species.

Kakisawa, H., *Tetrahedron,* 1965, **21**, 3091.

Grevillol 689

5-Tridecyl-1,3-benzenediol; 5-Tridecylresorcinol; Trifurcatol A_1

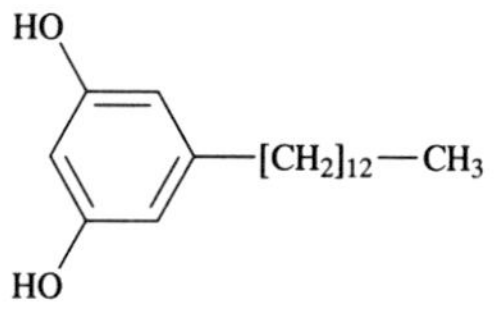

Alkylphenol

$C_{19}H_{32}O_2$ [5259-01-8] MW 292.46

Grevillea robusta and other Proteaceae.

Strong skin irritant, causing contact allergy. It inhibits 5-lipoxygenase, an enzyme of prostaglandin synthesis.

Ritchie, E., *Aust. J. Chem.,* 1965, **18**, 2015.

Grosshemin 690

Grossheimin

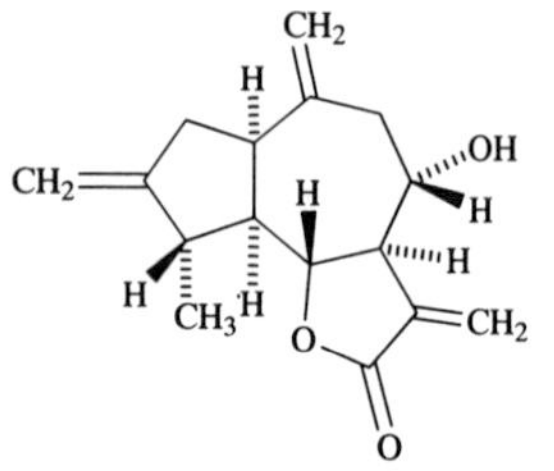

Sesquiterpenoid lactone (Guaianolide)

$C_{15}H_{18}O_4$ [22489-66-3] MW 262.31

Grossheimia macrocephala, *Amberboa lipii*, the artichoke, *Cynara scolymus*, *Chartolepsis intermedia* and *Venidium decurens* (Compositae).

Cytotoxic and antitumour properties. It is also an insect antifeedant.

Samek, Z., *Collect. Czech. Chem. Commun.,* 1972, **37**, 2611.

Guaiacol 691

2-Methoxyphenol; Methylcatechol

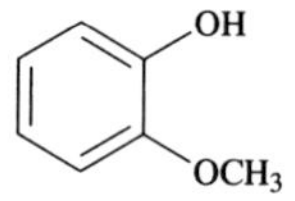

Phenol

$C_7H_8O_2$ [90-05-1] MW 124.14

In beechwood tar from *Betula* spp. (Betulaceae), in guaiac resin from *Guaiacum* spp. (Zygophyllaceae) and in various plant oils and saps, e.g. of celery seed, *Apium graveolens* (Umbelliferae) and of *Ruta montana* (Rutaceae).

Caustic. Earlier used medicinally to treat eczema and similar skin diseases. An expectorant in veterinary practice, mostly in the form of an ester.

Martindale, The Extra Pharmacopoeia, 30th edn, 1993, 881, The Pharmaceutical Press.

Guaiazulene 692

S-Guaiazulene; Azulon; Eucazulen; Kessazulen; Vaumigan

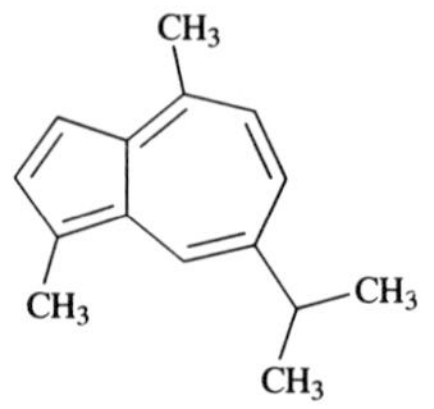

Sesquiterpenoid (Azulene)

$C_{15}H_{18}$ [489-84-9] MW 198.31

Blue oil produced during the steam distillation of chamomile, *Matricaria chamomilla* (Compositae) and guaiac wood, *Guaiacum officinale* or *Guaiacum sanctum* (Zygophyllaceae), from related substances in these plants.

Anti-inflammatory activity used to treat gastrointestinal disorders.

Šorm, F., *Collect. Czech. Chem. Commun.,* 1951, **16**, 168.

Gynocardin 693

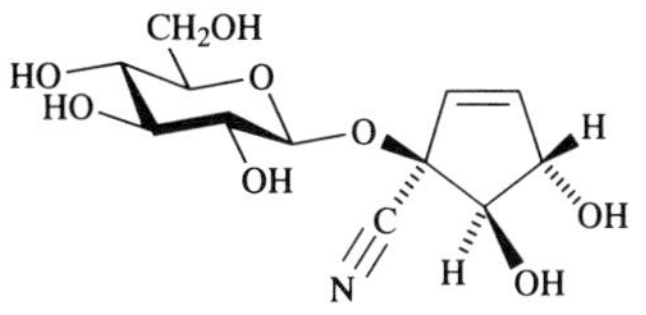

Cyanogenic glycoside

$C_{12}H_{17}NO_8$ [14332-17-3] MW 303.27

Seeds of *Gynocardia odorata* and the leaves and seeds of *Pangium edule* (Flacourtiaceae), the pericarps and immature seeds of *Carpotroche braziliensis* (also Flacourtiaceae).

Bound toxin, yielding cyanide on hydrolysis.

Coburn, R. A., *J. Org. Chem.,* 1966, **31**, 4312.

Gypsogenin 694

Albasapogenin; Astrantiagenin D; Githagenin; Gypsophilasapogenin

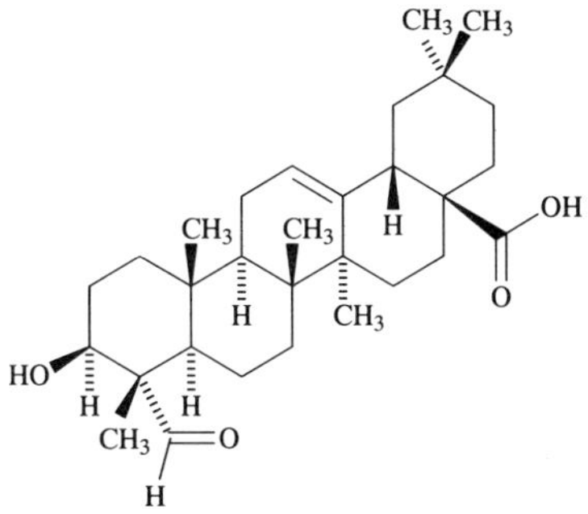

Triterpenoid

$C_{30}H_{46}O_4$ [639-14-5] MW 470.69

In glycosidic form in many plants, e.g. soapwort, *Saponaria officinalis*, *Gypsophila* spp. (Caryophyllaceae), *Swartzia* spp. (Leguminosae) and *Sideroxylon tomentosum* (Sapotaceae).

Probably toxic in both free state and in glycosidic combination. Suspected of causing livestock poisoning.

Ruzicka, L., *Helv. Chim. Acta,* 1937, **20**, 299.

Gyrocarpine 695

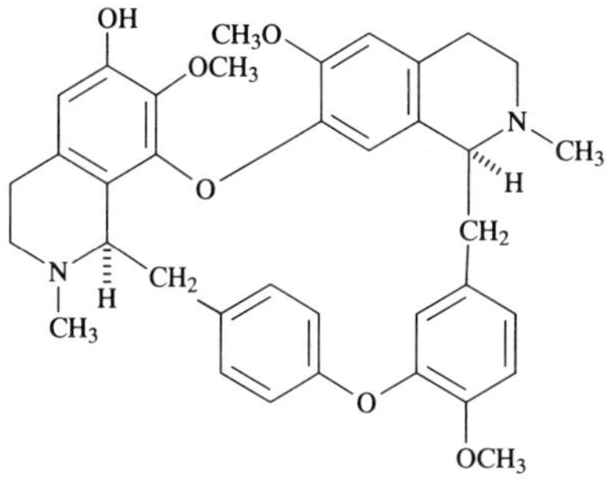

Bisbenzylisoquinoline alkaloid (Oxyacanthan)

$C_{37}H_{40}N_2O_6$ [102487-16-1] MW 608.73

Stembark of *Gyrocarpus americanus* (Hernandiaceae).

Antitrypanosomal activity against *Leishmania* species.

Chalandre, M. C., *J. Nat. Prod.,* 1986, **49**, 101.

H

Haematoxylin 696

Hydroxybrazilin; Hematoxylin; C.I. 75290; Natural Black 1

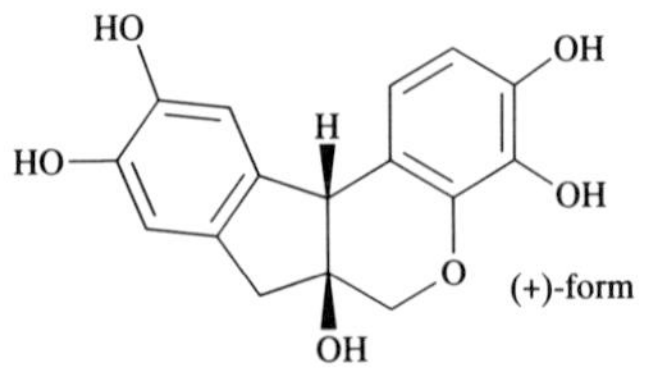

(+)-form

Phenol

$C_{16}H_{14}O_6$ [517-28-2] MW 302.28

Heartwood of *Haematoxylon campechianum* (Leguminosae).

Reddens on exposure to light. Used as a dye and as a stain in microscopy.

Morsingh, F., *Tetrahedron,* 1970, **26**, 281.

Haemocorin 697

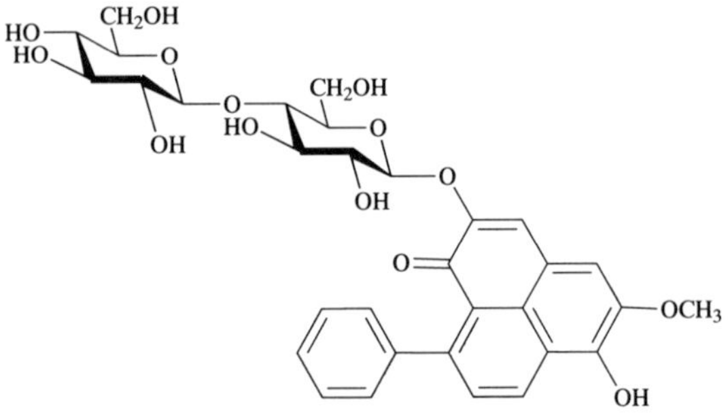

Phenolic glycoside

$C_{32}H_{34}O_{14}$ [11034-94-9] MW 642.61

In *Haemodorum corymbosum* (Haemodoraceae).

The aglycone shows antitumour and anti-inflammatory activities.

Cooke, R. G., *Aust. J. Chem.,* 1958, **11**, 230.

Haplophyllidine 698

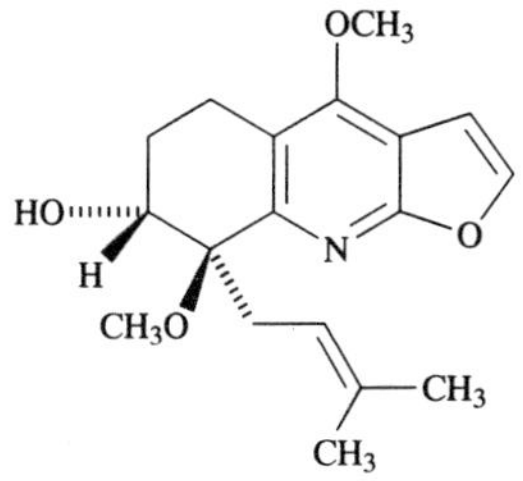

Furoquinoline alkaloid

$C_{18}H_{23}NO_4$ [18063-21-3] MW 317.38

Seeds of *Haplophyllum perforatum* and roots of *H. glabrinum* (Rutaceae).

Strong central nervous system depressant and hypnotic synergist.

Rózsa, Z., *Phytochemistry,* 1988, **27**, 2369.

Harmaline 699

3,4-Dihydroharmine; Harmidine

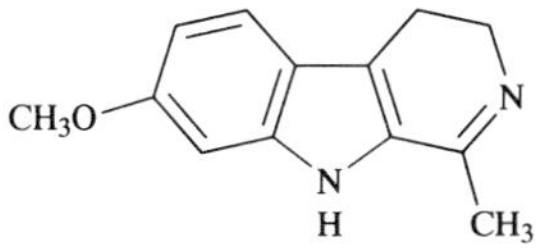

Indole alkaloid
(β-Carboline)

$C_{13}H_{14}N_2O$ [304-21-2] MW 214.27

Seeds of *Peganum harmala* (Zygophyllaceae), *Banisteria caapi* (Malpighaceae) and *Passiflora incarnata* (Passifloraceae).

Hallucinogenic. It causes ataxia and excitation tremors in animals. Was formerly used medically to treat Parkinson's disease.

Kermack, W. O., *J. Chem. Soc.,* 1921, 1602.

Harman 700

Aribine; Loturine; 1-Methyl-β-carboline; Passiflorin

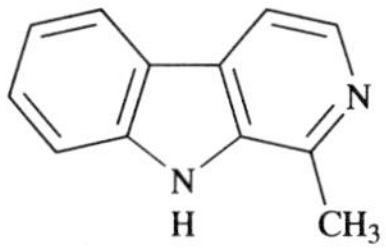

Indole alkaloid
(β-Carboline)

$C_{12}H_{10}N_2$ [486-84-0] MW 182.22

In the passion flower, *Passiflora incarnata* (Passifloraceae), *Singickia rubra* (Rubiaceae), *Zygophyllum fabago* and *Tribulus terrestris* (Zygophyllaceae) and *Symplocos racemosa* (Symplocaceae).

Motor depressant at low doses in animals but causes convulsions at high doses. Produces locomotor effects in sheep grazing on *Tribulus terrestris.*

Kermack, W. O., *J. Chem. Soc.,* 1921, 1602.

Harmine 701

Banisterine; Leucoharmine; Telepathine; Yageine

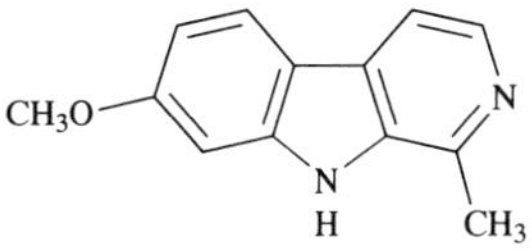

Indole alkaloid
(β-Carboline)

$C_{13}H_{12}N_2O$ [442-51-3] MW 212.25

Peganum harmala (Zygophyllaceae), *Banisteria caapi* (Malpighaceae) and the passion flower, *Passiflora incarnata* (Passifloraceae).

Central nervous system stimulant. It is hallucinogenic at high doses. The LD_{50} is 243 mg/kg body-weight when administered subcutaneously to the mouse.

Späth, E., *Ber. Dtsch. Chem. Ges.,* 1930, **63**, 120.

Harringtonine 702

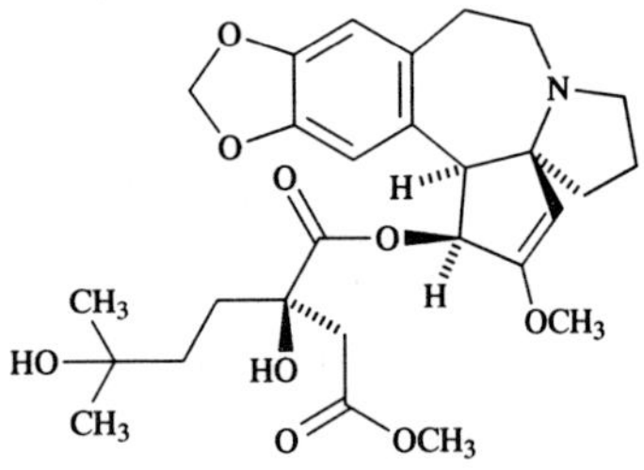

Cephalotaxus alkaloid

$C_{28}H_{37}NO_9$ [26833-85-2] MW 531.60

Cephalotaxus harringtonia, C. fortunei and *C. hainensis* (Cephalotaxaceae).

Shows antitumour activity and has been used clinically to treat acute myelocytic leukaemia.

Brossi, A., The Alkaloids, 1984, **23**, 157, Academic Press.

Harrisonin 703

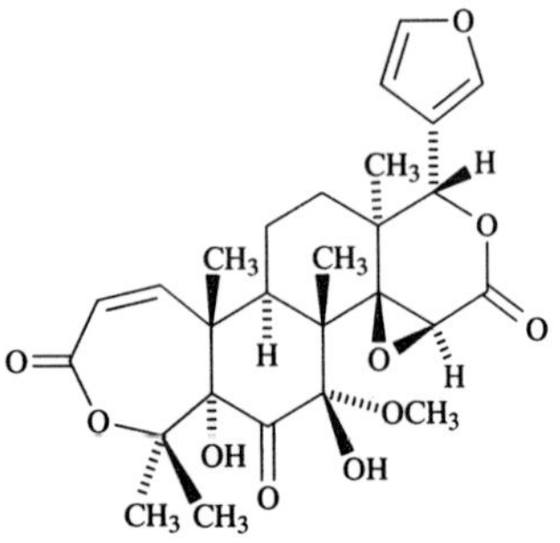

Nortriterpenoid
(Limonane)

$C_{27}H_{32}O_{10}$ [62026-30-6] MW 516.55

Roots of *Harrisonia abyssinica* (Simaroubaceae).

Insect antifeedant activity.

Kubo, I., *Heterocycles,* 1976, **5**, 485.

Hederagenin 3-*O*-arabinoside 704

Cauloside A; Tauroside B

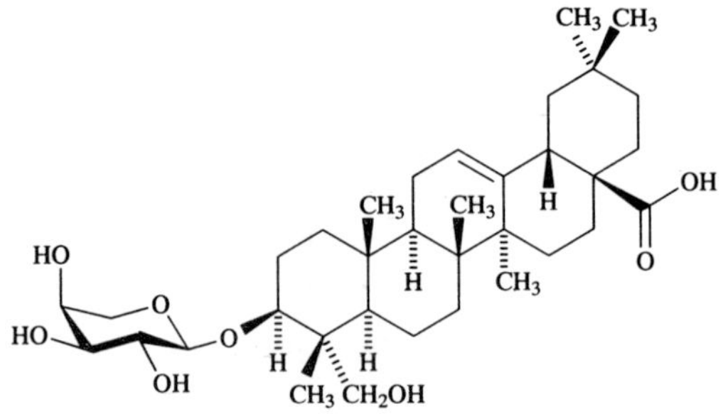

Triterpenoid saponin

$C_{35}H_{56}O_8$ [17184-21-3] MW 604.82

In ivy, *Hedera helix* (Araliaceae) and in *Lonicera nigra* (Caprifoliaceae).

Molluscicidal properties. The LD_{100} after 24 h for the snail *Biomphalaria glabrata* is 3 mg/l.

Hostettmann, K., *Helv. Chim. Acta,* 1980, **63**, 606.

Hederagenin 3-glucoside 705

Caulosaponin; Leontin; Vitalboside B

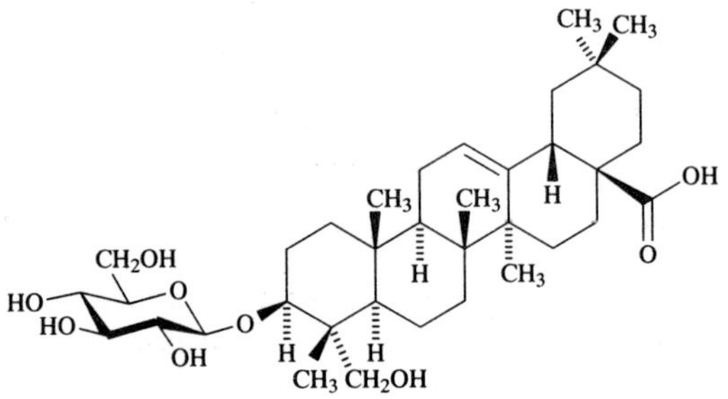

Triterpenoid saponin
(Oleanane)

$C_{36}H_{58}O_9$ [39776-12-0] MW 634.85

Roots of *Dolichos kilimandscharicus* (Leguminosae) and of *Calophyllum thalictroides* (Guttiferae).

Antifungal activity at a concentration of 5.0 μg/ml and molluscicidal at a concentration of 15 μg/l.

Aoki, T., *Phytochemistry,* 1976, **15**, 781.

α-Hederin 706

Helixin; Kalopanaxsaponin A; Kitzuta saponin K_6; Sapindoside A; Nepalin 2; Tauroside E; Hederoside B

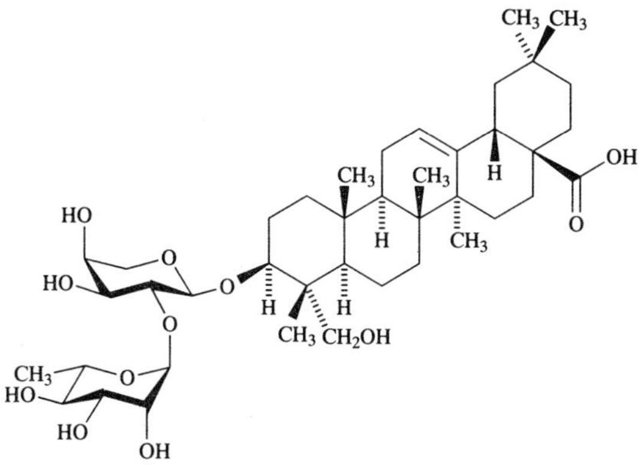

Triterpenoid saponin

$C_{41}H_{66}O_{12}$ [27013-91-8] MW 750.98

Ivy leaves, *Hedera helix* (Araliaceae) and root of *Kalopanax septemlobum* (Araliaceae).

Strong haemolytic properties. Also cancerostatic activity.

Tschesche, R., *Z. Naturforsch.*, 1965, **20B**, 708.

β-Hederin 707

Eleutheroside K; Prosapogenin CP_2; Tauroside C

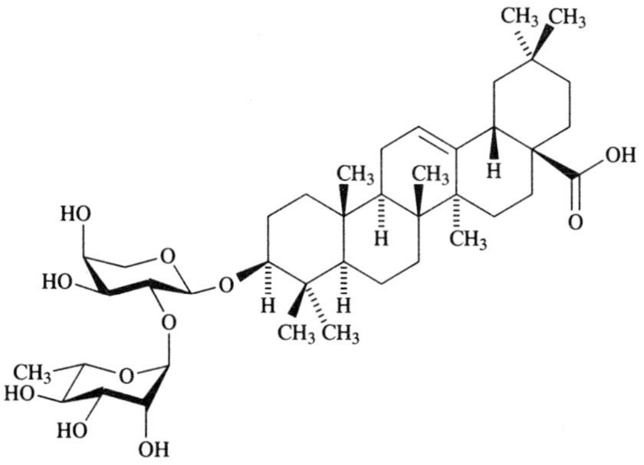

Triterpenoid saponin

$C_{41}H_{66}O_{11}$ [35790-95-5] MW 734.97

Ivy leaves, *Hedera helix* (Araliaceae).

Strong haemolytic properties.

Loloiko, A. A., *Chem. Nat. Compd.*, 1988, **24**, 320.

Helenalin 708

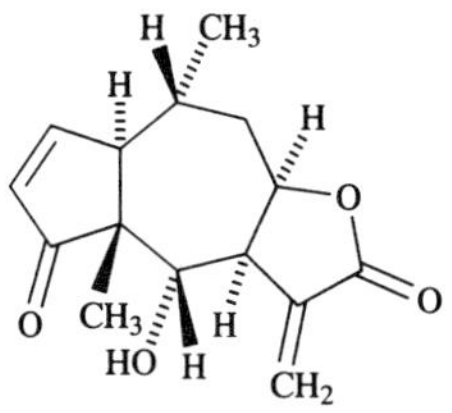

Sesquiterpenoid lactone (Pseudoguaianolide)

$C_{15}H_{18}O_4$ [6754-13-8] MW 262.31

In many *Helenium* spp. such as *H. autumnale* and *H. quadridentatum*; and in many other genera of the Compositae, such as *Anaphalis*, *Balduina* and *Gaillardia* spp..

Very poisonous to humans and other mammals. The oral LD_{50} in male mice is 92 mg/kg body-weight. Responsible for poisoning stock, e.g. of calves after feeding on *Helenium quadridentatum*. It causes paralysis of voluntary and cardiac muscles and fatal gastroenteritis, and is irritant to nose, eyes and stomach. It is also toxic to fish, molluscs and insects.

Lee, K. H., *J. Pharm. Sci.*, 1977, **66**, 1194.

Helianthoside A 709

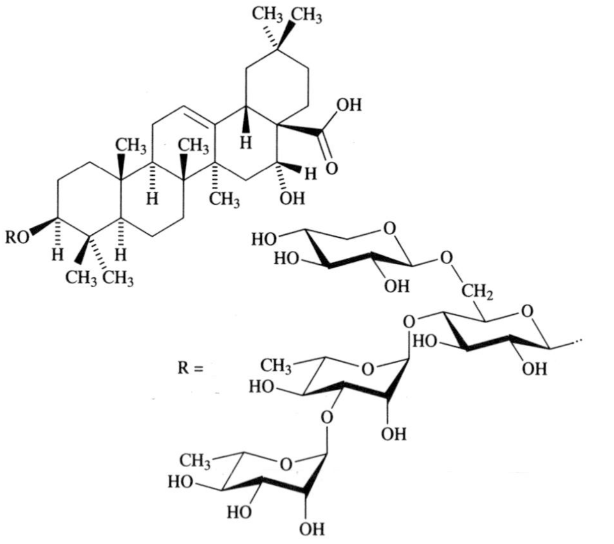

Triterpenoid saponin
(Oleanane)

$C_{53}H_{86}O_{21}$ [139164-70-8] MW 1059.25

Sunflower petals, *Helianthus annuus* (Compositae).

Bitter tasting. Haemolytic properties.

Bader, G., *Planta Med.,* 1991, **57**, 471.

Heliettin 710

Chalepin

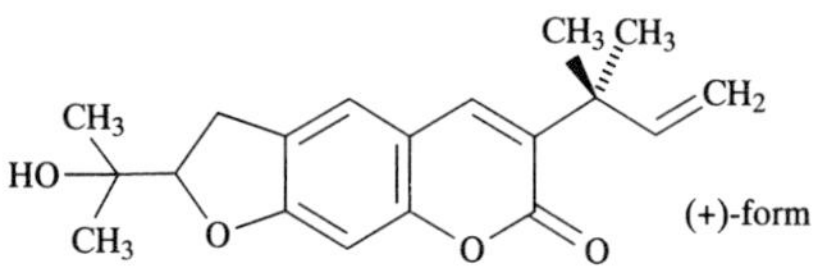

(+)-form

Furanocoumarin

$C_{19}H_{22}O_4$ [33054-89-6] MW 314.38

Bark of *Helietta longifolia* and aerial parts of *Ruta chalepensis* (Rutaceae).

Tumour inhibitor *in vitro.*

Pozzi, H., *Tetrahedron,* 1967, **23**, 1129.

Heliosupine 711

Cynoglossophine; 7-Angelyl-9-echimidinylheliotridine

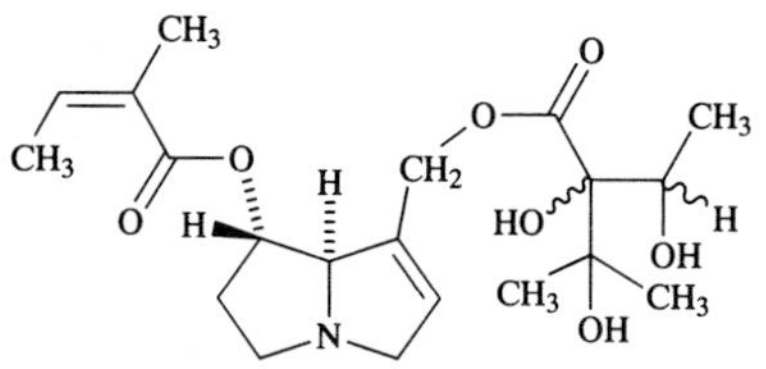

Pyrrolizidine alkaloid

$C_{20}H_{31}NO_7$ [32728-78-2] MW 397.47

In hound's tongue, *Cynoglossum officinale, C. australe* and *C. pictum*; in viper's bugloss, *Echium vulgare*; in *Heliotropium supinum, Symphytum asperum* and comfrey, *S. officinale* (Boraginaceae).

Hepatotoxic. Shows antitumour activity.

Cowley, H. C., *Aust. J. Chem.,* 1959, **12**, 694.

Heliotridine 712

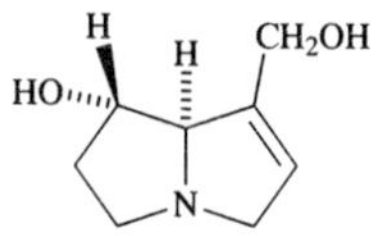

Pyrrolizidine alkaloid

$C_8H_{13}NO_2$ [520-63-8] MW 155.20

The most common necine base in American and Indian *Heliotropium* spp. (Boraginaceae), present both free and in ester form. Present in many other Boraginaceae, e.g. hound's tongue, *Cynoglossum officinale.*

Hepatotoxic. Responsible, in part, for calf, cattle and horse poisoning caused in North America from pastures containing hound's tongue.

Hart, D. J., *J. Org. Chem.,* 1985, **50**, 235.

Heliotrine 713

9-Heliotrylheliotridine

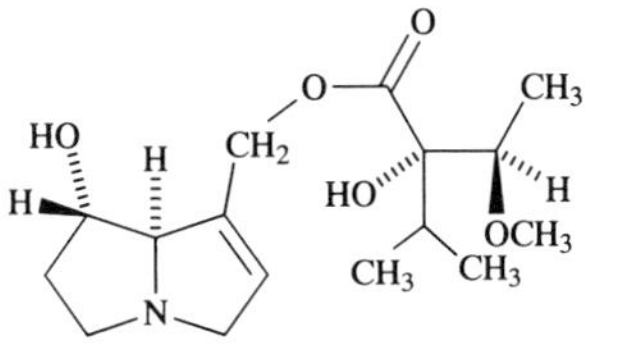

Pyrrolizidine alkaloid

$C_{16}H_{27}NO_5$ [303-33-3] MW 313.39

In heliotrope, *Heliotropium* spp. including *H. europaeum*, *H. arbainense*, *H. curassavicum*, *H. lasiocarpum* and *H. indicum* (Boraginaceae).

Hepatotoxic. Cause of calf poisoning in Israel, after consumption of *H. europaeum*. Human poisoning has also been recorded. One of four women taking *H. lasiocarpum* as a herbal tea died from liver damage.

Mattocks, A. R., Chemistry and Toxicology of Pyrrolizidine Alkaloids, 1986, Academic Press.

Hellebrigenin 3-acetate 714

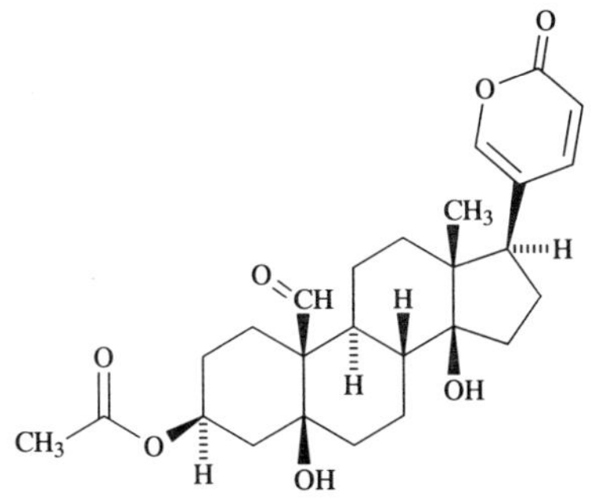

Bufadienolide

$C_{26}H_{34}O_7$ [4064-09-9] MW 458.55

Bersama abyssinica (Melianthaceae).

Tumour inhibitory properties.

Kupchan, S. M., *J. Org. Chem.*, 1969, **34**, 3894.

Hellebrin 715

Hellebrigenin 3-*O*-glucosylrhamnoside; Corelborin; Helborsid

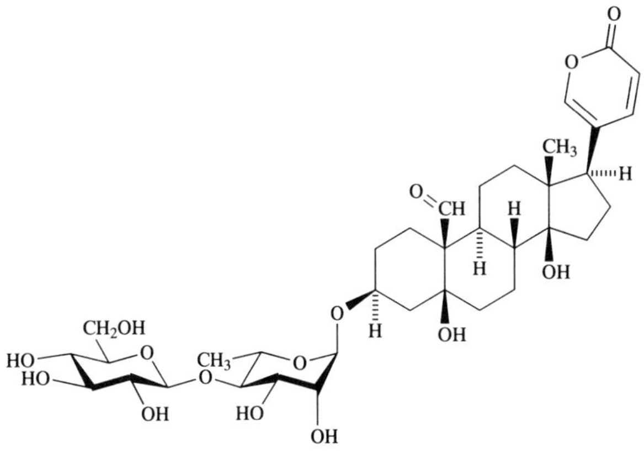

Bufadienolide

$C_{36}H_{52}O_{15}$ [13289-18-4] MW 724.80

Rhizomes of *Helleborus niger* and other *Helleborus* spp. (Ranunculaceae).

Very toxic to vertebrates. The LD_{50} is 0.85 μmol/kg body-weight when administered orally to guinea-pigs. In spite of this toxicity, there is some doubt whether hellebrin alone causes all the symptoms of animal poisoning by hellebores. Saponins and the ranunculin (q.v.) present seem to be the most active components.

Tschesche, R., *Z. Naturforsch.*, 1965, **20B**, 707.

Hellicoside 716

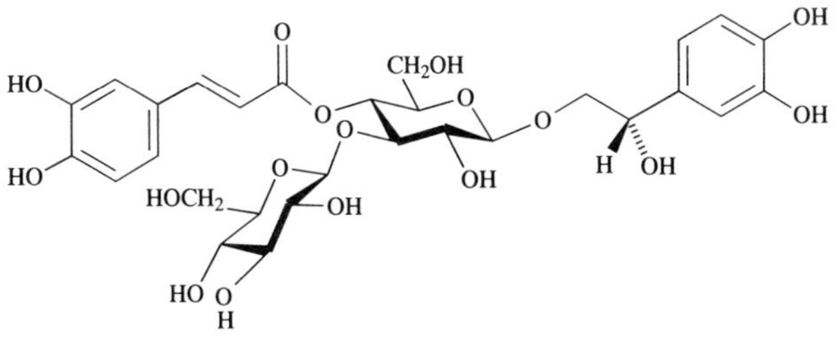

Phenylpropanoid

$C_{29}H_{36}O_{17}$ [132278-04-7] MW 656.59

Plantago asiatica (Plantaginaceae).

Inhibits cyclic adenosine monophosphate phosphodiesterate and 5-lipoxygenase. Has useful anti-inflammatory and anti-asthma activities.

Ravn, H., *Phytochemistry,* 1990, **29**, 3627.

Helveticoside 717

Strophanthidin 3-digitoxoside; Allioside A; Erysimin; Erysimotoxin; Alleoside

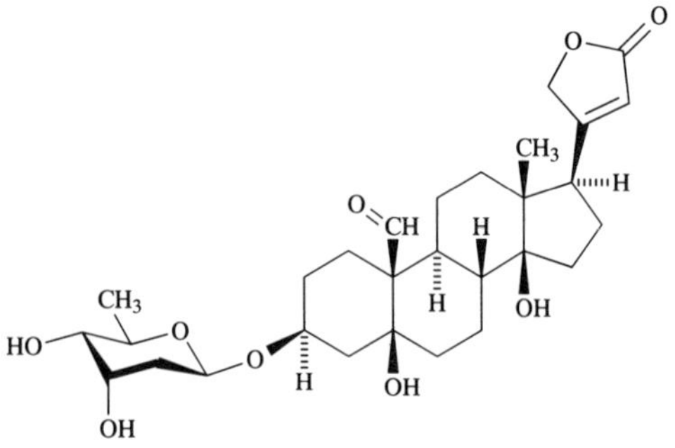

Cardenolide

$C_{29}H_{42}O_9$ [630-64-8] MW 534.65

Erysimum helveticum, *E. cheiranthoides* and *E. crepidifolium* (Cruciferae).

Very toxic to mammals. The LD_{50} is 0.104 mg/kg bodyweight when administered intravenously to cats.

Nagata, W., *Helv. Chim. Acta,* 1957, **40**, 41.

Heptadeca-1,9-diene-4,6-diyne-3,8-diol 718

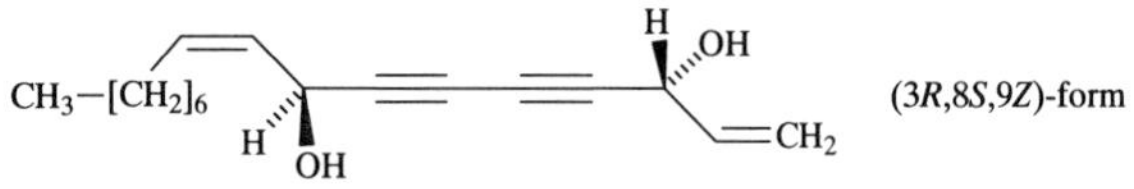

(3*R*,8*S*,9*Z*)-form

Polyacetylene

$C_{17}H_{24}O_2$ [30779-95-4] MW 260.38

Roots of *Angelica pubescens* (Umbelliferae). The (3*R*,8*S*,9*Z*-form is falcarindiol [55297-53-0].

Nematocidal activity.

Bentley, R. K., *J. Chem. Soc., C,* 1969, 685.

5-(Heptadec-12-enyl)resorcinol 719

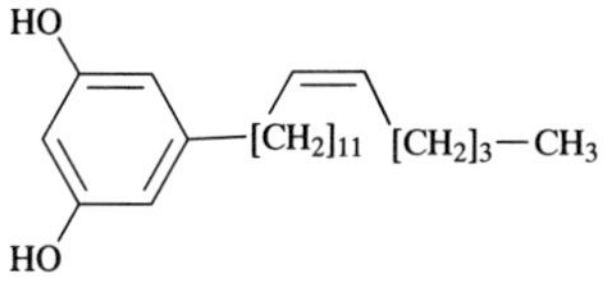

Phenol

$C_{23}H_{38}O_2$ [103462-06-2] MW 346.55

Unripe fruit of mango, *Mangifera indica* (Anacardiaceae).

Antifungal activity against 'black spot' mango disease organism *Alternaria alternata*.

Cojocaru, M., *Phytochemistry,* 1986, **25**, 1093.

Heritonin 720

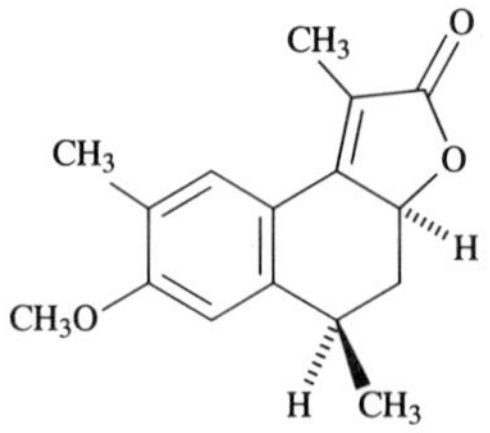

Aromatic lactone

$C_{16}H_{18}O_3$ [123914-48-7] MW 258.31

Roots of *Heritonia littoralis* (Sterculiaceae).

Piscicidal properties. The roots are used directly in the Philippines to kill fish.

Miles, D. H., *J. Nat. Prod.,* 1989, **52**, 896.

Hernandezine 721

Thalicsimine; Thaliximine

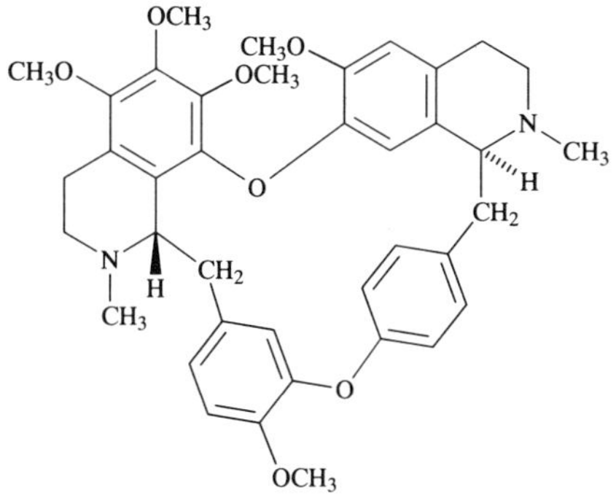

Bisbenzylisoquinoline alkaloid

$C_{39}H_{44}N_2O_7$ [6681-13-6] MW 652.79

Stephania hernandiifolia (Menispermaceae), *Thalictrum simplex* and *T. hernandezii* (Ranunculaceae).

Stephania plants are used directly as a fish poison. In rats, this alkaloid inhibits conditioned avoidance reactions and reflexes associated with movement and eating. It also has strong anti-inflammatory properties.

Padilla, J., *Tetrahedron,* 1962, **18**, 427.

Hernandulcin 722

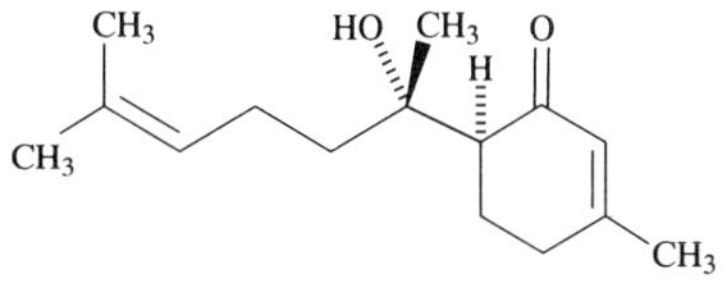

Sesquiterpenoid
(Bisabolane)

$C_{15}H_{24}O_2$ [95602-94-1] MW 236.35

Leaves and flowers of *Lippia dulcis* (Verbenaceae), a plant known to the Aztecs as "sweet herb".

It is 1 000 times sweeter than sucrose.

Compadre, C. M., *Science,* 1985, **227**, 417.

Heteratisine 723

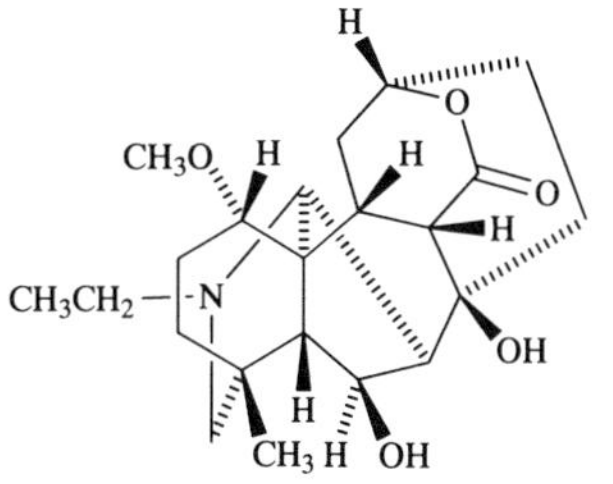

Diterpenoid alkaloid

$C_{22}H_{33}NO_5$ [3328-84-5] MW 391.51

Aconitum heterophyllum and *A. zeravschanicum* (Ranunculaceae).

Produces brief hypertension and disturbed respiration in mammals.

Aneja, R., *Tetrahedron Lett.,* 1964, 669.

Heterodendrin 724

Dihydroacacipetalin

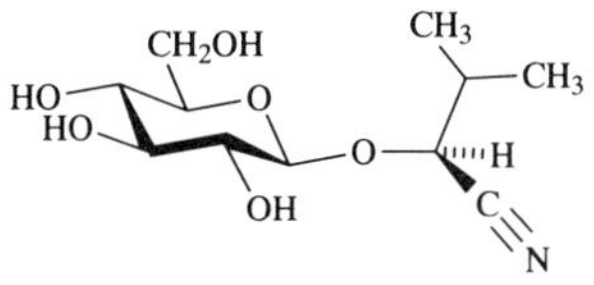

Cyanogenic glycoside

$C_{11}H_{19}NO_6$ [66465-22-3] MW 261.28

Leaves of South African *Acacia* spp. (Leguminosae), leaves of *Heterodendron oleaefolium* (Sapindaceae) and in leaves of *Sorbaria arborea* (Rosaceae). The epimer, (*R*)-epiheterodendrin [57103-47-6], occurs in *A. globulifera,* in dried leaves of barley, *Hordeum vulgare* (Gramineae) and in *Passiflora* spp. (Passifloraceae).

Both compounds are bound toxins, releasing cyanide on hydrolysis.

Hübel, W., *Phytochemistry,* 1975, **14**, 2723.

Hildecarpin 725

Isoflavonoid
(Pterocarpan)

$C_{17}H_{14}O_7$ [99624-64-3] MW 330.29

Roots of *Tephrosia hildebrandtii* (Leguminosae).

Insect antifeedant.

Lwande, W., *Insect Sci. Its Appl.,* 1985, **6**, 537.

Himbacine 726

Piperidine alkaloid

$C_{22}H_{35}NO_2$ [6879-74-9] MW 345.53

Bark of *Himantandra baccata* and *H. belgraveana* (Himantandraceae).

Antispasmodic properties.

Pinhey, J. T., *Aust. J. Chem.,* 1961, **14**, 106.

Hippeastrine 727

Trisphaerine; Trispherine

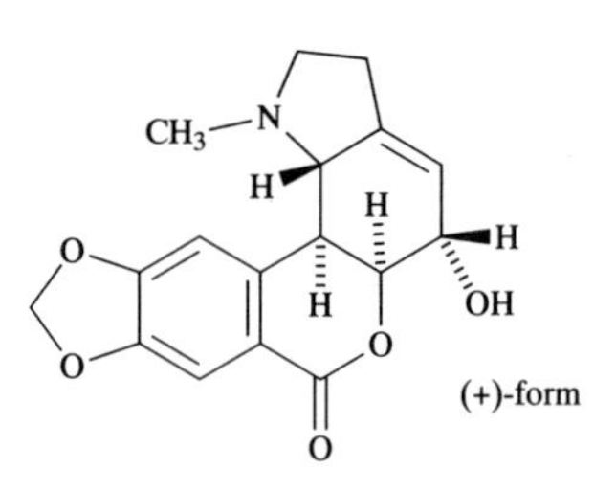

Lycoranan alkaloid

$C_{17}H_{17}NO_5$ [477-17-8] MW 315.33

Bulbs of *Clivia miniata*, *Crinum amabile*, *Hippeastrum* spp., *Lycoris radiata*, *Sternbergia lutea* and *Ungernia vvedenskyi* (Amaryllidaceae).

Lepidopteran feeding inhibitor.

Kitagawa, T., *J. Chem. Soc.,* 1959, 3741.

Hircinol 728

Phenanthrene

$C_{15}H_{14}O_3$ [41060-05-3] MW 242.27

Fungally-infected bulbs of *Loroglossum hircinum* (Orchidaceae) and constitutively present in tubers of *Dioscorea rotundata* (Dioscoreaceae).

Antifungal activity, inhibiting germ tube growth.

Fisch, M. H., *Phytochemistry,* 1973, **12**, 437.

Histamine 729

β-Aminoethylglyoxaline; 4-(2-Aminoethyl)imidazole; Ergamine; 4-Imidazoleethylamine; 1*H*-Imidazole-4-ethanamine; 2-(Imidazol-4-yl)ethylamine

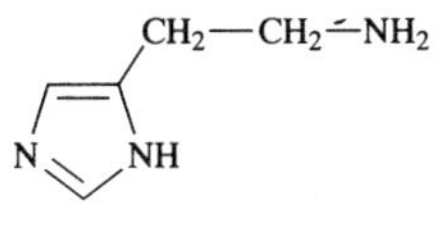

Amine

$C_5H_9N_3$ [51-45-6] MW 111.15

Widespread in nature, e.g. banana fruit, *Musa sapientum* (Musaceae), spinach, *Spinacea oleracea* (Chenopodiaceae), sundews, *Drosera* spp. (Droseraceae), pitcher plants, *Nepenthes* spp. (Nepenthaceae) and *Sarracenia* spp. (Sarraceniaceae); also in stinging hairs of the nettle, *Urtica dioica* (Urticaceae).

Vasodilator, irritant and bronchoconstrictor.

Silva, M. R. e, Handbook of Experimental Pharmacology, 1978, **18**, Springer.

Hiyodorilactone A 730

Eucannabinolide; Schkuhrin I; Hydroxychromolaenide

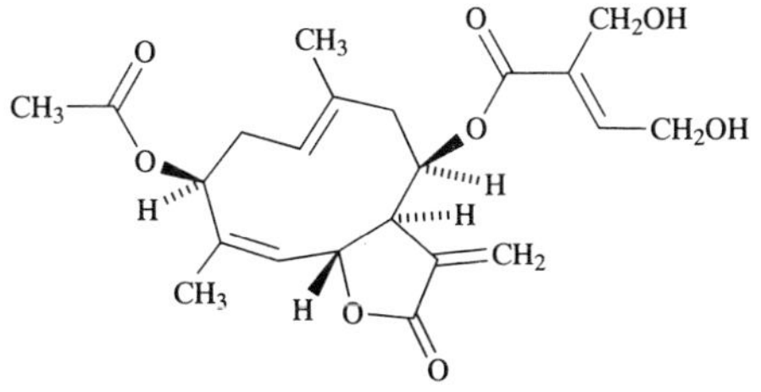

Sesquiterpenoid lactone (Germacranolide)

$C_{22}H_{28}O_8$ [38458-58-1] MW 420.46

Eupatorium sachalinense (Compositae).

Cytotoxic and antitumour properties.

Takahashi, T., *Chem. Lett.*, 1978, 1345.

L-Homoarginine 731

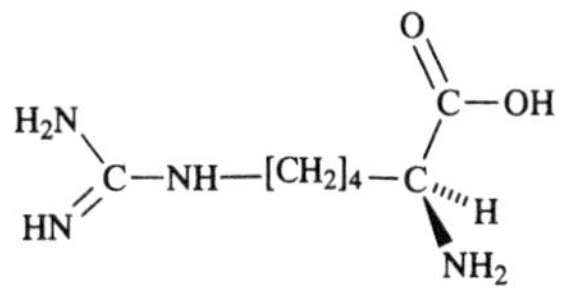

Amino acid

$C_7H_{16}N_4O_2$ [156-86-5] MW 188.23

In seeds of many *Lathyrus* spp., including *L. cicera* and *L. sativus* (Leguminosae).

Toxic in rats and some insects, with considerable variation in organism susceptibility. It presumably acts as a competitive inhibitor and antimetabolite of arginine (q.v.).

Bell, E. A., *Biochem. J.*, 1962, **85**, 91.

Hordenine 732

N,N-Dimethyltyramine; Eremursine; Anhaline; Cactine; Peyocactine

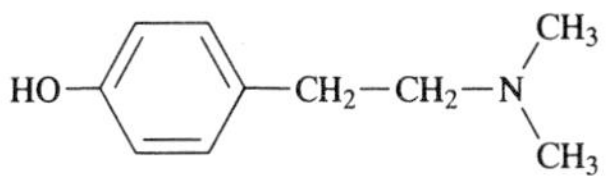

Aromatic amine

$C_{10}H_{15}NO$ [539-15-1] MW 165.24

Widespread in nature, e.g. in germinating barley, *Hordeum vulgare* and other Graminaeae, such as reed canary grass, *Phalaris arundinaea*; also in several Cactaceae, e.g. *Ariocarpus scapharostrus*.

Insect feeding inhibitor. Hypertensive in large doses, showing ephedrine-like activity.

Smith, T. A., *Phytochemistry*, 1977, **16**, 9.

Huratoxin 733

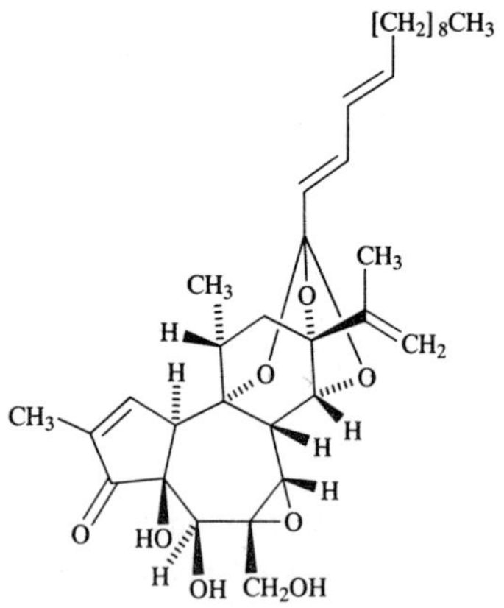

Diterpenoid
(Daphnetoxane)

$C_{34}H_{48}O_8$ [33465-16-6] MW 584.76

Latex of the hura tree, *Hura crepitans* (Euphorbiaceae).

Irritant and carcinogenic. The latex or seeds cause severe vomiting and diarrhoea when eaten. The latex has been used for poisoning fish.

Sakata, K., *Agric. Biol. Chem.,* 1971, **35**, 1084.

Hydrangenol 734

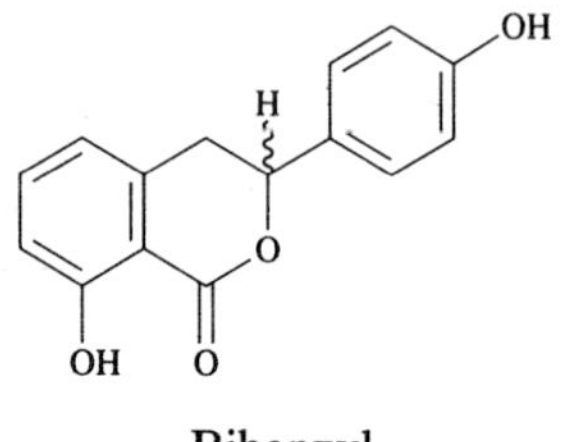

Bibenzyl

$C_{15}H_{12}O_4$ [480-47-7] MW 256.26

Hydrangea macrophylla var. *thunbergii* (Saxifragaceae).

Allergenic agent of *Hydrangea* plants.

Asahina, Y., *Ber. Dtsch. Chem. Ges.,* 1931, **64**, 1252.

15-Hydroperoxyabietic acid 735

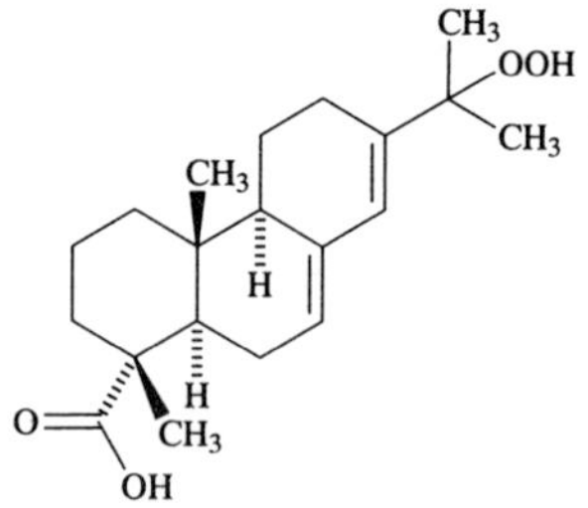

Diterpenoid
(Abietane)

$C_{20}H_{30}O_4$ [113903-96-1] MW 334.46

In the gum rosin (colophony) of *Pinus palustris* and other *Pinus* spp. (Pinaceae), obtained by distillation of the oleoresin.

One of several oxidation products of abietic and dehydroabietic acids (q.v.), which is a major contact allergen of gum rosin.

Karlberg, A. T., *J. Pharm. Pharmacol.,* 1988, **40**, 42.

Hydroquinidine 736

Hydrochinene; Hydroconchinene; Hydroconchinine; Quinotidine; Hydroquine; Hydroconquinine

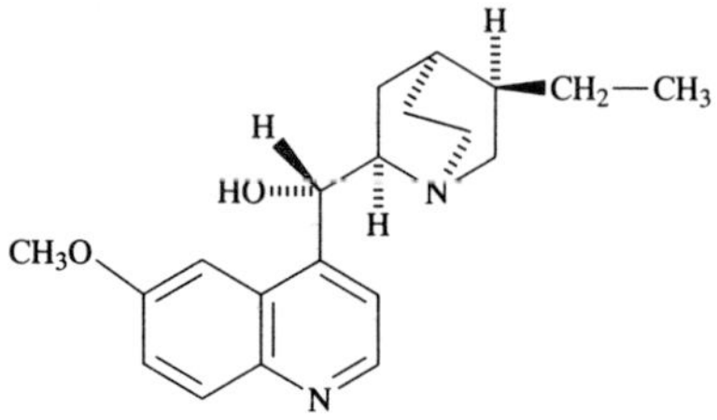

Quinoline alkaloid
(Cinchonan)

$C_{20}H_{26}N_2O_2$ [1435-55-8] MW 326.44

Bark of *Cinchona officinalis*, as a minor constituent compared to quinine (q.v.); also in *Remijia pedunculata* (also Rubiaceae); and in bark and leaves of *Aspidiosperma marcgravianum* (Apocynaceae).

Antimalarial properties. Toxic.

Grethe, G., *J. Am. Chem. Soc.,* 1978, **100**, 589.

Hydroquinone 737

Hydroquinol; Arctuvin; 1,4-Benzenediol; Pyrogentisic acid; Quinol; Eldoquin

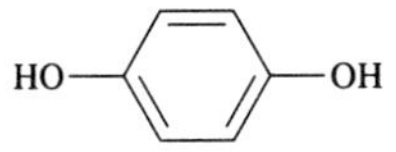

HO—⟨benzene ring⟩—OH

Phenol

$C_6H_6O_2$ [123-31-9] MW 110.11

Leaves of *Protea mellifera* (Protaceae), leaves of *Vaccinium vitis-idaea*, leaves of *Arbutus unedo* (both Ericaceae) and in leaf buds of pear, *Pyrus communis* (Rosaceae). Present also in bound form as the monoglucoside, arbutin (q.v.).

Antitumour, antimitotic and hypertensive properties. It is cytotoxic to rat hepatoma cells. Skin contact may cause dermatitis in man.

Karrer, W., Konstitution und Vorkommen der organischen Pflanzenstoffe, 1958, 87, Birkhäuser.

12α-Hydroxyamoorstatin 738

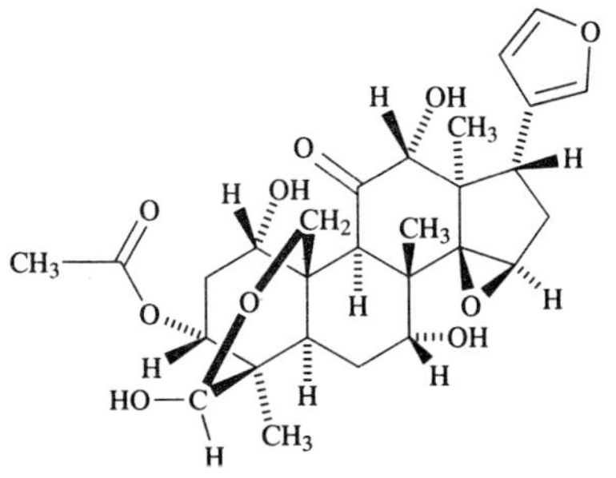

Nortriterpenoid
(Limonane)

$C_{28}H_{36}O_{10}$ [71590-47-1] MW 532.59

Seeds of *Aphanamixis grandifolia* (Meliaceae).

Cytotoxic and antileukaemic activities.

Polonsky, J., *Experientia,* 1979, **35**, 987.

8β-Hydroxyasterolide 739

Atractylenolide III

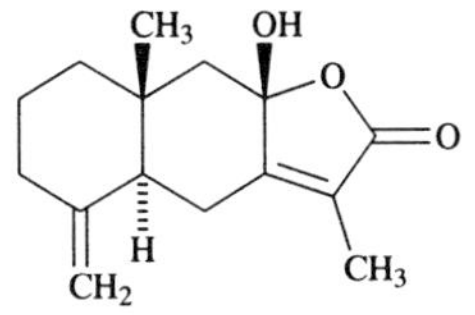

Sesquiterpenoid lactone
(Eudesmanolide)

$C_{15}H_{20}O_3$ [73030-71-4] MW 248.32

Trunk resin of *Trattinnickia aspera* (Burseraceae) and *Aster umbellatus* (Compositae).

Active principle of the *T. aspera* resin, which is used by white-nosed coatis to remove ectoparasites from their fur. It is presumably toxic or repellent to fleas, lice and ticks.

Urchida, M., *Chem. Pharm. Bull.,* 1980, **28**, 92.

1S-Hydroxy-α-bisabololoxide A acetate 740

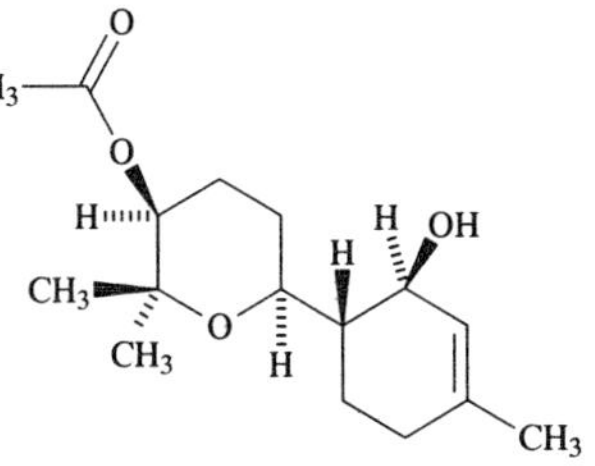

Sesquiterpenoid
(Bisabolane)

$C_{17}H_{28}O_4$ [119347-20-5] MW 296.38

Ambrosia abrotanum (Compositae).

Antiprotozoal activity.

Meriçli, A. H., *Planta Med.,* 1988, **54**, 463.

7-Hydroxycalamenene 741

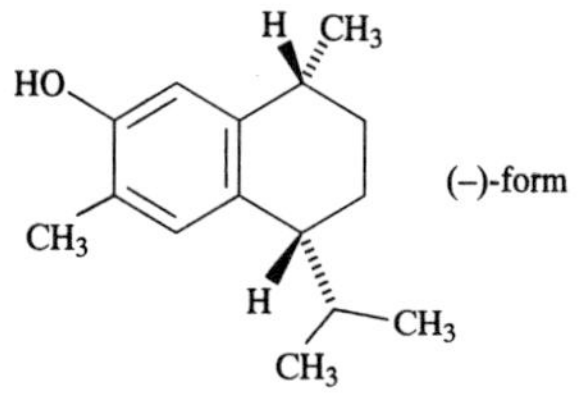

(–)-form

Sesquiterpenoid
(Cadinane)

$C_{15}H_{22}O$ [24406-03-9] MW 218.34
(+)-form [69126-74-5]

Wood of the lime tree, *Tilia europea* (Tiliaceae) infected with *Ganoderma applanatum*. Also constitutively present in the heartwood of the elm, *Ulmus thomasii* (Ulmaceae).

Antifungal; inhibitory to fungi at a concentration of 0.5 μg. The high concentration in infected lime (1% fresh-weight) indicates that it is effective *in vivo*.

Burden, R. S., *Phytochemistry,* 1983, **22**, 1039.

(+)-8-Hydroxycalamenene 742

(+)-2-Hydroxycalamenene

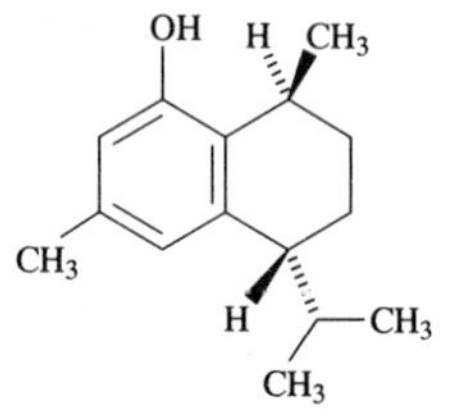

Sesquiterpenoid
(Cadinane)

$C_{15}H_{22}O$ [88642-92-6] MW 218.34

Seeds of *Dysoxylum alliaceum* and *D. acutangulum* (Meliaceae).

Piscicide. Major poisonous principle of *Dysoxylum* plants, which are used in Sumatra to kill fish.

Nishizawa, M., *Phytochemistry,* 1983, **22**, 2083.

11-Hydroxycanthin-6-one 743

Amarorine

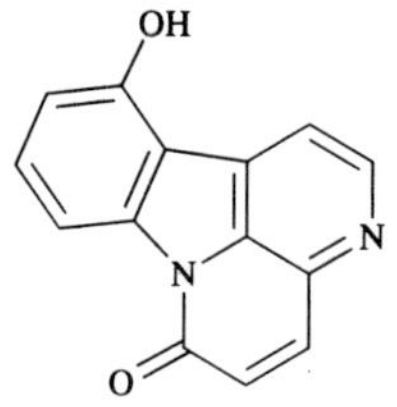

Indole alkaloid
(Canthinone)

$C_{14}H_8N_2O_2$ [75969-83-4] MW 236.23

Stembark of *Quassia kerstingii* and *Amaroria soulameoides* (Simaroubaceae).

Active against P388 lymphocytic leukaemia cell lines *in vitro*.

Clarke, P. J., *J. Chem. Soc., Perkin Trans. 1,* 1980, 1614.

Hydroxydictyodial 744

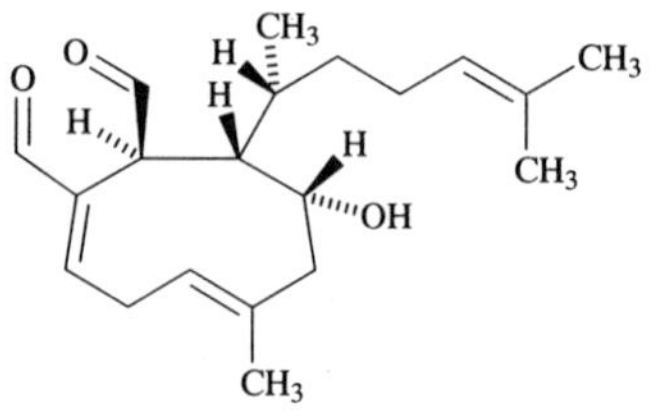

Diterpenoid

$C_{20}H_{30}O_3$ [89482-11-1] MW 318.46

Brown alga, *Dictyota spinulosa*.

Inhibits feeding of the omnivorous fish *Tilapia mossambica* at 10% concentration. Also antimicrobial.

Tanaka, J., *Chem. Lett.,* 1984, 231.

7-Hydroxyflavan 745

7-Hydroxy-2-phenylchroman

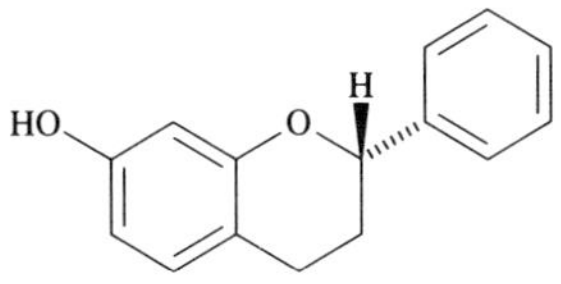

Flavan

$C_{15}H_{14}O_2$ [38481-95-7] MW 226.27

Fungally infected bulbs of the daffodil, *Narcissus pseudo-narcissus* (Amaryllidaceae).

Fungitoxic against *Botrytis cinerea*, with an ED_{50} of 22 μg/ml.

Coxon, D. T., *Phytochemistry,* 1980, **19**, 889.

3-Hydroxyheterodendrin 746

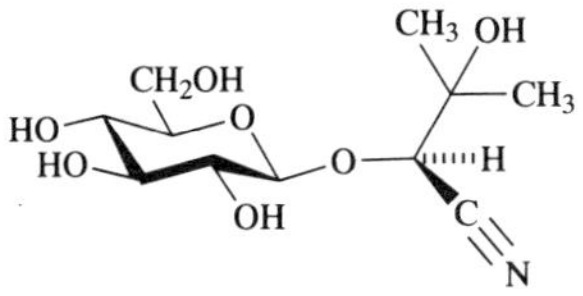

Cyanogenic glycoside

$C_{11}H_{19}NO_7$ [80750-13-6] MW 277.28

Pods of *Acacia sieberiana* var. *woodii* (Leguminosae).

Bound toxin, releasing poisonous cyanide on hydrolysis. The plant source is used as animal fodder in parts of Africa and this toxin, together with other cyanogenic glycosides present, could give rise to livestock poisoning.

Brimer, L., *Phytochemistry,* 1981, **20**, 2221.

13-Hydroxylupanine 747

13-Hydroxy-2-oxosparteine; Octalupine; Hydroxylupanine

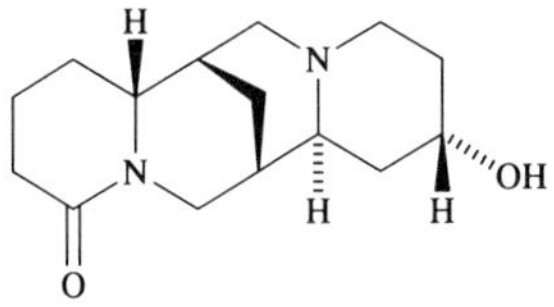

Quinolizidine alkaloid
(Sparteine)

$C_{15}H_{24}N_2O_2$ [15358-48-2] MW 264.37

In broom, *Cytisus scoparius, Genista cinerea, Thermopsis cinerea* and in several *Cadia* spp. and *Lupinus* spp. (Leguminosae). Also present with ester linkage at the 13-hydroxyl group.

Anti-arrythmic properties, also hypotensive.

Goosen, A., *J. Chem. Soc,* 1963, 3067.

N-(*p*-Hydroxyphenethyl)actinidine 748

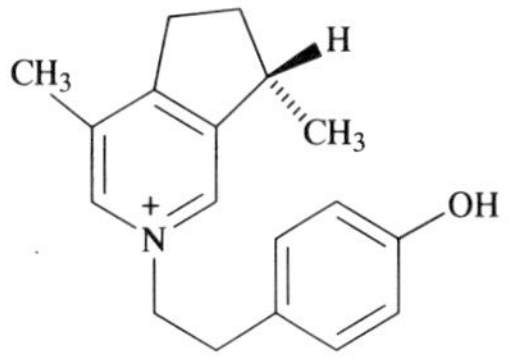

Monoterpenoid alkaloid

$[C_{18}H_{22}NO]^+$ [15794-92-0] MW 268.38

Roots of *Valeriana officinalis* (Valerianaceae).

Highly active inhibitor of cholinesterase activity.

Torssell, K., *Acta Chem. Scand.,* 1967, **21**, 53.

12β-Hydroxypregna-4,16-diene-3,20-dione 749

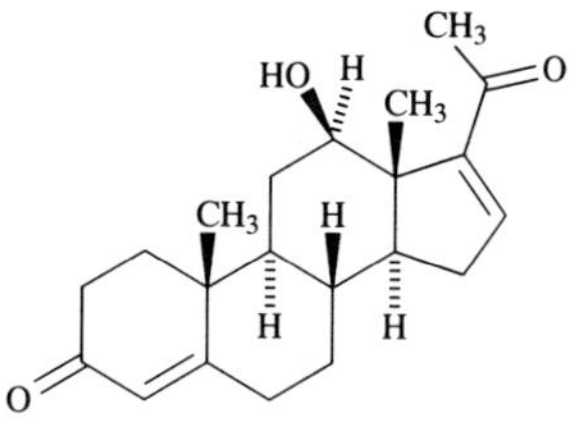

Phytosterol
(Pregnane)

$C_{21}H_{28}O_3$ [72959-46-7] MW 328.45

Gelsemium sempervirens (Loganiaceae).

Cytotoxic activity, with an ED_{50} of 0.7 μg/ml against P-388 cell lines.

Yamauchi, T., *Phytochemistry,* 1979, **18**, 1240.

6-Hydroxy-α-pyrufuran 750

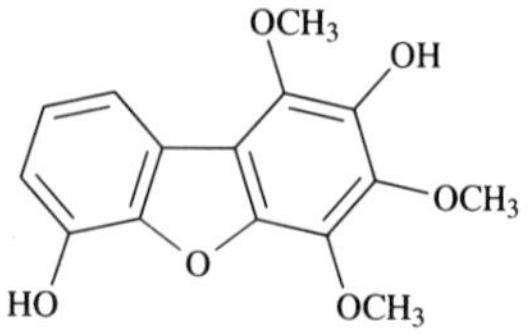

Dibenzofuran

$C_{15}H_{14}O_6$ [167278-43-5] MW 290.27

Sapwood of *Mespilus germanica* (Rosaceae) infected with *Nectria cinnabarina.*

Antifungal agent (phytoalexin), inhibiting spore germination at a concentration of 12 μg/ml.

Kokubun, T., *Phytochemistry,* 1995, **39**, 1039.

12a-Hydroxyrotenone 751

Rotenalone; Rotenolone; Rotenolone 1; Rotenolon 1

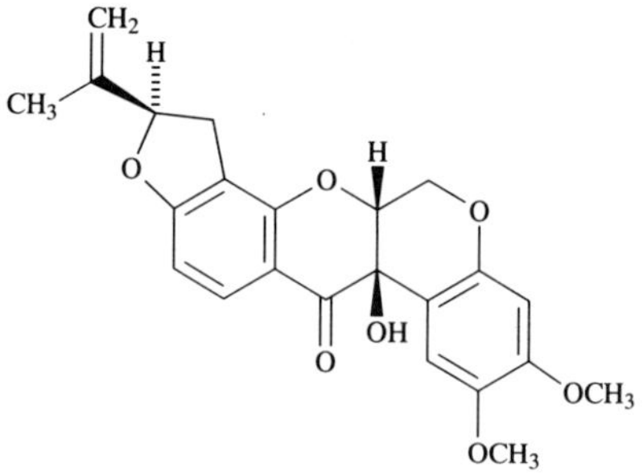

Isoflavonoid
(Rotenoid)

$C_{23}H_{22}O_7$ [54534-95-1] MW 410.42

Roots of *Derris urucu, Neorautanenia amboensis, Pachyrrhizus erosus* and *Tephrosia* spp. (Leguminosae).

Insecticidal activity.

Oberholzer, M. E., *Tetrahedron Lett.,* 1974, 2211.

6-Hydroxytremetone 752

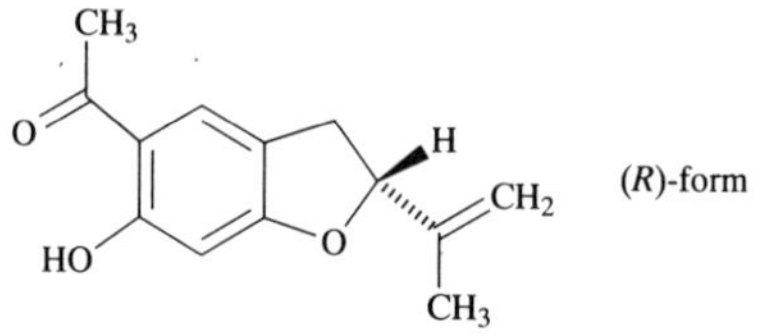

(*R*)-form

Benzofuran

$C_{13}H_{14}O_3$ [6906-88-3] MW 218.25
(*R*)-form [21491-62-3]
(*S*)-form [64234-07-7]

In many Compositae, e.g. *Encelia ventuorum, Hemizonia congesta* and *Ligularia intermedia.* Toxic principle of white snakeroot, *Eupatorium urticaefolium.*

Toxic to goldfish.

Bohlmann, F., *Chem. Ber.,* 1970, **103**, 90.

5-Hydroxy-L-tryptophan 753

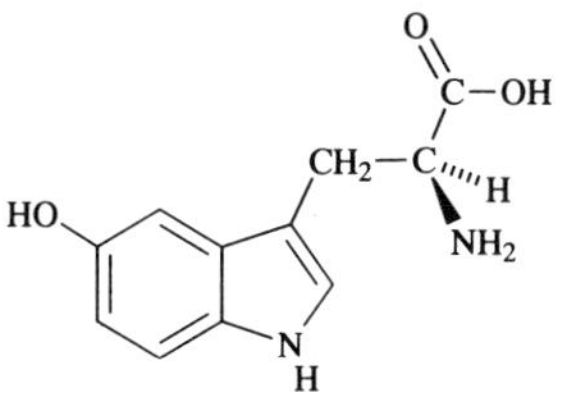

Amino acid

$C_{11}H_{12}N_2O_3$ [4350-09-8] MW 220.23

Seeds of *Mucuna pruriens* and *Griffonia simplicifolia* (Leguminosae).

Toxic to insects. Has antidepressant and anti-epileptic properties.

Bell, E. A., *Nature (London),* 1966, **210**, 529.

Hydroxyvernolide 754

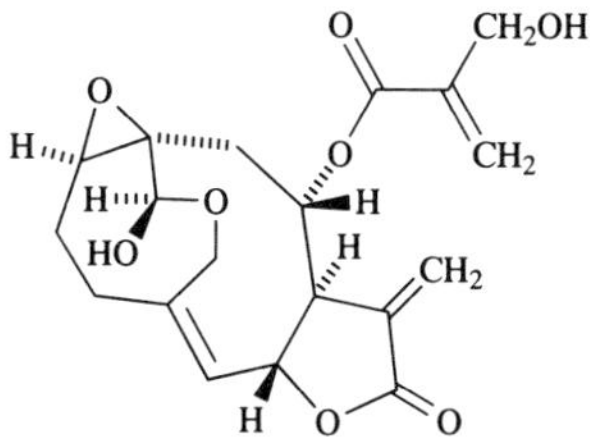

Sesquiterpenoid lactone
(Germacranolide)

$C_{19}H_{22}O_8$ [27470-84-4] MW 378.38

Vernonia amygdalina and *V. colorata* (Compositae).

Insect antifeedant.

Toubiana, R., *C. R. Acad. Sci., Ser. C,* 1969, **268**, 82.

Hymenoflorin 755

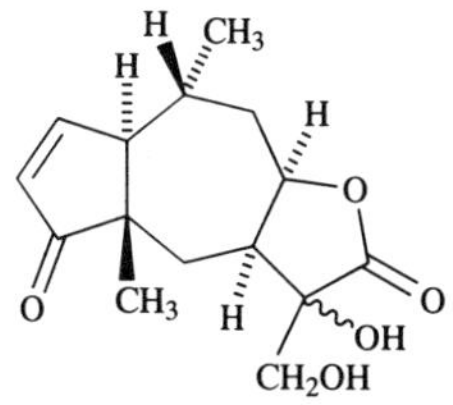

Sesquiterpenoid lactone
(Pseudoguaianolide)

$C_{15}H_{20}O_5$ [51292-63-8] MW 280.32

Hymenoxys grandiflora (Compositae).

Cytotoxic and antitumour properties.

Herz, W., *J. Org. Chem.,* 1974, **39**, 2013.

Hymenolin 756

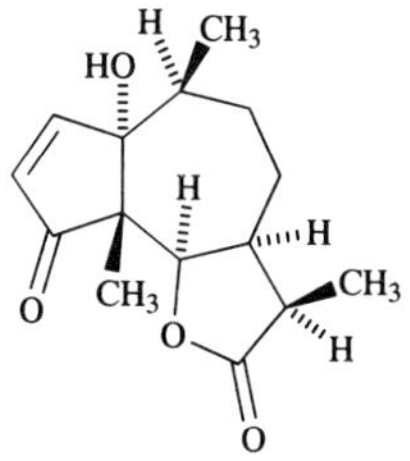

Sesquiterpenoid lactone
(Pseudoguaianolide)

$C_{15}H_{20}O_4$ [20555-05-9] MW 264.32

Hymenoclea salsola (Compositae).

Poisonous to mosquito larvae, *Aedes atropalpus*.

Toribio, F. P., *Phytochemistry,* 1968, **7**, 1623.

Hymenoxon 757

Hymenoxone

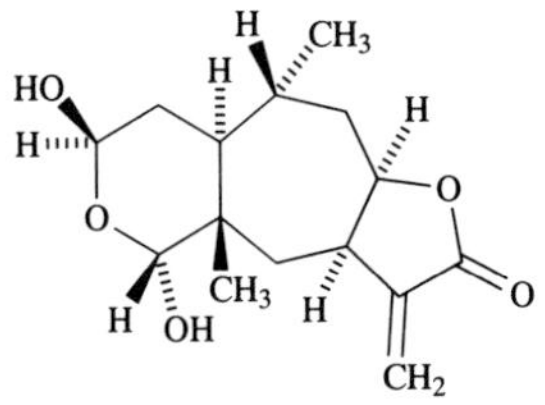

Sesquiterpenoid lactone
(Seco-pseudoguaianolide)

$C_{15}H_{22}O_5$ [57377-32-9] MW 282.34

In bitterweed, *Hymenoxys odorata*, in *H. richardsonii* and in *Dugaldia hoopesii* (Compositae). One of several epimers present together in these plants.

Highly toxic to mammals. Responsible for livestock poisoning, following grazing on bitterweed in North America. It is a potent stimulator of mass cell degranulation. The LD_{50} in sheep from a single oral dose is 75 mg/kg body-weight. The symptoms are identical whether the whole plant or the pure lactone is fed.

Pettersen, R. C., *J. Chem. Soc., Perkin Trans. 2,* 1976, 1399.

Hyoscine 758

6,7-Epoxytropine tropate; Scopine tropate; Scopolamine; Scopoline tropate; Oscine tropate; Scopoderm

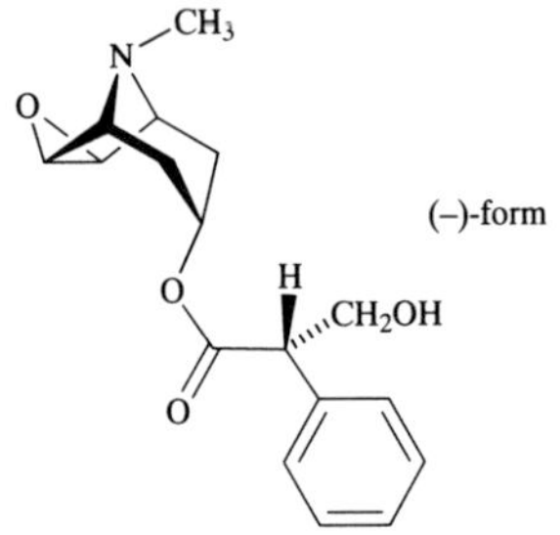

(–)-form

Tropane alkaloid

$C_{17}H_{21}NO_4$ [51-34-3] MW 303.36

In henbane, *Hyoscyamus niger*, *Datura metel*, *D. innoxia*, *Scopolia carniolica*, *Anthocercis viscosa* and *A. fasciculata* (Solanaceae). Also a minor alkaloid in deadly nightshade, *Atropa belladonna*.

Less toxic than hyoscyamine (q.v.). Anticholinergic, with both central and peripheral actions. It is used to treat motion sickness. Also it is employed as a pre-operative medication to sedate and induce anaesthesia.

Dobó, P., *J. Chem. Soc.,* 1959, 3461.

Hyoscyamine 759

Daturine; Duboisine; (*S*)-Tropine tropate

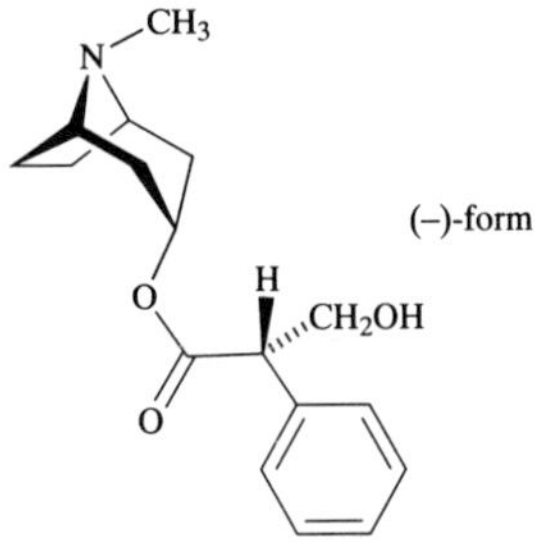

(–)-form

Tropane alkaloid

$C_{17}H_{23}NO_3$ [101-31-5] MW 289.37

In henbane, *Hyoscyamus niger*, Egyptian henbane, *H. muticus*, deadly nightshade, *Atropa belladonna*, thornapple, *Datura stramonium* and other Solanaceae. The highest concentrations are present in the roots and the fruits (especially the seeds).

Well known poison, causing human fatalities. It is anticholinergic with actions similar to but more potent than those of atropine (q.v.) which is the racemic form of hyoscyamine. It produces central nervous system depression, followed by stimulation. It is still in medical use as a premedication before operations.

Bremner, J. B., *Aust. J. Chem.,* 1968, **21**, 1369.

Hypaconitine 760

3-Deoxymesaconitine; Japaconitine B_1; Japaconitine C_1

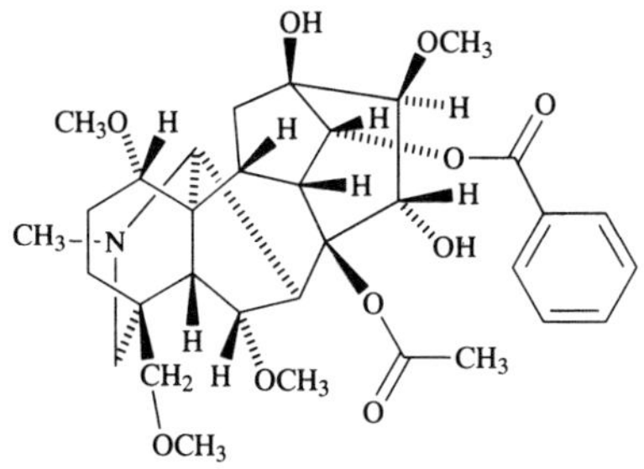

Diterpenoid alkaloid
(Aconitane)

$C_{33}H_{45}NO_{10}$ [6900-87-4] MW 615.72

In many aconite species, e.g. *Aconitum callianthum*, *A. carmichaeli* and *A. napellus* (Ranunculaceae).

Potent and quick-acting poison, slowing the heart rate and lowering the blood pressure. It has anti-inflammatory properties.

Tsuda, Y., *Justus Liebigs Ann. Chem.*, 1964, **680**, 88.

Hypaphorine 761

Tryptophan betaine; Trimethyltryptophan betaine

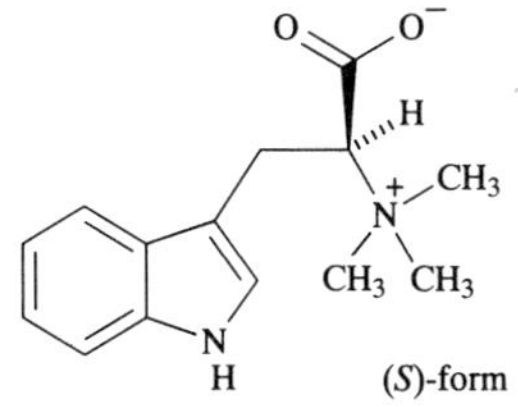

(*S*)-form

Indole alkaloid

$C_{14}H_{18}N_2O_2$ [487-58-1] MW 246.31

Seeds of *Erythrina hypaphorus* and of *Pterocarpus officinalis* (Leguminosae).

Convulsive poison. It is feeding deterrent to the seed-eating rodent *Liomys salvini*.

Folkers, K., *J. Am. Chem. Soc.*, 1939, **61**, 1232.

Hyperbrasilone 762

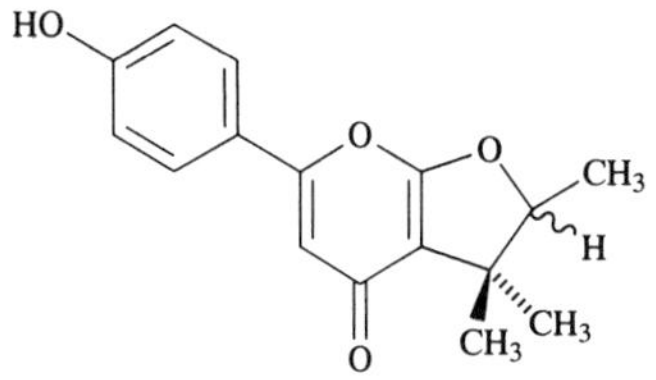

Pyrone

$C_{16}H_{16}O_4$ [158991-19-6] MW 272.30

Stems and roots of *Hypericum brasiliense* (Gutteriferae).

Antifungal; inhibitory to fungal growth at a concentration of 3 μg/ml.

Rocha, L., *Phytochemistry,* 1994, **36**, 1381.

Hypericin 763

Hypericum red; Mycoporphyrin

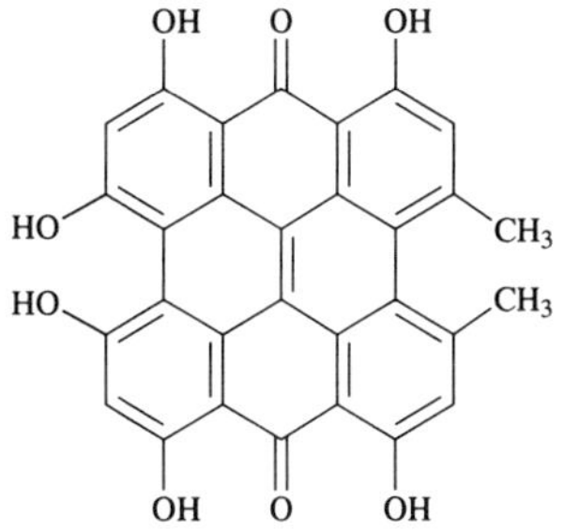

Bianthraquinone

$C_{30}H_{16}O_8$ [548-04-9] MW 504.45

In leaves and other parts of *Hypericum perforatum* (St. John's wort) and many other *Hypericum* spp. (Hypericaceae).

Causes facial eczema in sheep grazing on *Hypericum* plants; the photogenic disease produced is called 'hypericism'. It has photosensitising and antidepressant activity in many mammals. Also possesses antiretroviral properties.

Brockmann, H., *Naturwissenschaften,* 1951, **37**, 540.

L-Hypoglycin 764

Hypoglycin A; Hypoglycine

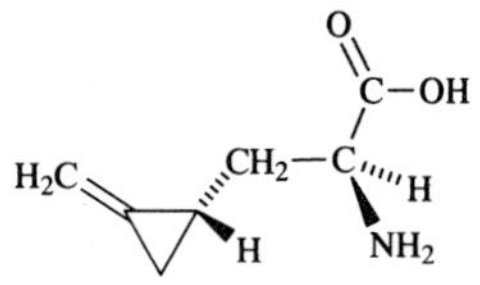

Amino acid

$C_7H_{11}NO_2$ [156-56-9] MW 141.17

In the arillus of the fruits of the ackee, *Blighia sapida* (Sapindaceae) and in *Billia hippocastanum* (Hippocastanaceae).

Very toxic and hypoglycaemic. It is responsible for a "vomiting sickness", characterised by violent retching, vomiting, convulsions and coma (sometimes fatal). The ackee fruit is eaten in the West Indies, where the sickness can reach epidemic proportions at certain times of the year. While the unripe fruit contains 1% fresh weight of toxin, the ripe fruit has only 0.0012% and is safe to eat.

Greenstein, J. P., Chemistry of the Amino Acids, 1961, **3**, 2742, Wiley.

Hyrcanoside 765

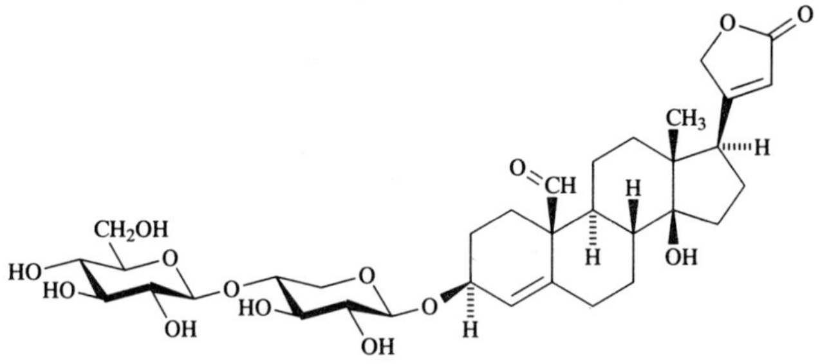

Cardenolide

$C_{34}H_{48}O_{14}$ [15001-93-1] MW 680.75

Leaf of crown vetch, *Coronilla varia*, of *C. hyrcana* and of *Securigera securidaca* (Leguminosae).

Toxin. Crown vetch is poisonous to livestock and hyrcanoside is the most likely cause, although the plant also contains toxic nitropropionic acid derivatives.

Bagirov, R. B., *Chem. Nat. Compd.,* 1966, **2**, 202.

I

Ibogaine 766

12-Methoxyibogamine

Indole alkaloid
(Ibogamine)

$C_{20}H_{26}N_2O$ [83-74-9] MW 310.44

Tabernanthe iboga and *Voacanga thouarsii* (Apocynaceae).

Central nervous system activity. It is a hallucinogen and has anticonvulsive properties.

Bartlett, M. F., *J. Am. Chem. Soc.,* 1958, **80**, 126.

Ichthyotherol 767

Cunaniol; Ichthyothereol

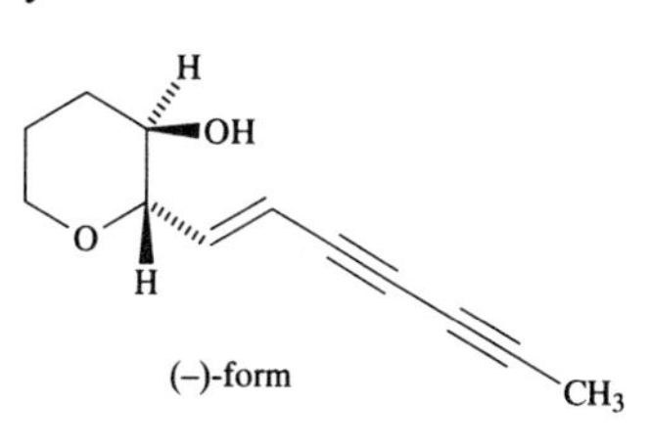

Polyacetylene

$C_{14}H_{14}O_2$ [2294-61-3] MW 214.26

In roots of *Dahlia coccinea* and in leaves of *Ichthyothere terminalis* and *Clibadium sylvestre* (Compositae).

Extremely toxic. Used as a fish poison.

Cascon, S. C., *J. Am. Chem. Soc.,* 1965, **87**, 5237.

Iforrestine 768

Miscellaneous alkaloid

$C_{14}H_{12}N_4O_3$ [125287-08-3] MW 284.27

Isotropis forrestii (Leguminosae).

Nephrotoxic in sheep. Consumption of between 500 and 1 000 g of plant material is sufficient to cause death.

Colegate, S. M., *Aust. J. Chem.,* 1989, **42**, 1249.

Ilexolide A 769

Triterpenoid saponin (Ursane)

$C_{35}H_{54}O_7$ [85344-31-6] MW 586.81

Roots of *Ilex pubescens* and leaves of *I. chinensis* (Aquifoliaceae).

Cardiac activity.

Zeng, L., *Chem. Abstr.,* 1984, **101**, 226825.

Imperatorin 770

Marmelosin; Ammidin; 8-Isoamylenoxypsoralen; Marmelide

Furocoumarin

$C_{16}H_{14}O_4$ [482-44-0] MW 270.28

Commonly present in roots and seeds of plants of the Umbelliferae, especially in *Angelica*, *Heracleum* and *Pastinaca* spp.. Also in seeds of *Citrus meyeri* and *Aegle marmelos* (Rutaceae) and in leaves of wild strawberries, *Fragaria* spp. (Rosaceae).

Toxic to toads and also has piscicidal properties. It has some antimutagenic activity.

Murray, R. D. H., The Natural Coumarins, 1982, 415, Wiley.

Indaconitine 771

15-Deoxyaconitine

Diterpenoid alkaloid (Aconitane)

$C_{34}H_{47}NO_{10}$ [4491-19-4] MW 629.75

First isolated from *Aconitum chasmanthum* and subsequently found in *A. ferox* and *A. falconeri* (Ranunculaceae).

Very poisonous, especially towards small animals.

Klasek, A., *J. Nat. Prod.,* 1972, **35**, 55.

Indicine 772

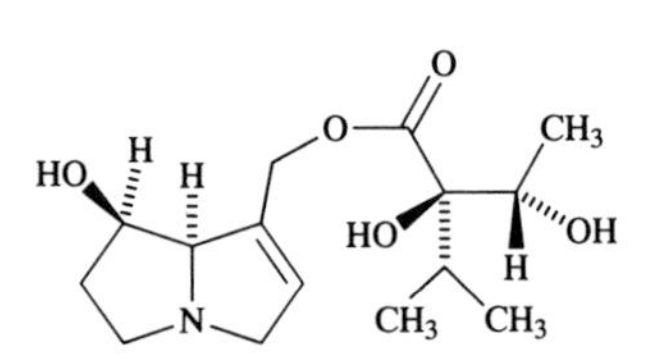

Pyrrolizidine alkaloid

$C_{15}H_{25}NO_5$ [480-82-0] MW 299.37

In the heliotrope, *Heliotropium indicum* and *H. amplexicaule* (Boraginaceae).

Poisonous, but less toxic in animals than other co-occurring pyrrolizidide alkaloids. The *N*-oxide [41708-76-3] has been used in clinical trials as an anticancer agent.

Mattocks, A. R., *J. Chem. Soc.*, 1961, 5400.

L-Indospicine 773

L-2-Amino-6-amidinohexanoic acid

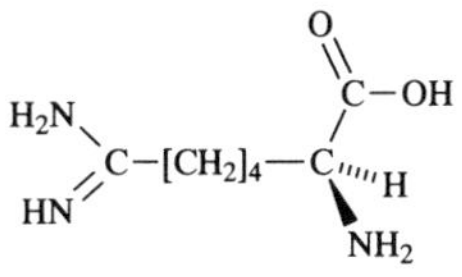

Amino acid

$C_7H_{15}N_3O_2$ [16377-00-7] MW 173.22

Leaves and seeds of *Indigofera* spp. including *I. spicata* and *I. linnaei* (Leguminosae).

Teratogenic, abortifacient and hepatotoxic properties. It causes cleft palate and dwarfism in rat foetuses, and liver damage in sheep, rabbits and cows. Is responsible for 'Birdsville horse disease' produced in Australia following the consumption of *Indigofera linnaei*. Dogs fed on contaminated horsemeat are also affected.

Culvenor, C. C. J., *Aust. J. Chem.*, 1971, **24**, 371.

Ingenol 3,20-dibenzoate 774

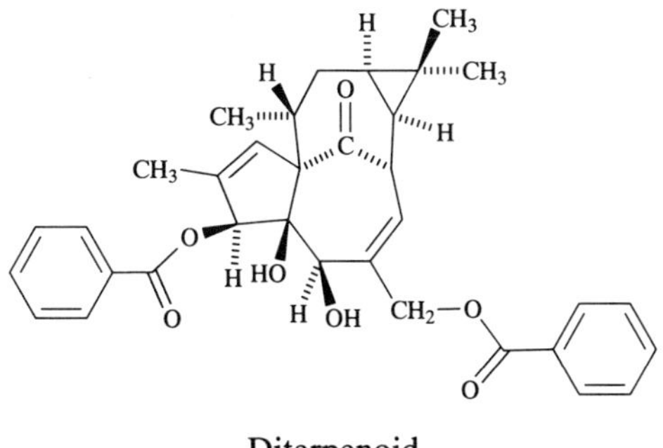

Diterpenoid
(Ingenane)

$C_{34}H_{36}O_7$ [59086-90-7] MW 556.66

Euphorbia esula (Euphorbiaceae).

Skin irritant and co-carcinogenic. It exhibits antileukaemic activity *in vivo*.

Kupchan, S. M., *Science*, 1976, **191**, 571.

Integerrimine 775

Squalidine; Alkaloid S-D

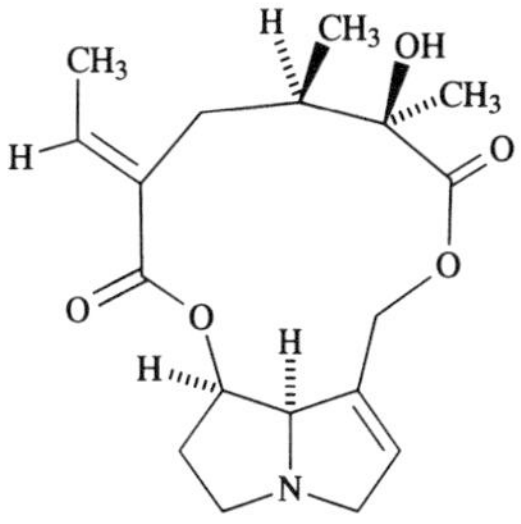

Pyrrolizidine alkaloid
(Senecionan)

$C_{18}H_{25}NO_5$ [480-79-5] MW 335.40

Senecio integerrimus, *S. squalidus*, *S. alpinus*, *S. brasiliensis*, *Cacalia hastata* and *Petasites hybridus* (Compositae) and in *Crotalaria brevifolia* and *C. incarnata* (Leguminosae).

Toxic.

Warren, F. L., *Fortschr. Chem. Org. Naturst*, 1955, **12**, 198.

Intermedine 776

9-(+)-Trachelanthylretronecine

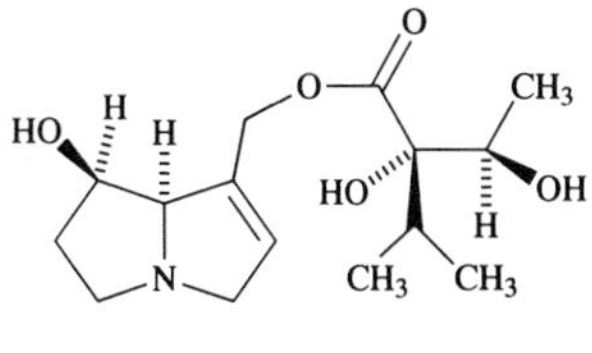

Pyrrolizidine alkaloid

$C_{15}H_{25}NO_5$ [10285-06-0] MW 299.37

In Russian comfrey, *Symphytum uplandicum*, *Amsinckia intermedia*, *A. lycopsoides* and *Trichodesma africana* (Boraginaceae) and in *Conoclinum coelastinum* (Compositae).

Hepatotoxic. Russian comfrey is sometimes eaten as a salad. Although the alkaloid content is low, this is dangerous because intermedine is a cumulative poison.

Culvenor, C. C. J., *Aust. J. Chem.*, 1966, **19**, 1955.

Inulicin 777

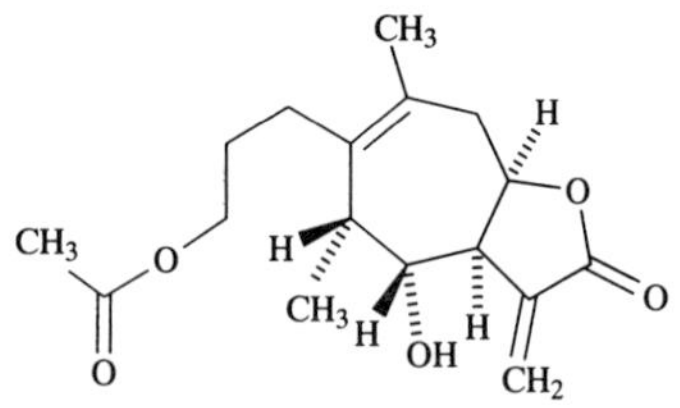

Sesquiterpenoid lactone
(Seco-pseudoguaianolide)

$C_{17}H_{24}O_5$ [33627-41-7] MW 308.37

Inula japonica (Compositae).

Stimulant of the central nervous system and of the smooth muscles of the intestine. Anti-ulcer properties.

Kiseleva, E. Y., *Chem. Nat. Compd.*, 1971, **7**, 254.

Ipomeamaronol 778

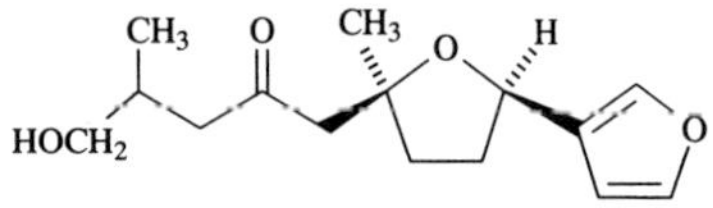

Furanosesquiterpenoid

$C_{15}H_{22}O_4$ [26767-96-4] MW 266.34

Produced in roots of sweet potato, *Ipomoea batatas* (Convolvulaceae) following damage or infection with the storage disease fungus *Ceratocystis fimbriata*.

Lung toxin. It can accumulate in damaged or infected swe potatoes and cause toxic effects when the potatoes are eaten. The LD_{50} is 266 mg/kg body-weight when injected intraperitoneally into mice.

Yang, D. T. C., *Phytochemistry,* 1971, **10**, 1653.

Ipomoeamarone 779

(+)-Ngaione

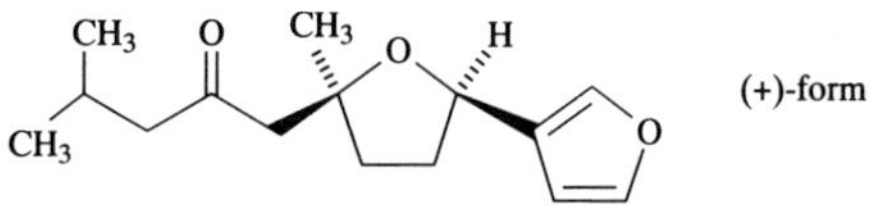

(+)-form

Furanosesquiterpenoid

$C_{15}H_{22}O_3$ [494-23-5] MW 250.34

Produced in roots of sweet potato, *Ipomoea batatas* (Convolvulaceae) following damage or infection with the fungus *Ceratocytis fimbriata*, a storage disease of sweet potato.

Lung toxin. It can accumulate in damaged sweet potatoes during storage and cause toxic effects when the potatoes are eaten. The LD_{50} is 200 mg/kg body-weight, when injected intraperitoneally in mice.

Birch, A. J., *Chem. Ind. (London),* 1954, 902.

Iridomyrmecin 780

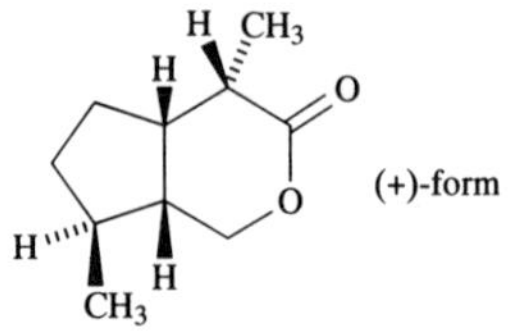

(+)-form

Monoterpenoid
(Iridane)

$C_{10}H_{16}O_2$ [485-43-8] MW 168.24

Actinidia polygama (Actinidiaceae).

Insecticidal activity.

Jaeger, R. H., *Tetrahedron Lett.,* 1959, (15), 14.

Isatidine 781

Retrorsine *N*-oxide

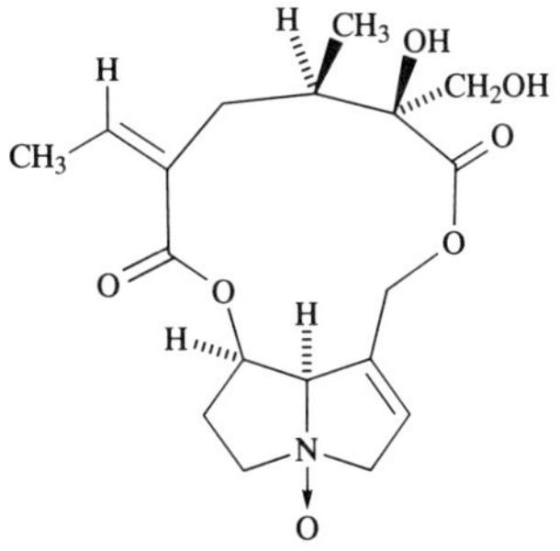

Pyrrolizidine alkaloid
(Senecionan)

$C_{18}H_{25}NO_7$ [15503-86-3] MW 367.40

Senecio longilobus and other *Senecio* spp. (Compositae).

Hepatotoxic, causing veno-occlusive disease in humans. Carcinogenic in livestock. It is far less toxic than retrorsine (q.v.) except when given orally, when the *N*-oxide is converted by gut enzymes to retrorsine base.

Molyneux, R. J., *Phytochemistry,* 1982, **21**, 439.

Isoajmaline 782

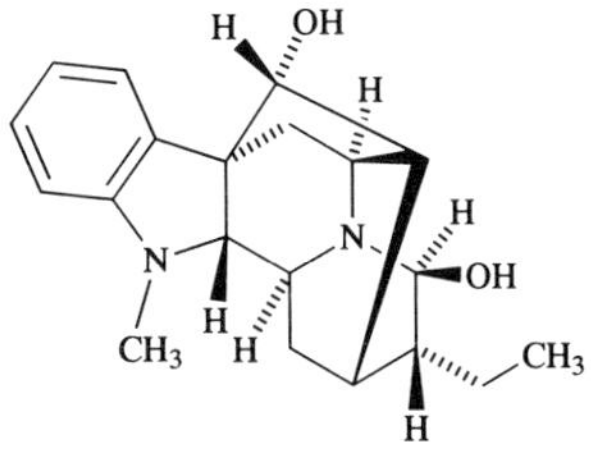

Indole alkaloid
(Ajmalan)

$C_{18}H_{26}N_2O_2$ [6989-79-3] MW 302.42

Roots of *Rauwolfia serpentina* (Apocynaceae).

Causes central nervous system stimulation followed by depression. It decreases blood pressure and also stimulates uterine contractions.

Mashimo, K., *Tetrahedron,* 1970, **26**, 803.

Isoalantolactone 783

Isohelenin

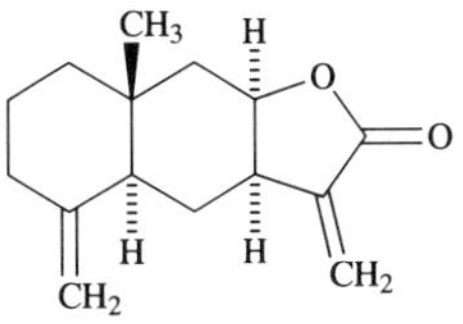

Sesquiterpenoid lactone
(Eudesmanolide)

$C_{15}H_{20}O_2$ [470-17-7] MW 232.32

In many *Inula* spp., e.g. in elecampane, *I. helenium*, in *Liatris cylindrica* and in *Telekia speciosa* (Compositae).

Antifeedant and affects the development of insects. It has been used as a vermifuge.

Marshall, J. A., *J. Org. Chem.,* 1964, **29**, 3727.

Isoaucuparin 784

3,5-Dimethoxy-4′-biphenylol

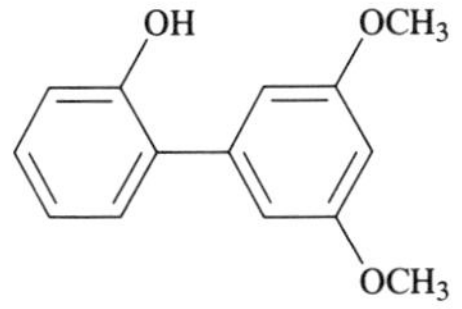

Biphenyl

$C_{14}H_{14}O_3$ [168301-25-5] MW 230.26

Sapwood of rowan, *Sorbus aucuparia* (Rosaceae) infected with *Nectria cinnabarina*.

Antifungal agent (phytoalexin).

Kokubun, T., *Phytochemistry,* 1995, **40**, 57.

Isobruceine A 785

Isobrucein A

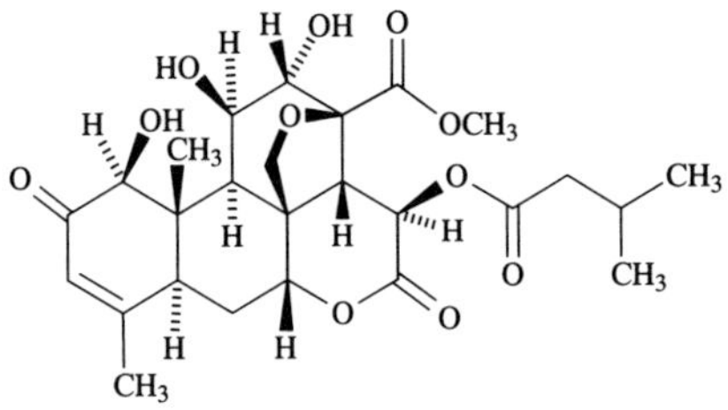

Nortriterpenoid
(Quassane)

$C_{26}H_{34}O_{11}$ [57629-50-2] MW 522.55

Soulamea tomentosa and *Brucea amarissima* (Simaroubaceae).

Insect antifeedant, e.g. against the southern armyworm *Spodoptera eridania*. It shows *in vivo* antileukaemic properties.

Polonsky, J., *Experientia,* 1975, **31**, 1113.

Isobutyrylmallotochromene 786

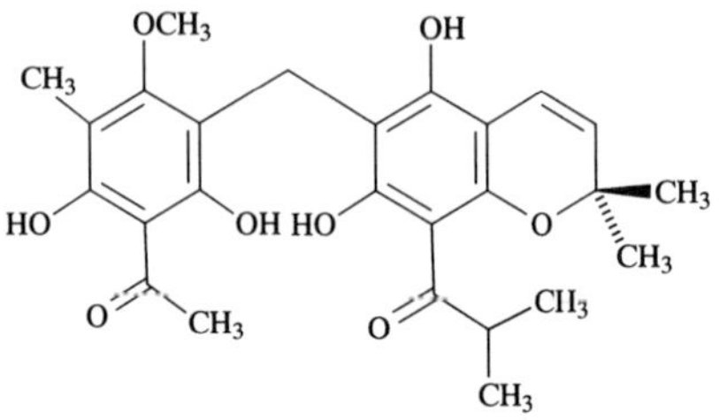

Chromene

$C_{26}H_{30}O_8$ [116964-16-0] MW 470.52

The pericarp of *Mallotus japonicus* (Euphorbiaceae).

Cytotoxic activity against KB cell lines.

Fujita, A., *J. Nat. Prod.,* 1988, **51**, 708.

Isochondrodendrine 787

Isobebeerine; Isendryl; Isodendril

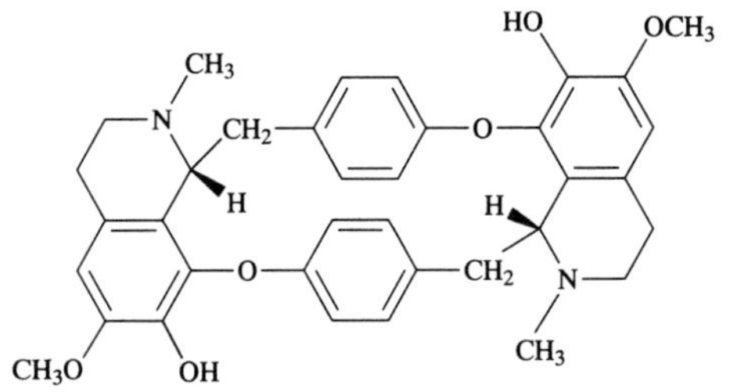

Bisbenzylisoquinoline alkaloid
(Cycleanan)

$C_{36}H_{38}N_2O_6$ [477-62-3] MW 594.71

In pareira, *Chondrodendron tomentosum*, *Sciadotenia toxifera* (both Menispermaceae), *Guatteria megalophylla* (Annonaceae) and *Heracleum wallichi* (Umbelliferae).

Muscle relaxant properties. The juice of plants containing this alkaloid are constituents of 'curare', the arrow poison used in South America.

Jeffreys, J. A. D., *J. Chem. Soc.,* 1956, 4451.

Isococculidine 788

O-Methylisococculine

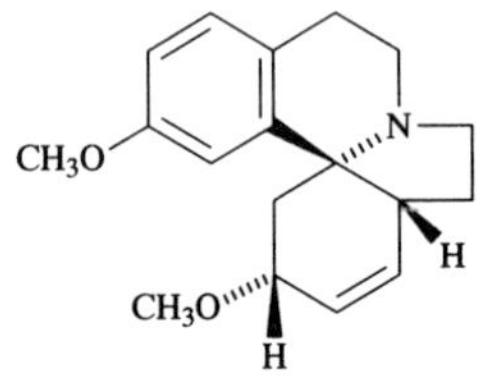

Erythrina alkaloid

$C_{18}H_{23}NO_2$ [60229-91-6] MW 285.39

Leaves of *Cocculus laurifolius* (Menispermaceae).

Neuromuscular blocking agent. Toxic.

Jain, S., *Indian J. Chem., Sect. B,* 1987, **26B**, 308.

Isocorydine 789

Artabotrine; Luteanine

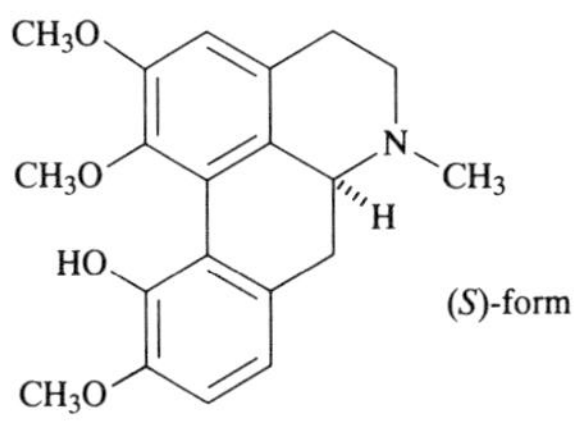

(*S*)-form

Benzylisoquinoline alkaloid
(Aporphine)

$C_{20}H_{23}NO_4$ [475-67-2] MW 341.41

Present in at least 12 plant families, including *Corydalis cava* (Fumariaceae), *Arbabotrys suaveolens* (Annonaceae), *Papaver* spp. (Papaveraceae) and *Phoebe clemensii* (Lauraceae).

Toxic to mammals. LD_{50} intraperitoneally in rats is 10.9 mg/kg body-weight. It is a sedative at low doses but cataleptic at high doses. It has anti-adrenergic properties.

Schlittler, E., *Helv. Chim. Acta,* 1952, **35**, 111.

Isocryptomerin 790

Hinokiflavone 7-methyl ether

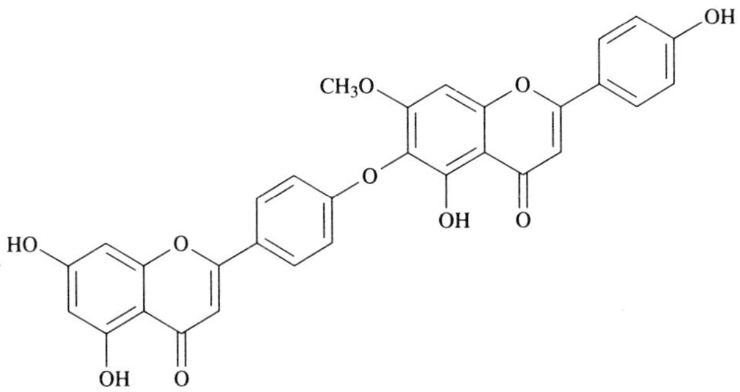

Biflavonoid

$C_{31}H_{21}O_{10}$ [20931-58-2] MW 552.49

Leaves of *Selaginella willdenowii* (Selaginellaceae).

Cytotoxic activity against the human cancer cell lines HT-1080 and Lu 1.

Miura, H., *Chem. Pharm. Bull.,* 1967, **15**, 232.

Isodiospyrin 791

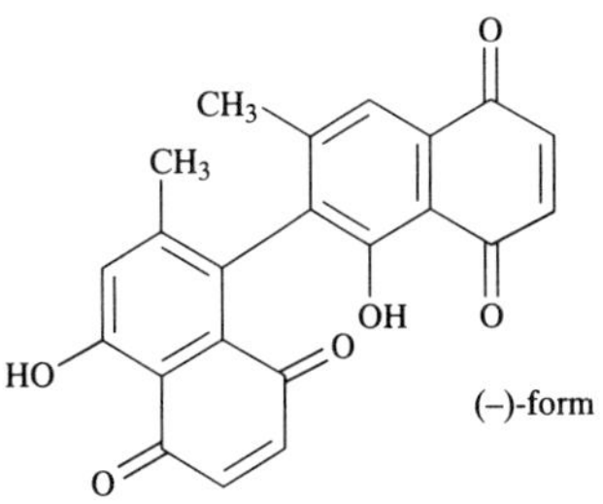

(–)-form

Binaphthoquinone

$C_{22}H_{14}O_6$ [20175-84-2] MW 374.35

Root and stem bark of *Diospyros usamabariensis* (Ebenaceae); also in other *Diospyros* spp. and in some of the related *Euclea* spp..

Molluscicide. It has cytotoxic properties.

Thomson, R. H., Naturally Occurring Quinones, 3rd edn, 1987, 172, Chapman and Hall.

Isodomedin 792

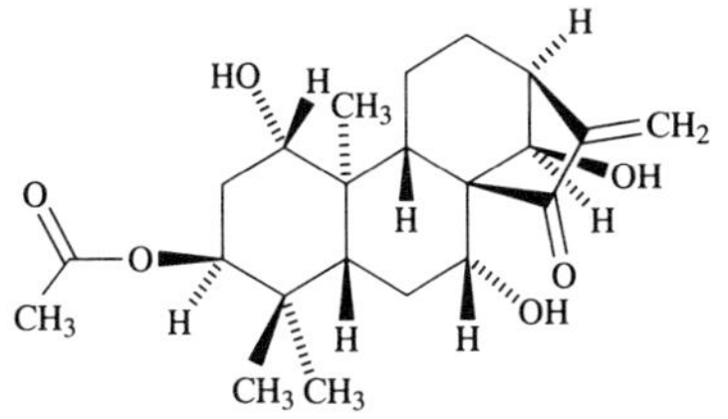

Diterpenoid
(Kaurane)

$C_{22}H_{32}O_6$ [39388-61-9] MW 392.49

Leaves of *Isodon shikokianus* var. *intermedius* (Labiatae).

Cytotoxic activity. Insect antifeedant. Is a bitter principle.

Kubo, I., *J. Chem. Soc., Chem. Commun.,* 1977, 555.

Isodomoic acid A 793

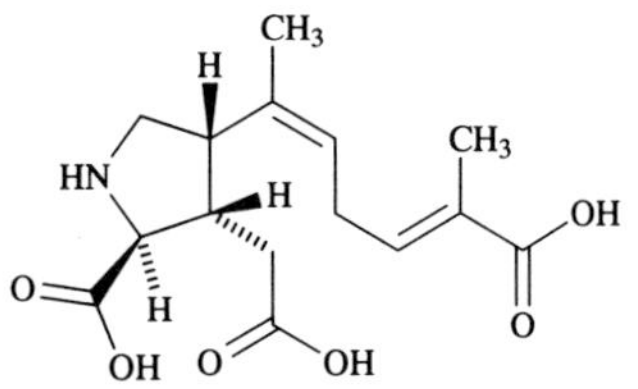

Amino acid

$C_{15}H_{21}NO_6$ [101899-44-9] MW 311.33

Red alga, *Chondria armata*.

Insecticidal.

Maeda, M., *Chem. Pharm. Bull.,* 1986, **34**, 4892.

Isodomoic acid B 794

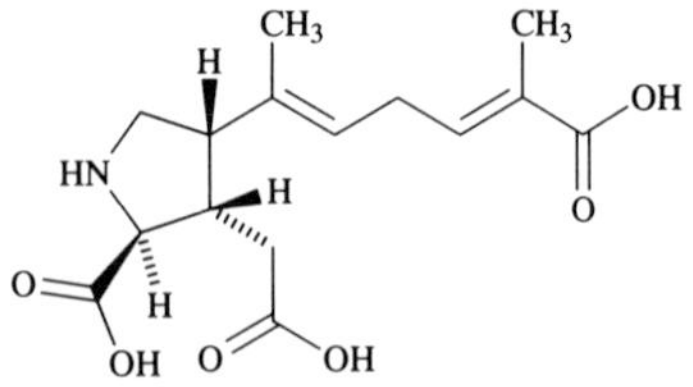

Amino acid

$C_{15}H_{21}NO_6$ [101877-25-7] MW 311.33

Red alga, *Chondria armata*.

Insecticidal, especially to the American cockroach.

Maeda, M., *Chem. Pharm. Bull.,* 1986, **34**, 4892.

Isoelemicin 795

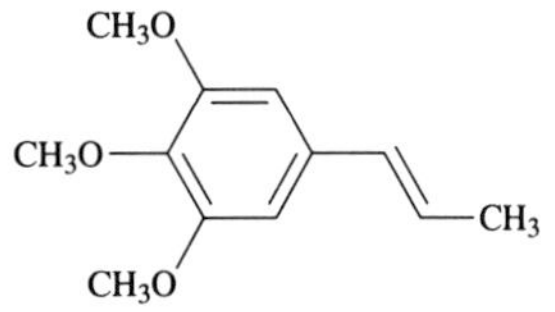

Phenylpropanoid

$C_{12}H_{16}O_3$ [5273-85-8] MW 208.26

Oil of nutmeg, *Myristica fragrans* (Myristicaceae), oil of *Backhousia myrtifolia* (Myrtaceae), *Libanotis transcaucasia* and *Diplolophium buchanani* (Umbelliferae).

Hypnotic activity. Toxic to mosquito larvae and antifungal.

Shulgin, A. T., *Naturwissenschaften,* 1964, **51**, 360.

Isofebrifugine 796

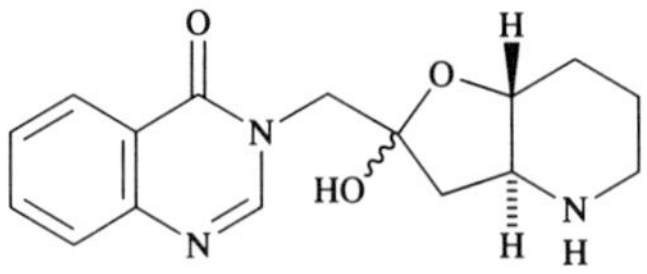

Quinazoline alkaloid

$C_{16}H_{19}N_3O_3$ [32434-44-9] MW 301.35

Dichroa febrifuga and *Hydrangea umbellata* (Saxifragaceae).

Toxic to moths.

Koepfli, J. B., *J. Am. Chem. Soc.,* 1950, **72**, 3323.

Isohelenol 797

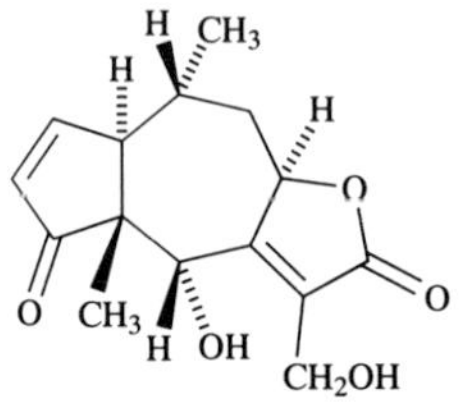

Sesquiterpenoid lactone
(Pseudoguaianolide)

$C_{15}H_{18}O_5$ [71013-32-6] MW 278.30

Helenium microcephalum (Compositae).

Cytotoxic and antitumour properties.

Sims, D., *J. Nat. Prod.,* 1979, **42**, 282.

Isomontanolide 798

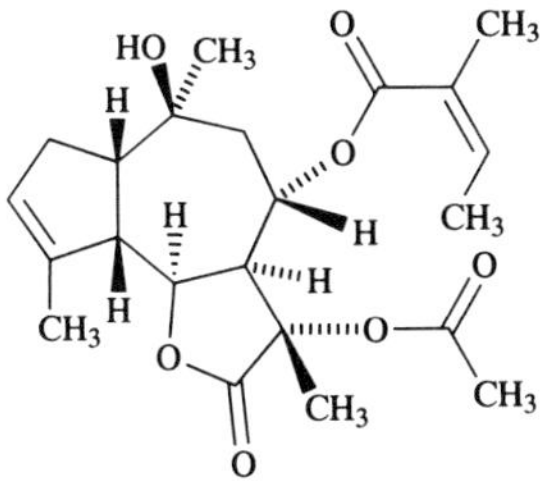

Sesquiterpenoid lactone
(Guaianolide)

$C_{22}H_{30}O_7$ [38114-47-5] MW 406.48

Fruit of *Laserpitium siler* (Umbelliferae).

Insect antifeedant.

Holub, M., *Collect. Czech. Chem. Commun.*, 1972, **37**, 1186.

Isoobtusilactone A 799

Borbonol 2

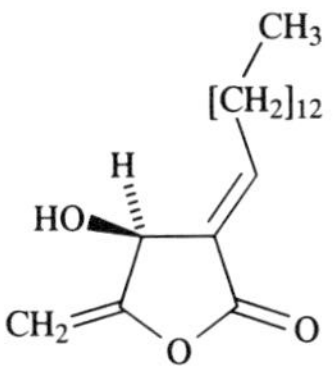

Aliphatic lactone

$C_{19}H_{32}O_3$ [56522-16-8] MW 308.46

Roots and stems of *Persea borbonia* and other *Persea* spp. (Lauraceae); co-occurs with three related lactones.

Antifungal, active against *Phytophthora cinnamomi* at a concentration of 1 μg/ml.

Niwa, M., *Chem. Lett.*, 1975, 655.

6-Isopentenylnaringenin 800

6-Prenylnaringenin

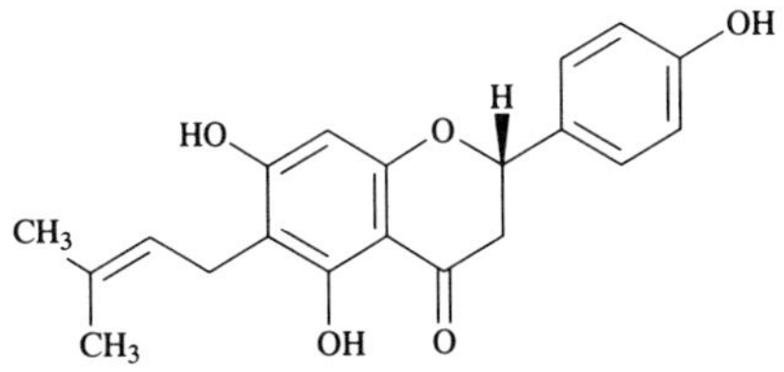

Flavonoid
(Flavanone)

$C_{20}H_{20}O_5$ [68236-13-5] MW 340.38

Wood resin of hops, *Humulus lupulus* (Cannabidaceae).

Antifungal properties.

Mizobuchi, S., *Agric. Biol. Chem.*, 1984, **48**, 2771.

Isopimpinellin 801

4,9-Dimethoxypsoralen

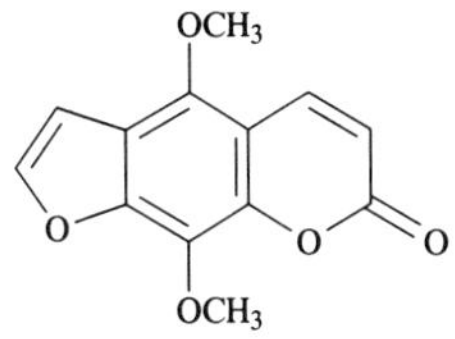

Furanocoumarin

$C_{13}H_{10}O_5$ [482-27-9] MW 246.22

In Umbelliferae genera, such as *Pimpinella* (roots), *Angelica* (roots), *Heracleum* (roots), *Pastinaca* (seeds), *Ferula* (seeds) and *Seseli* spp. (seeds). Also present in the Rutaceae in lime oil, *Citrus aurantifolia* and in the Compositae in *Trichocline incana*.

Toxic to *Schistosoma*-carrying snails and also to fish. Has tuberculostatic activity against *Myobacterium tuberculosis*.

Wessely, F., *Monatsh. Chem.*, 1932, **59**, 161.

Isosafrole 802

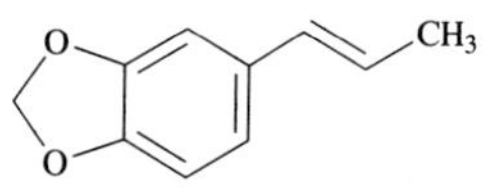

Phenylpropanoid

$C_{10}H_{10}O_2$ [120-58-1] MW 162.19

Essential oil of *Ligusticum acutilobum* (Umbelliferae), leaves of *Murraya koenigii* (Rutaceae), and ylang-ylang, *Cananga odorata* (Annonaceae).

Like safrole (q.v.), it is moderately toxic to humans. It is an inductor of cytochrome P450 and stimulates liver regeneration.

Robinson, R., *J. Chem. Soc.*, 1927, 2489.

Isotenulin 803

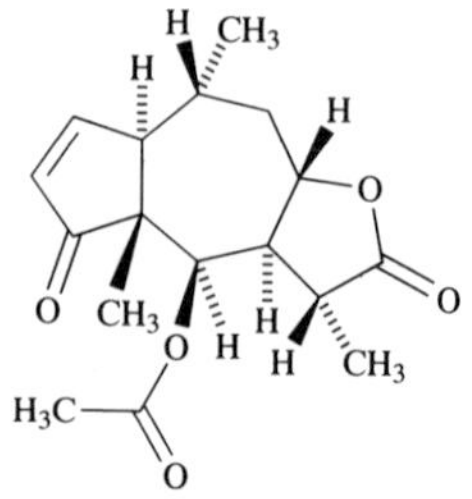

Sesquiterpenoid lactone
(Pseudoguaianolide)

$C_{17}H_{22}O_5$ [10092-04-3] MW 306.36

Helenium arizonicum and *H. bigelovii* (Compositae).

Weak analgesic properties when injected subcutaneously.

Barton, D. H. R., *J. Chem. Soc.*, 1956, 142.

Isothebaine 804

1-Hydroxy-2,11-dimethoxyaporphine

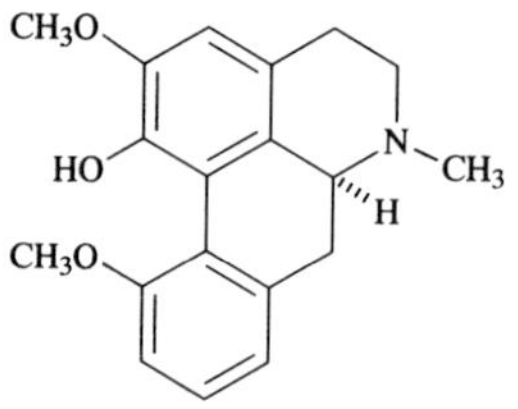

Aporphine alkaloid

$C_{19}H_{21}NO_3$ [568-21-8] MW 311.38

Papaver orientale and *P. pseudo-orientale* (Papaveraceae).

Poisonous to man. The LD_{50} in humans is 26 mg/kg bodyweight. It depresses respiration and heart rate and decreases motor activity. It has some useful properties, being anti-inflammatory and analgesic in animals.

Battersby, A. R., *J. Chem. Soc.*, 1965, 4550.

Isoxanthohumol 805

7,4′-Dihydroxy-5-methoxy-8-prenylflavanone

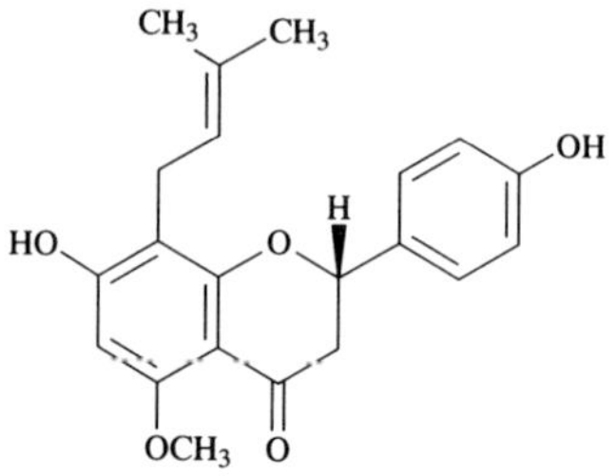

Flavonoid
(Flavanone)

$C_{21}H_{22}O_5$ [70872-29-6] MW 354.40

Hard resin of hops, *Humulus lupulus* (Cannabidaceae).

Antifungal activity.

Verzele, M., *Bull. Soc. Chim. Belg.*, 1957, **66**, 452.

Ivalin 806

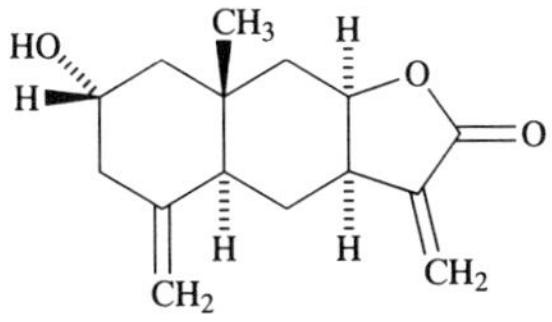

Sesquiterpenoid lactone
(Eudesmanolide)

$C_{15}H_{20}O_3$ [5938-03-4] MW 248.32

Iva imbricata, *I. microcephala* and in *Carpesium*, *Inula*, *Wedelia* and *Zaluzania* spp. (Compositae).

Toxic to mammals.

Herz, W., *J. Org. Chem.*, 1962, **27**, 905.

J

Jacobine 807

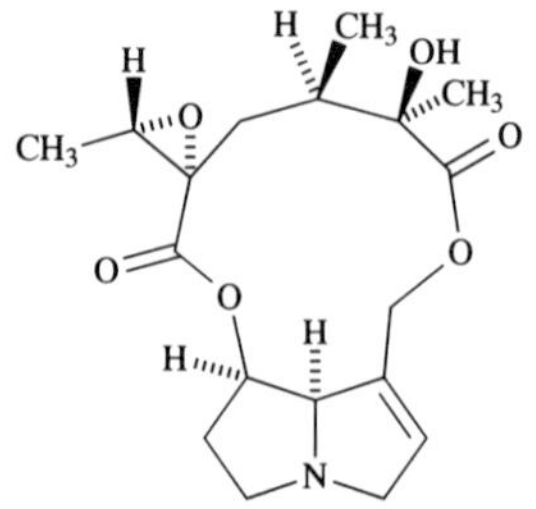

Pyrrolizidine alkaloid
(Senecionan)

$C_{18}H_{25}NO_6$ [6870-67-3] MW 351.40

Tansy ragwort, *Senecio jacobaea*, *S. alpinus*, *S. cineraria* and *S. incarnus* (Compositae).

Toxic. It is a potent inducer of hepatic epoxide hydrolase and is anticholinergic in rats.

Masamune, S., *J. Am. Chem. Soc.,* 1960, **82**, 5253.

Jatrophatrione 808

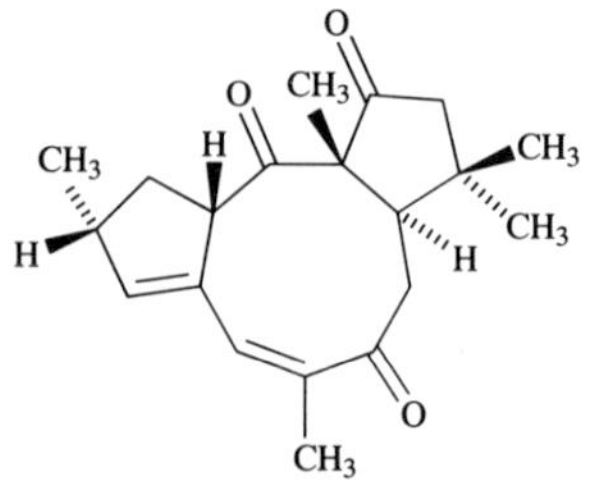

Diterpenoid
(Jatrophane B)

$C_{20}H_{26}O_3$ [58298-76-3] MW 314.42

Jatropha macrorhiza (Euphorbiaceae).

Antitumour activity.

Torrance, S. J., *J. Org. Chem.,* 1976, **41**, 1855.

Jatrophone 809

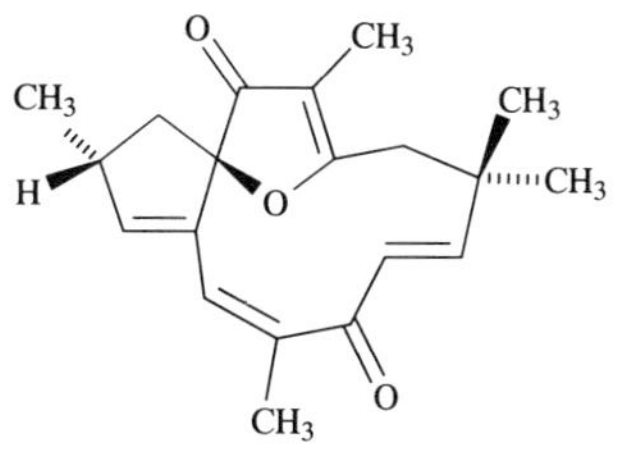

Diterpenoid
(Jatrophane A)

$C_{20}H_{24}O_3$ [29444-03-9] MW 312.41

Jatropha gossypiifolia (Euphorbiaceae).

Antitumour properties.

Kupchan, S. M., *J. Am. Chem. Soc.,* 1976, **98**, 2295.

Jatrorrhizine 810

Jateorrhizine; Jatrorhizine; Neprotine

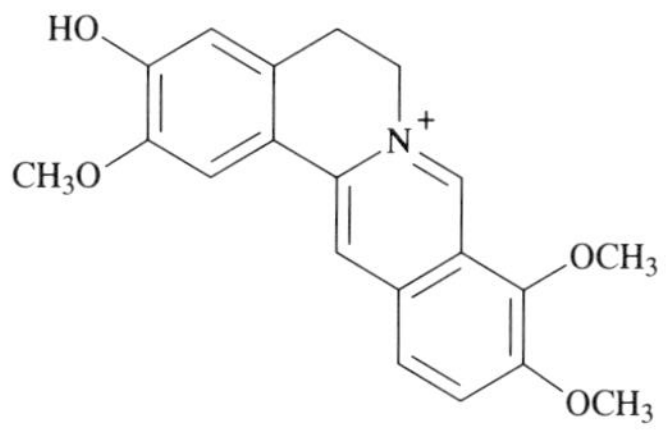

Benzylisoquinoline alkaloid
(Berbine)

$[C_{20}H_{20}NO_4]^+$ [3621-38-3] MW 338.38

Calumba root, *Jateorrhiza palmata* (Menispermaceae) and in many of the *Berberis*, *Mahonia* (Berberidaceae), *Thalictrum* (Ranunculaceae) and *Michelia* spp. (Magnoliaceae).

Sedative and hypotensive in animals.

Späth, E., *Ber. Dtsch. Chem. Ges.,* 1925, **58**, 1939.

Jervine 811

Jervanin-11-one

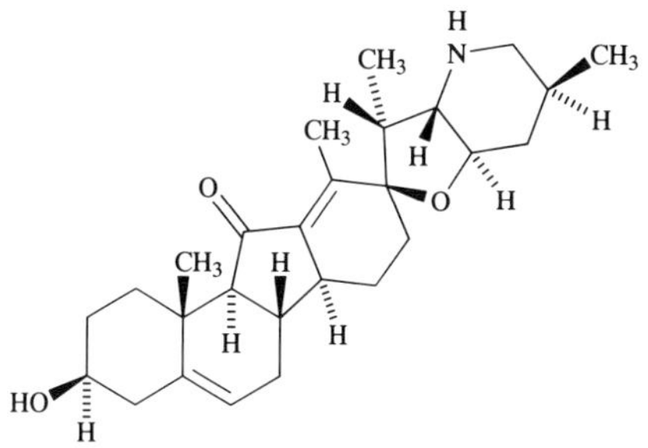

Steroidal alkaloid
(Jervanine)

$C_{27}H_{39}NO_3$ [469-59-0] MW 425.61

In rhizomes of *Veratrum album*, *V. grandiflorum*, *V. viride* and other *Veratrum* spp. (Liliaceae).

Toxin of *Veratrum album*, which is teratogenic in sheep and poisonous to humans. The fatal dose is 20 mg. As little as 1 to 2 g of dried root could kill, while the leaf also contains dangerous amounts.

Kupchan, S. M., *J. Am. Chem. Soc.,* 1968, **90**, 2730.

Jesaconitine 812

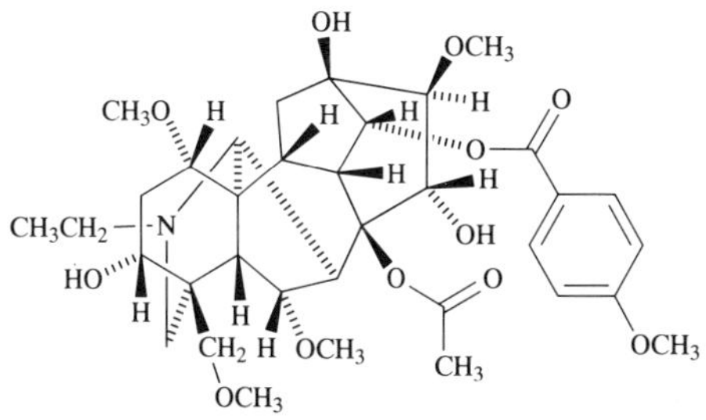

Diterpenoid alkaloid
(Aconitane)

$C_{35}H_{49}NO_{12}$ [16298-90-1] MW 675.77

Aconitum fischeri, *A. subcuneatum* and *A. sachalinense* (Ranunculaceae).

Highly toxic to mammals, affecting blood pressure, heart rate and respiration. Small doses increase the contractions of the mammalian intestine.

Keith, L. H., *J. Org. Chem.,* 1968, **33**, 2497.

Jodrellin B 813

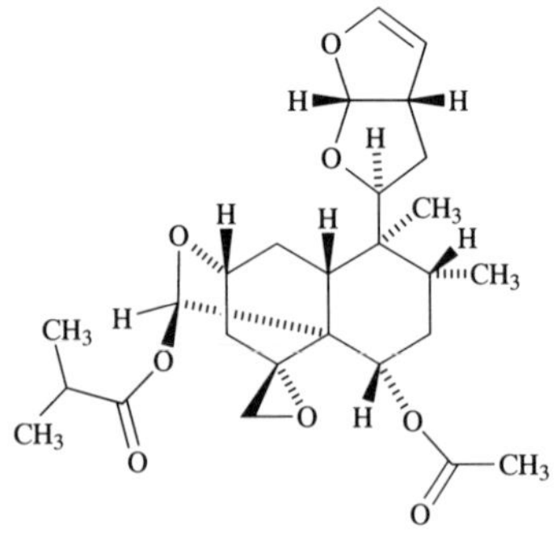

Diterpenoid
(Clerodane)

$C_{26}H_{36}O_8$ [124901-81-1] MW 476.57

Aerial parts of *Scutellaria violacea* and *S. woronowii* (Labiatae).

Antifungal activity against *Fusarium oxysporum*. Insect antifeedant.

Anderson, J. C., *Tetrahedron Lett.,* 1989, **30**, 4737.

Juglone 814

Nucin; Regianin; Natural Brown 7;
5-Hydroxy-1,4-naphthoquinone

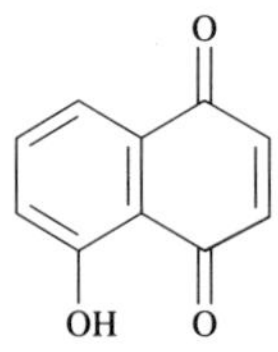

Naphthoquinone

$C_{10}H_6O_3$ [481-39-0] MW 174.16

Stem bark of black walnuts, *Juglans nigra*, *J. regia* and *Carya ovata*, leaves and nuts of *C. illinoensis* (Juglandaceae) and *Lomatia* spp. (Proteaceae).

Molluscicide. Sedating activity in mammals and fish. Feeding deterrent to the barkbeetle, *Scolytus multistriatus*. In spite of its toxic properties, it is not responsible for 'laminitis' in horses, which is produced by ingestion of black walnut tissue; the active agent is still unknown.

Thomson, R. H., Naturally Occurring Quinones, 3rd edn, 1987, 154, Chapman & Hall.

Justicidin A 815

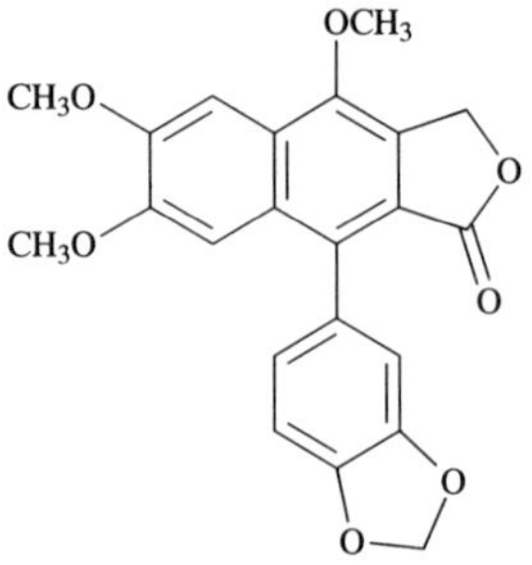

Lignan

$C_{22}H_{18}O_7$ [25001-57-4] MW 394.38

Justicia procumbens and *J. hayatai* var. *decumbens* (Acanthaceae).

Toxic to fish; also cytotoxic activity.

Okigawa, M., *Tetrahedron,* 1970, **26**, 4301.

Justicidin B 816

Dehydrocollinusin

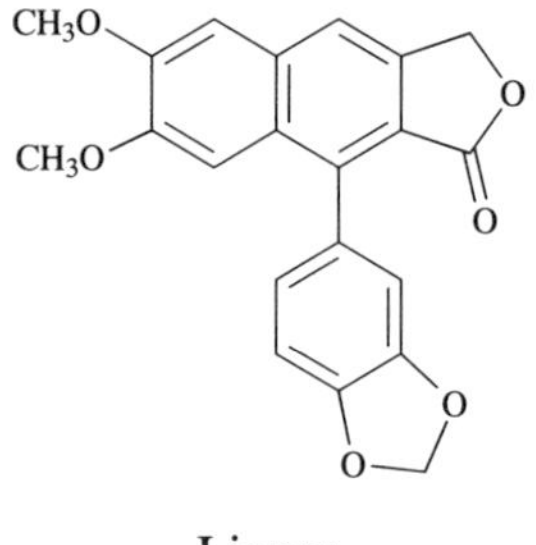

Lignan

$C_{21}H_{16}O_6$ [17951-19-8] MW 364.35

Phyllanthus acuminatus (Euphorbiaceae), *Justicia hayatai* (Acanthaceae), *Haplophyllum tuberculatum* (Rutaceae) and *Sesbania drummondii* (Leguminosae).

Piscicidal activity. Also has antiviral properties.

Abdullaev, N. D., *Chem. Nat. Compd.,* 1987, **23**, 63.

Juvabione 817

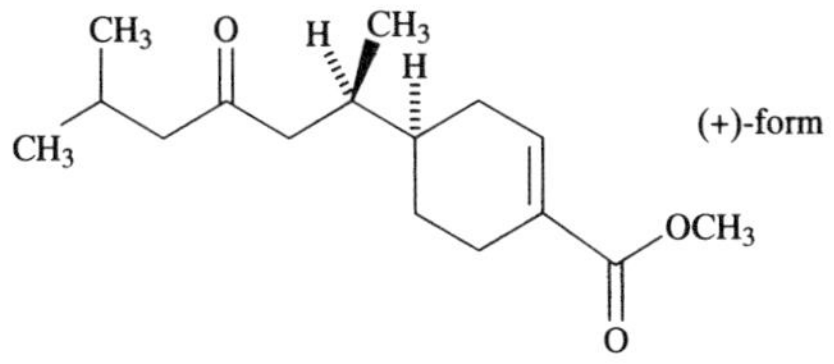

Sesquiterpenoid
(Bisabolane)

$C_{16}H_{26}O_3$ [17904-27-7] MW 266.38

Wood of balsam fir, *Abies balsamea* (Pinaceae) and of some other *Abies* spp..

Insect juvenile hormone activity, interfering with metamorphosis and preventing maturation, with fatal consequences.

Bowers, W. S., *Science,* 1966, **154**, 1020.

Juvadecene 818

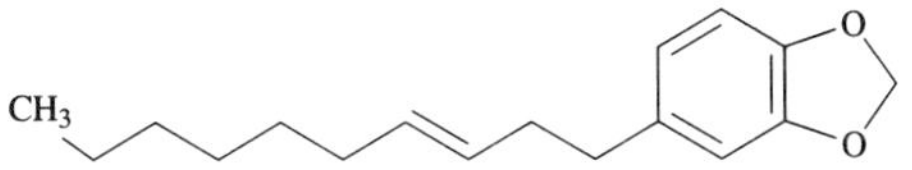

Phenylalkene

$C_{17}H_{24}O_2$ [89139-89-9] MW 260.38

Root of *Macropiper excelsum* (Piperaceae).

Insecticidal.

Nishida, R., *Arch. Insect Biochem. Physiol.,* 1983, **1**, 17.

Juvenile hormone III 819

JH III

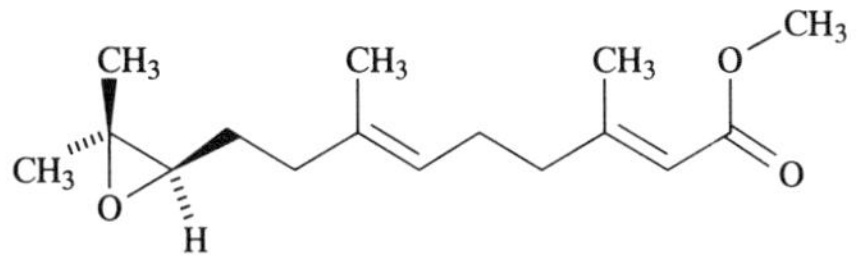

Sesquiterpenoid

$C_{16}H_{26}O_3$ [22963-93-5] MW 266.38

Grasshopper's cyperus, *Cyperus iria* (Cyperaceae).

Insect hormonal activity. It causes sterility in female grasshoppers eating *Cyperus iria*.

Toong, Y. C., *Nature (London),* 1988, **333**, 170.

Juvocimene 1 820

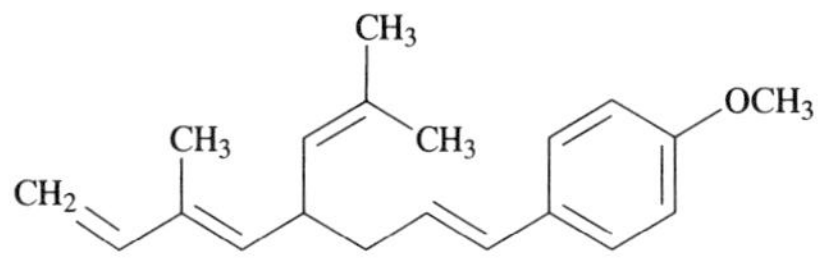

Monoterpenoid-phenylpropanoid

$C_{20}H_{26}O$ [75539-64-9] MW 282.43

Sweet basil, *Ocimum basilicum* (Labiatae). It co-occurs, with juvocimene 2 [75539-63-8], the closely related epoxide.

Insecticidal. A juvenile hormone mimic, it upsets normal metamorphosis in insect larvae.

Bowers, W. S., *Science,* 1980, **209**, 1030.

K

Kaempferol 3-(2,4-di-*p*-coumarylrhamnoside) 821

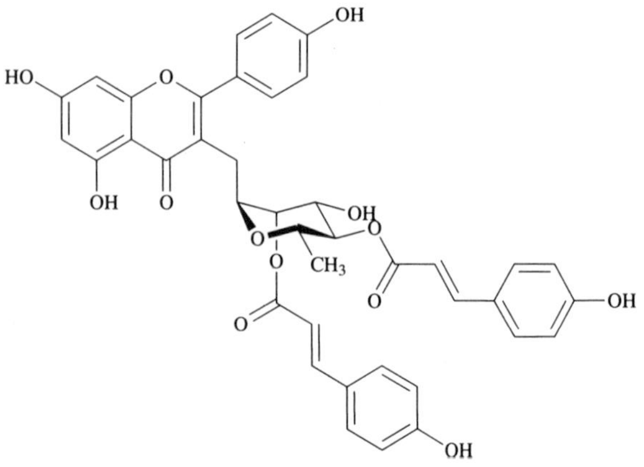

Flavonol glycoside

$C_{39}H_{32}O_{14}$ [163434-73-9] MW 724.67

Leaves of *Pentachondra pumila* (Epacridaceae).

Strong antibacterial activity against *Staphylococcus aureus*.

Bloor, S. J., *Phytochemistry,* 1995, **38**, 1033.

Kansuinine B 822

Kansuinin B

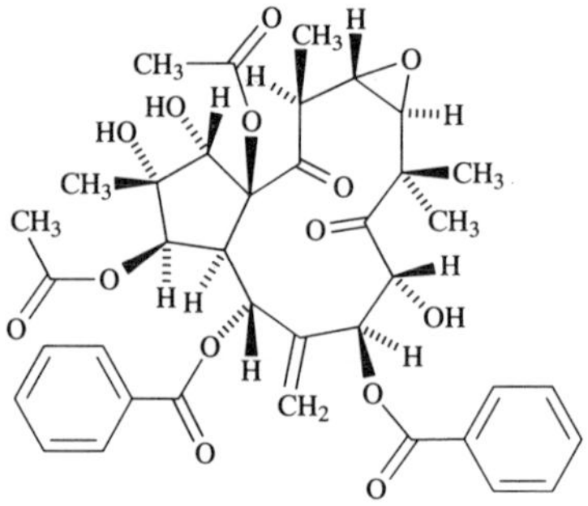

Diterpenoid
(Jatrophane A)

$C_{38}H_{42}O_{14}$ [57685-46-8] MW 722.74

Euphorbia kansui (Euphorbiaceae).

Highly toxic, with analgesic properties.

Hirata, Y., *Pure Appl. Chem.,* 1975, **41**, 175.

Karakin 823

1,2,6-Tris(3-nitropropanoyl)-β-D-glucoside

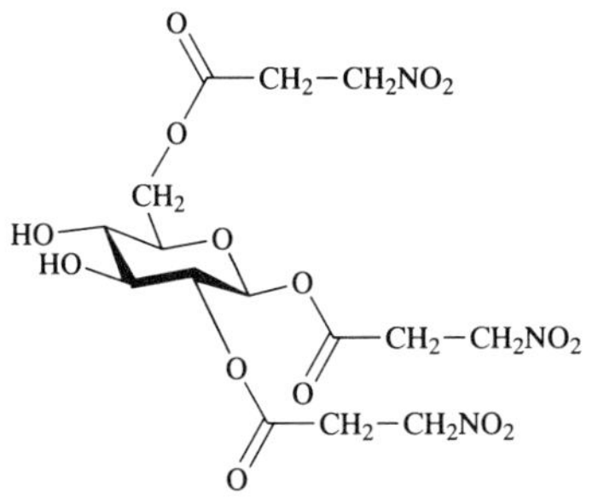

Nitro compound

$C_{15}H_{21}N_3O_{15}$ [1400-11-9] MW 483.34

Astragalus canadensis, *A. cibarius*, *A. falcatus* and *A. flexuosus* (Leguminosae). Its 1-epimer, coronillin [63368-43-4], occurs in aerial parts of *Coronilla varia* (also Leguminosae).

Toxin causing methaemoglobinaemia. May be responsible for poisoning livestock feeding on *Astragalus* plants and other legumes.

Harlow, M. C., *Phytochemistry,* 1975, **14**, 1421.

Karakoline 824

Karacoline; Carmichaeline

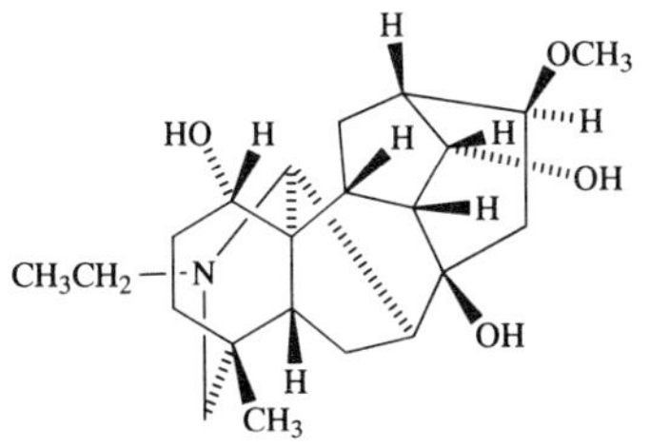

Diterpenoid alkaloid
(Aconitane)

$C_{22}H_{35}NO_4$ [39089-30-0] MW 377.52

Tubers of *Aconitum carmichaeli*, *A. karakolicum* and in *Delphinium pentagynum* (Ranunculaceae).

Acute toxicity is about twice as potent as that of talatizamine (q.v.) in mice. Intraperitoneal injection into mice produces muscular weakness, and death from respiratory depression. Intravenous administration to cats or dogs lowers the blood pressure for 15-20 min.

Sultankhodzhaev, M. N., *Chem. Nat. Compd.,* 1973, **9**, 194.

Karwinskione 825

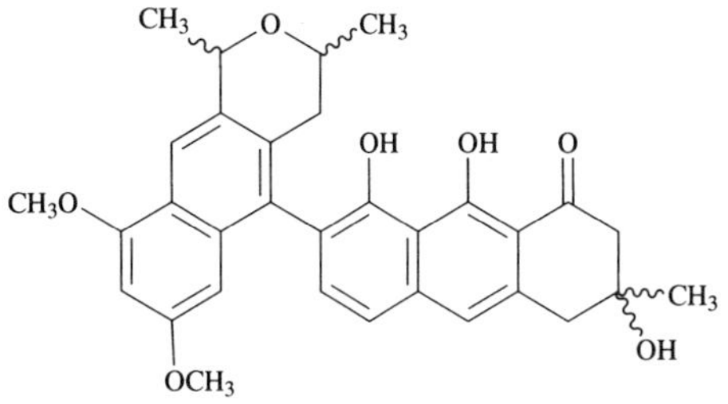

Anthracene

$C_{32}H_{32}O_7$ [59481-46-8] MW 528.60

Fruits of *Karwinskia humboldtiana* (Rhamnaceae), accompanied by tullidinol [56678-09-2], the 10′-hydroxy derivative.

The fruits of *Karwinskia* are edible, but the seeds contain these two neurotoxins, which can cause motor paralysis.

Dominguez, X. A., *Rev. Latinoam. Quim.,* 1976, **7**, 46.

Kievitone 826

Phaseolus substance II; Vignatin

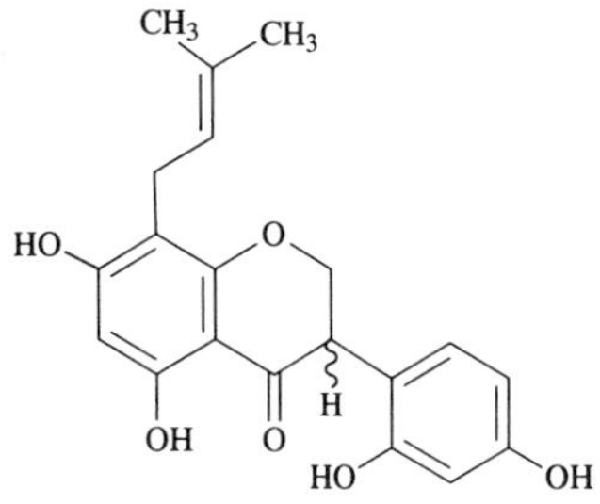

Isoflavonoid
(Isoflavanone)

$C_{20}H_{20}O_6$ [40105-60-0] MW 356.38

In fungally-infected leaves of cowpea, *Vigna unguiculata* (Leguminosae).

Antifungal properties (phytoalexin). Causes lysis of red blood cells.

Smith, D. A., *Physiol. Plant Pathol.*, 1973, **3**, 293.

Kigelinone 827

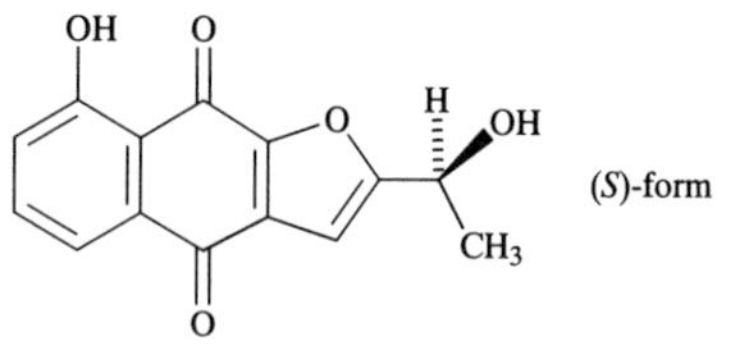

(*S*)-form

Naphthoquinone

$C_{14}H_{10}O_5$ [80931-34-6] MW 258.23

Wood of *Kigelia pinnata* and *Crescentia cujete* and stembark of *Tabebuia cassinoides* (Bignoniaceae).

Cytotoxic properties.

Fujimoto, Y., *J. Chem. Soc., Perkin Trans. 1*, 1991, 2323.

Kokusaginine 828

6,7-Dimethoxydictamnine

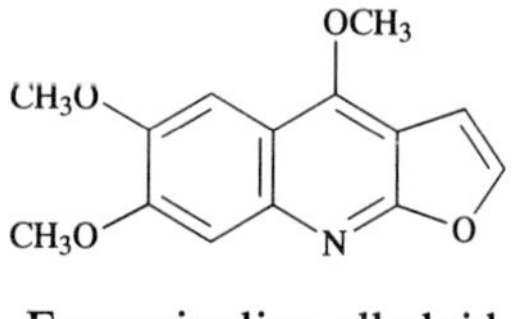

Furoquinoline alkaloid

$C_{14}H_{13}NO_4$ [484-08-2] MW 259.26

Common in the Rutaceae, e.g. in *Euodia*, *Orixa*, *Haplophyllum*, *Melicope* and *Acronychia* spp..

Moderate insect antifeedant properties. This alkaloid enhances noradrenaline and dopamine levels in the mouse brain.

Anet, F. A. L., *Aust. J. Sci. Res., Ser. A*, 1952, **5**, 412.

Kuwanone G 829

Kuwanon G; Moracenin B; Albanin F

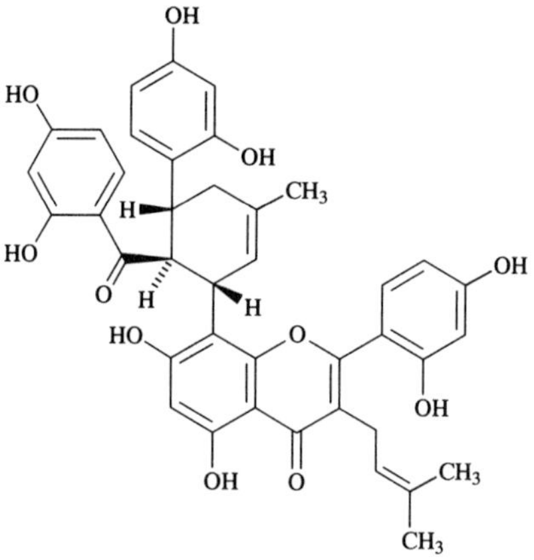

Flavone

$C_{40}H_{36}O_{11}$ [75629-19-5] MW 692.72

Rootbark of mulberry, *Morus alba* (Moraceae).

Hypotensive activity. It lowers blood pressure in rabbits when administered intravenously at a dose of 1.0 mg/kg body-weight.

Oshima, Y., *Tetrahedron Lett.*, 1980, **21**, 3381.

L

Labriformidin 830

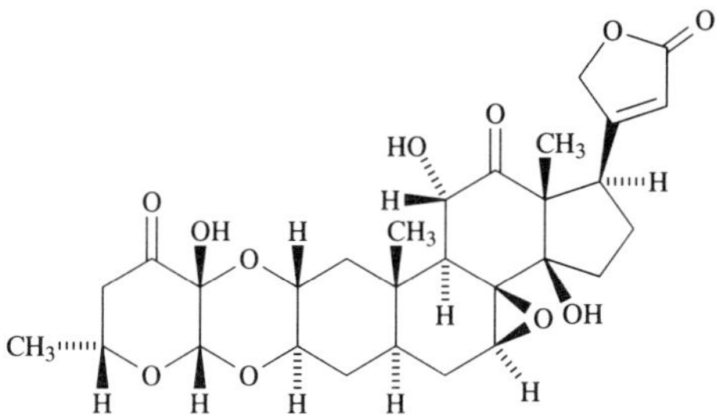

Cardenolide

$C_{29}H_{36}O_{11}$ [66419-08-7] MW 560.60

Leaves of the milkweeds, *Asclepias labriformis* and *A. eriocarpa* (Asclepiadaceae).

Very toxic to vertebrates. The LD_{50} is 3.1 mg/kg body-weight when administered intraperitoneally to male Swiss Webster mice.

Cheung, H. T. A., *J. Chem. Soc., Perkin Trans. 1,* 1980, 2169.

Labriformin 831

Labriformine

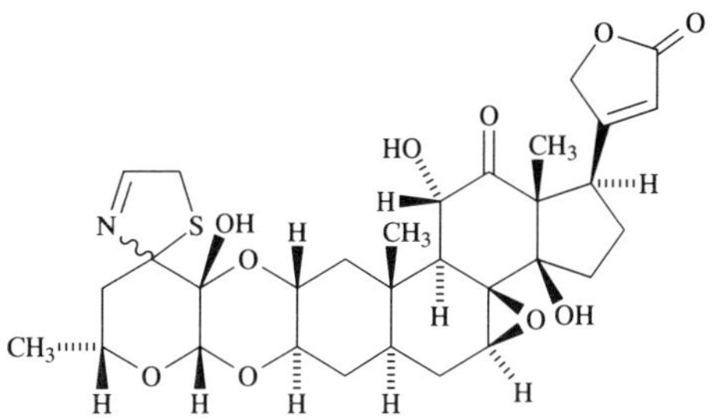

Cardenolide

$C_{31}H_{39}NO_{10}S$ [66419-07-6] MW 617.72

Leaves of the milkweeds, *Asclepias labriformis* and *A. eriocarpa* (Asclepiadaceae).

Very toxic to vertebrates. LD_{50} is 9.2 mg/kg body-weight when administered intraperitoneally in male Swiss Webster mice.

Fonseca, G., *J. Nat. Prod.,* 1991, **54**, 860.

Lachnophyllum lactone 832

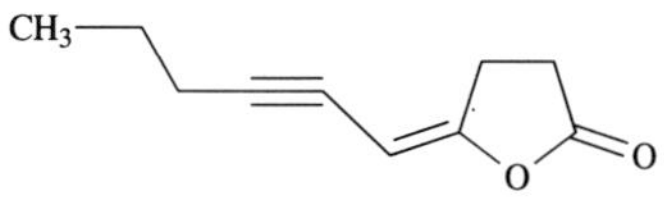

(E)-isomer

Acetylene

$C_{10}H_{10}O_2$ [23251-67-4] MW 162.19
(E)-isomer [81112-93-8]
(Z)-isomer [81122-95-4]

Leaf of *Baccharis pedunculata* (Compositae).

High toxicity (LD_{50} 2 μg/ml) against human keratinocytes. Antifungal activity against both human and plant pathogenic fungi.

Bohlmann, F., *Chem. Ber.*, 1965, **98**, 2236.

Lacinilene C 7-methyl ether 833

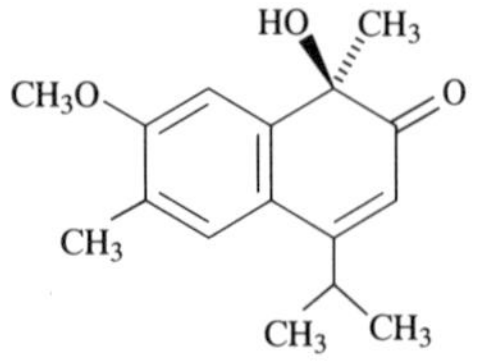

Sesquiterpenoid
(Cadinane)

$C_{16}H_{20}O_3$ [56362-72-2] MW 260.33

Formed in frost-killed bracts of cotton, *Gossypium hirsutum* (Malvaceae) and present in cotton dust.

Has *in vitro* chemotactic activity towards polymorphonuclear leukocytes. It has been implicated in the onset of byssinosis, an illness associated with workers inhaling cotton dust.

Stipanovic, R. D., *Phytochemistry,* 1975, **14**, 1041.

Lactucin 834

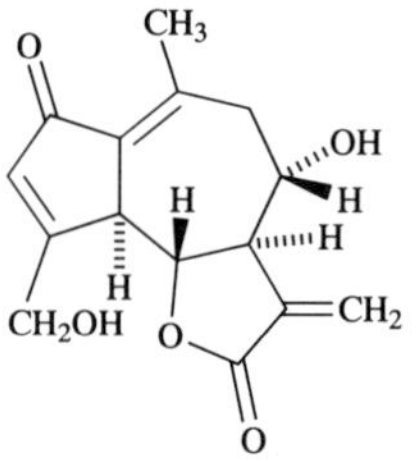

Sesquiterpenoid lactone
(Guaianolide)

$C_{15}H_{16}O_5$ [1891-29-8] MW 276.29

Chicory root and leaf, *Cichorium intybus*, and wild species of lettuce, *Lactuca canadensis*, *L. serriola* and *L. virosa* (Compositae).

Bitter principle. Cytotoxic and antitumour activities.

Bachelor, F. W., *Can. J. Chem.,* 1973, **51**, 3626.

Lactucopicrin 835

Intibin; Intybin

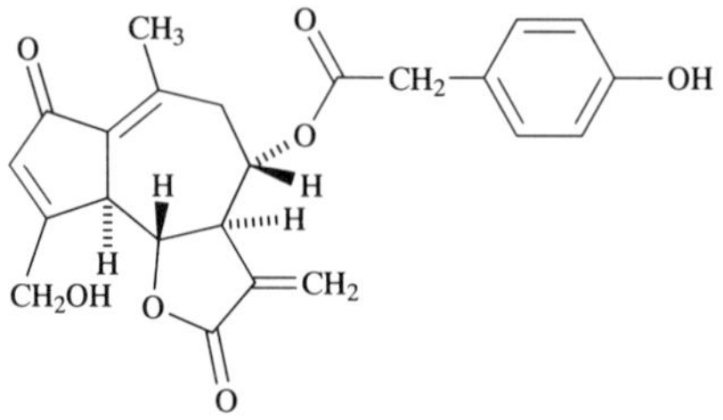

Sesquiterpenoid lactone
(Guaianolide)

$C_{23}H_{22}O_7$ [65725-11-3] MW 410.42

Root and leaf of chicory, *Cichorium intybus* and in wild species of lettuce, *Lactuca canadensis*, *L. serriola* and *L. virosa* (Compositae).

Bitter tasting. Has antifeedant activity against locusts. Hypoglycaemic properties.

Michl, H., *Monatsh. Chem.,* 1960, **91**, 500.

Lanceolatin B 836

Furano[4,5:7,8]flavone

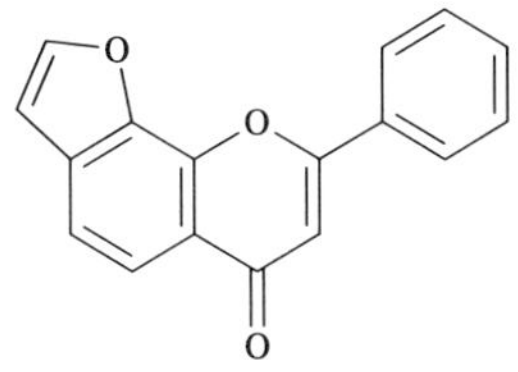

Furanoflavone

$C_{17}H_{10}O_3$ [482-00-8] MW 262.26

Seeds of *Pongamia glabra* and roots of *Derris mollis*. Also obtained from *Millettia ovalifolia*, *Tephrosia purpurea* and *T. lanceolata* (all Leguminosae).

Toxic principle of *Pongamia* seeds. Shown to be poisonous to chickens.

Rangawami, S., *Curr. Sci.,* 1955, **24**, 13.

Lanceotoxin A 837

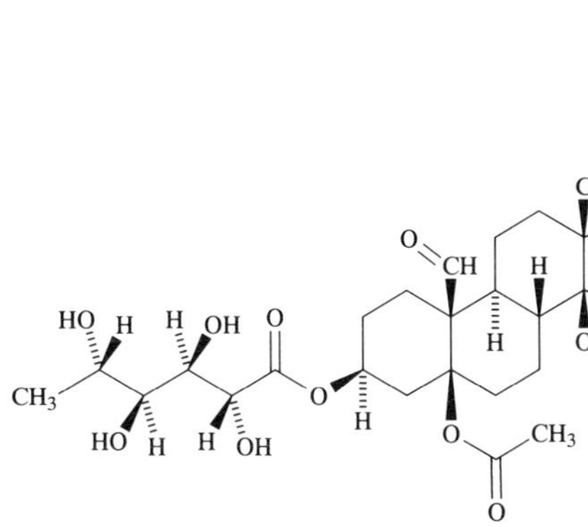

Bufadienolide

$C_{32}H_{44}O_{12}$ [93771-82-5] MW 620.69

Aerial parts of the succulent *Kalanchoe lanceolata* (Crassulaceae).

With lanceotoxin B (q.v.), responsible for the livestock poisoning caused by animals eating *Kalachoe lanceolata* plants. Also produces a chronic paralytic condition in small livestock known as 'krimpsiekte'.

Anderson, L. A. P., *J. Chem. Soc., Perkin Trans. 1,* 1984, 1573.

Lanceotoxin B 838

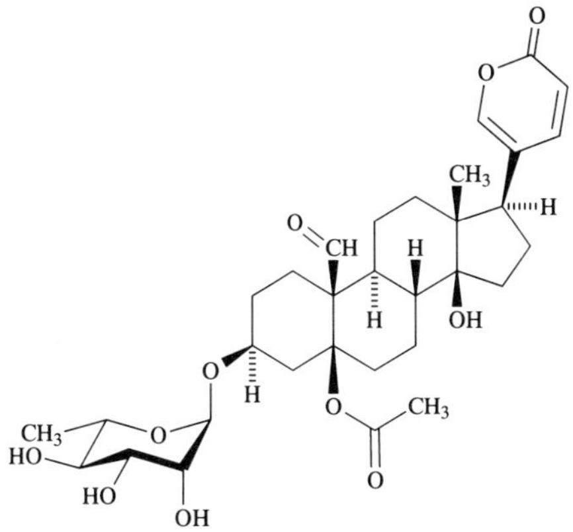

Bufadienolide

$C_{32}H_{44}O_{11}$ [93802-98-3] MW 604.69

Aerial parts of *Kalanchoe lanceolata* (Crassulaceae).

With lanceotoxin A (q.v.), responsible for the livestock poisoning caused by animals eating *Kalanchoe lanceolata* plants.

Anderson, L. A. P., *J. Chem. Soc., Perkin Trans. 1,* 1984, 1573.

Lantadene A 839

Rehmannic acid

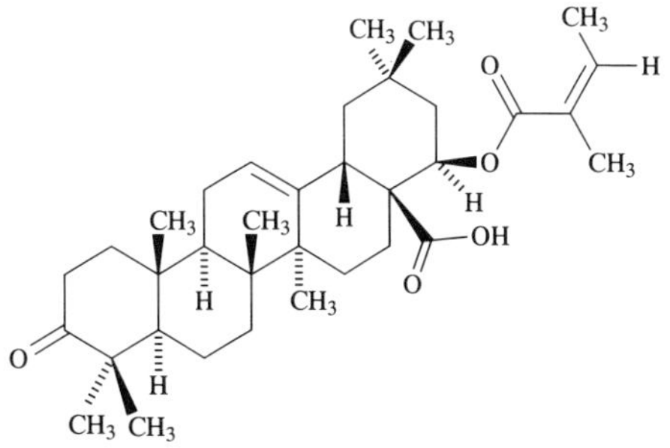

Triterpenoid
(Oleanane)

$C_{35}H_{52}O_5$ [467-81-2] MW 552.79

Aerial parts of some forms of the weedy ornamental *Lantana camara* (Verbenaceae). Also in the fruit of *Lippia rehmanni* (also Verbenaceae).

Lantana poisoning occurs commonly in sheep and cattle in northeastern Australia and in India. In sheep, lantadene A shows an oral toxicity of 60 mg/kg body-weight but an intravenous toxicity of 2 mg/kg body-weight.

Hart, N. K., *Aust. J. Chem.,* 1976, **29**, 655.

Lantadene B 840

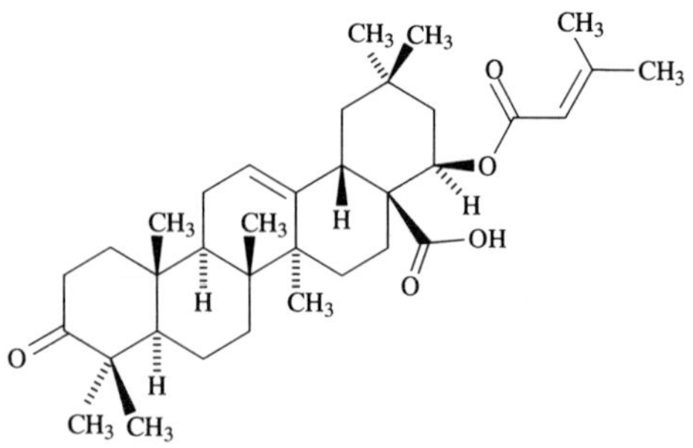

Triterpenoid
(Oleanane)

$C_{35}H_{52}O_5$ [467-82-3] MW 552.79

Aerial parts of some forms of the woody ornamental *Lantana camara* (Verbenaceae).

Responsible with lantadene A (q.v.) for the poisoning of cattle and sheep caused by lantana grazing.

Hart, N. K., *Aust. J. Chem.,* 1976, **29**, 655.

Lapachenole 841

Lapachenol; Lapachonone

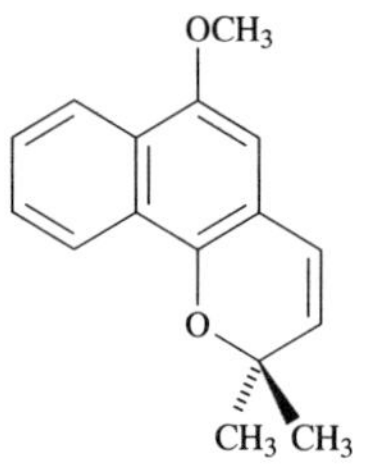

Phenolic

$C_{16}H_{16}O_2$ [573-13-7] MW 240.30

The essential oil of *Lippia graveolens* (Verbenaceae) and heartwood of *Tabebuia avellanedae* and *Tectona grandis* (Bignoniaceae).

Carcinogenic. It could be responsible for the antifertility activity of *Lippia graveolens* extracts.

Livingstone, R., *J. Chem. Soc.,* 1956, 3701.

Lapachol 842

Lapachic acid; Greenhartin; Taiguic acid; Tecomin

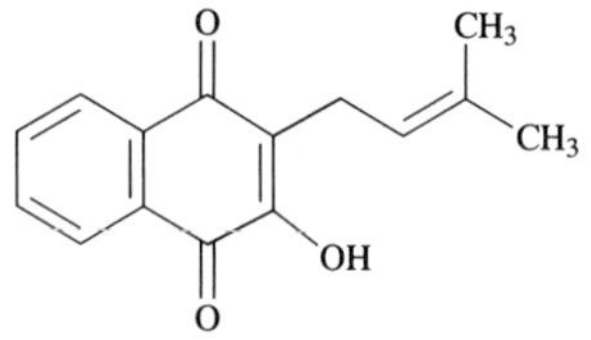

Naphthaquinone

$C_{15}H_{14}O_3$ [84-79-7] MW 242.27

Heartwood of *Haplophragma adenophyllum* and *Tabebuia rosea*, root of *Kigelia pinnata* (Bignoniaceae), heartwood of *Hibiscus tiliaceus* (Malvaceae), wood of *Diphysa robinoides* (Leguminosae) and root of *Conospermum teretifolium* (Proteaceae).

It inhibits respiratory processes and at high doses is cytotoxic. It has antitumour activity in the Walker 256 tumour cell system.

Thomson, R. H., Naturally Occurring Quinones, 3rd edn, 1987, 139, Chapman & Hall.

β-Lapachone 843

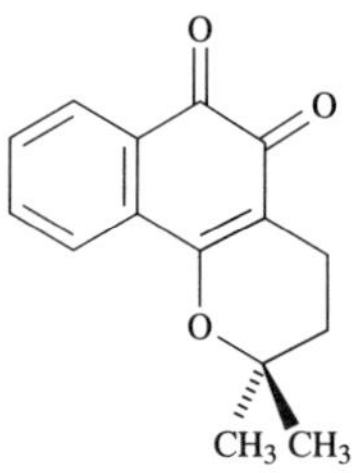

Naphthoquinone

$C_{15}H_{14}O_3$ [4707-32-8] MW 242.27

Heartwood of *Haplophragma adenophyllum*, *Tabebuia avellanedae* and *Phyllarthron comorense* (Bignoniaceae) and root of *Tectona grandis* (Verbenaceae).

Antitumour properties. It is active against the enzyme reverse transcriptase.

Thomson, R. H., Naturally Occurring Quinones, 3rd edn, 1987, 142, Chapman & Hall.

Lappaconitine 844

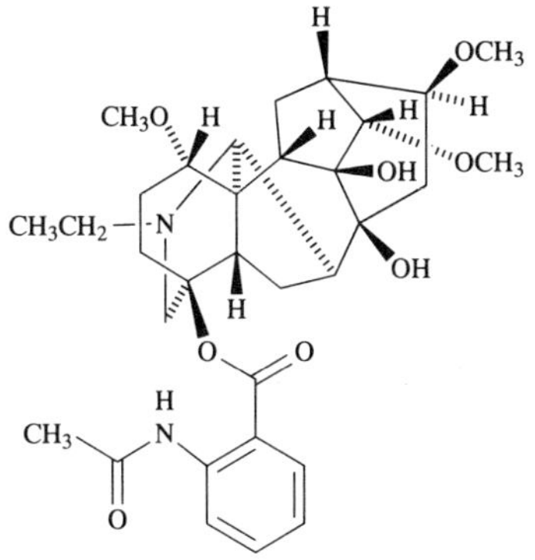

Diterpenoid alkaloid
(Aconitane)

$C_{32}H_{44}N_2O_8$ [32854-75-4] MW 584.71

Tubers and aerial parts of *Aconitum excelsum*, *A. orientale*, *A. ranunculaefolium* and *A. septentrionale* (Ranunculaceae).

Highly toxic, producing respiratory paralysis and having a direct action upon the heart, often terminating in ventricular fibrillation.

Khaimova, M., *Chem. Nat. Compd.*, 1970, **6**, 598.

(+)-Lariciresinol 845

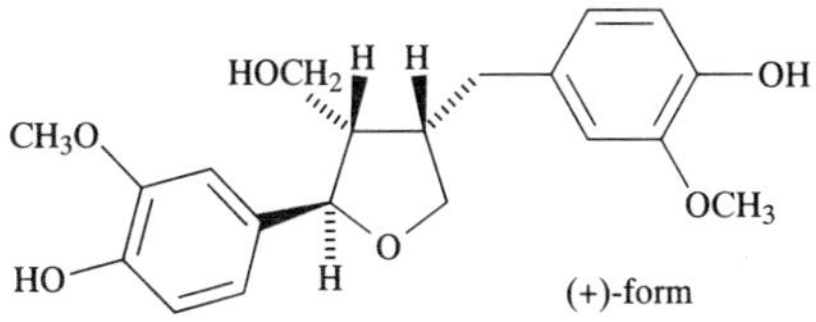

(+)-form

Lignan

$C_{20}H_{24}O_6$ [27003-73-2] MW 360.41

(−)-form [83327-19-9]
(±)-form [105367-81-5]

Wikstroemia elliptica (Thymelaeceae).

Cytotoxic activity, with an ED_{50} of 0.38 μg/ml against P-388 cells.

Badawi, M. M., *J. Pharm. Sci.*, 1983, **72**, 1285.

Laserolide 846

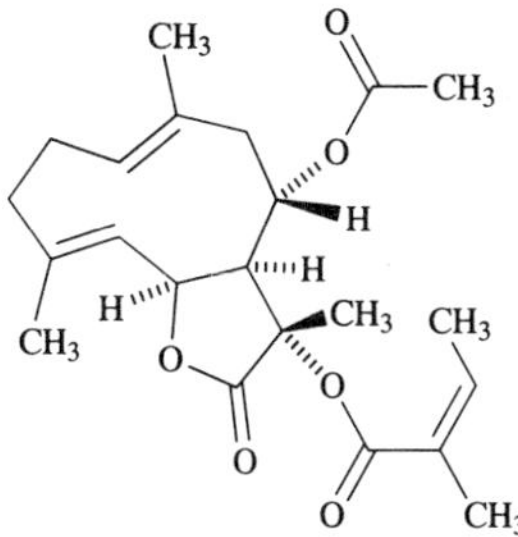

Sesquiterpene lactone
(Germacranolide)

$C_{22}H_{30}O_6$ [26560-24-7] MW 390.48

Roots of *Laser trilobium* (Umbelliferae).

Insect antifeedant.

Holub, M., *Collect. Czech. Chem. Commun.*, 1970, **35**, 284.

Lasiocarpine 847

Lassiocarpine; 7-Angelyl-9-lasiocarpylheliotridine

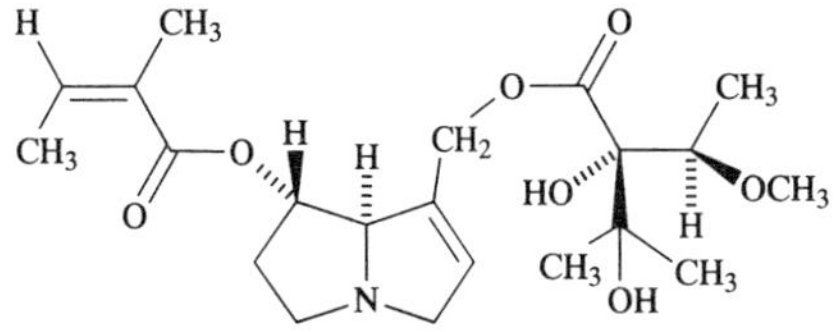

Pyrrolizidine alkaloid

$C_{21}H_{33}NO_7$ [303-34-4] MW 411.50

Heliotropium lasiocarpum, *H. hirsutum*, *H. arbainense* and *H. europaeum*, *Lappula intermedia*, comfrey, *Symphytum officinale* and *S. aspermum* (all Boraginaceae).

Hepatotoxic. It is also carcinogenic when tested on rat liver, skin and intestine.

Culvenor, C. C. J., *Aust. J. Chem.,* 1954, **7**, 277.

Lathodoratin 848

5,7-Dihydroxy-3-ethylchromone

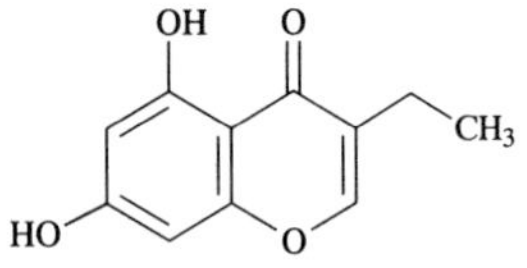

Chromone

$C_{11}H_{10}O_4$ [76693-50-0] MW 206.20

In fungally-infected leaves of *Lathyrus odoratus* and *L. hirsutus* (Leguminosae); accompanied in *L. odoratus* by the 7-methyl ether, methyllathodoratin [76690-64-7].

Antifungal agent (phytoalexin).

Robeson, D. J., *Phytochemistry,* 1980, **19**, 2171.

Lathyrol 849

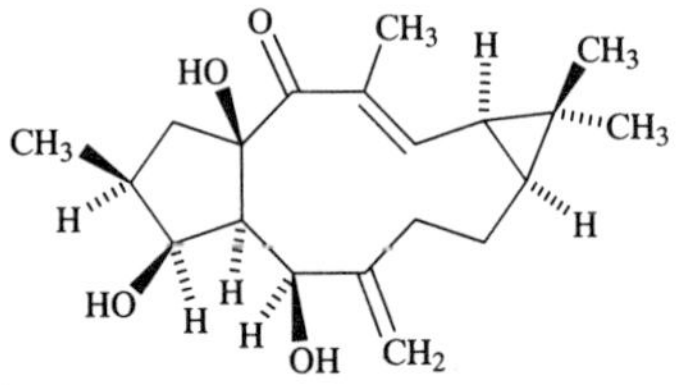

Diterpenoid
(Lathyrane)

$C_{20}H_{30}O_4$ [34420-19-4] MW 334.46

Seed oil of *Euphorbia lathyrus* (Euphorbiaceae).

Irritant and cocarcinogen.

Adolf, W., *Experientia,* 1971, **27**, 1393.

(−)-Laudanidine 850

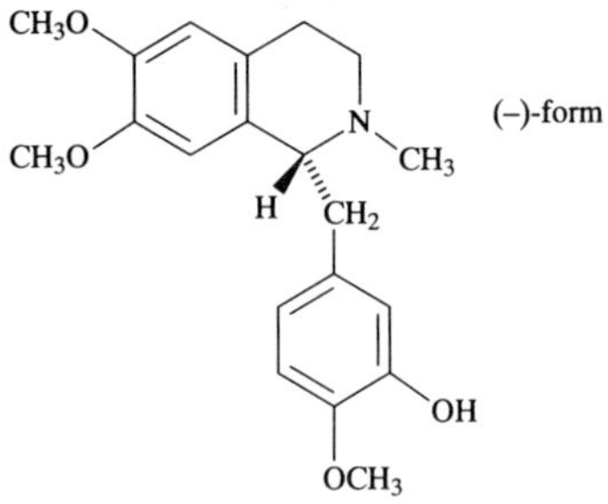

Benzylisoquinoline alkaloid

$C_{20}H_{25}NO_4$ [301-21-3] MW 343.42
(+)-form [3122-95-0]
(±)-form [85-64-3]

The (−)-form is in opium, the dried latex of *Papaver somniferum* (Papaveraceae). The (+)-form occurs in *Machitus obovatifolia* (Lauraceae) and in *Thalictrum dasycarpum* (Ranunculaceae). The racemic form, laudanine, is in *P. somniferum* and in *Xylopia pancheri* (Annonaceae).

Toxic. In animals, large doses have a strychnine-like effect, causing convulsions and paralysis.

Corrodi, H., *Helv. Chim. Acta,* 1956, **39**, 889.

Laudanosine 851

Laudanine methyl ether

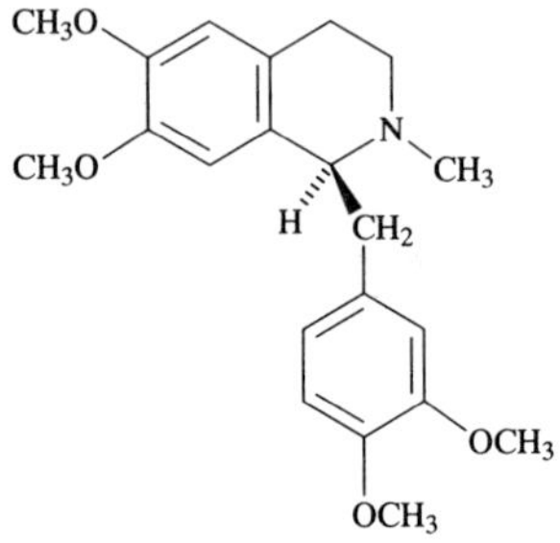

Benzylisoquinoline alkaloid

$C_{21}H_{27}NO_4$ [2688-77-9] MW 357.45

A minor alkaloid of opium, the dried latex of *Papaver somniferum* (Papaveraceae).

Strong tetanic poison. It depresses blood pressure and causes convulsions in animals.

Craig, J. C., *Tetrahedron,* 1966, **22**, 1335.

Lawsone 852

Isojuglone

Naphthoquinone

$C_{10}H_6O_3$ [83-72-7] MW 174.16

Leaves of *Lawsonia inermis* and *L. alba* (Lythraceae) and in *Impatiens balsamina* (Balsaminaceae).

Fungitoxic activity; minimal effective dose is 1000 p.p.m. Used as a dye, cosmetic and UV screen in therapy. Lawsone is the principle of the ancient 'henna' dye found on the nails of Egyptian mummies.

Tripathi, R. D., *Experientia,* 1978, **34**, 51.

Ledol 853

Ledum camphor

Sesquiterpenoid
(Aromadendrane)

$C_{15}H_{26}O$ [577-27-5] MW 222.37

Leaves of wild rosemary, *Ledum palustre* (Ericaceae).

Produces intoxicating and narcotic effects in humans. It brings about central stimulation, followed by subsequent paralysis.

Dolejš, L., *Collect. Czech. Chem. Commun.,* 1960, **25**, 1837.

Leiokinine A 854

Quinoline alkaloid

$C_{14}H_{17}NO_2$ [132587-64-4] MW 231.29

Leaves of *Esenbeckia leiocarpa* (Rutaceae).

Antifeedant to both monkeys and Lepidoptera.

Nakatsu, T, *J. Nat. Prod.,* 1990, **53**, 1508.

Lemmatoxin 855

Oleanoglycotoxin B

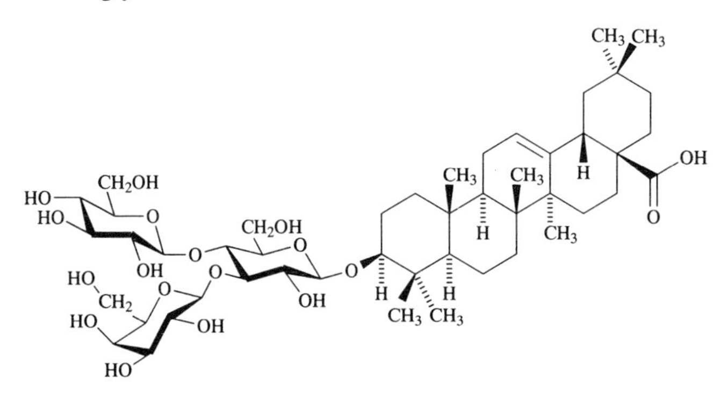

Triterpenoid saponin
(Oleanane)

$C_{48}H_{78}O_{18}$ [53043-29-1] MW 943.14

Berries of *Phytolacca dodecandra* (Phytolaccaceae), known as 'endod' in Ethiopia.

Molluscicidal, with an LD_{50} of 1.5 mg/litre for the snail *Biomphalaria glabrata*. It is active against human spermatozoa at concentrations of 50 mg/litre.

Parkhurst, R. M., *Can J. Chem.,* 1974, **52**, 702.

Lettucenin A 856

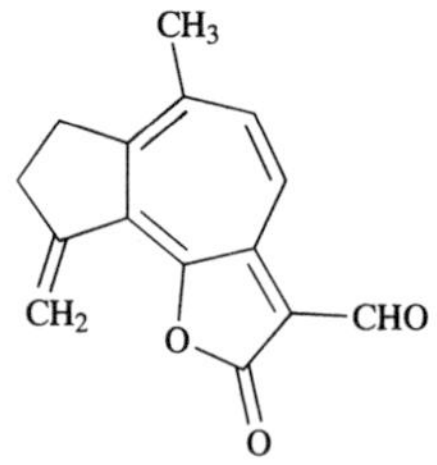

Sesquiterpenoid lactone
(Guaianolide)

$C_{15}H_{12}O_3$ [97915-46-3] MW 240.26

Leaves of lettuce, *Lactuca sativa* (Compositae) infected with *Pseudomonas cichorii.*

Antifungal agent (phytoalexin).

Takasugi, M., *J. Chem. Soc., Chem. Commun.,* 1985, 621.

Leurosidine 857

Vinrosidine; VRD; 4′α-Vincaleukoblastine

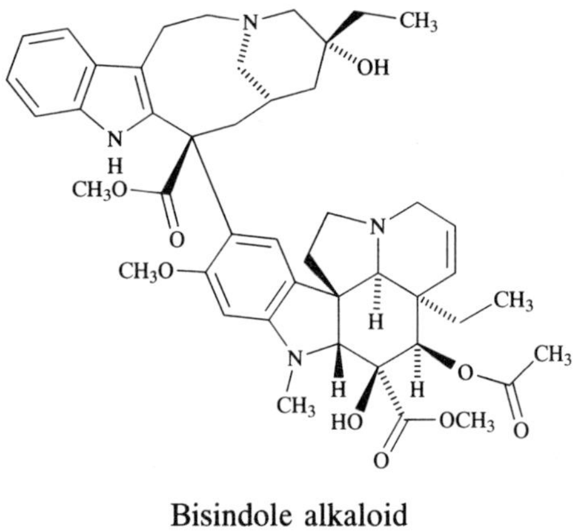

Bisindole alkaloid

$C_{46}H_{58}N_4O_9$ [15228-71-4] MW 810.99

Madagascar periwinkle, *Vinca rosea* (Apocynaceae).

Antineoplastic agent, but produces toxic side effects in man.

Neuss, N., *Tetrahedron Lett.,* 1967, 811.

Leurosine 858

Vinleurosine; VLR

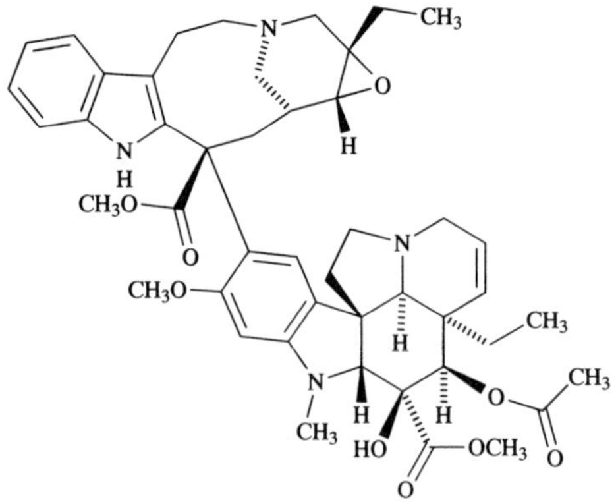

Bisindole alkaloid

$C_{46}H_{56}N_4O_9$ [23360-92-1] MW 808.97

Madagascar periwinkle, *Vinca rosea* and various *Catharanthus* spp. (Apocynaceae).

Antineoplastic agent.

Neuss, N., *J. Am. Chem. Soc.,* 1959, **81**, 4754.

Liatrin 859

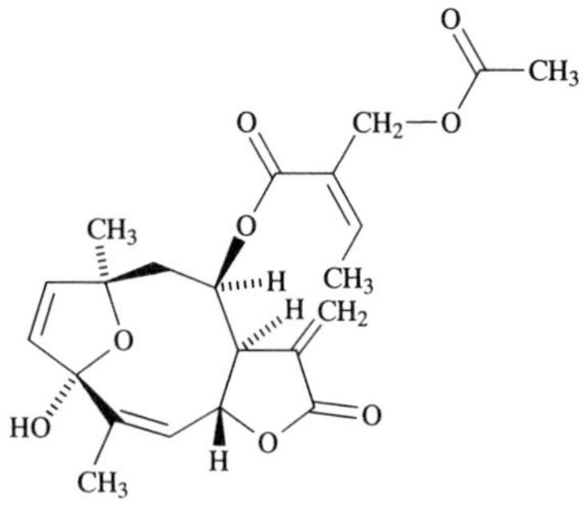

Sesquiterpenoid lactone
(Germacranolide)

$C_{22}H_{26}O_8$ [34175-79-6] MW 418.44

Blazing star, *Liatris chapmanii* (Compositae).

Cytotoxic and antitumour properties.

Kupchan, S. M., *J. Org. Chem.,* 1973, **38**, 1853.

Ligulatin B 860

Incanin

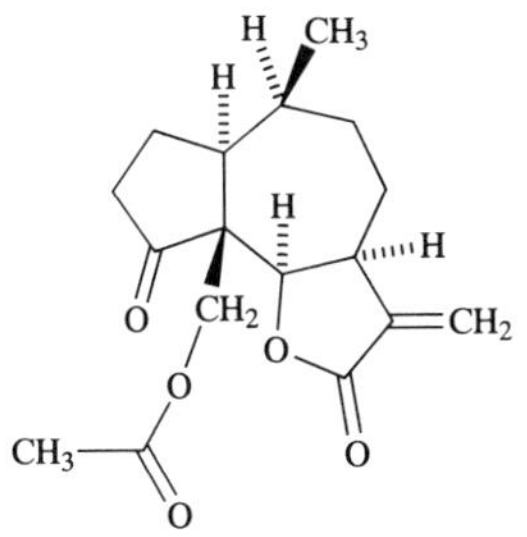

Sesquiterpenoid lactone
(Pseudoguaianolide)

$C_{17}H_{22}O_5$ [31299-06-6] MW 306.36

Parthenium ligulatum, *P. incanum*, *P. schottii* and *P. tomentosum* (Compositae).

Insect antifeedant and toxic to some insect species.

Rodriguez, E., *Phytochemistry,* 1971, **10**, 1145.

Limonene 861

Cajeputene; Cinene; Kautschin

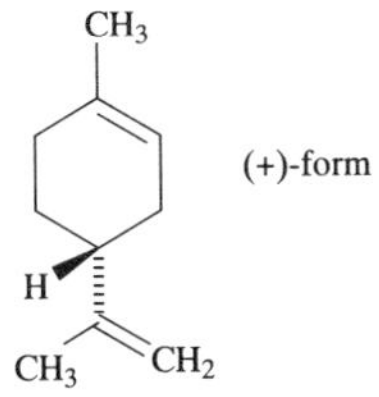

(+)-form

Monoterpenoid
(Menthane)

$C_{10}H_{16}$ [138-86-3] MW 136.24
(+)-form [5989-27-5]
(−)-form [5989-54-8]

(+)-Limonene is the major constituent of the oil of citrus fruits, such as lemon, tangerine and orange (Rutaceae). It is also in the oils of orange flowers, neroli, caraway and dill. (−)-Limonene is present in oil of needles and twigs of *Abies alba* (Pinaceae) and in mint oils, from *Mentha* spp. (Labiatae). The racemic form is dipentene (q.v.).

Skin irritant, with expectorant and sedative properties.

Thomas, A. F., *Nat. Prod. Rep.,* 1989, **6**, 291.

Limonin 862

Citrolimonin; Dictamnolactone; Evodin; Obaculactone

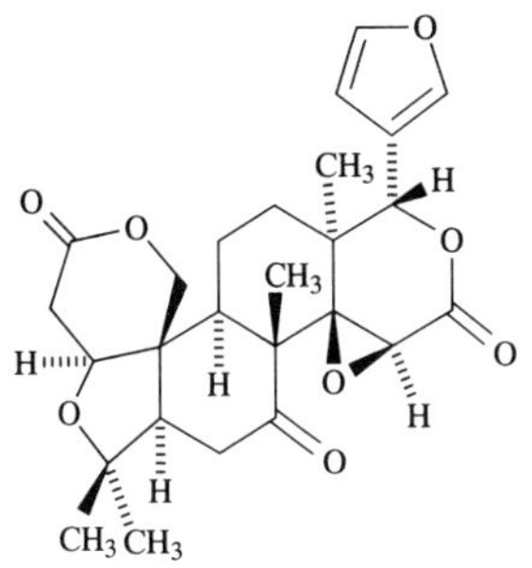

Nortriterpenoid
(Limonan)

$C_{26}H_{30}O_8$ [1180-71-8] MW 470.52

Citrus spp., especially in navel and valencia oranges and in lemon (Rutaceae), principally in the peel.

Bitter principle. Causes delayed bitterness in many *Citrus* fruit juices.

Barton, D. H. R., *J. Chem. Soc.,* 1961, 255.

Linamarin 863

Manihotoxine; Phaseolunatin

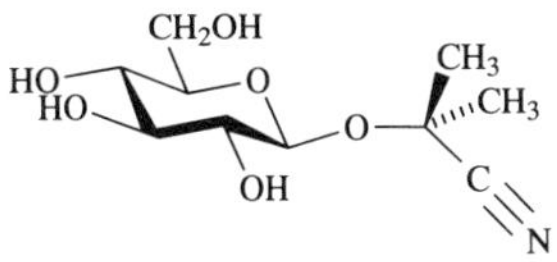

Cyanogenic glycoside

$C_{10}H_{17}NO_6$ [554-35-8] MW 247.25

Clover, *Trifolium repens*, birdsfoot trefoil, *Lotus corniculatus* and other legume pasture plants, *Acacia* spp. (Leguminosae), seedlings of flax, *Linum usitatissimum* (Linaceae), cassava, *Manihot esculentum* (Euphorbiaceae), *Passiflora* spp. (Passifloraceae) and several other families. It co-occurs with lotaustralin (q.v.).

Bitter-tasting toxin, releasing cyanide on hydrolysis or when plant tissue is damaged. It is the major toxin of cassava. Although acute poisoning from cassava consumption is rare, serious long term effects occur in people, due to the thiocyanate formed in the body during detoxification.

Olafsdottir, E. S., *Phytochemistry,* 1988, **28**, 127.

Linifolin A 864

Sesquiterpenoid lactone
(Pseudoguaianolide)

$C_{17}H_{20}O_5$ [5988-99-8] MW 304.34

Sneezeweeds, *Helenium linifolium, H. alternifolium, H. aromaticum, H. plantagineum* and *H. scorzoneraefolia* (Compositae).

Insecticidal. Cytotoxic and antitumour properties.

Herz, W., *J. Org. Chem.,* 1968, **33**, 2780.

Linustatin 865

Cyanogenic glycoside

$C_{16}H_{27}NO_{11}$ [72229-40-4] MW 409.39

Seed meal of flax, *Linum usitatissimum* (Linaceae), where it co-occurs with neolinustatin (q.v.). Also present in some *Passiflora* spp. (Passifloraceae).

Toxin, releasing cyanide on hydrolysis or tissue damage.

Smith, C. R., *J. Org. Chem.,* 1980, **45**, 507.

Lipiferolide 866

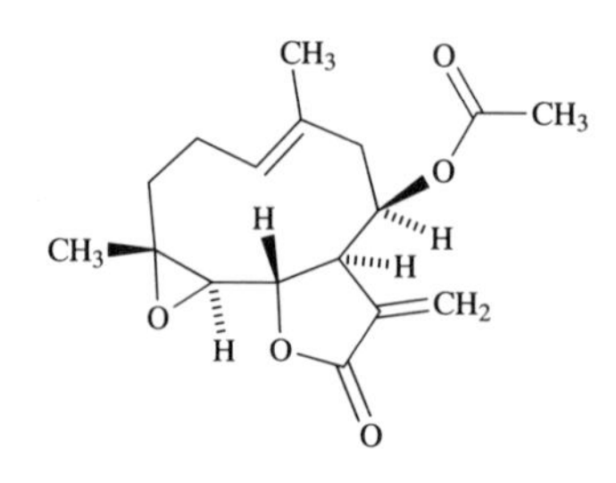

Sesquiterpenoid lactone
(Germacranolide)

$C_{17}H_{22}O_5$ [41059-80-7] MW 306.36

Tulip tree, *Liriodendron tulipifera* (Magnoliaceae).

Insect antifeedant. Has cytotoxic and antitumour properties.

Doskotch, R. W., *Phytochemistry,* 1975, **14**, 769.

Liriodenine 867

Spermatheridine

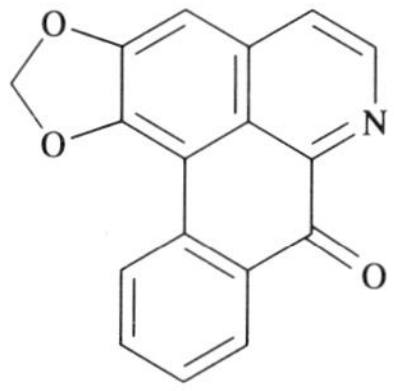

Benzylisoquinoline alkaloid
(Aporphine)

$C_{17}H_9NO_3$ [475-75-2] MW 275.26

Widely occurring in 13 plant families, including the Magnoliaceae (in *Liriodendron tulipifera* and *Magnolia obovata*).

Cytotoxic *in vitro* to human nasopharyngeal carcinoma cells.

Bick, I. R. C., *Tetrahedron Lett.,* 1964, 1629.

Lithospermic acid 868

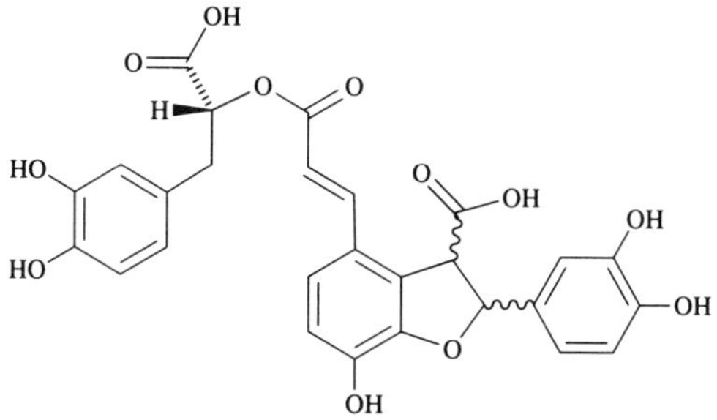

Benzofuran

$C_{27}H_{22}O_{12}$ [28831-65-4] MW 538.46

Roots of *Lycopus europaeus, L. virginicus, Lithospermum ruderale, L. officinale, Symphytum officinale, Anchusa officinale* and *Echium vulgare* (Boraginaceae).

Reported to have contraceptive properties in man, but this remains to be confirmed.

Kelley, C. J., *J. Org. Chem.,* 1976, **41**, 449.

Lobelanidine 869

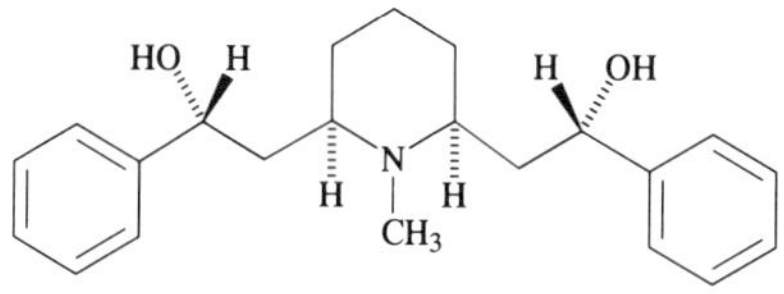

Piperidine alkaloid

$C_{22}H_{24}NO_2$ [552-72-7] MW 334.44

Lobelia inflata, L. hassleri, Isotoma longiflora and *Laurentia longiflora* (Campanulaceae) and *Sedum acre* (Crassulaceae).

Poisonous.

Wieland, H., *Justus Liebigs Ann. Chem.,* 1929, **473**, 83.

(−)-Lobeline 870

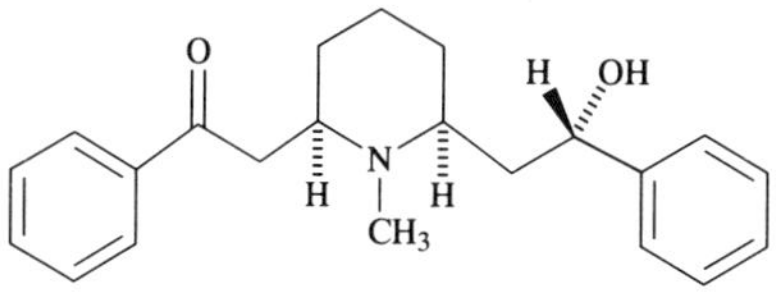

(−)-form

Piperidine alkaloid

$C_{22}H_{27}NO_2$ [90-69-7] MW 337.46

Lobelia inflata, L. nicotianaefolia and *L. hassleri*; also in the seed of *Campanula medium* (also Campanulaceae). The racemic form is lobelidine [134-65-6].

Causes nausea and hence is used in preparations to discourage tobacco smoking. It has a stimulating effect on respiration and has been used clinically to treat asthma. It also stimulates the secretion of catecholamine and thus increases the blood pressure. Lobelidine is an analeptic.

Wieland, H., *Justus Liebigs Ann. Chem.,* 1929, **473**, 83.

Loganin 871

Loganoside

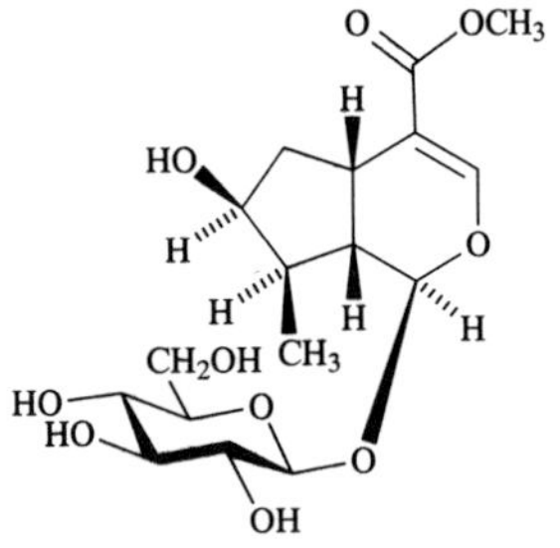

Monoterpenoid
(Iridane)

$C_{17}H_{26}O_{10}$ [18524-94-2] MW 390.39

Fruits of *Strychnos nux-vomica* (Loganaceae), rhizomes of *Menyanthes trifoliata* (Menyanthaceae), bark of *Hydrangea* spp. (Saxifragaceae) and *Vinca rosea* (Apocynaceae).

Bitter principle, toxic to birds. Has laxative activity. Crude drugs containing loganin are used as a bitter tonic.

Battersby, A. R., *J. Chem. Soc., C,* 1969, 721.

Lokundjoside 872

Bipindogenin 3-rhamnoside; Cuspidoside

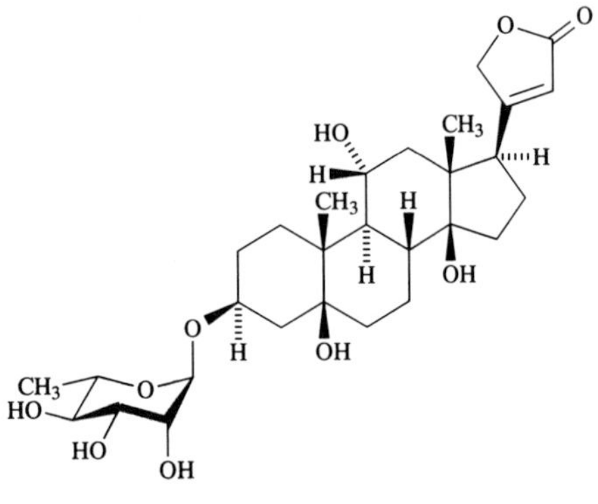

Cardenolide

$C_{29}H_{44}O_{10}$ [6869-51-8] MW 552.66

In seeds of *Strophanthus divaricatus*, *S. thollonii*, *S. sarmentosus* and other *Strophanthus* spp. (Apocynaceae) and leaves and flowers of lily-of-the-valley, *Convallaria majalis* (Liliaceae).

Highly poisonous to vertebrates.

Chen, R. F.., *Phytochemistry,* 1987, **26**, 2351.

Lophophorine 873

N-Methylanhalonine

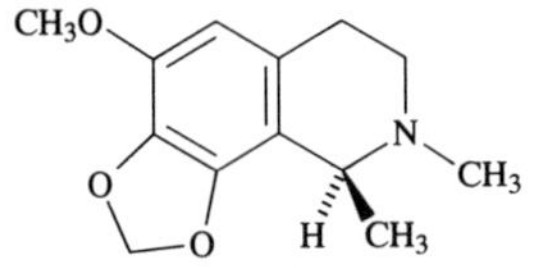

Isoquinoline alkaloid

$C_{13}H_{17}NO_3$ [17627-78-0] MW 235.28

Peyote (mescal buttons), *Lophophora williamsii* (Cactaceae).

Toxic, the LD_{50} in rabbits intravenously is 15-20 mg/kg body-weight. It is a respiratory stimulant and causes convulsions.

Späth, E., *Ber. Dtsch. Chem. Ges.,* 1935, **68**, 501.

Loroglossol 874

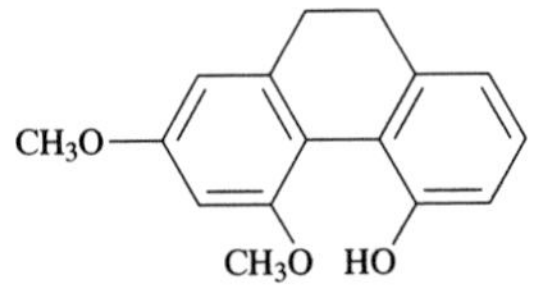

Phenanthrene

$C_{16}H_{16}O_3$ [41060-06-4] MW 256.30

Fungally-infected bulbs of *Loroglossum hircinum* (Orchidaceae).

Antifungal agent (phytoalexin).

Ward, E. W. B., *Phytopathology,* 1975, **65**, 632.

Lotaustralin 875

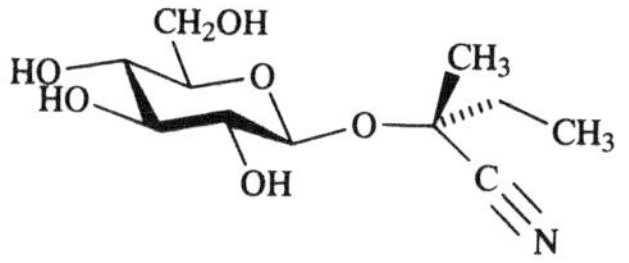

Cyanogenic glycoside

$C_{11}H_{19}NO_6$ [534-67-8] MW 261.28

Lotus corniculatus, *L. australis*, *Trifolium repens* and other legume herbs, *Haloragis erecta* (Haloragidaceae), *Linum usitatissimum* (Linaceae), *Passiflora* spp. (Passifloraceae) and some *Triticum* spp., including *T. monococcum* (Gramineae). The epimer, (*S*)-epilotaustralin [55758-42-4], occurs in *Triticum dicoccum* and some *Passiflora* spp..

Bitter-tasting toxin, releasing cyanide on hydrolysis. Activity similar to the co-occurring linamarin (q.v.).

Olafsdottir, E. S., *Phytochemistry,* 1988, **28**, 127.

Lubimin 876

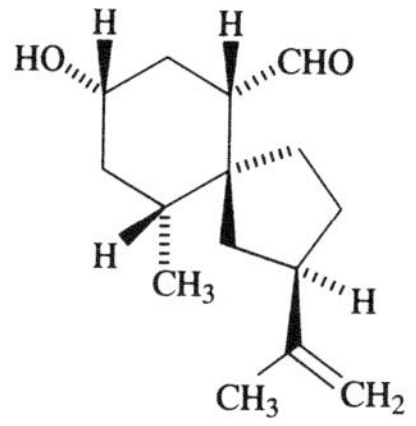

Sesquiterpenoid
(Vetispirane)

$C_{15}H_{24}O_2$ [35951-50-9] MW 236.35

Tubers of potato, *Solanum tuberosum*, infected with *Phytophthora infestans* and fruits of egg plant, *S. melongena* (Solanaceae) infected with *Monilinia fruticola*.

Antifungal agent (phytoalexin).

Birnbaum, G. I., *Can. J. Chem.,* 1977, **55**, 1619.

Lucumin 877

Lucuminoside; Prunasin xyloside

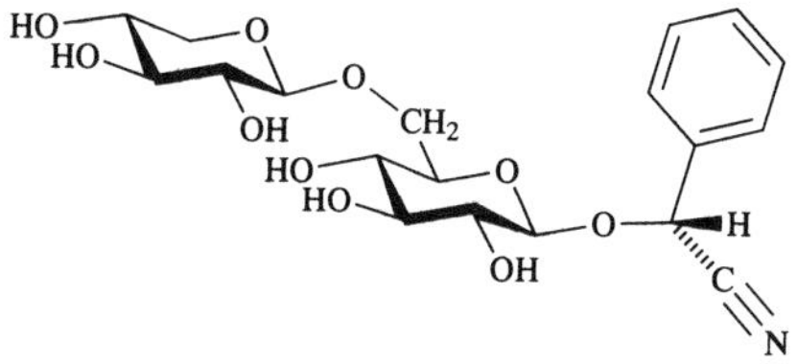

Cyanogenic glycoside

$C_{19}H_{25}NO_{10}$ [1392-28-5] MW 427.41

Seeds of *Calocarpum sapota* (Sapotaceae). The epimer, (*S*)-epilucumin [89460-01-5], is present in *Anthemis altissima* (Compositae), together with the two cyanogenic glycosides, *Anthemis* glycosides A and B (q.v.).

Bound toxin, releasing cyanide on hydrolysis or tissue damage.

Eyjólfsson, R., *Acta Chem. Scand.,* 1971, **25**, 1898.

Ludovicin A 878

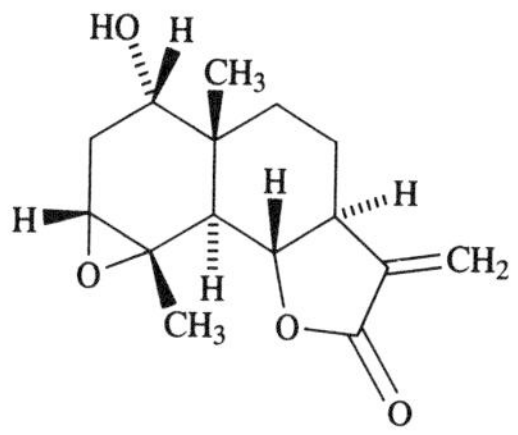

Sesquiterpenoid lactone
(Eudesmanolide)

$C_{15}H_{20}O_4$ [22740-13-2] MW 264.32

Artemisia ludoviciana var. *mexicana* (Compositae).

Cytotoxic and antitumour properties.

Lee, K. H., *Phytochemistry,* 1970, **9**, 403.

Lunacrine 879

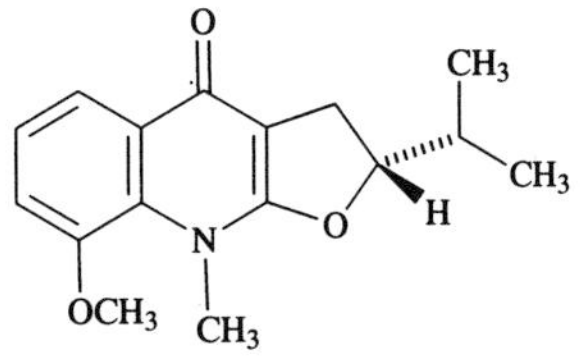

Furoquinoline alkaloid

$C_{16}H_{19}NO_3$ [82-40-6] MW 273.33

Bark of *Lunasia costulata* and *L. amara* (Rutaceae).

Toxic, with LD_{50} of 80 mg/kg body-weight intravenously in mice. It is a hypotensive agent, causing transient fall in blood pressure in cats.

Goodwin, S., *J. Am. Chem. Soc.*, 1959, **81**, 3065.

Lunamarine 880

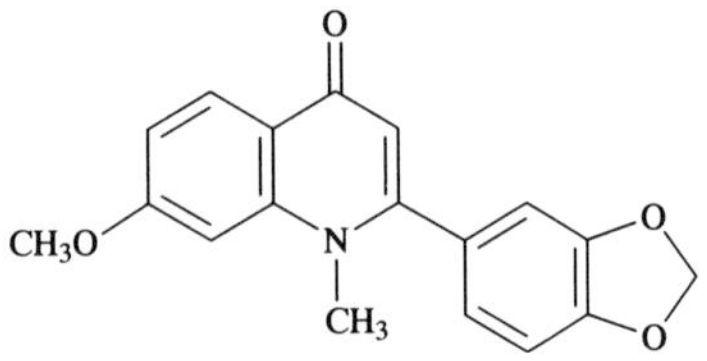

Quinoline alkaloid

$C_{18}H_{15}NO_4$ [483-52-3] MW 309.32

Lunasia amara (Rutaceae).

Shows hypotensive activity, causing a transient fall in blood pressure in cats. Also a weak smooth muscle stimulant, affecting isolated rabbit intestine and uterus.

Goodwin, S., *J. Am. Chem. Soc.*, 1959, **81**, 6209.

Lupanine 881

2-Oxo-11α-sparteine

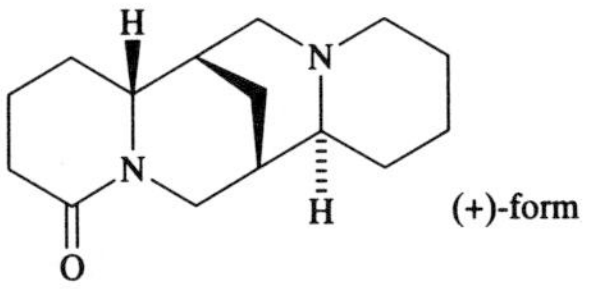

(+)-form

Quinolizidine alkaloid
(Sparteine)

$C_{15}H_{24}N_2O$ [550-90-3] MW 248.37
(−)-form [486-88-4]
(±)-form [4356-43-8]

Present in over 12 genera of the Leguminosae, especially in *Lupinus* spp. and in broom, *Cytisus scoparius*. Can occur in both the (+)- and (−)-forms, as well as the racemate (±).

Toxic, being responsible for cattle poisoning caused by grazing on *Lupinus* species in North America. It is anti-arrhythmic, hypotensive and hypoglycaemic.

Galinovsky, F., *Monatsh. Chem.*, 1955, **86**, 1014.

Lupeol 882

Fagasterol; Monogynol B; β-Viscol

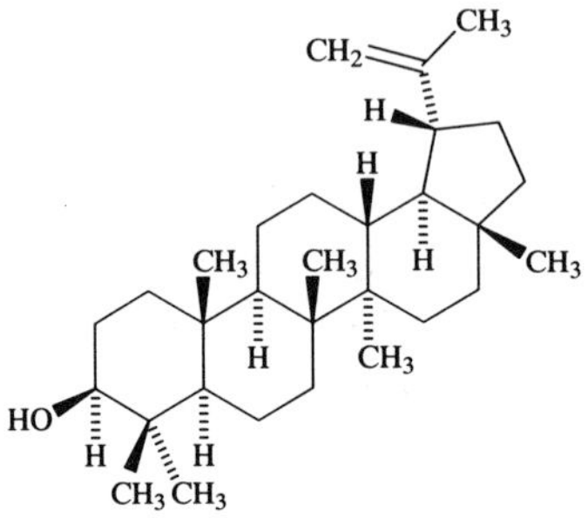

Triterpenoid
(Lupane)

$C_{30}H_{50}O$ [545-47-1] MW 426.73

First isolated from seed of the yellow lupin, *Lupinus luteus* (Leguminosae). A rich source (2.25% dry-weight present) is the bark of *Phyllanthus emblica* (Euphorbiaceae). Also present in the bark of many species in the Apocynaceae and Leguminosae.

Antitumour properties, active against the Walker carcinoma 256 tumour system. It shows antihyperglycaemic and hypotensive activities.

Karrer, W., Konstitution und Vorkommen der organischen Pflanzenstoffe, 2nd ed., 1981, **2, Part 1**, 472, Birkhäuser.

Lupeol acetate 883

Triterpenoid
(Lupane)

$C_{32}H_{52}O_2$ [1617-68-1] MW 468.76

Cordia buddleioides (Boraginaceae) and the date palm, *Phoenix dactylifera* (Palmae).

Antihyperglycaemic and anti-ulcer properties.

Ames, T. R., *J. Chem. Soc.*, 1951, 450.

Lupinine 884

Quinolizidine alkaloid

$C_{10}H_{19}NO$ [486-70-4] MW 169.27

Lupinus luteus and *L. palmeri* (Leguminosae) and *Anabasis aphylla* (Chenopodiaceae).

Orally toxic to mammals. It is also an insect antifeedant and a growth inhibitor for the grasshopper, *Melanoplus bivittatus*.

Cookson, R. C., *Chem. Ind. (London)*, 1953, 337.

Lupulone 885

β-Lupulic acid

Phenolic ketone

$C_{26}H_{38}O_4$ [468-28-0] MW 414.59

Hop cones, *Humulus lupulus* (Moraceae).

Mildly toxic with an LD_{50} in rats of 1.8 mg/kg body-weight. A bitter principle, contributing a bitter taste to beer.

Riedl, W., *Chem. Ber.*, 1956, **89**, 1863.

Luteolinidin 886

5,7,3′,4′-Tetrahydroxyflavylium

Flavonoid
(Anthocyanidin)

$[C_{15}H_{11}O_5]^+$ [1154-78-5] MW 271.25

Fungally-infected leaves of sugar cane, *Saccharum officinarum* and of *Sorghum bicolor* (Gramineae). Constitutively present in glycosidic form as an orange-red flower pigment in *Rechsteineria cardinalis* (Gesneriaceae).

Antifungal activity. Orange-red pigment.

Nicholson, R. L., *Proc. Natl Acad. Sci., U.S.A.*, 1987, **84**, 5520.

Luteone 887

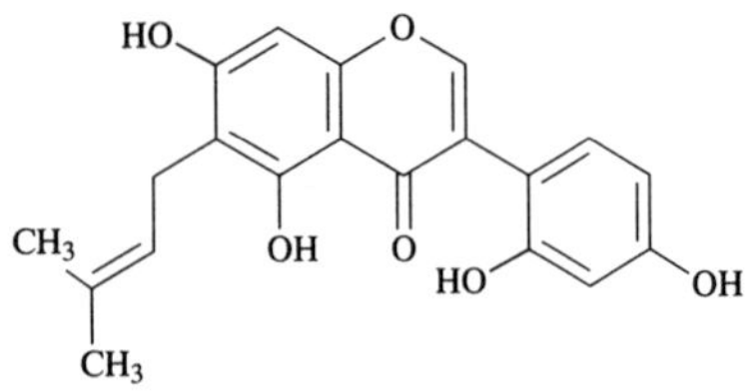

Isoflavonoid
(Isoflavone)

$C_{20}H_{18}O_6$ [41743-56-0] MW 354.36

Fungally-infected leaves of *Laburnum anagyroides*; constitutively present in leaves and fruit of *Lupinus albus* and other *Lupinus* spp. (Leguminosae).

Antifungal properties.

Harborne, J. B., *Phytochemistry,* 1976, **15**, 1485.

Lycaconitine 888

N-Succinylanthranoyllycoctonine

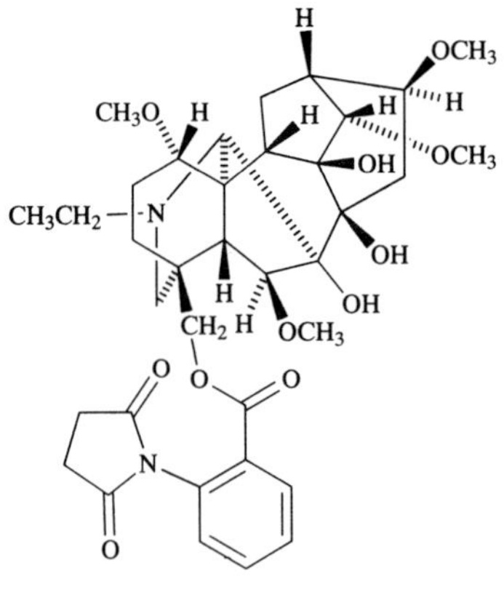

Diterpenoid alkaloid
(Aconitane)

$C_{36}H_{48}N_2O_{10}$ [25867-19-0] MW 668.78

Aconitum lycoctonum and *Delphinium cashmirianum* (Ranunculaceae).

Appears to be less toxic than most aconitine-type alkaloids. It has little effect on the heart rate of anaesthetised rabbits in doses of up to 5 mg/kg body-weight intravenously; doses above 0.25 mg/kg body-weight, however, produce a small rise in blood pressure.

Shamma, M., *J. Nat. Prod.*, 1979, **42**, 615.

Lycoctonine 889

Delsine; Royline

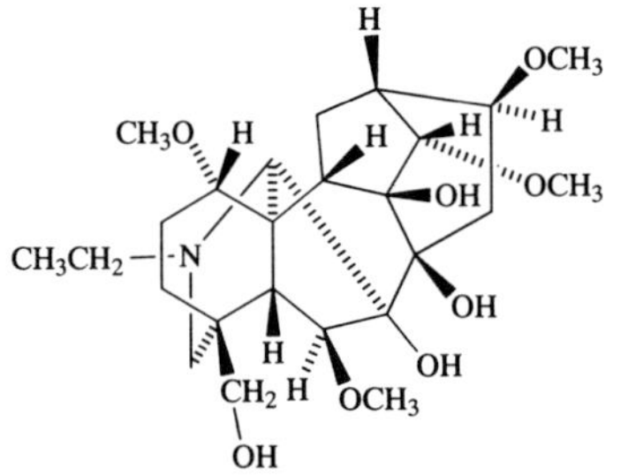

Diterpenoid alkaloid
(Aconitane)

$C_{25}H_{41}NO_7$ [26000-17-9] MW 467.60

Aconitum lycoctonum and *Delphinium consolida* (Ranunculaceae) and *Inula royleana* (Compositae).

Produces a sudden fall in blood pressure when given intravenously to cats in doses of 2-5 mg/kg body-weight. There is a hypotensive response in anaesthetised cats with intravenous doses of 5-15 mg/kg body-weight.

Edwards, O. E., *Can. J. Chem.*, 1956, **34**, 1315.

Lycopodine 890

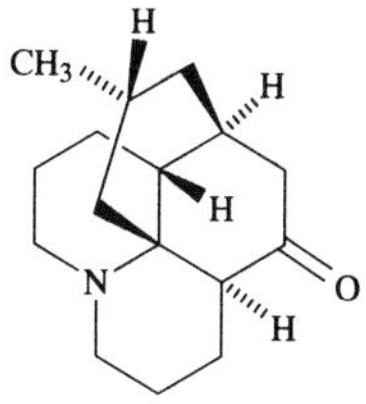

Lycopodium alkaloid
(Lycopodane)

$C_{16}H_{25}NO$ [466-61-5] MW 247.38

Lycopodium complanatum and *L. clavatum* (Lycopodiaceae).

Toxin. Causes paralysis in frogs. It produces uterine contractions in rabbits, rats and guinea-pigs. It is employed as a treatment for skin disorders in herbal medicine.

Harrison, W. A., *Can. J. Chem.*, 1961, **39**, 2086.

Lycopsamine 891

9-Viridiflorylretronecine

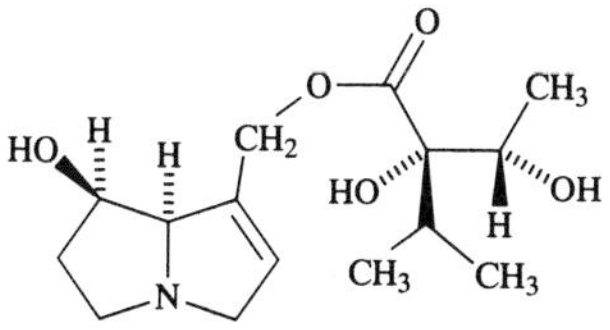

Pyrrolizidine alkaloid

$C_{15}H_{25}NO_5$ [10285-07-1] MW 299.37

Parsonsia eucalyptophylla and *P. straminea* (Apocynaceae), *Amsinckia hispida* and *A. intermedia*, *Anchusa arvensis*, *Heliotropium steudneri*, Russian comfrey, *Symphytum uplandicum* and borage, *Borago officinale* (Boraginaceae). Also present in *Eupatorium compositifolium* (Compositae). It is a stereoisomer of intermedine (q.v.).

Hepatotoxic. Consumption of Russian comfrey as a salad is to be discouraged, because this is a cumulative poison.

Culvenor, C. C. J., *Aust. J. Chem.*, 1966, **19**, 1955.

Lycorenine 892

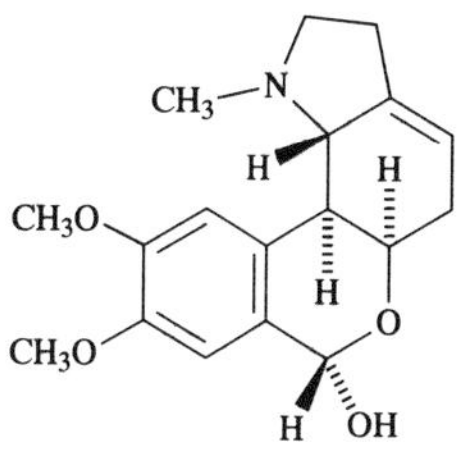

Amaryllidaceae alkaloid
(Lycoranan)

$C_{18}H_{23}NO_4$ [477-19-0] MW 317.38

Bulbs of *Lycoris radiata* and also in many other species of the Amaryllidaceae.

Insect antifeedant.

Mizukami, S., *Tetrahedron,* 1960, **11**, 89.

Lycoricidine 893

Margetine

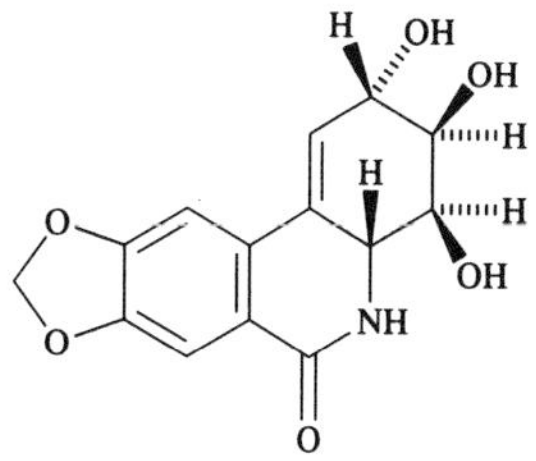

Amaryllidaceae alkaloid
(Narciclasine)

$C_{14}H_{13}NO_6$ [19622-83-4] MW 291.26

Lycoris radiata and *L. sanguinea* and also many other species of Amaryllidaceae.

Insect antifeedant. Has cytotoxic properties.

Okamoto, T., *Chem. Pharm. Bull.,* 1968, **16**, 1860.

Lycorine 894

Narcissine; Galanthidine

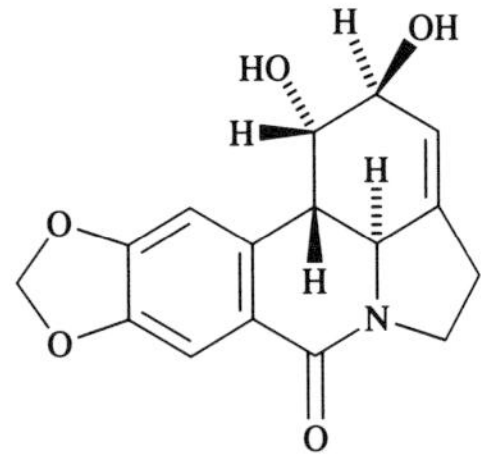

Amaryllidaceae alkaloid
(Galanthan)

$C_{16}H_{17}NO_4$ [476-28-8] MW 287.32

In bulbs of *Lycoris radiata* and of daffodils, *Narcissus* spp.; also widespread elsewhere in the Amaryllidaceae.

Highly toxic, frequently responsible for accidental poisoning by daffodil bulbs. The LD_{50} in dogs is 41 mg/kg body-weight. It also exhibits antiviral activity against poliomyelitis, coxsackie and herpes type I viruses.

Brossi, A., The Alkaloids, 1985, **25**, 1, Academic Press

Lyngbyatoxin A 895

Teleocidin A_1

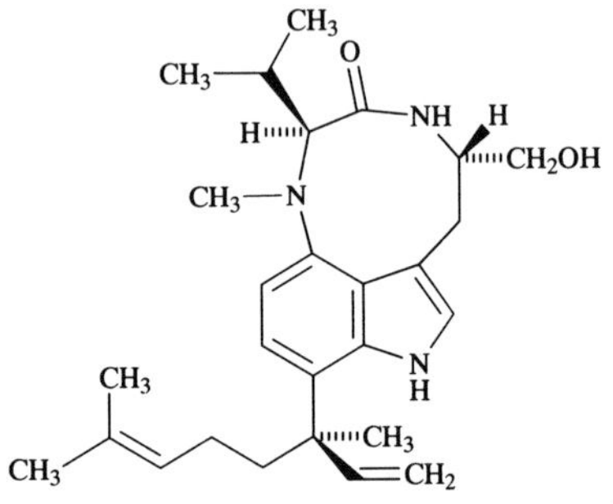

Indole alkaloid

$C_{27}H_{39}N_3O_2$ [70497-14-2] MW 437.63

Produced by shallow water isolates of the marine tropical and subtropical species of cyanobacteria, *Lyngbia majuscula*. Associated with contact dermatitis known as swimmers' itch. It has inflammatory, vesicatory and tumour promoting activity. Minimum lethal dose 0.3 mg/kg body-weight in mice (intraperitoneal). As little as 0.15 μg/ml in sea water is toxic to some species of fish.

Moore, R. E., *Pure Appl. Chem.*, 1982, **54**, 1919.

Lysergol 896

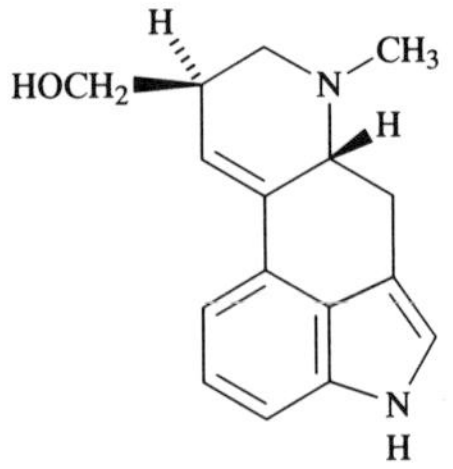

Indole alkaloid
(Ergoline)

$C_{16}H_{18}N_2O$ [602-85-7] MW 254.33

Seed and leaves of *Ipomoea parasitica* (Convolvulaceae). Also in seed of *Argyreia*, *Strictocardia*, *Rivea* and other *Ipomoea* spp. (Convolvulaceae).

Toxic alkaloid, with hallucinogenic effects.

Amor-Prats, D., *Chemoecology*, 1993, **4**, 55.

M

(−)-Maackiain 897

Inermin; Demethylpterocarpin

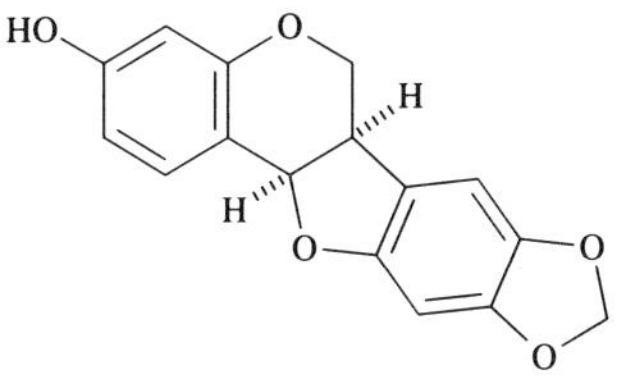

Isoflavonoid
(Pterocarpan)

$C_{16}H_{12}O_5$ [2035-15-6] MW 284.27
(+)-form [23513-53-3]
(±)-form [19908-48-6]

In fungally-infected leaves of chickpea, *Cicer arietinum* (Leguminosae). Constitutively present in heartwood of *Maackia amurensis* (also Leguminosae).

Antifungal properties.

Shibata, S., *Chem. Pharm. Bull.,* 1968, **11**, 771.

Macoline 898

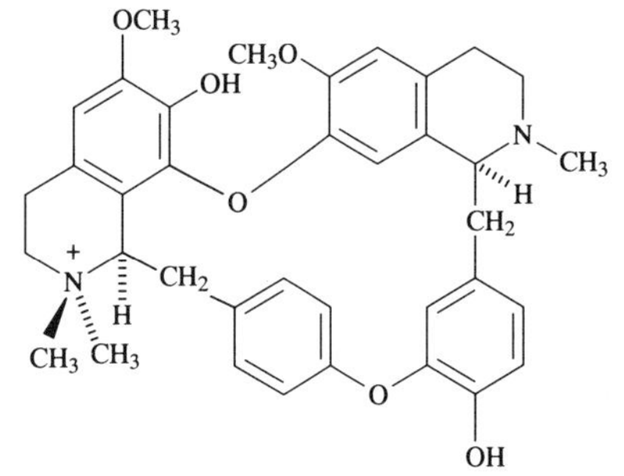

Bisbenzylisoquinoline alkaloid
(Oxyacanthan)

$[C_{37}H_{41}N_2O_6]^+$ [66216-59-9] MW 609.74

Stemwood of *Abuta grisebachii* (Menispermaceae).

The juice of the plant was mixed into a type of curare arrow poison by South American Indians. In line with this use, the alkaloid has muscle relaxant properties.

Galeffi, C., *Farmaco, Ed. Sci.,* 1977, **32**, 853.

Macrocarpamine 899

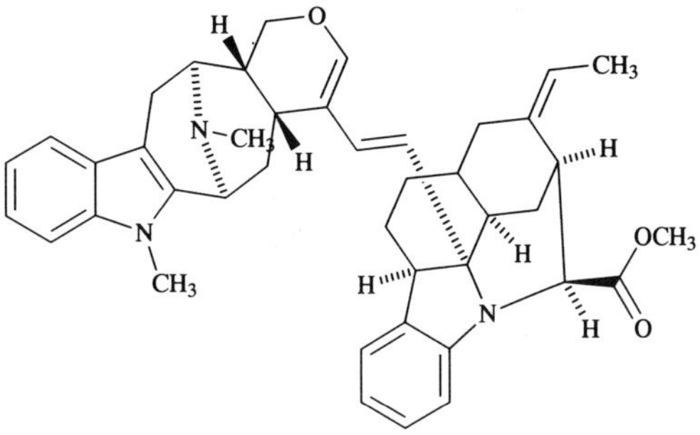

Bisindole alkaloid

$C_{41}H_{46}N_4O_3$ [66408-46-6] MW 642.84

Bark of *Alstonia macrophylla* (Apocynaceae).

Antiprotozoal activity.

Mayerl, F., *Helv. Chim. Acta,* 1978, **61**, 337.

Macrozamin 900

Methylazoxymethanol primeveroside

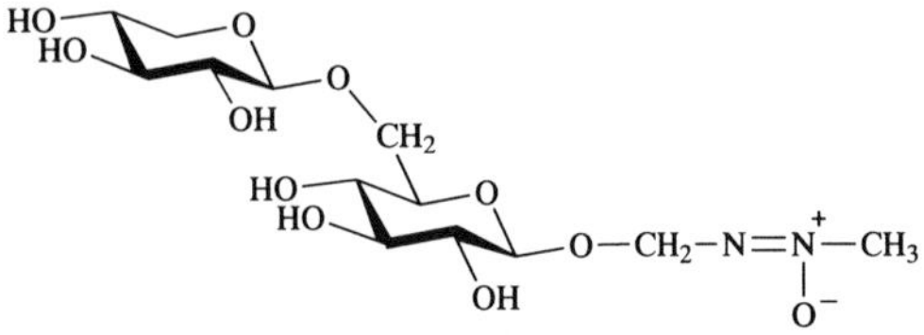

Azoxy compound

$C_{13}H_{24}N_2O_{11}$ [6327-93-1] MW 384.34

Macrozamia spiralis (Zamiaceae), *Cycas media* and some *Encephalartos* spp. (Cycadaceae).

Toxic agent. Responsible for toxicity of this plant to livestock. It is both hepatotoxic and carcinogenic.

Langley, B. W., *J. Chem. Soc.,* 1951, 2309.

Maesanin 901

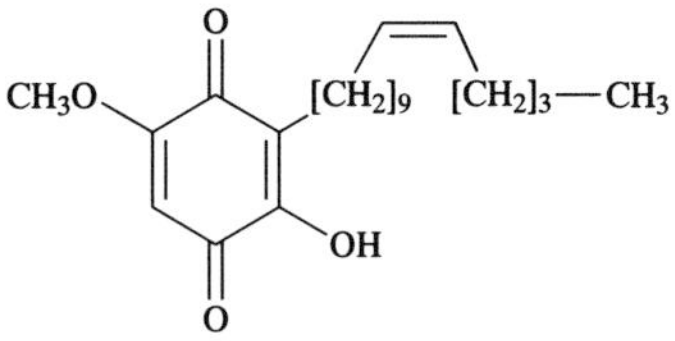

Benzoquinone

$C_{22}H_{34}O_4$ [82380-21-0] MW 362.51

Fruit of *Maesa lanceolata* (Myrsinaceae).

Antimicrobial agent. A hot water extract of the fruit is used locally in Africa to prevent cholera.

Kubo, I., *Tetrahedron Lett.,* 1983, **24**, 3825.

Magnolol 902

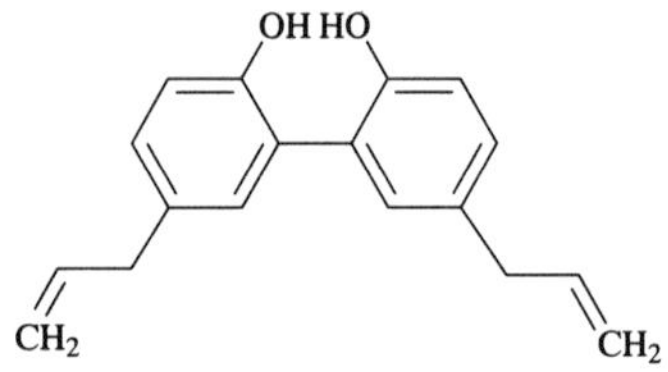

Neolignan

$C_{18}H_{18}O_2$ [528-43-8] MW 266.34

Twigs of *Cercidiphyllum japonicum* (Cercidiphyllaceae) infected with *Fusarium solani.* Also present constitutively in bark of *Magnolia officinalis* (Magnoliaceae) and roots of *Sassafras randaiense* (Lauraceae).

Antifungal and antibacterial agent.

Takasugi, M., *Phytochemistry,* 1986, **25**, 2751.

Makisterone B 903

Callinecdysone B

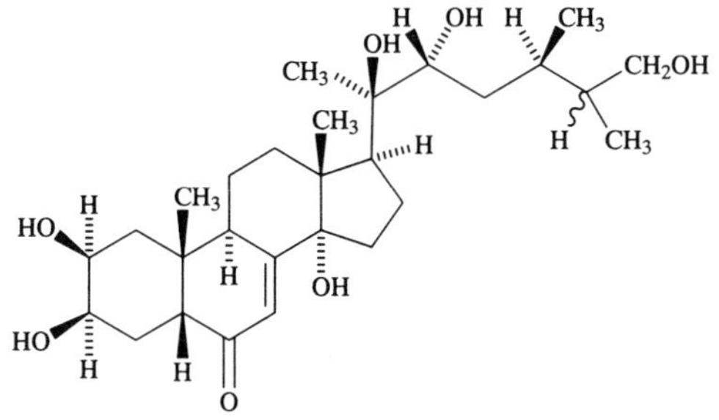

Steroid
(Cholestane)

$C_{28}H_{46}O_7$ [20512-31-6] MW 494.67

Ajuga chamaepitys (Labiatae) and seeds of *Diploclisia glaucescens* (Menispermaceae).

Potentially toxic to insects, because of its moulting hormone activity. It is an effective growth inhibitor and abortifacient when fed to females of the tsetse fly, *Glossina morsitans morsitans*. Relatively non-toxic in mammals.

Imai, S., *Tetrahedron Lett.,* 1968, 3887.

Mallotochromene 904

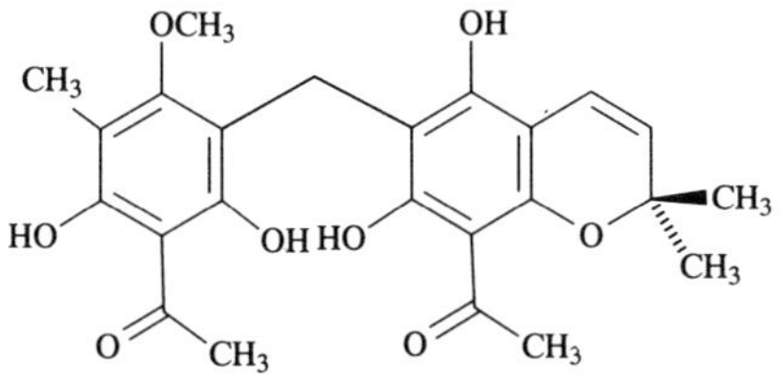

Chromene

$C_{24}H_{26}O_8$ [98569-62-1] MW 442.47

Pericarp of *Mallotus japonicus* (Euphorbiaceae).

Cytotoxic, with antileukaemic activity against KB cell lines.

Arisawa, M., *J. Nat. Prod.,* 1985, **48**, 455.

Mallotophenone 905

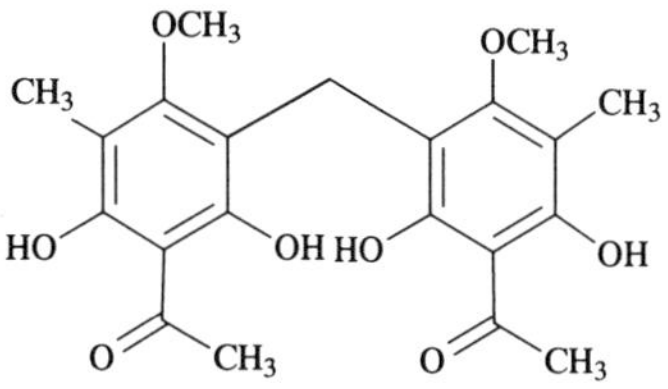

Phenolic ketone

$C_{21}H_{24}O_8$ [98569-63-2] MW 404.42

Pericarp of *Mallotus japonicus* (Euphorbiaceae).

Cytotoxic against mouse leukaemia L-5178Y cells and the KB cell system.

Arisawa, M., *J. Nat. Prod.,* 1985, **48**, 455.

Malonic acid 906

Methanedicarboxylic acid; Propanedioic acid

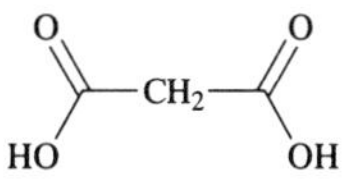

Organic acid

$C_3H_4O_4$ [141-82-2] MW 104.06

Occasionally accumulates in plant tissues, in leaf, e.g. parsley, *Apium graveolens* (Umbelliferae), in stem, e.g. bean, *Phaseolus coccineus* (Leguminosae), in root, e.g. beet, *Beta vulgaris* (Chenopodiaceae) or seed, e.g. barley, *Hordeum vulgare* (Gramineae).

A moderately strong acid, and a strong irritant, causing damage to mucous membranes and skin. The LD_{50} is 300 mg/kg body-weight intraperitoneally in mice.

Bentley, L. E., *Nature (London),* 1952, **170**, 847.

Malvalic acid 907

8,9-Methylene-8-heptadecenoic acid

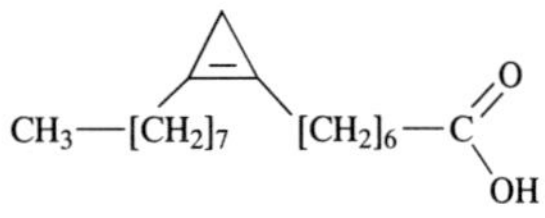

$CH_3—[CH_2]_7$ $[CH_2]_6—C$ O OH

Fatty acid

$C_{18}H_{32}O_2$ [503-05-9] MW 280.45

Seed oil of *Hibiscus syriacus* and of other members of the Malvaceae. Also in the seed oil of *Gnetum gnemon* (Gnetaceae).

Causes adverse effects in animals. The seeds of *Gnetum* are processed into chips in Java and consumption of such chips could be hazardous to humans.

Craven, B., *Nature (London),* 1959, **183**, 676.

Mammeisin 908

Mammea-compound A/AA

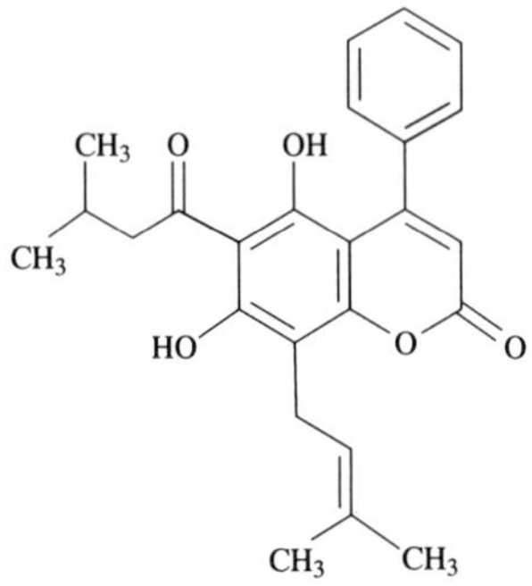

CH_3 O OH
CH_3
HO O O
CH_3 CH_3

4-Phenylcoumarin

$C_{25}H_{26}O_5$ [18483-64-2] MW 406.48

Mammea africana and *M. thwailesii* (Guttiferae). One of a number of related 4-phenylcoumarins present in these plants.

Antitumour properties.

Finnegan, R. A., *J. Org. Chem.,* 1961, **26**, 1180.

Mancinellin 909

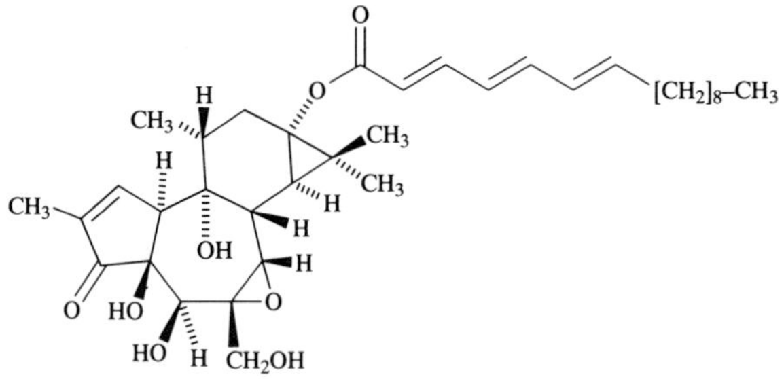

O
H O $[CH_2]_8$–CH_3
CH_3 CH_3
H CH_3
CH_3 H
OH H
H
O HO O
HO H CH_2OH

Diterpenoid
(Tigliane)

$C_{36}H_{52}O_8$ [57672-76-1] MW 612.80

Hippomane mancinella (Euphorbiaceae).

Highly toxic. It is also an irritant and a cocarcinogen.

Adolf, W., *Tetrahedron Lett.,* 1975, 1587.

Mansonone C 910

Cadalene-2,3-quinone

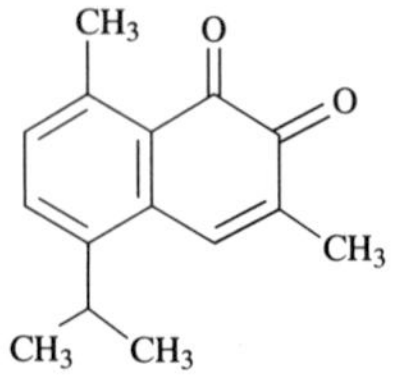

CH_3 O
O
CH_3
CH_3 CH_3

Sesquiterpenoid quinone
(Cadinane)

$C_{15}H_{16}O_2$ [5574-34-5] MW 228.29

Heartwood of *Mansonia altissima* (Sterculaceae) and of the elm, *Ulmus glabra* (Ulmaceae). This is one of ten or more co-occurring mansonones in these plants.

Sawdust containing mansonone C causes sneezing, vertigo and eczema.

Marini Bettòlo, G. B., *Tetrahedron Lett.,* 1965, 4857.

Maytansine 911

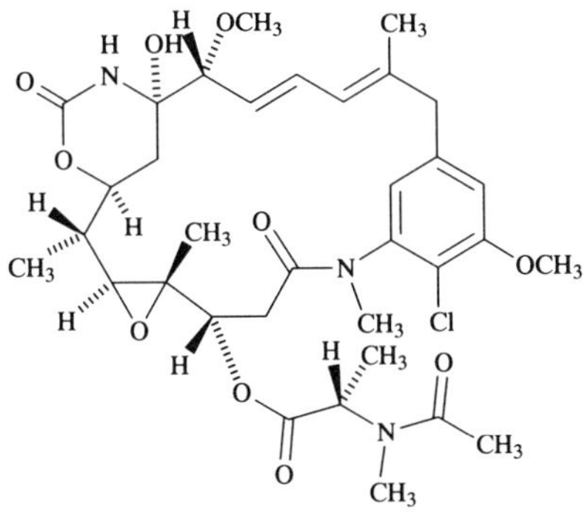

Maytansine alkaloid

$C_{34}H_{46}ClN_3O_{10}$ [35846-53-8] MW 692.21

Fruit of *Maytenus ovatus* and *M. senata* (Celastraceae). The stem and wood of *Putterlickia verrucosa* (Celastraceae) is a particularly rich source (12 mg/kg dry-weight).

Antileukaemic and cytotoxic properties. This is a very potent anticancer agent, but its harmful side-effects have so far prevented its clinical use.

Kupchan, S. M., *J. Am. Chem. Soc.*, 1972, **94**, 1354.

Mecambrine 912

Fugapavine

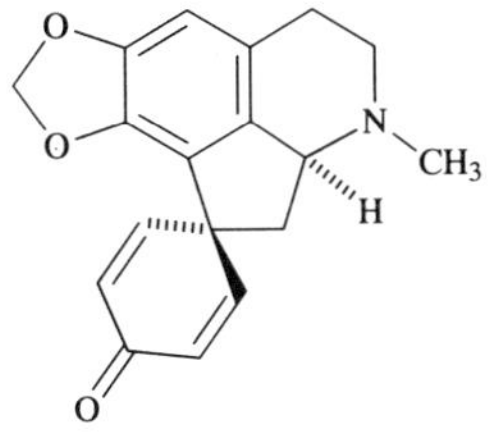

Benzylisoquinoline alkaloid
(Proaporphine)

$C_{18}H_{17}NO_3$ [1093-07-8] MW 295.34

Papaver fugax, *P. dubium* and the Welsh poppy, *Meconopsis cambrica* (Papaveraceae).

Large doses are toxic, causing convulsions. The LD_{50} in mice is 4.1 mg/kg body-weight. It increases blood-pressure, stimulates respiration and produces bradycardia in animals.

Slavík, J., *Collect. Czech. Chem. Commun.*, 1965, **30**, 914.

Medicagenic acid 3-glucoside 913

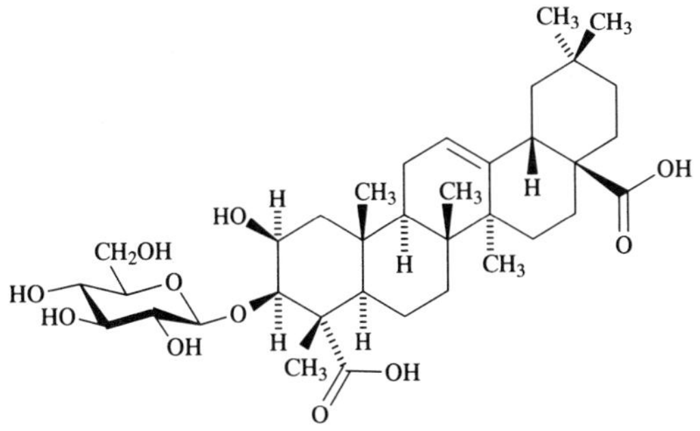

Triterpenoid saponin
(Oleanane)

$C_{36}H_{56}O_{11}$ [49792-23-6] MW 664.83

Roots of *Dolichos kilimandscharicus* and of *Medicago sativa* (Leguminosae).

Antifungal activity at a concentration of 5.0 μg/ml and molluscicidal at a concentration of 25 μg/l.

Morris, R. J., *J. Org. Chem.*, 1961, **26**, 1241.

Medicagenic acid 3-triglucoside 914

Medicogenic acid 3-triglucoside

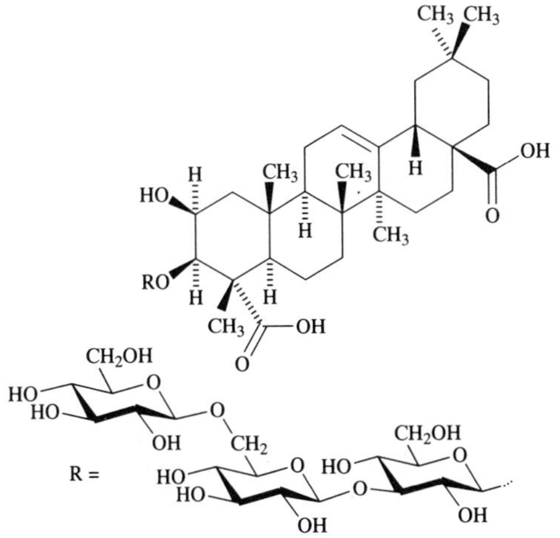

Triterpenoid saponin
(Oleanane)

$C_{48}H_{76}O_{21}$ [41162-94-1] MW 989.12

One of eleven glycosides of medicagenic acid present in the roots and aerial parts of alfalfa or lucerne, *Medicago sativa* (Leguminosae).

The saponins of lucerne have antinutritional properties and care should be exercised in using lucerne as an animal feed. They also have haemolytic properties.

Assa, Y., *Biochim. Biophys. Acta,* 1973, **307**, 83.

(−)-Medicarpin 915

Demethylhomopterocarpin

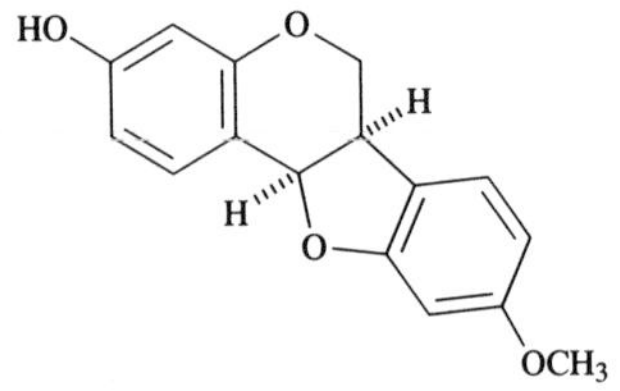

Isoflavonoid
(Pterocarpan)

$C_{16}H_{14}O_4$ [32383-76-9] MW 270.28
(+)-form [33983-39-0]
(±)-form [33983-40-1]

In fungally-infected leaves of *Medicago sativa* and *Trifolium pratense*; constitutively present in heartwood of *Andira inermis* and *Dalbergia variabilis* (Leguminosae).

Antifungal agent.

Smith, D. G., *Physiol. Plant Pathol.,* 1971, **1**, 41.

Megaphone 916

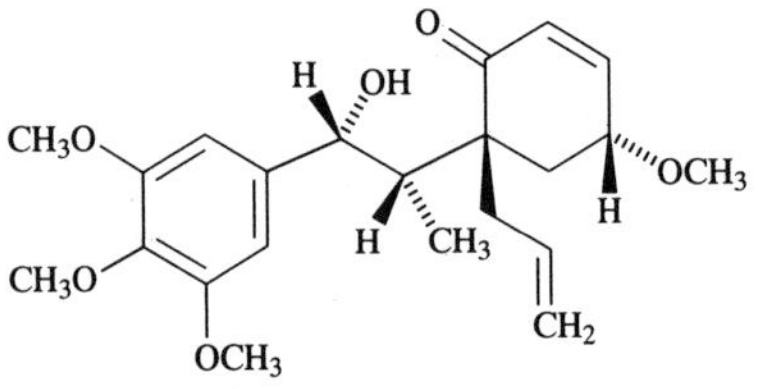

Neolignan

$C_{22}H_{30}O_6$ [64332-37-2] MW 390.48

Roots of *Aniba megaphylla* (Lauraceae).

Cyotoxic properties.

Kupchan, S. M., *J. Org. Chem.,* 1978, **43**, 586.

Melampodin A 917

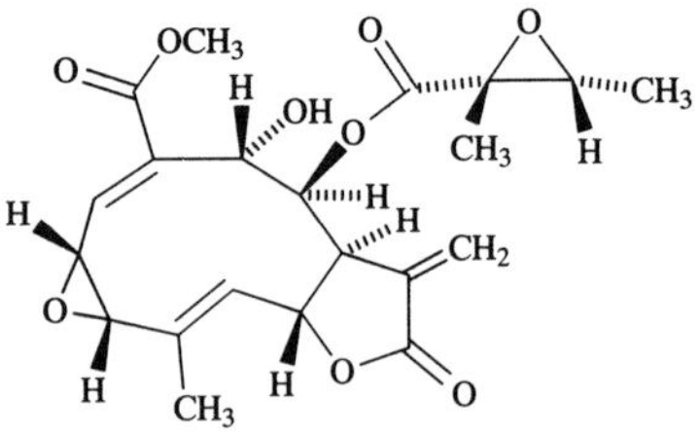

Sesquiterpenoid lactone
(Germacranolide)

$C_{21}H_{24}O_9$ [35852-26-7] MW 420.42

Melampodium heterophyllum and *M. leucanthum* (Compositae).

Insect antifeedant, which also affects insect growth and reduces insect survival.

Dominguez, X. A., *Phytochemistry,* 1981, **20**, 1431.

Melampodinin 918

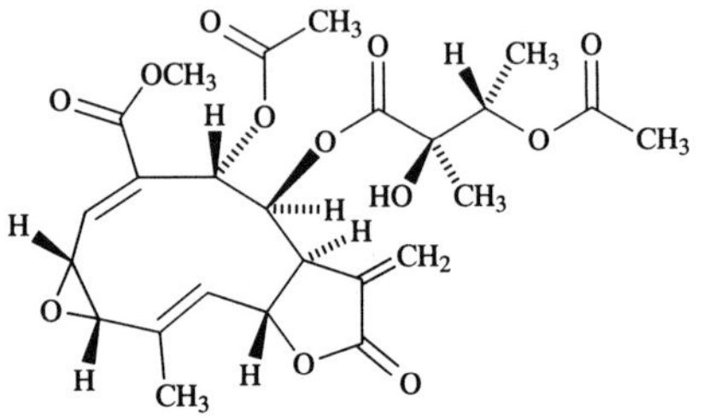

Sesquiterpenoid lactone
(Germacranolide)

$C_{25}H_{30}O_{12}$ [60295-53-6] MW 522.52

Melampodium americanum, M. diffusum and *M. longipes* (Compositae).

Insect antifeedant, which also affects insect growth and development.

Fischer, N. H., *J. Org. Chem.,* 1976, **41**, 3956.

Mellitoxin 919

4-Hydroxytutin; Hyenanchin; Ienancin

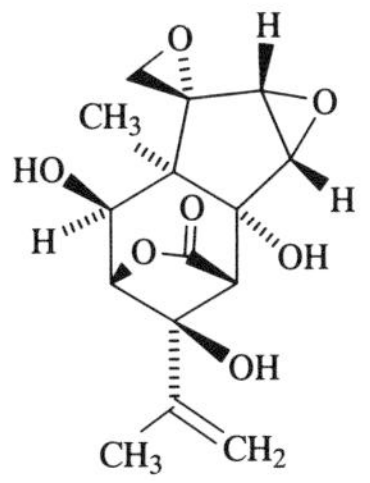

Sesquiterpenoid lactone
(Tutinanolide)

$C_{15}H_{18}O_7$ [3484-46-6] MW 310.30

Fruit of *Toxicodendron globosum* (Euphorbiaceae) and in *Coriaria arborea* (Coriariaceae).

Toxic principle in honey, causing extreme excitation of the central nervous system. First discovered in honey as a metabolite of the passion vine hopper, *Scolypopa australis*, which feeds on the tutin-containing *Coriaria arborea*.

Hodges, R., *Tetrahedron Lett.*, 1964, 371.

Melochinine 920

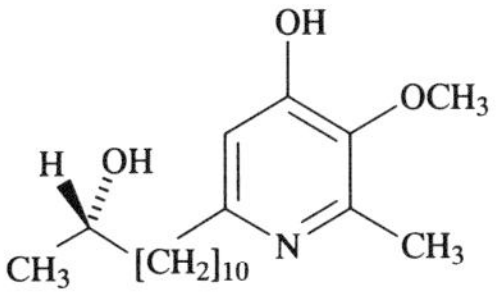

Pyridine alkaloid

$C_{19}H_{33}NO_3$ [70001-21-7] MW 323.48

Leaves of *Melochia pyramidata* (Sterculiaceae).

Causes paralysis in cattle, following the ingestion of *Melochia* plants. When administered to laboratory animals, it produces a variety of symptoms which suggest that the action of the alkaloid is at the cell membrane.

Medina, E., *Chem. Ber.*, 1979, **112**, 376.

Menthol 921

Mentol; Peppermint camphor; Menthacamphor

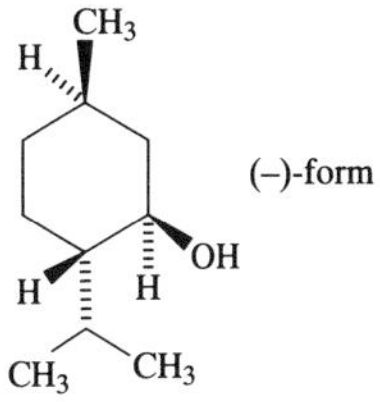

(–)-form

Monoterpenoid
(Menthane)

$C_{10}H_{20}O$ [2216-51-5]
(±)-form [15356-70-4] MW 156.27

Free and in ester form as the major constituent of peppermint oil, *Mentha piperita* and in other *Mentha* spp. (Labiatae). Present in many other plant essential oils in lesser amounts.

Can give rise to contact dermatitis in humans. Widely used to relieve symptoms of bronchial and nasal congestion.

Martindale, The Extra Pharmacopoeia, 30th edn, 1993, 1386, The Pharmaceutical Press.

Mesaconitine 922

3α-Hydroxyhypaconitine

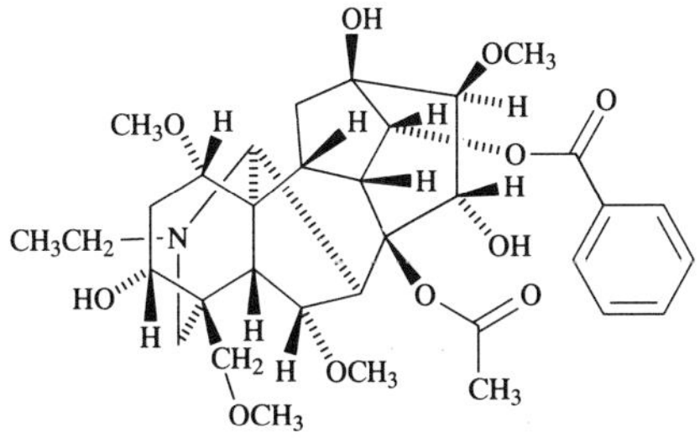

Diterpenoid alkaloid
(Aconitane)

$C_{33}H_{45}NO_{11}$ [2752-64-9] MW 631.72

Major alkaloid in certain subspecies of *Aconitum napellus*, the traditional source of aconitine; widely present in other *Aconitum* spp. (Ranunculaceae).

A potent poison, with similar activity to aconitine (q.v.).

Pelletier, S. W., *J. Am. Chem. Soc.*, 1976, **98**, 2626.

Mescaline 923

Mezcaline

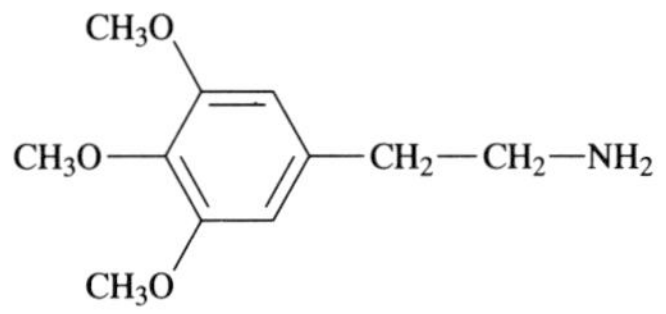

Aromatic amine

$C_{11}H_{17}NO_3$ [54-04-6] MW 211.26

Flowering heads of peyote, or mescal buttons, *Lophophora williamsii* and in other cacti, e.g. *Trichocereus pachanoi* (Cactaceae).

Toxic orally, with psychomimetic properties. It is a central nervous system depressant and hallucinogenic in high doses. one of the minor drugs of addiction.

Smith, T. A., *Phytochemistry,* 1977, **16**, 9.

Mesembrenone 924

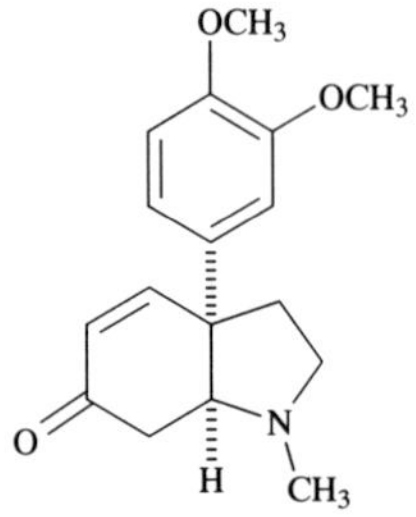

Indole alkaloid

$C_{17}H_{21}NO_3$ [80287-15-6] MW 287.36

Sceletium expansum, *S. tortuosum* and *S. anatomicum* (Aizoaceae).

Narcotic, cocaine-like stimulant. *Sceletium* leaves are chewed by Namaqualand bushmen and fermented to form a preparation known as "channa".

Jeffs, P. W., *J. Org. Chem.,* 1970, **35**, 3512.

Mesembrine 925

Mesembranone

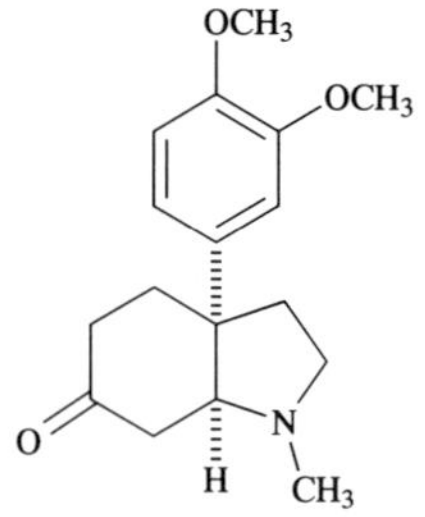

Indole alkaloid

$C_{17}H_{23}NO_3$ [24880-43-1] MW 289.37

Sceletium expansum, *S. tortuosum* and *S. anatomicum* (Aizoaceae).

Narcotic stimulant. *Sceletium* leaves are chewed by Namaqualand bushmen and fermented to form a preparation known as "channa".

Popelak, A., *Naturwissenschaften,* 1960, **47**, 231.

Mesembrinol 926

Mesembranol

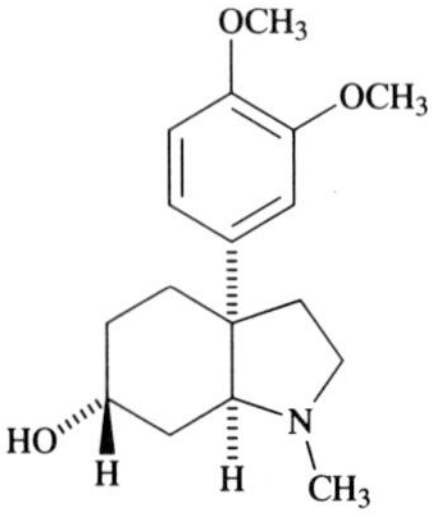

Indole alkaloid

$C_{17}H_{25}NO_3$ [23544-42-5] MW 291.39

Sceletium expansum, *S. tortuosum* and *S. anatomicum* (Aizoaceae).

Narcotic stimulant. *Sceletium* leaves are chewed by Namaqualand bushmen and fermented to form a preparation known as "channa".

Jeffs, P. W., *J. Am. Chem. Soc.,* 1969, **91**, 3831.

4′-Methoxyaucuparin 927

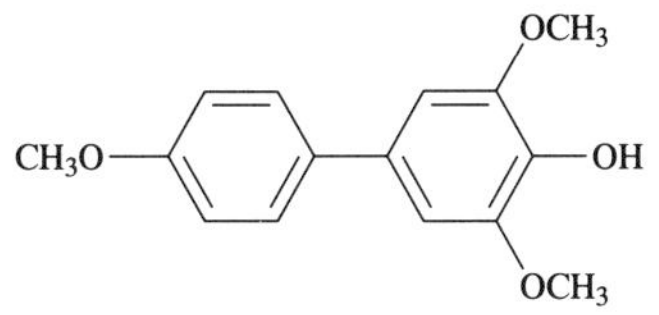

Biphenyl

$C_{15}H_{16}O_4$ [54961-04-5] MW 260.29

Sapwood of rowan, *Sorbus aucuparia* (Rosaceae) infected with *Nectria cinnabarina*.

Antifungal agent (phytoalexin).

Watanabe, K., *Agric. Biol. Chem.*, 1990, **54**, 1861.

p-Methoxycinnamic acid ethyl ester 928

Ethyl *p*-methoxycinnamate

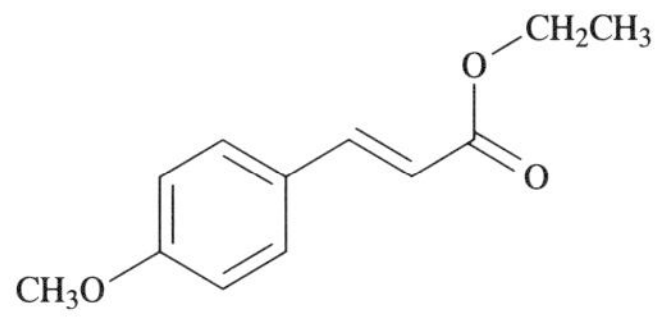

Phenylpropanoid

$C_{12}H_{14}O_3$ [1929-30-2] MW 206.24

Rhizomes of *Kaempferia galangae* and of *Hedychium spicatum* (Zingiberaceae).

Cytotoxic against HeLa cells.

Kosuge, T., *Chem. Pharm. Bull.*, 1985, **33**, 5565.

5-Methoxy-*N,N*-dimethyltryptamine 929

O-Methylbufotenine

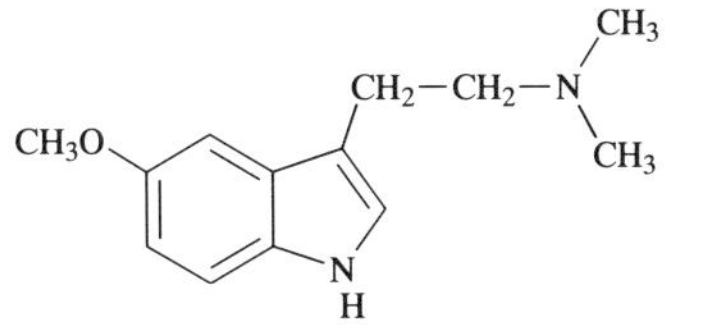

Aromatic amine

$C_{13}H_{18}N_2O$ [1019-45-0] MW 218.30

Pasture grasses *Phalaris arundinacea* and *P. tuberosa* (Gramineae) and in *Desmodium pulchellum* (Leguminosae).

Toxin. It is partly responsible for the toxic condition in sheep known as "Phalaris staggers" when the concentration in the grasses is high. It is psychotomimetic.

Baxter, C., *Phytochemistry,* 1972, **11**, 2767.

6-Methoxymellein 930

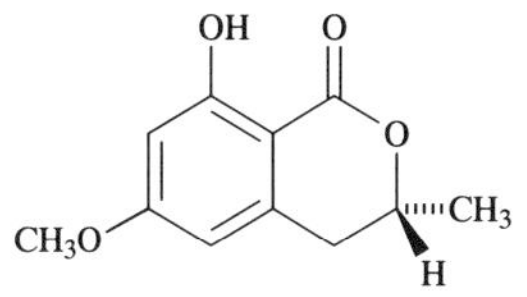

Isocoumarin

$C_{11}H_{12}O_4$ [13410-15-6] MW 208.21

Carrot root, *Daucus carota* infected with *Ceratocystis fimbriata*.

Antifungal agent (phytoalexin).

Govindachari, T. R., *Phytochemistry,* 1971, **10**, 1603.

2-Methoxy-1,4-naphthoquinone 931

Lawsone methyl ether

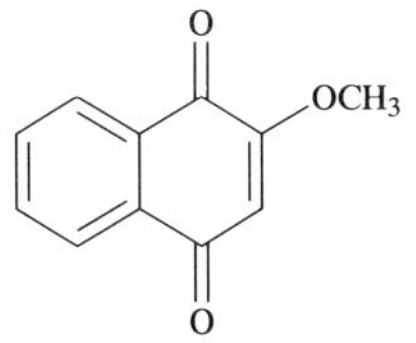

Naphthoquinone

$C_{11}H_8O_3$ [2348-82-5] MW 182.18

Leaf of *Swertia calycina* (Gentianaceae) and *Impatiens glandulifera* (Balsaminaceae).

Strong antifungal activity, effective against *Cladosporium cucumerinum* at a concentration of 0.1 μg/ml.

Little, J. E., *J. Biol. Chem.*, 1948, **174**, 335.

5-*O*-Methylalloptaeroxylin 932

Methylallopteroxylin; Perforatin A

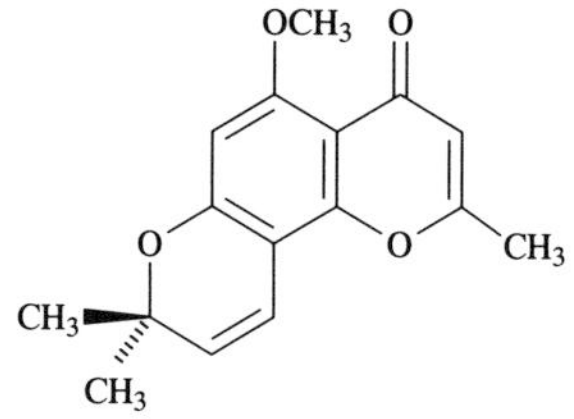

Chromone

$C_{16}H_{16}O_4$ [35930-31-5] MW 272.30

Leaves and wood of *Ptaeroxylon obliquum* and in *Cedrolopsis grevei* (Ptaeroxylaceae), *Neochamaelea pulverulenta* (Cneoraceae) and *Harrisonia perforata* (Simaroubaceae).

The powdered wood of *Ptaeroxylon* is pungent and irritating, causing violent sneezing. This plant is used in Southern Africa as a traditional medicine.

Dean, F. M., *Phytochemistry,* 1971, **10**, 3221.

3-Methylamino-L-alanine 933

N^{β}-Methyl-L-α,β-diaminopropionic acid

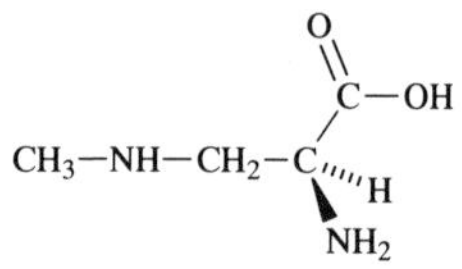

Amino acid

$C_4H_{10}N_2O_2$ [15920-93-1] MW 118.14

Seeds and leaves of the order Cycadales, including *Cycas circinalis.*

Growth retardant in rats when given orally, but toxic, causing convulsions and mortality when injected. It contributes to the toxicity of cycads in humans, although the main toxic principle present is cycasin (q.v.).

Vega, A., *Phytochemistry,* 1967, **6**, 759.

N-Methyl-2,6-bis(2-hydroxybutyl)-Δ^3-piperideine 934

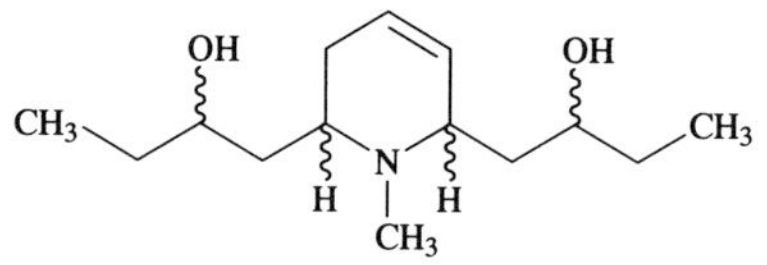

Piperidine alkaloid

$C_{13}H_{24}NO_2$ [105694-50-6] MW 226.34

In leaves of *Lobelia berlandieri* (Lobeliaceae).

One of three closely related piperideines presumed to be responsible for cattle poisoning caused by the consumption of this plant. Animals suffer from narcosis for up to 2 weeks and only recover if they are hand-fed and watered during that time.

Williams, H. J., *J. Agric. Food Chem.,* 1987, **35**, 19.

Methyl caffeate 935

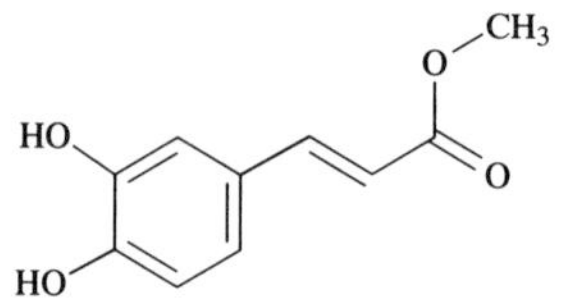

Phenylpropanoid

$C_{10}H_{10}O_4$ [3843-74-1] MW 194.19

Gaillardia pulchella, Tanacetum odessanum, Artemisia apiacea, Pseudostiffita kingii, Bedfordia solicina and *Gochnatra rusbyana* (Compositae), and *Scabiosa succisa* (Dipsacaceae).

Antitumour activity against Sarcoma 180.

Paler, M., *Planta Med.,* 1969, **17**, 139.

(+)-*N*-Methylconiine 936

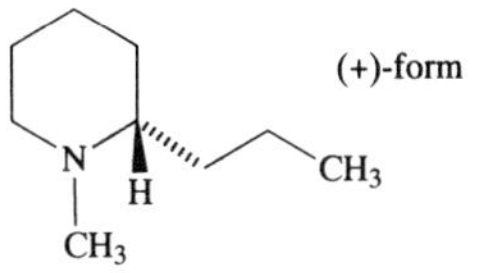

(+)-form

Piperidine alkaloid

$C_9H_{19}N$ [35305-13-6] MW 141.26

Minor alkaloid in leaf and seed of hemlock, *Conium maculatum* (Umbelliferae).

Toxic, with similar effects to coniine (q.v.).

Manske, R. H. F., The Alkaloids, 1950, **1**, 165, Academic Press.

S-Methyl-L-cysteine *S*-oxide 937

3-(Methylsulfinyl)alanine

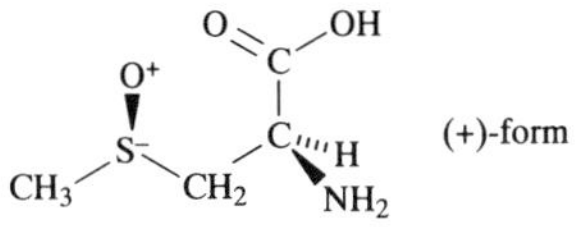

(+)-form

Amino acid

$C_4H_9NO_3S$ [32726-14-0] MW 151.19

Onion, *Allium cepa* (Alliaceae) and cabbages, *Brassica* spp. (Cruciferae).

Toxic in cattle fed cruciferous crops in excessive amounts. Is degraded in the ruminant to dimethylsulfide, which is the true toxin.

Barnsley, E. A., *Tetrahedron,* 1968, **24**, 3747.

α-(Methylenecyclopropyl)glycine 938

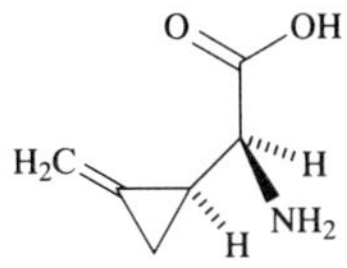

Amino acid

$C_6H_9NO_2$ [2517-07-9] MW 127.14

Seeds of *Litchi sinensis* (Sapindaceae), *Billia hippocastanum* (Hippocastanaceae) and the sycamore, *Acer pseudoplatanus* (Aceraceae).

Hypoglycaemic, with similar toxicity to hypoglycin (q.v.).

Gray, D. O., *Biochem. J.,* 1962, **82**, 385.

N-Methylflindersine 939

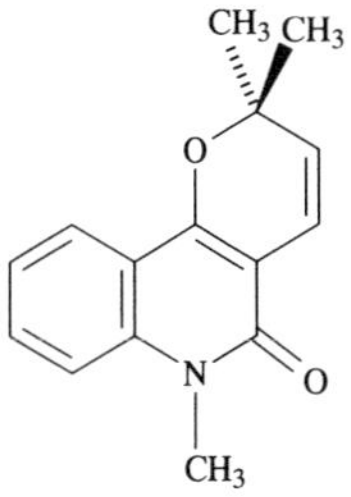

Quinoline alkaloid

$C_{15}H_{15}NO_2$ [50333-13-6] MW 241.29

Ptelea trifoliata, *Xylocarpus granatum*, *Atalantia roxburghiana* and *Fagara chalybaea* (Rutaceae).

Antifeedant activity against beetles.

Brown, R. F. C., *Aust. J. Chem.,* 1954, **7**, 348.

Methylillukumbin A 940

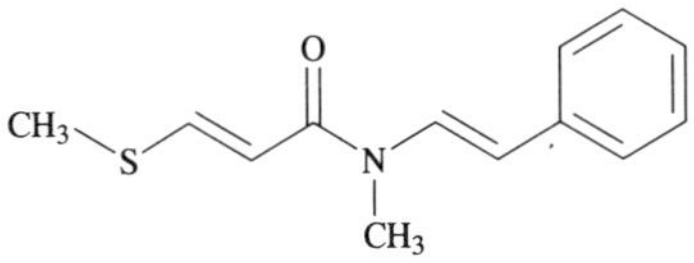

Phenylpropanoid
(Sulfur compound)

$C_{13}H_{15}NOS$ [152175-17-2] MW 233.33

Leaves of *Glycosmis mauritania* (Rutaceae).

Antifungal activity against *Cladosporium cladosporioides*.

Hinterberger, S., *Tetrahedron,* 1994, **50**, 6279.

Methylisoeugenol 941

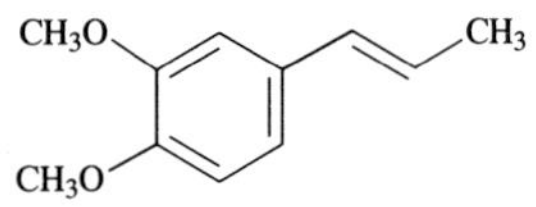

Phenylpropanoid

$C_{11}H_{14}O_2$ [6379-72-2] MW 178.23

Essential oil of the roots of *Asarum europaeum* (Aristolochiaceae) and of *Acorus calamus* (Araceae) and in the leaf wax of *Daucus carota* (Umbelliferae).

Moderately toxic to animals. Has expectorant, spasmolytic and antihistaminic properties. It is used as a local anaesthetic.

Naves, Y. R., *Bull. Soc. Chim. Fr.*, 1959, 1233.

Methyl linolenate 942

Methyl α-linolenate

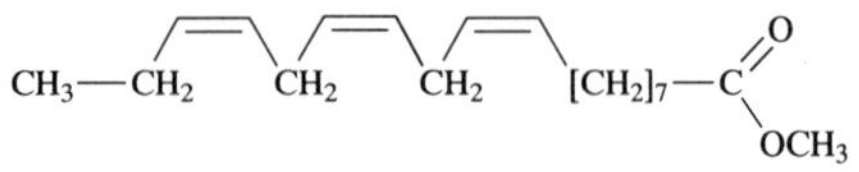

Fatty acid

$C_{19}H_{32}O_2$ [301-00-8] MW 292.46

Peanut, *Arachis hypogaea* (Leguminosae) infected with *Puccinia arachidus*.

Antifungal agent (phytoalexin).

Subba Row, P. V., *Oleagineux,* 1988, **43**, 173.

Methyl-lycaconitine 943

Delartine; Delsemidine

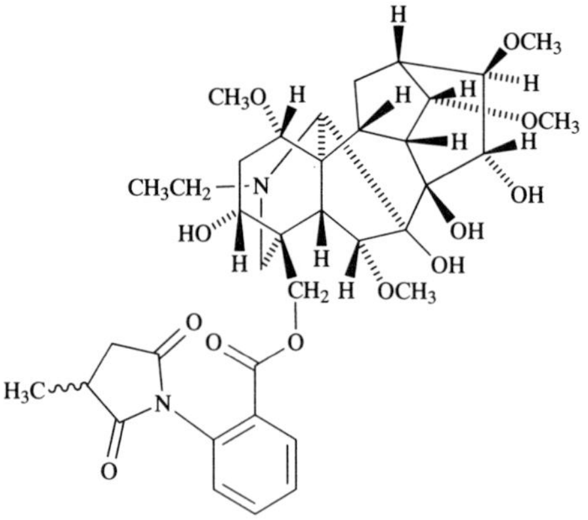

Diterpenoid alkaloid (Aconitane)

$C_{37}H_{50}N_2O_{10}$ [21019-30-7] MW 682.81

Roots of *Delphinium elatum*, aerial parts of *D. nuttallianum*, *D. barbeyi* and other *Delphinium* spp. (Ranunculaceae).

Potent neuromuscular poison in mammals with classical curariform activity. Cause of cattle poisoning in North America, following consumption of *Delphinium barbeyi* and *Delphinium nuttallianum*. The LD_{50} for total alkaloids of *Delphinium barbeyi* is 25 to 40 mg/kg body-weight for an oral dose to cattle.

Kuzovkov, A. D., *J. Gen. Chem. USSR,* 1959, **29**, 2746.

Methyl 2-*trans*,8-*cis*-matricarate 944

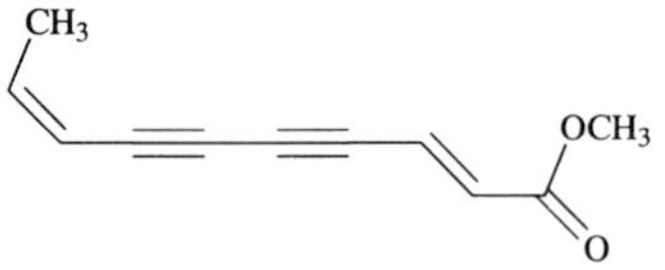

Polyacetylene

$C_{11}H_{10}O_2$ [23180-59-8] MW 174.20

Roots of *Erigeron philadelphicus* (Compositae), where it co-occurs with the 2-*cis*,8-*cis*-isomer [25019-41-4].

Both the 2-*trans*,8-*cis*- and the 2-*cis*,8-*cis*-forms of this acetylene are nematocidal, causing mortality at a concentration of 3.0 m[>g/ml.

Kimura, Y., *Agric. Biol. Chem.,* 1981, **45**, 2915.

Methyl mercaptan 945

Methanethiol; Methyl thioalcohol

CH_3-SH

Sulfur compound

CH_4S [74-93-1] MW 48.11

Roots of the radish, *Raphanus sativus* (Cruciferae) and present as a trace component in many plant volatiles.

Odour of rotten cabbage. Poisonous gas, if inhaled at a significant concentration. Used as a pesticide.

Koolhaas, D. R., *Biochem. Z.,* 1931, **230**, 446.

N-Methylmescaline 946

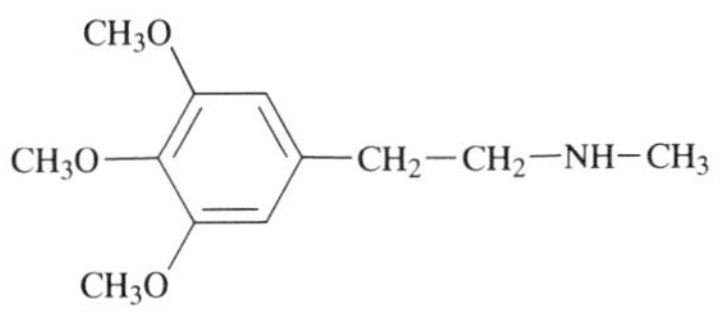

Aromatic amine

$C_{12}H_{19}NO_3$ [4838-96-4] MW 225.29

Cacti such as *Lophophora williamsii* (Cactaceae) and *Alhagi pseudoalhagi* (Leguminosae).

Psychotomimetic, similar in activity to mescaline.

Späth, E., *Ber. Dtsch. Chem. Ges.,* 1937, **70B**, 1446.

30-Methyloscillatoxin D 947

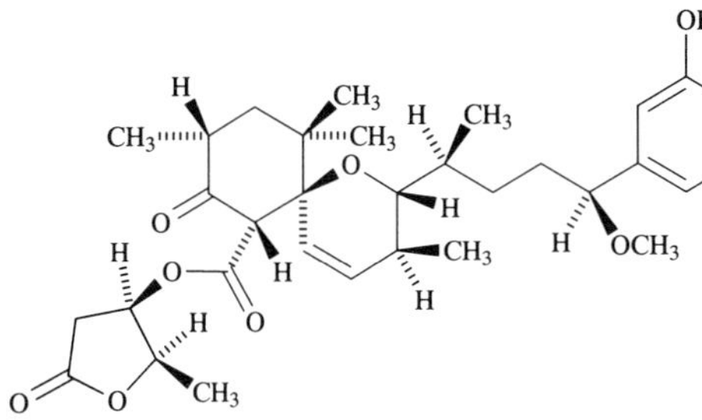

Phenolic bislactone

$C_{32}H_{44}O_8$ [95069-54-8] MW 556.70

Produced by a mixed culture of *Schizothrix calcicola* and *Oscillatoria nigroviridis*.

Associated with severe contact dermatitis known as swimmers' itch.

Entzeroth, M., *J. Org. Chem.,* 1985, **50**, 1255.

O^4-Methylptelefolonium 948

Ptelefolonium

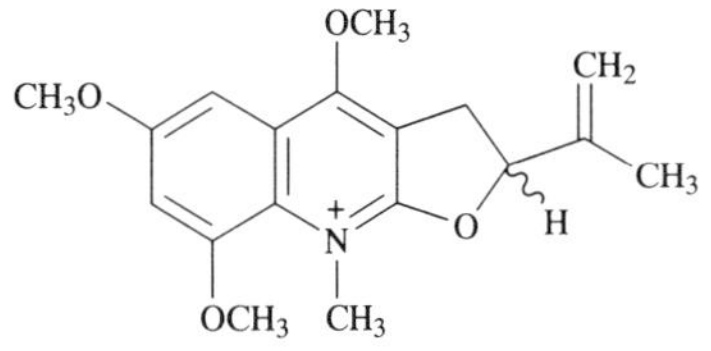

Quinoline alkaloid

$[C_{18}H_{22}NO_4]^+$ [52768-97-5] MW 316.38

Leaves of *Ptelea trifoliata* (Rutaceae).

Cytotoxic to animal tumours.

Reisch, J., *Phytochemistry,* 1973, **12**, 2552.

4′-*O*-Methylpyridoxine 949

Ginkgotoxin

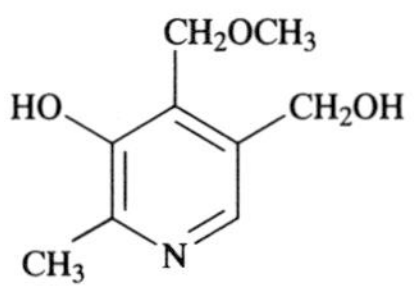

Pyridine alkaloid

$C_9H_{13}NO_3$ [1464-33-1] MW 183.21

Seeds of *Ginkgo biloba* (Ginkgoaceae), seeds and pods of *Albizia versicolor* and *A. tanganyicensis* (Leguminosae).

Toxic agent, with antivitamin B_6 activity. LD_{50} in humans is *ca* 11 mg/kg body-weight by oral route. Causes hypersensitivity, intermittent tetanic convulsions and mortality in South African cattle who consume *Albizia* pods.

Wada, K., *Chem. Pharm. Bull.,* 1988, **36**, 1779.

4′-*O*-Methylpyridoxine 5′-acetate 950

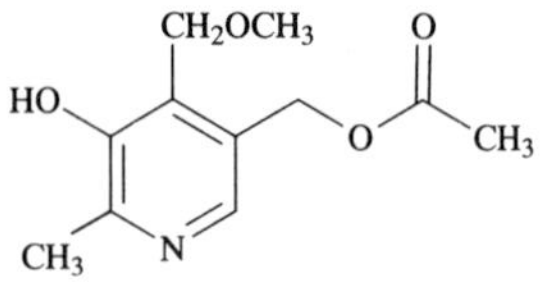

Pyridine alkaloid

$C_{11}H_{15}NO_4$ [82470-47-1] MW 225.24

Seeds and pods of *Albizia tanganyicensis* (Leguminosae).

Neurotoxic, causing cattle poisoning in South Africa along with 4′-methoxypyridoxine.

Steyn, P. S., *S. Afr. J. Chem.,* 1987, **40**, 191.

Methylripariochromene A 951

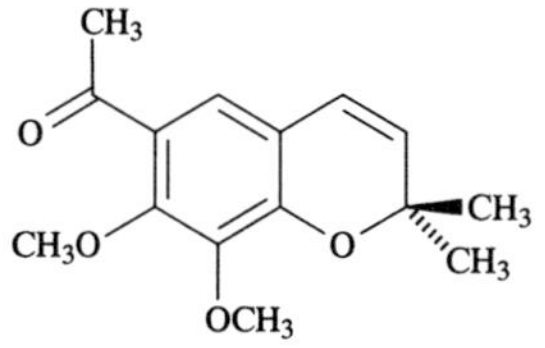

Chromene

$C_{15}H_{18}O_4$ [20819-46-9] MW 262.31

Roots of *Eupatorium riparium* and *Stevia serrata* (Compositae).

Antifungal activity against *Colletotrichum gloeosporioides.*

Taylor, D. R., *Phytochemistry,* 1971, **10**, 1665.

Se-Methyl-L-selenocysteine 952

(*R*)-2-Amino-3-selenomethylpropanoic acid; Selenomethylselenocysteine

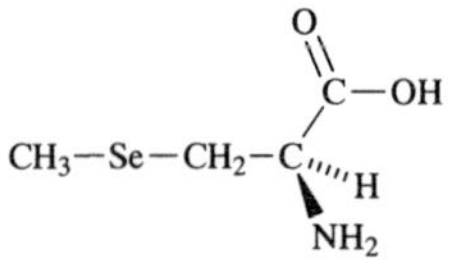

Amino acid

$C_4H_9NO_2Se$ [26046-90-2] MW 182.08

Seeds of accumulating plants growing on seleniferous soils, including *Astragalus bisulcatus* (Leguminosae) and *Oonopsis condensata* (Compositae).

Produces selenosis in livestock, in a similar manner to selenocystathione (q.v.) and is responsible for the often fatal 'blind staggers' in farm animals grazing on seleniferous plants.

Nigam, S. N., *Biochim. Biophys. Acta,* 1969, **192**, 185.

Mexicanin E 953

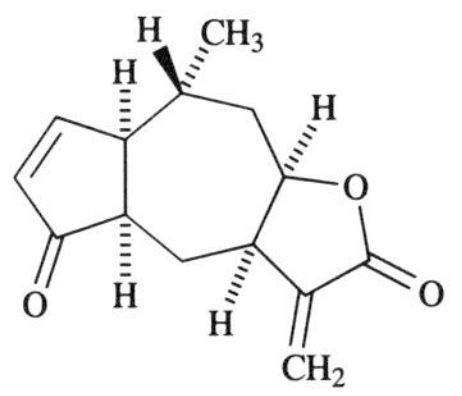

Sesquiterpenoid lactone
(Norpseudoguaianolide)

$C_{14}H_{16}O_3$ [5945-40-4] MW 232.28

Helenium mexicanum and other *Helenium* spp. (Compositae).

Highly toxic to mammals. It shows cytotoxic and antitumour properties.

Romo, J., *Tetrahedron,* 1963, **19**, 2317.

Mexicanin I 954

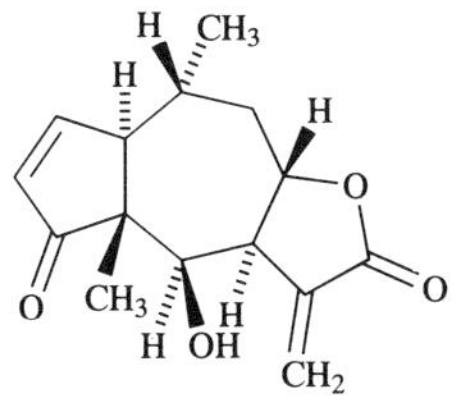

Sesquiterpenoid lactone
(Pseudoguaianolide)

$C_{15}H_{18}O_4$ [5945-41-5] MW 262.31

Sneezeweed, *Helenium autumnale*, several other *Helenium* spp., *Gaillardia pinnatifida* and *Hymenoxys linearis* (Compositae).

Cytotoxic and antitumour properties.

Domínguez, E., *Tetrahedron,* 1963, **19**, 1415.

Mezerein 955

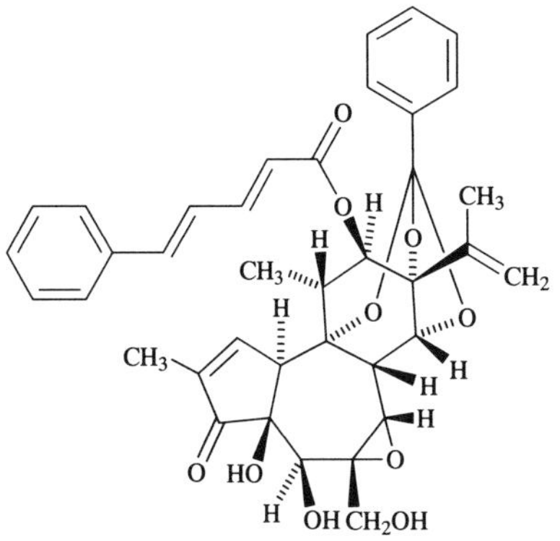

Diterpenoid
(Daphnane)

$C_{38}H_{38}O_{10}$ [34807-41-5] MW 654.71

Mezereon, *Daphne mezereum* (Thymelaeaceae), mainly in the fruits (0.04%).

Antitumour activity. Very active skin irritant. The 'mouse-ear inflammation' unit is 0.2 μg mezerein per ear. Responsible for human poisoning from mezereon. Two or three fruits can be fatal in a child.

Ronlán, A., *Tetrahedron Lett.,* 1970, 4261.

Michelenolide 956

Costunolide diepoxide

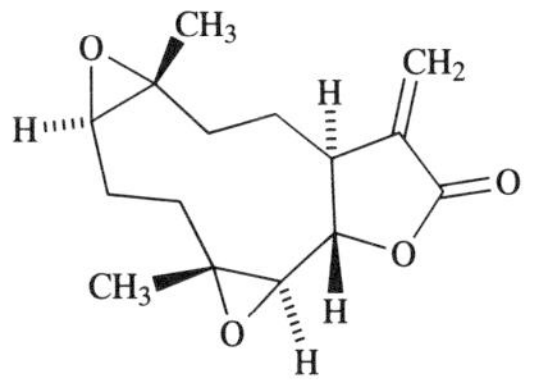

Sesquiterpenoid lactone
(Germacranolide)

$C_{15}H_{20}O_4$ [66392-96-9] MW 264.32

Rootbark of *Michelia compressa* (Magnoliaceae).

Cytotoxic and antitumour properties.

Ogura, M., *Phytochemistry,* 1978, **17**, 957.

Micheliolide 957

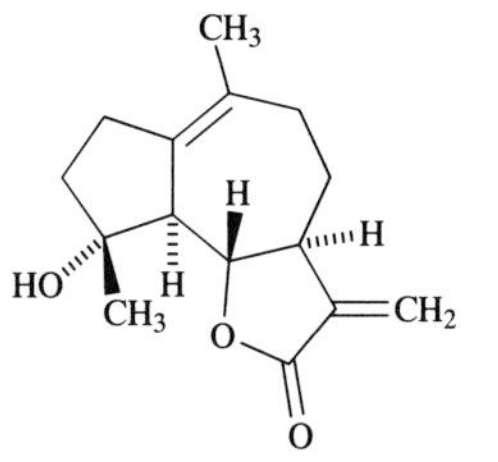

Sesquiterpenoid lactone (Guaianolide)

$C_{15}H_{20}O_3$ [68370-47-8] MW 248.32

Michelia compressa (Magnoliaceae).

Cytotoxic and antitumour properties.

Ogura, M., *Phytochemistry,* 1978, **17**, 957.

Michellamine B 958

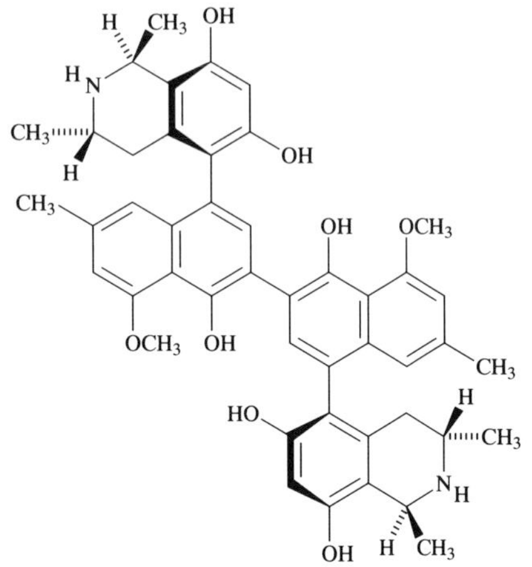

Bisisoquinoline alkaloid

$C_{46}H_{48}N_2O_8$ [137893-48-2] MW 756.90

Ancistrocladus korupensis (Ancistrocladaceae).

Potent antiviral activity, especially against HIV.

Manfredi, K. P., *J. Med. Chem.,* 1991, **34**, 3402.

Microcystin LA 959

Cyanoginosin LA; Fast Death Factor; Toxin BE-4; Aeruginosin

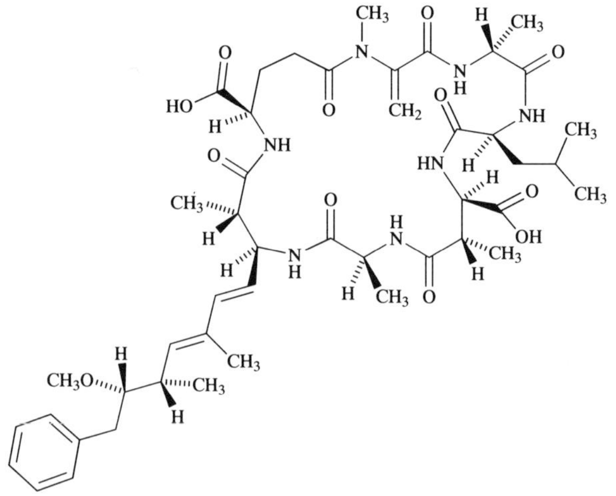

Cyclic heptapeptide

$C_{46}H_{67}N_7O_{12}$ [96180-79-9] MW 910.08

One of more than 40 microcystins produced by some strains of certain *Anabaena*, *Microcystis*, *Nostoc* and *Oscillatoria* spp.. *Microcystis aeruginosa* has been most intensively studied. Freshwater blooms of these organisms cause sporadic outbreaks of illness and death in wild and domestic animals.

A hepatotoxin, released from cyanobacterial cells in the stomach, absorbed from the ileum and concentrated via bile acid carriers by hepatocytes leading eventually to intrahepatic haemorrhage and liver failure. Intraperitoneal LD_{50} for the mouse is 50-100 μg/kg body-weight.

Falconer, I. R., Algal Toxins in Seafoods and Drinking Water, 1993, 187, Academic Press.

Microcystin LR 960

Cyanoginosin LR; Toxin BE-2; Toxin T17

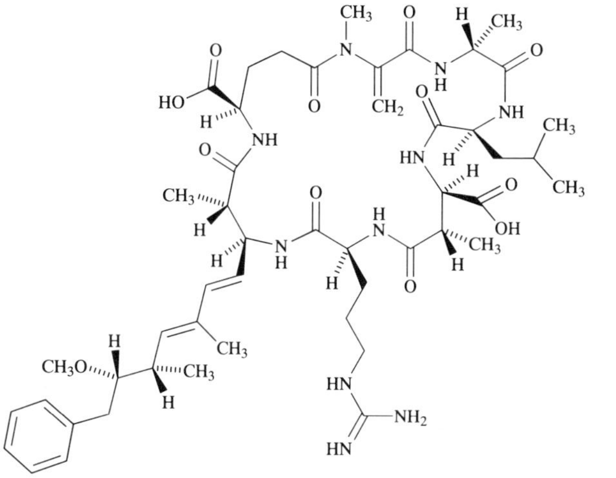

Cyclic heptapeptide

$C_{49}H_{74}N_{10}O_{12}$ [101043-37-2] MW 995.18

One of the more frequently produced microcystins associated with freshwater species of cyanobacteria such as *Microcystis aeruginosa*, *M. viridis* and *M. wesenbergii*.

An hepatotoxin causing intrahepatic haemorrhage which usually precedes death. LD_{50} 60-70 mg/kg body-weight intraperitoneally in the mouse.

Carmichael, W. W., *J. Appl. Bacteriol.*, 1992, **72**, 445.

Microhelenin-A 961

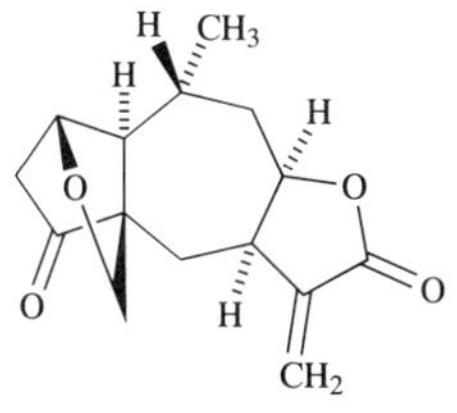

Sesquiterpenoid lactone
(Pseudoguaianolide)

$C_{15}H_{18}O_4$ [61490-63-9] MW 262.31

Helenium microcephalum (Compositae).

Cytotoxic and antitumour properties.

Lee, K. H., *J. Pharm. Sci.*, 1976, **65**, 1410.

Microhelenin C 962

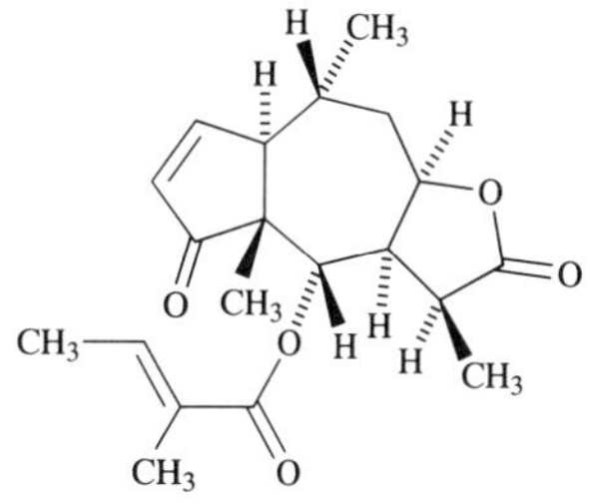

Sesquiterpenoid lactone
(Pseudoguaianolide)

$C_{20}H_{26}O_5$ [63569-07-3] MW 346.42

Helenium microcephalum (Compositae).

Cytotoxic and antitumour properties.

Lee, K. H., *Phytochemistry*, 1977, **16**, 393.

Microlenin 963

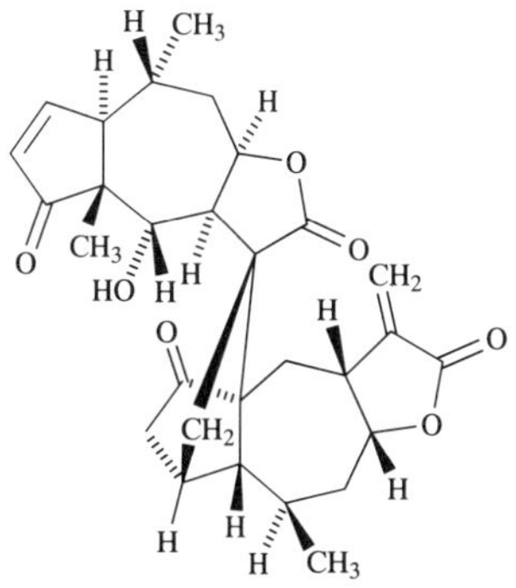

Bis-sesquiterpenoid lactone
(Pseudoguaianolide and norpseudoguaianolide)

$C_{29}H_{34}O_7$ [60622-41-5] MW 494.58

Helenium microcephalum (Compositae).

Cytotoxic and antitumour properties.

Lee, K. H., *J. Chem. Soc., Chem. Commun.*, 1976, 341.

Micromelin 964

Micromelumin

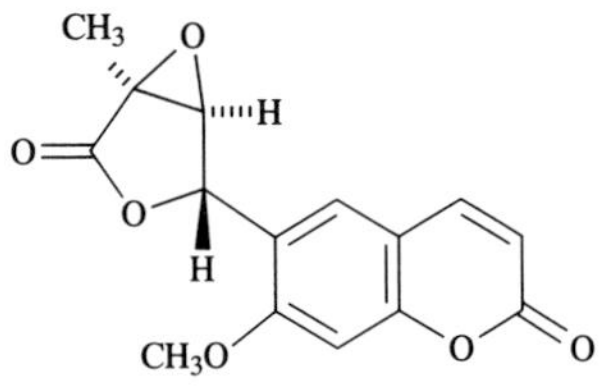

Coumarin

$C_{15}H_{12}O_6$ [15085-71-9] MW 288.26

Bark of *Caesaria graveolens* (Samydraceae), stem and leaf of *Micromelum integerrimum* and *M. minutum* (Rutaceae).

Antitumour properties.

Lamberton, J. A., *Aust. J. Chem.,* 1967, **20**, 973.

Milliamine L 965

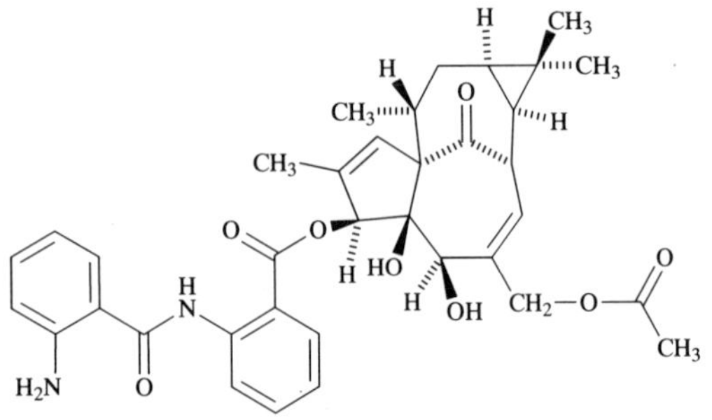

Diterpenoid alkaloid
(Ingenane)

$C_{35}H_{39}N_2O_8$ [152135-61-0] MW 615.70

Latex of *Euphorbia milii* var. *hislopii* (Euphorbiaceae).

Molluscicidal at 4 nM concentration. The most powerful molluscicide against *Biomphalaria glabrata* known to date.

Zani, C. L., *Phytochemistry,* 1993, **34**, 89.

L-Mimosine 966

Leucaenine; Leucaenol; Leucenine; Leucenol

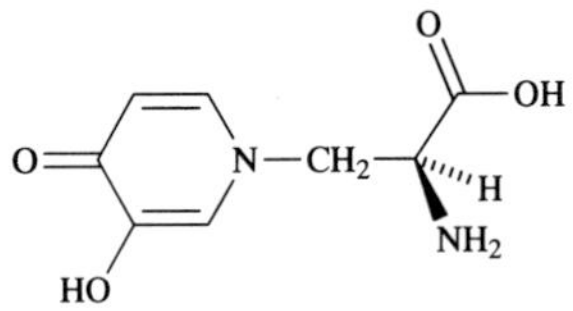

Amino acid

$C_8H_{10}N_2O_4$ [500-44-7] MW 198.18

Seeds and foliage of the jumbie bean, *Leucaena leucophylla* and of *Mimosa pudica* (Leguminosae). In *Leucaena*, the leaf concentration varies between 2 and 5% dry-weight, while the seed contains as much as 9% dry-weight.

Depilatory in horses, sheep and swine, and animals can become completely bald. It is goitrogenic in calves produced by heifers fed on *Leucaena*. In ruminants, goitrogenic effects occur, if 3-hydroxy-4(1*H*)-pyridone [1121-23-9], (a progoitrogen and an intermediate in detoxification) is not further metabolised. Green shoots, pods and ripe seed of *Leucaena* are used as human food, and there is a similar risk from the goitrogenic effects of the above detoxification product.

Adams, R., *J. Am. Chem. Soc.,* 1949, **71**, 705.

Miotoxin C 967

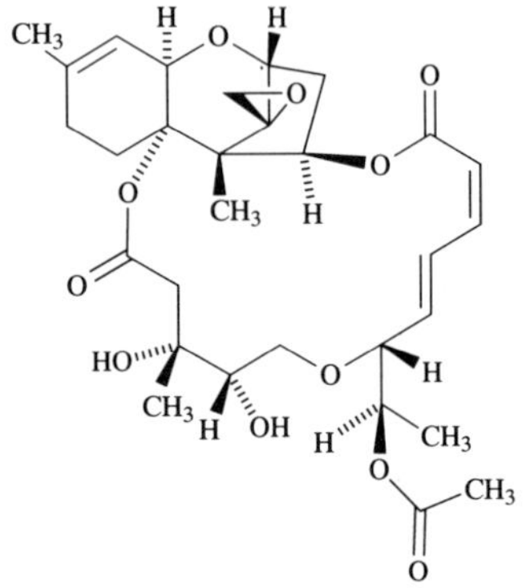

Sesquiterpenoid
(Trichothecane)

$C_{31}H_{42}O_{11}$ [93633-91-1] MW 590.67

Major toxin of aerial parts of *Baccharis megapotamica* (Compositae); also present in mio-mio, *B. cordifolia*.

Cause of livestock poisoning (see under roridin A).

Jarvis, B. B., *Phytochemistry,* 1991, **30**, 789.

Miroestrol 968

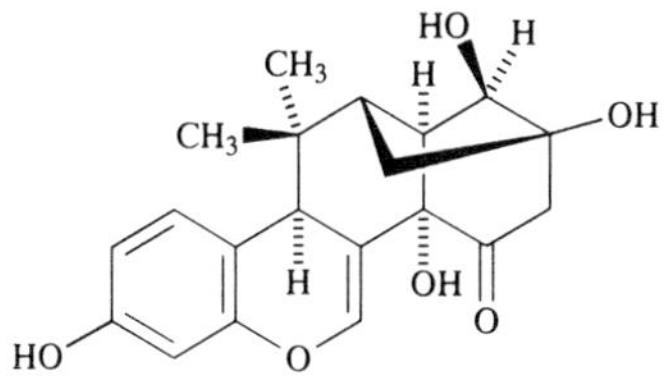

Steroid mimic

$C_{20}H_{24}O_6$ [2618-41-9] MW 360.41

Roots of *Pueraria mirifica* (Leguminosae).

Oestrogenic activity. The root extract has been used by pregnant women in Burma and Thailand to procure an abortion. Three times the activity of diethylstilboestrol when taken orally.

Bounds, D. G., *J. Chem. Soc.,* 1960, 3696.

Miserotoxin 969

3-Nitro-1-propyl-β-D-glucopyranoside

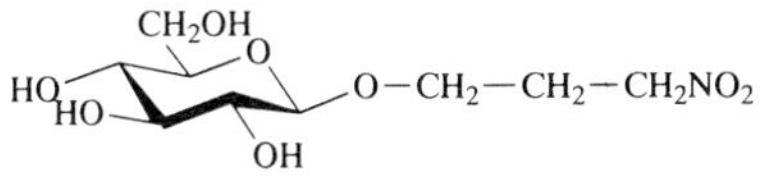

Nitro compound

$C_9H_{17}NO_8$ [24502-76-9] MW 267.24

Astragalus miser var. *oblongifolia*, *A. atropubescens*, *A. pterocarpus*, *A. tetrapleurus* and *A. toanus* (Leguminosae).

Bound toxin, yielding 3-Nitro-1-propanol [25182-84-7] on hydrolysis, which causes methaemoglobinaemia. Responsible for livestock poisoning in North America.

Stermitz, F. R., *Phytochemistry,* 1972, **11**, 1117.

Mistletoe lectins 970

Lectin I, composed of 4 chains, M_r 115 000
Lectin II, composed of 2 chains, M_r 60 000
Lectin III, composed of 2 chains, M_r 50 000

Glycoproteins MW 100 000-460 000

Vegetative parts of mistletoe, *Viscum album* (Viscaceae). See also viscumin.

Highly toxic to many animals, including Man. They react with human erythrocytes without specificity to blood groups A, B and O. Sugar specificities are; lectin I, D-galactose; lectin II, D-galactose and *N*-acetyl-D-galactosamine; lectin III, *N*-acetyl-D-galactosamine.

Franz, H., *Biochem. J.,* 1981, **195**, 481.

Mitragynine 971

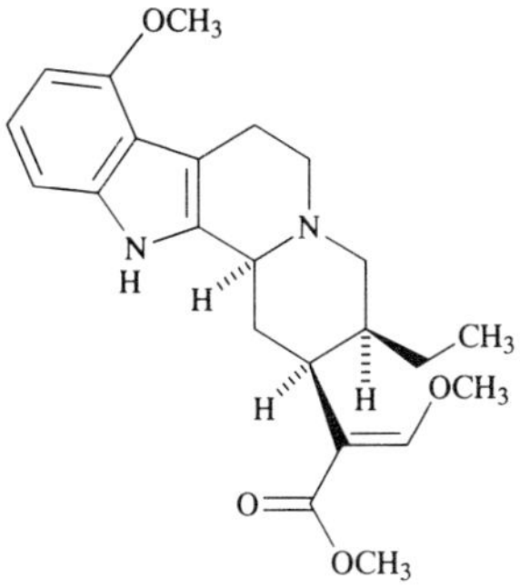

Indole alkaloid
(Corynan)

$C_{23}H_{30}N_2O_4$ [4098-40-2] MW 398.50

Mitragyna speciosa (Rubiaceae).

Central nervous system depressant; antitussive and analgesic activities. Leaves of *Mitragyna* are sometimes smoked for the narcotic action.

Joshi, B. S., *Chem. Ind. (London),* 1963, 573.

Molephantin 972

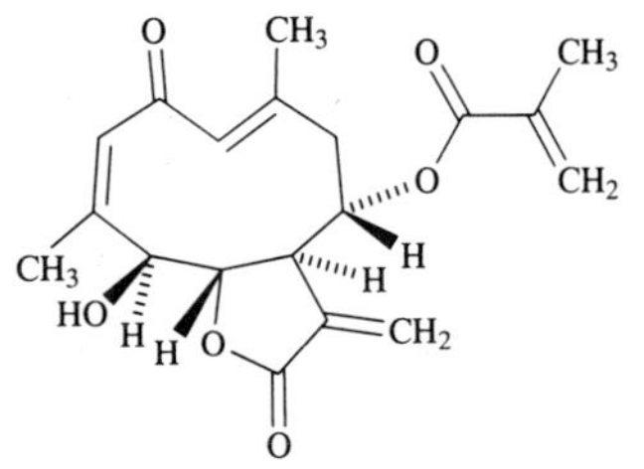

Sesquiterpenoid lactone
(Germacranolide)

$C_{19}H_{22}O_6$ [50656-66-1] MW 346.38

Elephantopus mollis (Compositae).

Cytotoxic and antitumour properties.

Lee, K. H., *J. Pharm. Sci.,* 1980, **69**, 1050.

Molephantinin 973

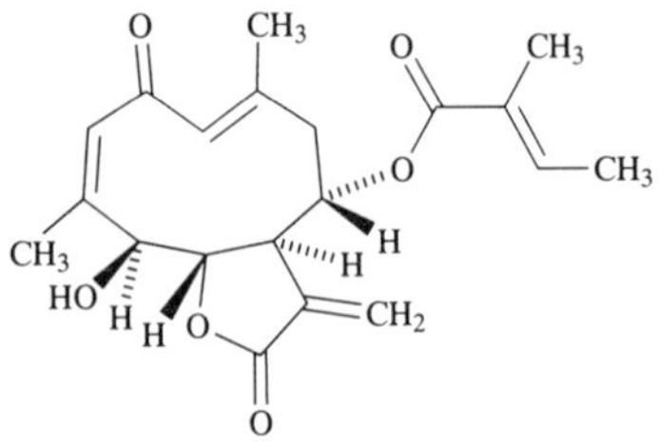

Sesquiterpenoid lactone
(Germacranolide)

$C_{20}H_{24}O_6$ [56221-98-8] MW 360.41

Elephantopus mollis (Compositae).

Cytotoxic and antitumour properties.

Lee, K. H., *J. Pharm. Sci.,* 1980, **69**, 1050.

Mollugogenol A 974

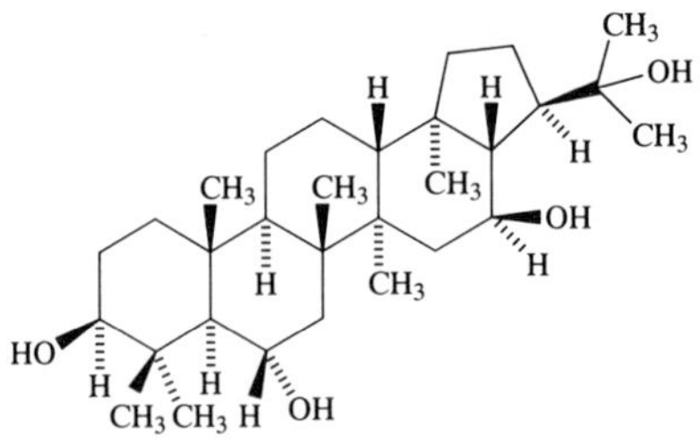

Triterpenoid
(Isohopane)

$C_{30}H_{52}O_4$ [22550-76-1] MW 476.74

Aerial parts of *Mollugo pentaphylla* and *M. hirta* (Molluginaceae).

Antifungal, active against *Cladosporium cucumerinum* at a concentration of 1.5 μg/ml.

Chakrabarti, P., *Tetrahedron,* 1969, **25**, 3301.

Momilactone A 975

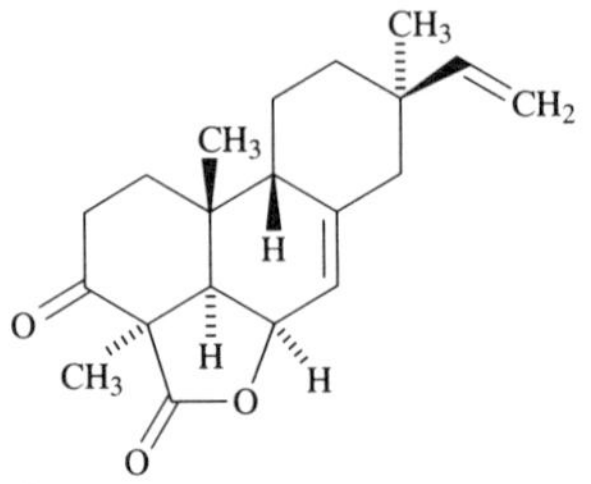

Diterpenoid
(Pimarane)

$C_{20}H_{26}O_3$ [51415-07-7] MW 314.42

Occurs, together with smaller amounts of the closely related momilactone B [51415-08-8], in leaves of rice, *Oryza sativa* (Gramineae) infected with *Cladosporium cucumerinum*, Both compounds occur constitutively in rice husks.

Antifungal agent.

Kato, T., *Tetrahedron Lett.,* 1973, 3861.

Momordin 976

Protein
(Lectin) MW 115 000

Seeds of bitter gourd, *Momordica charantia* (Cucurbitaceae).

Toxin, which inhibits protein synthesis.

Lin, J. Y., *Toxicon,* 1978, **16**, 653.

Monocrotaline 977

Crotaline

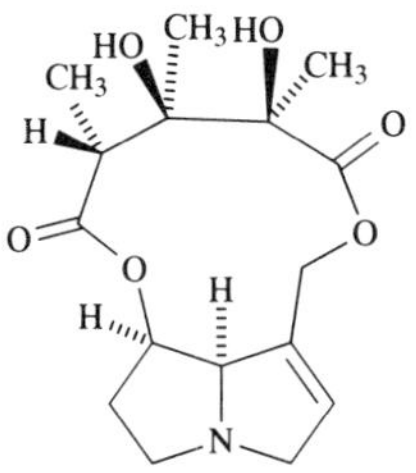

Pyrrolizidine alkaloid
(Norcrotalanan)

$C_{16}H_{23}NO_6$ [315-22-0] MW 325.36

Crotalaria crispata, C. paulina, C. quinquefolia, C. stipularia (Leguminosae) and *Lindelofia spectabilis* (Boraginaceae).

Hepatocarcinogenic and pneumotoxic. This is a causative agent of accidental poisoning in humans due to contamination of bread by alkaloid-containing material.

Adams, R., *J. Am. Chem. Soc.,* 1952, **74**, 5612.

Monofluoroacetic acid 978

Fluoroacetic acid

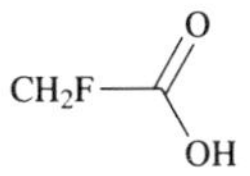

Organic acid

$C_2H_3FO_2$ [144-49-0] MW 78.04

First isolated in gifblaar, the South African plant *Dichapetalum cymosum* (Dichapetalaeae) and since detected in 13 other *Dichapetalum* spp.. Young leaves of *D. cymosum* contain 2500 μg dry-weight, while leaves of *D. braunii* have up to 8000 μg dry-weight. In Australia, *Acacia georginea* and many *Gastrolobium* and *Oxylobium* spp. (all Leguminosae) contain toxic levels. It is also present in the Brazilian plant *Palicourea marcgravii* (Rosaceae).

It is highly poisonous to humans and to farm animals. The fatal dose in humans is between 2 and 5 mg/kg body-weight. It stops respiration by blocking the Krebs tricarboxylic acid cycle, through the formation of fluorocitrate. Toxicity may be more closely connected with the inhibition of citrate transport. Farm animals are regularly poisoned by plants containing this toxin. However, some herbivorous animals in Australia (e.g. kangaroos) have evolved a resistance to fluoroacetate.

Harper, D. B., *Nat. Prod. Rep.,* 1994, **11**, 123.

Monotropitoside 979

Gaultherin

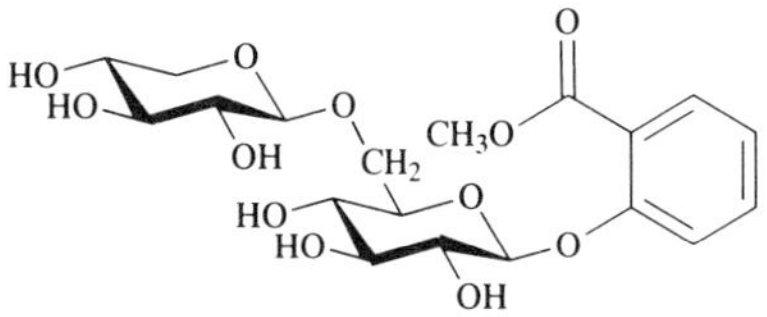

Phenolic acid
(Glycoside)

$C_{19}H_{26}O_{12}$ [490-67-5] MW 446.41

Checkerberry, *Gaultheria procumbens* (Ericaceae).

Steam distillation of the leaves gives oil of wintergreen, which is 96-99% methyl salicylate, derived by hydrolysis of monotropitoside. The oil has therapeutic value as an antirheumatic. Poisoning has occurred in children after ingestion of the pure oil.

Robertson, A., *J. Chem. Soc.,* 1931, 1881.

Montanine 980

O^2-Methylpancracine

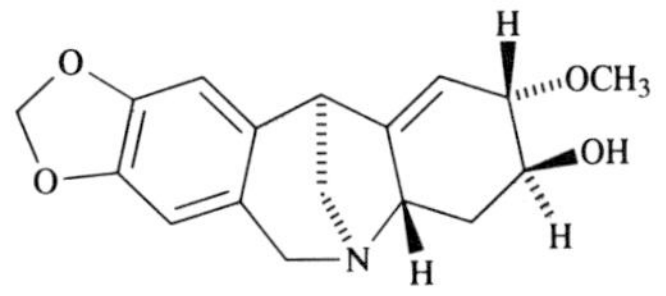

Amaryllidaceae alkaloid
(Pancracine)

$C_{17}H_{19}NO_4$ [642-52-4] MW 301.34

Bulbs of *Haemanthus montanus*, *H. amarylloides* and *H. coccineus* (Amaryllidaceae).

Poisonous, with an LD_{50} of 42 mg/kg body-weight intravenously in dogs. It is a weak hypotensive and convulsive agent.

Inubushi, Y., *J. Org. Chem.*, 1960, **25**, 2153.

Moracin A 981

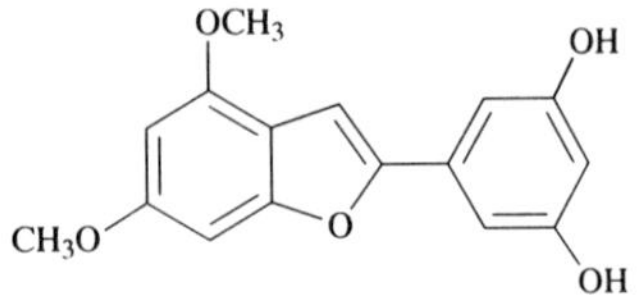

Benzofuran

$C_{16}H_{14}O_5$ [67259-17-0] MW 286.28

Together with a number of other related benzofurans in mulberry shoots, *Morus alba* (Moraceae) infected by *Fusarium solani*.

Antifungal agent (phytoalexin).

Takasugi, M., *Tetrahedron Lett.*, 1978, 797.

Moroidin 982

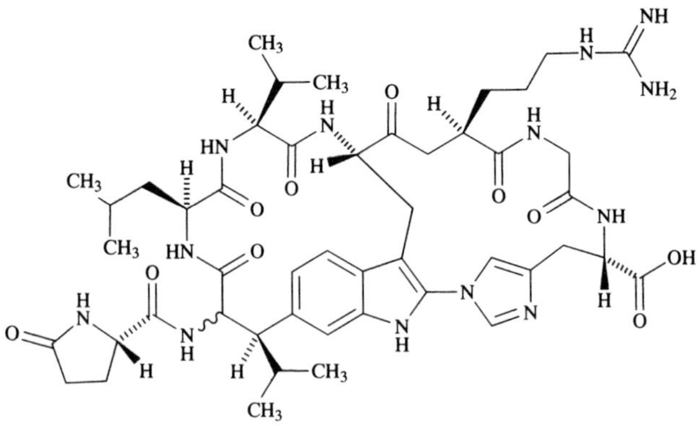

Cyclic peptide

$C_{47}H_{66}N_{14}O_{10}$ [104041-75-0] MW 987.13

In stinging hairs on leaves of *Dendrocnide moroides* (*Laportea moroides*) (Urticaceae).

One of the most active components of the stinging hairs of this plant and responsible for the long duration of the effects of the sting. An intense pain is caused in humans and farm animals, which normally lasts for 12 h but may recur for up to 6 months. In Australia horses have been killed from contact with this plant.

Kahn, S. D., *J. Org. Chem.*, 1989, **54**, 1901.

Morphine 983

Morphia

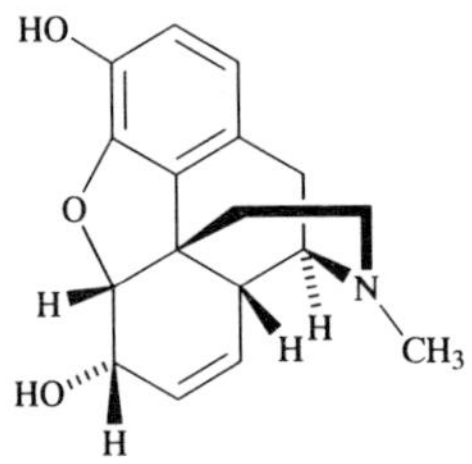

Benzylisoquinoline alkaloid
(Morphinan)

$C_{17}H_{19}NO_3$ [57-27-2] MW 285.34

Opium, the dried latex of the seed pod of *Papaver somniferum* (Papaveraceae).

Powerful analgesic, narcotic and sedative. Prolonged use leads to habituation. It is used extensively for pain relief, especially in terminal medical care. The lethal dose in humans lies between 1 and 10 mg and fatalities occur occasionally. The drug of addiction is the diacetate, heroin [561-27-3].

Muhtadi, F. J., *Anal. Profiles Drug Subst.*, 1988, **17**, 259.

Mukaadial 984

6,9-Dihydroxy-7-drimene-11,12-dial

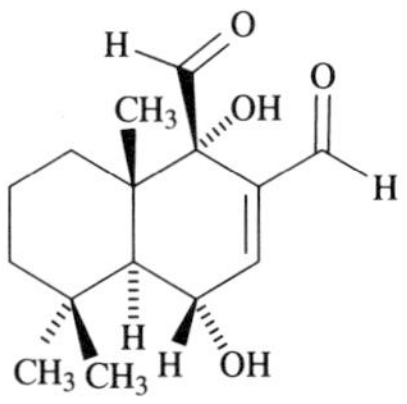

Sesquiterpenoid
(Drimane)

$C_{15}H_{22}O_4$ [87420-14-2] MW 266.34

Warburgia stuhlmannii, *W. ugandensis* and *Canella winterana* (Canellaceae).

Molluscicidal properties.

Kubo, J., *Chem. Lett.*, 1983, 979.

Multiflorine 985

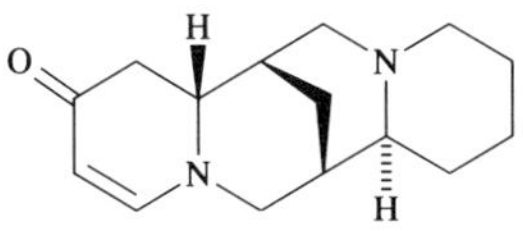

Quinolizidine alkaloid
(Sparteine)

$C_{15}H_{22}N_2O$ [529-80-6] MW 246.35

Cadia ellisiana, *Lupinus multiflorus* and white lupin, *L. albus* (Leguminosae).

Central nervous system depressant.

Comin, J., *Aust. J. Chem.*, 1959, **12**, 468.

Multigilin 986

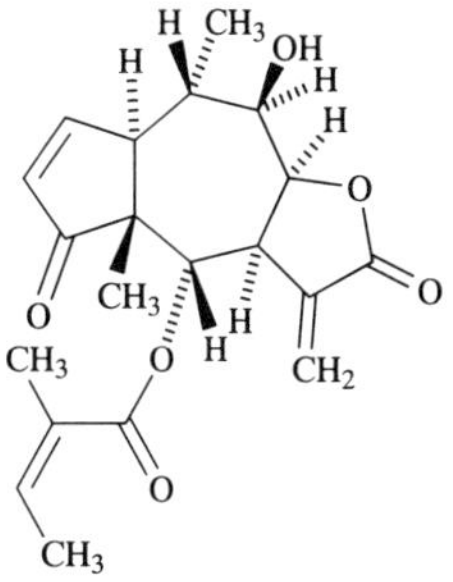

Sesquiterpenoid lactone
(Pseudoguaianolide)

$C_{20}H_{24}O_6$ [64937-25-3] MW 360.41

Baileya multiradiata (Compositae).

Cytotoxic and antitumour properties.

Pettit, G. R., *J. Org. Chem.*, 1978, **43**, 1092.

Multiradiatin 987

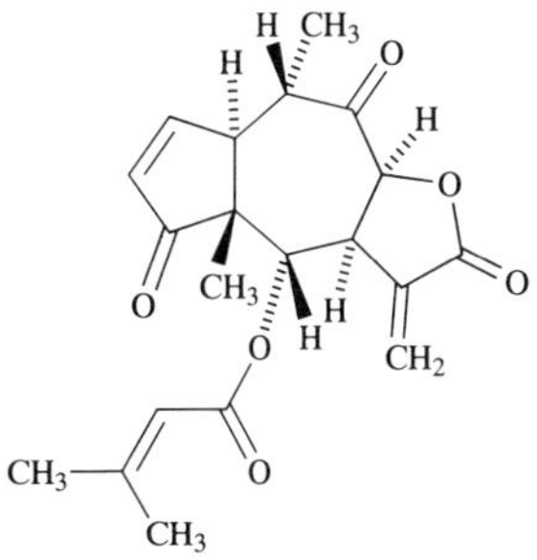

Sesquiterpenoid lactone
(Pseudoguaianolide)

$C_{20}H_{22}O_6$ [58262-52-5] MW 358.39

Baileya multiradiata (Compositae).

Cytotoxic and antitumour properties.

Pettit, G. R., *J. Org. Chem.*, 1978, **43**, 1092.

Multistatin 988

Sesquiterpenoid lactone
(Pseudoguaianolide)

$C_{20}H_{22}O_6$ [64937-26-4] MW 358.39

Baileya multiradiata (Compositae).

Cytotoxic and antitumour properties.

Pettit, G. R., *J. Org. Chem.,* 1978, **43**, 1092.

Musaroside 989

Sarmutogenin 3-digitaloside

Cardenolide

$C_{30}H_{44}O_{10}$ [465-97-4] MW 564.67

Strophanthus divaricatus (Apocynaceae).

Toxic to vertebrates.

Richter, R., *Helv. Chim. Acta,* 1954, **37**, 76.

Muzigadial 990

Canellal

Sesquiterpenoid
(*abeo*-Drimane)

$C_{15}H_{20}O_3$ [66550-09-2] MW 248.32

Bark of the 'muziga' tree, *Warburgia salutaris* and *Canella winterana* (Canellaceae).

Strong antifeedant against the armyworms *Spodoptera littoralis* and *S. exempta*.

Kubo, I., *Tetrahedron Lett.,* 1977, 4553.

Mycosinol 991

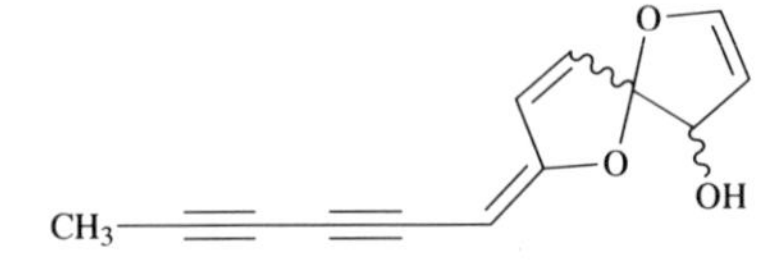

Polyacetylene

$C_{13}H_{10}O_3$ [111768-19-5] MW 214.22

Coleostephus myconis (Compositae) infected with *Botrytis cinerea*. Also present constitutively in roots of *Santolina oblongifolia* (Compositae).

Antifungal agent.

Marshall, P. S., *Phytochemistry,* 1987, **26**, 2493.

Myomontanone 992

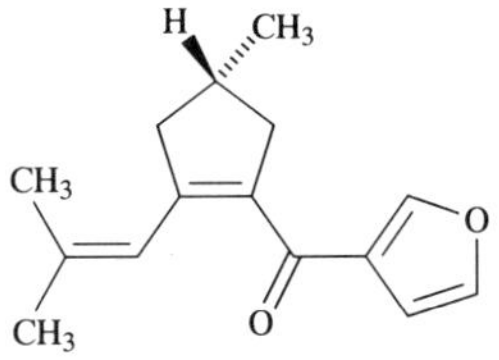

Sesquiterpenoid

$C_{15}H_{10}O_2$ [86989-09-5] MW 232.32

Leaves of *Myoporum montanum* (Myoporaceae).

Toxic to livestock feeding on *Myoporum* causing lung damage. In mice, it causes a typical interstitial pneumonia; the fatal dose is an intraperitoneal injection of 10 mg/kg body-weight.

Métra, P. L., *Tetrahedron Lett.,* 1983, **24**, 1749.

Myricoside 993

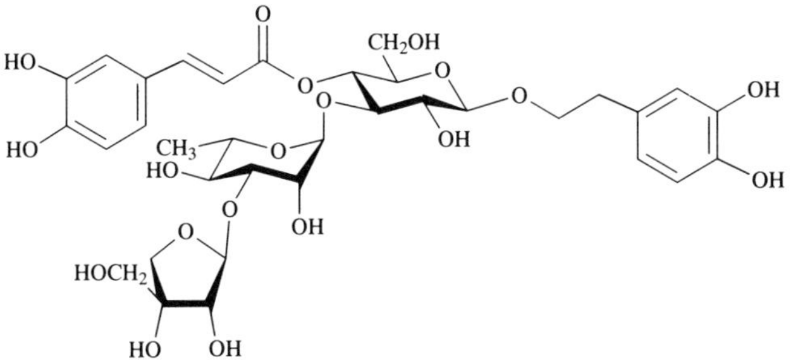

Phenylpropanoid

$C_{33}H_{43}O_{19}$ [76076-04-5] MW 743.69

Roots of *Clerodendron myricoides* (Verbenaceae).

Antifeedant properties against the African armyworm.

Cooper, R., *J. Am. Chem. Soc.,* 1980, **102**, 7953.

Myristicin 994

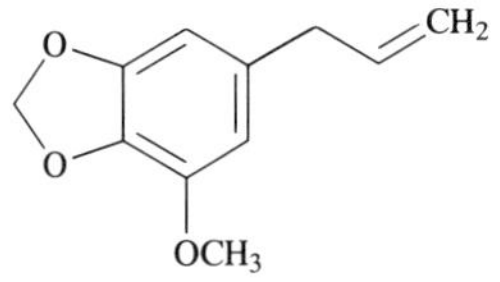

Phenylpropanoid

$C_{11}H_{12}O_3$ [607-91-0] MW 192.21

Seed oil of nutmeg, *Myristica fragrans* (Myristicaceae), essential oil of the wood of *Cinnamomum glanduliferum* (Lauraceae), leaf of *Apium graveolens*, fruit of *Petroselinum crispum* (Umbelliferae) and essential oil of *Orthodon* spp. (Labiatae).

Reputed to be psychotomimetic and probably toxic if taken in quantity. It shows synergistic activity with insecticides such as xanthotoxin (q.v.). It inhibits monoaminoxidase and rabbit platelet aggregation *in vitro*.

Shulgin, A. T., *Nature (London),* 1966, **210**, 380.

N

Nagilactone C 995

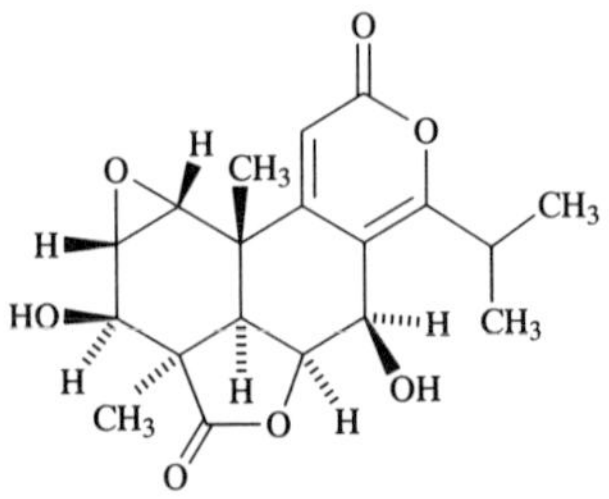

Diterpenoid
(Oxatotarane)

$C_{19}H_{22}O_7$ [24338-53-2] MW 362.38

Podocarpus nagi and other *Podocarpus* spp. (Podocarpaceae).

Antitumour properties. Larvicide.

Hayashi, Y., *Tetrahedron Lett.*, 1968, 2071.

Nandinin 996

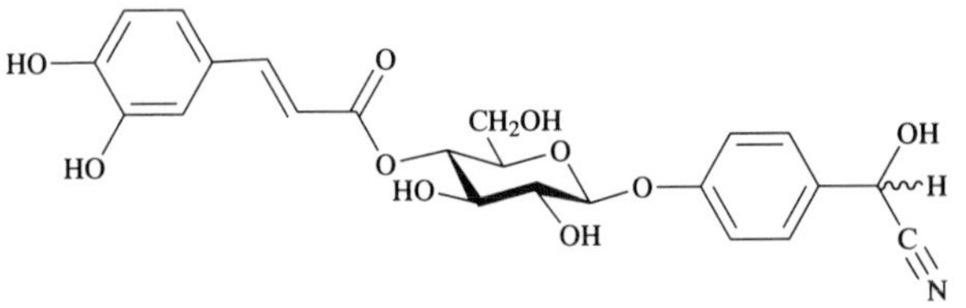

Cyanogenic glycoside

$C_{23}H_{23}NO_{10}$ [91919-94-7] MW 473.44

Heavenly bamboo, *Nandina domestica* (Berberidaceae), principally in the young leaves.

Toxin, giving poisonous cyanide on hydrolytic breakdown.

Olechno, J. D., *Phytochemistry*, 1984, **23**, 1784.

Napelline 997

Luciculine

Diterpenoid alkaloid
(Napellane)

$C_{22}H_{33}NO_3$ [5008-52-6] MW 359.51

Aconitum napellus and several other *Aconitum* spp. (Ranunculaceae).

Produces cardiovascular effects in cats, including brief lowering of the blood pressure and disturbed respiration.

Okamoto, T., *Chem. Pharm. Bull.,* 1965, **13**, 1270.

Naphtho[1,2-*b*]furan-4,5-dione 998

o-Naphthoquinone

$C_{12}H_6O_3$ [32358-83-1] MW 198.18

Wound tissue of the grey mangrove, *Avicennia marina* (Verbenaceae) infected by a *Phytophthora* species.

Antifungal agent (phytoalexin).

Sutton, D. C., *Phytochemistry,* 1985, **24**, 2877.

Narceine 999

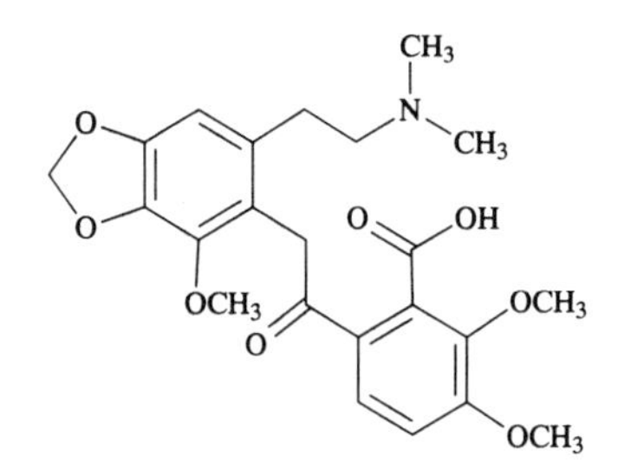

Seco-benzylsoquinoline alkaloid

$C_{23}H_{27}NO_8$ [131-28-2] MW 445.47

Opium, dried latex of seed pod, *Papaver somniferum* (Papaveraceae).

Antitussive, but without analgesic activity (c.f. codeine). It stimulates respiration, lowers blood pressure and stimulates intestinal peristalsis in animals.

Manske, R. H. F., The Alkaloids, 1954, **4**, 167, Academic Press.

Narciclasine 1000

Lycoricidinol

Amaryllidaceae alkaloid
(Phenanthridine)

$C_{14}H_{13}NO_7$ [29477-83-6] MW 307.26

Haemanthus kalbreyeri, *Lycoris longituba* and many *Narcissus* spp. (Amaryllidaceae).

Potent antifeedant to Lepidoptera. Antitumour agent, exerting an antimitotic effect by terminating protein synthesis in eukaryotic cells. The LD_{50} subcutaneously in mice is 5 mg/kg body-weight.

Fuganti, C., *J. Chem. Soc., Chem. Commun.,* 1972, 239.

α-Narcotine 1001

Noscapine

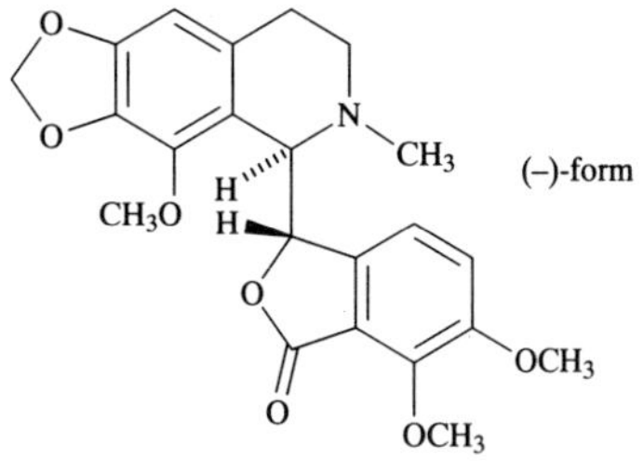

Phthalideisoquinoline alkaloid

$C_{22}H_{23}NO_7$ [128-62-1] MW 413.43

Opium from *Papaver somniferum* (Papaveraceae). It readily epimerises to the artefactual (−)-β-narcotine [3860-46-6] during its isolation.

Antitussive and spasmolytic properties.

Al-Yahya, M. A., *Anal. Profiles Drug Subst.*, 1982, **11**, 407.

Narcotoline 1002

Desmethylnarcotine

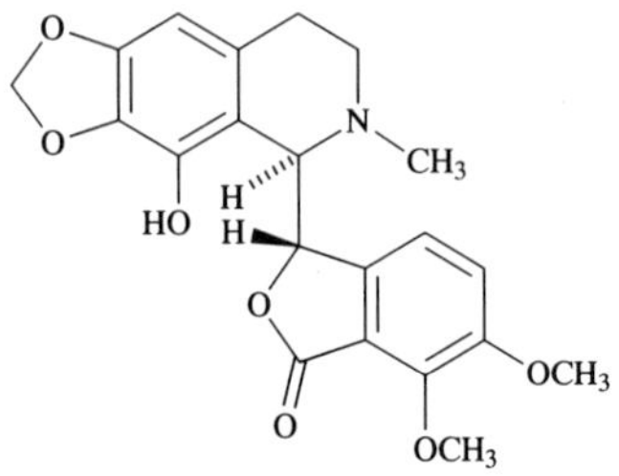

Phthalideisoquinoline alkaloid

$C_{21}H_{21}NO_7$ [521-40-4] MW 399.40

Opium from *Papaver somniferum* (Papaveraceae).

Respiratory stimulant and spasmolytic properties.

Battersby, A. R., *J. Chem. Soc.*, 1965, 1087.

Naringin 1003

Naringenin 7-neohesperidoside

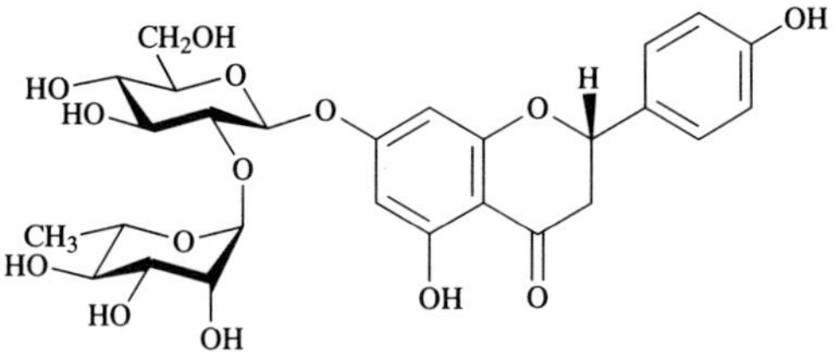

Flavanone glycoside

$C_{27}H_{32}O_{14}$ [10236-47-2] MW 580.54

Grapefruit, *Citrus paradisi* and other citrus fruits (Rutaceae), *Ceterach officinarum* and *Adiantum* spp. (Adiantaceae) and oregano, *Oreganum vulgare* (Labiatae).

Bitter-tasting, with one-fifth the bitterness of quinine on a molar basis. Shows antiperoxidative activity.

Hattori, S., *J. Am. Chem. Soc.*, 1952, **74**, 3614.

Narwedine 1004

Galanthaminone

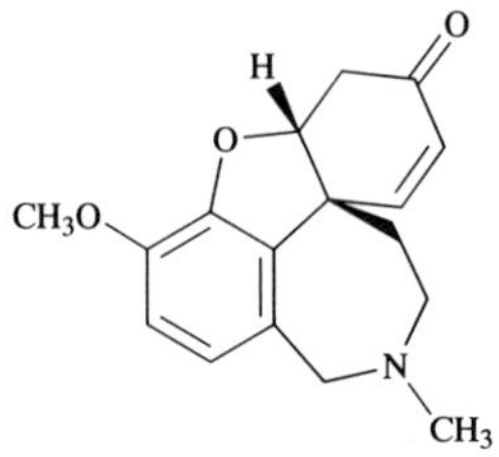

Amaryllidaceae alkaloid
(Galanthaman)

$C_{17}H_{19}NO_3$ [510-77-0] MW 285.34

Bulbs of the snowdrop, *Galanthus nivalis* and of *Lycoris guangxiensis*, *Ungernia severtzovii*, *U. victoris* and *U. vvedenskyi* (Amaryllidaceae).

Increases the amplitude and frequency of respiratory movements and increases the amplitude and decreases the frequency of cardiac contractions. It also potentiates the analgesic effects of morphine.

Boit, H. G., *Chem. Ber.*, 1957, **90**, 2197.

Neoisostegane 1005

Lignan

$C_{23}H_{26}O_7$ [87084-98-8] MW 414.46

Steganotaenia araliacea (Umbelliferae).

Cytotoxic and antitumour properties.

Taafrout, M., *Tetrahedron Lett.*, 1983, **24**, 2983.

Neolinustatin 1006

Cyanogenic glycoside

$C_{17}H_{29}NO_{11}$ [72229-42-6] MW 433.42

Seed of flax, *Linum usitatissimum* (Linaceae) where it co-occurs with linustatin (q.v.). Also present in some *Passiflora* spp. (Passifloraceae).

Bound toxin, releasing poisonous cyanide on hydrolysis.

Smith, C. R., *J. Org. Chem.*, 1980, **45**, 507.

Neopine 1007

β-Codeine

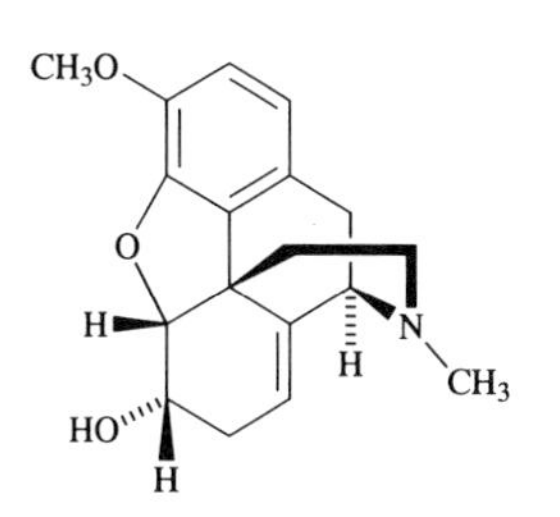

Benzylisoquinoline alkaloid
(Morphinan)

$C_{18}H_{21}NO_3$ [467-14-1] MW 299.37

Opium from *Papaver somniferim*, and also in *P. bracteatum* (Papaveraceae).

Analgesic and spasmolytic activities.

Manske, R. H. F., The Alkaloids, 1952, **2**, 1, Academic Press.

Neoquassin 1008

Nigakihemiacetal B; Simalikahemiacetal A

Nortriterpenoid
(Quassane)

$C_{22}H_{30}O_6$ [76-77-7] MW 390.48

Wood of *Quassia amara* and of several *Picrasma* spp. (Simaroubaceae).

Very bitter taste, with insecticidal activity.

Valenta, Z., *Tetrahedron,* 1962, **18**, 1433.

Neosaxitoxin 1009

Aphantoxin I; 5-Hydroxysaxitoxin

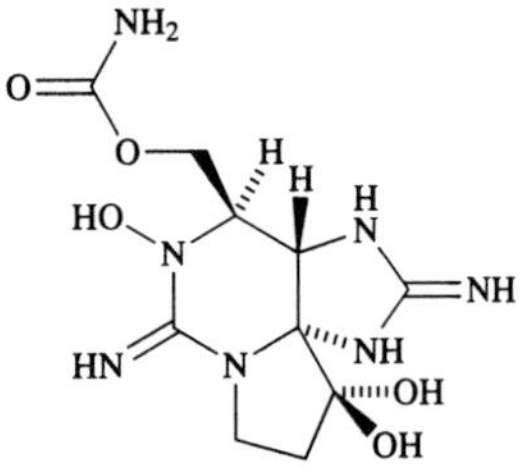

Cyclic bisguanidine alkaloid

$C_{10}H_{17}N_7O_5$ [64296-20-4] MW 315.29

Isolated from some strains of *Aphanizomenon flos-aquae* found in New Hampshire, USA.

A neurotoxin acting as a sodium channel blocking agent thus inhibiting the transmission of nerve impulses leading to death by respiratory failure.

Mahmood, N. A., *Toxicon,* 1986, **24**, 175.

Nepetalactone 1010

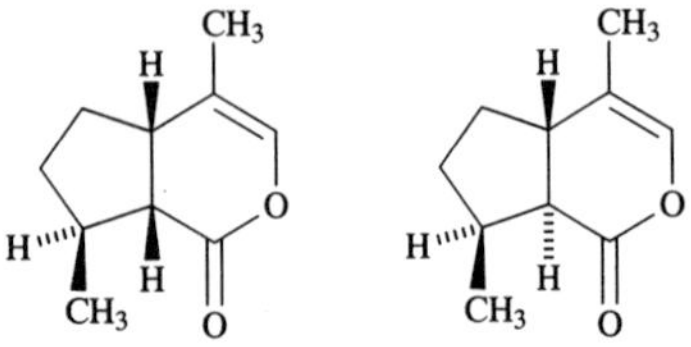

cis-trans-isomer *trans-cis*-isomer

Monoterpenoid
(Iridane)

$C_{10}H_{14}O_2$ MW 166.22

cis-trans-form [21651-62-7]
trans-cis-form [17257-15-7]

A mixture of *cis-trans* and *trans-cis* isomers in the volatile oil of catnip, *Nepeta cataria* (Labiatae).

Insect repellent. Has a potent ability to excite cats and other Felidae such as lions and jaguars.

De Pooter, H. L., *Phytochemistry,* 1987, **26**, 2311.

Nicotine 1011

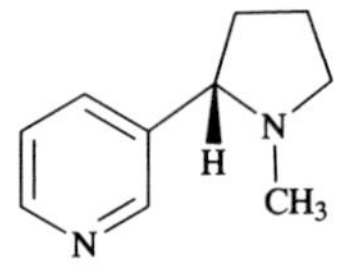

Pyridine alkaloid

$C_{10}H_{14}N_2$ [54-11-5] MW 162.23

In quantity in leaves of *Nicotiana tabacum* and most *Nicotiana* spp. (Solanaceae). Present as a trace component in *Asclepias syriaca* (Asclepiadaceae), *Lycopodium* spp. (Lycopodiaceae), *Equisetum arvense* (Equisetaceae) and *Sedum acre* (Crassulaceae).

Highly toxic, causing respiratory paralysis; the fatal dose in humans is about 50 mg. Very poisonous to most insects and widely used as an insecticide. Nicotine is the addictive component of tobacco, and has tranquillising properties.

Enzell, C. R., *Fortschr. Chem. Org. Naturst.,* 1977, **34**, 1.

Nilotin 1012

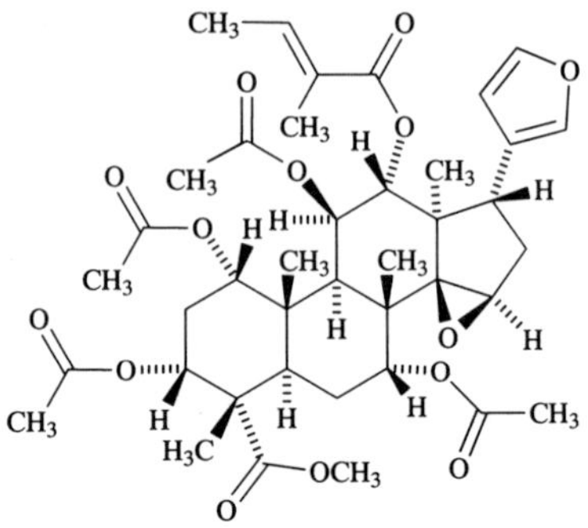

Nortriterpenoid
(Meliacane)

$C_{40}H_{52}O_{14}$ [165689-30-5] MW 756.84

Root bark of *Turraea nilotica* (Meliaceae).

Insect antifeedant, equally active with limonin (q.v.) against the colorado beetle, giving 50% reduction in feeding at 7 μg/ml.

Bentley, M. D., *J. Nat. Prod.,* 1995, **58**, 748.

Nitidine 1013

Angolinine

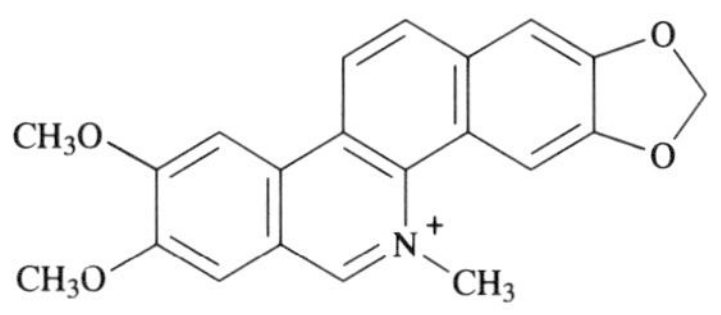

Benzylisoquinoline alkaloid
(Benzophenanthridine)

$[C_{21}H_{18}NO_4]^+$ [6872-57-7] MW 348.38

Bark of prickly ash, *Zanthoxylum americanum*, *Z. clava-herculis* and in various other *Zanthoxylum* and *Fagara* spp. (Rutaceae).

Antitumour properties, but too toxic for clinical use.

Arthur, H. R., *J. Chem. Soc.,* 1959, 1840.

Niveusin C 1014

Annuithrin

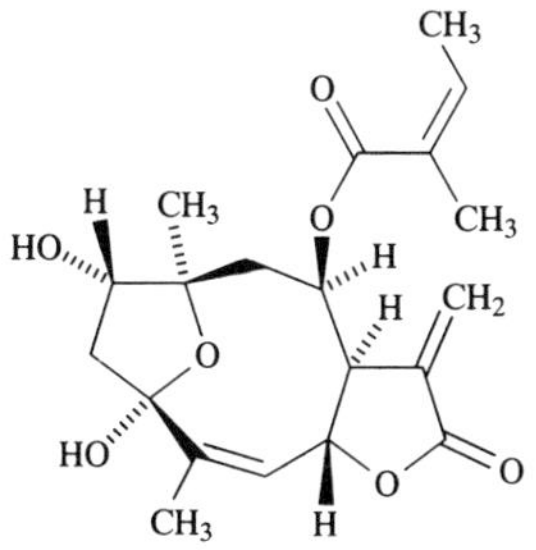

Sesquiterpenoid lactone
(Germacranolide)

$C_{20}H_{26}O_7$ [75680-27-2] MW 378.42

Helianthus niveus, sunflower, *H. annuus* and *H. maximiliani* (Compositae).

Cytotoxic and antitumour properties.

Herz, W., *Phytochemistry,* 1981, **20**, 93.

Nobilin 1015

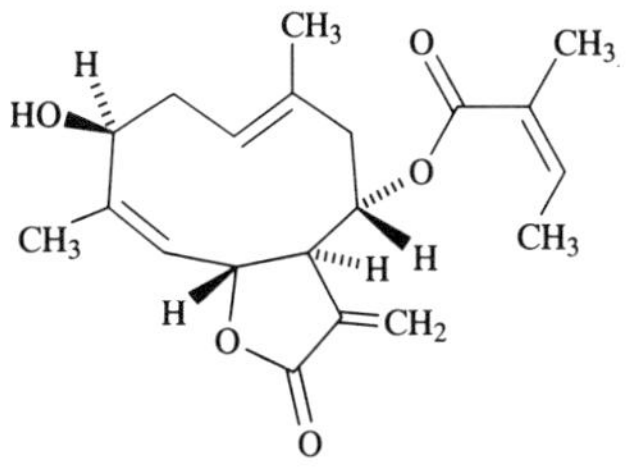

Sesquiterpenoid lactone
(Germacranolide)

$C_{20}H_{26}O_5$ [31824-11-0] MW 346.42

Chamomile, *Chamaemelum nobile* (Compositae).

Cytotoxic and antitumour properties.

Holub, M., *Collect. Czech. Chem. Commun.,* 1977, **42**, 1053.

Nodularin 1016

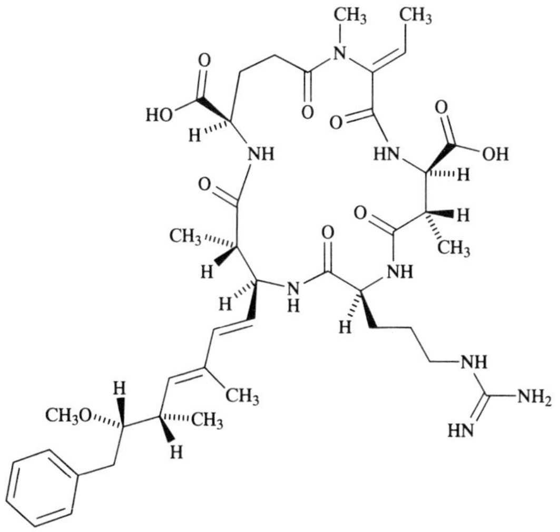

Cyclic pentapeptide

$C_{41}H_{60}N_8O_{10}$ [118399-22-7] MW 824.98

Produced by some strains of *Nodularia spumigena*, a filamentous cyanobacterium forming blooms in brackish water environments and responsible for illness and death in animals.

A hepatotoxin with a similar biological activity to the cyclic heptapeptides (*see* Microcystin).

Falconer, I. R., Algal Toxins in Seafoods and Drinking Water, 1993, 187, Academic Press.

Nomilin 1017

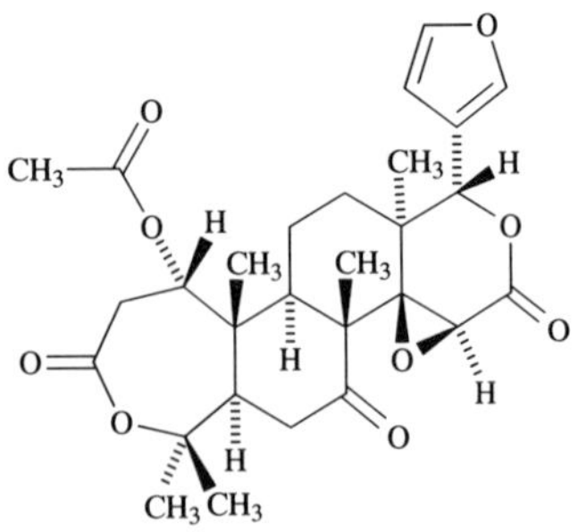

Nortriterpenoid
(Limonane)

$C_{28}H_{34}O_9$ [1063-77-0] MW 514.57

Fruits of cultivated *Citrus* spp. (Rutaceae).

About twice as bitter as limonin (q.v.), it contributes to the delayed bitterness of *Citrus* fruit juices. It shows insect antifeedant activity.

Barton, D. H. R., *J. Chem. Soc.*, 1961, 255.

Nordihydroguaiaretic acid 1018

NDGA; 3,3′,4,4′-Tetrahydroxylignan

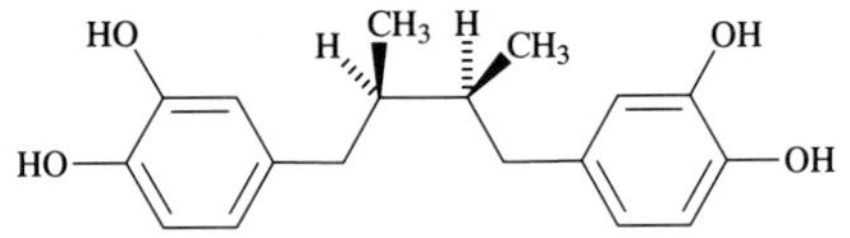

Lignan

$C_{18}H_{22}O_4$ [500-38-9] MW 302.37

Resinous exudate of creosote bush, *Larrea tridentata* and of other *Larrea* spp. and in wood resin of *Guaiacum sanctum* and *G. officinale* (all Zygophyllaceae). The concentrations in the first source range from 4 to 12% dry-weight.

Toxic to animals. Additions of 2% to laboratory rat diet produce kidney and caecal lesions, while addition of 3% causes mortality. Was once used as an antioxidant in fats and oils, but is now banned because it is suspected of inducing kidney cysts in humans.

McKecknie, J. S., *J. Chem. Soc, B,* 1969, 699.

Norerythrostachaldine 1019

19-Oxonorcassaidine

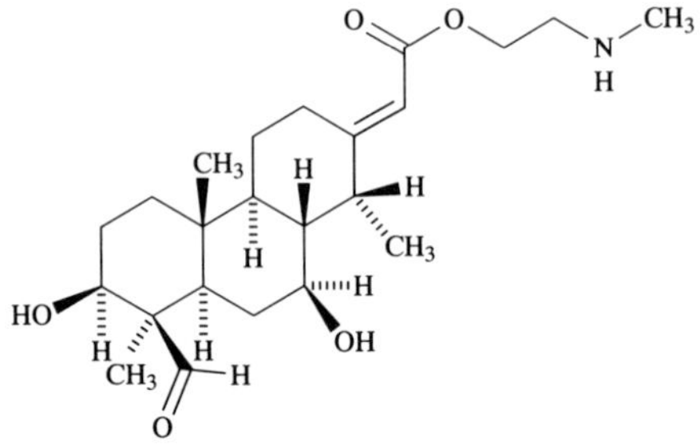

Diterpenoid alkaloid

$C_{23}H_{37}NO_5$ [55729-25-4] MW 407.55

Bark of *Erythrophleum chlorostachys* (Leguminosae).

Highly cytotoxic against human nasopharyngeal cancer cells.

Loder, J. W., *Aust. J. Chem.*, 1975, **28**, 651.

Norharman 1020

2-Carboline; β-Carboline

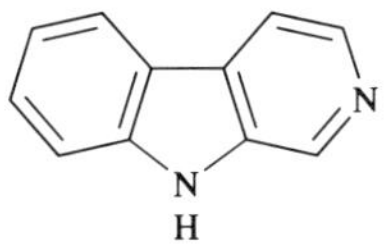

Indole alkaloid
(β-Carboline)

$C_{11}H_8N_2$ [244-63-3] MW 168.20

Tribulus terrestris (Zygophyllaceae), *Chrysophyllum lacourtianum* (Sapotaceae), *Catharanthus roseus* (Apocynaceae) and *Lolium perenne* (Gramineae).

Responsible, with harman (q.v.), for causing locomotor effects in sheep grazing on *Tribulus terrestris*. Also a tumour promoter.

Yomosa, K., *Agric. Biol. Chem.*, 1987, **51**, 921.

Norobtusifolin 1021

2-Hydroxychrysophanol

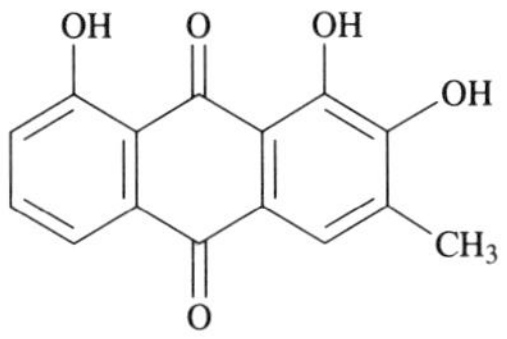

Anthraquinone

$C_{15}H_{10}O_5$ [58322-78-4] MW 270.24

Roots of *Myrsine africana* (Myrsinaceae).

Cytotoxic against three human tumour cell lines.

Takido, M., *Chem. Pharm. Bull.*, 1958, **6**, 397.

31-Noroscillatoxin B 1022

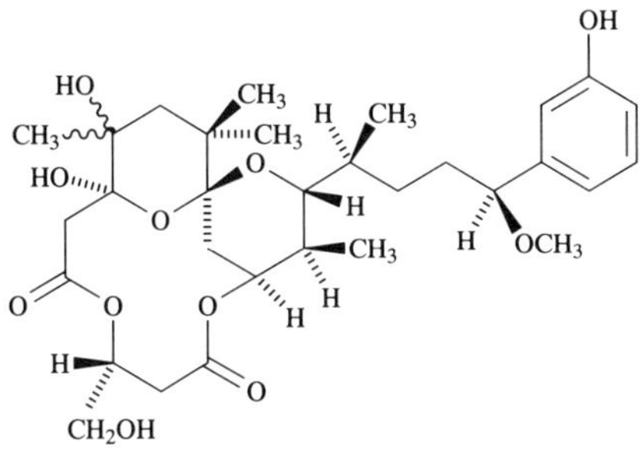

Phenolic bislactone

$C_{31}H_{44}O_{10}$ [95098-07-0] MW 576.68

Produced by a mixed culture of *Schizothrix calcicola* and *Oscillatoria nigroviridis*.

Associated with severe contact dermatitis known as swimmers' itch.

Entzeroth, M., *J. Org. Chem.*, 1985, **50**, 1255.

Nortrachelogenin 1023

Pinopalustrin; Wikstromol

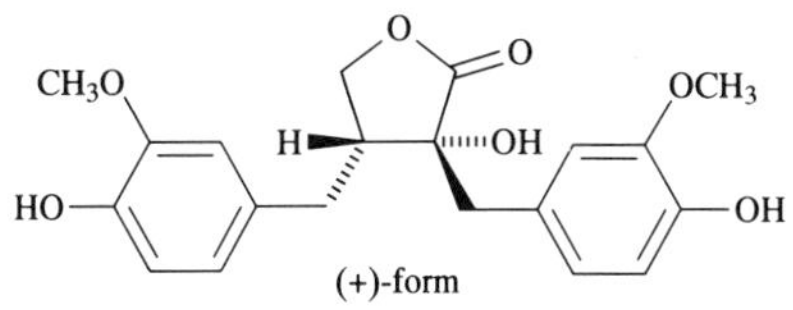

(+)-form

Lignan

$C_{20}H_{22}O_7$ [34444-37-6] MW 374.39
(+)-form [61521-74-2]
(−)-form [34444-37-6]

The (−)-form, known as pinopalustrin, is present in *Trachelospermum asiaticum* var. *intermedium* (Apocynaceae) and in *Pinus palustrus* (Pinaceae). The (+)- form, known as wikstromol, occurs in *Passerina vulgaris* and *Wikstroemia indica* (Thymeleaceae).

The (−)-form has cytotoxic and antileukaemic properties.

Nishibe, S., *Phytochemistry,* 1971, **10**, 2231.

Nudicauline 1024

Andersoline

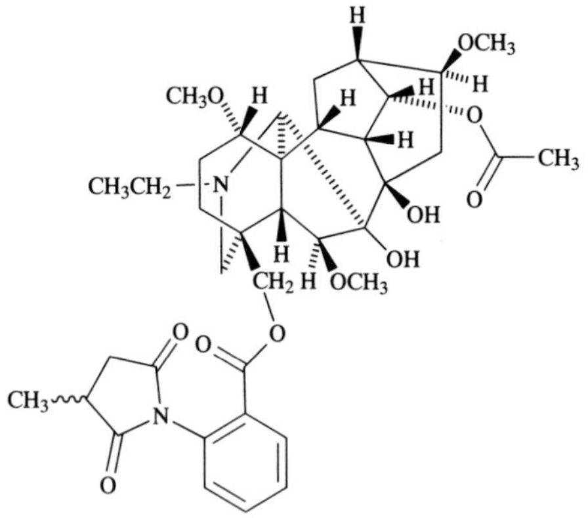

Diterpenoid alkaloid
(Aconitane)

$C_{38}H_{50}N_2O_{11}$ [99815-83-5] MW 710.82

Delphinium andersonii and *D. nudicaule* (Ranunculaceae).

One of several poisonous larkspur alkaloids causing livestock deaths in Canada and the U.S.A. The intravenous LD_{50} in mice is 2.7 mg/kg body-weight.

Pelletier, S. W., *Heterocycles,* 1988, **27**, 2387.

Ochrolifuanine A 1025

Indole alkaloid

$C_{29}H_{34}N_4$ [35527-46-9] MW 438.62

Ochrosia lifuana, *O. miana*, *O. confusa* and *Dyera costulata* (Apocynaceae). One of the four stereoisomeric forms of this alkaloid in these plants.

The root barks of these plants have been used as an ingredient of arrow poisons.

Peube-Locou, N., *C. R. Acad. Sci., Ser. C,* 1971, **273**, 905.

Odoratol 1026

Dihydrochalcone

$C_{17}H_{18}O_5$ [94943-12-1] MW 302.33

Cotyledons of sweet pea, *Lathyrus odoratus* infected with *Phytophthora megasperma*; constitutively present in heartwood of *Pterocarpus angolensis* (Leguminosae).

Antifungal activity.

Fuchs, A., *Phytochemistry,* 1984, **23**, 2199.

Oenanthotoxin 1027

Enanthotoxin

Polyacetylene

$C_{17}H_{22}O_2$ [20311-78-8] MW 258.36

Chiefly in the roots of water hemlock, *Oenanthe crocata* (Umbelliferae).

Toxic principle of the water hemlock, one of the most dangerous plants in the U.K. flora. Known as the 'five-finger death', due to the shape of the root, it has been responsible for fatalities in man and in cattle. Oenanthotoxin reversibly inhibits the sodium flux and action potentials in animal cell membranes.

Bohlmann, F., *Chem. Ber.,* 1955, **88**, 1245.

Oestrone 1028

Estrone; Estrol; Folliculin

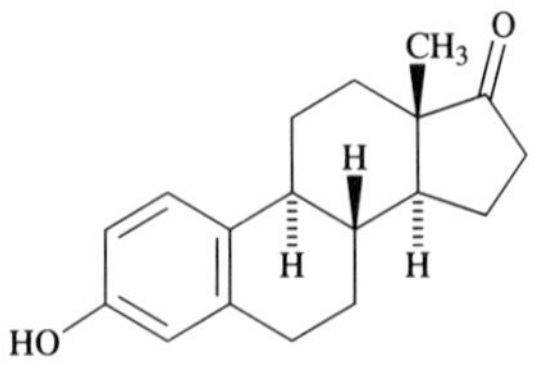

Steroid
(Estrane)

$C_{18}H_{22}O_2$ [53-16-7] MW 270.37

Pollen and seed of the date palm, *Phoenix dactylifera* (Palmae) and seed of pomegranate, *Punica granatum* (Punicaceae).

Oestrogenic activity.

Both, D., *Anal. Profiles Drug Subst.,* 1983, **12**, 135.

Ohchinolide B 1029

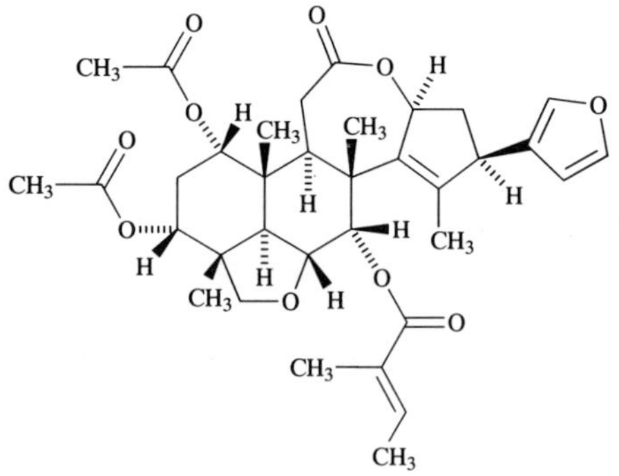

Nortriterpenoid
(Secomeliacane)

$C_{35}H_{44}O_{10}$ [71902-49-3] MW 624.73

Melia azedarach, the neem tree (Meliaceae).

Antifeedant against armyworms.

Kraus, W., *Chem. Ber.,* 1981, **114**, 267.

Ohioensin-A 1030

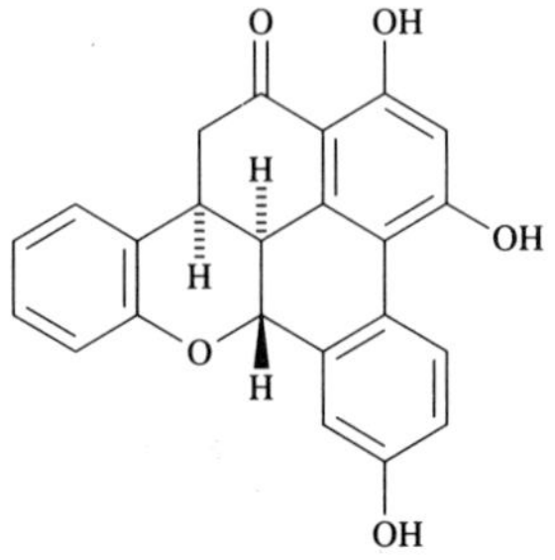

Phenanthrene

$C_{23}H_{16}O_5$ [121353-47-7] MW 372.38

The moss *Polytrichum ohioense* (Polytrichaceae).

Cytotoxic and antitumour properties.

Zheng, G. Q., *J. Am. Chem. Soc.,* 1989, **111**, 5500.

Okadaic acid 1031

PL Toxin II

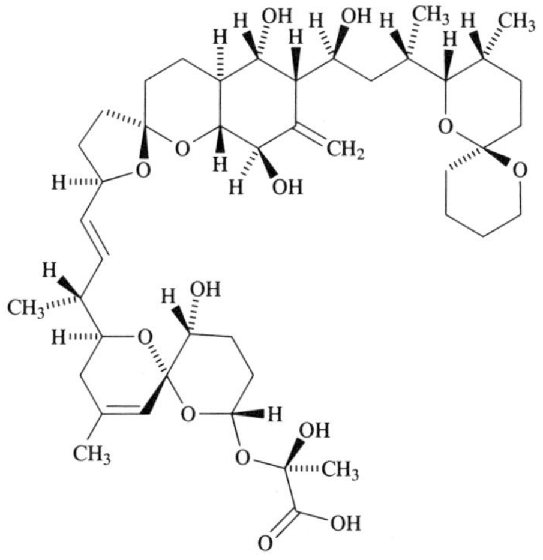

Polyether

$C_{44}H_{68}O_{13}$ [78111-17-8] MW 805.01

Produced by the alga *Prorocentrum lima* and dietarily sequestered by fish.

Mammalian toxin. The LD_{50} intraperitoneally in mice is 192 mg/kg body-weight. Okadaic acid is cytotoxic to human epidermoid carcinoma cells.

Tachibana, K., *J. Am. Chem. Soc.*, 1981, **103**, 2469.

Oleanoglycotoxin-A 1032

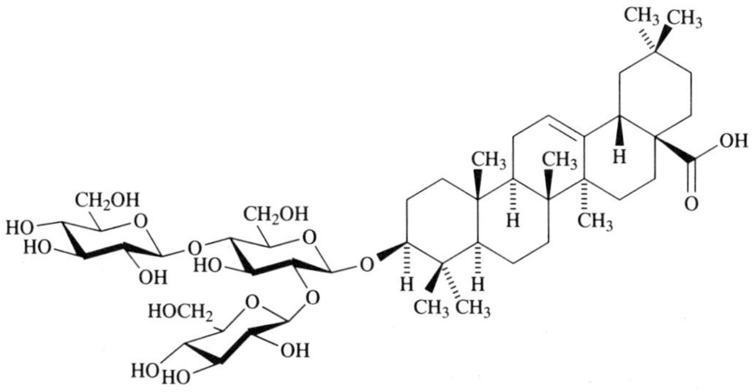

Triterpenoid saponin
(Oleanane)

$C_{48}H_{78}O_{18}$ [50657-29-9] MW 943.14

Berry of *Phytolacca dodecandra* (Phytolaccaceae).

Molluscicidal, with a LD_{100} of 6 mg/litre against the snail *Biomphalaria glabrata*. Active against human spermatozoa at a concentration of 50 mg/litre.

Domon, B., *Helv. Chim. Acta*, 1984, **67**, 1310.

Olivacine 1033

Guatambuinine

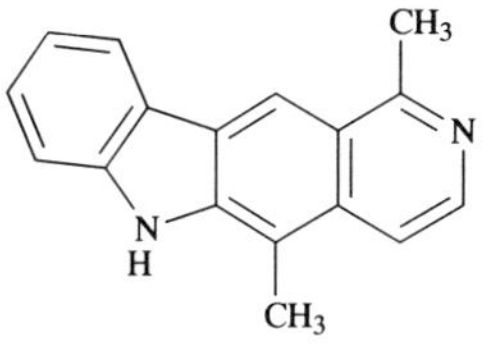

Indole alkaloid
(Pyridocarbazole)

$C_{17}H_{14}N_2$ [484-49-1] MW 246.31

Aspidosperma nigricans (Apocynaceae).

Strong cytotoxicity against human carcinomas. Antitrypanosomal *in vitro* against *Trypanosoma cruzi*.

Marini-Bettolo, G. B., *Helv. Chim. Acta*, 1959, **42**, 2146.

Onopordopicrin 1034

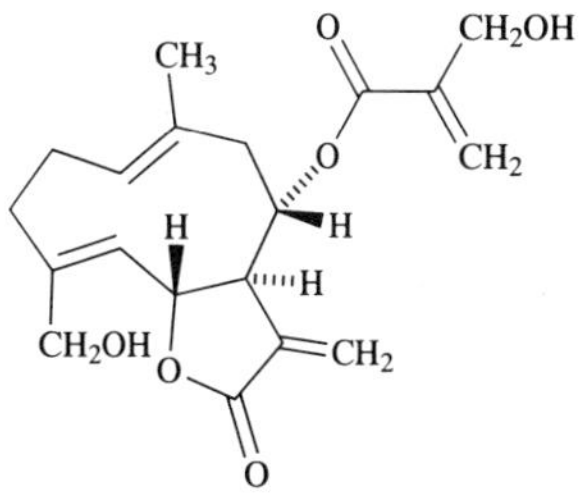

Sesquiterpenoid lactone
(Germacranolide)

$C_{19}H_{24}O_6$ [19889-00-0] MW 348.40

The Scotch thistle, *Onopordum acanthium*, the lesser burdock, *Arctium minus* and *Berkheya speciosa* (Compositae).

Insect antifeedant. Cytotoxic and antitumour properties.

Droździ, B., *Collect. Czech. Chem. Commun.*, 1968, **33**, 1730.

Orbicuside A 1035

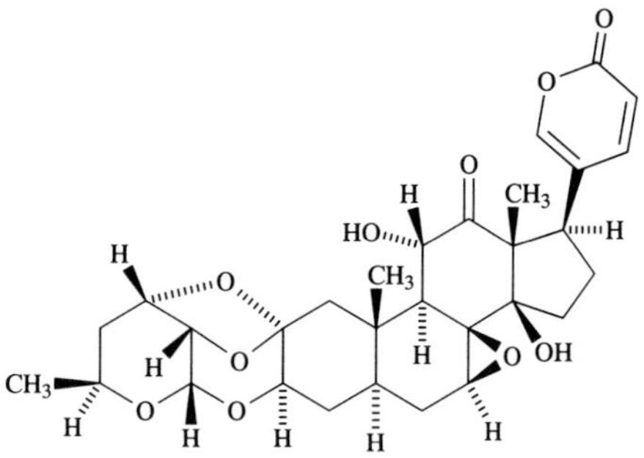

Bufadienolide

$C_{30}H_{36}O_{10}$ [105801-16-19] MW 556.61

One of three closely related bufadienolides present in aerial parts of *Cotyledon arbiculata* (Crassulaceae).

Toxic to livestock.

Steyn, P. S., *J. Chem. Soc., Perkin Trans. 1,* 1986, 1633.

Orchinol 1036

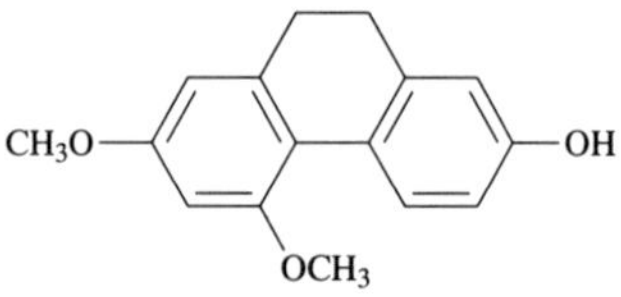

Phenanthrene

$C_{16}H_{16}O_3$ [41060-20-2] MW 256.30

Military orchid bulb, *Orchis militaris* (Orchidaceae) infected with *Rhizoctonia repens.*

Antifungal agent (phytoalexin).

Hardegger, E., *Helv. Chim. Acta,* 1963, **46**, 1354.

Orientin 1037

Luteolin 8-*C*-glucoside

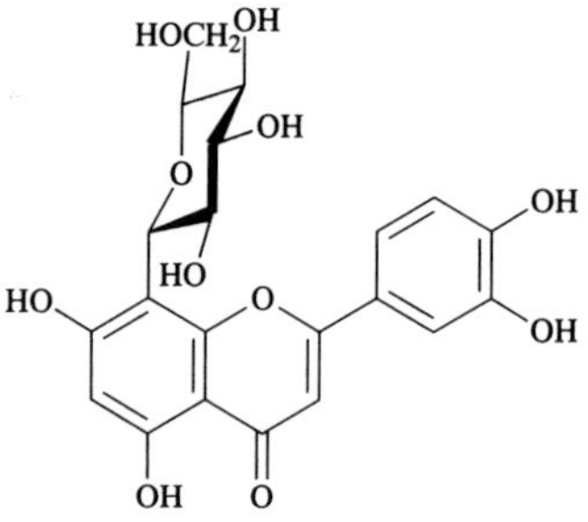

Flavone *C*-glycoside

$C_{21}H_{20}O_{11}$ [28608-75-5] MW 448.38

Widespread in nature, including millet, *Pennisetum americanum* (Gramineae) and *Polygonum orientale* (Polygonaceae).

Is an inhibitor of thyroid peroxidase. It is responsible, together with other phenolics present in the grain, for the goitrogenic and antithyroid activity of millet.

Karrer, W., Konstitution und Vorkommen der organischen Pflanzenstoffe, 2nd ed., part 1, 1981, 862, Birkhäuser.

Orizabin 1038

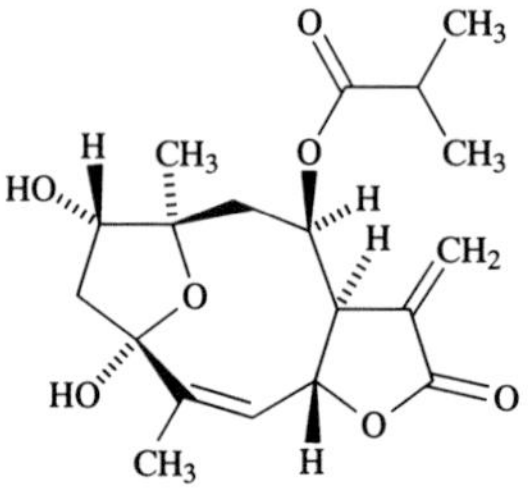

Sesquiterpenoid lactone (Germacranolide)

$C_{19}H_{26}O_7$ [34367-14-1] MW 366.41

Tithonia tubaeformis and *T. tagitiflora* (Compositae).

Cytotoxic and antitumour properties.

Ortega, A., *Rev. Latinoam. Quim.,* 1971, **2**, 38.

Oryzalexin A 1039

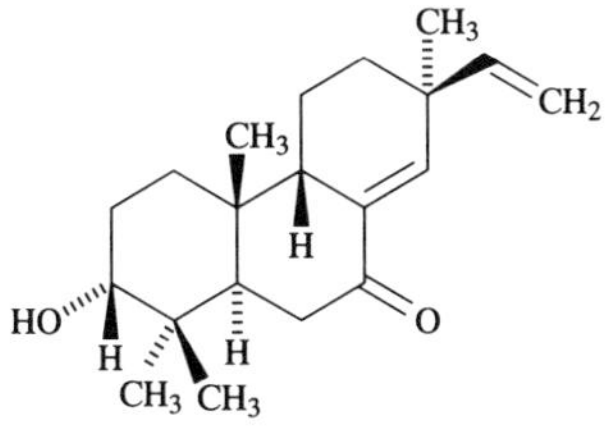

Diterpenoid
(Pimarane)

$C_{20}H_{30}O_2$ [85394-31-6] MW 302.46

In leaves of rice, *Oryza sativa* (Gramineae) infected with blast disease *Pyricularia oryzae*, together with several other related oryzalexins.

Antifungal agent (phytoalexin).

Kono, Y., *Agric. Biol. Chem.*, 1984, **48**, 253.

Oscillatoxin A 1040

31-Nordebromoaplysiatoxin

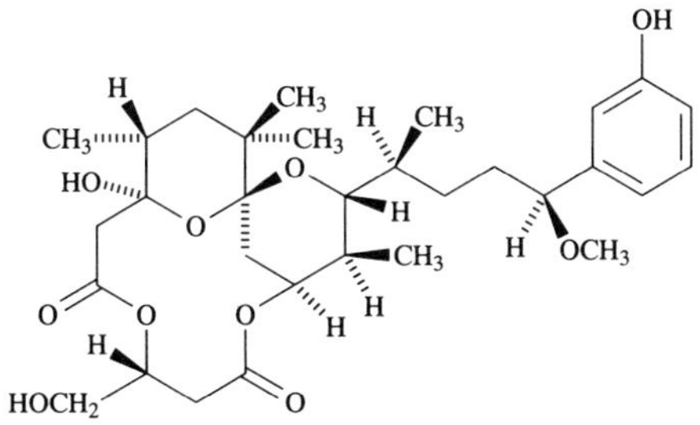

Phenolic bislactone

$C_{31}H_{46}O_{10}$ [66671-95-2] MW 578.70

Produced by mixed cultures of *Schizothrix calcicola* and *Oscillatoria nigroviridis*.

Associated with severe contact dermatitis known as swimmers' itch. Minimum lethal dose in mice 0.2 mg/kg bodyweight.

Moore, R. E., *Pure Appl. Chem.*, 1982, **54**, 1919.

Oscillatoxin B1 1041

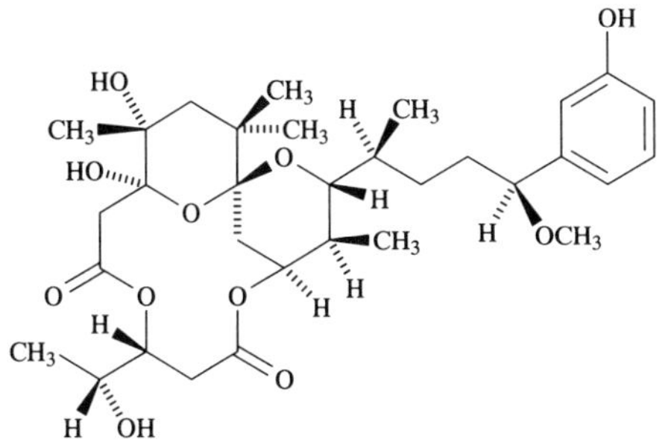

Phenolic bislactone

$C_{32}H_{46}O_{10}$ [95189-16-5] MW 590.71

Produced by mixed cultures of *Schizothrix calcicola* and *Oscillatoria nigroviridis*.

Associated with severe contact dermatitis known as swimmers' itch.

Moore, R. E., *Pure Appl. Chem.*, 1982, **54**, 1919.

Oscillatoxin B2 1042

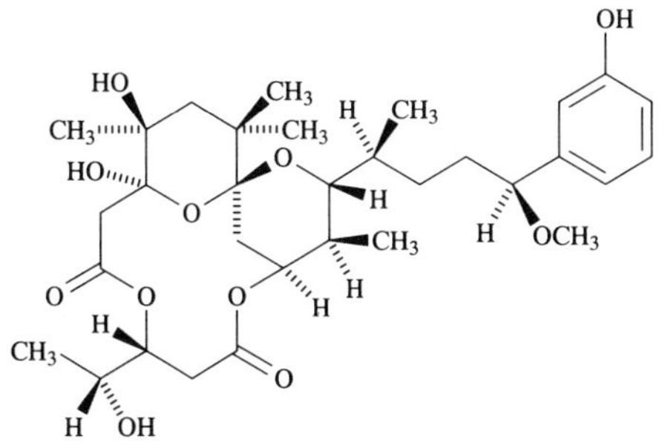

Phenolic bislactone

$C_{32}H_{46}O_{10}$ [95189-17-6] MW 590.71

Produced by mixed culture of *Schizothrix calcicola* and *Oscillatoria nigroviridis*.

Associated with severe contact dermatitis known as swimmers' itch.

Entzeroth, M., *J. Org. Chem.*, 1985, **50**, 1255.

Oscillatoxin D 1043

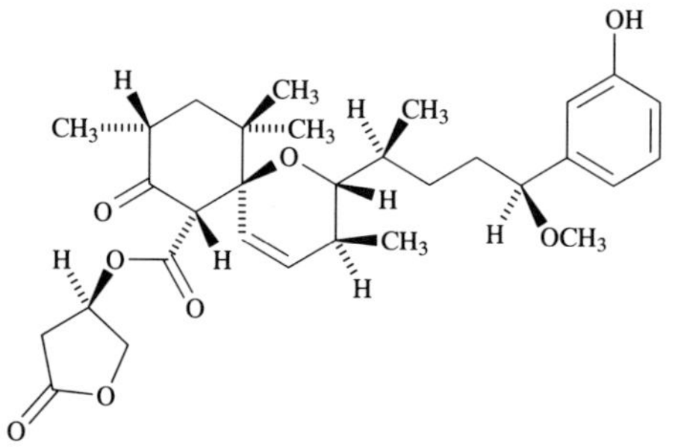

Phenolic lactone

$C_{31}H_{42}O_8$ [95069-53-7] MW 542.67

Produced by a mixed culture of *Schizothrix calcicola* and *Oscillatoria nigroviridis*.

Associated with severe contact dermatitis known as swimmers' itch.

Entzeroth, M., *J. Org. Chem.*, 1985, **50**, 1255.

Ostruthin 1044

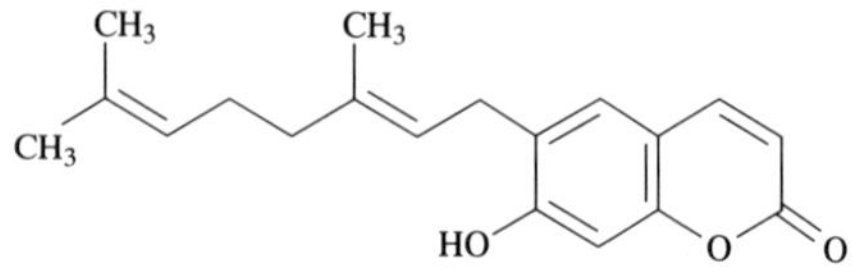

Coumarin

$C_{19}H_{22}O_3$ [148-83-4] MW 298.38

Rhizomes of *Peucedanum ostruthium* (Umbelliferae), leaves and twigs of *Eriostemon tomentellus* and in *Luvunga eleutherandra* (Rutaceae).

Piscicidal agent. Also, has antimalarial properties.

Nikovov, G. K., *Chem. Nat. Compd.*, 1968, **4**, 268.

Otonecine 1045

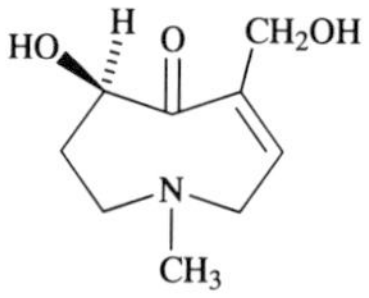

Pyrrolizidine alkaloid

$C_9H_{15}NO_3$ [6887-34-9] MW 185.22

Present, together with related esters such as senkirkine (q.v.), in many *Senecio* spp., including *S. jacobaea* and *S. kirkii* (Compositae).

Hepatotoxin.

Culvenor, C. C. J., *Aust. J. Chem.*, 1967, **20**, 801.

Ouabain 1046

Oubagenin 3-rhamnoside; *g*-Strophanthin; Acocantherin; Gratibain; Gratus strophanthin; Astrobain; Purostrophan

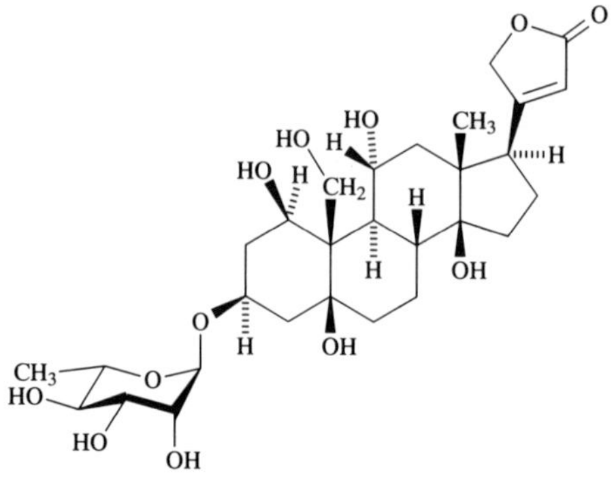

Cardenolide

$C_{29}H_{44}O_{12}$ [630-60-4] MW 584.66

Leaves of *Acokanthera ouabaio* and seeds of *Strophanthus gratus* (Apocynaceae).

Very toxic to vertebrates. The LD_{50} intravenously in cats is 0.11 mg/kg body-weight and in rats 14 mg/kg body-weight. Widely used in Kenya as an arrow poison. The most effective poisons are obtained from *Acokanthera* plants containing ouabain as the main cardenolide. Is in clinical use as a cardiotonic.

Hauschild-Rogat, P., *Helv. Chim. Acta,* 1967, **50**, 2299.

Ovatifolin 1047

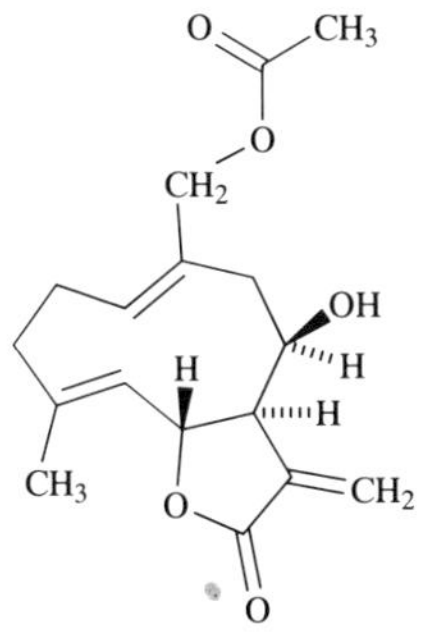

Sesquiterpene lactone
(Germacranolide)

$C_{17}H_{22}O_5$ [50886-56-1] MW 306.36

Leaves and stems of *Podanthus ovatifolius* and *P. mitiqui* (Compositae).

Cytotoxic and antitumour properties.

Gnecco, S., *Phytochemistry,* 1973, **12**, 2469.

Ovatine 1048

(15*β*)-Veatchine acetate

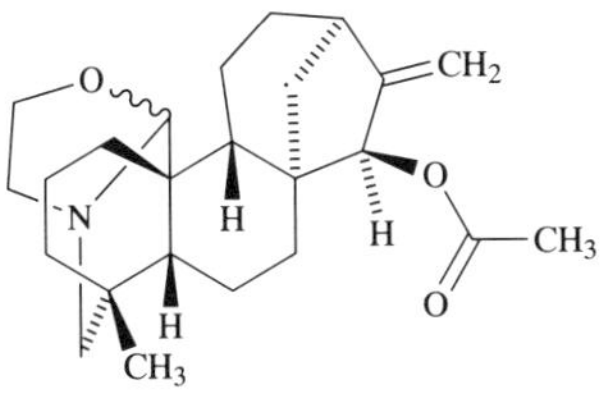

Diterpenoid alkaloid
(Veatchane)

$C_{24}H_{35}NO_3$ [68719-14-2] MW 385.55

Garrya ovata var. *lindheimeri* (Garryaceae).

Crude extracts of the bark and leaves of *Garrya ovata* show antitumour activity *in vitro* due to ovatine and a related alkaloid, lindheimerine [68831-67-4].

Pelletier, S. W., *J. Org. Chem.,* 1981, **46**, 1840.

Oxalic acid 1049

Ethanedioic acid

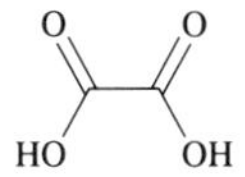

Organic acid

$C_2H_2O_4$ [144-62-7] MW 90.04

Wide occurrence as crystalline occlusions in cell vacuoles, known as 'raphides', as the insoluble calcium salt, e.g. in *Philodendron* and *Dieffenbachia* spp. (Araceae). Also occurs either as the free acid or the soluble potassium salt, in spinach, *Spinacia oleracea* (Chenopodiaceae), rhubarb, *Rheum rhaponticum* (Polygonaceae), *Oxalis* spp. (Oxalidaceae), *Kochia scoparia* (Caryophyllaceae), *Amaranthus reflexus* (Amaranthaceae) and *Mesembryanthemum nodiflorum* (Aizoaceae).

Poisonous when taken in the soluble form (e.g. from rhubarb leaves), causing paralysis of the nervous system. It is corrosive to the skin and precipitates blood calcium. Livestock poisoning has occurred following the consumption of *Kochia*, *Amaranthus* and *Mesembryanthemum* species. The calcium oxalate raphides can also be dangerous, e.g. in *Dieffenbachia* leaves. The sharp ends produce injuries which can cause death in humans, especially children.

Karrer, W., Konstitution und Vorkommen der organischen Pflanzenstoffe, 2nd ed., 1981, 195, Birkhäuser.

N-(3-Oxobutyl)cytisine 1050

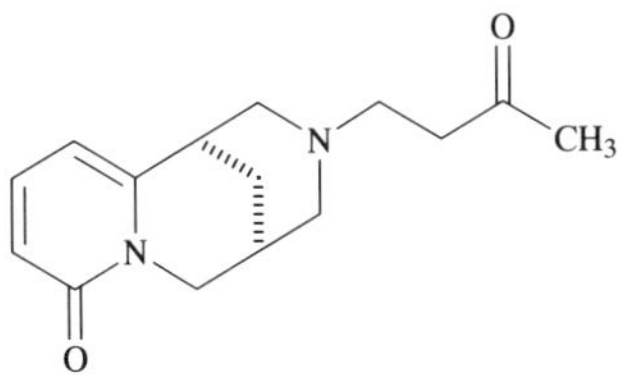

Quinolizidine alkaloid
(Cytisine)

$C_{15}H_{20}N_2O_2$ [64408-08-8] MW 260.34

Aerial parts of *Echinosophora koreenis* (Leguminosae).

Toxic. LD_{50} in mice is 71 mg/kg body-weight.

Murakoshi, I., *Phytochemistry,* 1977, **16**, 1460.

7-Oxodehydroabietic acid 1051

7-Ketodehydroabietic acid

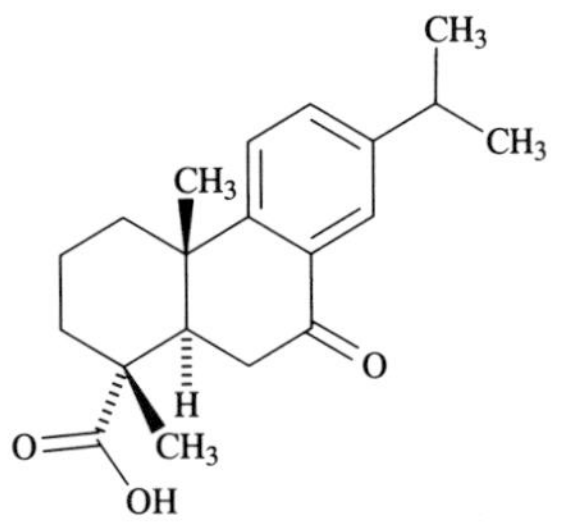

Diterpenoid
(Abietane)

$C_{20}H_{26}O_3$ [18684-55-4] MW 314.42

Surface wax of needles of *Pinus radiata* (Pinaceae).

Antifungal activity against *Dothistroma pini*.

Franich, R. A., *Physiol. Plant Pathol.*, 1983, **23**, 183.

2-Oxo-13-epimanool 1052

2-Keto-13-epimanool

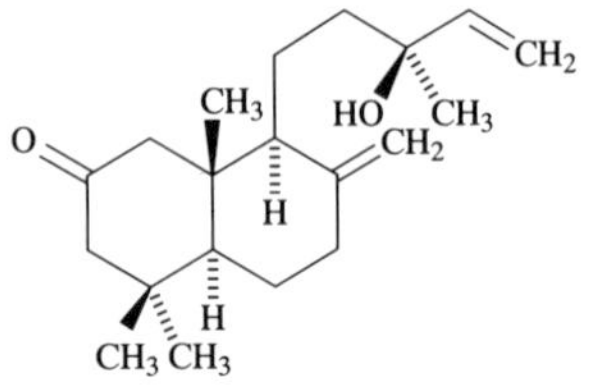

Diterpenoid
(Labdane)

$C_{20}H_{32}O_2$ [86561-13-9] MW 304.47

Leaf wax of *Nicotiana glutinosa* (Solanaceae).

Toxic to tobacco powdery mildew, *Erysiphe cichoracearum*.

Cohen, Y., *Physiol. Plant Pathol.*, 1983, **22**, 143.

Oxyayanin A 1053

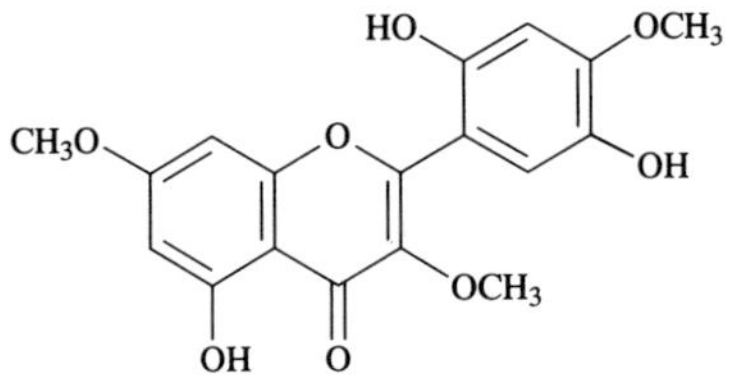

Flavonol

$C_{18}H_{16}O_8$ [549-17-7] MW 360.32

Ayan or Nigerian satinwood, *Distemonanthus benthamianus* (Leguminosae).

Allergenic properties. It causes contact allergic skin rashes in coffin workers handling ayan wood.

King, F. E., *J. Chem. Soc.*, 1954, 4587.

Oxyayanin B 1054

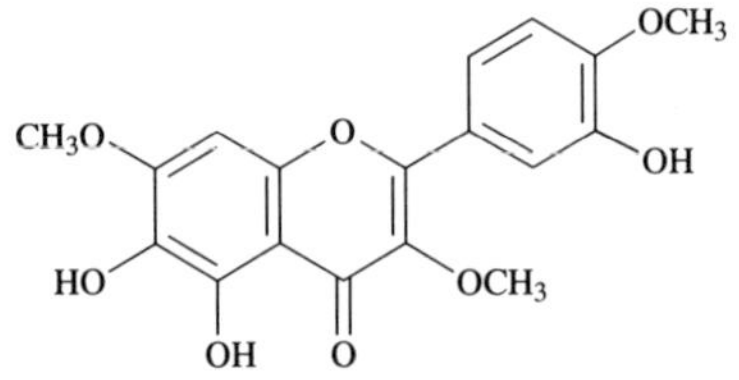

Flavonol

$C_{18}H_{16}O_8$ [548-74-3] MW 360.32

Ayan wood, *Distemonanthus benthamianus* (Leguminosae).

Allergenic properties. Causes contact allergic skin rashes in coffin workers handling ayan wood.

King, F. E., *J. Chem. Soc.*, 1954, 4589.

(+)-Oxyfrullanolide 1055

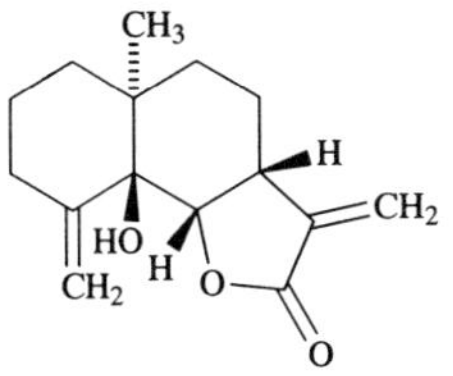

Sesquiterpenoid lactone
(Eudesmanolide)

$C_{15}H_{20}O_3$ [62870-71-7] MW 248.32

The liverworts *Frullania tamarisci*, *F. dilotata* and *F. nisquallensis*.

Causes allergenic contact dermatitis in those handling these liverworts.

Asakawa, Y., *Bull. Soc. Chim. Fr.*, 1976, 1465.

P

Pachyrrhizone 1056

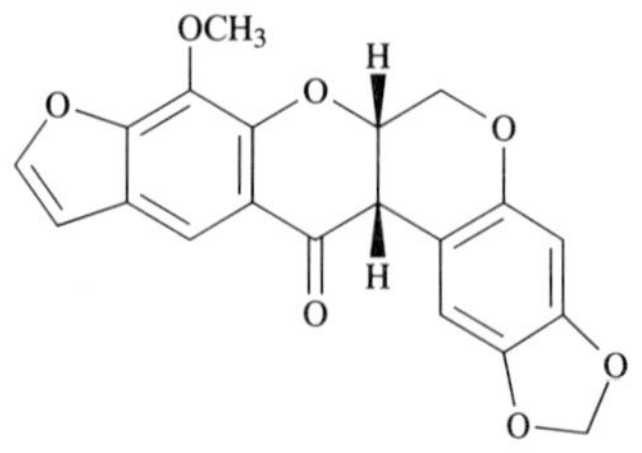

Rotenoid

$C_{20}H_{14}O_7$ [478-09-1] MW 366.33

Seeds of *Pachyrrhizus erosus* (Leguminosae).

Insecticidal properties.

Bickel, H., *Helv. Chim. Acta,* 1953, **36**, 664.

Pachysandrine A 1057

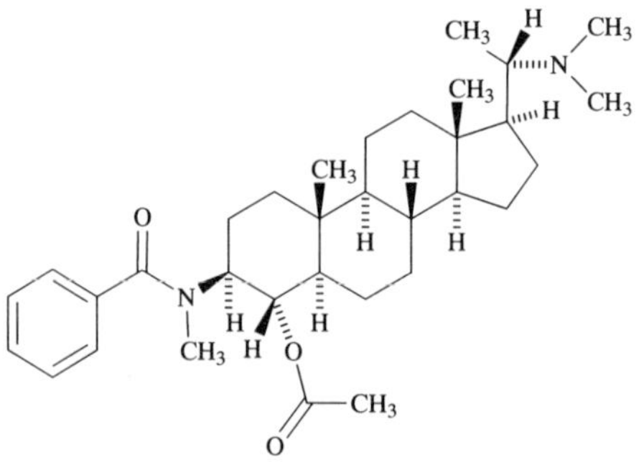

Steroid alkaloid
(Pregnane)

$C_{33}H_{50}N_2O_3$ [6878-28-3] MW 522.77

Japanese spurge, *Pachysandra terminalis* (Buxaceae). One of 23 alkaloids present in this plant.

Toxic.

Tomita, M., *Tetrahedron Lett.,* 1964, 1053.

Palmatine 1058

Berbericinine; Calystigine; Gindarinine

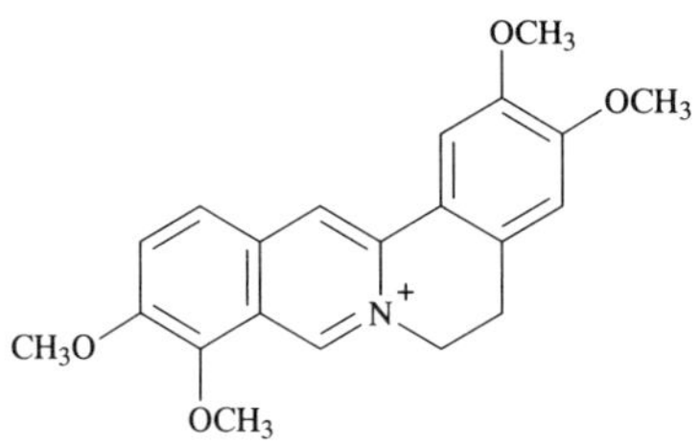

Benzylisoquinoline alkaloid
(Berbine)

$[C_{21}H_{22}NO_4]^+$ [3486-67-7] MW 352.41

Calumba root, *Jateorrhiza palmata* (Menispermaceae), *Berberis* and *Mahonia* spp. (Berberidaceae) and *Papaver* spp. (Papaveraceae).

Anti-arrhythmic, anticholinesterase and analgesic properties in experimental animals.

Späth, E., *Monatsh. Chem.,* 1928, **50**, 341.

12-*O*-Palmitoyl-16-hydroxyphorbol 13-acetate 1059

Croton factor F_1; Welensalifactor F_1;
12-Hexadecanoyl-16-hydroxyphorbol 13-acetate

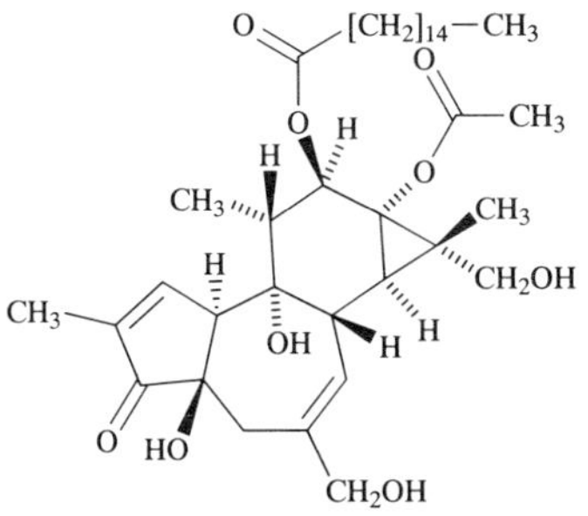

Diterpenoid
(Tigliane)

$C_{38}H_{60}O_9$ [67492-53-9] MW 660.89

Fruit of *Aleurites fordii* and leaves of welensali, *Croton flavens* (Euphorbiaceae).

Welensali leaves are commonly used in Curacao to prepare a herbal tea and this is one of the causes of high incidences of cancer of the oesophagus. This and several related diterpenoids present are the active co-carcinogenic principles. Piscicidal activity against kellie fish, *Orizias leptipes*.

Hecker, E., *Bot. J. Linn. Soc.,* 1987, **94**, 197.

Palustrine 1060

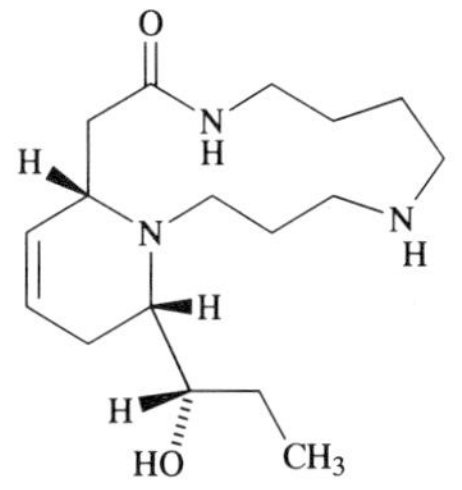

Macrocyclic alkaloid

$C_{17}H_{31}N_3O_2$ [22324-44-3] MW 309.45

Marsh horsetail, *Equisetum palustre* (Equisetaceae); the alkaloid content undergoes considerable fluctuations, from 96 to 302 mg/kg dry-weight. Also present in *E. arvense*, *E. limosum*, *E. silvaticum* and *E. ramossissimum*.

Livestock poisoning from horsetail is due to this toxic principle. Symptoms in the horse include excitability, reeling gait, falling over and eventually death from exhaustion.

Natsume, M., *Chem. Pharm. Bull.,* 1984, **32**, 3789.

Palustrol 1061

1-Aromadendranol

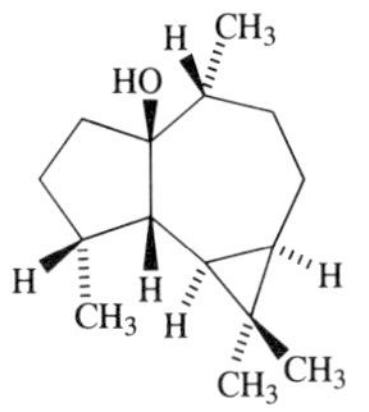

Sesquiterpenoid
(Aromadendrane)

$C_{15}H_{26}O$ [5986-49-2] MW 222.37
(+)-form [36948-71-7]

Leaf of wild rosemary, *Ledum palustre* (Ericaceae).

Together with ledol (q.v.), is responsible for intoxicating and narcotic effects of this plant. Dried branches are used as an insecticide.

Dolejš, L., *Collect. Czech. Chem. Commun.*, 1961, **26**, 811.

Pancratistatin 1062

Amaryllidaceae alkaloid
(Narciclasine)

$C_{14}H_{15}NO_8$ [96203-70-2] MW 325.28

Bulbs of *Pancratium littorale* and *Zephyranthes grandiflora* (Amaryllidaceae).

Anticancer activity when tested against lymphocytic P388 leukaemia, PS system and M5076 ovary sarcoma.

Pettit, G. R., *J. Nat. Prod.*, 1986, **49**, 995.

Papaverine 1063

Benzylisoquinoline alkaloid

$C_{20}H_{21}NO_4$ [58-74-2] MW 339.39

Opium, *Papaver somniferum* (Papaveraceae) and *Rauwolfia serpentina* (Apocynaceae).

Smooth muscle relaxant, cerebral vasodilator and antitussive. It is used medicinally as a spasmolytic and in cough medicines. The LD_{50} intravenously in mice is 25 mg/kg body-weight.

Manske, R. H. F., The Alkaloids, 1954, **4**, 29, Academic Press.

Parasorbic acid 1064

Hexenolactone; 5-Hydroxy-2-hexenoic acid lactone; Sorbic oil

Lactone

$C_6H_8O_2$ [10048-32-5] MW 112.13

Fruit of the rowan tree, *Sorbus aucuparia* and of other *Sorbus* spp. (Rosaceae).

Mildly toxic, with an LD_{50} intraperitoneally in mice of 750 mg/kg body-weight, and also carcinogenic.

Crombie, L., *J. Chem. Soc., C,* 1968, 2852.

Paravallarine 1065

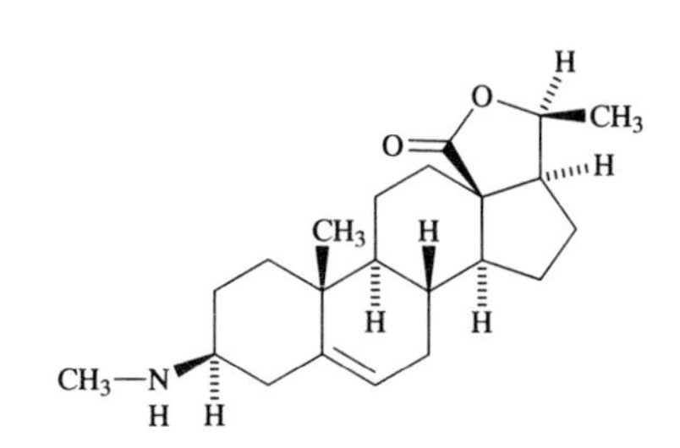

Steroid alkaloid
(Pregnane)

$C_{22}H_{33}NO_2$ [510-31-6] MW 343.51

Paravallaris microphylla (Apocynaceae).

Fish poison.

Le Men, J., *Bull. Soc. Chim. Fr.*, 1960, 860.

Parillin 1066

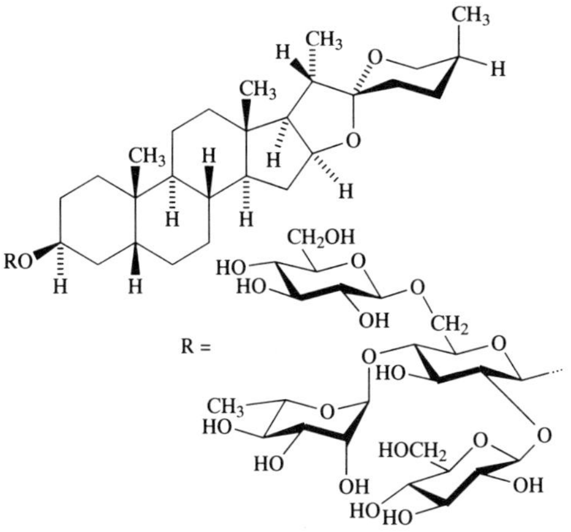

Triterpenoid saponin
(Spirostan)

$C_{51}H_{84}O_{22}$ [19507-61-5] MW 1049.31

Roots of sarsaparilla, *Smilax aristolochiaefolia* (Liliaceae).

Strong haemolytic activity; also has some cancerostatic activity.

Tschesche, R., *Justus Liebigs Ann. Chem.*, 1966, **699**, 212.

Parthenin 1067

Parthenicin

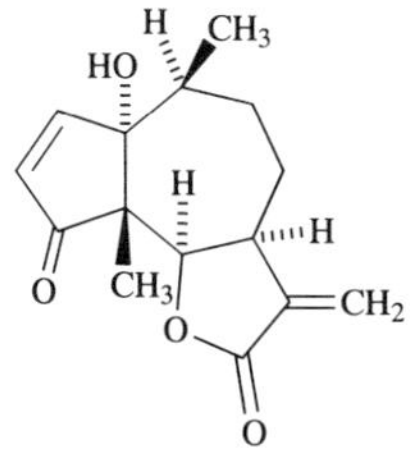

Sesquiterpenoid lactone
(Pseudoguaianolide)

$C_{15}H_{18}O_4$ [508-59-8] MW 262.31

Parthenium hysterophorus, *Ambrosia psilostachya* and *Iva nevadensis* (Compositae).

Parthenin is the main allergen responsible for the allergic contact dermatitis caused by the weed *Parthenium hysterophorus*. The dermatitis produced in man is occasionally serious enough to cause death. It is a direct cardiac depressant in dogs, is toxic to cattle and has molluscicidal properties. It is both antifeedant and toxic to insects.

Herz, W., *J. Am. Chem. Soc.*, 1962, **84**, 2601.

Parthenolide 1068

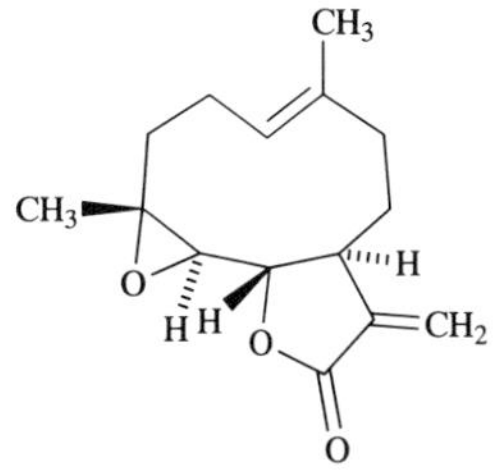

Sesquiterpenoid lactone
(Germacranolide)

$C_{15}H_{20}O_3$ [20554-84-1] MW 248.32

Major lactone in the glands on the leaf surface of feverfew, *Tanacetum parthenium*; also present in *Ambrosia* and *Arctotis* spp. (Compositae). Occurs in *Michelia champaca* and *M. lanuginosa* (Magnoliaceae).

Cytotoxic and antitumour properties. It is regarded as the main active principle of feverfew, which is used as a plant drug in the treatment of migraine.

Govindachari, T. R., *Tetrahedron,* 1965, **21**, 1509.

Passicapsin 1069

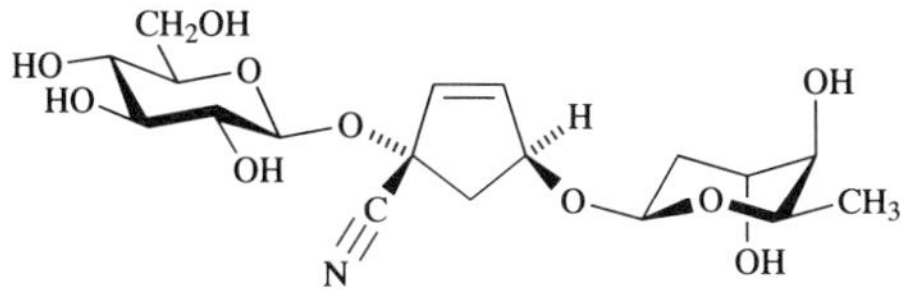

Cyanogenic glycoside

$C_{18}H_{27}NO_{10}$ [82829-54-7] MW 417.41

Leaves of *Passiflora capsularis* (Passifloraceae).

Bound toxin, releasing poisonous cyanide on hydrolysis.

Olafsdottir, E. S., *Acta Chem. Scand.*, 1989, **43**, 51.

Causal agent of shellfish poisoning.

Yasumoto, T., *Tetrahedron*, 1985, **41**, 1019.

Paucin 1070

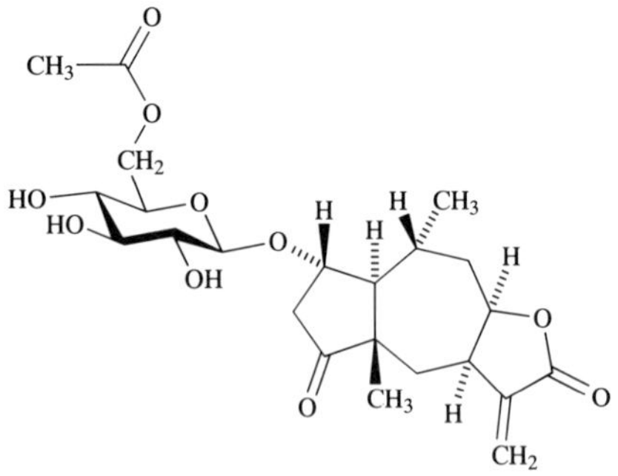

Sesquiterpenoid lactone
(Pseudoguaianolide)

$C_{23}H_{32}O_{10}$ [26836-43-1] MW 468.50

Baileya pauciradiata, *B. pleniradiata*, and several *Hymenoxys* spp. (Compositae).

Cytotoxic and antitumour properties.

Waddell, T. G., *Tetrahedron Lett.*, 1969, 515.

Pectenotoxin 1 1071

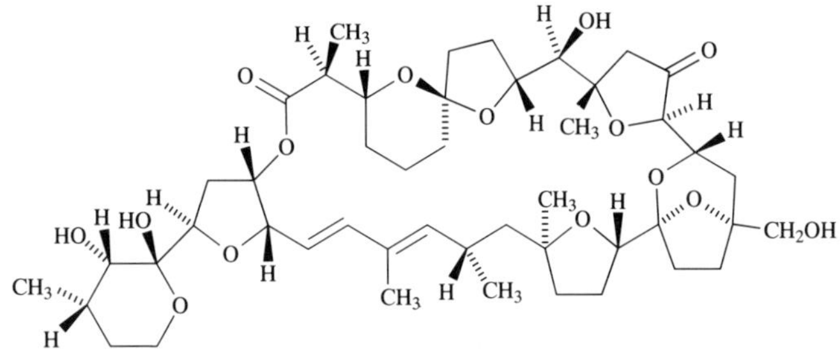

Polyether

$C_{47}H_{70}O_{15}$ [97564-90-4] MW 875.06

One of five almost identical toxins produced by the alga *Dinophysis acuminata* and dietarily sequestered by scallops.

Pedunculagin 1072

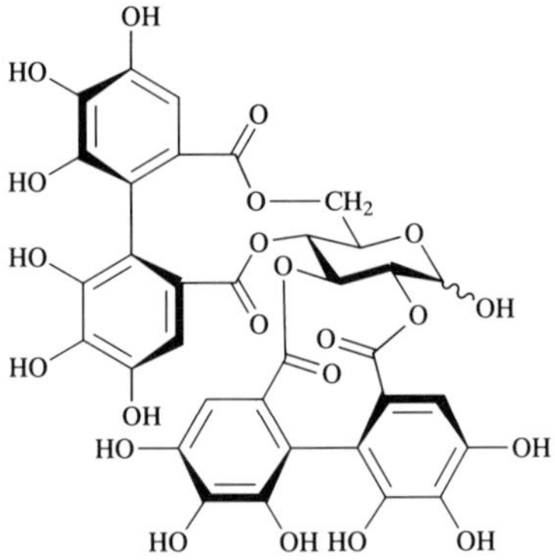

Ellagitannin

$C_{34}H_{24}O_{22}$ [7045-42-3] MW 784.55

Quercus spp. (Fagaceae) and of *Rubus* spp. (Rosaceae), *Casuarina stricta* (Casuarinaceae), *Stachyurus praecox* (Stachyuraceae) and *Camellia japonica* (Theaceae).

Contributes to the toxicity of *Quercus* leaves to livestock. Has *in vitro* antihepatotoxic activity due to inhibitory effects on the enzyme glutamine-pyruvic transaminase.

Okuda, T., *J. Chem. Soc., Perkin Trans. 1*, 1983, 1765.

Peganine 1073

Linarine; Vasicine

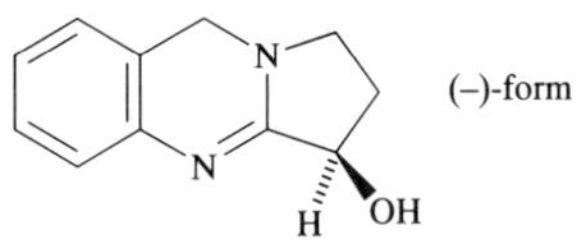

(–)-form

Quinoline alkaloid

$C_{11}H_{12}N_2O$ [6159-55-3] (±)-form [6159-56-4] MW 188.23

The (−)-form is found in seeds of *Peganum harmala* (Zygophyllaceae), the leaves of *Adhatoda vasica* (Acanthaceae) and the roots of *Sida cordifolia* (Malvaceae). The racemic (±)-form occurs in *Anisotes sessiliflorus* (Acanthaceae).

Causes hypotension and stimulates respiration. Is an abortifacient in mammals and also has anthelmintic properties.

Späth, E., *Ber. Dtsch. Chem. Ges.,* 1936, **69**, 384.

(−)-Pelletierine 1074

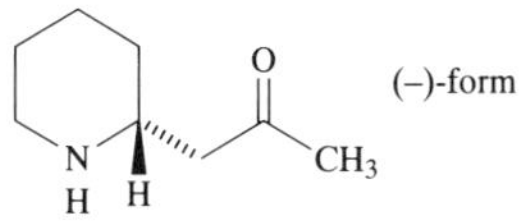

(−)-form

Piperidine alkaloid

$C_8H_{15}NO$ [4396-01-4] MW 141.21

Pomegranate, *Punica granatum* (Punicaceae), *Duboisia myoporoides* (Solanaceae) and *Sedum acre* (Crassulaceae).

Highly toxic to tapeworms and has been used clinically as an anthelmintic.

Drillien, G., *Bull. Soc. Chim. Fr.,* 1963, 2393.

α-Peltatin 1075

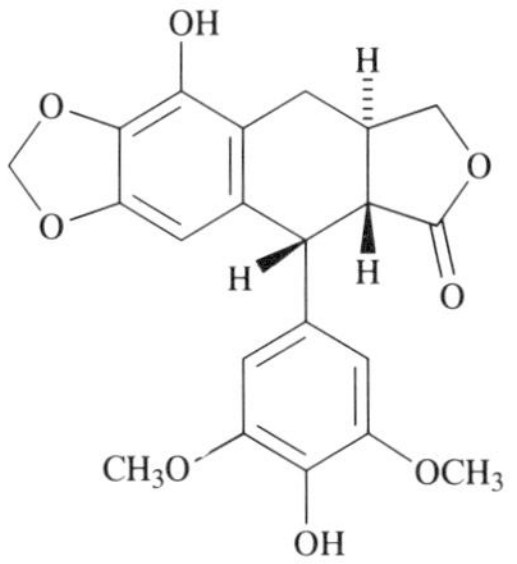

Lignan

$C_{21}H_{20}O_8$ [568-53-6] MW 400.39

Resin of *Podophyllum peltatum* (Podophyllaceae).

Antitumour properties.

Hartwell, J. L., *J. Am. Chem. Soc.,* 1950, **72**, 246.

β-Peltatin A methyl ether 1076

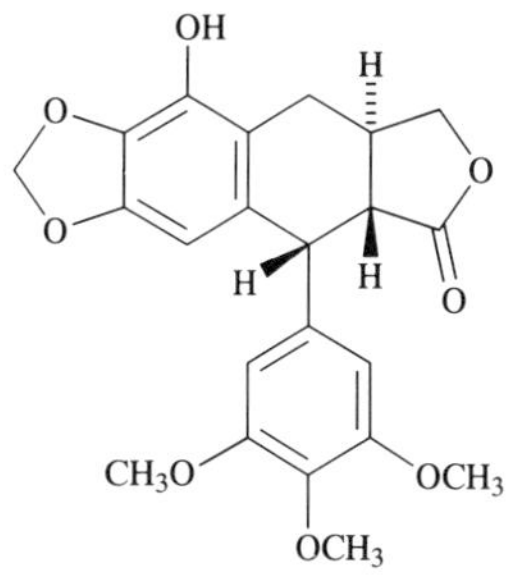

Lignan

$C_{23}H_{24}O_8$ [23978-65-6] MW 428.44

Podophyllum resin, *Podophyllum peltatum* (Podophyllaceae), needles of *Juniperus sabina* (Cupressaceae) and *Bursera fagaroides* (Burseraceae).

Antitumour properties.

Hartwell, J. L., *J. Am. Chem. Soc.,* 1952, **74**, 6285.

Perilla ketone 1077

1-(3-Furanyl)-4-methylpentan-1-one

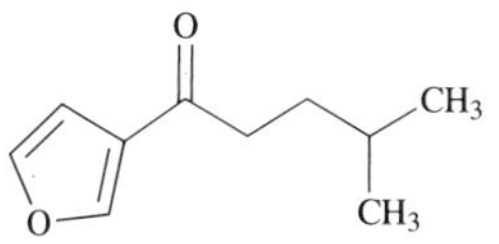

Monoterpenoid
(Furanoketone)

$C_{10}H_{14}O_2$ [553-84-4] MW 166.22

In the essential oil of *Perilla frutescens* (Labiatae).

Lung toxin causing damage and death to livestock feeding on *Perilla* plants.

Koezuka, Y., *Planta Med.,* 1985, 480.

1-Peroxyferolide 1078

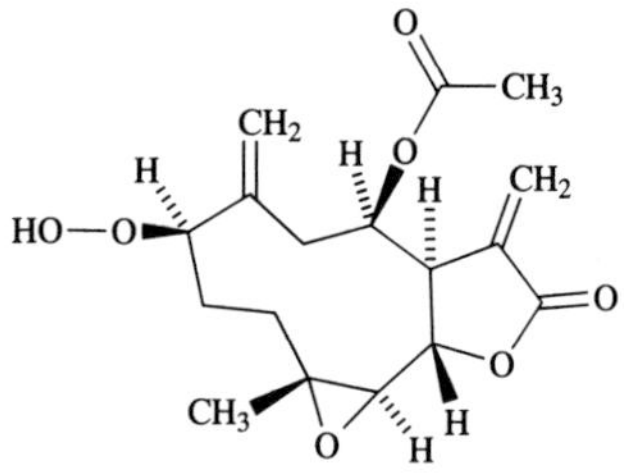

Sesquiterpenoid lactone
(Germacranolide)

$C_{17}H_{22}O_7$ [61228-73-7] MW 338.36

Tulip tree, *Liriodendron tulipifera* (Magnoliaceae).

Insect antifeedant.

Doskotch, R. W., *J. Chem. Soc., Chem. Commun.,* 1976, 402.

Peruvoside 1079

Cannogenin 3-gulomethyloside

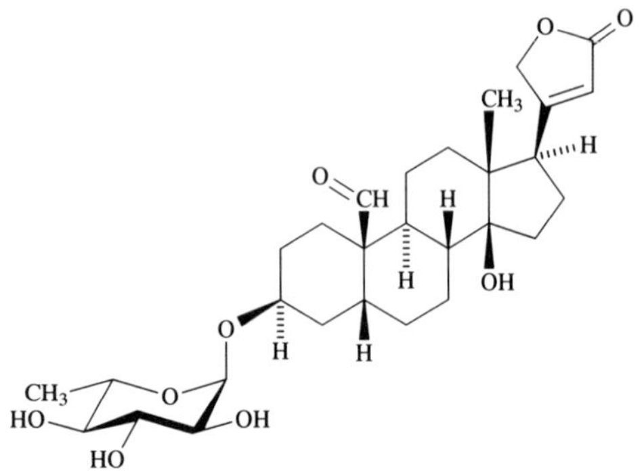

Cardenolide

$C_{29}H_{42}O_9$ [1182-87-2] MW 534.65

Yellow oleander, *Thevetia peruviana* (Apocynaceae).

Toxic to mammals and other vertebrates. Seeds of yellow oleander are particularly dangerous and they are used in India to poison cattle and to commit murders. Peruvoside is used medicinally for cardiac insufficiency.

Bloch, R., *Helv. Chim. Acta,* 1960, **43**, 652.

Petasin 1080

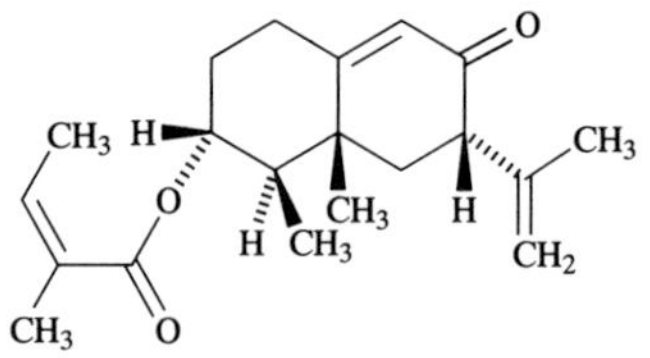

Sesquiterpenoid
(Eremophilane)

$C_{20}H_{28}O_3$ [26577-85-5] MW 316.44

Leaf and root of butterbur, *Petasites hybridus* (Compositae).

Spasmolytic properties, 14 times more active than the alkaloid papaverine (q.v.).

Herbst, D., *J. Am. Chem. Soc.,* 1960, **82**, 4337.

Petasitenine 1081

Fukinotoxin

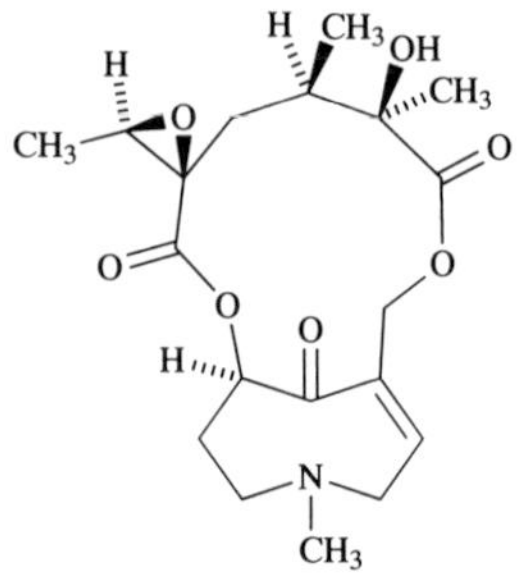

Pyrrolizidine alkaloid
(Crotalanan)

$C_{19}H_{27}NO_7$ [60102-37-6] MW 381.43

Petasites japonicus and *P. hybridus* (Compositae).

Hepatotoxic and carcinogenic in vertebrates.

Yamada, K., *Chem. Lett.,* 1976, 461.

Pfaffic acid 1082

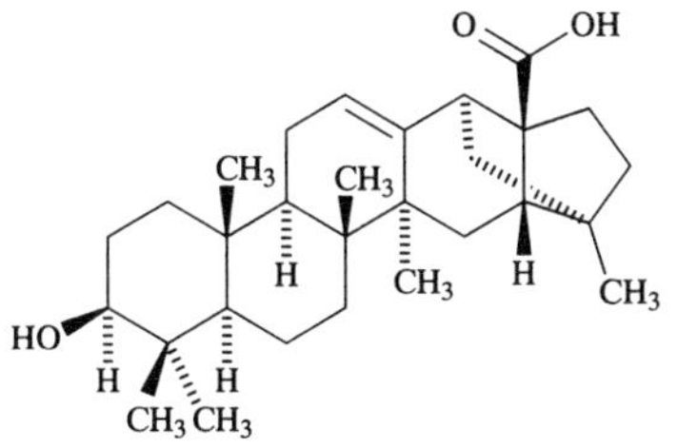

Triterpenoid
(Cyclonoroleanane)

$C_{29}H_{44}O_3$ [86432-14-6] MW 440.67

Roots of Brazil ginseng, *Pfaffia paniculata* (Amaranthaceae)

Antitumour properties, inhibiting the growth of tumour cells at concentrations of 4 to 6 μg/ml.

Takemoto, T., *Tetrahedron Lett.,* 1983, **24**, 1057.

Pfaffoside A 1083

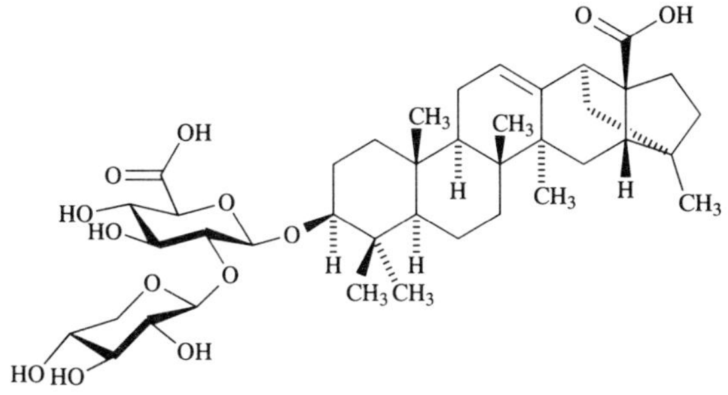

Triterpenoid
(Cyclonoroleanane)

$C_{40}H_{60}O_{13}$ [90745-17-8] MW 748.91

Root of Brazil gingseng, *Pfaffia paniculata* (Amaranthaceae).

Antitumour properties, inhibiting the growth of tumour cells at concentrations of 30 to 50 μg/ml.

Nishimoto, N., *Phytochemistry,* 1984, **23**, 139.

Phaeantharine 1084

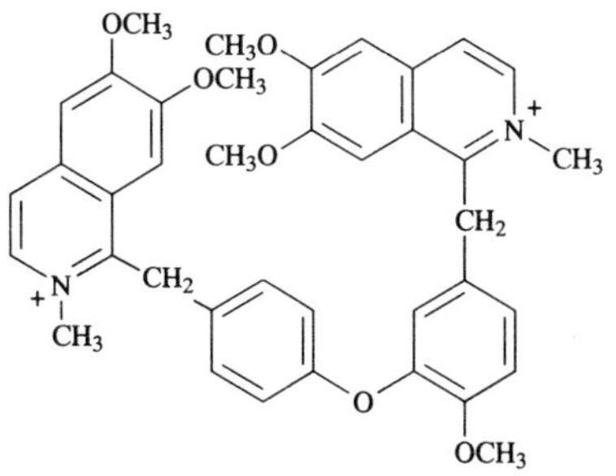

Bisbenzylisoquinoline alkaloid

$[C_{39}H_{40}N_2O_6]^{2+}$ [27670-80-0] MW 632.76

Bark of *Phaeanthus ebracteolatus* (Annonaceae).

Anticancer properties.

Van Beek, T. A., *J. Nat. Prod.,* 1983, **46**, 226.

Phantomolin 1085

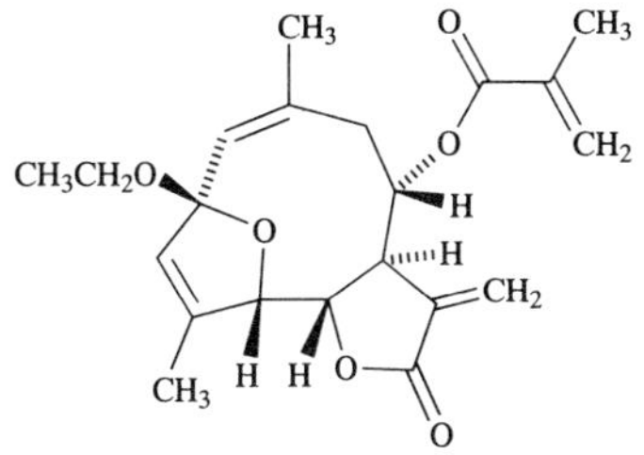

Sesquiterpenoid lactone
(Germacranolide)

$C_{21}H_{26}O_6$ [55306-08-6] MW 374.43

Elephantopus mollis (Compositae).

Cytotoxic and antitumour properties.

McPhail, A. T., *Tetrahedron Lett.,* 1974, 2739.

Phasin 1086

PHA

Protein
(Lectin)
[1392-87-6] MW 126000

Seed of the common bean, *Phaseolus vulgaris* (Leguminosae).

Toxin of raw beans, with clinical symptoms similar to ricin (q.v.). Vegetarians and children, who eat the raw beans, are liable to suffer poisoning. The toxin is denatured by heating and the cooked beans are perfectly safe.

Manen, J. F., *Planta,* 1982, **155**, 328.

α-Phellandrene 1087

p-Mentha-1,5-diene

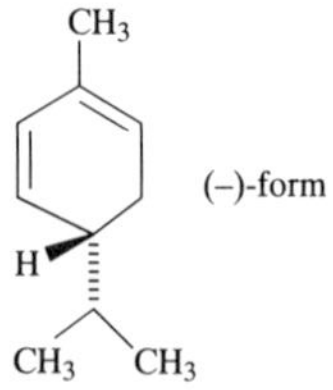

(−)-form

Monoterpenoid
(Menthane)

$C_{10}H_{16}$ [99-83-2] MW 136.24
(−)-form [4221-98-1]
(+)-form [2243-33-6]

The (−)-form occurs widely in essential oils, e.g. in oils of eucalypt (Myrtaceae). The (+)-form is present in oil of fennel, *Foeniculum vulgare* (Umbelliferae) and oil of black pepper fruit, *Piper nigrum* (Piperaceae).

Irritating to, and absorbed through, the skin. Ingestion can cause vomiting and/or diarrhoea.

Karrer, W., Konstitution und Vorkommen der organischen Pflanzenstoffe, 2nd ed., 1981, 26, Birkhäuser.

β-Phellandrene 1088

p-Mentha-1(7),2-diene

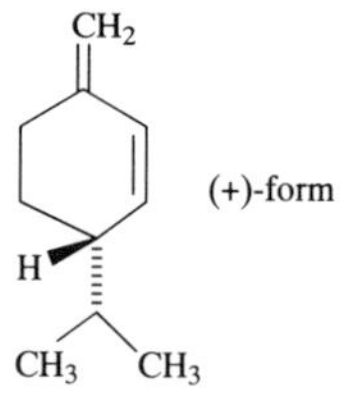

(+)-form

Monoterpenoid
(Menthane)

$C_{10}H_{16}$ [555-10-2] MW 136.24
(+)-form [6153-17-9]
(−)-form [6153-16-8]

Common constituent of many essential oils. The (−)-form occurs in the oils of *Abies*, *Picea* and *Pinus* spp. (Pinaceae), whereas the (+)-form is the main constituent of the seed oil of the water dropwort, *Oenanthe aquatica* (Umbelliferae).

Expectorant.

Karrer, W., Konstitution und Vorkommen der organischen Pflanzenstoffe, 2nd ed., 1981, 27, Birkhäuser.

Phenethylamine 1089

β-Phenylethylamine; *β*-Aminoethylbenzene; 1-Amino-2-phenylethane; Benzeneethanamine

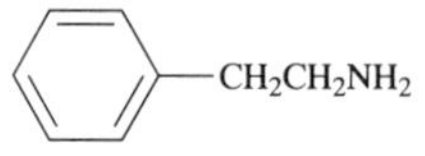

$-CH_2CH_2NH_2$

Aromatic amine

$C_8H_{11}N$ [64-04-0] MW 121.18

Oil of bitter almonds, *Prunus dulcis* var. *amara* (Rosaceae), fruit of banana, *Musa sapientum* (Musaceae) and red algae (species of Rhodophyceae).

Fishy odour. Skin irritant and sensitiser.

Smith, T. A., *Phytochemistry,* 1977, **16**, 9.

Phenol 1090

Carbolic acid; Phenylic acid; Hydroxybenzene

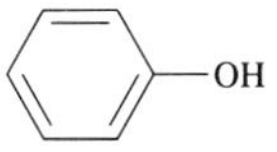

Phenol

C_6H_6O [108-95-2] MW 94.11

Essential oil of *Perovskia angustifolia* (Labiatae), fruit of *Paedera chinensis* (Rubiaceae), *Elscholtzia nipponica* (Labiatae), wood of *Populus tremuloides* (Salicaceae) and *Paeonia albiflora* (Paeoniaceae).

Toxic to humans, with caustic effects on the skin. Aqueous solutions are used as a topical anaesthetic.

Karrer, W., Konstitution und Vorkommen der organischen Pflanzenstoffe, 2nd ed., 1981, 72, Birkhäuser.

2-Phenylethanol 1091

Phenethyl alcohol; Benzeneethanol

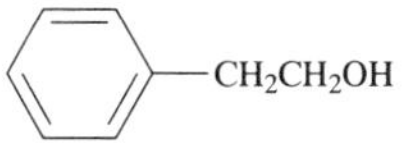

Aromatic alcohol

$C_8H_{10}O$ [60-12-8] MW 122.17

In essential oils of plants, e.g. in rose oil, *Rosa rugosa* (Rosaceae).

Moderately toxic.

Karrer, W., Konstitution und Vorkommen der organischen Pflanzenstoffe, 2nd ed., 1981, 78, Birkhäuser.

1-Phenylhepta-1,3,5-triyne 1092

Phenylheptatriyne

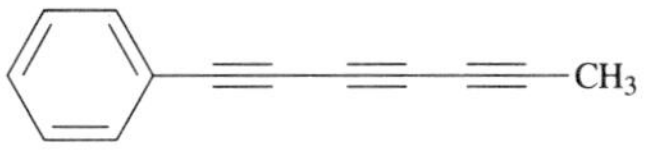

Polyacetylene

$C_{13}H_8$ [4300-27-0] MW 164.21

Coreopsis grandiflora, *Dahlia* and *Bidens* spp. (Compositae).

Phototoxic agent and insect antifeedant.

Sörensen, J. S., *Acta Chem. Scand.*, 1958, **12**, 765.

Phloridzin 1093

Phoretin 2′-glucoside

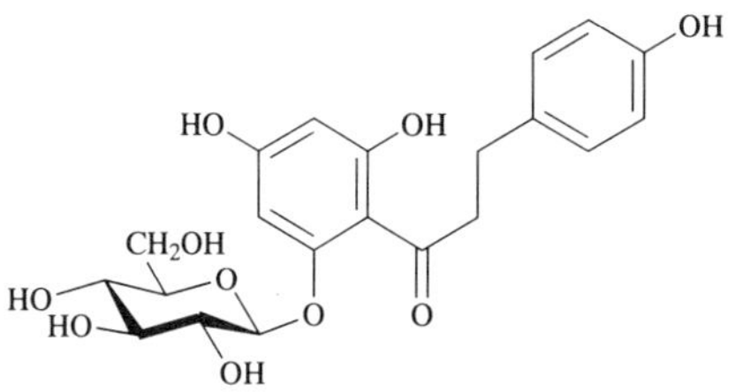

Dihydrochalcone

$C_{21}H_{24}O_{10}$ [60-81-1] MW 436.42

Leaves and skins of apple, *Malus domestica* (Rosaceae), *Kalmia latifolia*, *Pieris japonica* and *Rhododendron* spp. (Ericaceae) and *Symplocos* spp. (Symplocaceae).

Bitter tasting glycoside. When taken orally, it causes glycosuria by interfering with the tubular readsorption of glucose in the kidney. It shows feeding deterrent activity to aphids.

Farkas, L., *Chem. Ber.*, 1965, **98**, 2926.

Phloroglucinol trimer 1094

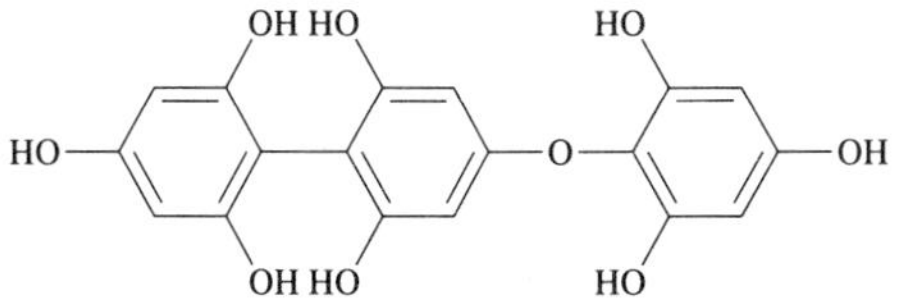

Phenol

$C_{18}H_{14}O_9$ [58878-18-5] MW 374.30

Brown algae, *Fucus vesiculosus* and *Ascophyllum nodosum*.

Antifeedant to the marine snail *Littorina littorea*.

Geiselman, J. A., *J. Chem. Ecol.*, 1981, **7**, 1115.

Phoratoxin 1095

Lys-Ser-Cys-Cys-Pro-Thr-Thr-Thr-Ala-Arg-
-Asn-Ile-Tyr-Asn-Thr-Cys-Arg-Phe-Gly-Gly-
-Gly-Ser-Arg-Pro-Val-Cys-Ala-Lys-Leu-Ser-
-Gly-Cys-Lys-Ile-Ile-Ser-Gly-Thr-Lys-Cys-
-Asp-Ser-Gly-Trp-Asn-His

Protein

[37339-68-7] MW 4881

Californian mistletoe, *Phoradendron tomentum* (Viscaceae). Contains 46 amino acid residues.

Toxic to heart muscle.

Mellstrand, S. T., *Acta Pharm. Suec.,* 1974, **11**, 367.

Phorbol 12-tiglate 13-decanoate 1096

Diterpenoid
(Tigliane)

$C_{35}H_{52}O_8$ [59086-92-9] MW 600.79

Seed oil of *Croton tiglium* (Euphorbiaceae).

Irritant, with antileukaemic activity.

Clarke, E., *Z. Krebsforsch.,* 1965, **676**, 192.

Phrymarolin I 1097

Lignan

$C_{24}H_{24}O_{11}$ [38303-95-6] MW 488.45

Roots of *Phryma leptostachya* (Phrymaceae).

Insecticidal activity, especially effective when combined with pyrethrin (q.v.) or the pesticide sevin. The roots are used directly in Kyushu, Japan to exterminate houseflies.

Taniguchi, E., *Agric. Biol. Chem.,* 1972, **36**, 1497.

Phyllanthin 1098

Lignan

$C_{24}H_{34}O_6$ [10351-88-9] MW 418.53

Leaves of *Phyllanthus niruri* (Euphorbiaceae).

Feeding deterrent, with a bitter taste.

Row, L. R., *Tetrahedron,* 1966, **22**, 2899.

Phyllanthostatin A 1099

Lignan

$C_{29}H_{30}O_{13}$ [119767-19-0] MW 586.55

Roots of *Phyllanthus acuminatus* (Euphorbiaceae).

Cytostatic activity.

Pettit, G. R., *J. Nat. Prod.,* 1988, **51**, 1104.

Physoperuvine 1100

1-Hydroxytropane

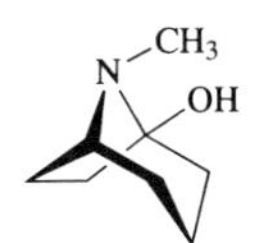

Tropane alkaloid

$C_8H_{15}NO$ [60723-27-5] MW 141.21

Cape gooseberry, *Physalis peruviana* (Solanaceae).

Moderately toxic.

McPhail, A. T., *Tetrahedron,* 1984, **40**, 1661.

Physostigmine 1101

Eserine; Physosterine; Physostol

Indole alkaloid

$C_{15}H_{21}N_3O_2$ [57-47-6] MW 275.35

Major alkaloid of the calabar bean, the seed of *Physostigma venenosum* (Leguminosae).

Anticholinesterase agent. It has wide-ranging parasympathetic activity when taken internally and can be used to counteract the effects of anticholinergics such as the alkaloid atropine (q.v.). Large doses, however, can be fatal. The main clinical use is in a form of eye drops as a miotic.

Manske, R. H. F., The Alkaloids, 1968, **10**, 383, Academic Press.

Physovenine 1102

Indole alkaloid

$C_{14}H_{18}N_2O_3$ [6091-05-0] MW 262.31

Calabar bean, the seed of *Physostigma venenosum* (Leguminosae).

Anticholinesterase agent. It is similar in activity to physostigmine (q.v.), but it is not in clinical use.

Robinson, B., *J. Chem. Soc.,* 1964, 1503.

Phytolaccagenin 1103

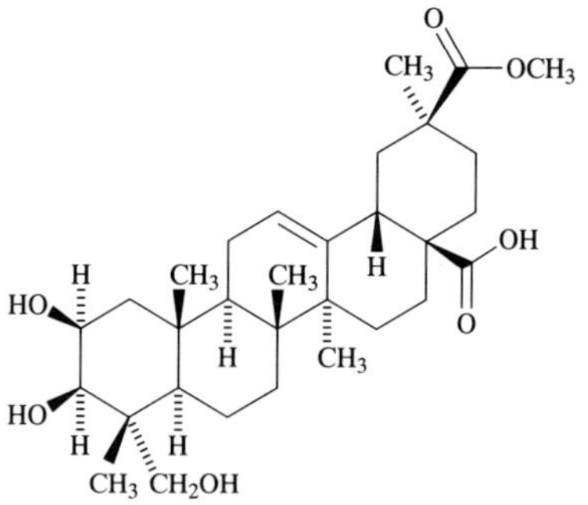

Triterpenoid sapogenin
(Oleanane)

$C_{31}H_{48}O_7$ [1802-12-6] MW 532.72

In glycosidic form in pokeberry, *Phytolacca americana* (Phytolaccaceae). The xylosylglucoside is called phytolaccoside E [65497-07-6].

Phytolaccatoxin and other glycosides of phytolaccagenin are toxic to humans. The leaves and roots are the most likely sources of poisoning. The concentrations of toxin in the attractive black berries are too low to cause any problems to adults but could be fatal to children.

Stout, G. H., *J. Am. Chem. Soc.,* 1964, **86**, 957.

Phytuberin 1104

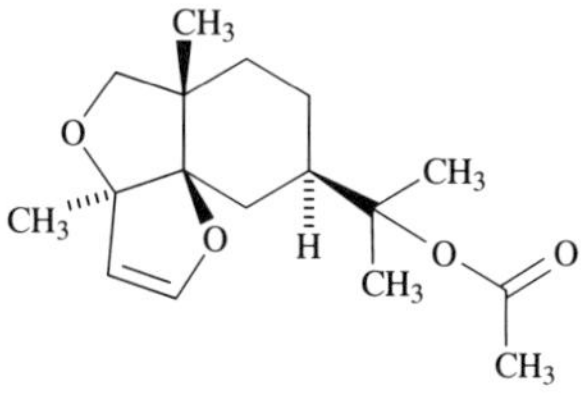

Sesquiterpenoid
(Secoeudesmane)

$C_{17}H_{26}O_4$ [37209-50-0] MW 294.39

Occurs, together with the unacylated parent alcohol, phytuberol [56857-64-8] in tubers of potato, *Solanum tuberosum* (Solanaceae) infected with *Phytophthora infestans.*

Antifungal agent (phytoalexin).

Coxon, D. T., *J. Chem. Soc., Perkin Trans. 1,* 1977, 53.

Piceatannol 1105

3,4,3′,5′-Tetrahydroxystilbene; Astringenin

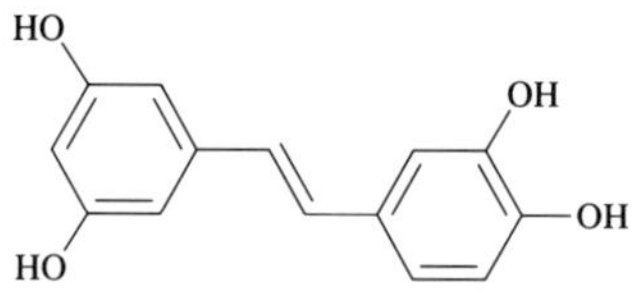

Stilbenoid

$C_{14}H_{12}O_4$ [10083-24-6] MW 244.25

Heartwood of *Picea* and *Pinus* spp. and of *Laburnum anagyroides* (Leguminosae).

Fungitoxin and fish poison.

King, F. E., *J. Chem. Soc.,* 1956, 4477.

Picrotin 1106

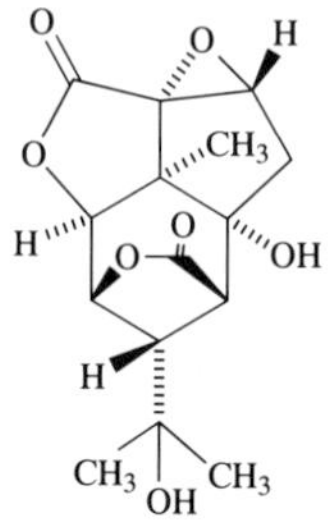

Sesquiterpenoid lactone
(Tutinanolide)

$C_{15}H_{18}O_7$ [21416-53-5] MW 310.30

'Fish berries', the drupes of *Anamirta paniculata* (Menispermaceae).

Causes extreme excitation of the central nervous system. Has been used to stun fish and also to treat skin diseases.

Taylor, W. I., Cyclopentanoid Terpene Derivatives, 1969, 147, Marcel Dekker.

Picrotoxinin 1107

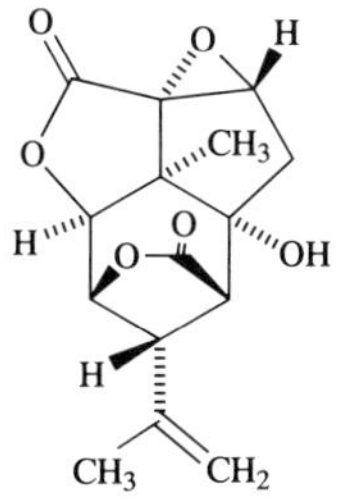

Sesquiterpenoid lactone
(Tutinanolide)

$C_{15}H_{16}O_6$ [17617-45-7] MW 292.29

'Fish berries', the drupes of *Anamirta paniculata* (Menispermaceae).

As a mixture with picrotin (q.v.), it produces extreme excitation of the central nervous system. The mixture has been used to stun fish and also to treat skin diseases.

Porter, L. A., *Chem. Rev.*, 1967, **67**, 441.

Pilocarpine 1108

Syncarpine

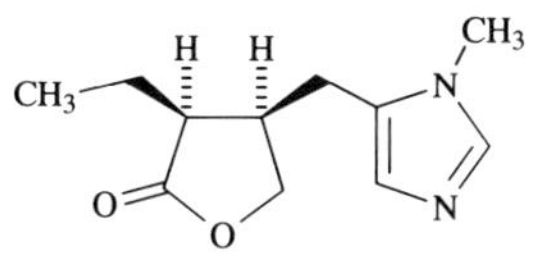

Imidazole alkaloid

$C_{11}H_{16}N_2O_2$ [92-13-7] MW 208.26

Pilocarpus microphyllus and other *Pilocarpus* spp. (Rutaceae).

Stimulates the parasympathetic nerve endings, increasing thereby salivatory, gastric and lachrymal secretions. It is used in the treatment of glaucoma and of hepatitis.

Al-Badr, A. A., *Anal. Profiles Drug Subst.*, 1983, **12**, 385.

Pilosine 1109

Carpidine

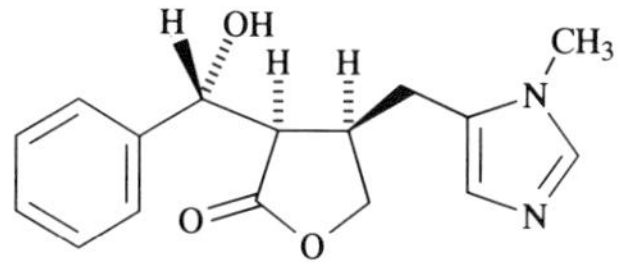

Imidazole alkaloid

$C_{16}H_{18}N_2O_3$ [13640-28-3] MW 286.33

Pilocarpus microphyllus (Rutaceae).

Narcotic to mammals. Similar to pilocarpine (q.v.) in its pharmacological properties.

Link, H., *Helv. Chim. Acta,* 1974, **57**, 2199.

α-Pinene 1110

2-Pinene; Pinene

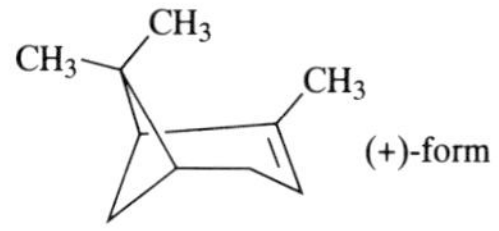

(+)-form

Monoterpenoid
(Pinane)

$C_{10}H_{16}$ [80-56-8] MW 136.24
(+)-α form [7785-70-8]
(−)-α form [7785-26-4]

Constituent of many essential oils. Notably present in oil of turpentine, from the oleoresin of *Pinus palustris* and other *Pinus* spp. (Pinaceae). Also in leaf oils of many Pinaceae, Cupressaceae, Myrtaceae, Labiatae and Rutaceae.

Irritant. It can cause skin eruption, delirium, ataxia and kidney damage.

Banthorpe, D. V., *Chem. Rev.*, 1966, **66**, 643.

β-Pinene 1111

Nopinene

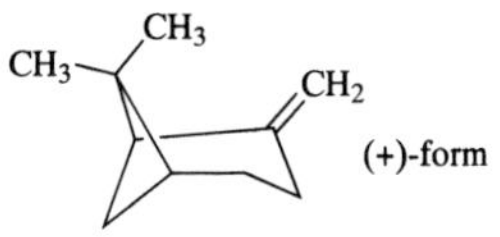

Monoterpenoid
(Pinane)

$C_{10}H_{16}$ [127-91-3] MW 136.24

In oil of turpentine and most other essential oils containing α-pinene (q.v.), but is present in smaller amounts. Present in oil of cumin, from fruits of *Cuminum cyminum* (Umbelliferae).

Irritant. It can cause skin eruption, delirium, ataxia and kidney damage.

Banthorpe, D. V., *Chem. Rev.,* 1966, **66**, 643.

Pinocembrin 1112

5,7-Dihydroxyflavanone

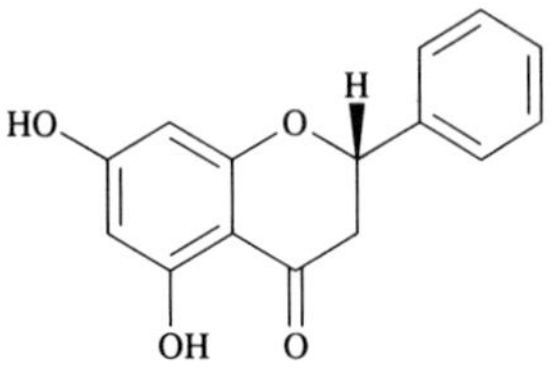

Flavanone

$C_{15}H_{12}O_4$ [480-39-7] MW 256.26
(±)-form [61490-55-9]

Widespread in nature; in leaf glands of *Populus deltoides*, heartwood of *Pinus cembra* (Pinaceae), in *Prunus* spp. (Rosaceae) and *Helichrysum* spp. (Compositae).

Antifungal activity, e.g. against *Melampsora medusae*; antibacterial activity, e.g. against *Bacillus subtilis* at a concentration of 3 μg/ml.

Lindstedt, G., *Acta Chem. Scand.,* 1951, **5**, 121.

Pinosylvin methyl ether 1113

5-Methoxy-3-stilbenol; Pinosylvin monomethyl ether

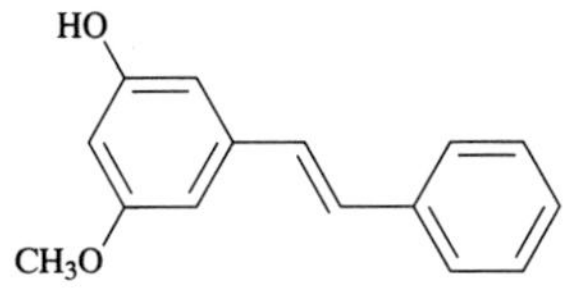

Stilbenoid

$C_{15}H_{14}O_2$ [35302-70-6] MW 226.27

Wood of sixty *Pinus* spp. (Pinaceae) and in *Alnus sieboldiana* and *A. crispa* (Betulaceae).

Antifeedant to the snowshoe hare, *Lepus americanus.*

Lindstedt, G., *Acta Chem. Scand.,* 1949, **3**, 755.

Pipercide 1114

Retrofractamide B

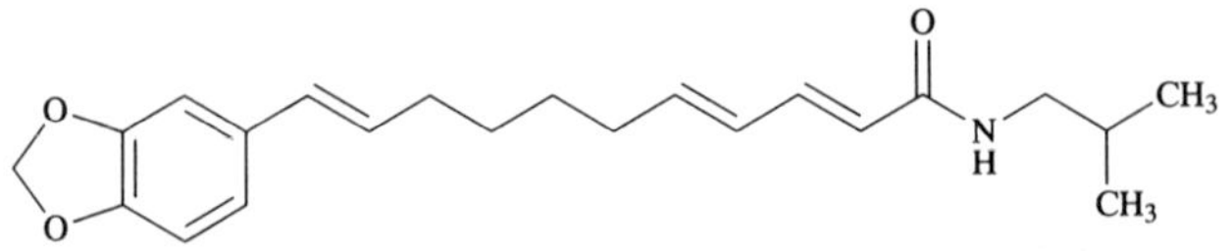

Aliphatic amide

$C_{22}H_{28}NO_3$ [54794-74-0] MW 355.48

Fruits of *Piper nigrum* (Piperaceae).

Insecticidal.

Miyakado, M., *Agric. Biol. Chem.,* 1079, **43**, 1609.

Pisatin 1115

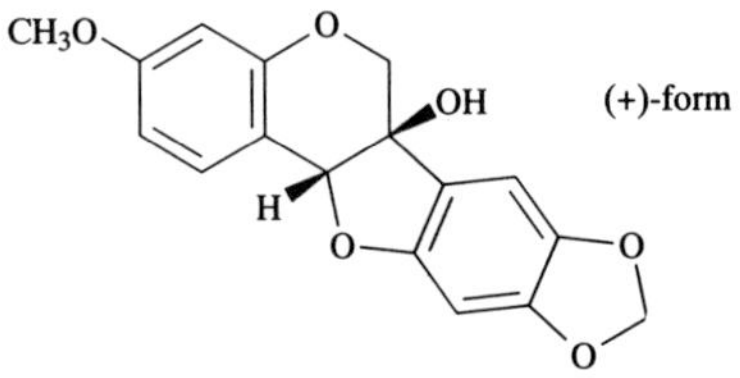

Isoflavonoid
(Pterocarpan)

$C_{17}H_{14}O_6$ [469-01-2] MW 314.29

(−)-form [20186-22-5]
(±)-form [3187-47-1]

Fungally affected leaves and pods of pea, *Pisum sativum* (Leguminosae).

Antifungal activity (phytoalexin). Causes lysis in red blood cells and inhibits respiration in rat liver mitochondria.

Perrin, D. R., *J. Am. Chem. Soc.,* 1962, **84**, 1919.

Pisiferic acid 1116

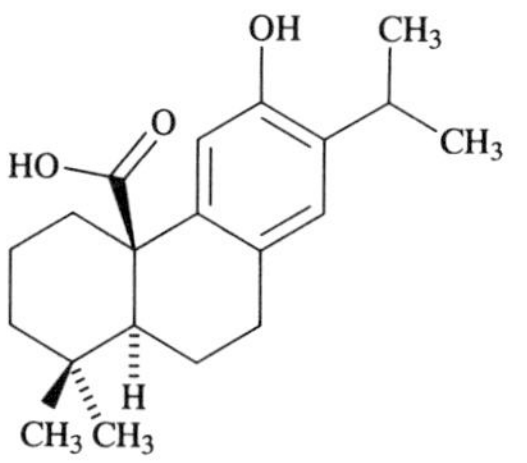

Diterpenoid
(Abietane)

$C_{20}H_{28}O_3$ [67494-15-9] MW 316.44

Chamaecyparis pisifera (Cupressaceae).

Fungitoxic to the rice pathogen *Pyricularia oryzae* at a concentration of 25 μg/ml.

Fukui, H., *Agric. Biol. Chem.,* 1978, **42**, 1419.

Pithecolobine 1117

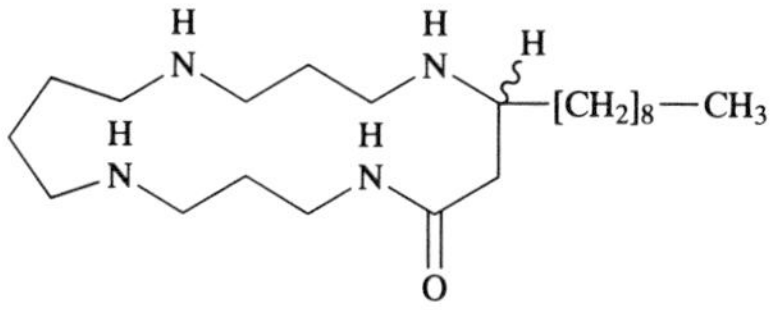

Macrocyclic alkaloid

$C_{22}H_{46}N_4O$ [22368-82-7] MW 382.63

Major component of a mixture of analagous alkaloids in *Samanea saman* (Leguminosae).

Toxic to vertebrates.

Wiesner, K., *Can. J. Chem.,* 1968, **46**, 3617.

Plagiochiline A 1118

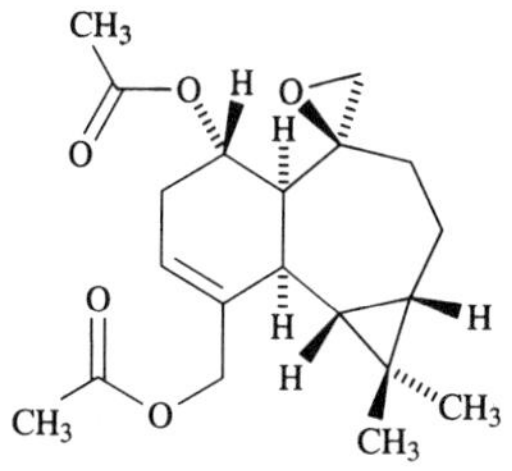

Sesquiterpenoid
(Secoaromadendrane)

$C_{19}H_{26}O_6$ [67779-73-1] MW 350.41

The liverworts *Plagiochila yokogurensis* and *P. hattoriana.*

Strongly pungent taste. Toxic to fish, causing death at a concentration of 0.4 p.p.m. within 240 min. Also an effective insect antifeedant.

Asakawa, Y., *Tetrahedron Lett.,* 1978, 1553.

Pleniradin 1119

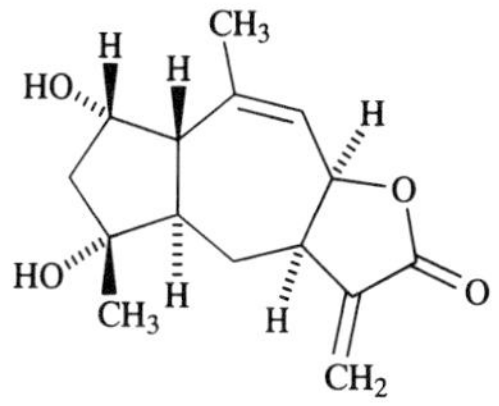

Sesquiterpenoid lactone
(Guaianolide)

$C_{15}H_{20}O_4$ [25941-24-6] MW 264.32

Baileya pleniradiata (Compositae).

Cytotoxic and antitumour properties.

Yoshitake, A., *Phytochemistry,* 1969, **8**, 1753.

Plenolin 1120

Dihydrohelenalin

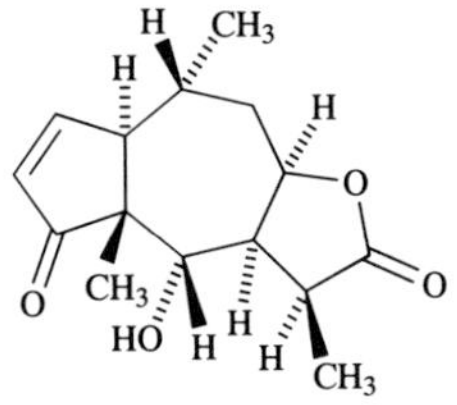

Sesquiterpenoid lactone
(Pseudoguaianolide)

$C_{15}H_{20}O_4$ [34257-95-9] MW 264.32

Baileya pleniradiata and sneezeweed, *Helenium autumnale* (Compositae).

Cytotoxic and antitumour properties.

Waddell, T. G., *Phytochemistry,* 1969, **8**, 2371.

Plicatic acid 1121

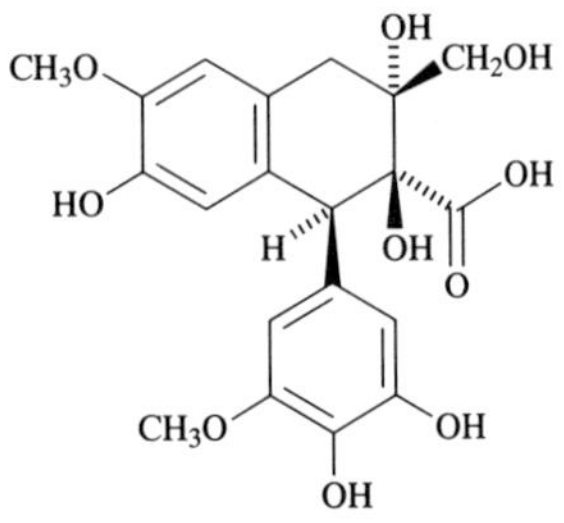

Lignan

$C_{20}H_{22}O_{10}$ [16462-65-0] MW 422.39

Sawdust of the gymnosperm tree, *Thuja plicata* (Cupressaceae).

Allergen of the sawdust, causing asthma and rhinitis in wood workers.

Gardner, J. A. F., *Can. J. Chem.,* 1966, **44**, 52.

Plumbagin 1122

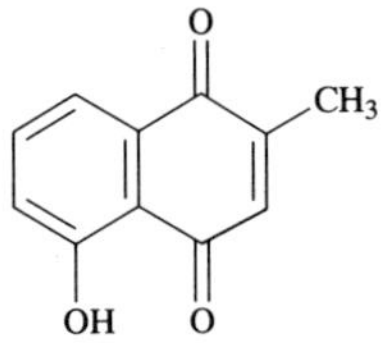

Naphthoquinone

$C_{11}H_8O_3$ [481-42-5] MW 188.18

Present in bound form in roots of *Plumbago europaea* (Plumbaginaceae) and *Dionaea muscipula* and *Drosera rotundifolia* (Droseraceae). Present in the free state in *Aristea*, *Sisyrynchium* and *Sparaxis* spp. (Iridaceae), in bark of *Diospyros* spp. (Ebenaceae) and root bark of *Pera ferruginea* (Euphorbiaceae).

Active molluscicidal agent and insect antifeedant. When rubbed on the human skin, produces a 'bruised' purple appearance. It is cytotoxic at high doses and immunostimulating at low doses. It has an irritating odour and causes sneezing.

Thomson, R. H., Naturally Occurring Quinones, 2nd edn, 1971, 228, Academic Press.

Plumericin 1123

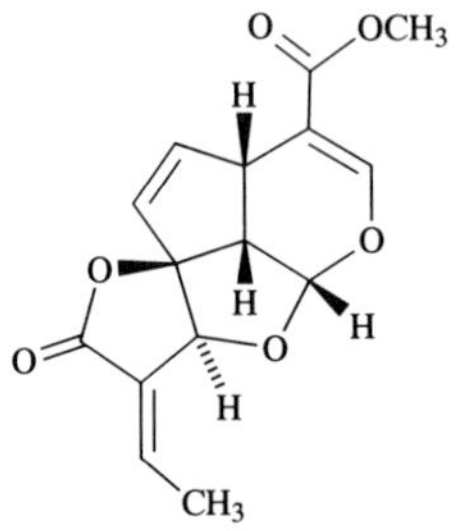

Monoterpenoid
(Iridoid lactone)

$C_{15}H_{14}O_6$ [77-16-7] MW 290.27

In the roots of *Plumeria multiflora* and *P. rubra* (Apocynaceae) and in the leaves of *Duroia hirsuta* (Rubiaceae).

Toxic to molluscs and to algae, and is also cytotoxic.

Page, J. E., *Experientia,* 1994, **50**, 840.

Podecdysone B 1124

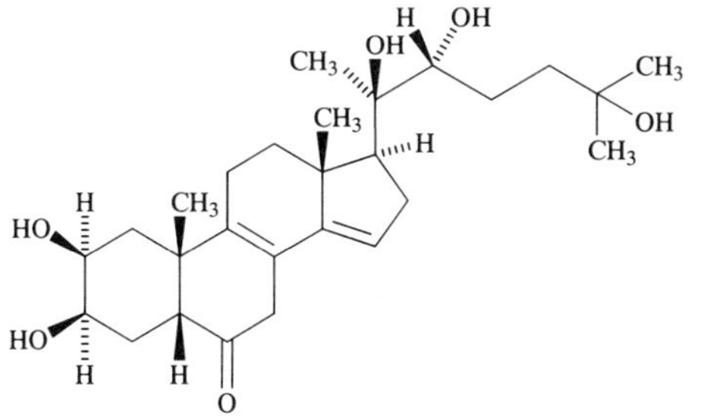

Steroid
(Cholestane)

$C_{27}H_{42}O_6$ [22612-27-7] MW 462.63

Bark of *Podocarpus elatus* (Podocarpaceae).

Phytoecdysteroid, with insect moulting hormone activity.

Galbraith, M. N., *J. Chem. Soc., Chem. Commun.,* 1969, 402.

Podolide 1125

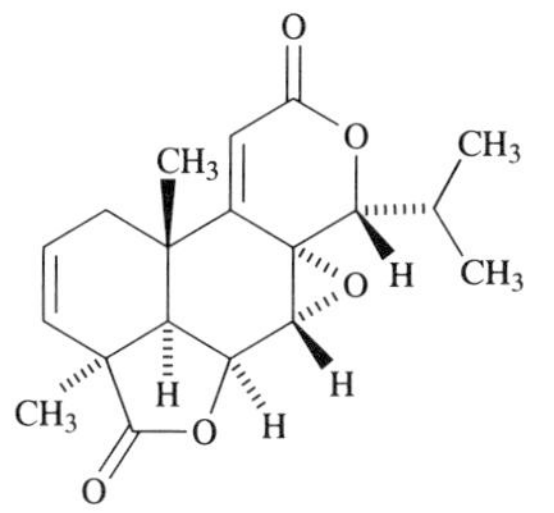

Diterpenoid
(Oxatotarane)

$C_{19}H_{22}O_5$ [55786-36-2] MW 330.38

Podocarpus gracilior (Podocarpaceae).

Antitumour properties.

Kupchan, S. M., *Experientia,* 1975, **31**, 137.

Podophyllotoxin 1126

Podophyllinic acid lactone

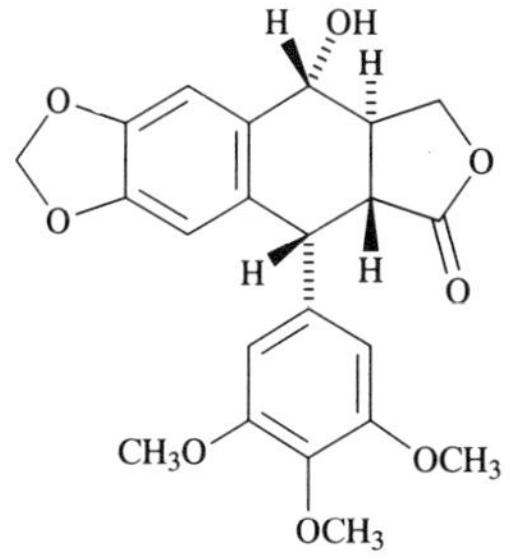

Lignan

$C_{22}H_{22}O_8$ [518-28-5] MW 414.41

Rhizomes of *Podophyllum peltatum*, *P. hexandrum* and *P. pleianthum*, roots of *Diphylleia grayi* (Podophyllaceae). Also present in needles of *Juniperus sabina*, shoots of *J. virginiana* and needles of *Callitris drummondii* (Cupressaceae).

Antitumour, antimitotic and antiviral properties. Highly toxic orally.

Petcher, T. J., *J. Chem. Soc., Perkin Trans. 2,* 1973, 288.

Podophyllotoxone 1127

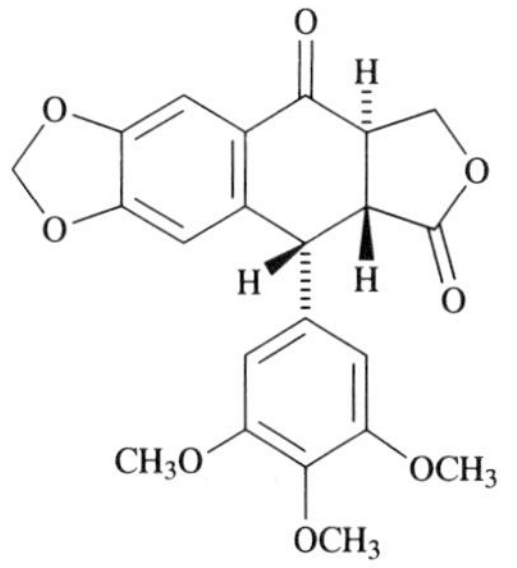

Lignan

$C_{22}H_{22}O_8$ [477-49-6] MW 414.41

Rhizomes of *Podophyllum hexandrum* and *P. peltatum* (Podophyllaceae).

Cytotoxic properties.

Gensler, W. J., *J. Am. Chem. Soc.,* 1955, **77**, 3674.

Polhovolide 1128

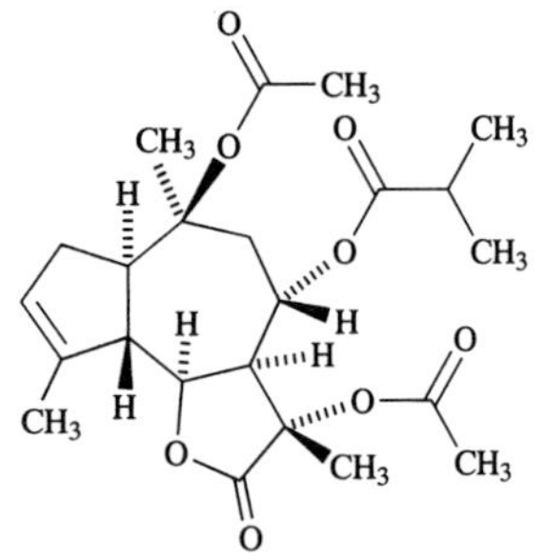

Sesquiterpenoid lactone
(Guaianolide)

$C_{23}H_{32}O_8$ [68799-88-2] MW 436.50

Root of *Laserpitium siler* (Umbelliferae).

Insect antifeedant, useful against storage pests.

Holub, M., *Coll. Czech. Chem. Commun.*, 1978, **43**, 2471.

Polycavernoside A 1129

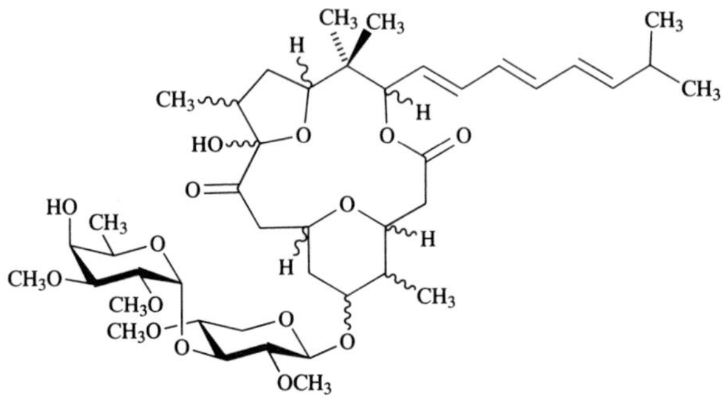

Macrolide glycoside

$C_{43}H_{68}O_{15}$ [146644-24-8] MW 825.00

Red alga, *Polycavernosa tsudai*.

One of two closely similar structures which produced an outbreak of human illness and death following the ingestion of this red alga.

Yotsu-Yamashita, M., *J. Am. Chem. Soc.*, 1993, **115**, 1147.

Polygodial 1130

Tadeonal

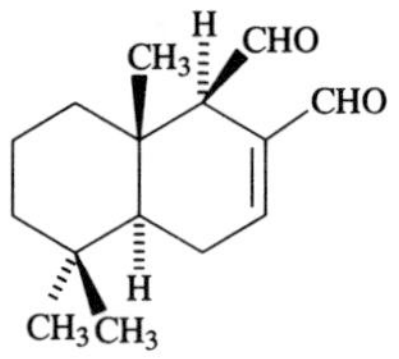

Sesquiterpenoid
(Drimane)

$C_{15}H_{22}O_2$ [6754-20-7] MW 234.34

Water pepper, *Polygonum hydropiper* (Polygonaceae), *Drymis lanceolata* (Winteraceae) and bark of *Warburgia stuhlmanii* (Canellaceae).

Sharp peppery taste. Molluscicide and insect antifeedant.

Barnes, C. S., *Aust. J. Chem.*, 1962, **15**, 322.

Polypodine B 1131

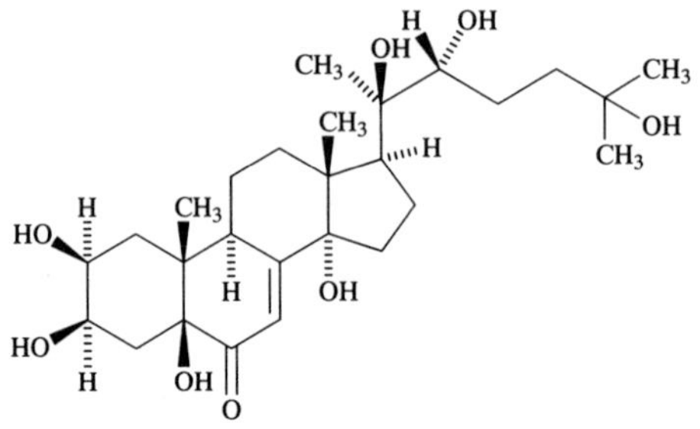

Steroid
(Cholestane)

$C_{27}H_{44}O_8$ [18069-14-2] MW 496.64

Polypodium vulgare (Polypodiaceae), *Lychnis fulgens* (Caryophyllaceae), *Ajuga reptans* (Labiatae), roots of *Pfaffia iresinoides* (Amaranthaceae) and of *Rhaponticum carthamoides* (Compositae).

Phytoecdysteroid, which disrupts the growth and development of insect larvae.

Jizba, J., *Tetrahedron Lett.*, 1967, 5139.

Ponasterone A 1132

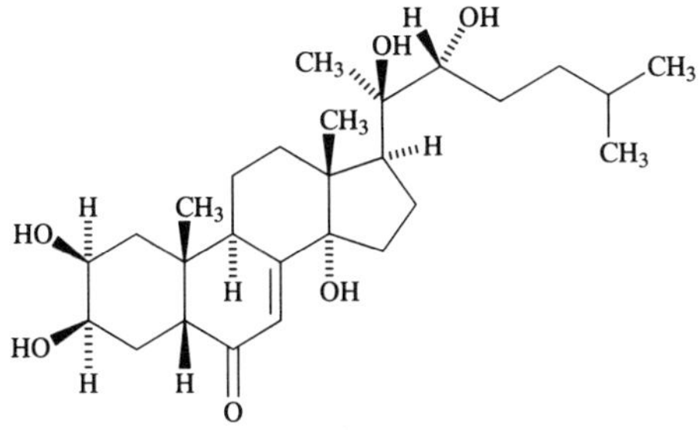

Steroid
(Cholestane)

$C_{27}H_{44}O_6$ [13408-56-5] MW 464.64

Leaves of *Podocarpus nakaii* (Podocarpaceae).

Phytoecdysteroid, with insect moulting hormone activity. Inhibits egg production in newly emerged female houseflies and greatly inhibits larval development in insects that have been tested.

Nakanashi, K., *J. Chem. Soc., Chem. Commun.,* 1966, 915.

Prangolarine 1133

Oxypeucedanin; Prangolarin

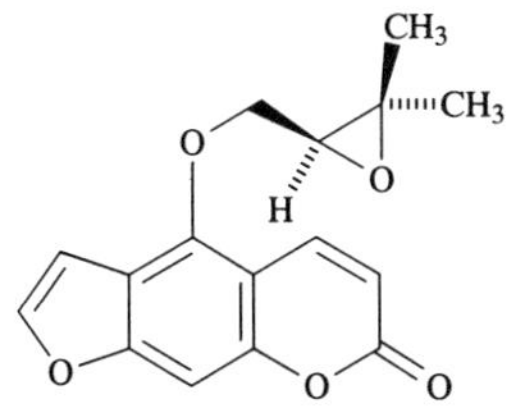

Furocoumarin

$C_{16}H_{14}O_5$ [737-52-0] MW 286.28
(+)-form [3173-02-2]
(−)-form [2609-73-6]
(±)-form [28919-33-7]

Root of *Peucedanum palustre*, *Prangos pabularia* and *Angelica archangelica*, leaf of *Diplolophium buchanani* and fruit of *Cymopterus longipes* (Umbelliferae).

Fish poison. At a concentration of 25 p.p.m. has larvicidal activity against *Aedes aegypti*.

Marston, A., *J. Nat. Prod.,* 1995, **58**, 128.

Precocene 1 1134

7-Methoxy-2,2-dimethylchromene

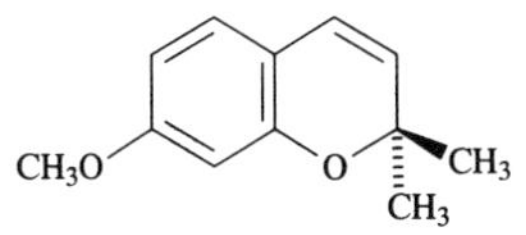

Chromene

$C_{12}H_{14}O_2$ [17598-02-6] MW 190.24

Aerial parts of *Ageratina aromatica* (Compositae).

Insecticidal activity. Acts as an antijuvenile hormone, causing precocious metamorphosis.

Bowers, W. S., *Science,* 1976, **193**, 542.

Precocene 2 1135

Ageratochromene; 6,7-Dimethoxy-2,2-dimethylchromene

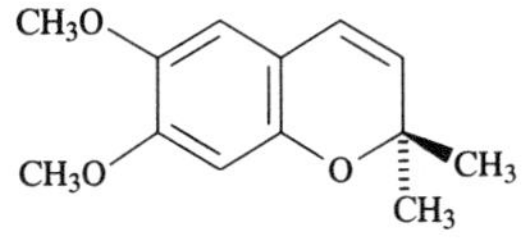

Chromene

$C_{13}H_{16}O_3$ [644-06-4] MW 220.27

Ageratum houstonianum, *Ageratina aromatica* and *Senecio longifolius* (Compositae).

Insecticidal, acting as an antijuvenile hormone and causing precocious metamorphosis.

Bowers, W. S., *Science,* 1976, **193**, 542.

Prenylbenzoquinone 1136

Prenyl-1,4-benzoquinone

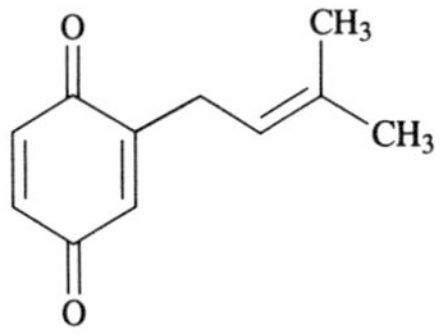

Benzoquinone

$C_{11}H_{12}O_2$ [5594-02-5] MW 176.22

Fruit, leaf and stem of *Phagnalon sordidum* (Compositae).

The contact allergen of *Phagnalon* species, producing painful skin rashes.

Howard, B. M., *Tetrahedron Lett.,* 1979, 4449.

Prenyl caffeate 1137

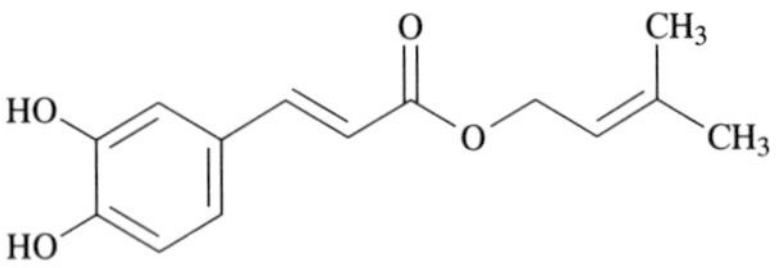

Phenylpropenoid

$C_{14}H_{16}O_4$ [118971-61-2] MW 248.28

Buds of *Populus* spp. (Salicaceae) and used by bees in the manufacture of propolis.

One of the major contact allergens in bee propolis.

Hashimoto, T., *Z. Naturforsch.,* 1988, **43C**, 470.

5′-Prenylhomoeriodictyol 1138

Sigmoidin B 4′-methyl ether

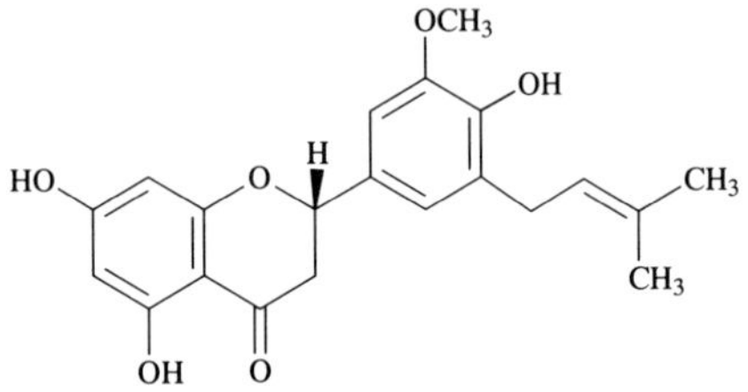

Flavanone

$C_{21}H_{22}O_6$ [114340-00-1] MW 370.40

Stem bark of *Erythrina berteroana* (Leguminosae).

Antifungal activity against *Cladosporium cucumerinum.*

Maillard, M., *Planta Med.,* 1989, **55**, 281.

Pretazettine 1139

Isotazottine

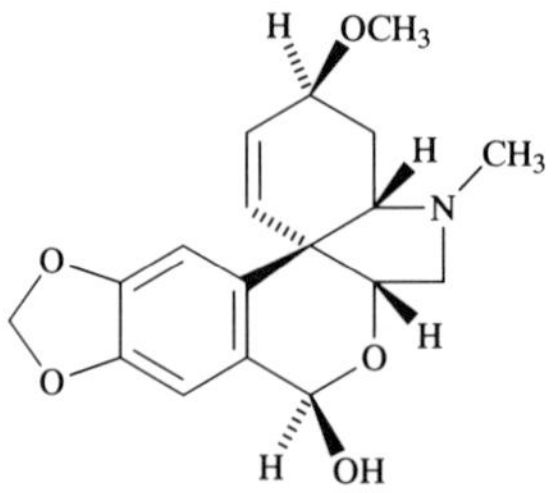

Amaryllidaceae alkaloid
(Tazettine)

$C_{18}H_{21}NO_5$ [17322-84-8] MW 331.37

Leucojum aestivum, *Narcissus tazetta*, *Pancratium biflorum*, *Lycoris radiata* and *Zephyranthus carinatus* (Amaryllidaceae).

Antitumour activity, inhibiting HeLa cell growth as well as protein synthesis.

Wildman, W. C., *J. Am. Chem. Soc.,* 1967, **89**, 5514.

Primetin 1140

5,8-Dihydroxyflavone

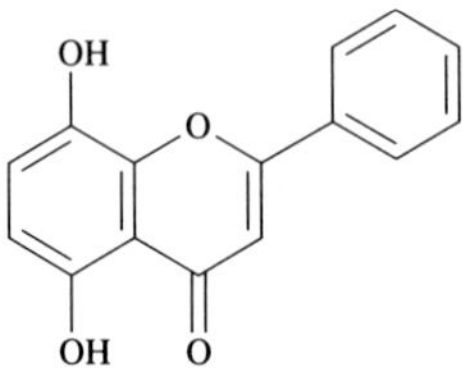

Flavone

$C_{15}H_{10}O_4$ [548-58-3] MW 254.24

External farina on leaves and stalks of *Primula modesta* and *P. mistassinica* (Primulaceae).

Contact allergen.

Harborne, J. B., *Phytochemistry,* 1971, **10**, 472.

Primin 1141

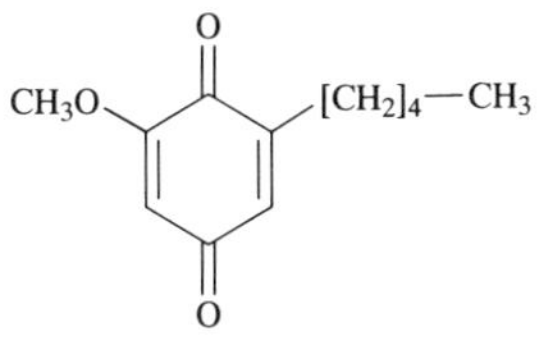

Benzoquinone

$C_{12}H_6O_3$ [15121-94-5] MW 198.18

Glandular hairs on leaves of *Primula obconica*, *P. elatior*, *Anagallis hirtella*, *Dionysia aretioides* and *Glaux maxima* (Primulaceae). Also present in root bark of *Miconia* spp. (Melastomataceae).

Responsible for the skin irritation produced by handling *Primula obconica* plants. Active as a molluscicide and as an insect feeding deterrent.

Marini-Bettolo, G. B., *Gazz. Chim. Ital.,* 1971, **101**, 41.

Pristimerin 1142

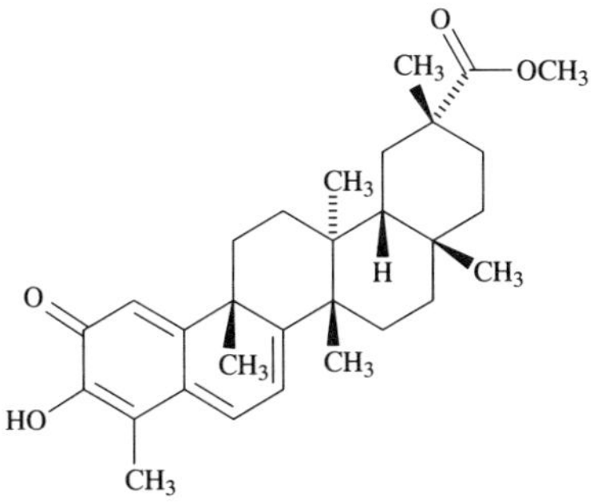

Triterpenoid
(Friedelane)

$C_{30}H_{40}O_4$ [1258-84-0] MW 464.65

Roots of *Pristimera indica*, *Catha edulis*, *Schaefferia cuneifolia* and *Maytenus* spp. (Celastraceae).

Toxin, but also potent antitumour properties.

Harada, R., *Tetrahedron Lett.,* 1962, 603.

Proacaciberin 1143

Proacacipetalin 6′-arabinoside

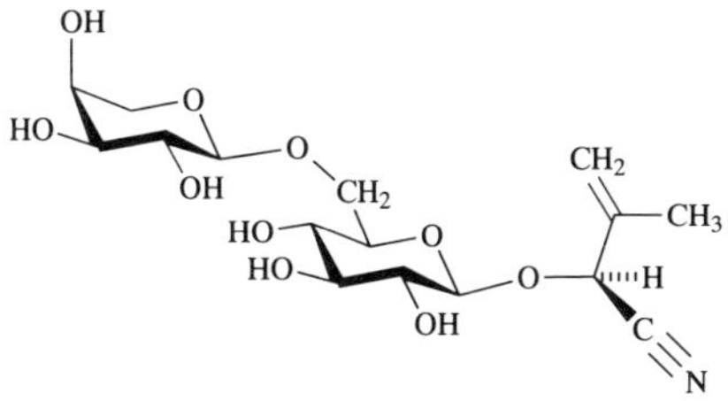

Cyanogenic glycoside

$C_{16}H_{25}NO_{10}$ [79197-21-0] MW 391.38

Pods of *Acacia sieberiana* (Leguminosae), where it co-occurs with proacacipetalin (q.v.).

Bound toxin, releasing cyanide on hydrolysis.

Nartley, F., *Phytochemistry,* 1981, **20**, 1311.

Proacacipetalin 1144

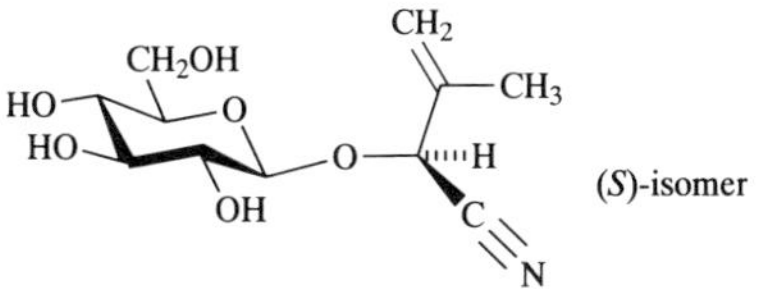

(*S*)-isomer

Cyanogenic glycoside

$C_{11}H_{17}NO_6$ [66871-89-4] MW 258.26

Many of the *Acacia* spp. (Leguminosae). The epimer (*R*)-epiproacacipetalin, [66871-88-3], occurs in *Acacia globulifera*.

Bound toxin, releasing cyanide poison on hydrolysis.

Butterfield, C. S., *Phytochemistry,* 1975, **14**, 993.

Progoitrin 1145

Glucorapiferin; (2*R*)-2-Hydroxybut-3-enylglucosinolate

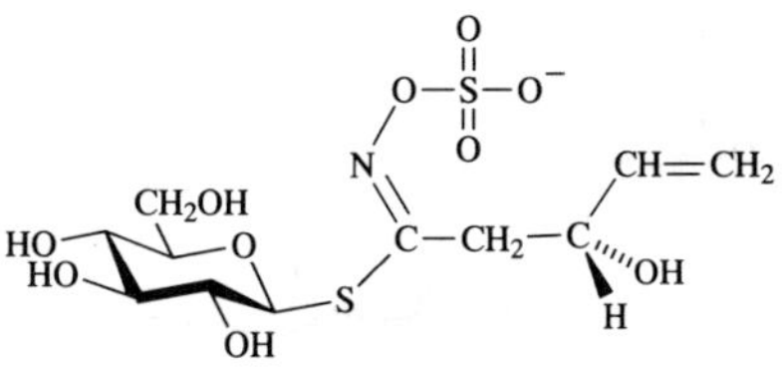

Glucosinolate

$[C_{11}H_{18}NO_{10}S_2]^-$ [585-95-5] MW 388.40

Present in most cultivated brassica crops, including *Brassica napus*, brown mustard, *B. juncea* and cabbage, *B. oleracea* var. *capitata* (Cruciferae). The epimer, epigoitrin, [1072-93-1], occurs in *Crambe abyssinica* (Cruciferae).

Flavour component. It is the precursor of (*R*)-5-vinyl-2-oxazolidinethione, 'goitrin', [500-12-9], which is responsible for the bitter taste of some frozen *Brassica* vegetables and which is goitrogenic. Both goitrin and epigoitrin are toxic in animals.

Greer, M. A., *Arch. Biochem. Biophys.*, 1962, **99**, 369.

Propanethial *S*-oxide 1146

Thiopropanal *S*-oxide

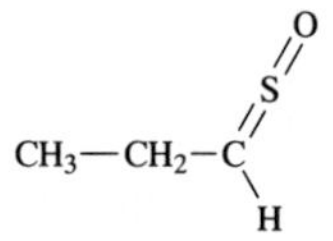

Sulfur compound

C_3H_6OS [32157-29-2] MW 90.15

Onion, *Allium cepa* (Alliaceae).

Major lachrymatory agent of onions.

Wilkens, W. F., *Chem. Abstr.*, 1964, **61**, 9771

Prostratin 1147

12-Deoxyphorbol 13-acetate; Stillingia factor S_7

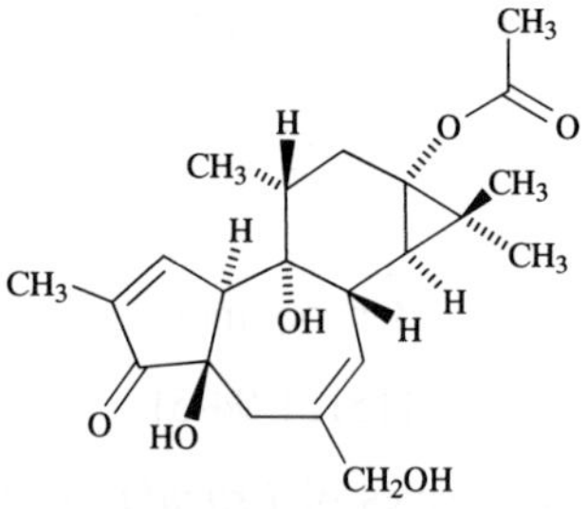

Diterpenoid
(Tigliane)

$C_{22}H_{30}O_6$ [60857-08-1] MW 390.48

In stemwood of *Homalanthus nutans* (Euphorbiaceae) and *Pimelea prostrata* (Thymelaeaceae).

Toxic; cytostatic and antiviral agent, active *in vitro* against the HIV-1 virus. Unlike most other phorbol esters isolated from the Euphorbiaceae, it shows no tumour-promoting activity.

McCormick, I. R. N., *Tetrahedron Lett.*, 1976, 1735.

Proteacin 1148

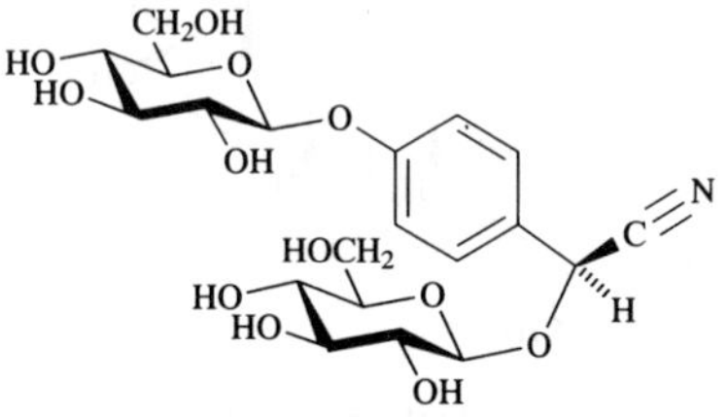

Cyanogenic glycoside

$C_{20}H_{27}NO_{12}$ [22660-96-4] MW 473.43

Seed kernels of bitter Macadamia nuts, *Macadamia ternifolia* (Proteaceae).

Bound toxin, releasing poisonous cyanide on hydrolysis.

Swenson, W. K., *Phytochemistry*, 1989, **28**, 821.

Protoveratrine A 1149

Protoalba

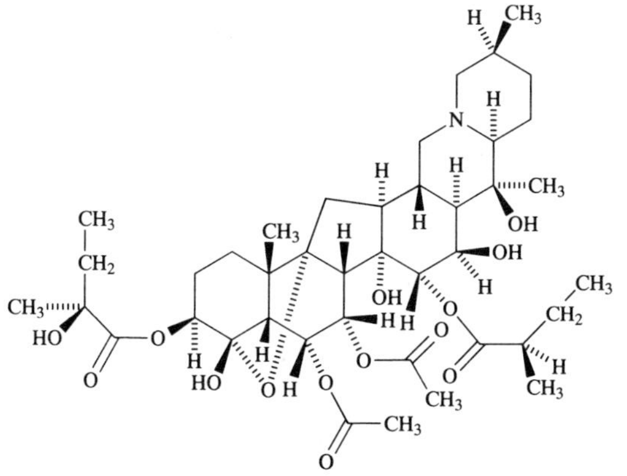

Steroidal alkaloid
(Cevane)

$C_{41}H_{63}NO_{14}$ [143-57-7] MW 793.95

Root and leaf of white hellebore, *Veratrum alba* (Liliaceae).

Highly toxic alkaloid. The lethal dose in man is about 20 mg, corresponding to 1-2 g dried root. It causes burning in the mouth, numbness and then vomiting and diarrhoea, followed by respiratory difficulties. The plant can be mistaken for gentian, sometimes with fatal consequences.

Kupchan, S. M., *J. Am. Chem. Soc.,* 1960, **82**, 2252.

Protoveratrine B 1150

Neoprotoveratrine; Veratetrine

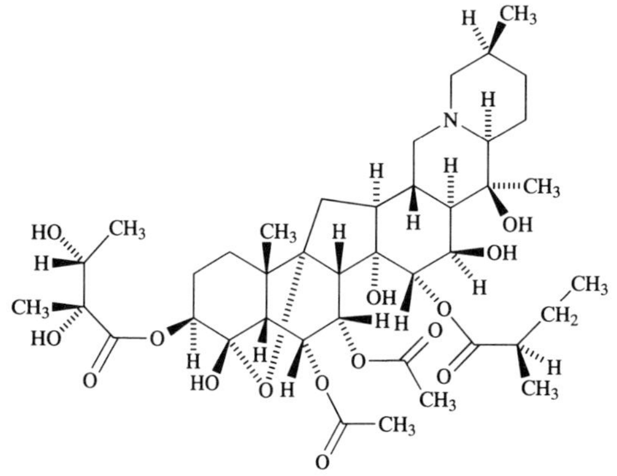

Steroidal alkaloid
(Cevane)

$C_{41}H_{63}NO_{15}$ [124-97-0] MW 809.95

Root and leaf of white hellebore, *Veratrum alba* (Liliaceae), together with protoveratrine A (q.v.).

Highly toxic alkaloid. The lethal dose in man is about 20 mg, corresponding to 1-2 g dried roots. Symptoms of poisoning are described under protoveratrine A and both alkaloids are similar in their pharmacological properties.

Kupchan, S. M., *J. Am. Chem. Soc.,* 1960, **82**, 2252.

Provincialin 1151

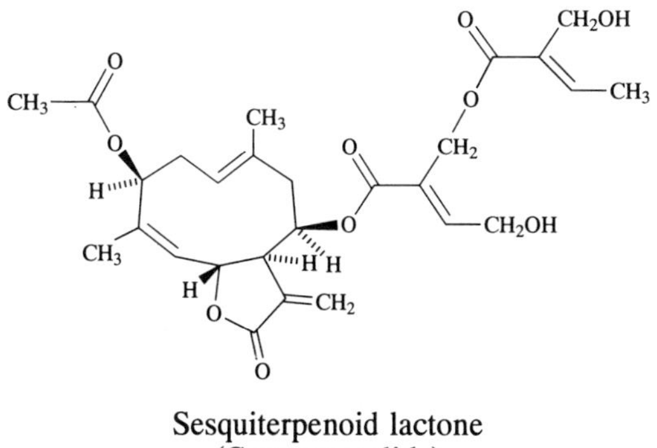

Sesquiterpenoid lactone
(Germacranolide)

$C_{27}H_{34}O_{10}$ [40328-96-9] MW 518.56

Aerial parts of blazing star, *Liatris provincialis* (Compositae).

Cytotoxic and antitumour properties.

Herz, W., *J. Org. Chem.,* 1973, **38**, 2485.

Prunasin 1152

Mandelonitrile glucoside

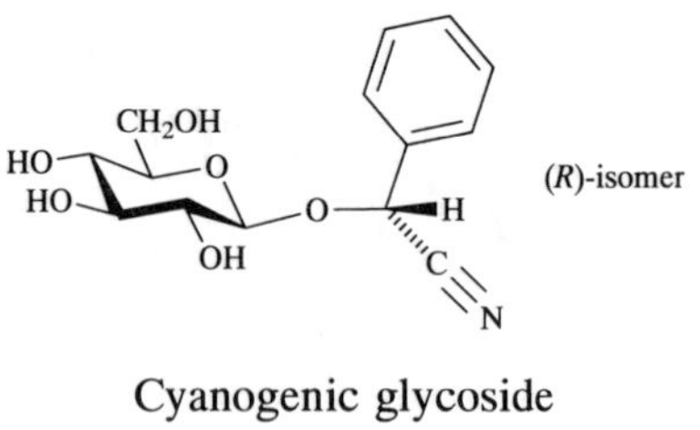

(*R*)-isomer

Cyanogenic glycoside

$C_{14}H_{17}NO_6$ [99-18-3] MW 295.29

Major sources are leaves and barks of cherry, *Prunus* spp. (Rosaceae). Also present in several other families, notably *Eucalyptus cladocalyx* (Myrtaceae) and *Holocalyx glaziovii* (=*H. balansae*) (Leguminosae). The epimer (*S*)-sambunigrin [99-19-4] occurs in elderberry, *Sambucus nigra* (Caprifoliaceae), *Acacia glaucescens* (Leguminosae) and *Ximenia americana* (Olacaceae).

Bound toxin, releasing poisonous cyanide on hydrolysis. Lethal dose in guinea pigs is 0.25 mg/kg body-weight orally. Livestock poisoning has been recorded, following grazing on *Holocalyx* in Brazil and on *Eucalyptus* in Australia.

Kofod, H., *Tetrahedron Lett.,* 1966, 1289.

Pseudaconitine 1153

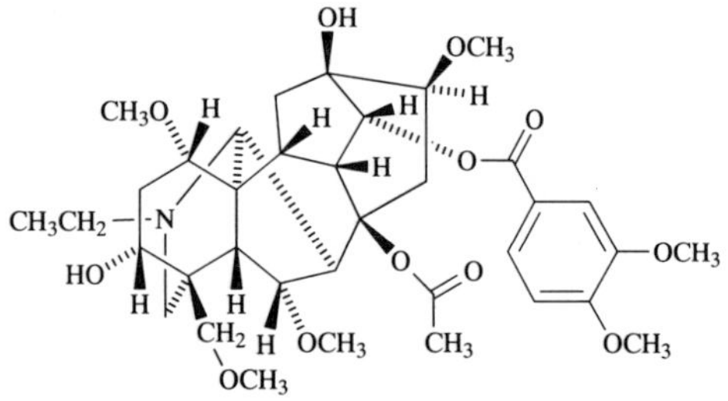

Diterpenoid alkaloid
(Aconitane)

$C_{36}H_{51}NO_{12}$ [127-29-7] MW 689.80

Tubers of *Aconitum ferox, A. falconeri* and *A. spictatum* (Ranunculaceae).

About twice as toxic in small animals and birds as aconitine (q.v.). It is more effective as a respiratory depressant than aconitine, but similar in its cardiovascular potency.

Klasek, A., *Lloydia,* 1972, **35**, 55.

Pseudoconhydrine 1154

ψ-Conhydrine; 5-Hydroxy-2-propylpiperidine

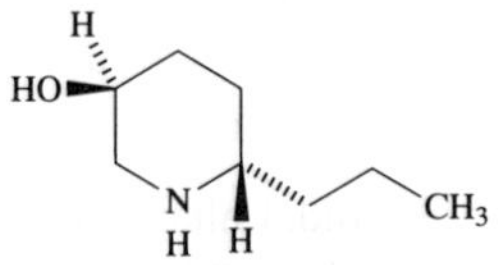

Piperidine alkaloid

$C_8H_{17}NO$ [144-55-6] MW 143.23

Minor component of hemlock, *Conium maculatum* (Umbelliferae).

Contributing, with coniine and γ-coniceine (q.v.), to the toxic properties of the hemlock plant.

Yanai, H. S., *Tetrahedron,* 1959, **6**, 103.

Pseudopurpurin 1155

ψ-Purpurin; Purpurin-3-carboxylic acid

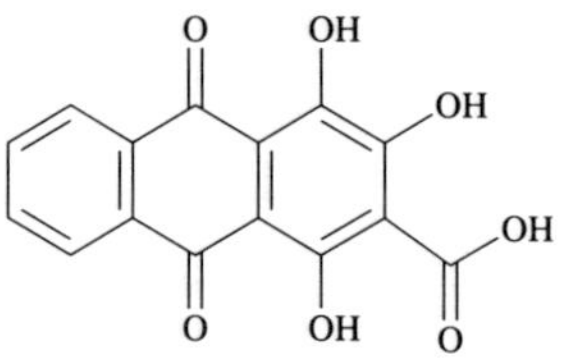

Anthraquinone

$C_{15}H_8O_7$ [476-41-5] MW 300.22

Roots of *Rubia tinctorum* and various *Galium* and *Asperula* spp. (Rubiaceae).

Genotoxic in the hamster fibroblast-mutagenicity assay.

Thomson, R. H., Naturally Occurring Quinones, 2nd edn, 1971, 408, Academic Press.

Pseudotropine 1156

ψ-Tropine; 3-Pseudotropanol; 3β-Tropanol

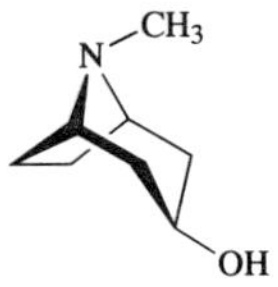

Tropane alkaloid

$C_8H_{15}NO$ [135-97-7] MW 141.21

Aerial parts of field bindweed, *Convolvulus arvensis* (Convolvulaceae), in *Scopolia carniolica* and other members of the Solanaceae.

Probable cause of poisoning in horses feeding on pastures containing field bindweed. The symptoms are intestinal fibrosis and vascular sclerosis of the small intestine. Toxic effects have been observed in mice fed bindweed. The LD_{50} of pseudotropine is 164 mg/kg body-weight in mice, when administered intravenously, but the oral toxicity may be more considerable.

Todd, F. G., *Phytochemistry,* 1995, **39**, 301.

Psoralen 1157

Ficusin

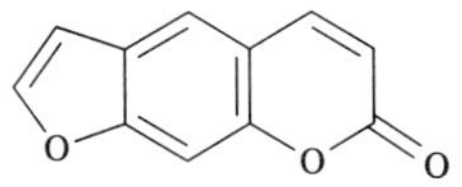

Furanocoumarin

$C_{11}H_6O_3$ [66-97-7] MW 186.17

Seeds of *Psoralea* spp. and *Coronilla glauca* (Leguminosae), essential oil of *Phebalium argenteum* (Rutaceae), in the fig, *Ficus carica* (Moraceae) and commonly present in root, leaf and seed in members of the Umbelliferae, including parsnip, and celery.

Phototoxin. In the presence of ultraviolet light, it causes impairment of DNA synthesis. Also is a photocarcinogen.

Murray, R. D. H., The Natural Coumarins, 1982, 176, Wiley.

Psorospermin 1158

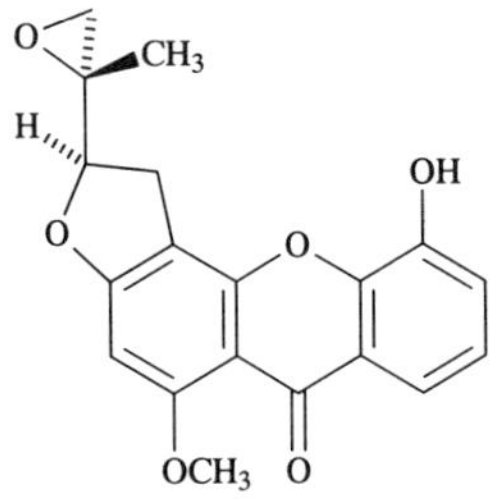

Xanthone

$C_{19}H_{16}O_6$ [74045-97-9] MW 340.33

Root of *Psorospermum febrifugum* (Guttiferae).

Antitumour and antileukaemic properties.

Kupchan, S. M., *J. Nat. Prod.,* 1980, **43**, 296.

Ptaeroglycol 1159

Cneorum chromone F; Pteroglycol

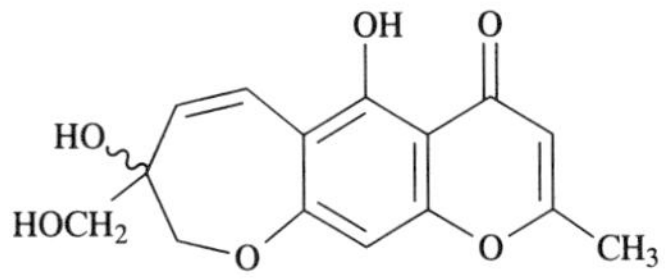

Chromone

$C_{15}H_{14}O_6$ [18836-12-9] MW 290.27

Heartwood of *Cedrelopsis grevei* and *Ptaeroxylon obliquum* (Ptaeroxylaceae) and in *Cneorum tricoccum* and *C. pulverulentum* (Cneoraceae).

Cytotoxic against HeLa cells.

Dean, F. M., *Tetrahedron Lett.,* 1967, 3459.

Ptaquiloside 1160

Aquilide A

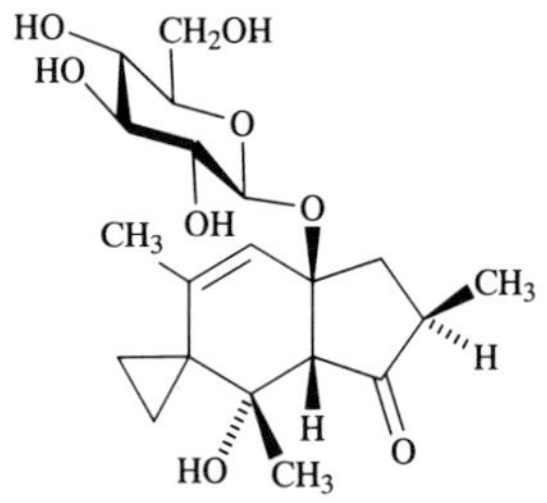

Sesquiterpenoid
(Norilludane)

$C_{20}H_{30}O_8$ [87625-62-5] MW 398.45

In bracken fern, *Pteridium aquilinum* (Pteridaceae). Also present in several other ferns including *Pteris cretica*, *Histiopteris incisa*, *Hypolepsis punctata*, *Dennstaedtia hirsuta*, *Pteridium esculentum* and *Cheilanthes sieberi* (all Pteridaceae). Bracken fronds (dry-weight) contain from 0.02 to 0.16% of ptaquiloside, while the rhizomes, (also dry-weight), have from 0.03 to 0.12%.

Potent carcinogen. Cause of death in cattle which have been browsing on bracken. Although young fronds of bracken are widely eaten in Japan, human poisoning is unlikely. The content in young fronds is very low and heat treatment causes slow decomposition.

Saito, K., *Phytochemistry,* 1990, **29**, 1475.

Pterosterone 1161

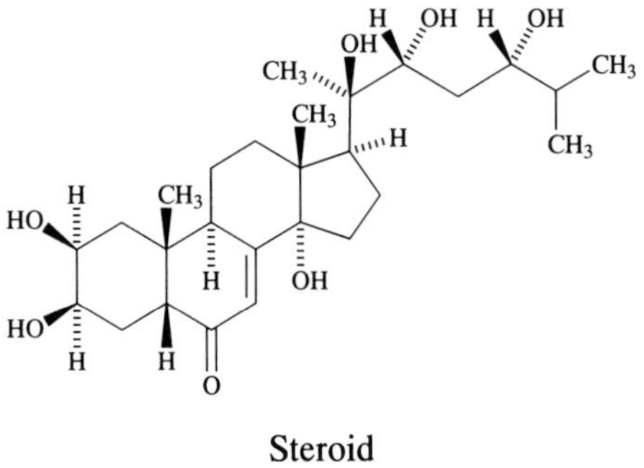

Steroid
(Cholestane)

$C_{27}H_{44}O_7$ [18089-44-6] MW 480.64

Roots of *Pfaffia iresinoides* (Amaranthaceae) and seeds of *Diploclisia glaucescens* (Menispermaceae).

Phytoecdysteroid, with insect moulting hormone activity, disrupting metamorphosis.

Takemoto, T., *Tetrahedron Lett.,* 1968, 375.

Pterostilbene 1162

Resveratrol 3,5-dimethyl ether;
4′-Hydroxy-3,5-dimethoxystilbene

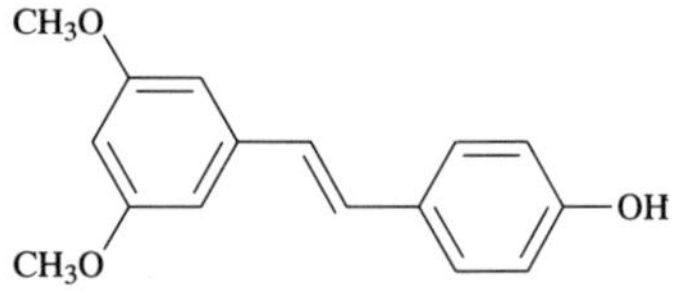

Stilbenoid

$C_{16}H_{16}O_3$ [537-42-8] MW 256.30

In vine leaves, *Vitis vinifera* (Vitaceae) infected with downy mildew, *Plasmopara viticola*. Also present constitutively in sandalwood, *Pterocarpus santalinus* (Leguminosae).

Antifungal agent. As a phytoalexin of the grape vine, it is a product in smaller amounts than ε-viniferin (q.v.) but its antifungal activity is much higher.

Langcake, P., *Phytochemistry,* 1979, **18**, 1025.

Pulegone 1163

β-Pulegone

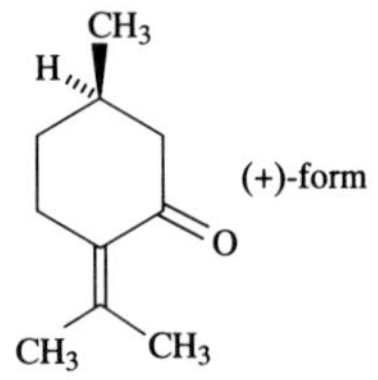

(+)-form

Monoterpenoid
(Menthane)

$C_{10}H_{16}O$ [89-82-7] MW 152.24
(−)-form [3391-90-0]

The main constituent of oils of pulegium and hedeoma, distilled from the leaves and flowering heads of *Mentha pulegium* and *Hedeoma pulegioides* respectively (Labiatae). Also occurs in other *Mentha* spp. oils.

Induces abortion in animals and in man. The use of pennyroyal tea, prepared from the leaves of *Mentha pulegium*, is preferable, since the pure pulegone is relatively toxic.

Simonsen, J. L., The Terpenes, 2nd edn, 1947, **1**, 370, Cambridge University Press.

Pulverochromenol 1164

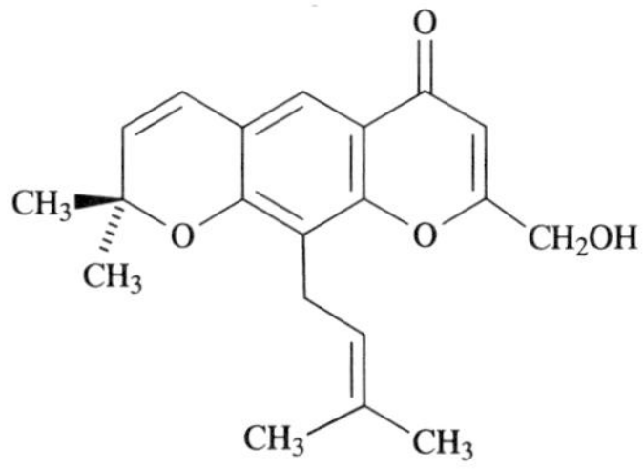

Chromone

$C_{20}H_{22}O_4$ [85394-12-3] MW 326.39

Cneorum tricoccum and *C. pulverulentum* (Cneoraceae).

Inhibitory to HeLa cells.

Gonzalez, A. G., *Planta Med.,* 1983, **47**, 56.

Punicalagin 1165

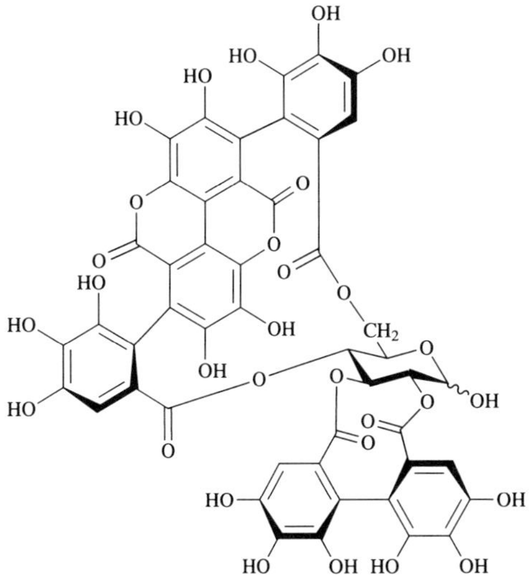

Hydrolysable tannin

$C_{48}H_{28}O_{30}$ [65995-63-3] MW 1084.73

In leaves of yellow-wood, *Terminalia oblongata* (Combretaceae) and of the pomegranate, *Punica granatum* (Punicaceae).

Hepatotoxic agent of *Terminalia* leaves, affecting browsing ruminants in Australia. Both cattle and sheep are poisoned and die after feeding on this plant. A second tannin, terminalin (q.v.), is also involved as a kidney toxin.

Mayer, W., *Justus Liebigs Ann. Chem.,* 1977, 1976.

Purothionin A-I 1166

β-Purothionin

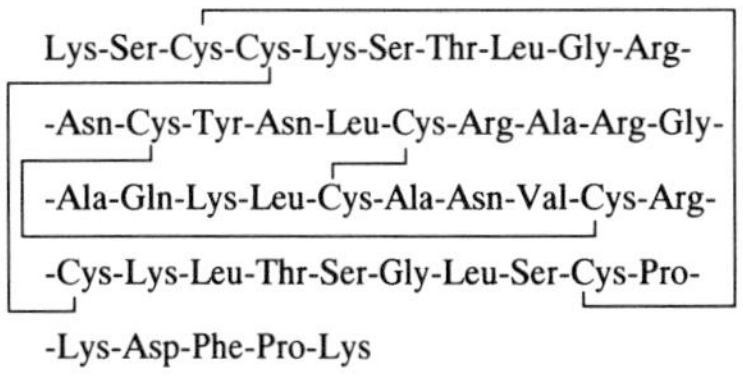

Lys-Ser-Cys-Cys-Lys-Ser-Thr-Leu-Gly-Arg-
-Asn-Cys-Tyr-Asn-Leu-Cys-Arg-Ala-Arg-Gly-
-Ala-Gln-Lys-Leu-Cys-Ala-Asn-Val-Cys-Arg-
-Cys-Lys-Leu-Thr-Ser-Gly-Leu-Ser-Cys-Pro-
-Lys-Asp-Phe-Pro-Lys

Protein

[58239-10-4] MW 4927

One of several closely related proteins present in wheat flour, prepared from hexaploid, tetraploid or diploid *Triticum* spp. (Gramineae). Contains 45 amino acid residues.

Toxic protein. It binds to cell membranes and inhibits sugar incorporation.

Ohtani, S., *J. Biochem. (Tokyo),* 1977, **82**, 753.

Purpurin 1167

Madder purple

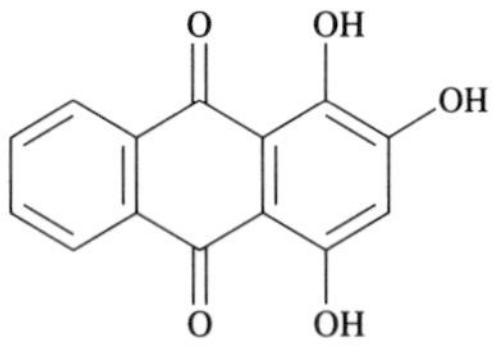

Anthraquinone

$C_{14}H_8O_5$ [81-54-9] MW 256.21

Roots of the madder plant, *Rubia tinctorum*, of *Asperula odorata* and *Relbunium hypocarpum* (Rubiaceae).

Genotoxic in the hamster fibroblast-mutagenicity assay. One of the pigments of the madder plant.

Thomson, R. H., Naturally Occurring Quinones, 2nd edn, 1971, 407, Academic Press.

Pyrethrin I 1168

Chrysanthemum monocarboxylic acid pyrethrolone ester

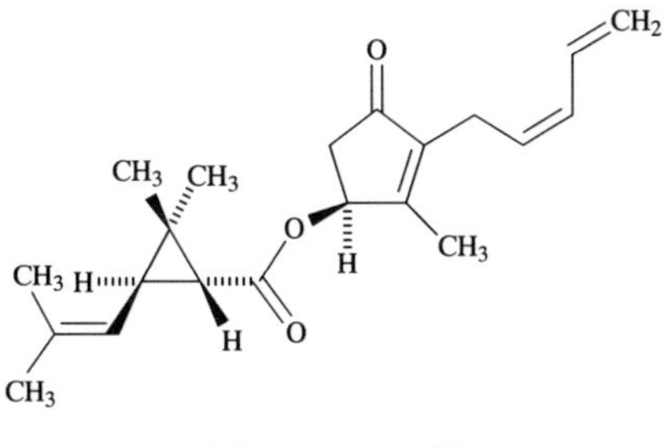

Monoterpenoid

$C_{21}H_{28}O_3$ [121-21-1] MW 328.45

Flowers of pyrethrum, *Tanacetum cinerariifolium* (Compositae).

Widely used as an insecticide. Toxic in mammals with an LD_{50} orally in rats of 1.2 g/kg body-weight. Can cause severe allergic dermatitis.

Crombie, L., *Pestic. Sci.,* 1980, **11**, 102.

Pyrethrin II 1169

Chrysanthemum dicarboxylic acid monomethyl ester pyrethrolone ester

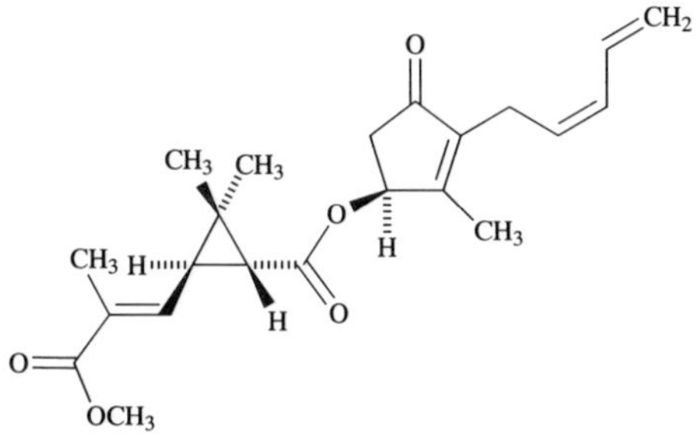

Monoterpenoid

$C_{22}H_{28}O_5$ [121-29-9] MW 372.46

Together with pyrethrin I, in flowers of pyrethrum, *Tanacetum cinerariifolium* (Compositae).

Widely used as an insecticide. Toxic in mammals with an LD_{50} orally in rats of 1.2 g/kg body-weight. Can cause severe allergic dermatitis.

Crombie, L., *Pestic. Sci.,* 1980, **11**, 102.

Pyrethrosin 1170

Chrysanthin

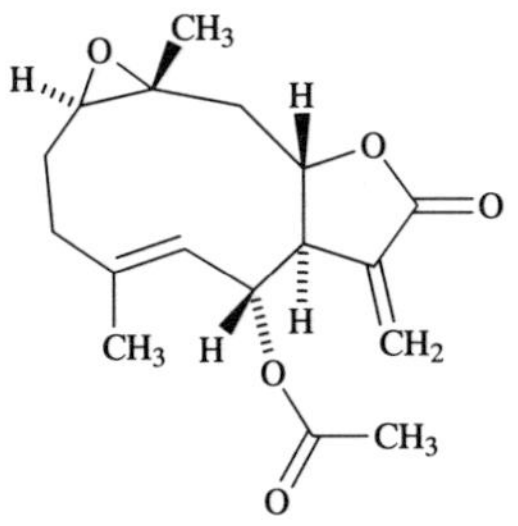

Sesquiterpenoid lactone
(Germacranolide)

$C_{17}H_{22}O_5$ [28272-18-6] MW 306.36

Flowers of pyrethrum, *Tanacetum cinerariifolium* and in *Anthemis cupaniana* (Compositae).

Molluscicide. Causes allergic contact dermatitis in humans.

Gabe, E. J., *J. Chem. Soc., Chem. Commun.,* 1971, 559.

Pyrogallol 1171

Pyrogallic acid; 1,2,3-Trihydroxybenzene

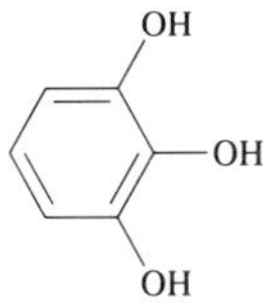

Phenol

$C_6H_6O_3$ [87-66-1] MW 126.11

Fruit of *Ceratonia siliqua*, root of *Statice gmelinii*, leaf of *Phyllanthus reticulatus* (Euphorbiaceae) and leaf of *Rosa* spp. (Rosaceae).

Toxic, when present in any quantity. Easily absorbed through the skin. Causes kidney and liver damage.

Tanenbaum, S. W., *Biochim. Biophys. Acta,* 1958, **28**, 21.

Pyrrole-3-carbamidine 1172

Brunfelsamidine

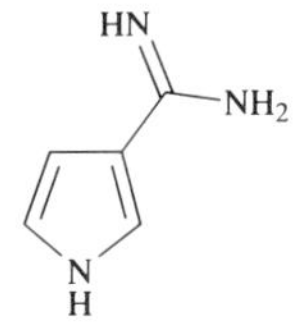

Pyrrole alkaloid

$C_5H_7N_3$ [97744-98-4] MW 109.13

Whole plant of *Nierembergia hippomanica* and root bark of *Brunfelsia grandiflora* var. *schultesii* (Solanaceae).

Lethal principle of *Nierembergia*, which is poisonous to cattle, sheep, goats, horses and rabbits. Death is preceded by symptoms of diarrhoea, midriasis, locomotor ataxia, weakened heart action and strong convulsions.

Buschi, C. A., *Phytochemistry,* 1987, **26**, 863.

α-Pyrufuran 1173

1,3,4-Trimethoxy-2-dibenzofuranol

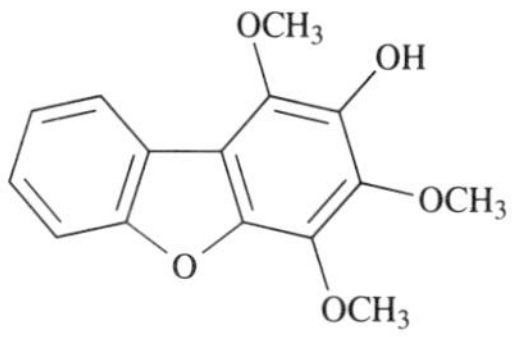

Dibenzofuran

$C_{15}H_{14}O_5$ [88256-05-7] MW 274.27

Sapwood of an infected pear tree, *Pyrus communis* (Rosaceae).

Antifungal agent (phytoalexin).

Kemp, M. S., *J. Chem. Soc., Perkin Trans. 1,* 1983, 2267.

Q

Quassimarin 1174

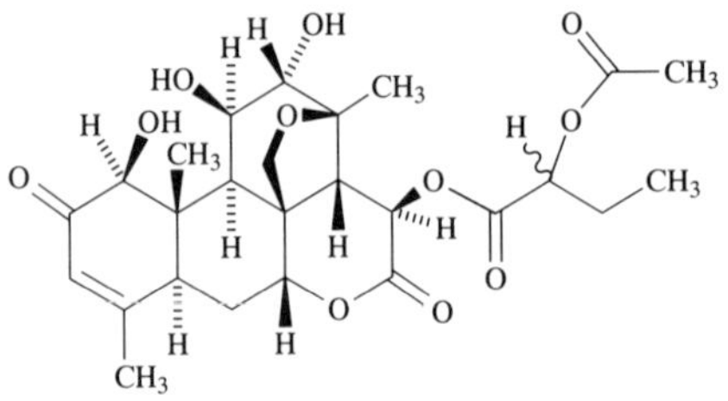

Nortriterpenoid
(Quassane)

$C_{27}H_{36}O_{11}$ [59938-97-5] MW 536.58

Surinam quassia wood, *Quassia amara* (Simaroubaceae).

Antileukaemic properties, with good *in vivo* activity.

Kupchan, S. M., *J. Org. Chem.,* 1976, **41**, 3481.

Quassin 1175

Nigakilactone D; Quassiin

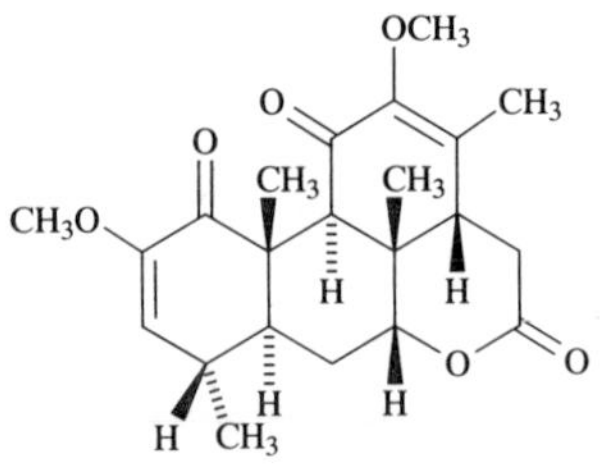

Nortriterpenoid
(Quassane)

$C_{22}H_{28}O_6$ [76-78-8] MW 388.46

Surinam quassia wood, *Quassia amara* and in wood of several *Picrasma* spp. (Simaroubaceae).

Very bitter tasting, with a bitterness threshold of 1:60000. The wood is used as a febrifuge, insecticide, vermicide and as a bitter tonic. Quassin has antifertility activity, inhibiting testosterone secretion of rat Leydig cells.

Valenta, Z., *Tetrahedron,* 1962, **18**, 1433.

Quinine 1176

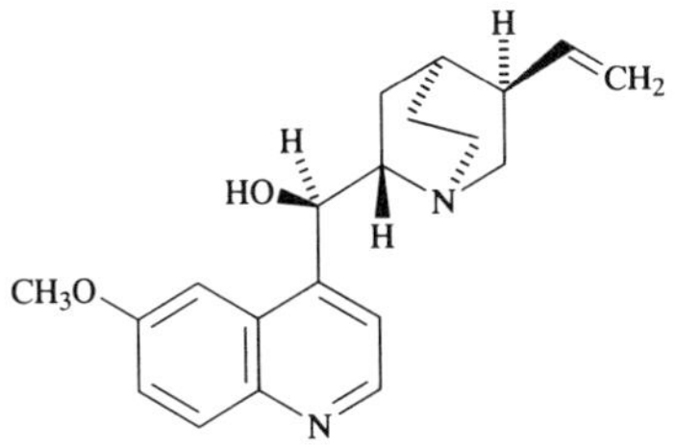

Quinoline alkaloid
(Cinchonan)

$C_{20}H_{24}N_2O_2$ [130-95-0] MW 324.42

Bark of *Cinchona officinalis* and other *Cinchona* spp., also in *Remijia pedunculata* (Rubiaceae). It co-occurs with its stereoisomer quinidine [56-54-2].

One of the bitterest substances known, with significant bitterness to humans at a molar concentration of 1×10^{-5}. Well known as an effective antimalarial drug used against *Plasmodium falcicarpum*. It has a relatively low toxicity with an LD_{50} of 115 mg/kg body-weight intraperitoneally in mice.

Muhtadi, F. J., *Anal. Profiles Drug Subst.,* 1983, **12**, 547.

Quinquangulin 1177

7-Methylrubrofusarin

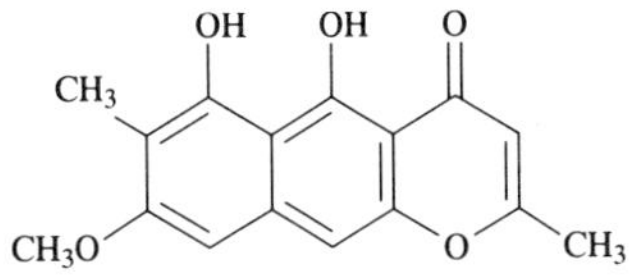

Chromone

$C_{16}H_{14}O_5$ [64894-58-2] MW 286.28

Root of *Cassia quinquangulata* (Leguminosae).

Cytotoxic in the P-388 lymphocytic leukaemia cell system. Red-brown pigment.

Ogura, M., *Lloydia,* 1977, **40**, 347.

R

Radiatin 1178

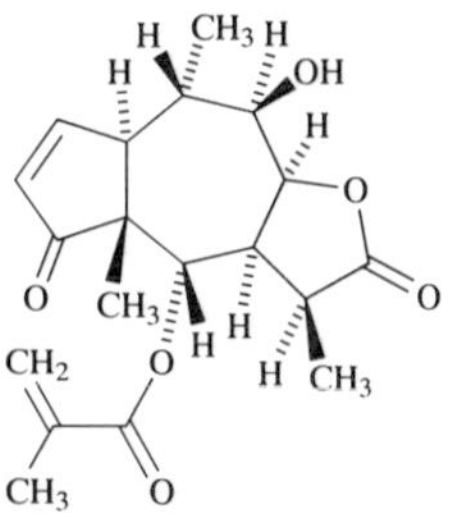

Sesquiterpenoid lactone
(Pseudoguaianolide)

$C_{19}H_{24}O_6$ [25873-31-8] MW 348.40

Baileya pleniradiata and *B. multiradiata* (Compositae).

Cytotoxic and antitumour properties.

Yoshitake, A., *Phytochemistry,* 1969, **8**, 1753.

Ragweed pollen allergen Ra5 1179

Allergen Ra5

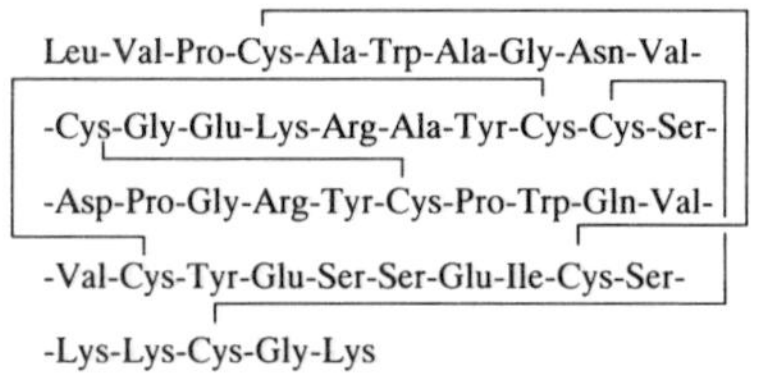

Leu-Val-Pro-Cys-Ala-Trp-Ala-Gly-Asn-Val-
-Cys-Gly-Glu-Lys-Arg-Ala-Tyr-Cys-Cys-Ser-
-Asp-Pro-Gly-Arg-Tyr-Cys-Pro-Trp-Gln-Val-
-Val-Cys-Tyr-Glu-Ser-Ser-Glu-Ile-Cys-Ser-
-Lys-Lys-Cys-Gly-Lys

Protein

[56092-26-3] MW 4979

A minor protein of ragweed, *Ambrosia eliator* (Compositae). Contains 45 amino acid residues.

Allergen.

Metzler, W. J., *Biochemistry,* 1992, **31**, 8697.

Ranunculin 1180

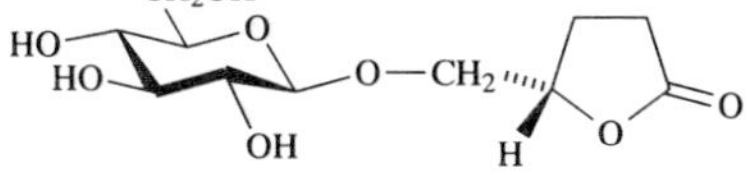

Lactone

$C_{11}H_{15}O_8$ [644-69-9] MW 275.24

In buttercups, *Ranunculus acris*, *R. scleratus*, *R. bulbosus*, *R. thora*, *R. flammula* and bur buttercup *R. falcatus*; also in *Anemone* and *Clematis* spp. and many other members of the Ranunculaceae.

Bitter tasting. It is converted, when fresh plants are bruised, to protoanemonin, [108-28-1], a vesicant oil with an acid taste. This causes an unpleasant subepidermal blistering of human skin. Protoanemonin is toxic to livestock. In central Utah 150 ewes were killed by eating bur buttercup, the lethal plant dosage being about 11 g wet-weight of plant per kilogram body-weight of sheep.

Nachman, R. J., *J. Agric. Food Chem.*, 1983, **31**, 1358.

Rapanone 1181

Oxaloxanthin

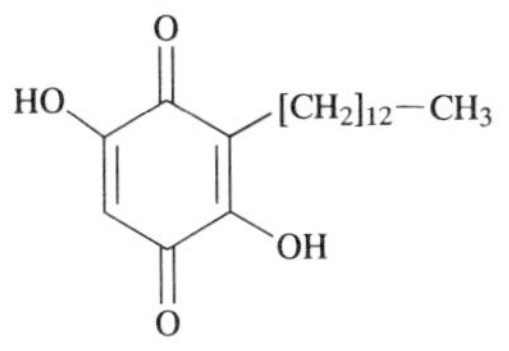

Benzoquinone

$C_{19}H_{30}O_4$ [573-40-0] MW 322.44

Bark and wood of *Rapanea maximowiczii* (Myrsinaceae), stem and twig of *Aegiceras corniculata* (Aegicerataceae), root of *Connarus monocarpus* (Connaraceae) and bulb of *Oxalis purpurata* (Oxalidaceae).

Anthelmintic activity.

Murthy, V. K., *Tetrahedron,* 1965, **21**, 1445.

Repin 1182

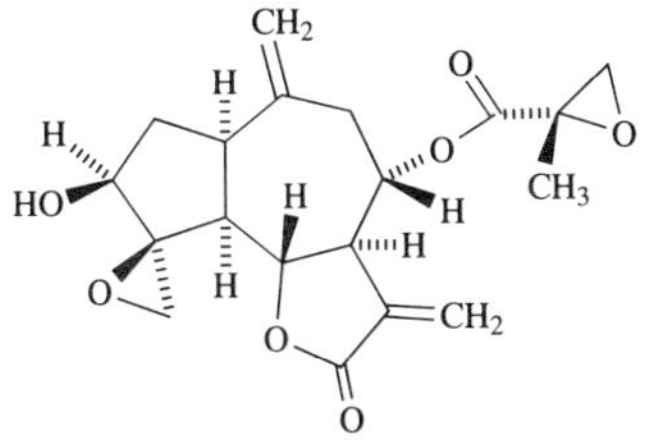

Sesquiterpenoid lactone
(Guaianolide)

$C_{19}H_{22}O_7$ [11024-67-2] MW 362.38

Centaurea hyrcanica and *Acroptilon repens* (Compositae).

Causes nervous disorder (equine nigropallidal encephalomalica) in horses.

Stevens, K.L., *J. Nat. Prod.,* 1990, **53**, 218.

Rescinnamine 1183

Reserpinine

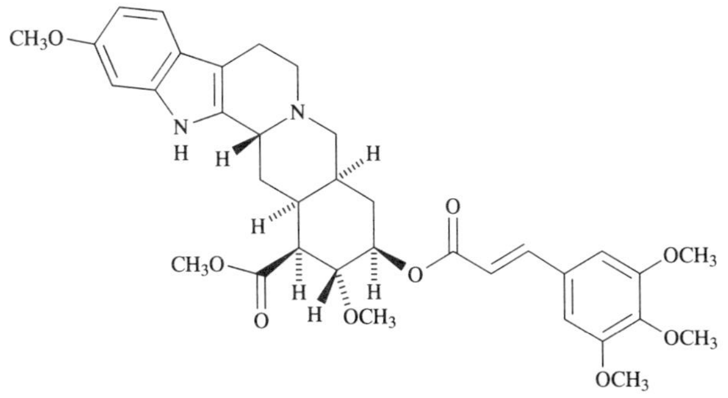

Indole alkaloid
(Yohimban)

$C_{35}H_{42}N_2O_9$ [84-34-4] MW 634.73

Rauwolfia serpentina, *R. vomitoria* and *R. nitida* (Apocynaceae).

Antihypertensive and tranquilliser.

Klohs, M. W., *J. Am. Chem. Soc.,* 1955, **77**, 2241.

Reserpine 1184

Indole alkaloid
(Yohimban)

$C_{33}H_{40}N_2O_9$ [50-55-5] MW 608.69

Rauwolfia serpentina and *R. vomitoria* (Apocynaceae).

Possibly carcinogenic. In clinical use as an antihypertensive drug and tranquilliser.

Manske, R. H. F., The Alkaloids, 1965, **8**, 287, Academic Press.

Resiniferatoxin 1185

Euphorbia factor RL_9

Diterpenoid
(Daphnane)

$C_{37}H_{40}O_9$ [57444-62-9] MW 628.74

Euphorbia resinifera and *E. poisonii* (Euphorbiaceae).

One of the most irritant diterpenoid esters of the family Euphorbiaceae, with co-carcinogenic properties.

Evans, F. J., *Phytochemistry,* 1976, **15**, 333.

Retronecine 1186

Pyrrolizidine alkaloid

$C_8H_{13}NO_2$ [480-85-3] MW 155.20

Senecio pseudo-orientalis and other *Senecio* spp. (Compositae), some *Crotalaria* spp. (Leguminosae) and several species in the Boraginaceae.

Hepatotoxic. It is dehydrogenated after ingestion to the related pyrrole, which is more toxic and binds to the DNA in the liver.

Geissman, T. A., *J. Org. Chem.,* 1962, **27**, 139.

Retrorsine 1187

β-Longilobine

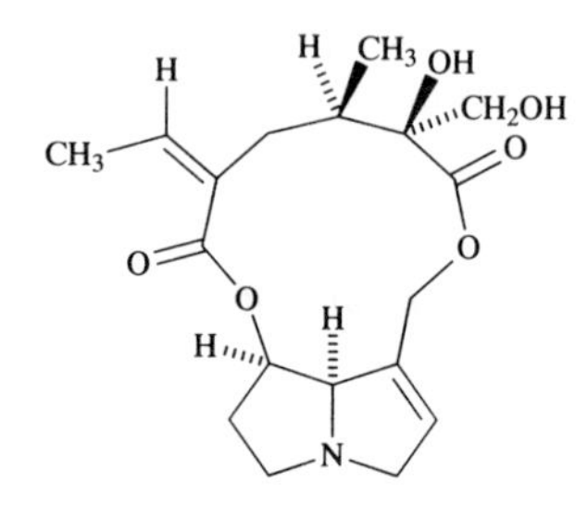

Pyrrolizidine alkaloid
(Senecionan)

$C_{18}H_{25}NO_6$ [480-54-6] MW 351.40

Senecio retrorsus, S. vulgaris, S. filaginoides, S. phillipicus, S. grisebachii (Compositae), *Crotalaria usaramoenensis* and *C. spartioides* (Leguminosae).

Hepatotoxic and pneumotoxic. Contributes to cattle poisoning after grazing on *Senecio* plants.

Leisegang, E. C., *J. Chem. Soc.,* 1950, 702.

Rhaphiolepsin 1188

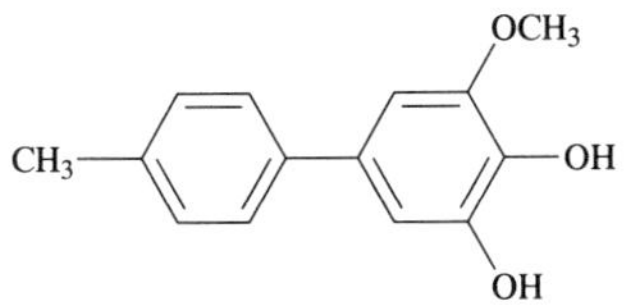

Biphenyl

$C_{14}H_{14}O_4$ [130364-26-4] MW 247.27

Fungally-infected leaves of *Raphiolepsis umbellata* (Rosaceae).

Antifungal agent (phytoalexin). Inhibits spore germination at a concentration of 10 μg/ml.

Watanabe, K., *Agric. Biol. Chem.,* 1990, **54**, 1861.

Rhaponticin 1189

Rhapontin

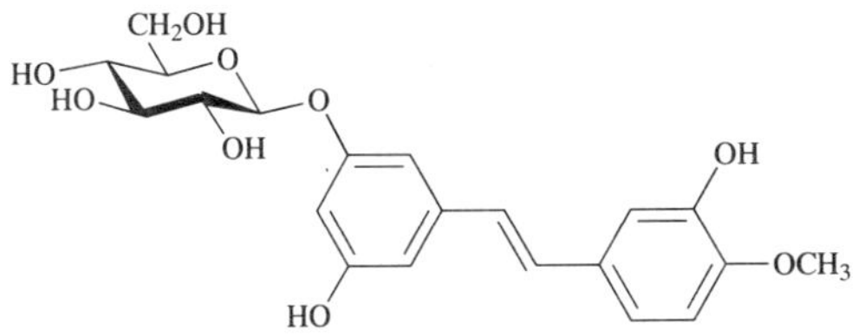

Stilbenoid

$C_{21}H_{24}O_9$ [155-58-8] MW 420.42

Bark of Sitka spruce, *Picea sitchensis* (Pinaceae).

Antifungal activity, toxic to *Sparassis crispa.*

Kawamura, S., *J. Pharm. Soc. Jpn,* 1938, **58**, 405.

Rhipocephalin 1190

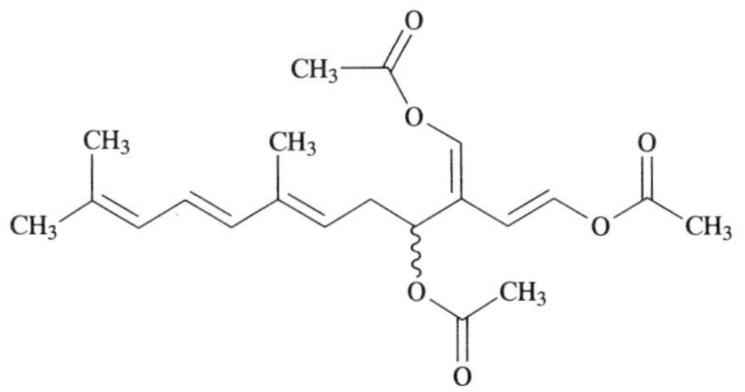

Sesquiterpenoid

$C_{21}H_{28}O_6$ [71135-78-9] MW 376.45

The Caribbean green alga *Rhipocephalus phoenix* (Codiaceae).

Toxic to pomacentrid fish and feeding deterrent to herbivorous Caribbean fish.

Sun, H. H., *Tetrahedron Lett.,* 1979, 685.

Rhipocephenal 1191

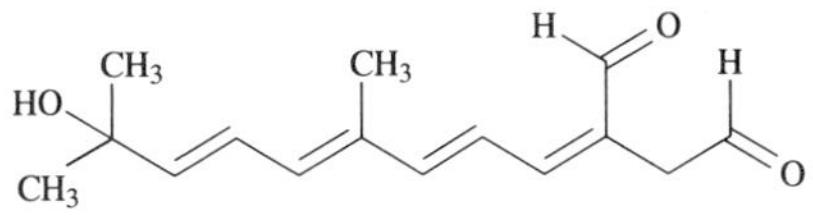

Sesquiterpenoid

$C_{15}H_{20}O_3$ [71135-77-8] MW 248.32

The Caribbean green alga *Rhipocephalus phoenix* (Codiaceae).

Toxic to pomacentrid fish and feeding deterrent to herbivorous Caribbean fish.

Sun, H. H., *Tetrahedron Lett.,* 1979, 685.

Rhodexin A 1192

Rohdexin A; Sarmentogenin 3-rhamnoside

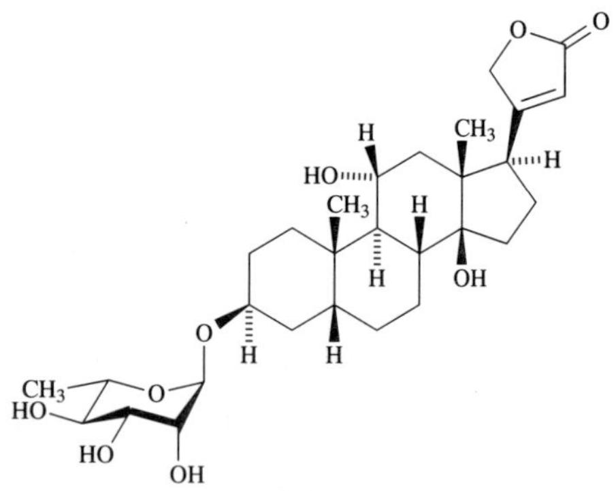

Cardenolide

$C_{29}H_{44}O_9$ [545-49-3] MW 536.66

Rhodea japonica and *Ornithogalum magnum* (Liliaceae).

Toxic to vertebrates.

Komisssarenko, N. F., *Chem. Nat. Compd.,* 1965, **1**, 120.

Rhododendrin 1193

Betuloside

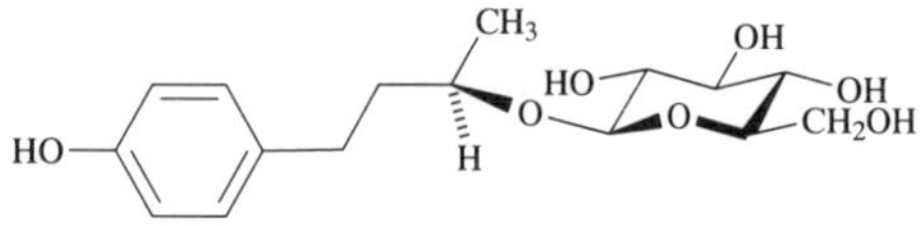

Phenol

$C_{16}H_{24}O_7$ [497-78-9] MW 328.36

Leaves of *Rhododendron chrysanthum*, *R. fauriae* and other *Rhododendron* spp. (Ericaceae) and also *Betula* spp. (Betulaceae).

Bitter tasting. It exhibits diuretic and diaphoretic properties.

Klischies, M., *Phytochemistry,* 1978, **17**, 1281.

Rhodojaponin IV 1194

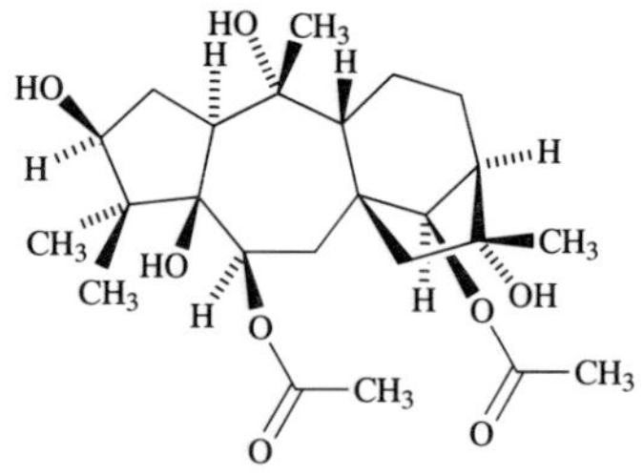

Diterpenoid
(Grayanotoxin)

$C_{24}H_{34}O_8$ [30460-34-5] MW 454.56

Rhododendron japonicum (Ericaceae).

Highly toxic.

Hikino, H., *Chem. Pharm. Bull.,* 1970, **18**, 2357.

ψ-Rhodomyrtoxin 1195

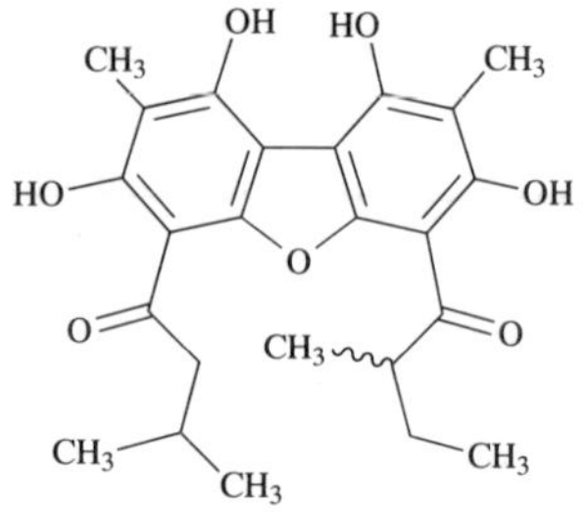

Dibenzofuran
(Pseudorhodomyrtoxin)

$C_{24}H_{28}O_7$ [24563-20-0] MW 428.48

Fruits of *Rhodomyrtus macrocarpa* (Myrtaceae).

Consumption of the fruits causes blindness and poisoning in livestock, but there is no proof that this compound is responsible. Nevertheless, it is toxic to mice and therefore could be the toxic agent.

Sargent, M. V., *J. Chem. Soc., Perkin Trans. 1,* 1983, 231.

Rhoeadine 1196

Rheadine

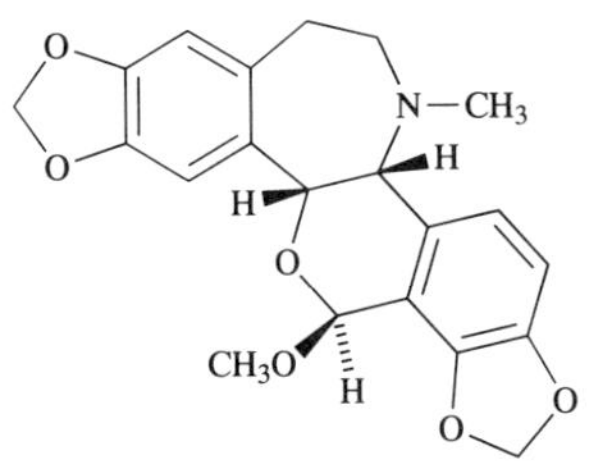

Isoquinoline alkaloid
(Rhoeadan)

$C_{21}H_{21}NO_6$ [2718-25-4] MW 383.40

Seed capsules of the corn poppy, *Papaver rhoeas* and present in other *Papaver* spp. (Papaveraceae).

Toxic, with an LD_{50} intraperitoneally in rats of 530 mg/kg body-weight. Large doses cause spasms in animals. *Papaver rhoeas* is used as a sedative and mild expectorant in folk medicine.

Šantavý, F., *Collect. Czech. Chem. Commun.*, 1965, **30**, 3479.

Ricin 1197

A mixture of four lectins:
two non-toxic agglutinins (RCA), RCL_I and RCL_{II}
(tetrameric, with two 30 000 and two 35 000 M_r units);
and two toxins, ricin D (RCL_{III}) and RCL_{IV}
(dimeric with a 30 000 and a 33 000 M_r unit).

Glycoprotein

Castor bean, *Ricinus communis* (Euphorbiaceae).

Ricin is one of the most toxic substances known. One mg of toxin can be isolated from 1 g of seed and a single seed of 0.25 g contains a lethal dose. It is very stable to proteolytic enzymes and hence is not destroyed when taken orally. Poisonings occur because the seed is attractive to children and is used decoratively in a necklace.

Montfort, W., *J. Biol. Chem.*, 1987, **262**, 5398.

Ricinine 1198

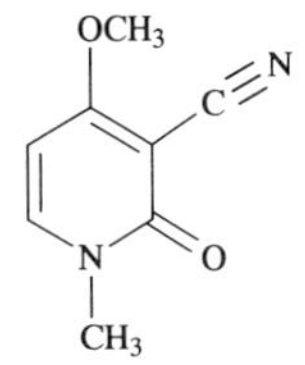

Piperidine alkaloid

$C_8H_8N_2O_2$ [524-40-3] MW 164.16

Seed and leaves of castor oil, *Ricinus communis* (Euphorbiaceae).

Highly toxic. May contribute to the toxicity of castor seed, but the major toxin is undoubtedly the protein ricin (q.v.). Also present in the leaves, and is suspected of being the cause of cattle poisoning in Brazil, following grazing on these leaves.

Robinson, T., *Phytochemistry,* 1978, **17**, 1903.

Riddelline 1199

Riddelliine; 18-Hydroxyseneciphylline

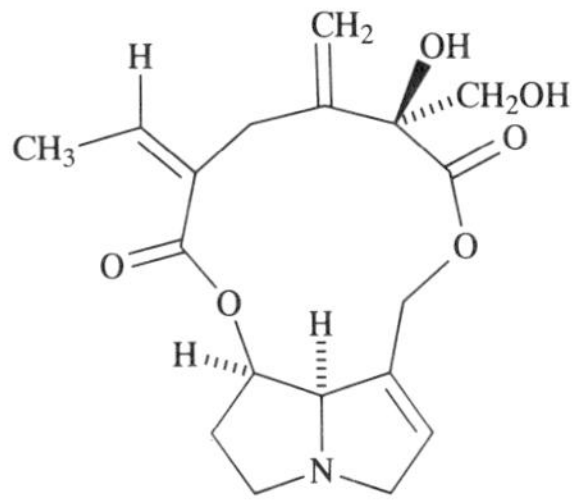

Pyrrolizidine alkaloid
(Senecionan)

$C_{18}H_{23}NO_6$ [23246-96-0] MW 349.38

Tansy ragwort, *Senecio vulgaris*, *S. riddellii*, *S. longilobus*, *S. aegypticus*, *S. ambrosioides* and *S. eremophilus* (Compositae) and *Crotalaria juncea* (Leguminosae). The alkaloid levels in *Senecio riddellii* are exceptionally high, with some plants containing 10-18% of the leaf dry-weight.

Hepatotoxic poison. *Senecio longilobus* has caused fatal veno-occlusive disease. It is anticholinergic in animals.

Adams, R., *J. Am. Chem. Soc.*, 1953, **75**, 4638.

Ridentin 1200

Ridentin A

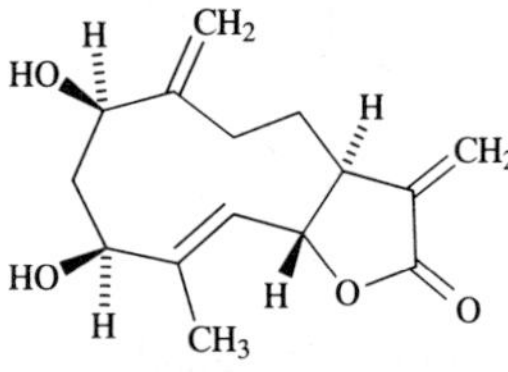

Sesquiterpenoid lactone
(Germacranolide)

$C_{15}H_{20}O_4$ [28148-84-7] MW 264.32

Artemisia tridentata, *A. cana* and *A. tripartita* (Compositae).

Cytotoxic and antitumour properties.

Irwin, M. A., *Phytochemistry,* 1969, **8**, 2009.

Rinderine 1201

9-(+)-Trachelanthylheliotridine

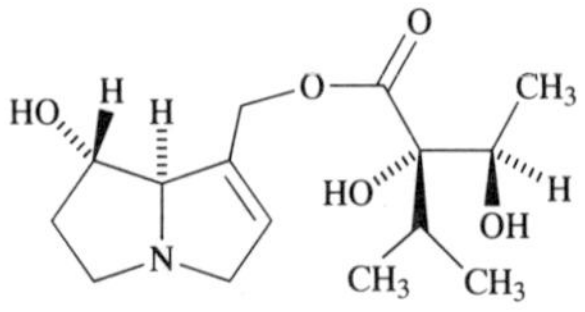

Pyrrolizidine alkaloid

$C_{15}H_{25}NO_5$ [6029-84-1] MW 299.37

Rindera baldschuanica and *Solenanthus circinatus* (Boraginaceae), *Eupatorium altissimum* and *E. cannabinum* (Compositae).

Hepatotoxic and pneumotoxic.

Akramov, S. T., *Chem. Abstr.,* 1962, **57**, 16676

Rishitin 1202

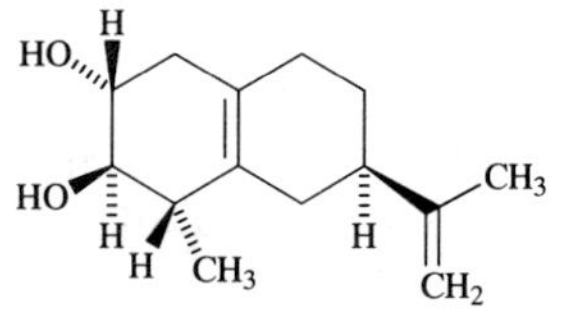

Sesquiterpenoid
(Noreudesmane)

$C_{14}H_{22}O_2$ [18178-54-6] MW 222.33

Tuber of potato, *Solanum tuberosum* (Solanaceae) infected with *Phytophthora infestans.*

Antifungal agent (phytoalexin). Also bactericide.

Bukhari, S. T. K., *J. Chem. Soc., C,* 1969, 1073.

Robin 1203

Robinia lectin

Protein
(Lectin)
[1393-13-1] MW 110000

Bark and seed of the false acacia, *Robinia pseudoacacia* (Leguminosae).

Haemagglutinating and mitogenic properties. Less toxic on oral administration than abrin (q.v.).

Hořejší, V., *Biochem. Biophys. Acta,* 1978, **532**, 98.

Rodiasine

1204

6′-*O*-Methylphlebicine

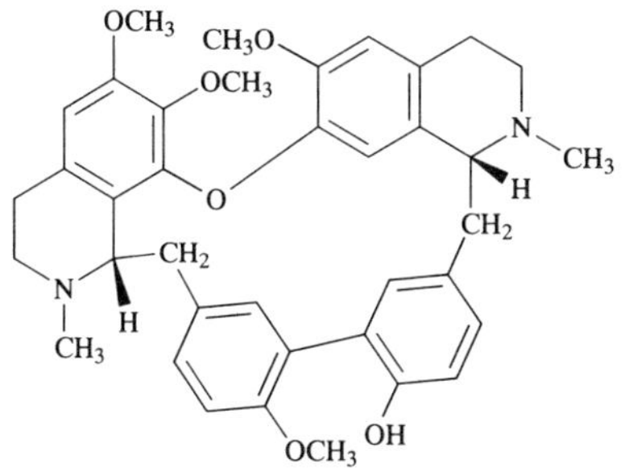

Bisbenzylisoquinoline alkaloid
(Rodiasan)

$C_{38}H_{42}N_2O_6$ [6391-64-6] MW 622.76

Bark and seed of *Ocotea venenosa* (Lauraceae).

Neuromuscular blocking agent, with muscle-relaxing activity. The plant has been used as a curare ingredient in South American arrow poisons.

Grundon, M. F., *J. Chem. Soc., C,* 1966, 1082.

Roridin A

1205

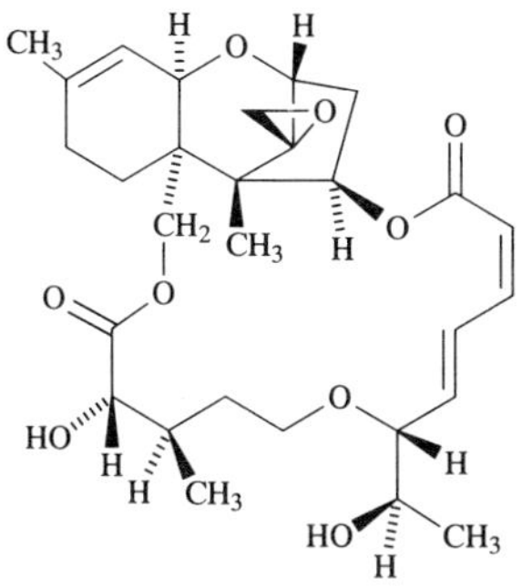

Sesquiterpenoid
(Trichothecane)

$C_{29}H_{40}O_9$ [14729-29-4] MW 532.63

Major component of seedcoats of female plants of *Baccharis coridifolia* (Compositae) and in aerial parts of male plants.

Antifungal, cytostatic and poisonous. Upper aerial parts of *B. coridifolia* are lethal to sheep at levels as low as 1-2 g/kg body-weight. This is a major source of stock poisoning in Brazil, Argentinia and Uruguay. Leaves, stems and roots are all toxic but the flowering tops are 4 to 8 times more toxic.

Jarvis, B. B., *J. Nat. Prod.,* 1982, **45**, 440.

Roridin E

1206

Satratoxin D

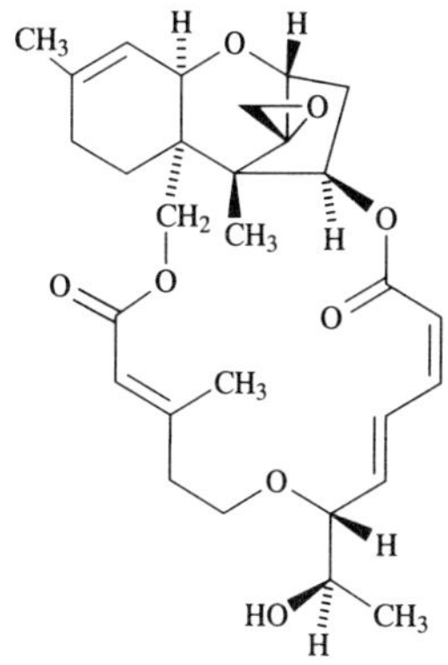

Sesquiterpenoid
(Trichothecane)

$C_{29}H_{38}O_8$ [16891-85-3] MW 514.61

Major toxin of seedcoats of female plants of *Baccharis coridifolia* (Compositae) and in aerial parts of male plants.

Antimicrobial and poisonous to livestock (see under roridin A).

Still, W. C., *J. Am. Chem. Soc.,* 1984, **106**, 260.

Rotenone 1207

Tubotoxin; Nicouline

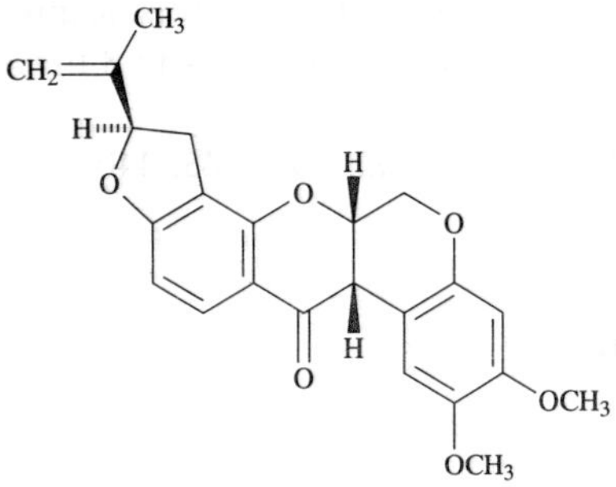

Isoflavonoid
(Rotenoid)

$C_{23}H_{22}O_6$ [83-79-4] MW 394.42

Major sources are the roots of *Derris elliptica* and *Piscidia erythrina*, but also recorded in over sixty other species of tropical Leguminosae. Also present in leaves of *Verbascum thapsus* (Scrophulariaceae).

Poisonous to fish and insects. The lethal dose in silkworms is 0.003 mg/kg body-weight. Widely used as an insecticide. However, it can be toxic to humans when inhaled rather than ingested. The LD_{50} intraperitoneally in mice is 2.8 mg/kg body-weight. It has been used as an arrow poison in Sumatra and for committing suicide in New Guinea.

Jacobson, M., Naturally Occurring Insecticides, 1971, 71, Marcel Dekker.

Rottlerin 1208

Mallotoxin

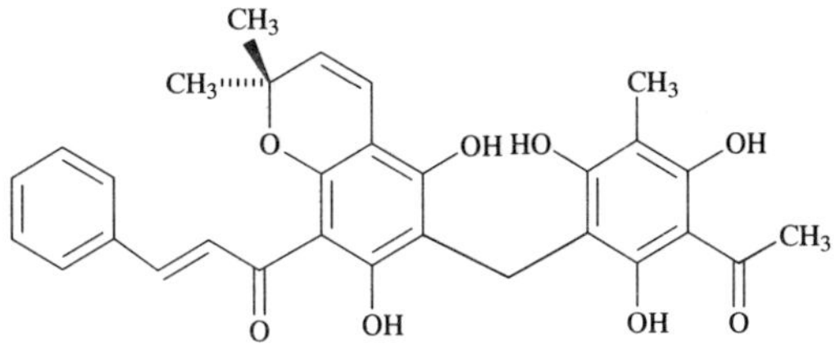

Chalcone

$C_{30}H_{28}O_8$ [82-08-6] MW 516.55

The fruit glands of *Mallotus philippensis* (Euphorbiaceae).

Toxic pinkish-brown pigment. The anthelmintic activity is used in veterinary practice. In India, it is used commercially as a dye for silk.

McGookin, A., *J. Chem. Soc.*, 1939, 1579.

Rubrofusarin 1209

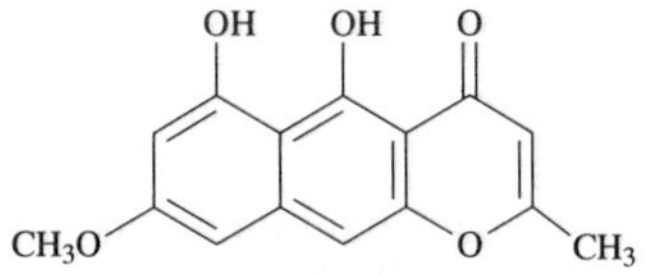

Chromone

$C_{15}H_{12}O_5$ [3567-00-8] MW 272.26

Root of *Cassia tora* and *C. quinquangulata* (Leguminosae).

Moderately cytotoxic in the P-388 lymphocytic leukaemia cell system. Acts as a depressant on the central nervous system in animals.

Stout, G. H., *Acta Crystallogr.*, 1961, **15**, 451.

Rugosin D 1210

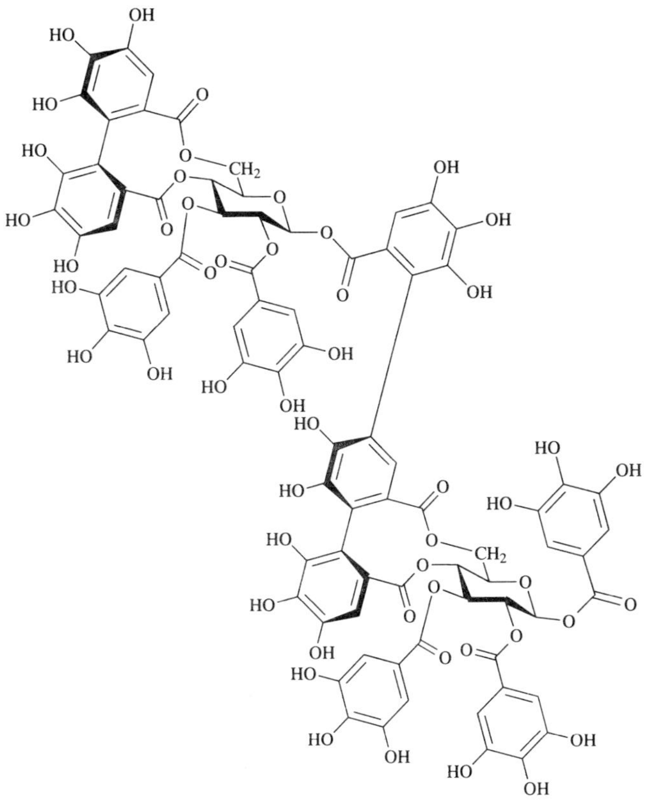

Ellagitannin

$C_{82}H_{58}O_{52}$	[84754-11-0]	MW 1875.33

Flower petals of *Rosa rugosa* and in *Filipendula ulmaria* (Rosaceae).

Antitumour activity against Sarcoma 180 in mice.

Okuda, T., *Chem. Pharm. Bull.*, 1982, **30**, 4234.

Rutamarin 1211

Chalepin acetate

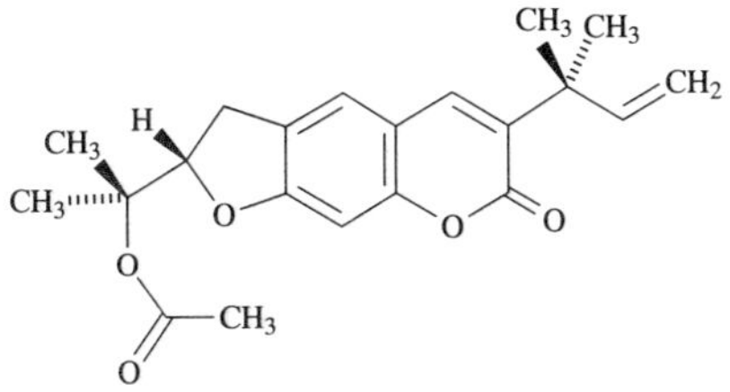

Furocoumarin

$C_{21}H_{24}O_5$	[14882-94-1]	MW 356.42

Leaves of *Boenninghausenia japonica*, heartwood of *Chloroxylon swietenia* and in *Ruta graveolens* (Rutaceae).

Antitumour properties against HeLa cells.

Reisch, J., *Acta Pharm. Suec.*, 1967, **4**, 179.

Ryanodine 1212

Ryanex; Ryanicide

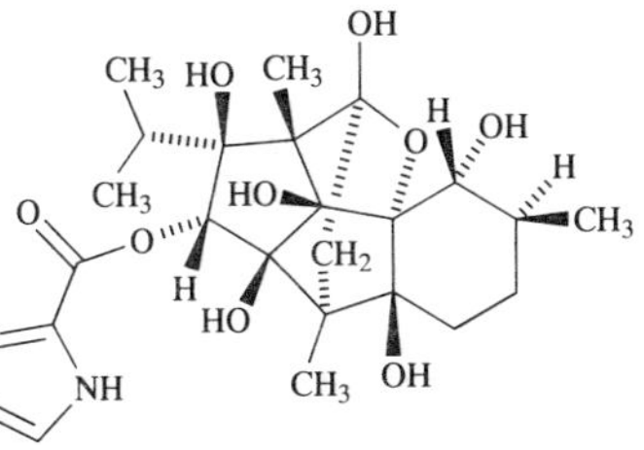

Diterpenoid alkaloid

$C_{25}H_{35}NO_9$	[15662-33-6]	MW 493.55

Ryania speciosa (Flacourtiaceae).

Insecticidal against the European cornborer. It inhibits the binding of calcium to muscle protein and retards circulation by vascular constriction.

Wiesner, K., *Adv. Org. Chem.*, 1972, **8**, 295.

S

Sabinol 1213

4(10)-Thujen-3-ol

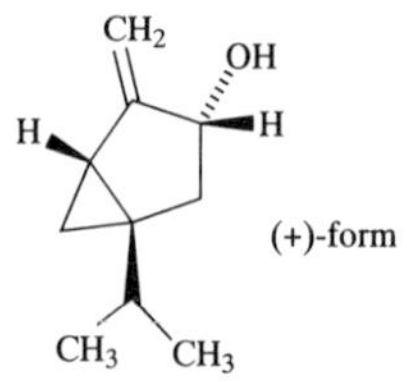

(+)-form

Monoterpenoid
(Thujane)

$C_{10}H_{16}O$ [471-16-9] MW 152.24

Free and as the sabinyl ester in oil of savin, from the fresh tops of *Juniperus sabina* (Cupressaceae). Also present in essential oil of other *Juniperus* spp..

Toxic. It is used as an emmenagogue and anthelmintic.

Bergqvist, M. S., *Arch. Kemi.,* 1964, **22**, 137.

Sacculatal 1214

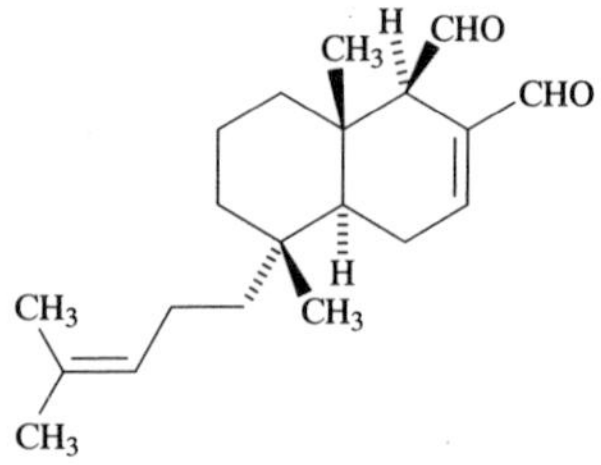

Diterpenoid

$C_{20}H_{30}O_2$ [64242-90-6] MW 302.46

The liverworts *Pellia endiviifolia* and *Trichocoleopsis sacculata*.

Persistent pungent taste. Piscicidal, causing death at a concentration of 0.4 p.p.m. within 120 min. Also skin irritant and tumour promotor.

Asakawa, Y., *Tetrahedron Lett.,* 1977, 1407.

Safrole 1215

Allylcatechol methylene ether; Shikimol

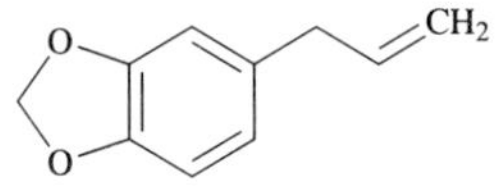

Phenylpropanoid

$C_{10}H_{10}O_2$ [94-59-7] MW 162.19

Sassafras albidum (Lauraceae), *Magnolia salicifolia* (Magnoliaceae), leaves of *Illicium religiosum* (Illiciaceae), bark of *Nemuaron humboldtii* (Atherospermataceae), *Ocimum basilicum* (Labiatae) and *Myristica fragrans* (Myristicaceae).

Moderately toxic to man and a low grade hepatocarcinogen. Anticonvulsant, DNA-binding and hypothermic properties. It is used as a topical antiseptic, a pediculicide and as a carminative.

Perkin, W. H., *J. Chem. Soc.,* 1927, 1663.

Safynol 1216

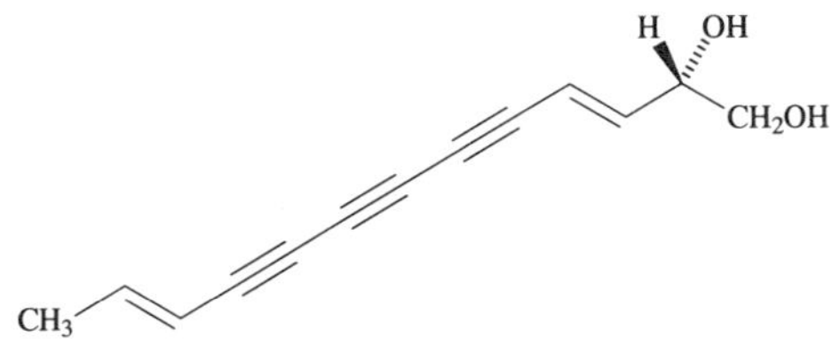

Polyacetylene

$C_{13}H_{12}O_2$ [27978-14-9] MW 200.24

Safflower, *Carthamus tinctorius* (Compositae) infected with *Phytophthora drechsleri.*

Antifungal agent (phytoalexin).

Allen, E. H., *Phytochemistry,* 1971, **10**, 1579.

Saikosaponin BK1 1217

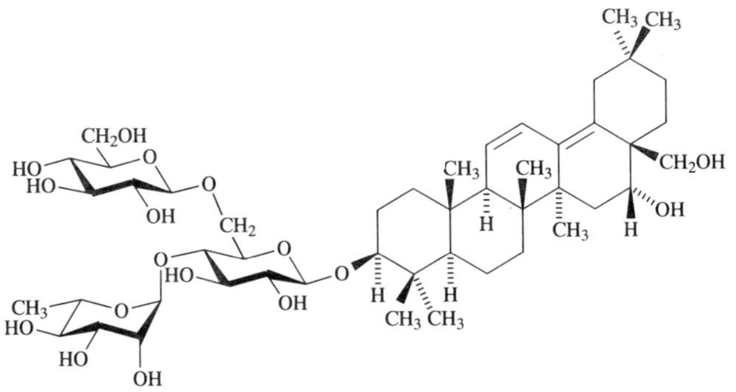

Triterpenoid saponin
(Oleanane)

$C_{48}H_{78}O_{17}$ [110352-77-7] MW 927.14

Bupleurum kummingense (Umbelliferae).

Antileukaemic activity *in vitro.*

Luo, S. Q., *Agric. Biol. Chem.,* 1987, **51**, 1515.

Sainfuran 1218

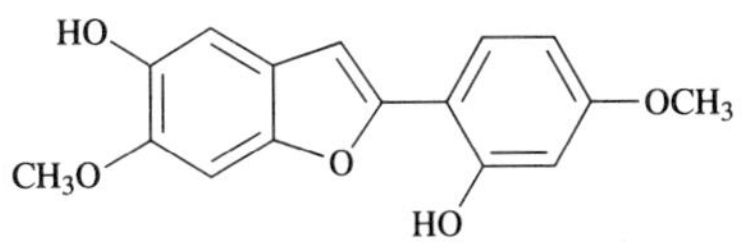

Benzofuran

$C_{16}H_{14}O_5$ [90664-32-7] MW 286.28

Roots of *Onobrychis viciifolia* and of *Hedysarum polybotris* (Leguminosae).

Insect antifeedant. Displays antifungal activity against *Cladosporium cladosporoides.*

Russell, G. B., *Phytochemistry,* 1984, **23**, 1417.

Sakuranetin 1219

Naringenin 7-methyl ether

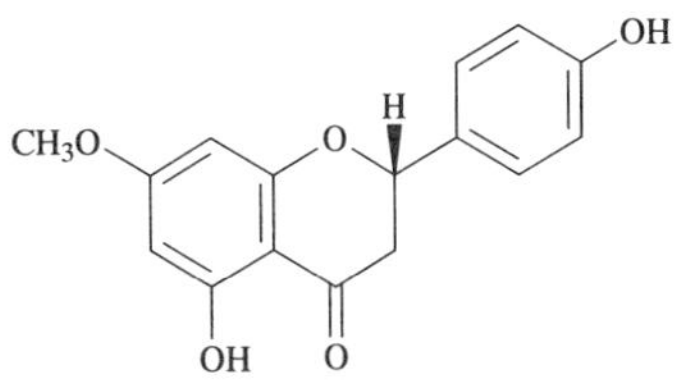

Flavanone

$C_{16}H_{14}O_5$ [2957-21-3] MW 286.28

Fungally infected leaves of rice, *Oryza sativa* (Gramineae) and constitutively present in leaf surface wax of blackcurrant leaves, *Ribes nigrum* (Grossulariaceae).

Antifungal activity. The ED_{50} against the rice pathogen *Pyricularia oryzas* is 30 p.p.m.

Kodama, O., *Phytochemistry,* 1992, **31**, 3807.

Sakuraso-saponin 1220

Triterpenoid saponin
(Oleanane)

$C_{60}H_{98}O_{27}$ [59527-84-3] MW 1251.42

Leaves of *Rapanea melanophloeos* (Myrsinaceae).

Antifungal activity, toxic to *Cladosporium cucumerinum* at 1 μg/ml concentration; also molluscicide at 3 p.p.m.

Kitagawa, I., *Chem. Pharm. Bull.*, 1980, **28**, 296.

Salannin 1221

Nortriterpenoid

$C_{34}H_{44}O_9$ [992-20-1] MW 596.72

Essential oil of *Azadirachta indica* (Meliaceae).

Insect antifeedant.

Henderson, R., *Tetrahedron*, 1968, **24**, 1525.

Salicylic acid 1222

2-Hydroxybenzoic acid

Phenolic acid

$C_7H_6O_3$ [69-72-7] MW 138.12

Free in the spadix of the voodoo lily, *Sauromatum guttatum* (Araceae) and as the methyl ester [119-36-8] in wintergreen leaves, *Gaultheria procumbens* (Ericacae) and bark of birch, *Betula lenta* (Betulaceae). The free acid also occurs in the leaves of 20 of 27 angiosperms tested, e.g. in pea, *Pisum sativum* (0.03 μg/g fresh weight) and in rice, *Oryza sativa* (37 μg/g fresh weight).

The acetyl derivative, aspirin [50-78-2] , is widely used as a mild pain killer. It undergoes hydrolysis *in vivo* and can attack the stomach lining in some people. The free acid is used in medicine as a topical keratolytic, but it can cause skin rashes in sensitive people.

Pierpoint, W. S., *Adv. Bot. Res.*, 1994, **20**, 163.

Salonitenolide 1223

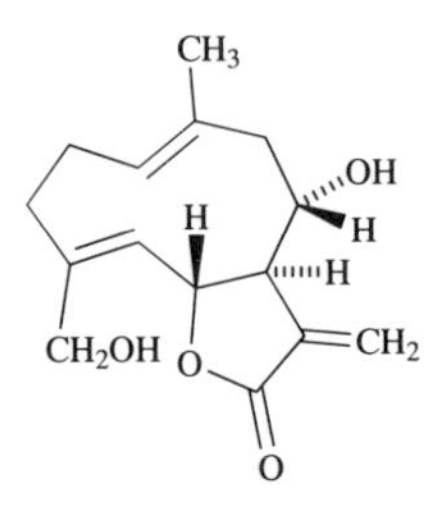

Sesquiterpenoid lactone
(Germacranolide)

$C_{15}H_{20}O_4$ [26931-94-2] MW 264.32

Flowers and seeds of *Centaurea salonitana*, the blessed thistle, *Cnicus benedictus*, *Berkheya speciosa* and *Jurinea maxima* (Compositae).

Cytotoxic and antitumour properties. It is also an insect antifeedant.

Yoshioka, H., *J. Chem. Soc., Chem. Commun.*, 1970, 148.

(−)-Salsoline 1224

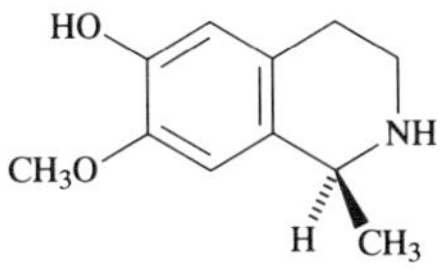

Isoquinoline alkaloid

$C_{11}H_{15}NO_2$ [89-31-6] MW 193.25

Salsola richteri (Chenopodiaceae).

Lethal to mice when given intravenously, but does not appear to be toxic orally.

Späth, E., *Ber. Dtsch. Chem. Ges.*, 1934, **67**, 1214.

(+)-Salutaridine 1225

Floriparine

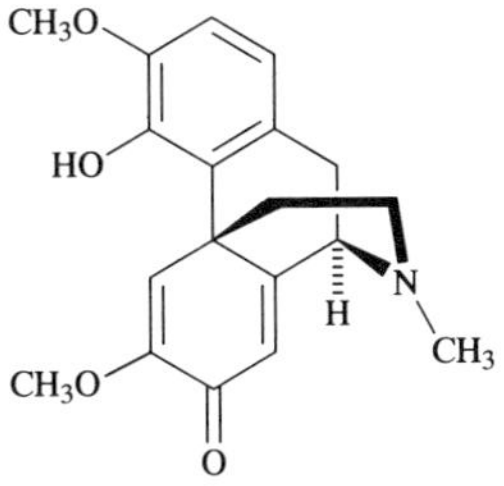

Isoquinoline alkaloid
(Morphinan)

$C_{19}H_{21}NO_4$ [1936-18-1] MW 327.38

Papaver somniferum, *P. orientale* and *P. bracteatum* (Papaveraceae), *Croton salutaris* and *C. balsamifera* (Euphorbiaceae).

Antitumour activity against Walker 256 carcinosarcoma.

Barton, D. H. R., *J. Chem. Soc.*, 1965, 2423.

Samaderin A 1226

Samaderine A

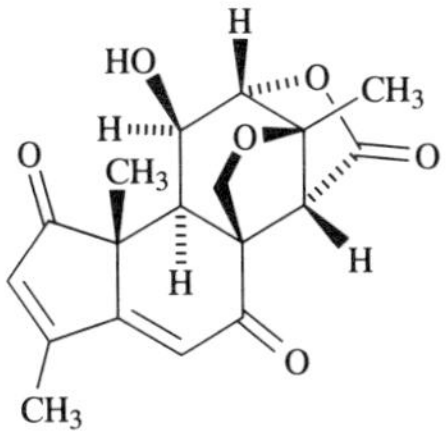

Nortriterpenoid
(Norquassane)

$C_{18}H_{18}O_6$ [64364-76-7] MW 330.34

Bark and seed of *Quassia indica* (Simaroubaceae), co-occurring with the related structures samaderins B and C [803-22-5 and 803-21-4 respectively].

A plant extract of *Quassia indica* is used as a vermifuge and as an insecticide. Samaderins A, B and C show antileukaemic activity.

Onan, K. D., *J. Chem. Res. (S),* 1978, 14.

Sanguinarine 1227

ψ-Chelerythrine; Pseudochelerythrine

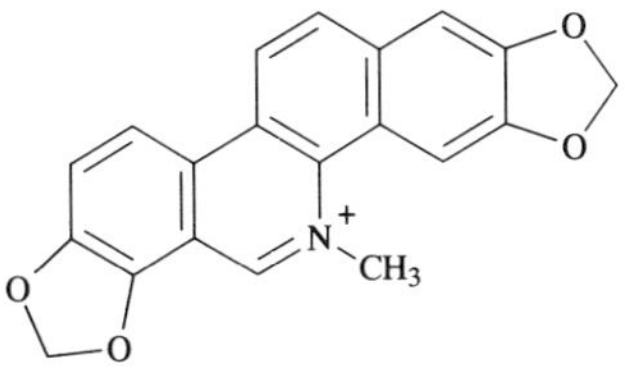

Isoquinoline alkaloid
(Benzophenanthridine)

$[C_{20}H_{14}NO_4]^+$ [2447-54-3] MW 332.34

Bloodroot, *Sanguinaria canadensis* and fumitory, *Fumaria officinalis* (Fumariaceae), *Papaver somniferum* and *Chelidonium majus* (Papaveraceae), *Zanthoxylum* spp. (Rutaceae) and *Pteridophyllum* spp. (Sapindaceae).

Toxic alkaloid, with an LD_{50} in mice orally of 19.4 mg/kg body-weight. It has a positive inotropic effect on the heart. An enzyme inhibitor, high doses cause glaucoma. It is used in dentifrice and mouthwashes because of its antiplaque activity.

Späth, E., *Ber. Dtsch. Chem. Ges.,* 1931, **64**, 2034.

Santamarin 1228

Santamarine; Balchanin

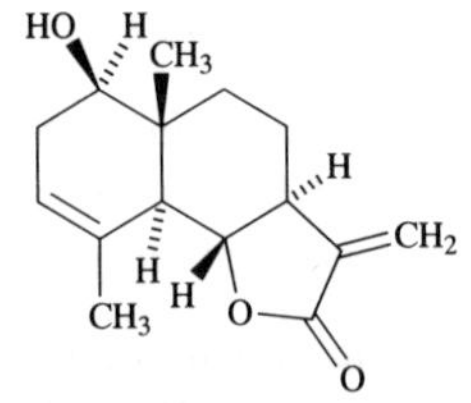

Sesquiterpenoid lactone
(Eudesmanolide)

$C_{15}H_{20}O_3$ [4290-13-5] MW 248.32

Feverfew, *Tanacetum parthenium*, *Ambrosia confertiflora* and several *Artemisia* spp. (Compositae), also present in *Michelia compressa* (Magnoliaceae).

Cytotoxic and antitumour properties.

Romo de Vivar, A., *Tetrahedron,* 1965, **21**, 1741.

α-Santonin 1229

Santonin

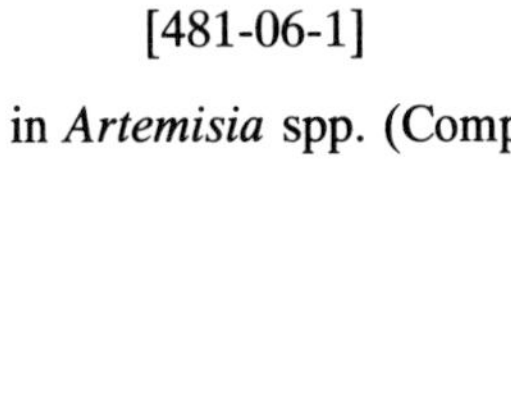

Sesquiterpenoid lactone
(Eudesmanolide)

$C_{15}H_{18}O_3$ [481-06-1] MW 246.31

Widely occurring in *Artemisia* spp. (Compositae).

Anthelmintic and ascaricidal properties; insect antifeedant. Also exhibits cytotoxic and antitumour properties. Has been used medicinally in the treatment of nervous complaints.

Asher, J. D. M., *J. Chem. Soc.,* 1965, 6041.

β-Santonin 1230

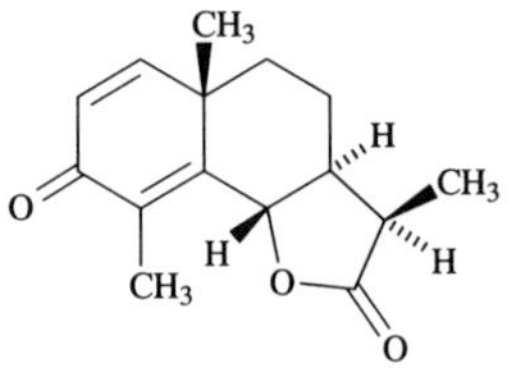

Sesquiterpenoid lactone
(Eudesmanolide)

$C_{15}H_{18}O_3$ [481-07-2] MW 246.31

Artemisia caerulescens, *A. cina* and *A. compacta* and other *Artemisia* spp. (Compositae).

Once used as a vermifuge, but because of its toxicity, it has been excluded from official use in many countries.

Cocker, W., *J. Chem. Soc.,* 1955, 4430.

Sapatoxin A 1231

Euphorbia factor Ti_1

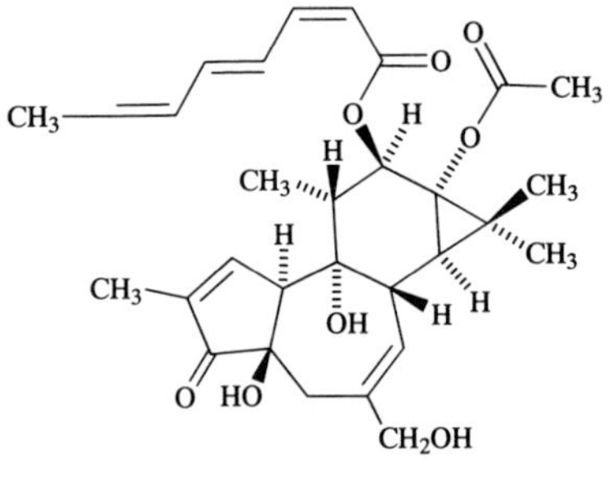

Diterpenoid
(Tigliane)

$C_{32}H_{44}O_7$ [82467-18-3] MW 540.70

Unripe fruits of *Sapium indicum* (Sapindaceae) and in *Euphorbia tirucalli* (Euphorbiaceae).

Toxic irritant. A minor toxin compared to sapintoxin A (q.v.) in the unripe *Sapium* fruits.

Taylor, S. E., *Phytochemistry,* 1982, **21**, 405.

Sapelin A 1232

Triterpenoid
(Euphorbane)

$C_{30}H_{50}O_4$ [26790-93-2] MW 474.72

Wood of *Entandophragma cylindricum* (Meliaceae).

Cytotoxic properties.

Chan, W. R., *J. Chem. Soc., C,* 1970, 311.

Sapintoxin A 1233

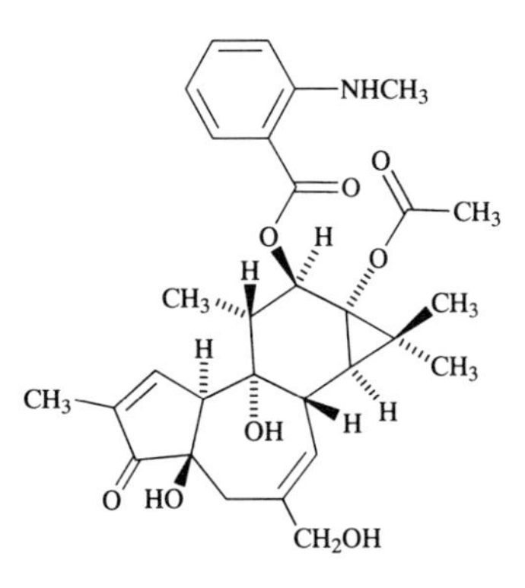

Diterpenoid
(Tigliane)

$C_{30}H_{37}NO_7$ [79083-69-5] MW 523.63

Unripe fruits of *Sapium indicum* (Sapindaceae).

Irritant. Fruit extracts induce erythema of the skin.

Taylor, S. E., *Experientia,* 1981, **37**, 681.

Sarmentoloside 1234

Sarmentologenin 3-(6-deoxy-α-L-taloside)

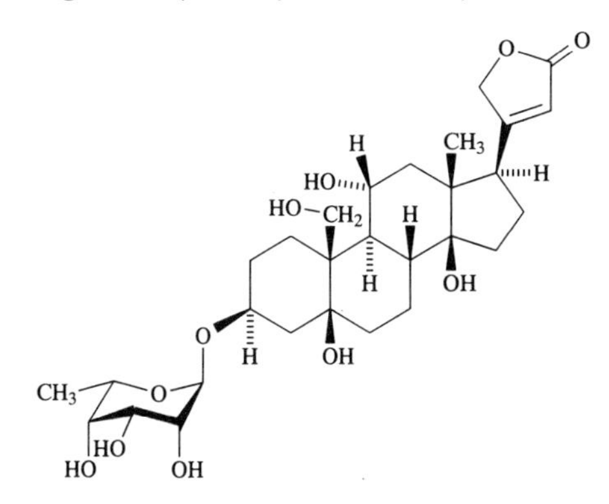

Cardenolide

$C_{29}H_{44}O_{11}$ [6847-59-2] MW 568.66

Strophanthus sarmentosus var. *senegambiae* and *S. divaricatus* (Apocynaceae).

Toxic to vertebrates.

Fechtig, B., *Helv. Chim. Acta,* 1959, **42**, 1448.

Sarmentosin epoxide 1235

Epoxysarmentosin

Cyanogenic glycoside

$C_{11}H_{17}NO_8$ [81907-02-0] MW 291.26

Sedum cepaea (Crassulaceae).

Bound toxin. This is an unusual cyanogenic glycoside, which has to undergo hydrolysis by an epoxyhydrolase before it will release its poisonous cyanide.

Nahrstedt, A., *Phytochemistry,* 1982, **21**, 107.

Sarothralin 1236

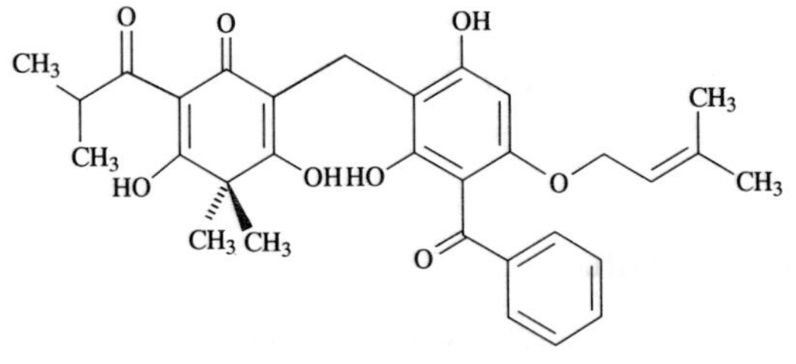

Phenolic ketone

$C_{31}H_{34}O_8$ [96624-40-7] MW 534.61

From the whole plant of *Hypericum japonicus* (Guttiferae).

Antimicrobial activity, inhibitory to *Bacillus cereus* and *Staphylococcus aureus*.

Ishigura, K., *J. Chem. Soc., Chem. Commun.*, 1985, 26.

Saupirin 1237

Saupirine

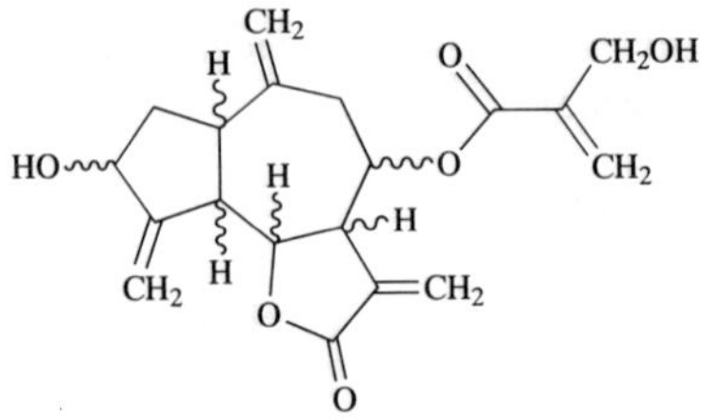

Sesquiterpenoid lactone
(Guaianolide)

$C_{19}H_{22}O_6$ [35932-39-9] MW 346.38

Flowers of *Saussurea pulchella* and *S. neopulchella* (Compositae).

Active against protozoa pathogenic to humans, i.e. *Entamoeba histolytica* and *Trichomonas vaginalis*.

Chugunov, P. V., *Chem. Nat. Compd.*, 1971, **7**, 706.

Saxitoxin 1238

Aphantoxin II; Saxitoxin hydrate

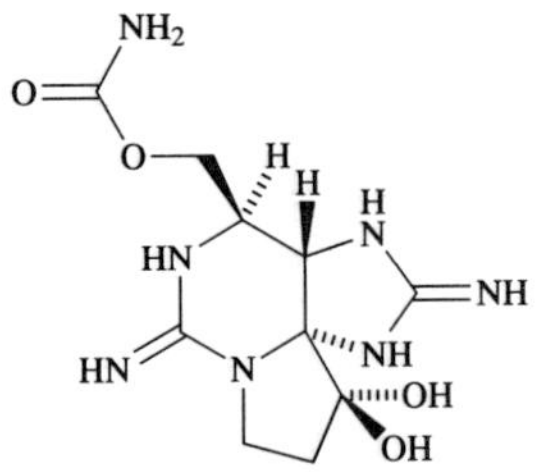

Alkaloid
(Cyclic bisguanide)

$C_{10}H_{17}N_7O_4$ [35523-89-8] MW 299.29

Produced by some strains of *Aphanizomenon flos-aquae* found in New Hampshire, USA, as well as blooms of *Anabaena circinalis* in rivers and water-storage reservoirs in Australia where formation of these blooms have caused the deaths of approximately 1600 cattle.

Saxitoxin had previously been isolated from dinoflagellates causing paralytic shellfish poisoning. It is a neurotoxin acting as a sodium channel blocking agent and death is usually by failure of the respiratory system.

Mahmood, N. A., *Toxicon*, 1986, **24**, 175.

Schizozygine 1239

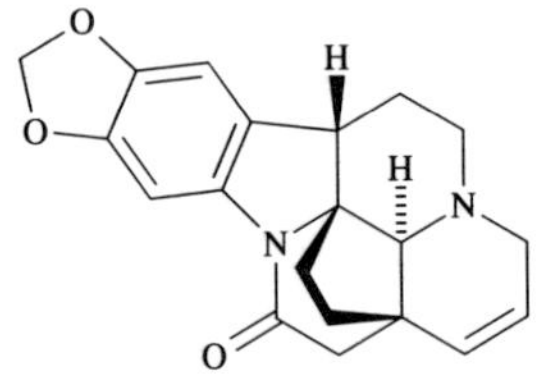

Indole alkaloid

$C_{20}H_{20}N_2O_3$ [2047-63-4] MW 336.39

Fruit of *Schizozygia caffaeoides* (Apocynaceae).

Very poisonous.

Renner, U., *Helv. Chim. Acta*, 1965, **48**, 308.

Scillaren A 1240

Scillarenin 3-glucosylrhamnoside; Glucoproscillaridin A; Transvaalin

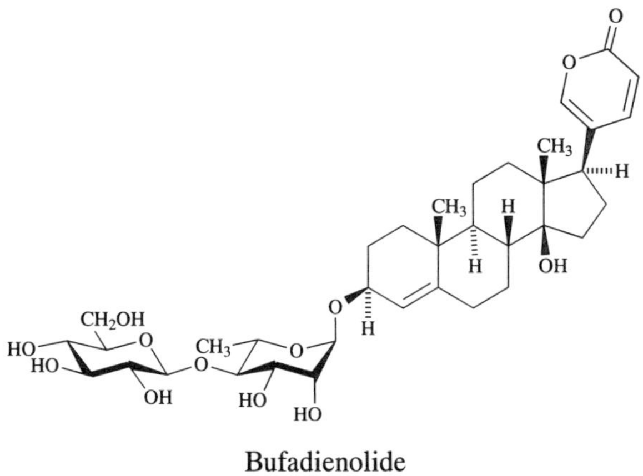

Bufadienolide

$C_{36}H_{52}O_{13}$ [124-99-2] MW 692.80

White form of squill or sea onion, *Urginea maritima* (Liliaceae).

Very toxic, with an LD_{50} intravenously in cats of 0.143 mg/kg body-weight. It has a very bitter taste and is used as a cardiotonic.

Stoll, A., *Helv. Chim. Acta,* 1951, **34**, 1431.

Scilliroside 1241

Scillirosidin 3-glucoside; Silmurin

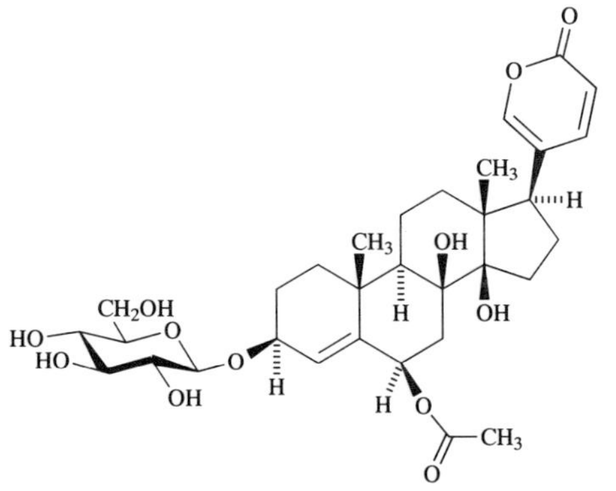

Bufadienolide

$C_{32}H_{44}O_{12}$ [507-60-8] MW 620.69

Red form of the squill or sea onion, *Urginea maritima* (Liliaceae).

Very toxic, with an LD_{50} orally in male rats of 0.7 mg/kg body-weight. It is an effective raticide.

v. Wartburg, A., *Helv. Chim. Acta,* 1959, **42**, 1620.

Sclareol 1242

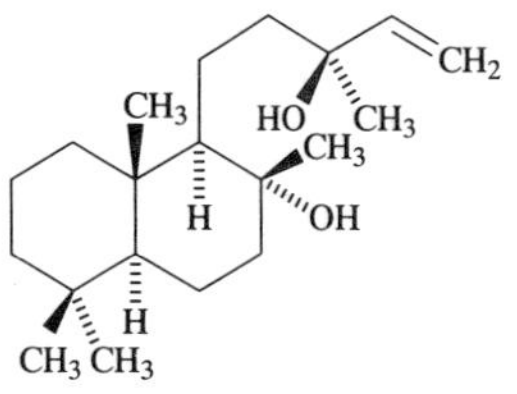

Diterpenoid
(Labdane)

$C_{20}H_{36}O_2$ [515-03-7] MW 308.50

Leaf surface wax of *Nicotiana glutinosa* (Solanaceae) and leaf of *Salvia sclarea* (Labiatae). In *Nicotiana* sp., is accompanied by the epimer, 13-episclareol [4630-08-4].

Antifungal. Both sclareol and episclareol at concentrations of 5-100 μg/ml inhibited the growth of 16 out of 18 fungal species.

Bailey, J. A., *J. Gen. Microbiol.,* 1974, **85**, 57.

Scorpioidin 1243

Scorpioidine

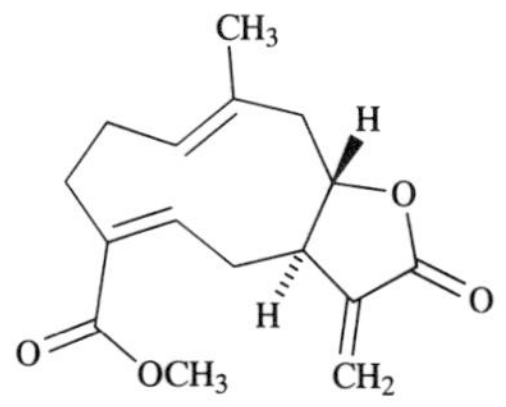

Sesquiterpenoid lactone
(Germacranolide)

$C_{16}H_{20}O_4$ [76045-40-4] MW 276.33

Aerial parts of *Vernonia scorpioides* (Compositae). Scorpioidine is also the name of a pyrrolidine alkaloid from *Myosotis scorpioides*.

Larvicidal and anthelmintic properties. Antifeedant to the locust at 0.1% concentration.

Drew, M. G. B., *J. Chem. Soc., Chem. Commun.,* 1980, 802.

Scullcapflavone II 1244

Neobaicalein

Flavone

$C_{19}H_{18}O_8$ [55084-08-7] MW 374.35

Root of *Scutellaria baicalensis* (Labiatae).

Cytotoxic properties.

Takido, M., *Yakugaku Zasshi,* 1975, **95**, 108.

Scytonemin A 1245

Cyclic peptide

$C_{71}H_{106}N_{12}O_{21}$ [112793-66-5] MW 1463.69

The cyanophyte *Scytonema* sp., obtained from a soil sample in the Marshall Islands.

Antibacterial and antifungal activity.

Helms, G. L., *J. Org. Chem.,* 1988, **53**, 1298.

Securinine 1246

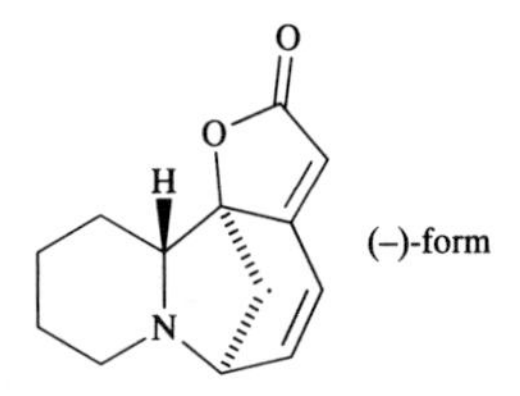

Securinega alkaloid

$C_{13}H_{15}NO_2$ [5610-40-2] MW 217.27

Leaves, roots and stems of *Securinega suffruticosa* and *Phyllanthus discoides* (Euphorbiaceae) and bark of *Securidaca longepedunculata* (Leguminosae).

It is a central nervous system stimulant with strychnine-like activity and increase of blood pressure. Used medically in the treatment of paralysis following an infectious disease.

Horii, Z., *Tetrahedron,* 1963, **19**, 2101.

L-Selenocystathionine 1247

Amino acid

$C_7H_{14}N_2O_4Se$ [2196-58-9] MW 269.16

Seeds of accumulating plants growing on soils rich in soluble selenium salts, which substitute for sulfur. These plants include *Stanleya pinnata* (Cruciferae), monkey nut, *Lecythis ollaria* (Lecythidaceae), *Astralagus* spp. and *Neptunia amplexicaulis* (Leguminosae).

Toxic in livestock, causing abdominal pain, nausea, vomiting, diarrhoea and, within a week or so, loss of scalp and body hair. It causes the acute selenium poisoning known as "blind staggers".

Horn, J., *J. Biol. Chem.,* 1941, **139**, 649.

Sempervirine 1248

Sempervirene; Sempervine

Indole alkaloid
(Yohimbane)

$C_{19}H_{16}N_2$ [6882-99-1] MW 272.35

Roots and rhizomes of the Carolina or yellow jessamine, *Gelsemium sempervirens* and *Mostuea buchholzii* (Loganiaceae).

Although it occurs in a poisonous plant, *Gelsemium*, it is not the major toxin, which is gelsemine (q.v.). However, it does possess antitumour properties.

Woodward, R. B., *J. Am. Chem. Soc.,* 1949, **71**, 379.

Senaetnine 1249

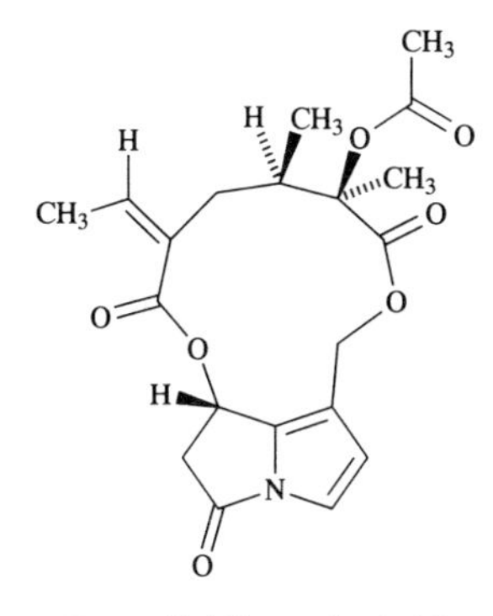

Pyrrolizidine alkaloid
(Senecionan)

$C_{20}H_{23}NO_7$ [64191-69-1] MW 389.41

Senecio aetnensis (Compositae).

Causes damage to pulmonary vascular tissue when fed to rats but does not produce the symptoms of hepatoxicity, as do most other pyrrolizidine alkaloids.

Bohlmann, F., *Chem. Ber.,* 1978, **111**, 3009.

Senampeline A 1250

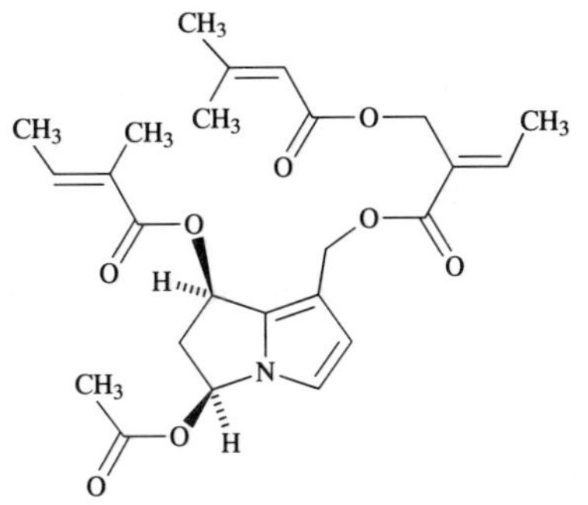

Pyrrolizidine alkaloid

$C_{25}H_{31}NO_8$ [62787-00-2] MW 473.52

Senecio aetnensis, *S. aucheri* and *S. pterophorus* (Compositae).

Suspected of being hepatotoxic, but this has yet to be proved beyond doubt.

Bohlmann, F., *Chem. Ber.*, 1977, **110**, 474.

Senecionine 1251

12-Hydroxysenecionane-11,16-dione

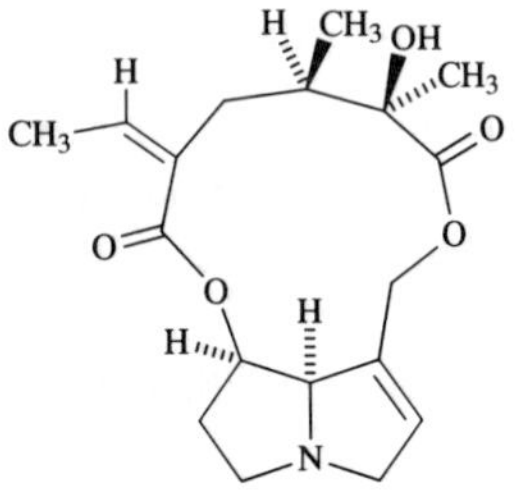

Pyrrolizidine alkaloid
(Senecionan)

$C_{18}H_{25}NO_5$ [130-01-8] MW 335.40

Common groundsel, *Senecio vulgaris*, tansy ragwort, *S. jacobaea* and many other *Senecio* spp. (Compositae).

Hepatotoxic, pneumotoxic and genotoxic. It is a cause of grazing toxicity in livestock. The LD_{50} orally in rats is 85 mg/kg body-weight.

Koekemoer, M. S., *J. Chem. Soc.*, 1955, 63.

Seneciphylline 1252

Jacodine; α-Longilobine

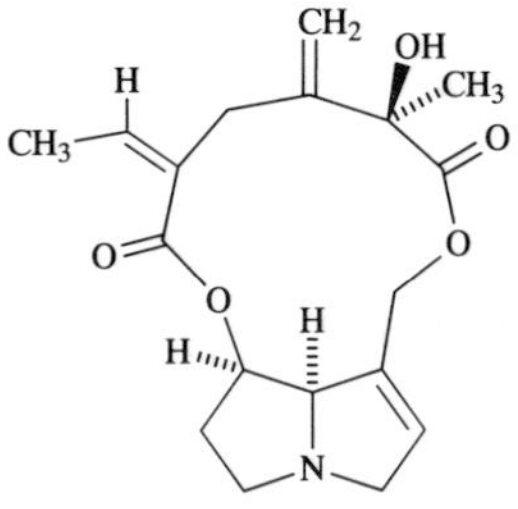

Pyrrolizidine alkaloid
(Senecionan)

$C_{18}H_{23}NO_5$ [480-81-9] MW 333.38

Senecio platyphyllus, *S. phillipicus* and many other *Senecio* spp. (Compositae); also present in *Crotalaria juncea* (Leguminosae).

Hepatotoxic, cardiotoxic and pneumotoxic. It is one of the *Senecio* alkaloids responsible for livestock poisoning, following grazing on these plants. The LD_{50} orally in rats is 77 mg/kg body-weight and such a dose will undoubtedly be reached, since it is a cumulative poison. For instance, liver cirrhosis developed in cattle in Switzerland following the ingestion of *Senecio alpinus* with an alkaloid content (including seneciphylline) of 0.3-0.4% dry-weight.

Warren, F. L., *Fortschr. Chem. Org. Naturst.*, 1955, **12**, 198.

Senecivernine 1253

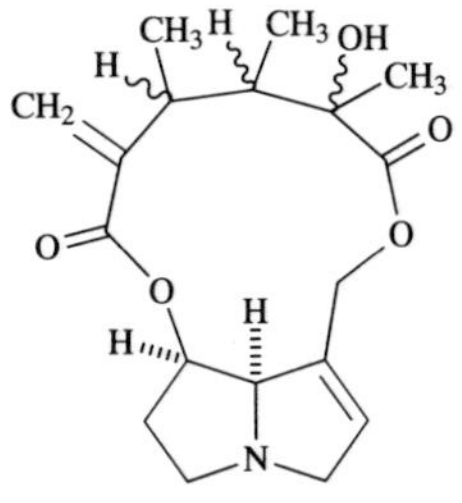

Pyrrolizidine alkaloid

$C_{18}H_{25}NO_5$ [72755-25-0] MW 335.40

Senecio vernalis and *S. seratophiloides* (Compositae).

Hepatotoxic.

Röder, E., *Planta Med.*, 1979, **37**, 131.

Senkirkine 1254

Renardine

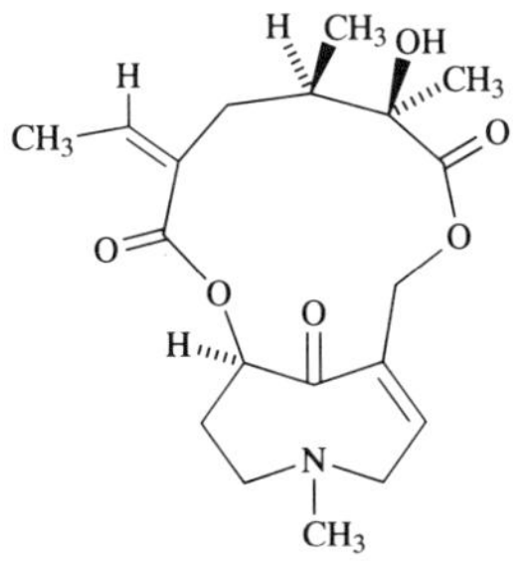

Pyrrolizidine alkaloid
(Senecionan)

$C_{19}H_{27}NO_6$ [2318-18-5] MW 365.43

Emilia sonchifolia, *Brachyglottis repanda*, *Farfugium japonicum*, *Petasites albus*, *P. hybridus*, *Senecio jacobaea*, *S. kirki*, *S. illinutus*, *S. renardii*, *S. verularis* and coltsfoot, *Tussilago farfara* (Compositae); also in *Crotalaria laburnifolia* (Leguminosae).

Hepatocarcinogenic in animals, and also hepatotoxic. Could be the cause of poisoning in humans, following the ingestion of herbal teas based on *Senecio* or *Crotalaria* leaves or of coltsfoot leaves as a herbal remedy. Contamination of cereal grain with seeds of plants containing pyrrolizidine alkaloids has occurred in India and Afghanistan, leading to poisoning epidemics with a high human mortility rate.

Briggs, L. H., *J. Chem. Soc.,* 1965, 2492.

Sennoside A 1255

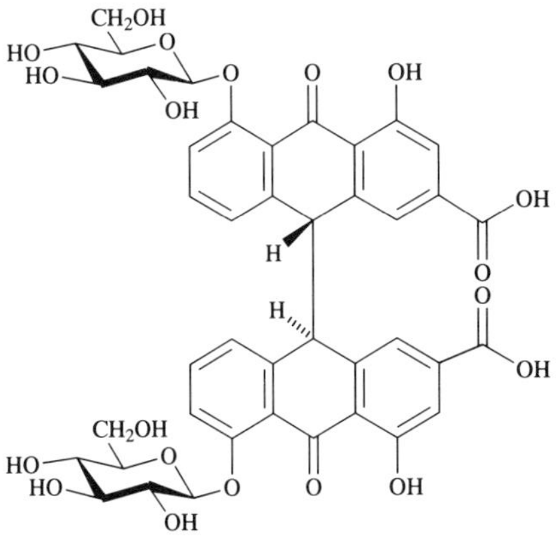

Bianthrone

$C_{42}H_{38}O_{20}$ [81-27-6] MW 862.75

Leaves of *Cassia senna*, fruits of *C. angustifolia* (Leguminosae) and rhizomes of *Rheum palmatum* (Polygonaceae).

Cathartic properties. Used medicinally in the treatment of chronic constipation and hence a component of various herbal remedies.

Stoll, A., *Helv. Chim. Acta,* 1950, **33**, 313.

Serpentine 1256

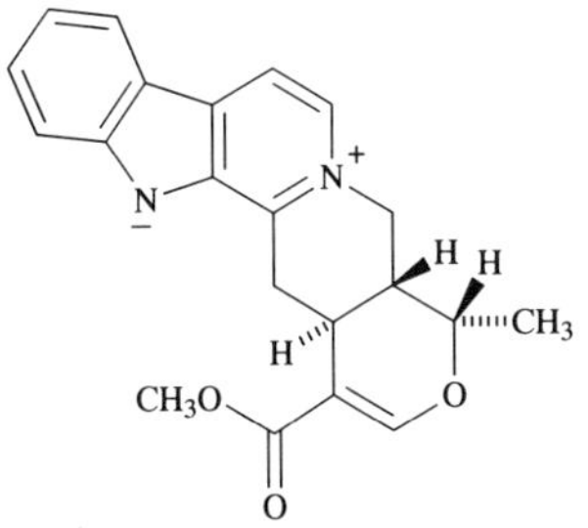

Indole alkaloid
(Oxayohimbane)

$C_{21}H_{20}N_2O_3$ [18786-24-8] MW 348.40

Rauwolfia serpentina, *R. beddomei*, *R. fruticosa*, *Vinca major* and *V. rosea* (Apocynaceae).

Antihypertensive drug. It also exhibits antitumour properties.

Schlittler, E., *Helv. Chim. Acta,* 1954, **37**, 1912.

Sesamol 1257

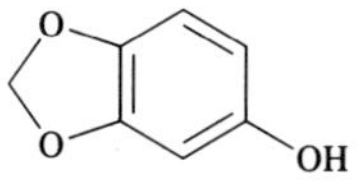

Phenol

$C_7H_6O_3$ [533-31-3] MW 138.12

Oil of *Sesamum indicum* (Pedaliaceae).

Causes allergic skin reactions in humans.

Böeseken, J., *Rec. Trav. Chim.,* 1936, **55**, 815.

Sesartemin 1258

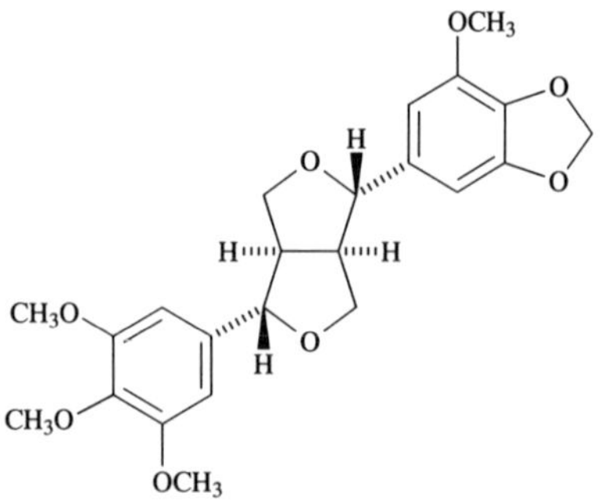

Lignan

$C_{23}H_{26}O_8$ [77394-27-5] MW 430.45

Bark of *Virola elongata* (Myristicaceae) and root of *Artemisia absinthium* (Compositae).

Inhibits the gut microsomal monooxygenase of *Ostrinia nubilalis* (corn borer). The bark resin of *V. elongata* is used as a hallucinogenic snuff and as an arrow poison by various Indian tribes.

Gregor, H., *Tetrahedron,* 1980, **36**, 3551.

Sesbanimide A 1259

Sesbanimide

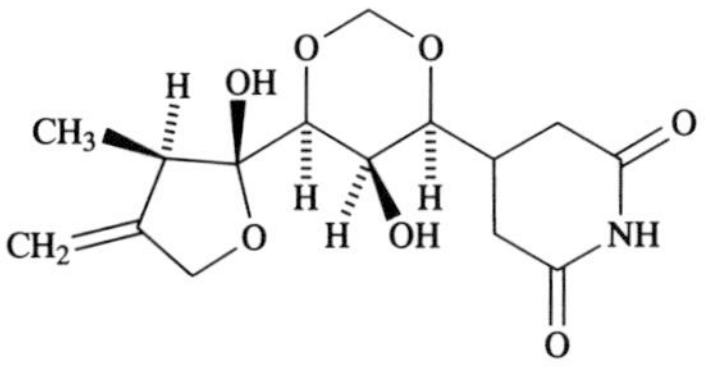

Piperidine alkaloid

$C_{15}H_{21}NO_7$ [85719-78-4] MW 327.33

Seeds of rattlebox, *Sesbania drummondii* and seeds of *S. punicea* (Leguminosae). It co-occurs with two closely related alkaloids sesbanimide B and C.

Sesbania has a long history of toxicity to livestock, but it is not known whether sesbanimide A is the active principle. The seeds are toxic to birds. Sesbanimide A shows antitumour properties, markedly active in the PS leukaemia and KB cell culture systems.

Powell, R. G., *Phytochemistry,* 1984, **23**, 2789.

Shikimic acid 1260

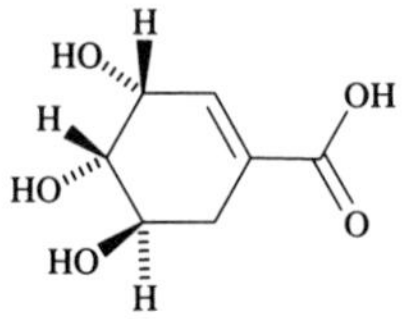

Organic acid

$C_7H_{10}O_5$ [138-59-0] MW 174.15

Fruit of *Illicium religiosum* (Magnoliaceae) and frond of bracken, *Pteridium aquilinum* (Pteridaceae); also a minor acid in several plant fruits, including gooseberry, cherry and strawberry. Universally present in plants in trace amounts.

Powerful mutagen. Once thought to be responsible for the carcinogenic effects of bracken on cattle grazing on this fern, but a potent carcinogen ptaquiloside (q.v.) has since been identified in bracken.

Dangschat, G., *Biochim. Biophys. Acta,* 1950, **4**, 199.

Shikodonin 1261

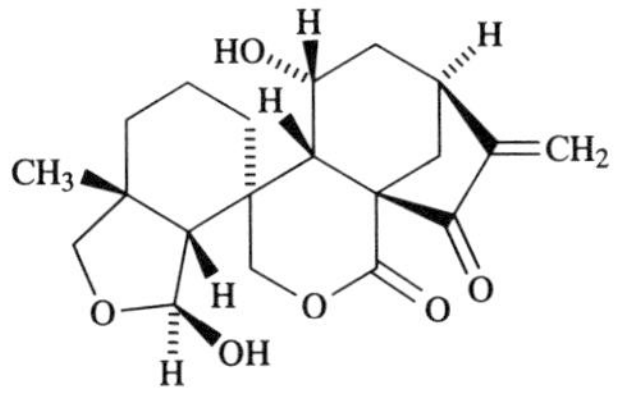

Diterpenoid
(Seco-*ent*-kaurane)

$C_{20}H_{26}O_6$ [66548-00-3] MW 362.42

Isodon shikokianus var. *intermedius* (Labiatae).

Antitumour and insecticidal properties.

Kubo, I., *J. Am. Chem. Soc.*, 1978, **100**, 628.

Shikonin 1262

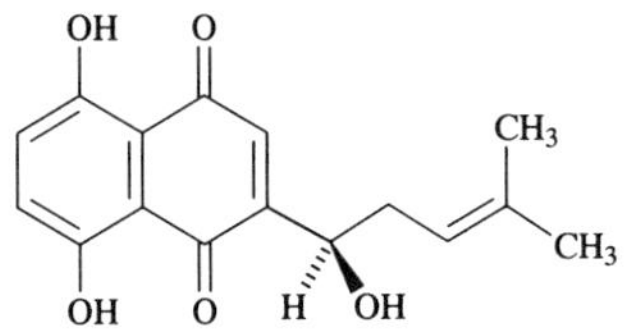

Naphthoquinone

$C_{16}H_{16}O_5$ [517-89-5] MW 288.30

Roots of *Lithospermum erythrorhizon*, in *Echium lycopsis* and *Onosma caucasicum* (Boraginaceae).

Red dye. Produced commercially in Japan from *Lithospermum* cell cultures, and is used medicinally, and for colouring lipsticks. Shows antitumour activity.

Arakawa, H., *Chem. Ind. (London)*, 1961, 947.

Shiromodiol diacetate 1263

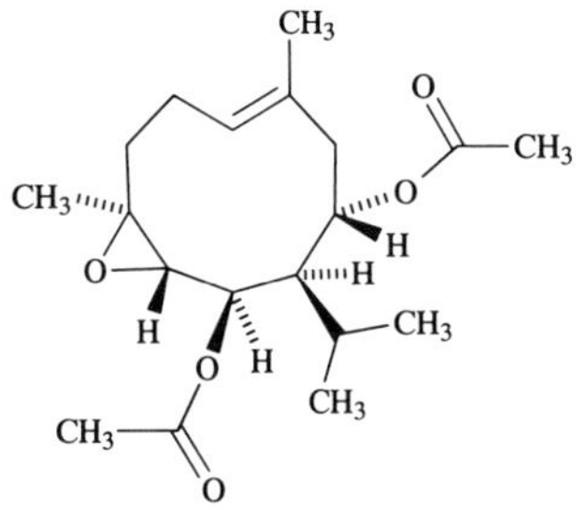

Sesquiterpenoid
(Germacrane)

$C_{19}H_{30}O_5$ [13095-59-9] MW 338.44

Leaves of *Lindera triloba* (Lauraceae).

Insect antifeedant.

McClure, R. J., *J. Chem. Soc., Chem. Commun.*, 1970, 128.

Simalikilactone D 1264

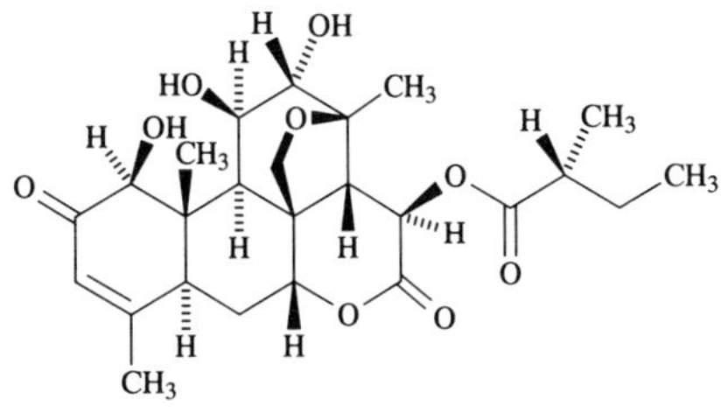

Nortriterpenoid
(Quassane)

$C_{25}H_{34}O_9$ [35321-80-3] MW 478.54

Quassia africana (Simaroubaceae).

Very potent antifeedant to the southern armyworm, at the level of 50 p.p.m. Also has amoebicidal, antimalarial and antileukaemic properties.

Tresca, J. P., *C. R. Acad. Sci., Ser. C*, 1971, **273**, 601.

Simplexin 1265

Wikstrotoxin D

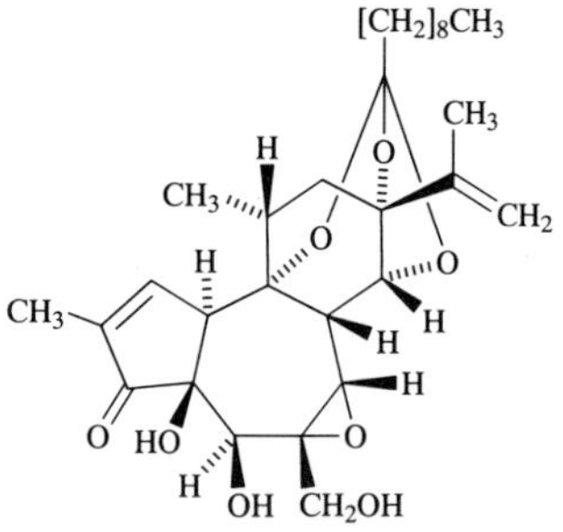

Diterpenoid
(Daphnetoxane)

$C_{30}H_{44}O_8$ [1404-62-2] MW 532.67

Pimelea simplex (Thymelaeaceae).

Poisonous, causing the 'St. George' disease in cattle, a cardiopulmonary syndrome resulting from regular grazing on the above plant.

Jolad, S. D., *J. Nat. Prod.,* 1983, **46**, 675.

Sinharine 1266

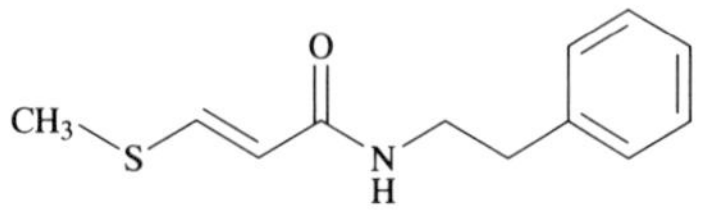

Phenylpropanoid
(Sulfur compound)

$C_{12}H_{15}NOS$ [142717-65-5] MW 221.32

Leaves of *Glycosmis cyanocarpa* (Rutaceae).

Antifungal activity against *Cladosporium cladosporioides.*

Johnson, W. M., *Aust. J. Chem.,* 1994, **47**, 751.

Sinomenine 1267

Cucoline

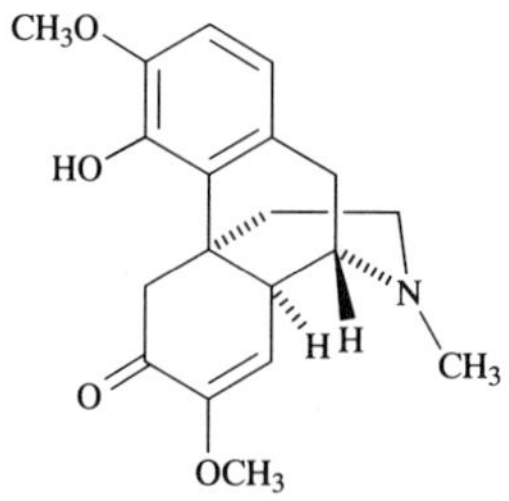

Isoquinoline alkaloid
(Morphinan)

$C_{19}H_{23}NO_4$ [115-53-7] MW 329.40

Sinomenium acutum (Menispermaceae).

Abortifacient in large doses. It has weak analgesic properties.

Goto, K., *Justus Leibigs Ann. Chem.,* 1931, **485**, 247.

Solacapine 1268

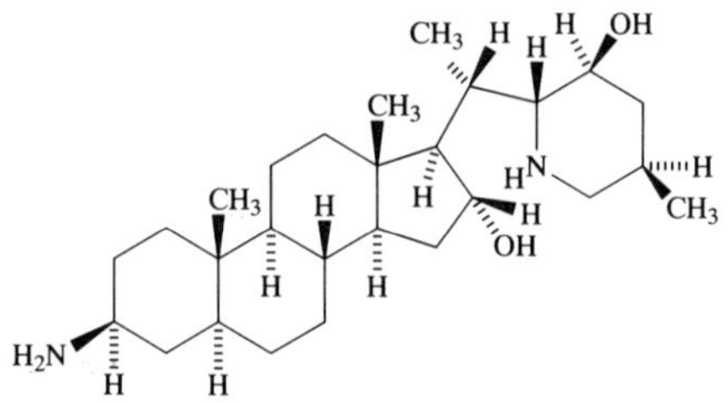

Steroid alkaloid
(Cholestane)

$C_{27}H_{48}N_2O_2$ [63785-22-8] MW 432.69

All parts of *Solanum pseudocapsicum* or Jerusalem cherry (Solanaceae).

Eating a few of the berries causes nausea, abdominal pains, dilation of the pupils and drowsiness. Toxicity is much less when taken orally as compared to intraperitoneal injection.

Chakravarty, A. K., *J. Chem. Soc., Perkin Trans. 1,* 1984, 467.

Solanidine 1269

Purapuridine; Solanidine T; Solatubine

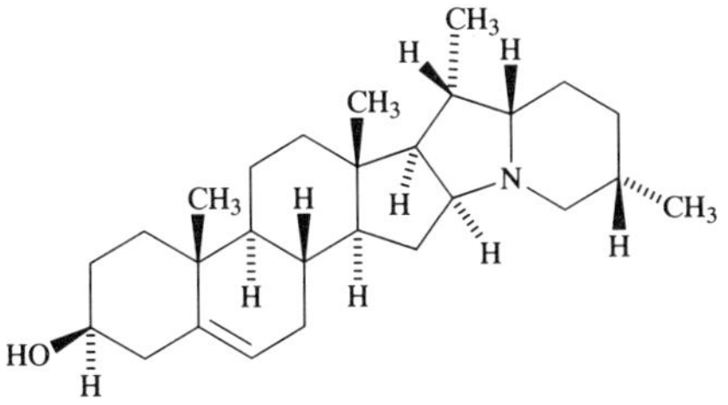

Steroid alkaloid
(Solanidane)

$C_{27}H_{43}NO$ [80-78-4] MW 397.64

The cultivated potato, *Solanum tuberosum* and in many wild *Solanum* spp., especially black nightshade, *S. nigrum* (Solanaceae). Potato sprouts can contain 0.008% solanidine.

One of the toxic constituents of the domestic potato, together with α-solanine (q.v.).

Manske, R. H. F., The Alkaloids, 1953, **3**, 247, Academic Press.

α-Solanine 1270

Solatunine

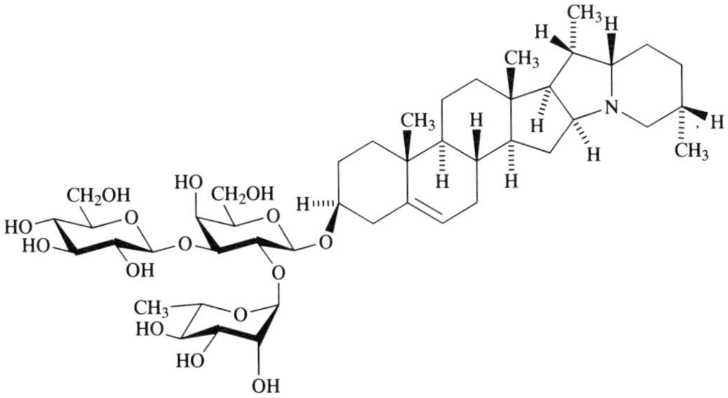

Steroid alkaloid
(Solanidane)

$C_{45}H_{73}NO_{15}$ [20562-02-1] MW 868.07

In domestic potato, *Solanum tuberosum*, where it occurs with the closely related glycosides, β-solanine [61877-94-9] and γ-solanine [511-37-5]. The alkaloid content is mainly in the leaves (0.5%), flowers (0.7%), fruits (1.0%) and sprouts (0.8-5.0% dry-weight). There are normally only trace amounts (7 mg/100g) in the tubers, but can reach toxic levels (35 mg/100g) in 'greened' tissue. α-Solanine also occurs in woody nightshade, *S. nigrum* and in the tomato, *Lycopersicon esculentum* (Solanaceae), with tomatine (q.v.).

Human toxicity symptoms include vomiting, diarrhoea, hallucination and coma. Oral ingestion of 2.8 mg/kg body-weight is toxic in man. Poisoning from potato tubers is relatively rare since cooking and peeling reduce alkaloid levels considerably. Farm animals can suffer serious poisoning if fed potato tops or sprouted potatoes.

Kuhn, R., *Angew. Chem.,* 1954, **66**, 639.

Solanocapsine 1271

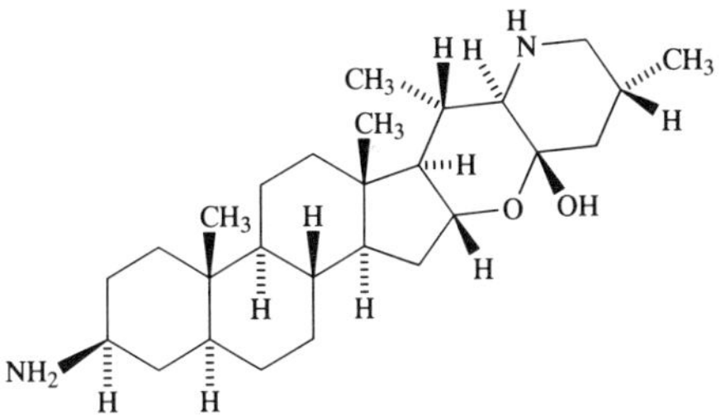

Steroid alkaloid
(Secosolanidan)

$C_{27}H_{46}N_2O_2$ [639-86-1] MW 430.67

All parts of Jerusalem cherry, *Solanum pseudocapsicum* (Solanaceae).

Eating a few of the berries causes nausea, abdominal pains, dilatation of the pupils and drowsiness. Toxicity is much less when taken orally as compared to intraperitoneal injection.

Ripperger, H., *Justus Liebigs Ann. Chem.,* 1969, **723**, 159.

Solasodine 1272

Purpapuridine; Solancarpidine; Solanidine-S

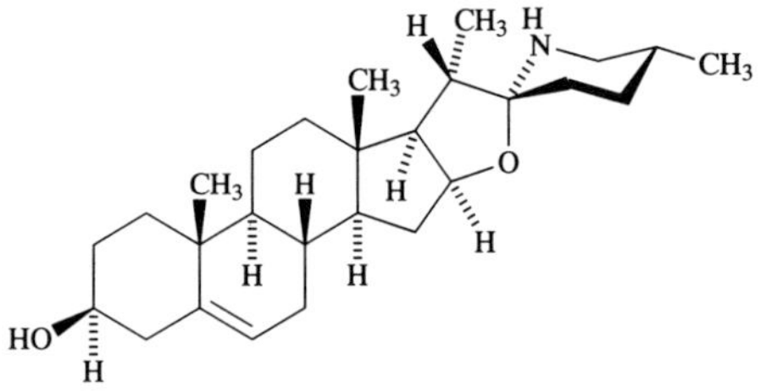

Steroid alkaloid
(Spirosolane)

$C_{27}H_{43}NO_2$ [126-17-0] MW 413.64

Widespread in *Solanum* spp., in glycosidic combination, with high concentrations in *S. laciniatum* (1-2% solasodine) and in *S. nigrum* (Solanaceae); occurs in the fruit of bittersweet, *S. dulcamara*, together with α- and β-soladulcine and tomatidenol [37337-73-8, 11093-43-9 and 546-40-7 respectively].

Teratogenic when fed to rats and guinea-pigs. Poisoning from fruits of bittersweet is relatively unlikely because the alkaloid content drops sharply during ripening, so that a fatal dose would require the consumption of many fruit.

Briggs, L. H., *J. Chem. Soc.*, 1950, 3013.

Solasonine 1273

Purapurine; Solanine-S; Solasodamine

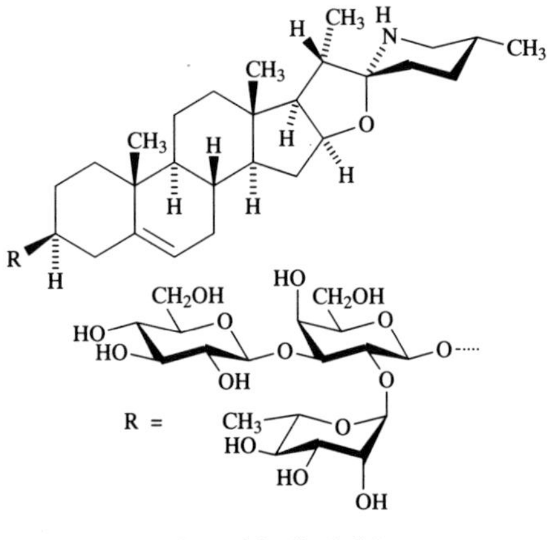

Steroid alkaloid
(Spirosolane)

$C_{45}H_{73}NO_{16}$ [19121-58-5] MW 884.07

Fruits of *Solanum aviculare*, *S. sodomeum*, *S. torvum*, *S. viarum S. xanthocarpum*, *S. incanum* and *S. melongena* (Solanaceae).

Potentially toxic, if several fruits are consumed. The fruit of *Solanum sodomeum* is known to be lethal to cockroaches.

Briggs, L. H., *J. Chem. Soc.*, 1963, 2848.

Solavetivone 1274

Katahdinone

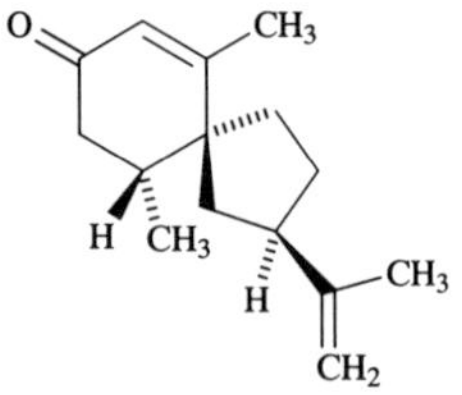

Sesquiterpenoid
(Vetispirane)

$C_{15}H_{22}O$ [54878-25-0] MW 218.34

Tubers of potato, *Solanum tuberosum* (Solanaceae) infected with *Phytophthora infestans.*

Antifungal agent (phytoalexin).

Anderson, R. C., *J. Chem. Soc., Chem. Commun.*, 1977, 27.

Songorine 1275

Napellonine; Bullatine G

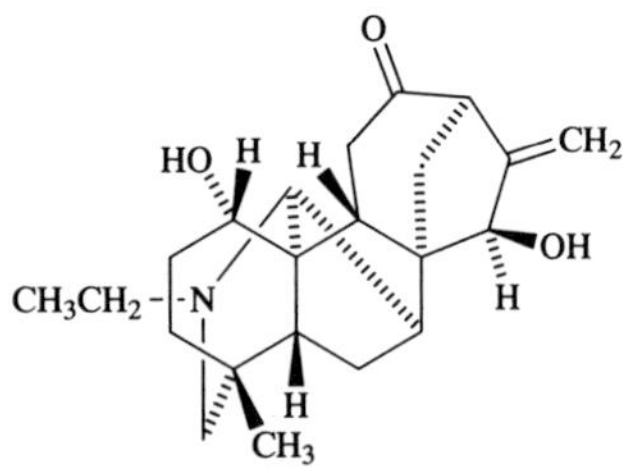

Diterpenoid alkaloid
(Napellane)

$C_{22}H_{31}NO_3$ [509-24-0] MW 357.49

Aconitum karakolicum, *A. soongaricum* and *A. monticola* (Ranunculaceae).

Acutely toxic to mice at relatively low doses (20 mg/kg body-weight), causing fall in blood pressure. Larger doses lead to a decrease in motor activity, to respiratory difficulties, tremor and clonicotonic convulsions.

Okamoto, T., *Chem. Pharm. Bull.,* 1965, **13**, 1270.

Soularubinone 1276

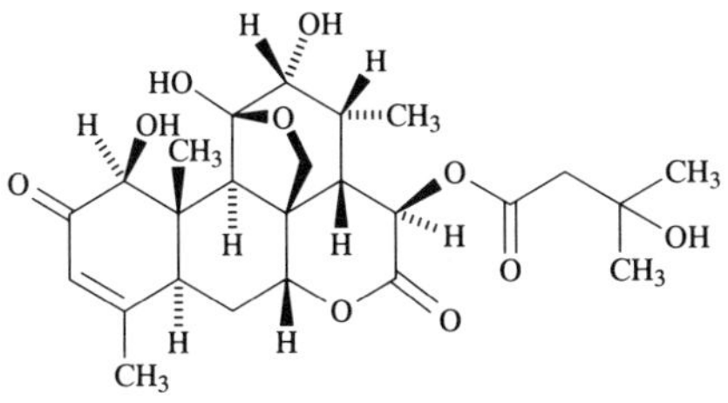

Nortriterpenoid
(Quassane)

$C_{25}H_{34}O_{10}$ [74156-49-3] MW 494.54

Leaves of *Soulamea tomentosa* (Simaroubaceae).

Antimalarial activity at a concentration as low as 0.006 μg/ml. Also antileukaemic properties.

Van Tri, M., *J. Nat. Prod.,* 1981, **44**, 279.

(−)-Sparteine 1277

Lupinidine; Pachycarpine

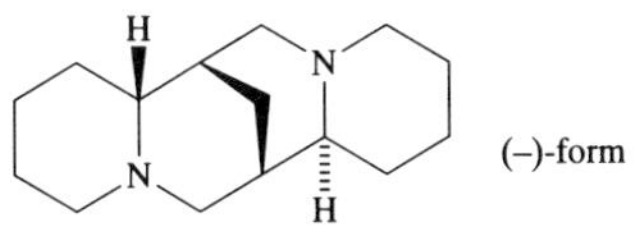

(−)-form

Quinolizidine alkaloid
(Sparteine)

$C_{15}H_{26}N_2$ [90-39-1] MW 234.38
(+)-form [492-08-0]

Commonly occurring in several legume genera, *Baptisia*, *Cytisus*, *Lupinus* and *Sarothamnus* spp. (Leguminosae).

Cause of livestock poisoning after grazing on wild lupin species, and also toxic to most insects. Has diuretic and hypoglycaemic properties.

Clemo, G. R., *J. Chem. Soc.,* 1933, 644.

Spatheliabischromene 1278

Cneorum chromone A

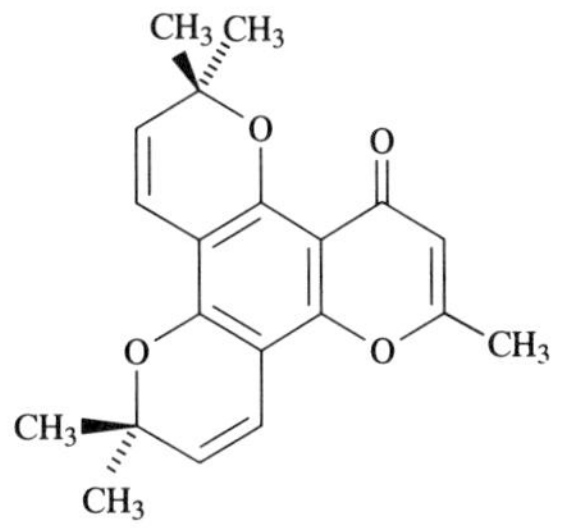

Chromone

$C_{20}H_{20}O_4$ [34411-93-3] MW 324.38

Cneorum tricoccum, *C. pulverulentum* (Cneoraceae), *Spathelia glabrescens* and *S. sorbifolia* (Rutaceae).

Active against HeLa cells.

González, A. G., *Phytochemistry,* 1974, **13**, 2305.

Sphondin 1279

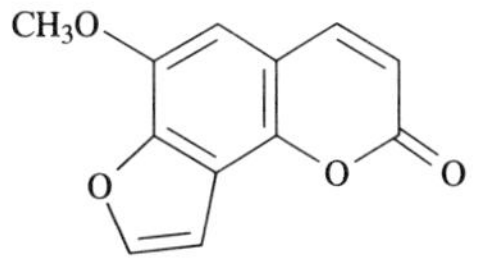

Furocoumarin

$C_{12}H_8O_4$ [483-66-9] MW 216.19

Seeds and leaves of wild parsnip, *Pastinaca sativa* (Umbelliferae).

Toxic to the parsnip webworm, *Depressaria pastinacella*.

Späth, E., *Ber. Dtsch. Chem. Ges.,* 1941, **74B**, 595.

Spicatin 1280

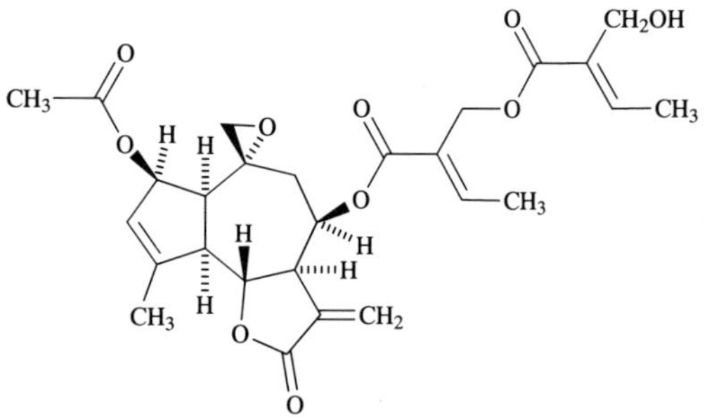

Sesquiterpenoid lactone
(Guaianolide)

$C_{27}H_{32}O_{10}$ [53142-46-4] MW 516.55

Liatris spicata, *L. pycnostachya* and *L. tenuifolia* (Compositae).

Cytotoxic and antitumour properties.

Herz, W., *J. Org. Chem.,* 1975, **40**, 199.

Spinoside A 1281

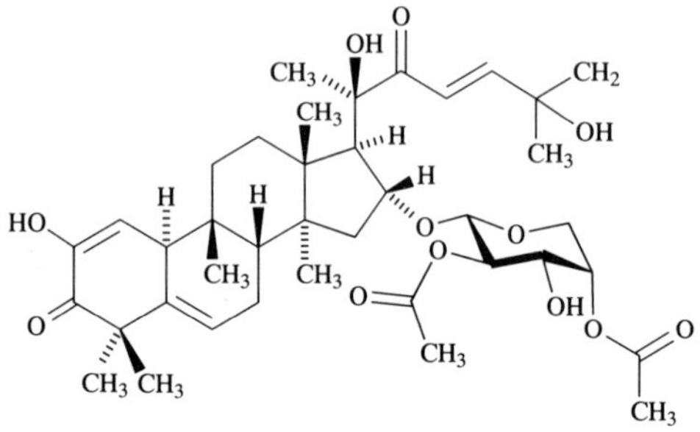

Triterpenoid
(Cucurbitane)

$C_{39}H_{56}O_{12}$ [119626-74-3] MW 716.87

Desfontainia spinosa (Desfontainiaceae), together with the closely related spinoside B [119626-75-4].

Spinoside A and spinoside B both show cytotoxic properties.

Reddy, K. S., *Phytochemistry,* 1988, **27**, 3781.

Spirobrassinin 1282

Spirobrassinine

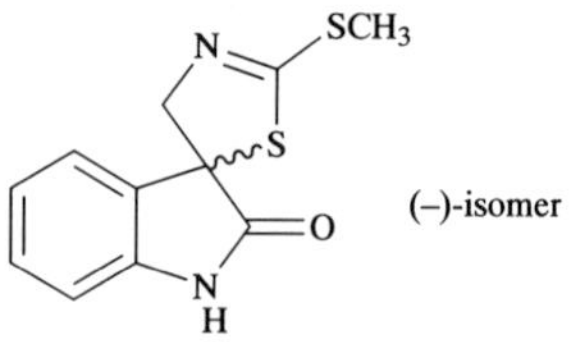

(−)-isomer

Indole

$C_{11}H_{10}N_2OS_2$ [113866-40-3] MW 250.35

One of nine similar indoles formed in fungally-infected cabbage leaves, *Brassica campestris* (Cruciferae).

Antifungal agent (phytoalexin).

Takasugi, M., *Chem. Lett.,* 1987, 1631.

Sporochnol A 1283

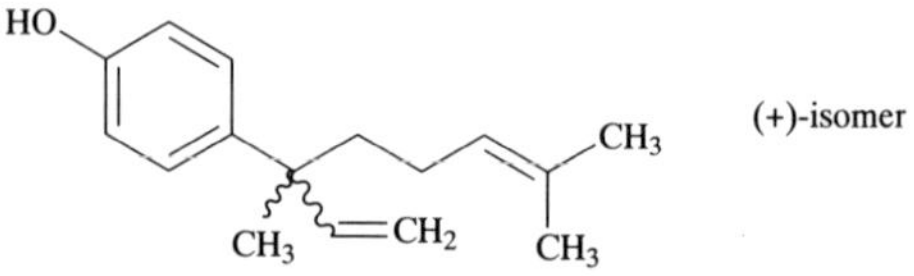

(+)-isomer

Phenol

$C_{16}H_{22}O$ [147821-59-8] MW 230.35

Brown alga, *Sporochnus bolleanus*.

Antifeedant to marine herbivores.

Shen, Y. C., *Phytochemistry,* 1993, **32**, 71.

Spruceanol 1284

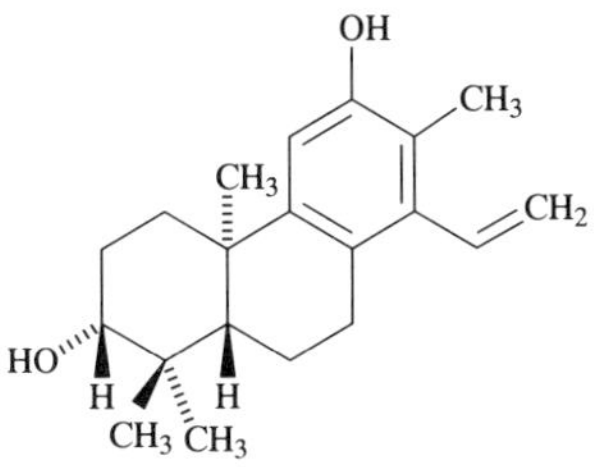

Diterpenoid
(Abeo-*ent*-pimarane)

$C_{20}H_{28}O_2$ [72963-56-5] MW 300.44

Root and root bark of *Micranda spruceana* (Euphorbiaceae).

Cytotoxic properties.

Gunasekera, S. P., *J. Nat. Prod.*, 1979, **42**, 658.

Steganacin 1285

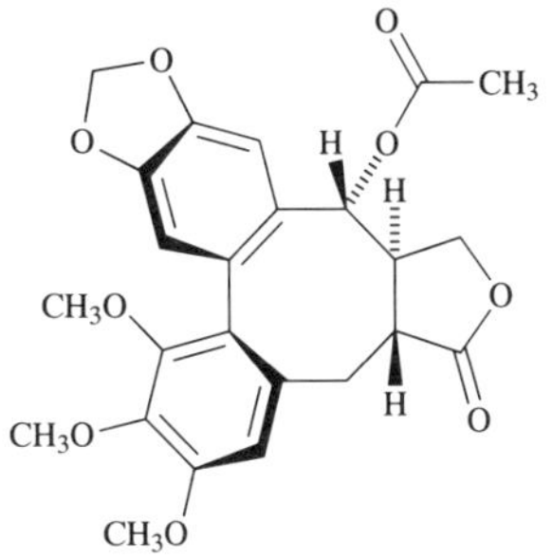

Lignan

$C_{24}H_{24}O_9$ [41451-68-7] MW 456.45

Wood, stem and bark of *Steganotaenia araliacea* (Umbelliferae).

Inhibits HeLa cell growth and possesses antileukaemic and antitumour properties.

Kupchan, S. M., *J. Am. Chem. Soc.*, 1973, **95**, 1335.

Sterculic acid 1286

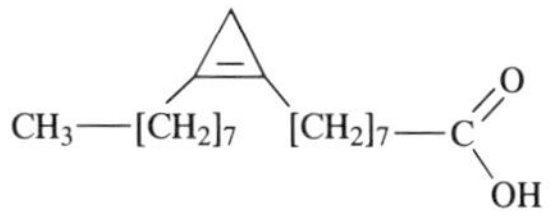

$CH_3—[CH_2]_7$ $[CH_2]_7—C(=O)OH$

Fatty acid

$C_{19}H_{34}O_2$ [738-87-4] MW 294.48

Major fatty acid (53%) in the seed oil of *Sterculia foetida* (Sterculiaceae). Occurs in seed oils of species in Malvaceae, Sterculiaceae, Tiliaceae and Bombacaceae. A minor constituent in cottonseed oil, *Gossypium hirsutum* (Malvaceae) and seed oil of *Gnetum gnemon* (Gnetaceae).

The sterculic acid component of cottonseed oil has a deleterious effect on livestock fed this oil. Seeds of *Gnetum* are processed into chips for human consumption in Java and the sterculic acid content could be a dietary hazard.

Nunn, J. R., *J. Chem. Soc.*, 1952, 313.

Stizophyllin 1287

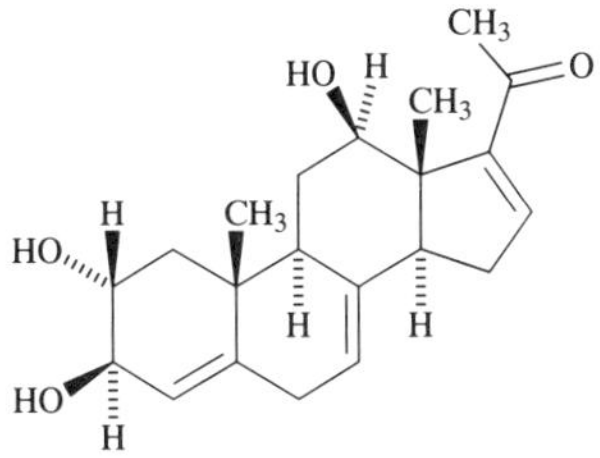

Phytosterol
(Pregnane)

$C_{21}H_{28}O_4$ [109237-00-5] MW 344.45

Stizophyllum riparium (Bignoniaceae).

Very high cytotoxicity, with an ED_{50} of 0.07 μg/ml against P-388 cell lines.

Duh, C. Y., *J. Nat. Prod.*, 1987, **50**, 63.

Stramonin-B 1288

Sesquiterpenoid lactone
(Pseudoguaianolide)

$C_{15}H_{18}O_4$ [65179-88-6] MW 262.31

Parthenium tomentosum var. *stramonium* (Compositae).

Cytotoxic and antitumour properties

Grieco, P. A., *J. Org. Chem.*, 1978, **43**, 4552.

k-Strophanthoside 1289

Strophanthin; *k*-Strophanthin;
Strophanthidin 3-diglucosylcymaroside; Combetin;
Eustrophinum

Cardenolide

$C_{41}H_{64}O_{19}$ [33279-57-1] MW 860.95

Seed of *Strophanthus kombé* and *S. arnotdianus* (Apocynaceae). The concentration in the seed of the latter plant is 0.49% dry-weight.

Very toxic to vertebrates, with an LD_{50} intravenously in rats of 15 mg/kg body-weight. The seeds of *Strophanthus kombé* have been used in Africa for preparing arrow poisons.

Reichstein, T., *Adv. Carbohydr. Chem.*, 1962, **17**, 65.

Strychnine 1290

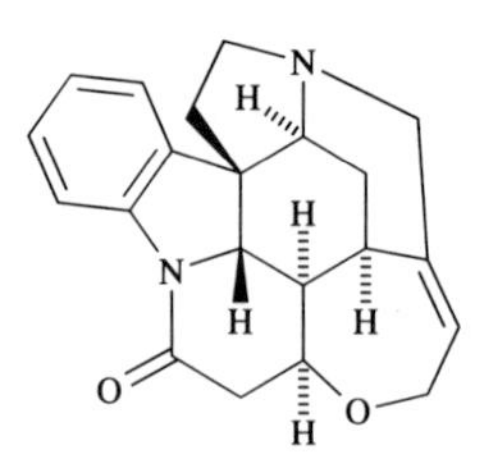

Indole alkaloid
(Strychnidine)

$C_{21}H_{22}N_2O_2$ [57-24-9] MW 334.42

Nux-vomica, the seed of *Strychnos nux-vomica*, ignatius beans, seed of *S. ignatii* and from other *Strychnos* spp. (Loganiaceae). The richest known source (6.6% dry-weight) is the bark of *S. icaja*.

Central nervous system and respiratory stimulant. Well known poison and used extensively as a rodenticide. Once used medically in low doses as a nervous tonic and appetite stimulant.

Robinson, R., *Prog. Org. Chem.*, 1952, **1**, 1.

Stypandrol 1291

Binaphthalene

$C_{26}H_{22}O_6$ [99305-33-6] MW 430.46

Blindgrass, *Stypandra imbricata* and *Dianella revoluta* (Liliaceae).

It is responsible for the toxicity of these plants after ingestion by sheep and goats, leading to paralysis and sometimes death. Those animals which survive intoxication are often blind as a side effect.

Colegate, S. M., *Aust. J. Chem.,* 1985, **38**, 1233.

Styraxin 1292

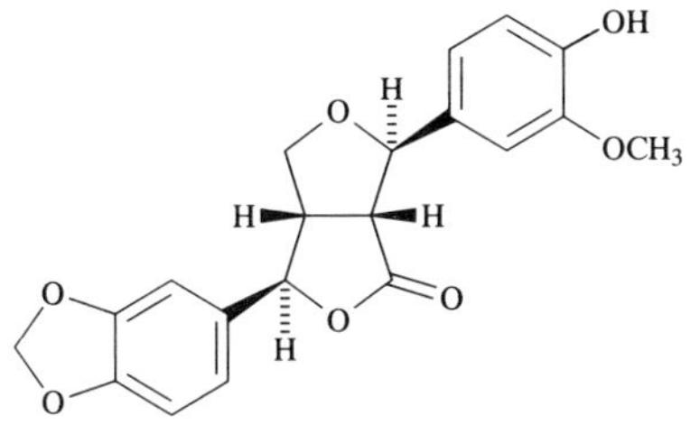

Lignan

$C_{20}H_{18}O_7$ [69742-32-1] MW 370.36

Aerial parts of *Styrax officinalis* (Styracaceae).

Antitumour properties. Used as a parasiticide in veterinary medicine.

Ulubelen, A., *Planta Med.,* 1978, **34**, 403.

Sumatrol 1293

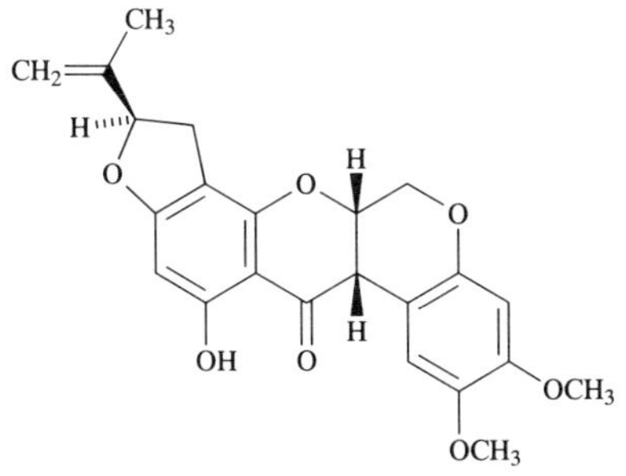

Isoflavonoid
(Rotenoid)

$C_{23}H_{22}O_7$ [82-10-0] MW 410.42

Root of *Derris malaccensis* and *Piscidia erythrina* (Leguminosae).

Insecticidal properties.

Crombie, L., *J. Chem. Soc.,* 1961, 5445.

Supinine 1294

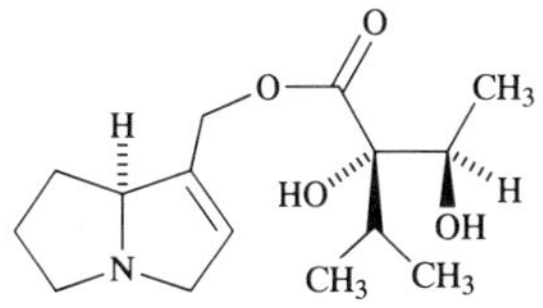

Pyrrolizidine alkaloid

$C_{15}H_{25}NO_4$ [551-58-6] MW 283.37

Heliotropium indicum, H. supinum, Tournefortia sarmentosa, T. zeylandicum (Boraginaceae), *Eupatorium cannabinum, E. serotinum* and *E. stoechadosum* (Compositae).

Hepatotoxic alkaloid. Has tumour-inhibiting activity.

Mattocks, A. R., *Nature (London),* 1968, **217**, 723.

Surangin B 1295

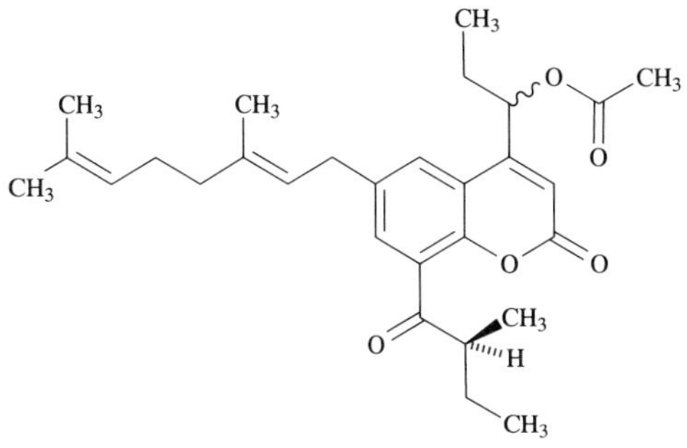

Coumarin

$C_{29}H_{38}O_7$ [28319-38-2] MW 498.62

Mammea longifolia (Guttiferae).

Insecticidal, especially toxic to mosquito larvae.

Crombie, L., *J. Chem. Soc., Perkin Trans. 1,* 1987, 345.

Swainsonine 1296

Indolizidine alkaloid

$C_8H_{15}NO_3$ [72741-87-8] MW 173.21

Swainsona canescens, *S. luteola*, *S. galagifolia*, *Oxytropis ochrocephala* and *O. kansuensis* (Leguminosae). The leaf content in the latter two plants is 0.012% and 0.21% dry-weight respectively.

Toxic to livestock feeding on *Swainsona* plants, causing symptoms similar to the genetic disorder mannosidosis, and eventually causing death. It is an α-mannosidase inhibitor. Toxic to goats feeding on *Oxytropis* plants, producing a variety of symptoms but eventually death by exhaustion.

Colegate, S. M., *Aust. J. Chem.*, 1979, **32**, 2257.

Symlandine 1297

7-Angelyl-9-(−)-viridiflorylretronecine

Pyrrolizidine alkaloid

$C_{20}H_{31}NO_6$ [74410-74-5] MW 381.47

Russian comfrey, *Symphytum* × *uplandicum* (Boraginaceae).

Toxicity not established, but suspected of being hepatotoxic.

Culvenor, C. C. J., *Aust. J. Chem.*, 1980, **33**, 1105.

Symphytine 1298

7-Tiglyl-9-(−)-viridoflorylretronecine

Pyrrolizidine alkaloid

$C_{20}H_{31}NO_6$ [22571-95-5] MW 381.47

Comfrey, *Symphytum officinale*, Russian comfrey, *S.* × *uplandicum*, *S. orientale* and the water forget-me-not, *Myosotis scorpiodes* (Boraginaceae).

Hepatocarcinogenic. The use of comfrey leaves as a salad constituent is not to be recommended because although the content of toxic alkaloid is low, it is a cumulative liver poison.

Furuya, T., *Phytochemistry,* 1971, **10**, 2217.

Synaptolepsis factor K_1 1299

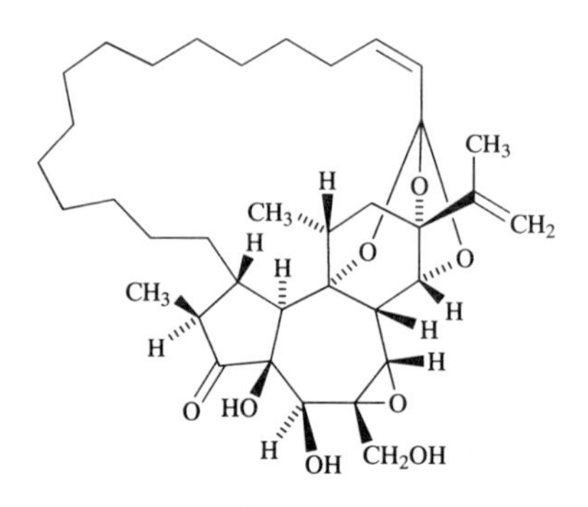

Diterpenoid
(Daphnetoxane)

$C_{36}H_{54}O_8$ [66268-94-8] MW 614.82

Leaves, stems and roots of *Synaptolepsis* spp. (Thymelaeaceae).

Highly irritant.

Zayed, S., *Tetrahedron Lett.,* 1977, 3481.

(+)-Syringaresinol 1300

Lirioresinol B

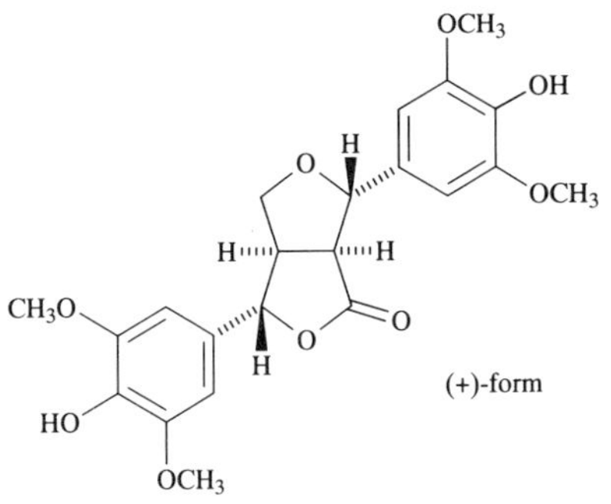

Lignan

$C_{22}H_{26}O_8$ [21453-69-0] MW 418.44
(−)-form [6216-81-5]
(±)-form [1177-14-6]

Wood of *Populus* spp. (Salicaceae) and in *Wikstroemia* spp. (Thymelaeaceae). The racemate occurs in the beech tree, *Fagus sylvatica* (Fagaceae).

Cytotoxic properties.

Bryan, R. F., *J. Chem. Soc., Perkin Trans. 2,* 1976, 341.

T

Tabernamine 1301

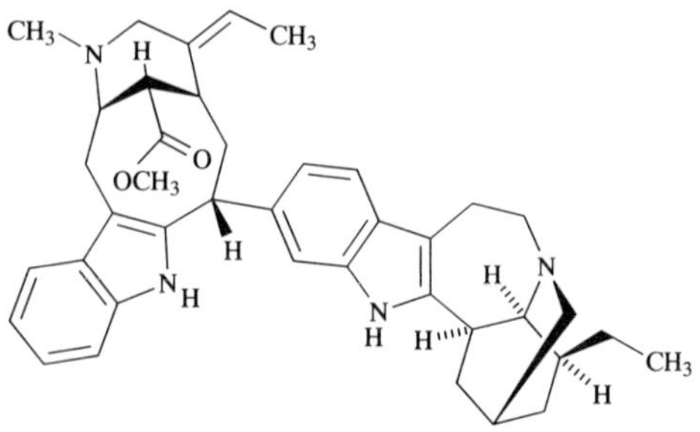

Bisindole alkaloid
(Ibogamine and vobasan)

$C_{40}H_{48}N_4O_2$ [59626-92-5] MW 616.85

Stem bark of *Tabernaemontana johnstonii* (Apocynaceae).

Cytotoxic to P-388 lymphocytic leukaemia cells *in vitro* and antitumour properties.

Kingston, D. G. I., *Tetrahedron Lett.,* 1976, 649.

Tabernanthine 1302

13-Methoxyibogamine

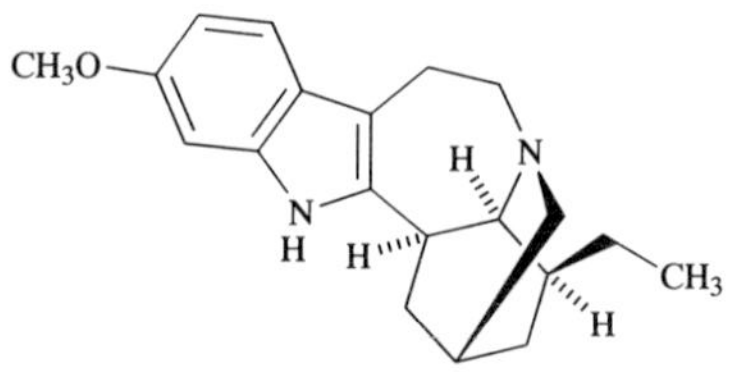

Indole alkaloid
(Ibogamine)

$C_{20}H_{26}N_2O$ [83-94-3] MW 310.44

Tabernanthe iboga and in *Conopharingia*, *Tabernaemontana* and *Stemmadenia* spp. (Apocynaceae).

Central nervous system stimulant.

Walls, F., *Tetrahedron,* 1958, **2**, 173.

Tagitinin F 1303

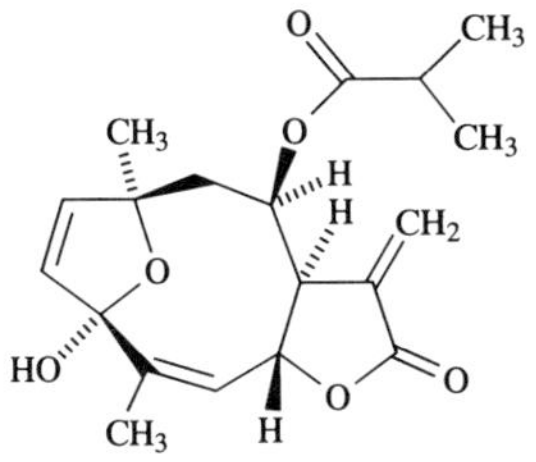

Sesquiterpenoid lactone
(Germacranolide)

$C_{19}H_{24}O_6$ [59979-57-6] MW 348.40

Tithonia tagitiflora and *T. diversifolia* (Compositae).

Cytotoxic and antitumour properties.

Pal, R., *Indian J. Chem., Sect. B,* 1977, **15B**, 208.

Tamaulipin-A 1304

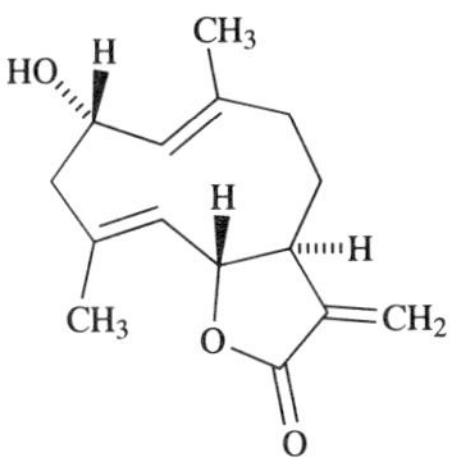

Sesquiterpenoid lactone
(Germacranolide)

$C_{15}H_{20}O_3$ [19888-11-0] MW 248.32

Ambrosia confertiflora and *A. dumosa* (Compositae).

Cytotoxic and antitumour properties.

Fischer, N. H., *Tetrahedron,* 1968, **24**, 4091.

Tangeretin 1305

Ponkanetin; 5,6,7,8,4′-Pentamethoxyflavone

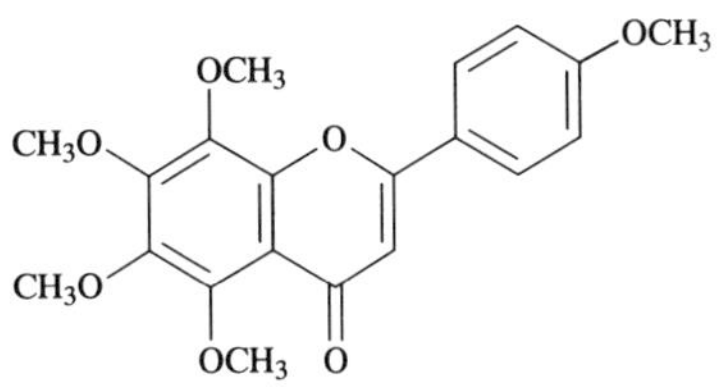

Flavone

$C_{20}H_{20}O_7$ [481-53-8] MW 372.37

Rind of *Citrus* spp. fruits.

When fed at the rate of 10 mg/kg body-weight each day to female rats during gestation, it caused the death of 83% of the offspring. It also inhibits the proliferation and invasion of malignant tumour cells *in vitro*.

Goldsworthy, L. J., *Chem. Ind. (London),* 1957, 47.

Taxifolin 1306

Dihydroquercetin; Distylin;
3,5,7,3′,4′-Pentahydroxyflavanone

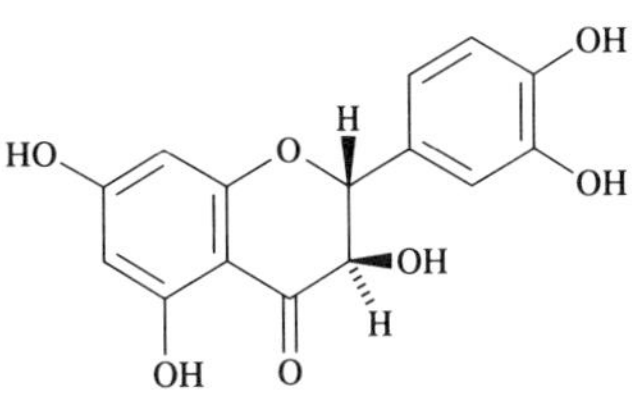

Dihydroflavonol

$C_{15}H_{12}O_7$ [480-18-2] MW 304.26

Widespread, especially in woody plants, e.g. in *Acacia catechu* (Leguminosae) and *Salix capraea* (Salicaceae).

Inhibits the growth of *Heliothis zea* larvae. Also has anti-inflammatory, antihepatotoxic and antioxidant properties.

Geissman, T. A., The Chemistry of Flavonoid Compounds, 1962, 575, Pergamon.

Taxine A 1307

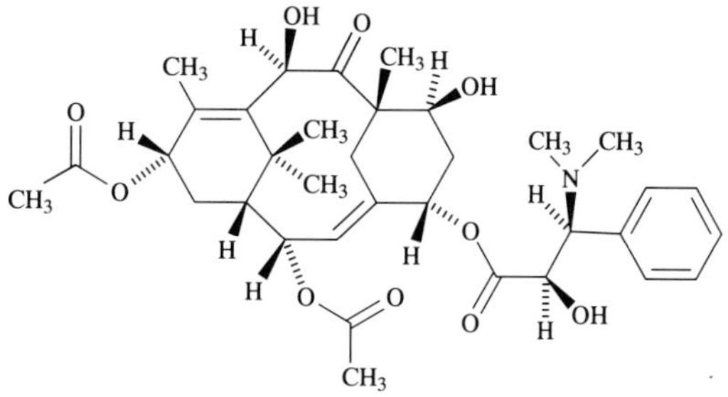

Diterpenoid alkaloid
(Taxane)

$C_{35}H_{47}NO_{10}$ [1361-49-5] MW 641.76

Leaf and all other parts, except the fruit aril, of the yew, *Taxus baccata* (Taxaceae).

A major alkaloid of the yew, responsible for cattle poisoning by yew leaves. Yew leaves have been used successfully for suicide attempts, because there is no known antidote. Symptoms include nausea, dizziness, abdominal pains and shallow breathing. Death occurs through respiratory paralysis with the heart in diastolic arrest.

Graf, E., *Justus Liebigs Ann. Chem.*, 1982, 376.

Taxiphyllin 1308

Phyllanthoside; Phyllanthin

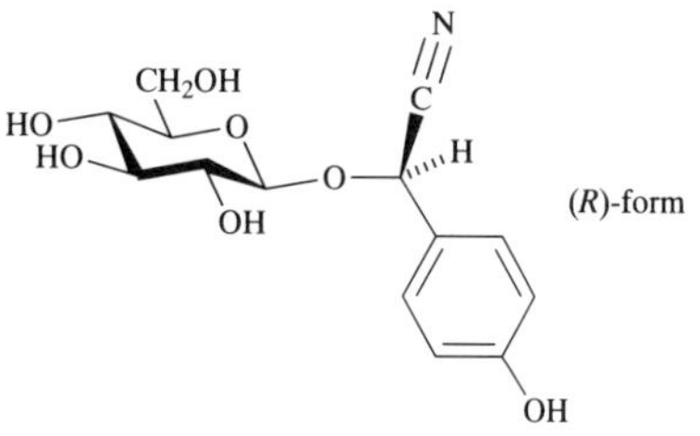

(*R*)-form

Cyanogenic glycoside

$C_{14}H_{17}NO_7$ [21401-21-8] MW 311.29

First isolated from *Phyllanthus gasstroemeri* (Euphorbiaceae), it occurs, like the *(S)*-epimer dhurrin (q.v.), in several Gramineae. It is relatively widespread in dicotyledons, occurring in the tulip tree, *Liriodendron tulipifera* (Magnoliaceae). Releases cyanide on hydrolysis. It occurs in young bamboo shoots and has been responsible for several cases of human cyanide poisoning.

Schwarzmaier, U., *Chem. Ber.*, 1976, **109**, 3250.

Taxodione 1309

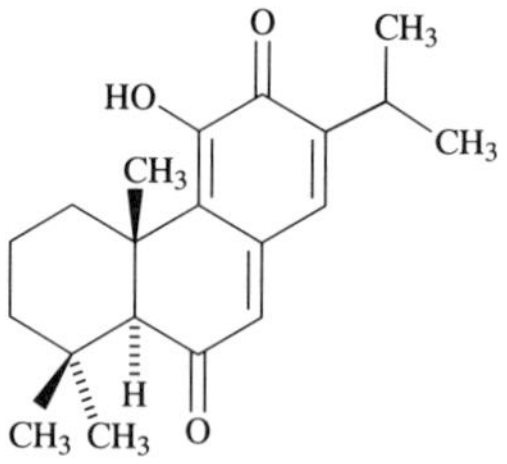

Diterpenoid
(Abietane)

$C_{20}H_{26}O_3$ [19026-31-4] MW 314.42

Taxodium distichum (Taxodiaceae).

Antitumour properties.

Kupchan, S. M., *J. Org. Chem.*, 1969, **34**, 3912.

Taxodone 1310

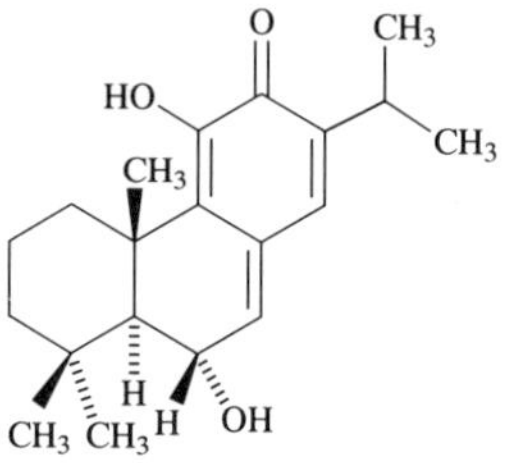

Diterpenoid
(Abietane)

$C_{20}H_{28}O_3$ [19039-02-2] MW 316.44

Swamp cypress, *Taxodium distichum* (Taxodiaceae).

Antitumour properties.

Kupchan, S. M., *J. Org. Chem.*, 1969, **34**, 3912.

Taxol 1311

Paclitaxel; Taxol A

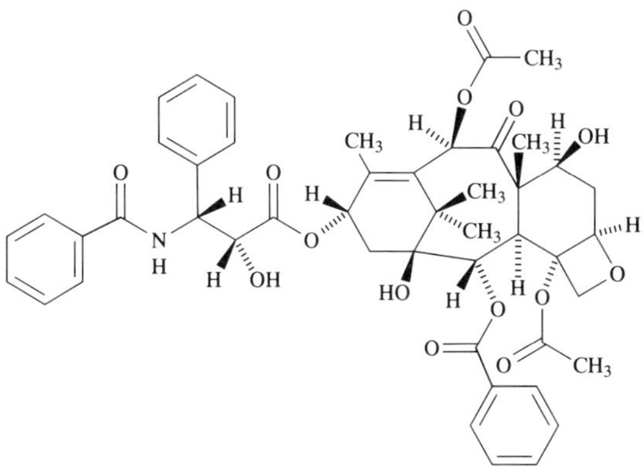

Diterpenoid alkaloid
(Taxane)

$C_{47}H_{51}NO_{14}$ [33069-62-4] MW 853.92

Stem bark of the Pacific yew, *Taxus brevifolia* and of the Japanese yew, *T. cuspidata*; also present in trace amounts in other *Taxus* spp., including needles of the common yew, *T. baccata* (Taxaceae).

Toxic, with an oral LD_{50} in the dog of 9 mg/kg body-weight; the major toxin of the common yew, however, is taxine A (q.v.). Has valuable antitumour activity and is currently used to treat ovarian and breast cancer patients. One of the most effective naturally occurring anticancer drugs available at the present time.

Wani, M. C., *J. Am. Chem. Soc.,* 1971, **93**, 2325.

Tecomine 1312

Tecomanine

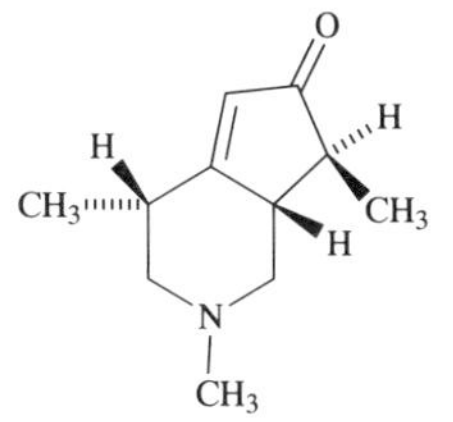

Monoterpenoid alkaloid

$C_{11}H_{17}NO$ [6878-83-7] MW 179.26

Tecoma stans (Bignoniaceae), *Calopogonium stans* and *C. fulva* (Leguminosae).

Hypoglycaemic in fasting rabbits when administered intravenously at 20 mg/kg body-weight or orally at 50 mg/kg body-weight. Used in Mexico for treating diabetes mellitus.

Dickinson, E. M., *Tetrahedron,* 1969, **25**, 1523.

Tecostanine 1313

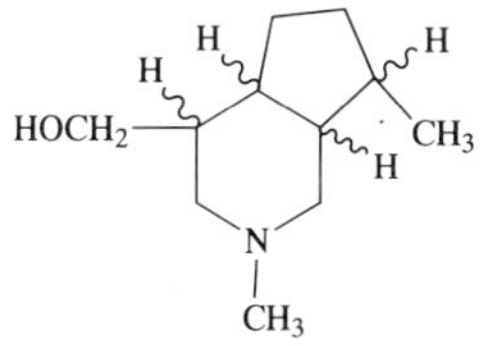

Monoterpenoid alkaloid

$C_{11}H_{21}NO$ [708-18-9] MW 183.29

Tecoma stans (Bignoniaceae).

Lowers blood sugar levels in experimental animals, e.g. rabbits, at similar doses to those of tecomine (q.v.).

Hammouda, Y., *Bull. Soc. Chim. Fr.,* 1963, 2802.

Telfairine 1314

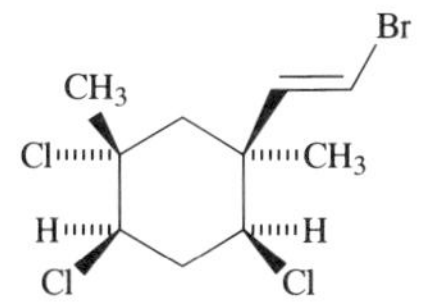

Halogenated cycloalkane

$C_{10}H_{14}BrCl_3$ [120163-22-6] MW 320.48

Red alga, *Plocamium telfairiae*.

Insecticidal.

Watanabe, K., *Phytochemistry,* 1989, **28**, 77.

Tenulin 1315

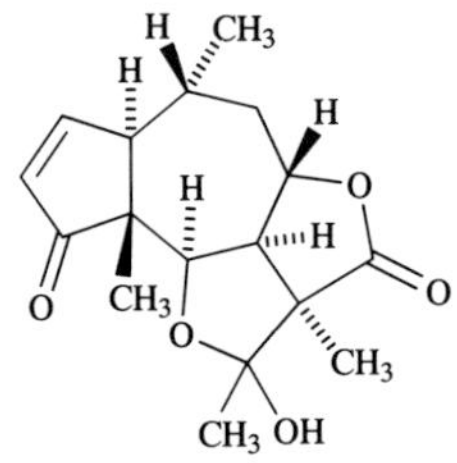

Sesquiterpenoid lactone
(Pseudoguaianolide)

$C_{17}H_{22}O_5$ [19202-92-7] MW 306.36

Many *Helenium* spp., including *H. tenuifolium*, *H. amarum*, *H. autumnale*, *H. elegans* and *H. puberulum* (Compositae).

Toxic to hamster, mouse and sheep. Barely toxic to cows, but tenulin can be found in the milk, where it affects the taste. It has cytotoxic and antitumour properties.

Herz, W., *J. Org. Chem.*, 1970, **40**, 2557.

Terminalin 1316

Gallagic acid dilactone; Gallagyldilactone

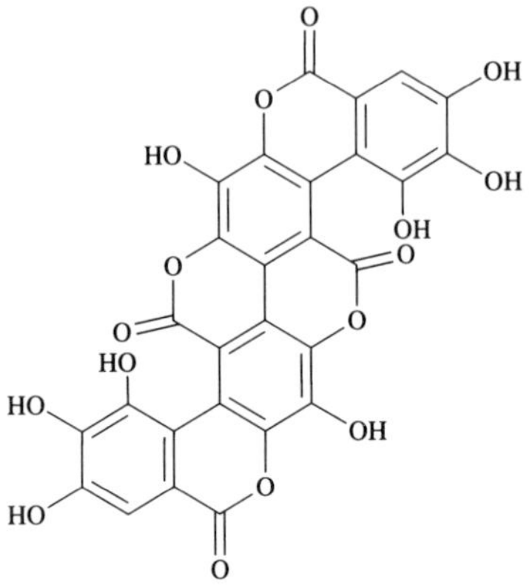

Condensed tannin

$C_{28}H_{10}O_{16}$ [155144-63-1] MW 602.38

In leaves of yellow wood, *Terminalia oblongata* (Combretaceae).

Major toxin, together with punicalagin (q.v.), responsible for poisoning and death of cattle and sheep in Queensland, following ingestion of this plant. It specifically produces kidney necrosis in both mice and sheep. It has a high toxicity (20 mg/kg body-weight) to male mice.

Oelrichs, P. B., Natural Toxins, 1994, **2**, 144, Iowa University Press.

α-Terthienyl 1317

2,2′:5′,2-Terthiophene; α-T

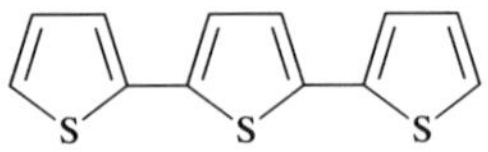

Thiophene

$C_{12}H_8S_3$ [1081-34-1] MW 248.39

Root and leaf of the marigold, *Tagetes erecta* (Compositae).

It is phototoxic and induces photodermatitis in humans. It is also a nematocide and insecticide.

Zechmeister, L., *J. Am. Chem. Soc.*, 1947, **69**, 273.

12-Tetradecanoylphorbol 13-acetate 1318

Croton factor A_1; Cocarcinogen A_1

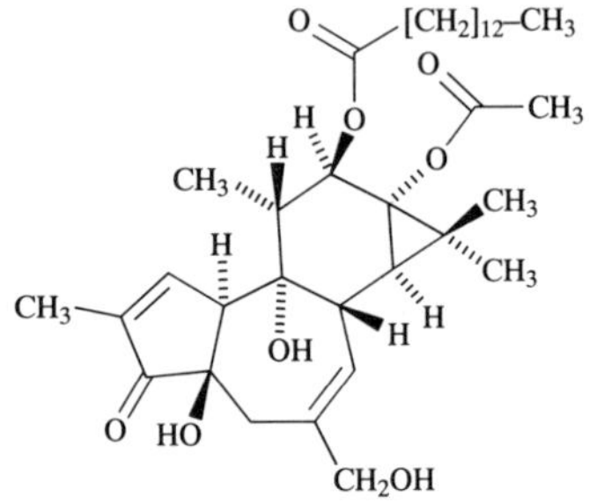

Diterpenoid
(Tigliane)

$C_{36}H_{56}O_8$ [16561-29-8] MW 616.84

Seed oil of *Croton tiglium* (Euphorbiaceae).

Irritant and cocarcinogen.

Libermann, C., *Nature (London)*, 1968, **217**, 563.

Tetradymol 1319

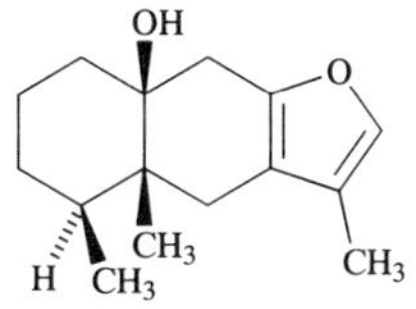

Sesquiterpenoid
(Eremophilane)

$C_{15}H_{22}O_2$ [52279-13-7] MW 234.34

Stem and flower bud of *Tetradymia glabrata* (Compositae).

Hepatotoxic agent, responsible for the death of sheep feeding on *T. glabrata*. The oral LD_{50} in mice is 250 mg/kg bodyweight.

Jennings, P. W., *J. Org. Chem.*, 1974, **39**, 3392.

Δ^1-Tetrahydrocannabinol 1320

Δ^9-Tetrahydrocannabinol; Dronabinol

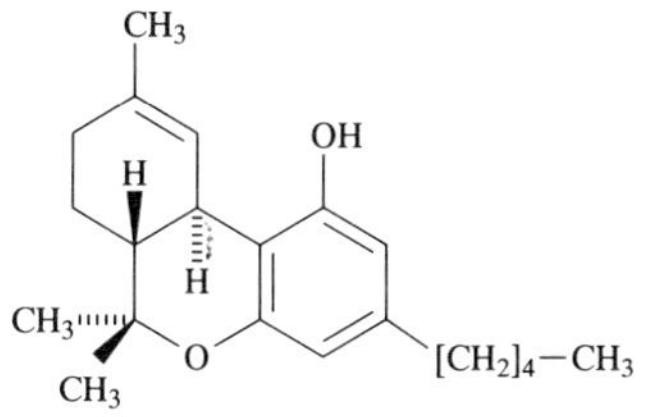

Cannabinoid

$C_{21}H_{30}O_2$ [1972-08-3] MW 314.47

Resin of *Cannabis sativa* (Cannabaceae) and in marihuana (dried tips of shoots of the same plant).

Active principle of marihuana. It has anti-inflammatory, anti-emetic and hallucinogenic properties. It may have long term toxic side effects in man, although this is still controversial. Is toxic to lepidopteran larvae.

Gaoni, Y., *J. Am. Chem. Soc.*, 1971, **93**, 217.

Tetrahydropalmatine 1321

Corydalis B; Gindarine; Caseanine

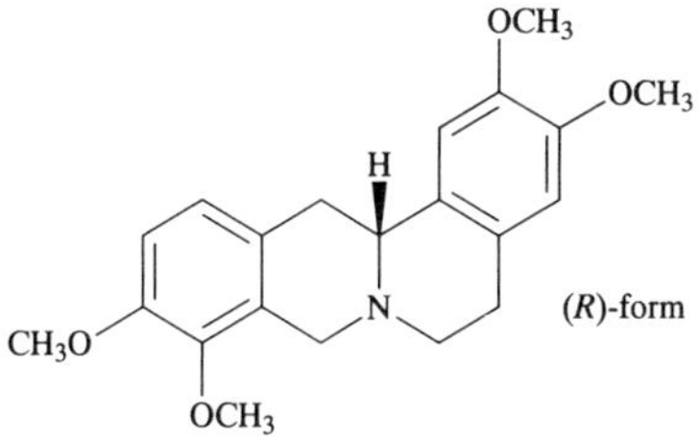

Isoquinoline alkaloid
(Berbine)

$C_{21}H_{25}NO_4$ [10097-84-4] MW 355.43
(*R*)-form [3520-14-7]
(*S*)-form [483-14-7]

Corydalis aurea (Fumariaceae); also present in *Stephania glabra* (Menispermaceae), *Berberis tinctoria* (Berberidaceae) and *Coptis tecta* (Ranunculaceae).

Cause of cattle poisoning in Canada following grazing on *C. aurea*.

Späth, E., *Ber. Dtsch. Chem. Ges.*, 1923, **56**, 875.

(+)-Tetrandine 1322

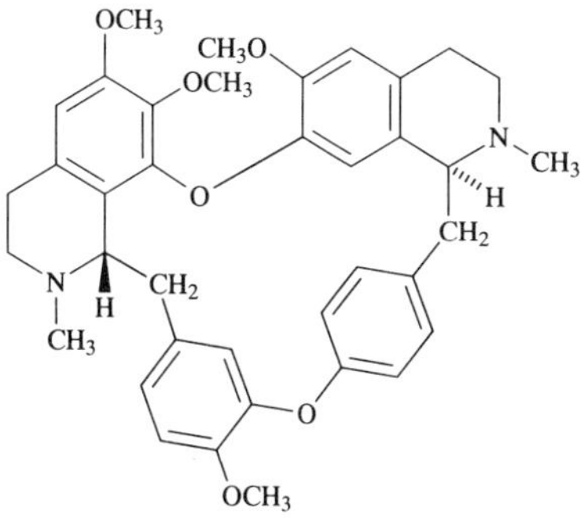

Bisbenzylisoquinoline alkaloid
(Berbaman)

$C_{38}H_{42}N_2O_6$ [518-34-3] MW 622.76
(±)-form [23495-89-8]

Stephania tetrandra, *S. discolor*, *Cyclea peltata* and *Cissampelos pareira* (Menispermaceae). The (−)-isomer is phaeanthine [1263-79-2].

Antimalarial activity. Also possesses analgesic, antipyretic and anti-inflammatory properties. The plant *C. pareira* is used in Uruguay for fertility control and treating snakebites. Chronic administration of tetrandrine to humans causes liver necrosis.

Brossi, A., The Alkaloids, 1985, **25**, 163, Academic Press.

Tetraneurin-A 1323

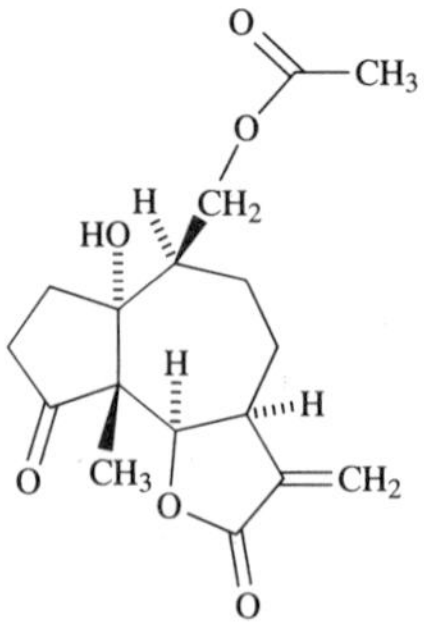

Sesquiterpenoid lactone
(Pseudoguaianolide)

$C_{17}H_{22}O_6$ [22621-72-3] MW 322.36

Parthenium alpinum, P. cineraceum, P. confertum, P. fruticosum and *P. hysterophorus* (Compositae).

One of several sesquiterpene lactones present in *Parthenium hysterophorus* responsible for the allergic contact dermatitis of this plant. Also is an antifeedant.

Rüesch, H., *Tetrahedron,* 1969, **25**, 805.

Tetraneurin-E 1324

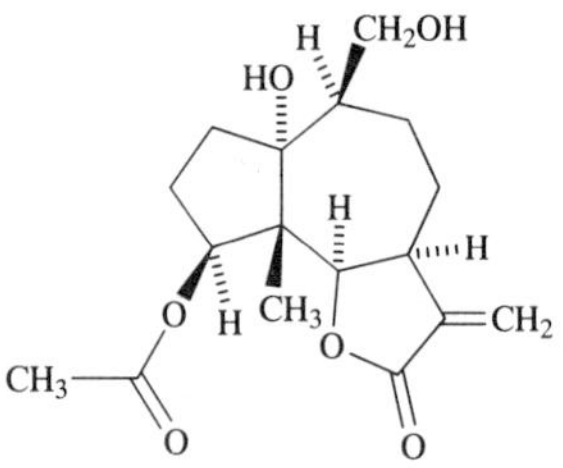

Sesquiterpenoid lactone
(Pseudoguaianolide)

$C_{17}H_{24}O_6$ [25383-30-6] MW 324.37

Parthenium confertum, P. fruticosum, P. hispidum, P. integrifolium and *P. lozanianum* (Compositae).

Inhibits the growth and development of insect larvae.

Yoshioka, H., *J. Org. Chem.,* 1976, **35**, 2888.

Thalicarpine 1325

Thaliblastine

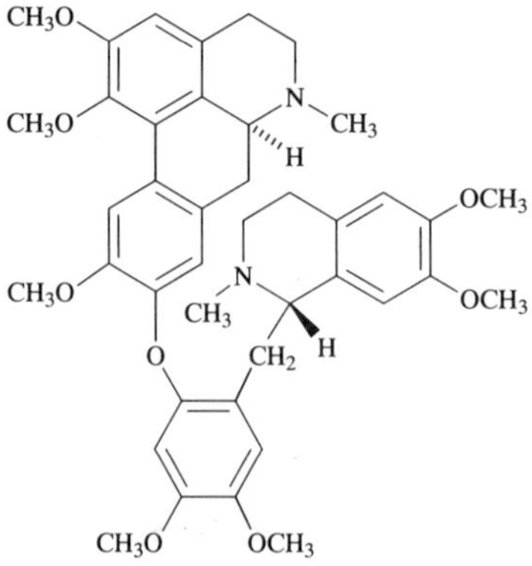

Benzylisoquinoline alkaloid
(Aporphine)

$C_{41}H_{48}N_2O_8$ [5373-42-2] MW 696.84

Thalictrum dasycarpum, T. flavum and *T. polygamum* (Ranunculaceae).

Toxic, with an LD_{50} subcutaneously in mice of 58.6 mg/kg body-weight. Has hypotensive, vasodilatory and antitumour properties.

Tomita, M., *Tetrahedron Lett.,* 1965, 4309.

Thalicoside A 1326

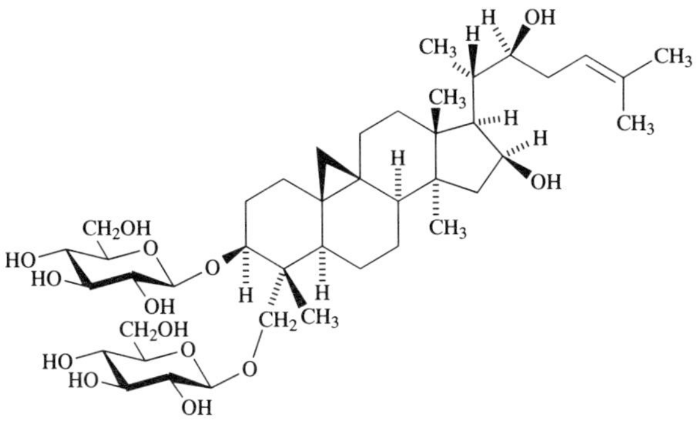

Triterpenoid saponin
(Cycloartane)

$C_{42}H_{70}O_{14}$ [93208-45-8] MW 799.01

Thalictrum minus (Ranunculaceae).

Antitumour activity.

Gromova, A. S., *Chem. Nat. Compd.,* 1984, **20**, 197.

Thalmine 1327

Talmine

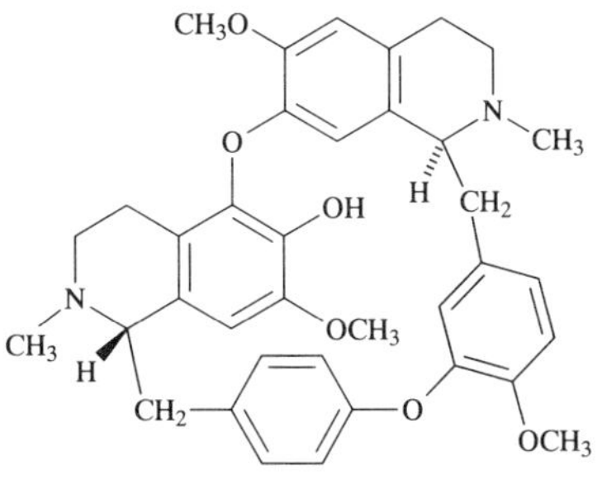

Bisbenzylisoquinoline alkaloid
(Thalman)

$C_{37}H_{40}N_2O_6$ [7682-65-7] MW 608.73

Thalictrum spp. (Ranunculaceae), including *T. minus* and *T. kuhistanicum.*

Has antitumour activity against ascites lymphoma in rats and mice. Also anti-inflammatory agent.

Telezhenetskaya, M. V., *Chem. Nat. Compd.,* 1966, **2**, 83.

Thalsimine 1328

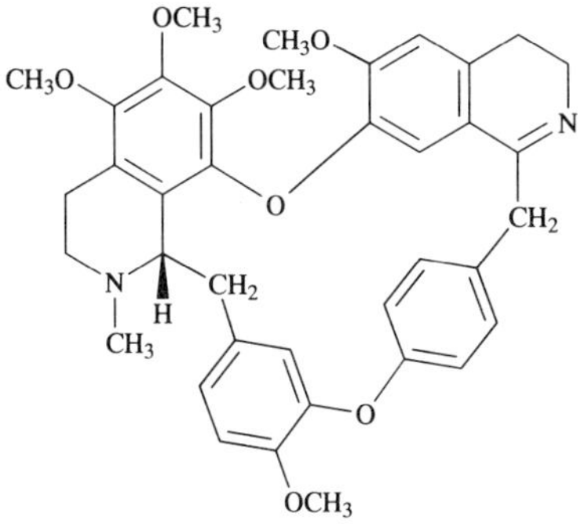

Bisbenzylisoquinoline alkaloid
(Berbaman)

$C_{38}H_{40}N_2O_7$ [5525-36-0] MW 636.74

Thalictrum simplex and *T. rugosum* (Ranunculaceae).

Affects the nervous system. It inhibits conditioned avoidance reactions and reflexes associated with eating and movement in rats and also reduces the time taken for dogs to run through a labyrinth.

Shamma, M., *J. Chem. Soc., Chem. Commun.,* 1966, 7.

Thapsigargin 1329

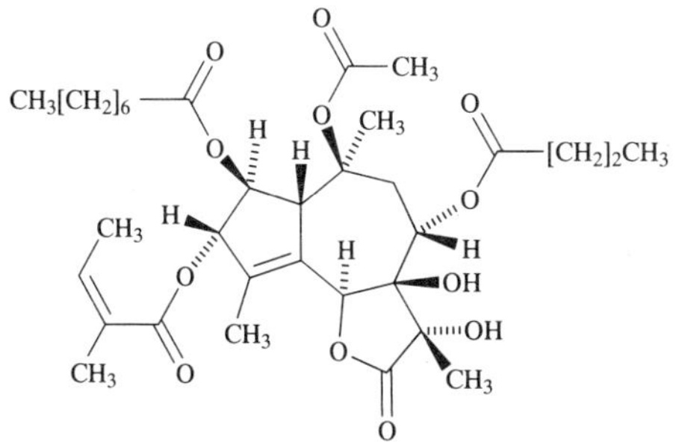

Sesquiterpenoid lactone
(Guaianolide)

$C_{34}H_{50}O_{12}$ [67526-95-8] MW 650.76

Thapsia garganica (Umbelliferae). The concentration of thapsigargin (plus the closely related thapsigargicin [67526-94-7]) varies from 0.2 to 1.2% dry-weight in the roots, and from 0.7 to 1.5% in ripe fruits. The content of the leaf is much lower (0.1%).

Causes a vigorous contact dermatitis, expressed as erythema, itching and small resiculae in humans. Thapsigargin is a non-cytotoxic histamine liberator. It is also a calcium-ATPase inhibitor, and is used in biochemical studies of calcium homeostasis.

Christensen, S. B., *J. Org. Chem.,* 1983, **48**, 396.

Theasaponin 1330

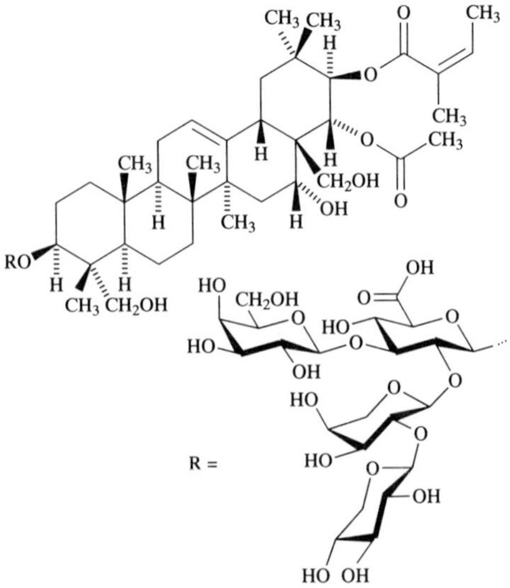

Triterpenoid saponin
(Oleanane)

$C_{59}H_{92}O_{27}$ [11055-93-9] MW 1233.36

Seeds of the tea plant, *Thea sinensis* (Theaceae).

Theasaponin is a mixture of glycoside diesters of the aglycones theasapogenol A, B and E and camelliagenin A and D [13844-22-9, 13844-01-4, 15399-41-4, 53227-91-1 and 25122-87-6, respectively]. Strong haemolytic properties. Also has anti-exudative and insecticidal activities.

Tschesche, R., *Justus Liebigs Ann. Chem.,* 1969, **721**, 209.

Thebaine 1331

Paramorphine

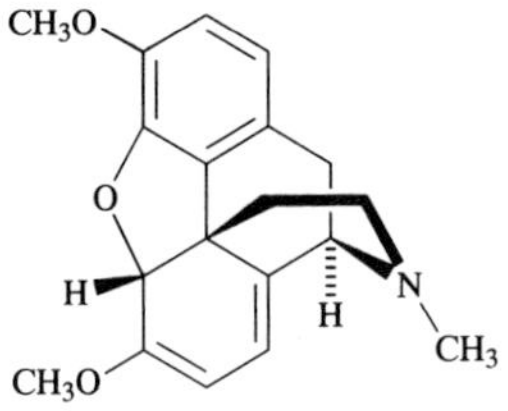

Alkaloid
(Morphinan)

$C_{19}H_{21}NO_3$ [115-37-7] MW 311.38

Minor alkaloid of opium, from the latex of *Papaver somniferum*; also present in greater amounts in some strains of *P. bracteatum* (Papaveraceae).

Weak narcotic and analgesic properties compared to the closely related morphine, but more toxic. Large doses of thebaine cause convulsions.

Bentley, K. W., Chemistry of the Morphine Alkaloids, 1954, 184, Oxford Univ. Press.

Thevetin A 1332

Cannogenin 3-gentiobiosylthevetoside

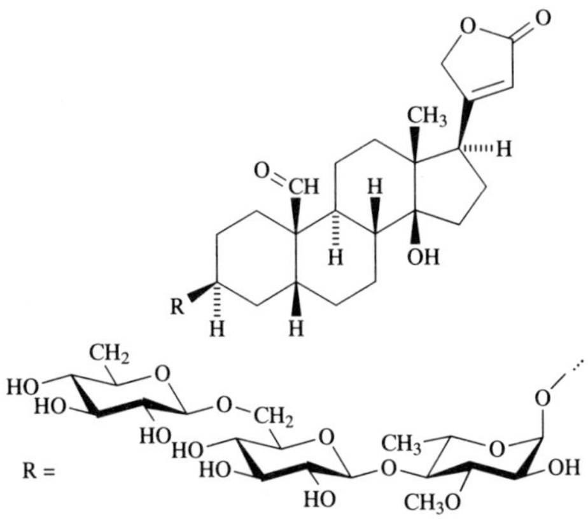

Cardenolide

$C_{42}H_{64}O_{19}$ [37933-66-7] MW 872.96

Thevetia neriifolia (Apocynaceae).

Very toxic to vertebrates.

Helfenberger, H., *Helv. Chim. Acta,* 1948, **31**, 1470.

Thevetin B 1333

Cerberoside; Digitoxigenin 3-gentiobiosylthevetoside

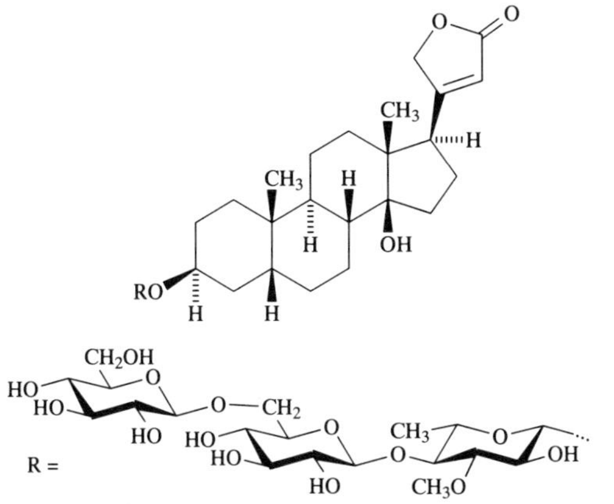

Cardenolide

$C_{35}H_{54}O_{14}$ [27127-79-3] MW 698.80

In *Cerbera odollam* and *Thevetia neriifolia* (Apocynaceae).

Toxic to vertebrates. Ground seed kernels of *Thevetia* have been used as a rat poison.

Tori, K., *Tetrahedron Lett.,* 1977, 717.

Thiarubrine A 1334

TR-A

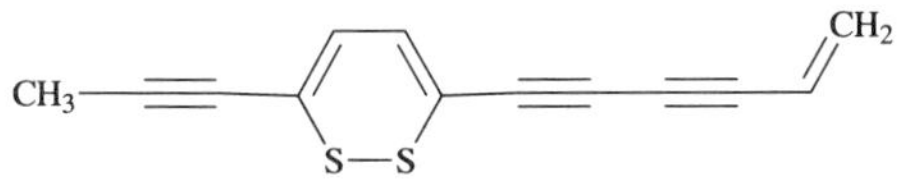

Polyacetylene

$C_{13}H_8S_2$ [63543-09-9] MW 228.34

Aspilia spp. (Compositae).

Phototoxic agent. Also is a nematocide.

Norton, R. A., *Phytochemistry,* 1985, **24**, 356.

α-Thujone 1335

(−)-Isothujone; (−)-β-Thujone; Thujan-3-one

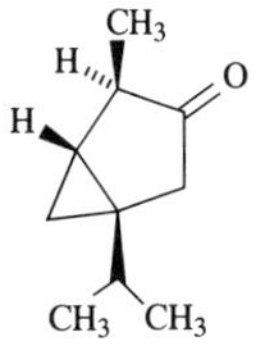

Monoterpenoid
(Thujane)

$C_{10}H_{16}O$ [546-80-5] MW 152.24

Thujone [1125-12-8] is present as a mixture of stereoisomers, α- and β-thujone [471-15-8], in many essential oils. Occurs in some concentration in oil of leaves of *Thuja occidentalis* (Cupressaceae), oil of tansy, *Tanacetum vulgare* and in oil of wormwood, *Artemisia absinthium* (both Compositae).

Potentially toxic, causing convulsions. Absinthe liqueurs are prepared from oleum Absinthii, which contains considerable amounts of thujone. Large doses are very toxic, leading to chronic poisoning. For this reason, the preparation of absinthe is banned in Germany and Switzerland. Oil of wormwood has been used as an anthelmintic.

Whittaker, D., *Chem. Rev.,* 1972, **72**, 305.

Thymol 1336

m-Thymol; Timol; 3-*p*-Cymenol; Thyme camphor; 6-Isopropyl-*m*-cresol

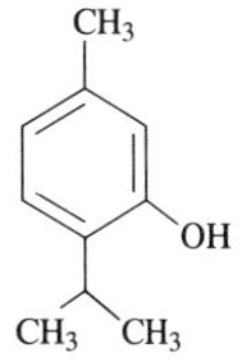

Monoterpenoid
(Menthane)

$C_{10}H_{14}O$ [89-83-8] MW 150.22

Common constituent of plant essential oils. Best sources are essential oils of *Thymus vulgaris* and *Monarda punctata* (Labiatae) and of seeds of *Carum copticum* (Umbelliferae).

Irritates the gastric mucosa. It is antiseptic, being 20 times more active than phenol. It is used for preserving biological specimens and in dentistry.

Martindale, The Extra Pharmacopoeia, 30th edn, 1993, 805, The Pharmaceutical Press.

Tigloidine 1337

3β-Tigloyloxytropane; Tiglylpseudotropeine

Tropane alkaloid

$C_{13}H_{21}NO_2$ [533-08-4] MW 223.32

Duboisia myoporoides and *Datura innoxia* (Solanaceae).

Central nervous system depressant. It has been used to treat muscular rigidity and Parkinson's disease.

Barger, G., *J. Chem. Soc.,* 1937, 1820.

Tiliacorine 1338

Bisbenzylisoquinoline alkaloid

$C_{36}H_{36}N_2O_5$ [27073-72-9] MW 576.69

Bark and root of *Tiliacora acuminata* and of *T. racemosa* (Menispermaceae).

Antimalarial activity against *Plasmodium falciparum in vivo.*

Bhakuni, D. S., *J. Chem. Soc., Perkin Trans. 1,* 1981, 2598.

Tinyatoxin 1339

Diterpenoid
(Daphnetoxane)

$C_{36}H_{38}O_8$ [58821-95-7] MW 598.69

Latex of tinya, *Euphorbia poisonii* (Euphorbiaceae).

Toxin, producing severe skin inflammation in humans.

Evans, F. J., *Phytochemistry,* 1976, **15**, 333.

Tomatidine 1340

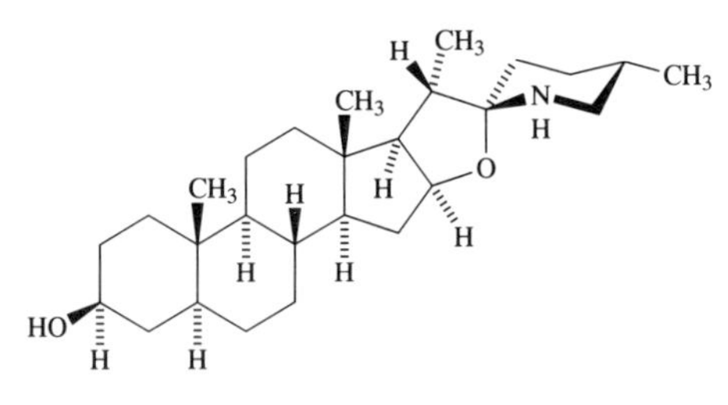

Steroid alkaloid
(Spirosolane)

$C_{27}H_{45}NO_2$ [77-59-8] MW 415.66

Roots of *Lycopersicon esculentum*, Rutgers tomato plant and *Solanum demissum.* Widely present in glycosidic form (e.g. as tomatine, q.v.) in other *Lycopersicon* spp. and *Solanum* spp. (Solanaceae).

Toxic alkaloid with cytotoxic activity. It is a cholinesterase inhibitor and clinically active against forms of dermatitis. It is a repellent to the Colorado potato beetle.

Sato, Y., *J. Am. Chem. Soc.*, 1957, **79**, 6089.

Tomatine 1341

α-Tomatine; Lycopericin

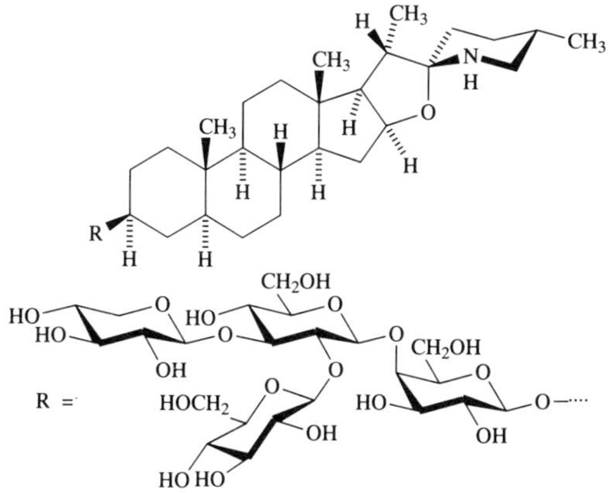

Steroid alkaloid
(Spirosolane)

$C_{50}H_{83}NO_{21}$ [17406-45-0] MW 1034.20

Leaves and fruits of the cultivated tomato plant, *Lycopersicon esculentum* (Solanaceae). Also in wild *Lycopersicon* spp. and in several *Solanum* spp..

Used as an insecticide, especially to repel the Colorado beetle. It precipitates steroids and has been proposed as an alternative to digitonin. It has antihistamine properties.

Kuhn, R., *Chem. Ber.*, 1957, **90**, 203.

Toonacilin 1342

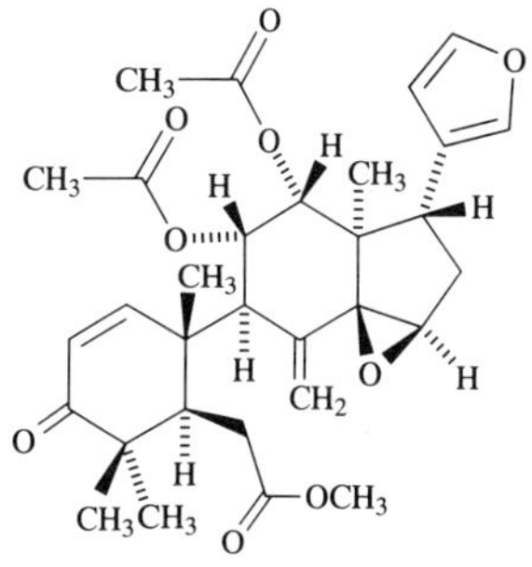

Nortriterpenoid
(Normeliacane)

$C_{31}H_{38}O_9$ [66610-71-7] MW 554.64

Bark of *Toona ciliata* (Simaroubaceae).

Insect antifeedant, e.g. against the Mexican bean beetle, *Epilachna varivestis*.

Kraus, W., *Angew. Chem., Int. Ed. Engl.*, 1978, **17**, 452.

Toxicarol 1343

α-Toxicarol

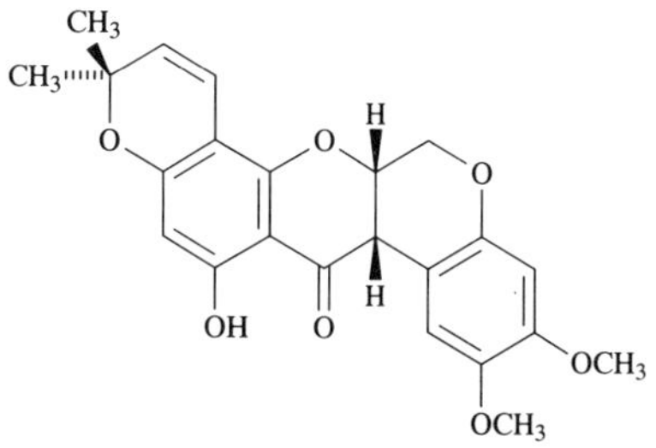

Isoflavonoid
(Rotenoid)

$C_{23}H_{22}O_7$ [82-09-7] MW 410.42

Roots of *Tephrosia toxicaria* and *Derris elliptica* (Leguminosae).

Poisonous to fish and insects.

Haller, H. L., *Chem. Rev.*, 1942, **30**, 33.

Toxiferine I 1344

C-Toxiferine I; Toxiferine V; Toxiferine XI

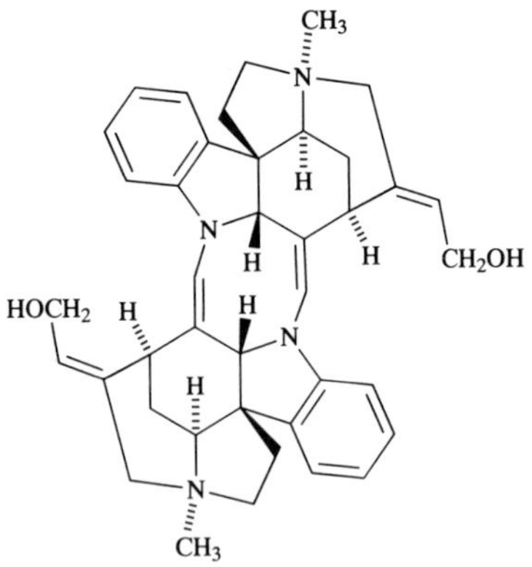

Bisindole alkaloid

$[C_{40}H_{46}N_4O_2]^{2+}$ [6888-23-9] MW 614.83

A component of calabash curare, from *Strychnos toxifera* and *S. froesii* (Loganiaceae).

Neuromuscular blocking agent, six to eight times more potent than tubocurarine (q.v.). Calabash curare is used in South America as an arrow poison.

Battersby, A. R., *J. Chem. Soc.*, 1960, 736.

Toxol 1345

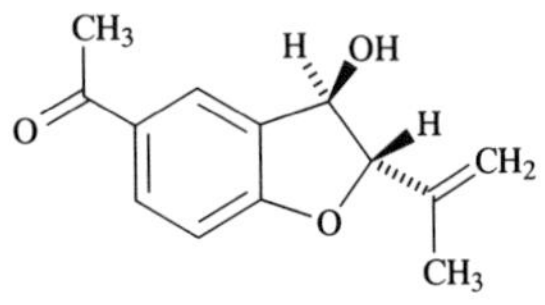

Benzofuran

$C_{13}H_{14}O_3$ [26296-56-0] MW 218.25

Haplopappus heterophyllus and *Morithamnus crassua* (Compositae).

Haplopappus heterophyllus is claimed to be responsible for causing milk sickness after human consumption of milk from cattle feeding on the plant. Toxol and tremetone (q.v.) are joint toxic factors. Toxol also has antitumour properties.

Zalkow, L. H., *Tetrahedron Lett.*, 1972, 2873.

Trachelogenin 1346

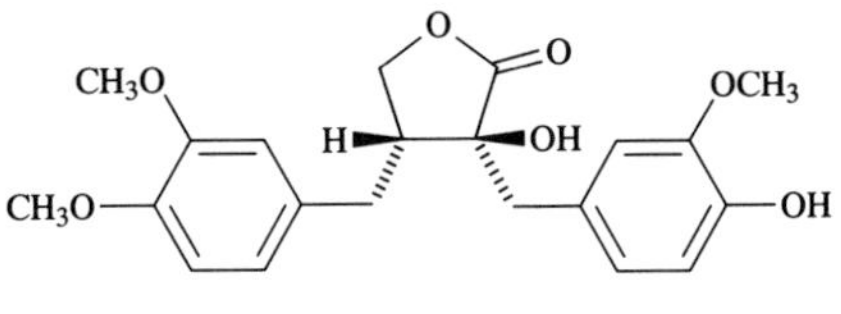

Lignan

$C_{21}H_{24}O_7$ [34209-69-3] MW 388.42

Trachelospermum asiaticum var. *intermedium* (Apocynaceae) and *Ipomoea cairica* (Convolvulaceae).

Cytotoxic activity in lymphoma cell systems.

Nishbe, S., *Phytochemistry*, 1971, **10**, 2231.

Tremetone 1347

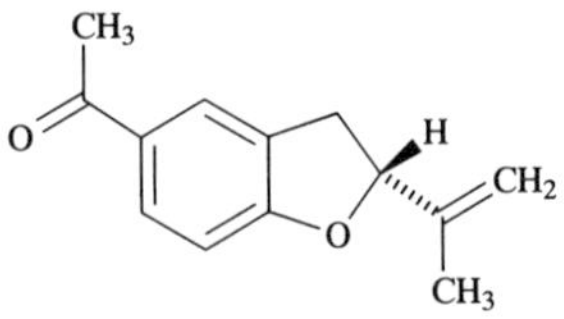

Benzofuran

$C_{13}H_{14}O_2$ [4976-25-4] MW 202.25

Eupatorium urticaefolium, *E. rugosum*, *Haplopappus heterophyllus* and *Ageratina*, *Brickellia*, *Liatris*, *Ligularia*, *Baccharis* and *Grindelia* spp. (Compositae).

It is jointly responsible, with dehydrotremetone (q.v.), for milk sickness in humans after consumption of milk from cattle feeding on *Eupatorium urticaefolium*. It is also toxic to goldfish.

Bonner, W. A., *Tetrahedron*, 1964, **20**, 1419.

Triangularine 1348

6-Angelyl-9-sarracinylretronecine

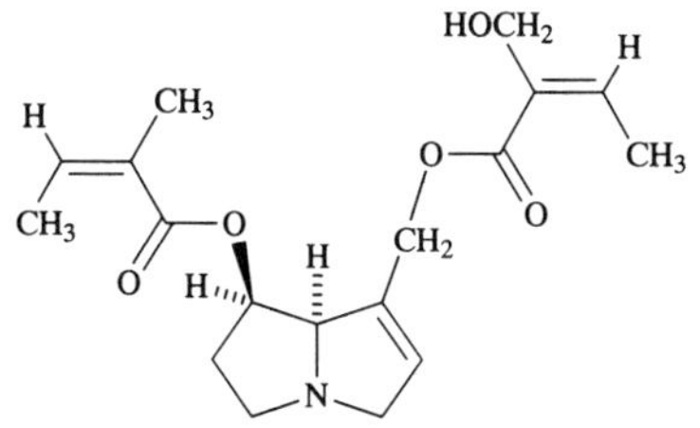

Pyrrolizidine alkaloid

$C_{18}H_{25}NO_5$ [87340-27-0] MW 355.40

Senecio triangularis (Compositae) and *Alkanna tinctoria* (Boraginaceae).

Toxicity not established, but suspected of being hepatotoxic.

Roitman, J. N., *Aust. J. Chem.,* 1983, **36**, 1203.

Trichilin A 1349

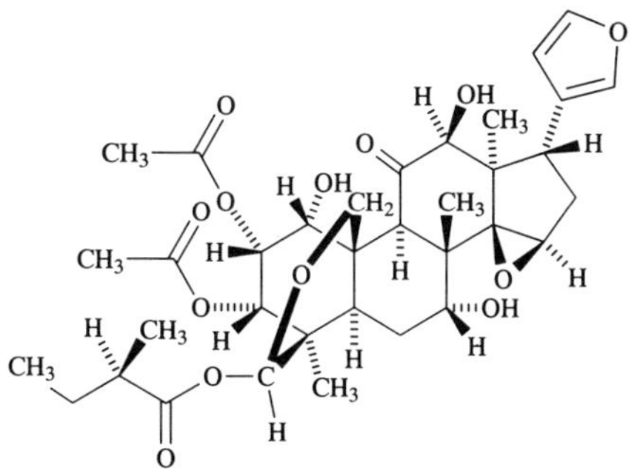

Nortriterpenoid
(Meliacane)

$C_{35}H_{46}O_{13}$ [77182-69-5] MW 674.74

Trichilia roka (Meliaceae).

Strong antifeedant activity, e.g. against the southern army-worm, *Spodoptera eridania*.

Nakatani, M., *J. Am. Chem. Soc.,* 1981, **103**, 1228.

Trichosanthin 1350

Protein
[60318-52-7] MW 24000

Tubers of *Trichosanthes kirilowii* and seeds of *Momordica charantia* (Cucurbitaceae).

Abortifacient, used clinically in China.

Wang, Y., *Pure Appl. Chem.,* 1986, **58**, 789.

Trichotomine 1351

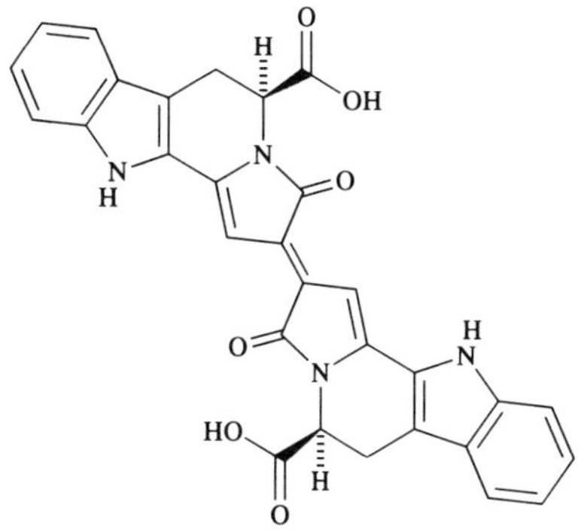

Bisindole alkaloid
(Indolizinoindole)

$C_{30}H_{20}N_4O_6$ [53472-14-3] MW 532.51

Fruit of *Clerodendron trichotomum* (Verbenaceae).

Hypotensive agent, with bronchodilator and sedative effects.

Iwadare, S., *Tetrahedron,* 1974, **30**, 4105.

Tricornine 1352

18-*O*-Acetyllycoctonine; Lycoctonine 18-acetate

Diterpenoid alkaloid
(Aconitane)

$C_{27}H_{43}NO_8$ [2871-60-3] MW 509.64

Delphinium tricorne (Ranunculaceae).

Low doses produce a small response in the rat to the phrenic nerve-diaphragm preparation; higher doses cause an initial enhancement followed by depression.

Pelletier, S. W., *Phytochemistry,* 1977, **16**, 1464.

2-Tridecanone 1353

Methyl undecyl ketone

Aliphatic ketone

$C_{13}H_{26}O$ [593-08-8] MW 198.35

Leaf trichomes of *Lycopersicon* spp. (Solanaceae). The concentration is 72 times greater in the wild tomato, *L. hirsutum* var. *glabratum* than in the domestic plant *L. esculentum*.

Insecticidal, toxic to lepidoptera and aphid pest insects.

Williams, W. G., *Science,* 1980, **207**, 888.

3-*trans*,11-*trans*-Trideca-1,3,11-triene-5,7,9-triyne 1354

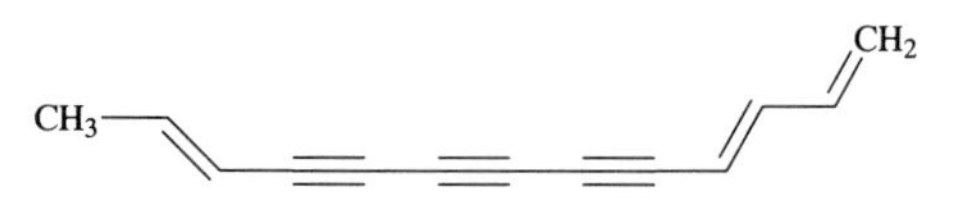

Polyacetylene

$C_{13}H_{10}$ [18668-89-8] MW 166.22

Flowers of *Carthamus tinctorius* (Compositae).

Nematocidal at a concentration of 1 μg/ml against the white-tip nematode, *Aphelenchoides besseyi*.

Kogiso, S., *Tetrahedron Lett.,* 1976, 109.

Tridec-1-ene-3,5,7,9,11-pentayne 1355

Polyacetylene

$C_{13}H_6$ [81900-91-6] MW 161.19

Roots of *Cirsium japonicum* (Compositae).

Nematocidal activity, inhibiting reproduction in *Bursaphelenchus xylophilus* at a concentration of 16 μg/ml.

Kawazu, K., *Agric. Biol. Chem.,* 1980, **44**, 903.

Triglochinin 1356

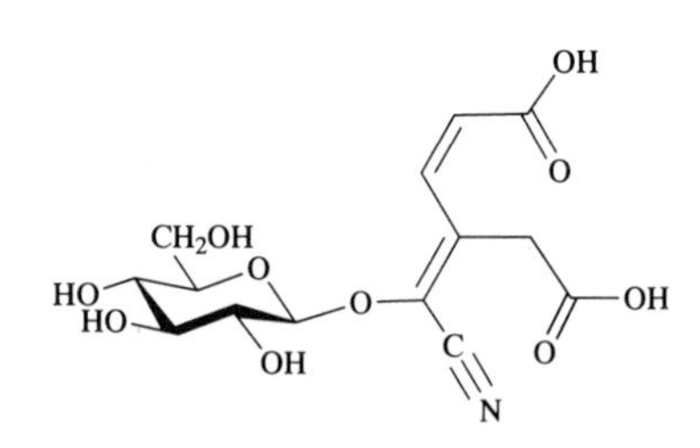

Cyanogenic glycoside

$C_{14}H_{17}NO_{10}$ [28876-11-1] MW 359.29

Arrow grass, *Triglochin maritima* (Juncaginaceae), *Campanula cochleariifolia* and *C. rotundifolia* (Campanulaceae).

Bound toxin, releasing cyanide on hydrolysis. Cattle and sheep poisoning have been recorded after the ingestion of arrow grass.

Eyjólfsson, R., *Phytochemistry,* 1970, **9**, 845.

Trilobine 1357

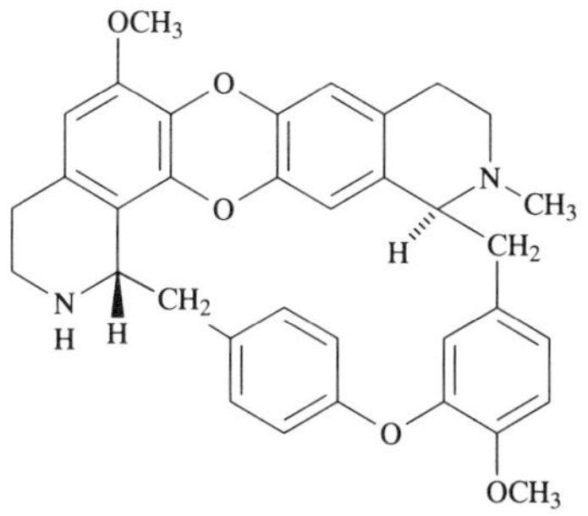

Bisbenzylisoquinoline alkaloid

$C_{35}H_{34}N_2O_5$ [6138-73-4] MW 562.67

Roots of *Cocculus trilobus* and *C. sarmentosus* and stems of *Anisocycla grandidieri* (Menispermaceae).

Toxic alkaloid, with an oral LD_{50} in man of about 500 mg/kg body-weight.

Faltis, F., *Ber. Dtsch. Chem. Ges.,* 1941, **74B**, 79.

Trilobolide 1358

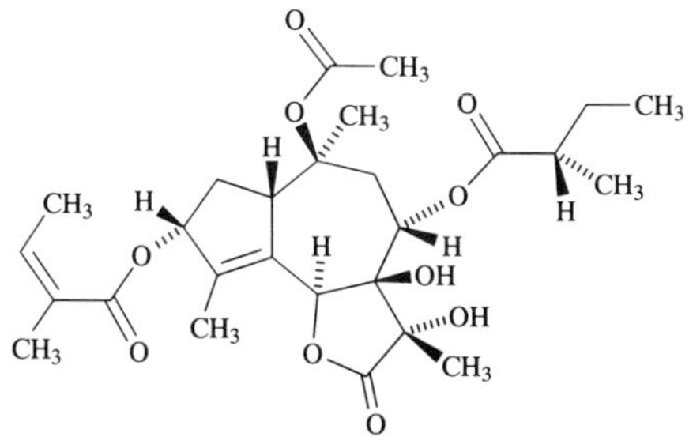

Sesquiterpenoid lactone (Guaianolide)

$C_{27}H_{38}O_{10}$ [50657-07-3] MW 522.59

Roots of *Laser trilobum* (Umbelliferae).

Cytotoxic and antitumour properties. It is an insect antifeedant.

Holub, M., *Collect. Czech. Chem. Commun.,* 1973, **38**, 1551.

2,4,5-Trimethoxystyrene 1359

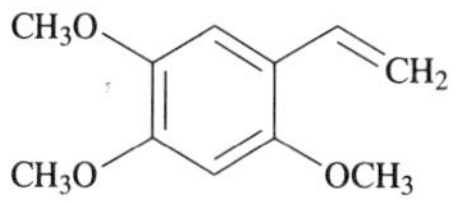

Phenylpropanoid

$C_{11}H_{14}O_3$ [17598-03-7] MW 194.23

Duguetia panamensis (Annonaceae).

Toxic to brine shrimps, with an LC_{50} of 8 p.p.m. Also cytotoxic.

Wang, Z. W., *J. Nat. Prod.,* 1988, **51**, 382.

4,5′,8-Trimethylpsoralen 1360

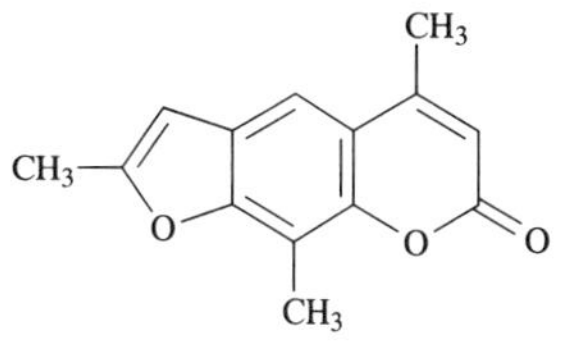

Furocoumarin

$C_{14}H_{12}O_3$ [3902-71-4] MW 228.25

Produced as a phytoalexin by celery plants, *Apium graveolens* (Umbelliferae) infected by the fungus *Sclerotinia sclerotiorum.*

Causes dermatitis in humans.

Scheel, L. D., *Biochemistry,* 1963, **2**, 1127.

Tripdiolide 1361

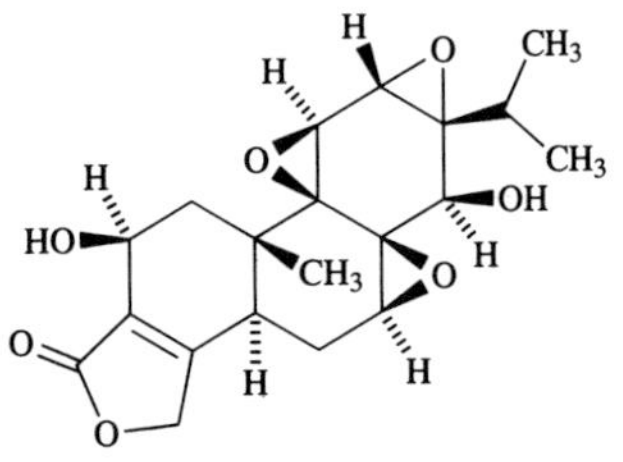

Diterpenoid
(*abeo*-Abietane)

$C_{20}H_{24}O_7$ [38647-10-8] MW 376.41

Tripteryginum wilfordii (Celastraceae).

Antitumour properties.

Kutney, J. P., *Can. J. Chem.*, 1981, **59**, 2677.

Tripteroside 1362

Norathyriol 6-glucoside

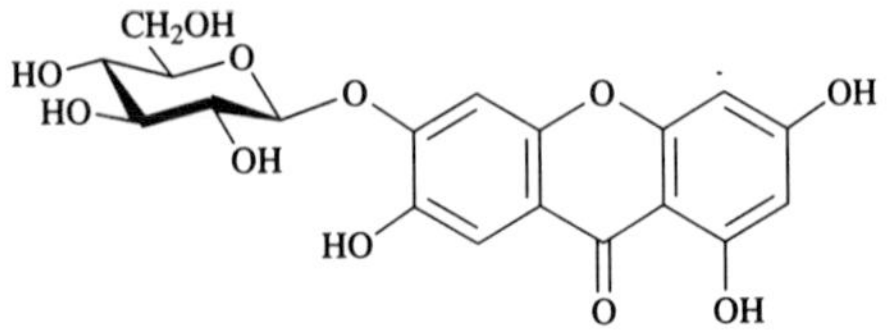

Xanthone

$C_{19}H_{18}O_{11}$ [82855-00-3] MW 422.35

Aerial parts of *Tripterospermum taiwanense* (Gentianaceae).

Central nervous system depressant in vertebrates.

Lin, C. N., *Phytochemistry,* 1989, **21**, 948.

Triptolide 1363

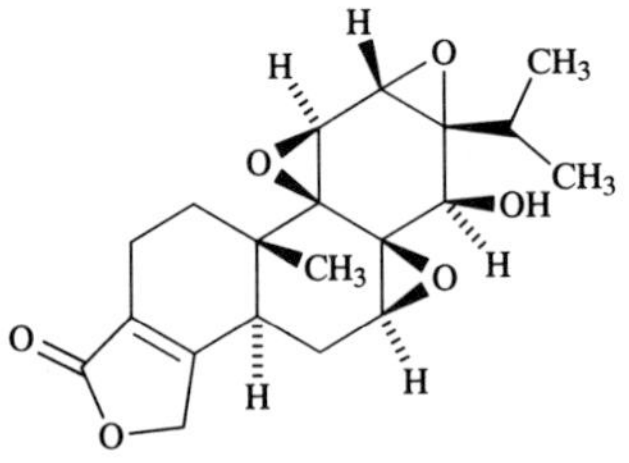

Diterpenoid
(*abeo*-Abietane)

$C_{20}H_{24}O_6$ [38748-32-2] MW 360.41

Tripterygium wilfordii (Celastraceae). Co-occurs with the 2β-hydroxy derivative, tripdiolide (q.v.), which has similar activity.

Growth inhibitor for insects. Has male contraceptive activity, but its most potent property is against the human cancer cells, L-1210 and p-388.

Kupchan, S. M., *J. Am. Chem. Soc.,* 1972, **94**, 7194.

Tropacocaine 1364

Tropacaine; Benzoylpseudotropeine;
Pseudotropine benzoate

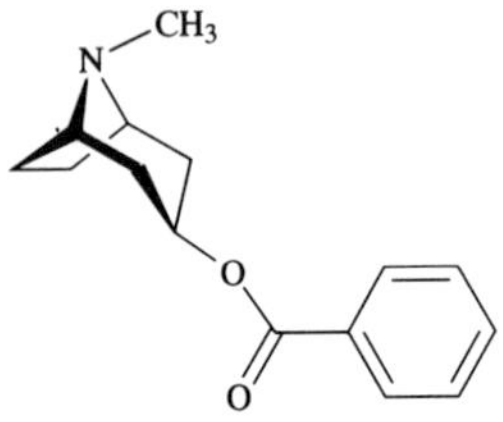

Tropane alkaloid

$C_{15}H_{19}NO_2$ [537-26-8] MW 245.32

Javanese and Peruvian coca leaves, *Erythroxylum truxillense* (Erythroxylaceae) and *Peripentadenia mearsii* (Elaeocarpaceae).

Poisonous.

Beyerman, H. C., *Rec. Trav. Chim.,* 1956, **75**, 1445.

Tropine 1365

Tropeine; 2,3-Dihydro-3α-hydroxytropidine; 1α*H*,5α*H*-Tropan-3α-ol

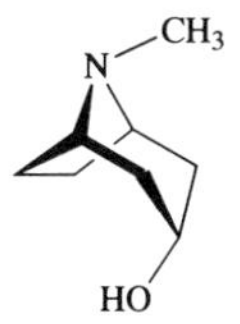

Tropane alkaloid

$C_8H_{15}NO$ [120-29-6] MW 141.21

Root and leaf of deadly nightshade, *Atropa belladonna* and in *Scopolia carniolica* (Solanaceae).

Highly toxic.

Hardeggeer, E., *Helv. Chim. Acta,* 1953, **36**, 1186.

L-Tryptophan 1366

α-Aminoindole-3-propionic acid

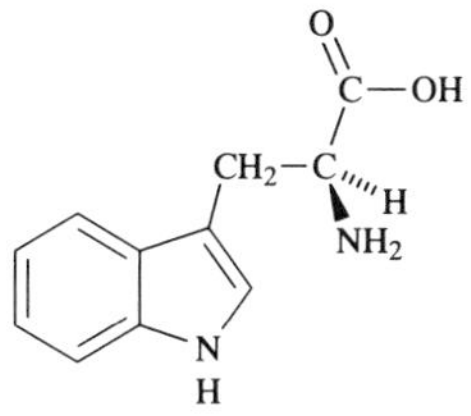

Amino acid

$C_{11}H_{12}N_2O_2$ [73-22-3] MW 204.23

As a protein amino acid, it is found in all plants.

High tryptophan levels may develop in some grasses. Cattle grazing on these grasses can develop acute pulmonary oedema and emphysema. This is due to ruminant microbial conversion of tryptophan to 3-methylindole, which is a pneumotoxin. Tryptophan is used clinically in humans to treat depression.

Greenstein, J. P., Chemistry of the Amino Acids, Part 3, 1961, 2316, Wiley.

Tsibulin 2 1367

4-Hexylcyclopenta-1,3-dione

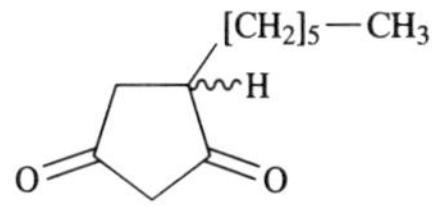

Aliphatic ketone

$C_{11}H_{18}O_2$ [126624-27-9] MW 182.26

Fungally-infected onion bulbs, *Allium sativa* (Alliaceae), co-occurring with the 4-octyl analogue, tsibulin 1 [126624-26-8].

Antifungal agent (phytoalexin).

Tverskoy, L., *Phytochemistry,* 1991, **30**, 799.

Tubeimoside I 1368

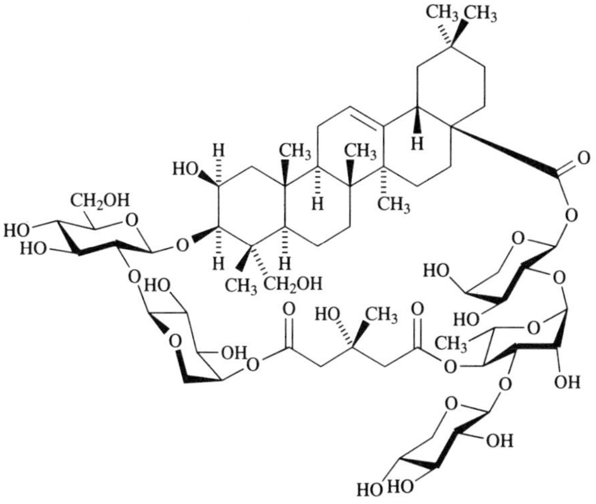

Triterpenoid saponin
(Oleanane)

$C_{63}H_{98}O_{29}$ [102040-03-9] MW 1319.45

Bulbs of *Bolbostemma paniculatum* (Cucurbitaceae).

Moderate antitumour properties.

Kasai, R., *Chem. Pharm. Bull.,* 1986, **34**, 3974.

(+)-Tubocurarine 1369

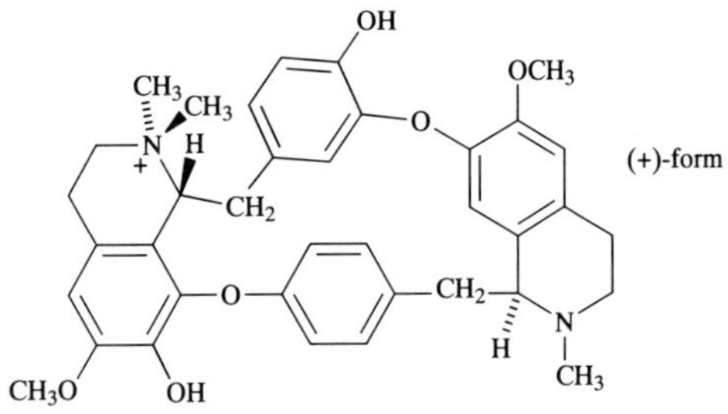

Bisbenzylisoquinoline alkaloid

$[C_{37}H_{42}N_2O_6]^+$ [57-95-4] MW 610.75

Pareira bark, *Chondrodendron tomentosum* and in other *Chondrodendron* spp. (Menispermaceae). The (−)-form also occurs in *C. tomentosum*, but is much less active than the (+)-form (illustrated).

A major ingredient of the South American arrow poison 'curare'. It is a skeletal muscle relaxant and is used to paralyse muscles during surgical operations.

Manske, R. H. F., The Alkaloids, 1977, **16**, 319, Academic Press.

Tubulosine 1370

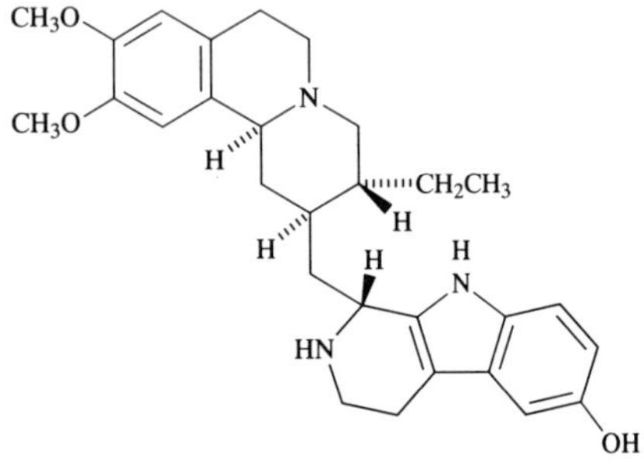

Indole alkaloid
(Tubulosan)

$C_{29}H_{37}N_3O_3$ [2632-29-3] MW 475.63

Pogonopus tubulosus, *Psychotria granadensis* and *Cephaelis ipecacuanha* (Rubiaceae).

Highly toxic, with antitumour activity. It is also amoebicidal.

Brauchli, P., *J. Am. Chem. Soc.*, 1964, **86**, 1895.

Tulipinolide 1371

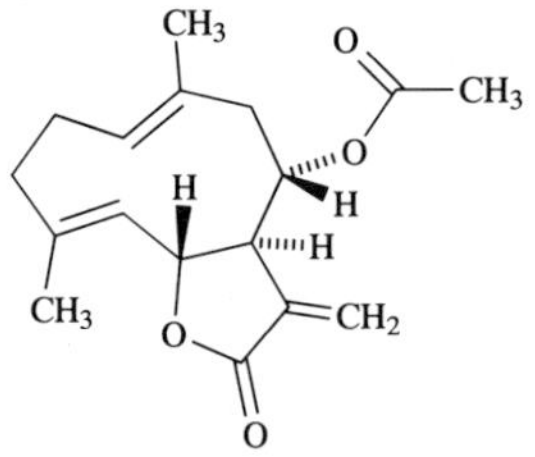

Sesquiterpenoid lactone
(Germacranolide)

$C_{17}H_{22}O_4$ [24164-12-3] MW 290.36

Root bark of the tulip tree, *Liriodendron tulipifera* (Magnoliaceae) and in aerial parts of *Ambrosia camphorata* and *A. dumosa* (Compositae).

Cytotoxic properties.

Doskotch, R. W., *J. Org. Chem.*, 1970, **35**, 1928.

Tuliposide A 1372

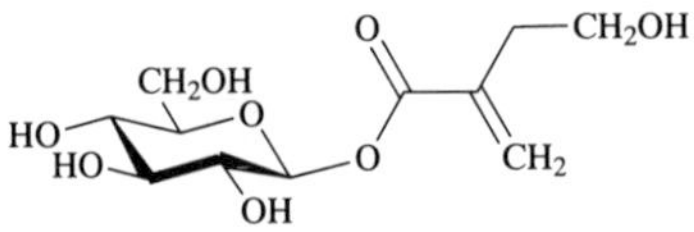

Aliphatic glucoside

$C_{11}H_{17}O_8$ [19870-30-5] MW 277.25

Tulip bulbs, *Tulipa hybrida* and widespread in other Liliaceae.

Bound toxin. Rearranges in damaged tulip bulbs to the lactone, tulipalin A [547-65-9], which is fungitoxic and allergenic. It is responsible, together with the lactone from tuliposide B (q.v.), for the skin disease caused in horticulturists by the frequent handling of tulip bulbs.

Tschesche, R., *Chem. Ber.*, 1969, **102**, 2057.

Tuliposide B 1373

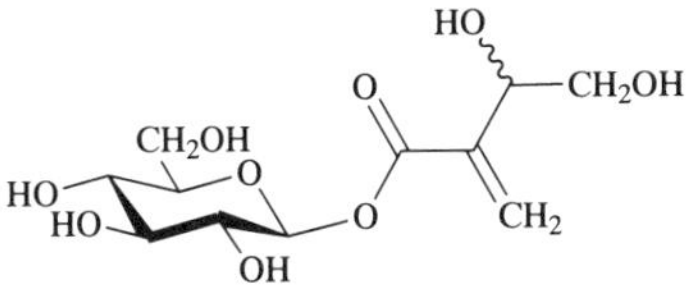

Aliphatic glucoside

$C_{11}H_{17}O_9$ [19870-33-8] MW 293.25

Tulip bulbs, *Tulipa hybrida* and widespread in other members of the Liliaceae.

Bound toxin. Rearranges in damaged tulip bulbs to the lactone tulipalin B [38965-80-9], which is fungitoxic and allergenic in humans. Responsible for the skin disease caused by frequent handling of tulip bulbs.

Tschesche, A., *Chem. Ber.,* 1969, **102**, 2057.

Tuliposide D 1374

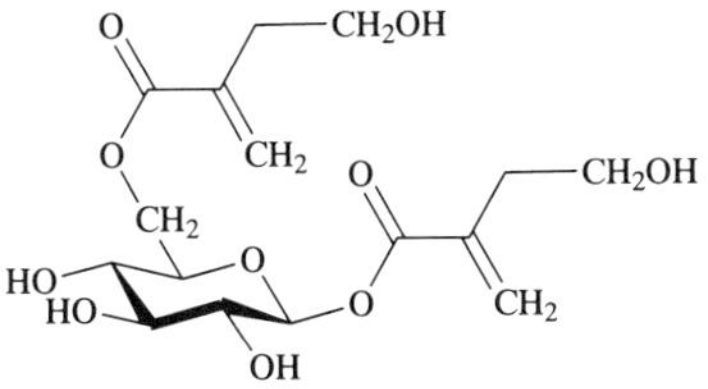

Glucose ester

$C_{16}H_{24}O_{10}$ [164991-88-2] MW 376.36

Stems, leaves and flowers of *Alstroemeria angustifolia*, *A. aurea*, *A. ligtu* and *A. revoluta* (Alstroemeriaceae).

Responsible, with tuliposide A (q.v.), for the contact dermatitis caused by handling *Alstroemeria* flowers.

Christensen, L. P., *Phytochemistry,* 1995, **38**, 1371.

Tullidinol 1375

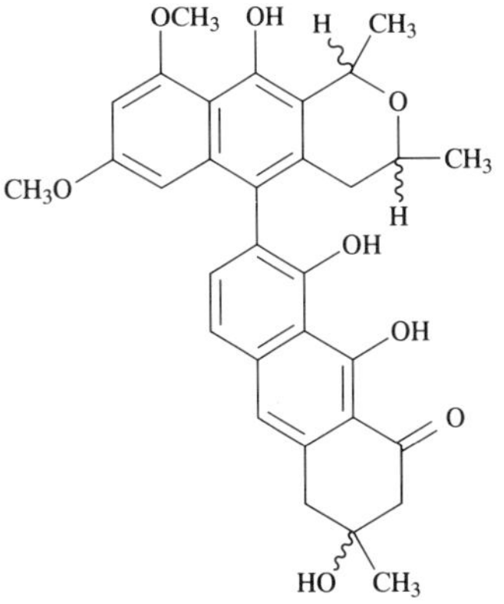

Anthracenone

$C_{32}H_{32}O_8$ [56678-09-2] MW 544.60

Fruit and root of *Karwinskia humboldtiana* (Rhamnaceae); also root of *K. umbelluta*, *K. subcordata*, *K. mollis* and *K. johnstonii*.

Neurotoxic agent of fruit, causing paralysis.

Guerrero, M., *Toxicon,* 1987, **25**, 565.

Turricolol E 1376

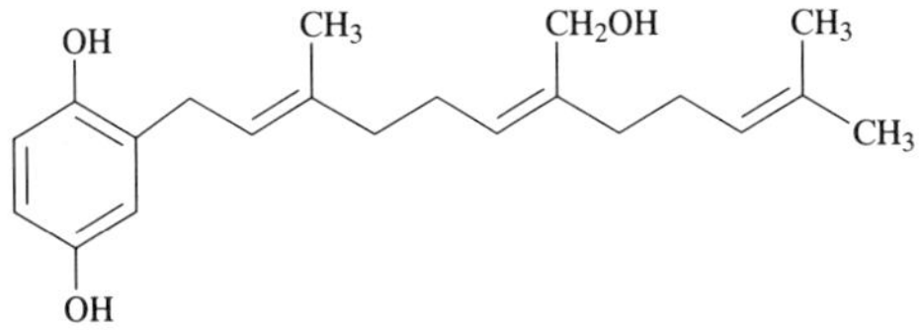

Phenolic

$C_{21}H_{30}O_3$ [101392-12-5] MW 330.47

Glandular secretions of *Turricula parryi* (Hydrophyllaceae).

Induces allergic skin reactions and dermatitis in humans.

Reynolds, G. W., *Planta Med.,* 1985, **40**, 494.

Tussilagine 1377

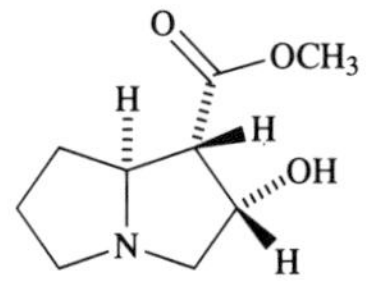

Pyrrolizidine alkaloid

$C_{10}H_{17}NO_3$ [80151-77-5] MW 199.25

Coltsfoot, *Tussilago farfara* (Compositae).

Toxicity yet to be established.

Röder, E., *Planta Med.,* 1981, **43**, 99.

Tutin 1378

2-Hydroxycoriamyrtin; Coriarin

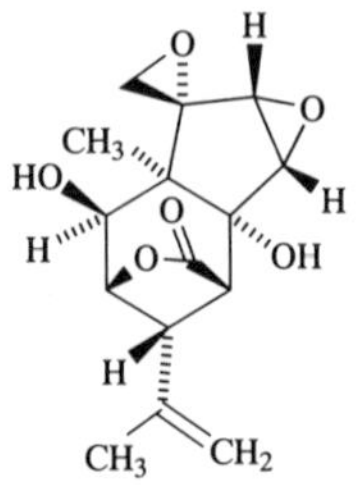

Sesquiterpenoid lactone (Tutinanolide)

$C_{15}H_{18}O_6$ [2571-22-4] MW 294.30

Coriaria angustissima and other *Coriaria* spp. (Coriariaceae); *Toxicodendrum capense* (Euphorbiaceae).

Produces extreme excitation of the central nervous system. It stimulates the respiratory, vasomotor and cardioinhibitory centres in the brain. It has caused the death of many cattle feeding on *Coriaria* species in New Zealand, because of these toxic effects.

Okuda, T., *Tetrahedron Lett.,* 1965, 2137.

Tyledoside A 1379

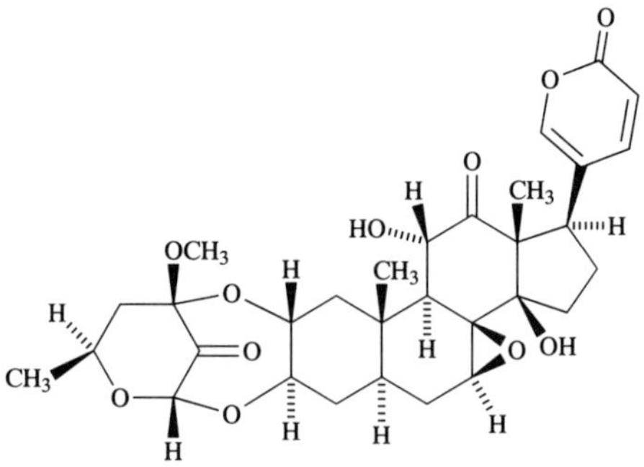

Bufadienolide

$C_{31}H_{38}O_{11}$ [102694-24-6] MW 586.64

One of six closely related bufadienolides present in aerial parts of *Tylecodon grandiflorus* (Crassulaceae).

Tyledoside A and the related tyledosides B, D, F, and G [102694-25-7, 102694-26-8, 102694-27-9 and 102694-28-0 respectively] are responsible for the toxicity to cattle of *Tylecodon grandiflorus.*

Steyn, P. S., *J. Chem. Soc., Perkin Trans. 1,* 1986, 429.

(−)-Tylocrebine 1380

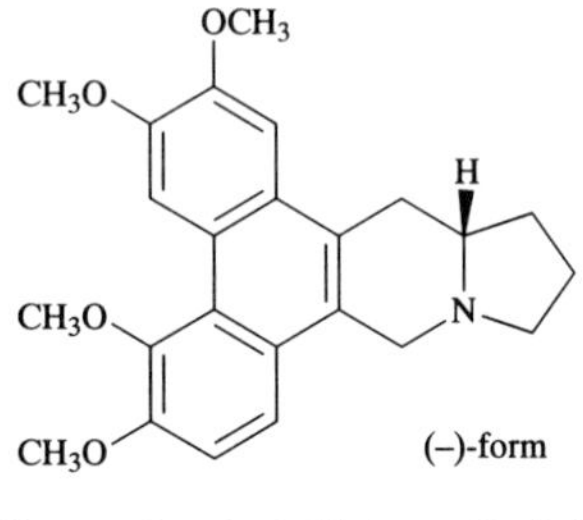

(−)-form

Phenanthroindolizine alkaloid

$C_{24}H_{27}NO_4$ [61302-92-9] MW 393.48
(+)-form [6879-02-3]

The (−)-form occurs in *Tylophora crebiflora* (Asclepiadaceae), while the (+)-form is present in the wild fig tree, *Ficus septica* (Moraceae).

Both forms are toxic and powerful vesicants. The (−)-form possesses antitumour activity in the mouse and has central nervous system toxicity in humans.

Gellert, E., *J. Chem. Soc.,* 1962, 1008.

Tylophorine 1381

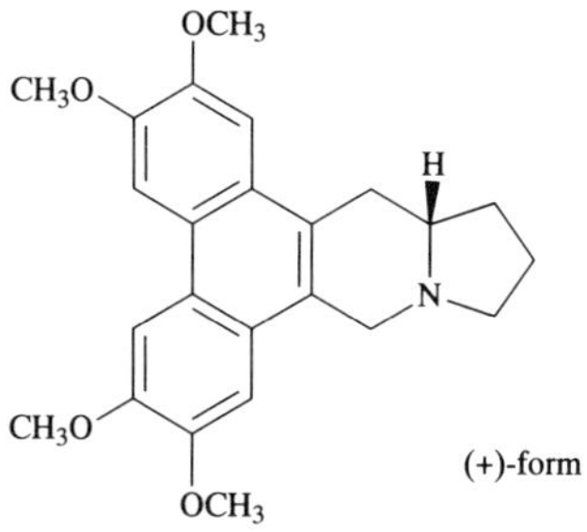

OCH₃

Phenanthroindolizine alkaloid

$C_{24}H_{27}NO_4$ [482-20-2] MW 393.48
(−)-form [111408-21-0]

Tylophora asthmatica, *Cynanchum vincetoxicum*, *Pergularia pallida* and *Vincetoxicum officinale* (Asclepiadaceae) and *Ficus septica* (Moraceae).

Highly toxic to frogs but shows low toxicity towards mice. It is a powerful vesicant, with weak antitumour properties.

Govindachari, T. R., *J. Chem. Soc., Perkin Trans. 1,* 1974, 1161.

U

Undulatone 1382

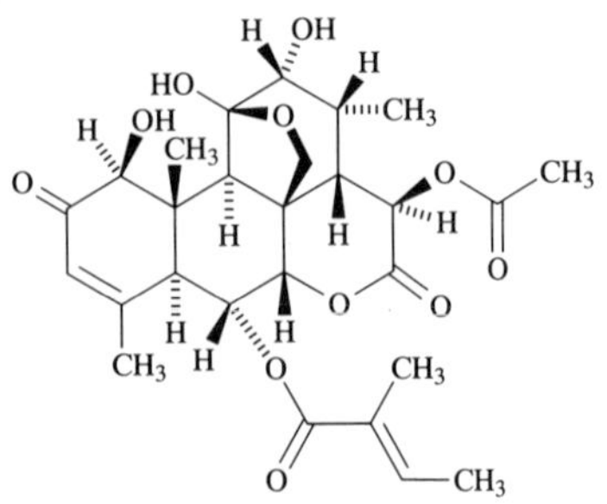

Nortriterpenoid
(Quassane)

$C_{27}H_{34}O_{11}$ [70993-77-0] MW 534.56

Root bark of *Hannoa undulata* (Simaroubaceae).

Antileukaemic properties.

Wani, M. C., *Tetrahedron,* 1979, **35**, 17.

Uplandicine 1383

7-Acetyl-9-echimidinylretronecine

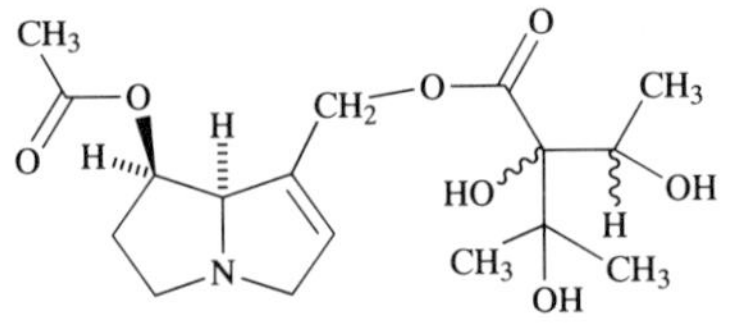

Pyrrolizidine alkaloid

$C_{17}H_{27}NO_7$ [74202-10-1] MW 357.40

Russian comfrey, *Symphytum* × *uplandicum* (Boraginaceae).

Toxicity not established, but suspected of being hepatotoxic.

Culvenor, C. C. J., *Aust. J. Chem.,* 1980, **33**, 1105.

Urechitoxin 1384

Oleandrin monoglucoside

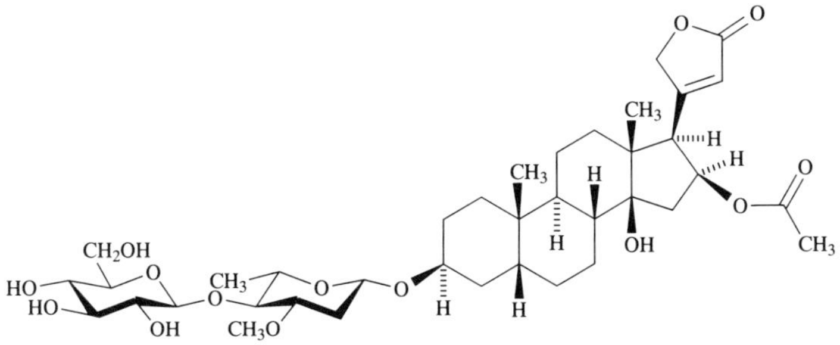

Cardenolide

$C_{38}H_{58}O_{14}$ [56774-61-9] MW 738.87

Leaves of *Nerium indicum* and of yellow nightshade, *Urechites suberecta* (Apocynaceae).

Major toxin of the leaf of yellow nightshade. Causes pain in the oral cavity, nausea, emesis, cramping and diarrhoea after ingestion.

Hassall, C. H., *J. Chem. Soc.*, 1951, 3193.

Ursiniolide A 1385

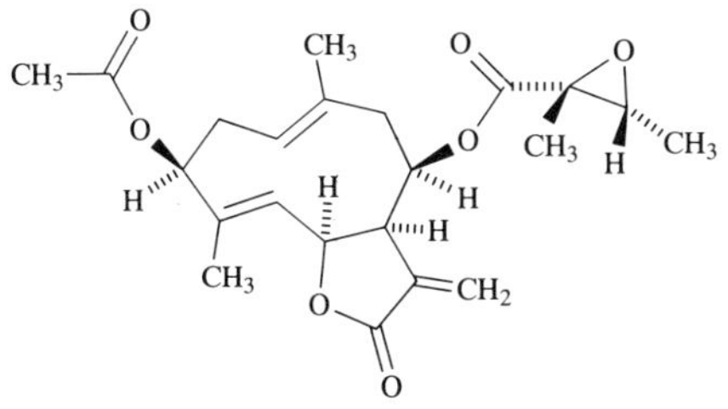

Sesquiterpenoid lactone
(Germacranolide)

$C_{22}H_{28}O_7$ [52677-96-0] MW 404.46

Ursinia anthemoides (Compositae).

Cytotoxic and antitumour properties.

Samek, Z., *Tetrahedron Lett.*, 1979, 2691.

Ursolic acid 1386

Malol; Malolic acid; Micromerol; Prunol; Urson

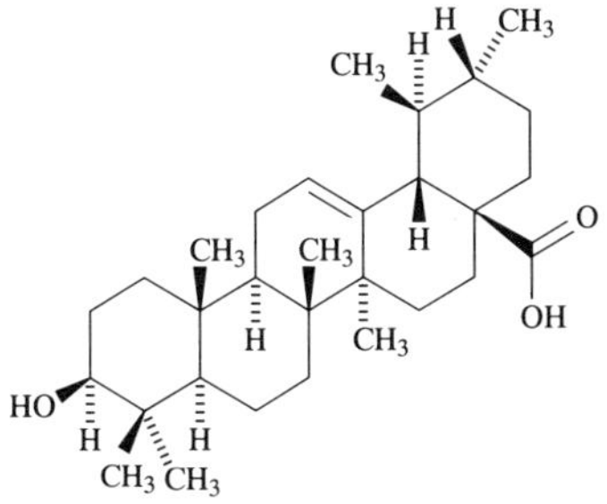

Triterpenoid
(Ursane)

$C_{30}H_{48}O_3$ [77-52-1] MW 456.71

Widespread in nature, in *Prunella vulgaris* (Labiatae), cranberry, *Vaccinium macrocarpon*, bearberry, *Arctostaphylos uva-ursi* (Ericaceae) and in waxy coating of apple, *Malus* spp. and pear skins, *Pyrus* spp. (Rosaceae).

Cytotoxic and antileukaemic properties.

Stout, G. H., *J. Org. Chem.*, 1963, **28**, 1259.

Urushiol III 1387

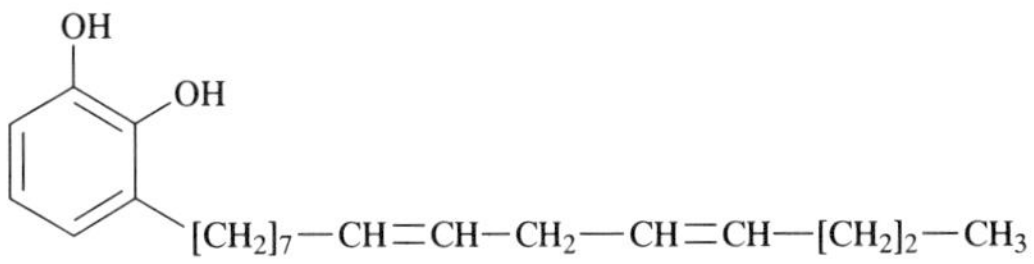

$[CH_2]_7-CH=CH-CH_2-CH=CH-[CH_2]_2-CH_3$

Phenolic

$C_{21}H_{32}O_2$ [492-91-1] MW 316.48

In the irritant oil of poison ivy, *Toxicodendron radicans* (Anacardiaceae). It is one of five closely related alkyl-catechols present in this plant.

Provokes heavy allergic skin reactions and inhibits arachidonic acid metabolism. It is estimated that some 2 million people can suffer each year from poison ivy allergy and about 70% of the population are sensitised. The urushiols are used positively as anti-allergic agents in hyposensitisation therapy.

Tyman, J. H. P., *Chem. Soc. Rev.*, 1979, **8**, 499.

Usambarensine 1388

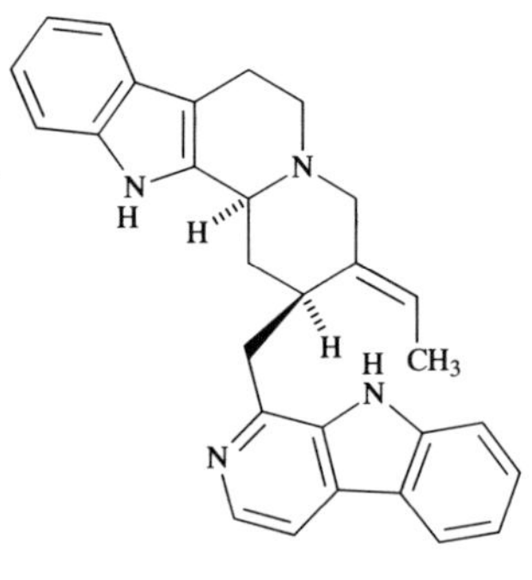

Indole alkaloid
(Tubulosan)

$C_{29}H_{28}N_4$ [36150-14-8] MW 432.57

Root bark of *Strychnos usambarensis* (Loganiaceae).

One of the active principles of *Strychnos* root bark, which is used as an arrow poison. It shows antimuscarinic effects on isolated rat intestinal muscle.

Angenot, L., *J. Pharm. Belg.*, 1971, **26**, 585.

Usambarine 1389

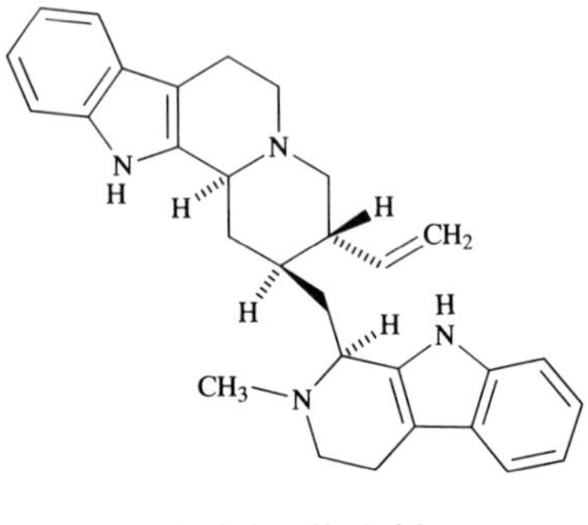

Indole alkaloid
(Tubulosan)

$C_{30}H_{34}N_4$ [35226-29-0] MW 450.63

Root bark of *Strychnos usambarensis* (Loganiaceae).

One of the active principles of *Strychnos* root bark, which is used as an arrow poison.

Koch, M., *C. R. Acad. Sci., Ser. C*, 1971, **273**, 753.

Usaramine 1390

Mucronatine; Usuramine

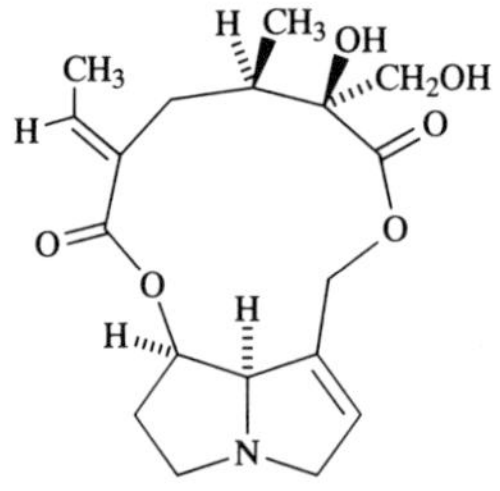

Pyrrolizidine alkaloid
(Senecionan)

$C_{18}H_{25}NO_6$ [15503-87-4] MW 351.40

Crotalaria brevifolia, *C. incana*, *C. intermedia*, *C. mucronata* and *C. usaramoensis* (Leguminosae).

Both hepatotoxic and pneumotoxic. It has hypotensive activity.

Culvenor, C. C. J., *Aust. J. Chem.*, 1966, **19**, 2127.

Uscharidin 1391

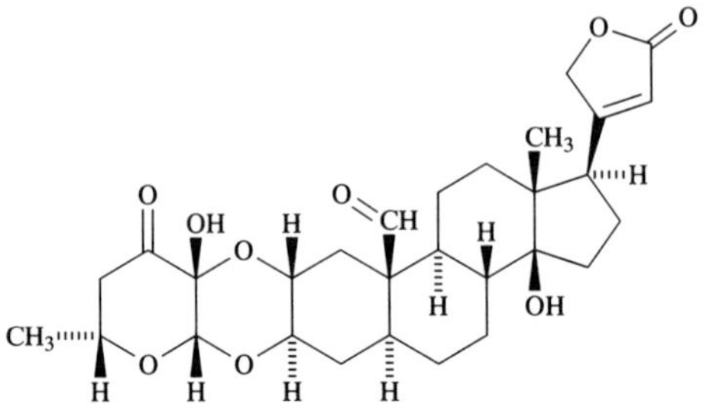

Cardenolide

$C_{29}H_{38}O_9$ [24321-47-9] MW 530.62

Calotropis procera (Asclepiadaceae).

Toxic to vertebrates. The intraperitoneal LD_{50} in male Swiss Webster mice is 11.8 mg/kg body-weight, while the intravenous LD_{50} in cats is 1.4 mg/kg body-weight.

Brüschweiler, F., *Helv. Chim. Acta*, 1969, **52**, 2276.

Uscharin 1392

Uscharine

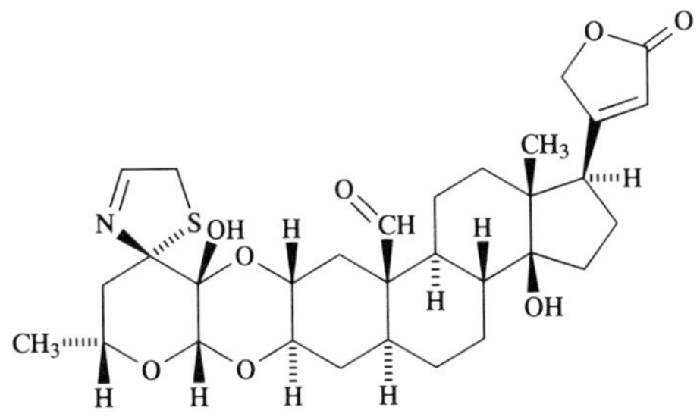

Cardenolide

$C_{31}H_{41}NO_8S$ [24211-81-2] MW 587.73

Calotropis procera (Asclepidiaceae).

Heart poison.

Cheung, H. T. A., *J. Chem. Soc., Perkin Trans. 1*, 1983, 2827.

Uvaretin 1393

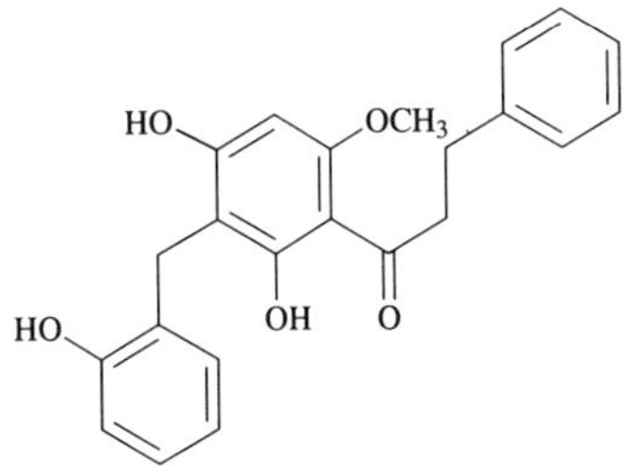

Flavonoid
(Dihydrochalcone)

$C_{23}H_{22}O_5$ [58449-06-2] MW 378.42

Roots of *Uvaria angolensis* and *U. chamae* (Annonaceae).

Cytotoxic properties.

Hufford, C. D., *J. Org. Chem.*, 1976, **41**, 1297.

V

Vanillic acid 1394

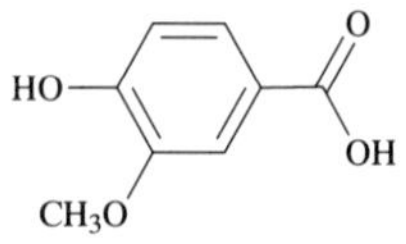

Phenolic acid

$C_8H_8O_4$ [121-34-6] MW 168.15

Fairly widespread in nature, e.g. in *Alnus japonica* (Betulaceae), *Melia azedarach* (Meliaceae) and *Rosa canina* (Rosaceae).

Anthelmintic and antisickling properties. *In vitro* tests suggest that it has anti-inflammatory activity.

Karrer, W., Konstitution und Vorkommen der organischen Pflanzenstoffe, 2nd ed., part 1, 1958, 160, Birkhäuser.

Vasicinol 1395

7-Hydroxypeganine

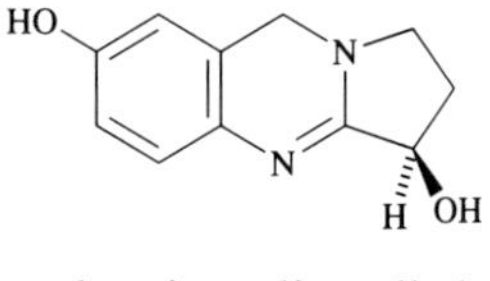

Pyrroloquinazoline alkaloid

$C_{11}H_{12}N_2O_2$ [5081-51-6] MW 204.23

Roots, leaves and seeds of *Adhatoda vasica* (Acanthaceae) and roots of *Sida cordifolia* (Malvaceae).

Has antifertility effects in mammals and is an antifeedant in insects. It is active as a histamine antagonist, a cardiac depressant and a transient hypotensive agent.

Bhatnagar, A. K., *Indian J. Chem.,* 1965, **3**, 524.

Vasicinone 1396

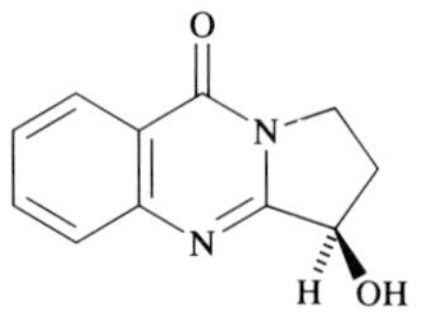

(–)-form

Pyrroloquinazoline alkaloid

$C_{11}H_{10}N_2O_2$ [486-64-6] MW 202.21
(±)-form [35387-16-7]

Seeds and aerial parts of *Peganum harmala*, leaves of *P. nigellastrum* (Zygophyllaceae), roots of *Sida cordifolia* (Malvaceae) and leaves of *Adhatoda vasica* (Acanthaceae).

Anthelmintic and antifertility properties. It is also an insect antifeedant.

Brossi, A., The Alkaloids, 1986, **29**, 99, Academic Press.

Veatchine 1397

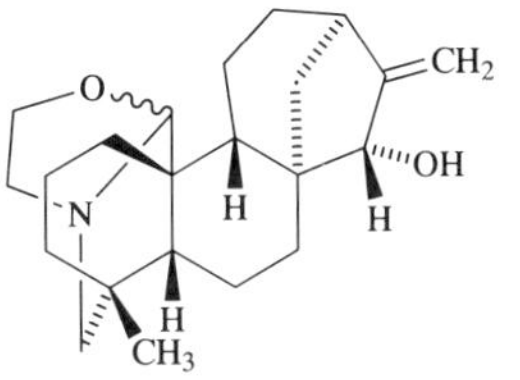

Diterpenoid alkaloid
(Veatchane)

$C_{22}H_{33}NO_2$ [76-53-9] MW 343.51

Garrya veatchii (Garryaceae). Garryfoline [509-30-8], the 15-epimer, occurs in *G. laurifolia* and *G. ovata* var. *lindheimeri* (Garryaceae).

Toxicity is relatively high, compared to the other diterpenoid alkaloids (c.f. aconitine). Toxic symptoms include gasping, convulsions and respiratory failure. Garryfoline is slightly less toxic than veatchine.

Wiesner, K., *Experientia,* 1955, **11**, 255.

Veratramine 1398

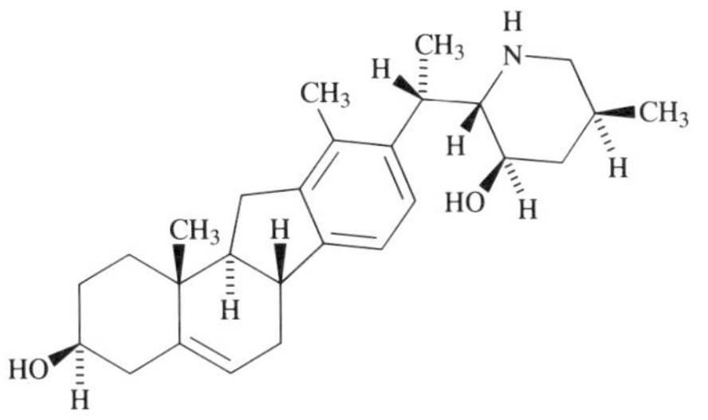

Steroid alkaloid
(Veratraman)

$C_{27}H_{39}NO_2$ [60-70-8] MW 409.61

Rhizomes of *Veratrum grandiflorum* and *V. viride* (Liliaceae).

Antihypertensive drug. *Veratrum viride* extracts have been used in veterinary practice as circulatory depressants, emetics and parasiticides.

Masamune, T., *Tetrahedron,* 1971, **27**, 3369.

Verbascoside 1399

Acteoside; Kusaginin

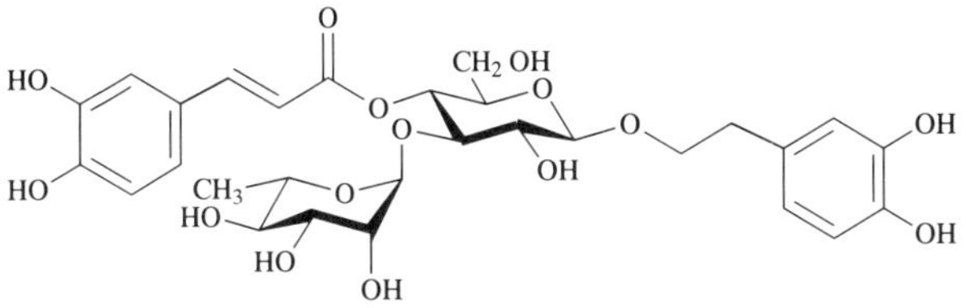

Phenylpropanoid

$C_{29}H_{36}O_{15}$ [61276-17-3] MW 624.60

Leaf of *Buddleja globosa*, *B. officinalis*, fruit of *Forsythia* spp. (Oleaceae) and *Verbascum sinuatum* (Scrophulariaceae).

Hypertensive properties. It also has antihepatotoxic and anti-inflammatory activities.

Birkofer, L., *Z. Naturforsch.,* 1968, **23B**, 1051.

Verbenalin 1400

Verbenaloside; Cornin

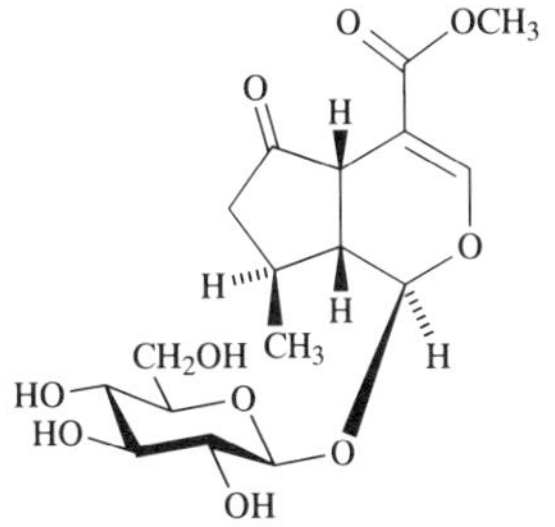

Monoterpenoid
(Iridane)

$C_{17}H_{24}O_{10}$ [548-37-8] MW 388.37

Verbena officinalis (Verbenaceae) and *Cornus* spp. (Cornaceae).

Activity on the uterus resembling that of the ergot alkaloids. Also shows laxative properties.

Büchi, G., *Tetrahedron,* 1962, **18**, 1049.

Verbenone 1401

Pin-2-en-4-one

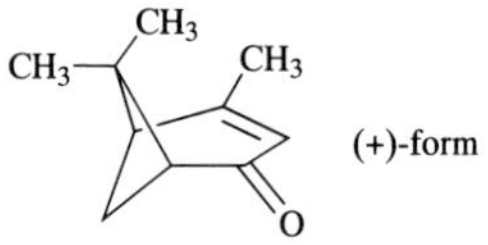

(+)-form

Monoterpenoid
(Pinane)

$C_{10}H_{14}O$ [18309-32-5] MW 150.22

Spanish verbena oil, *Verbena triphylla* (Verbenaceae).

Toxic to mammals. The LD_{50} intraperitoneally in mice is 250 mg/kg body-weight.

Banthorpe, D. V., *Chem. Rev.,* 1966, **66**, 643.

Vermeerin 1402

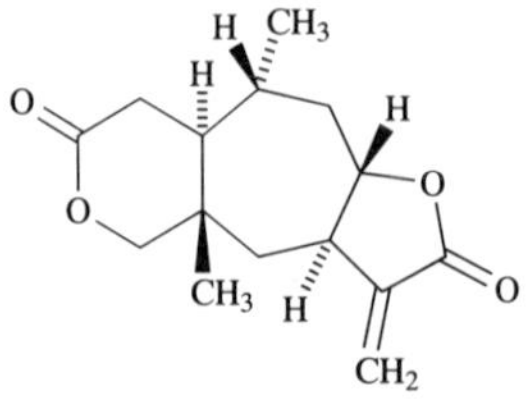

Sesquiterpenoid lactone
(Seco-pseudoguaianolide)

$C_{15}H_{20}O_4$ [16983-23-6] MW 264.32

Geigera aspera, G. africana, Hymenoxys anthemoides, H. richardsonii and *Psilostrophe villosa* (Compositae).

Toxic to mammals. It causes mass poisoning ("vomiting disease") among sheep in Africa after feeding on *Geigera* plants.

Anderson, L. A. P., *Tetrahedron,* 1967, **23**, 4153.

Vernadigin 1403

Strophadogenin 3-diginoside

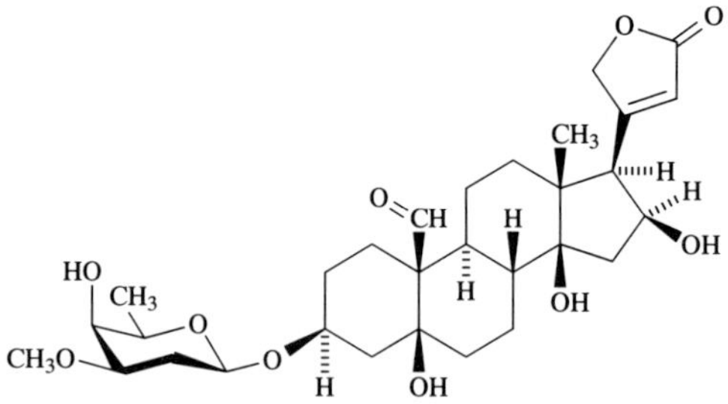

Cardenolide

$C_{30}H_{44}O_{10}$ [30285-47-3] MW 564.67

One of twenty cardenolides in yellow pheasant's eye, *Adonis vernalis* (Ranunculaceae).

Toxic to vertebrates. Human poisoning is not recorded and is unlikely, since these toxins are poorly absorbed and show little tendency to accumulate.

Poláková, A., *Chem. Ind. (London),* 1963, 1766.

Vernodalin 1404

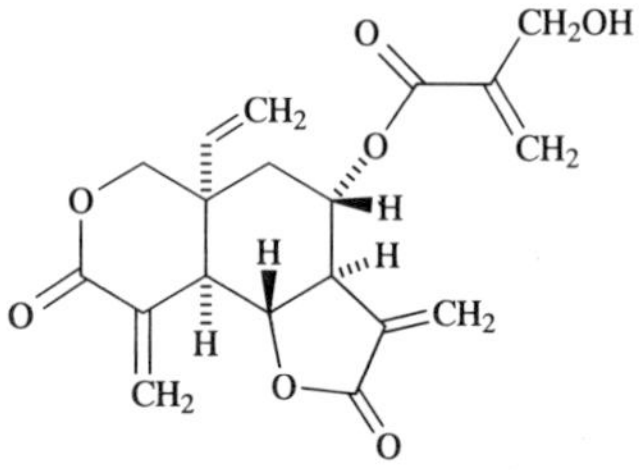

Sesquiterpenoid lactone
(Elemanolide)

$C_{19}H_{20}O_7$ [21871-10-3] MW 360.36

Ironweeds, *Vernonia amygdalina* and *V. guineensis* (Compositae).

Major toxin of *V. amygdalina*. It is an insect antifeedant, but also has cytotoxic and antitumour properties.

Kupchan, S. M., *J. Org. Chem.,* 1969, **34**, 3908.

Vernodalol 1405

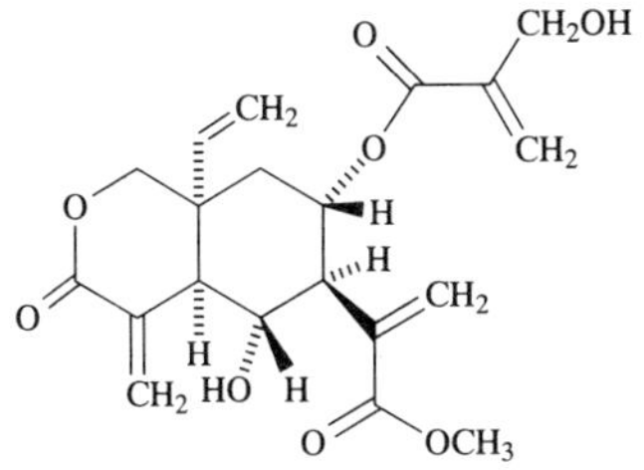

Sesquiterpenoid
(Elemane)

$C_{20}H_{24}O_8$ [65388-17-2] MW 392.41

Ironweeds, *Vernonia amygdalina* and *V. anthelmintica* (Compositae).

Insect antifeedant.

Asaka, Y., *Phytochemistry,* 1977, **16**, 1838.

Vernoflexin 1406

Vernoflexine; Zaluzanin C senecioate

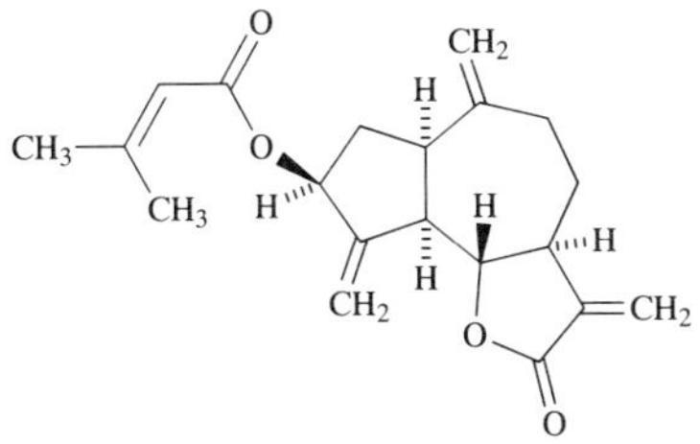

Sesquiterpenoid lactone
(Guaianolide)

$C_{20}H_{24}O_4$ [57576-43-9] MW 328.41

Vernonia flexuosa, *V. arkansana* and *V. chinensis* (Compositae).

Cytotoxic and antitumour properties

Kisiel, W., *Pol. J. Pharmacol. Pharm.,* 1975, **27**, 461.

Vernoflexuoside 1407

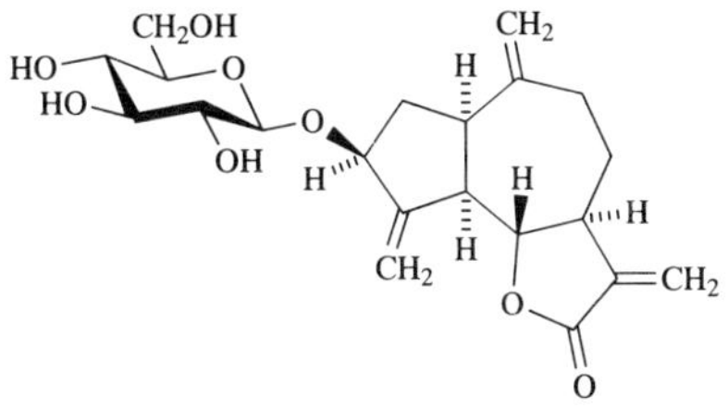

Sesquiterpenoid lactone
(Guaianolide)

$C_{21}H_{28}O_8$ [57576-33-7] MW 408.45

Vernonia flexuosa (Compositae).

Cytotoxic properties.

Kisiel, W., *Pol. J. Pharmacol. Pharm.,* 1975, **27**, 461.

Vernolepin 1408

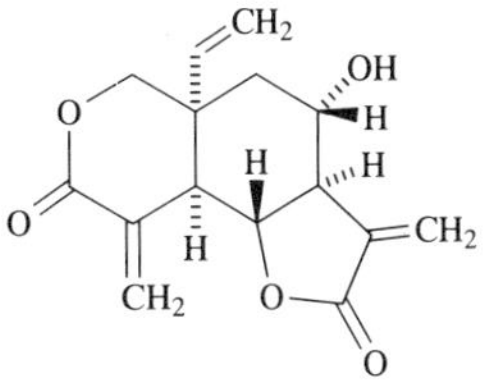

Sesquiterpenoid lactone
(Elemanolide)

$C_{15}H_{16}O_5$ [18542-37-5] MW 276.29

Vernonia guineensis and *V. hymenolepis* (Compositae).

Cytotoxic and antitumour properties.

Kupchan, S. M., *J. Org. Chem.,* 1969, **34**, 3903.

Vernolide 1409

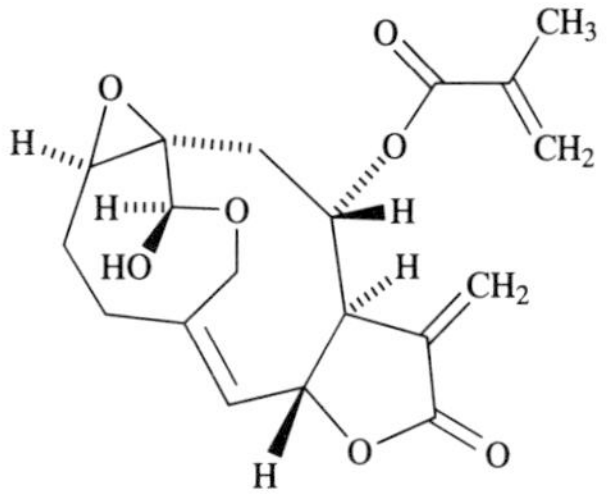

Sesquiterpenoid lactone
(Germacranolide)

$C_{19}H_{22}O_7$ [27428-86-0] MW 362.38

Leaves of *Vernonia amygdalina* and *V. colorata* (Compositae).

Together with vernodalin (q.v.), it constitutes the bitter and toxic principle of 'bitter leaf', *Vernonia amygdalina*. Goats are poisoned by eating the leaves. The toxicity is avoided by processing and leaching and the leaves are eaten locally as a vegetable in Africa. The plant is also used medicinally against intestinal parasites and chimpanzees have been observed to swallow the leaf juice for the same purpose.

Pascard, C., *Tetrahedron Lett.,* 1970, 4131.

Vernomenin 1410

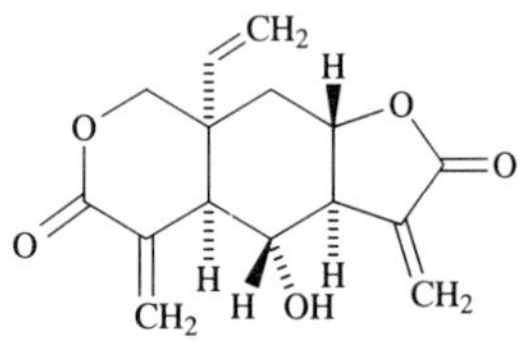

Sesquiterpenoid lactone
(Elemanolide)

$C_{15}H_{16}O_5$ [20107-26-0] MW 276.29

Vernonia hymenolepis (Compositae).

Cytotoxic and antitumour properties.

Kupchan, S. M., *J. Org. Chem.,* 1969, **34**, 3903.

Vernomygdin 1411

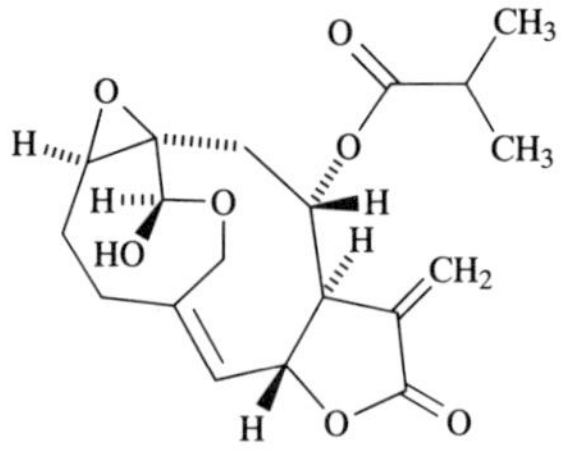

Sesquiterpenoid lactone
(Germacranolide)

$C_{19}H_{24}O_7$ [21871-14-7] MW 364.40

Vernonia amygdalina (Compositae).

Cytotoxic and antitumour properties.

Kupchan, S. M., *J. Org. Chem.,* 1969, **34**, 3908.

Vestitol 1412

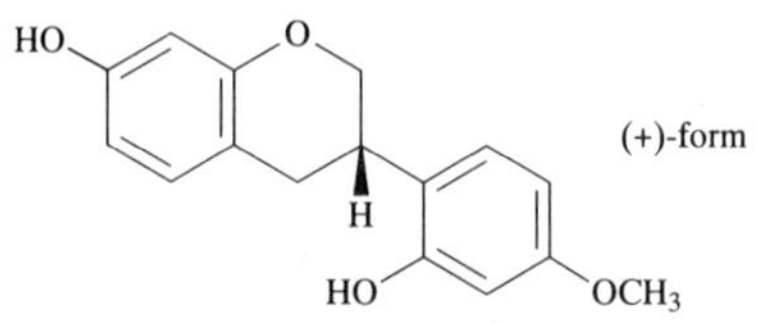

(+)-form

Isoflavonoid
(Isoflavan)

$C_{16}H_{16}O_4$ [20879-05-4] MW 272.30
(−)-form [35878-41-2]

In the (−)-form in lucerne, *Medicago sativa* and *Lotus* spp. (Leguminosae) infected with *Cladosporium fulvum*; constitutively present as the (+)-form in heartwood of *Dalbergia variabilis* (Leguminosae).

Antifungal agent, with antifeedant properties.

Kurosawa, K., *J. Chem. Soc., Chem. Commun.,* 1968, 1263.

Vicianin 1413

Mandelonitrile vicianoside

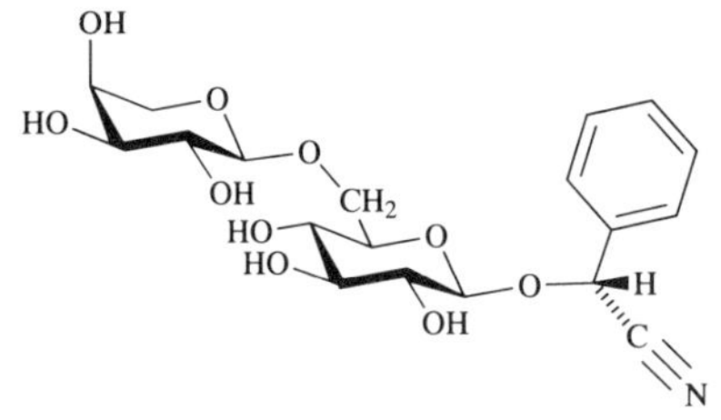

Cyanogenic glycoside

$C_{19}H_{25}NO_{10}$ [155-57-7] MW 427.41

Seeds of *Vicia angustifolia* and *V. sativa* (Leguminosae), subterranean parts of *Gerbera jamesonii* (Compositae) and fronds of *Davallia* spp. (Davalliaceae).

Bound toxin, releasing cyanide on hydrolysis or on tissue damage.

Chaudhury, D. N., *J. Chem. Soc.,* 1949, 2054.

Vignafuran 1414

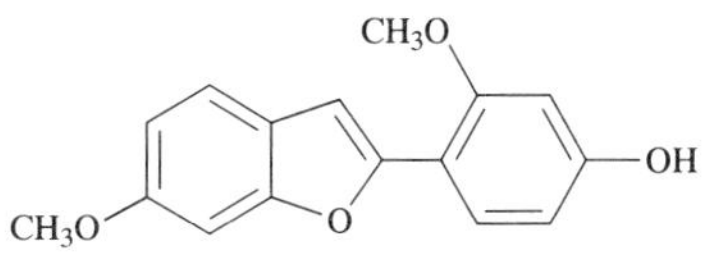

Benzofuran

$C_{16}H_{14}O_4$ [57800-41-6] MW 270.28

Fungally-infected hypocotyls of *Lablab niger* and leaves of cowpea, *Vigna unguiculata* (Leguminosae).

Antifungal agent (phytoalexin).

Preston, N. W., *Phytochemistry,* 1975, **14**, 1843.

Viguiestenin 1415

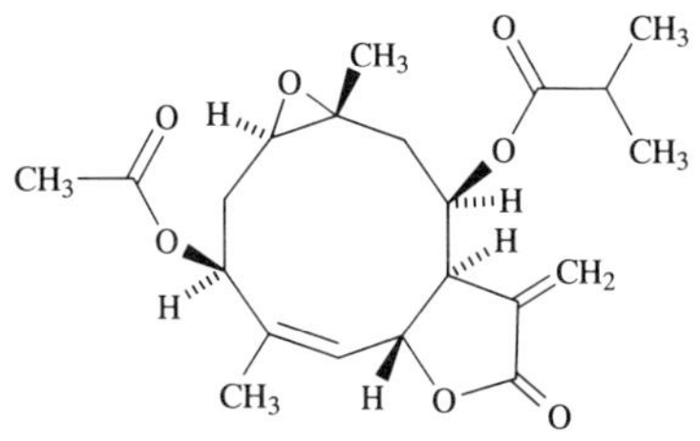

Sesquiterpenoid lactone
(Germacranolide)

$C_{21}H_{28}O_7$ [54153-71-8] MW 392.45

Viguiera stenoloba and *V. pinnatilobata* (Compositae).

Cytotoxic and antitumour properties.

Romo de Vivar, A., *Rev. Latinoam. Quim.,* 1978, **9**, 171.

Villalstonine 1416

Alkaloid B

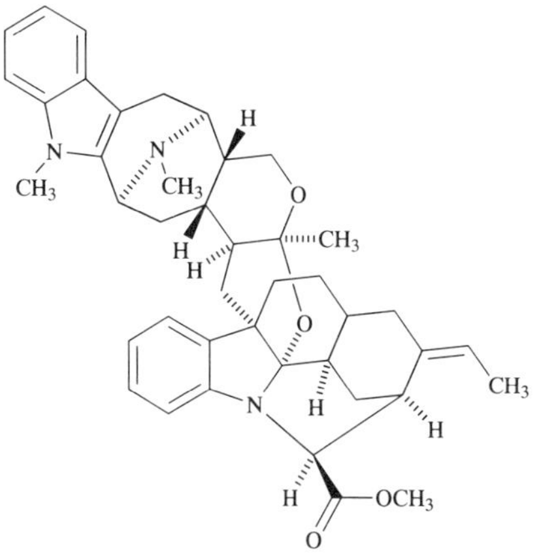

Bisindole alkaloid

$C_{41}H_{48}N_4O_4$ [2723-56-0] MW 660.86

Alstonia muelleriana, *A. macrophylla* and *A. spectabilis* (Apocynaceae).

Antiprotozoal activity.

Hesse, M., *Helv. Chim. Acta,* 1966, **49**, 1173.

Vinblastine 1417

Vincaleukoblastine; VLB

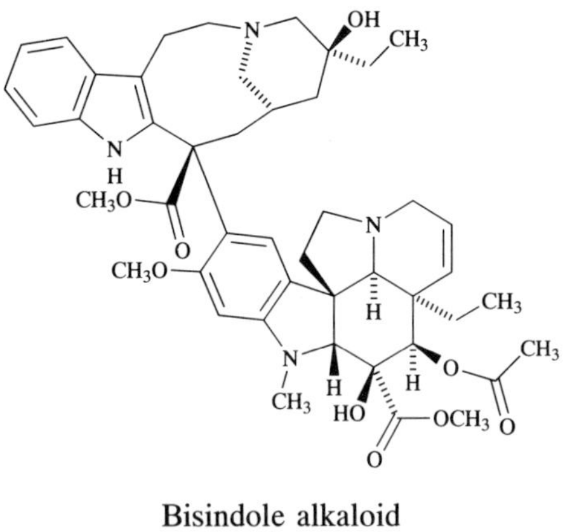

Bisindole alkaloid

$C_{46}H_{58}N_4O_9$ [865-21-4] MW 810.99

Leaves of the periwinkle, *Vinca rosea* (Apocynaceae).

Well known for its effective antitumour activity. It is in clinical use for treating leukaemia and Hodgkin's disease.

Neuss, N., *J. Am. Chem. Soc.,* 1964, **86**, 1439.

Vincamine 1418

Vincamarine; Minorine

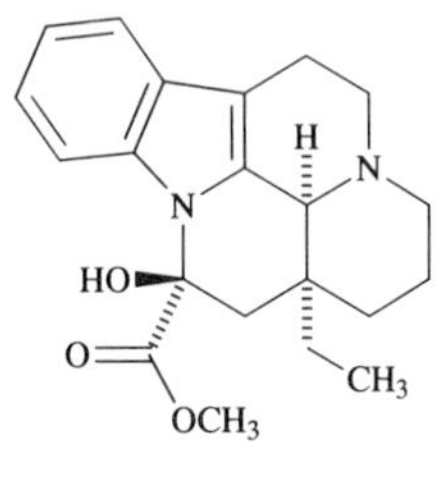

Indole alkaloid
(Eburnane)

$C_{21}H_{26}N_2O_3$ [1617-90-9] MW 354.45

Leaves of the greater periwinkle, *Vinca major* (Apocynaceae).

Vasodilatory properties. It is used as a drug, in a variety of disorders, to increase cerebral blood circulation.

Trojánek, J., *Collect. Czech. Chem. Commun.,* 1968, **33**, 2950.

Vincristine 1419

Leurocristine; 22-Oxovincaleukoblastine; VCR; LCR

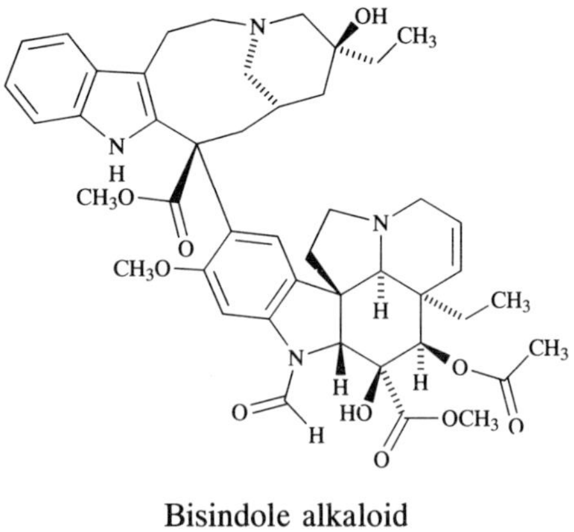

Bisindole alkaloid

$C_{46}H_{56}N_4O_{10}$ [57-22-7] MW 824.97

Madagascar periwinkle, *Vinca rosea* (Apocynaceae).

Well known antitumour agent. It is used clinically for acute childhood lymphocytic leukaemia.

Neuss, N., *J. Am. Chem. Soc.,* 1964, **86**, 1440.

ε-Viniferin 1420

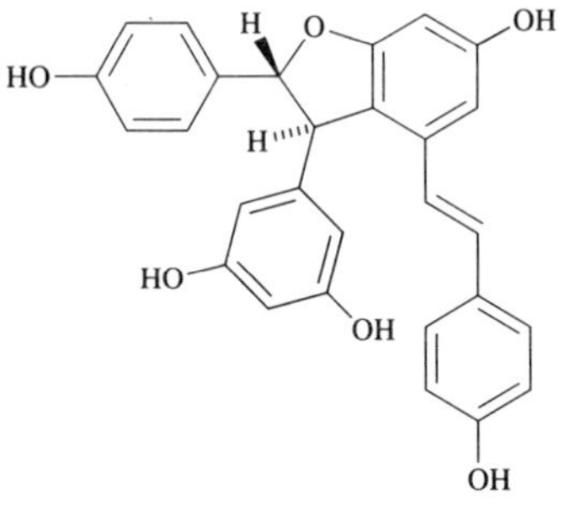

Bis-stilbenoid

$C_{28}H_{22}O_6$ [62218-08-0] MW 454.48

Fungally infected leaves of the grapevine, *Vitis vinifera* (Vitaceae).

Antifungal agent (phytoalexin).

Kurihara, H., *Agric. Biol. Chem.,* 1990, **54**, 1097.

Viscotoxin 1421

Lys-Ser-Cys-Cys-Pro-Asn-Thr-Thr-Gly-Arg-
-Asn-Ile-Tyr-Asn-Thr-Cys-Arg-Phe-Gly-Gly-
-Gly-Ser-Arg-Glu-Val-Cys-Ala-Ser-Leu-Ser-
-Gly-Cys-Lys-Ile-Ile-Ser-Ala-Ser-Thr-Cys-
-Pro-Ser-Tyr-Pro-Asp-Lys

Protein
[76822-96-3] MW 4834

Mistletoe, *Viscum album* (Viscaceae). Contains 46 amino acid residues.

Toxic to heart muscle and cytotoxic. It inhibits DNA synthesis.

Olson, T., *Acta Pharm. Suec.*, 1974, **11**, 381.

Viscumin 1422

Two chains with sulfide links
A chain M_r 29 000, B chain M_r 34 000

Protein
(Lectin) MW ~ 60 000

Mistletoe, *Viscum album* (Viscaceae).

Cytotoxic. It inhibits protein synthesis in cell free systems. Toxicity can be prevented by pretreatment with galactose, lactose and a calcium ionophore.

Olsnes, S., *J. Biol. Chem.*, 1982, **257**, 13271.

Visnagin 1423

Visnagidin; Visnacorin

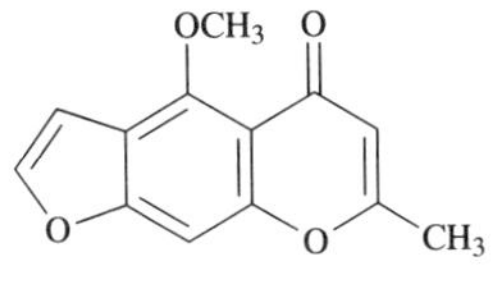

Furochromone

$C_{13}H_{10}O_4$ [82-57-5] MW 230.22

Fruits of *Ammi visnaga* (Umbelliferae).

Phototoxic.

Späth, E., *Chem. Ber.*, 1941, **74**, 1492.

Vitexin 1424

Apigenin 8-*C*-glucoside

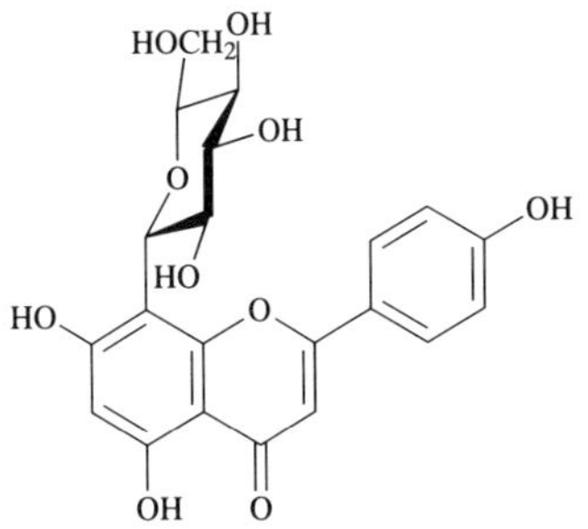

Flavonoid glucoside

$C_{21}H_{20}O_{10}$ [3681-93-4] MW 432.38

Widespread natural occurrence; good sources include wood of *Vitex lucens* (Verbenaceae) and grain of millet, *Pennisetum americanum* (Gramineae).

Potent inhibitor of thyroid peroxidase. It is responsible, with other phenolics present, for the goitrogenic and antithyroid activity of *Pennisetum americana* grain.

Horowitz, R. M., *Chem. Ind. (London)*, 1964, 498.

Voacamine 1425

Voacanginine

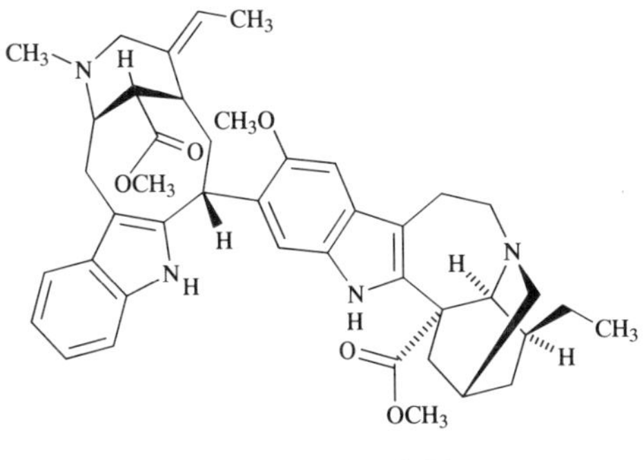

Bisindole alkaloid

$C_{43}H_{52}N_4O_5$ [3371-85-5] MW 704.91

Voacanga africana, *V. globosa*, *V. grandiflora* and *V. thouarsii*, *Tabernaemontana arborea*, *T. australis* and *T. oppositifolia*, and in *Hedranthera barteri* (Apocynaceae).

Cytotoxic properties.

Büchi, G., *J. Am. Chem. Soc.,* 1964, **86**, 4631.

Volkenin 1426

Barterin; Epitetraphyllin B

(1*R*,4*R*)-form

Cyanogenic glycoside

$C_{12}H_{17}NO_7$ [66575-40-4] MW 287.27

Barteria fistulosa (Passifloraceae). Volkenin, the (1*R*,4*R*)-form, co-occurs in the Passifloraceae with three epimers, tetraphyllin B [34323-07-4] (1*S*,4*S*)-form, taraktophyllin [110115-55-4] (1*R*,4*S*)-form and epivolkenin [109905-56-8] (1*S*,4*R*)-form.

Volkenin and its epimeric forms are bound toxins, releasing cyanide on hydrolysis or tissue damage.

Jaroszewski, J. W., *Acta Chem. Scand.,* 1987, **41B**, 410.

Vulgarin 1427

Judaicin; Tauremisin

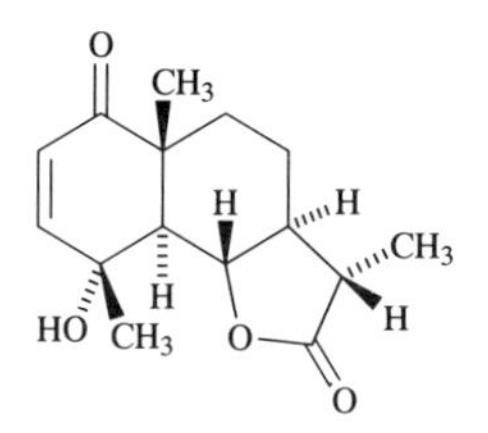

Sesquiterpenoid lactone
(Eudesmanolide)

$C_{15}H_{20}O_4$ [3162-56-9] MW 264.32

Mugwort, *Artemisia vulgaris*; in *A. judaica*, *A. taurica* and many other *Artemisia* spp. (Compositae).

Cytotoxic and antitumour properties.

Rybalko, K. S., *Collect. Czech. Chem. Commun.,* 1961, **26**, 2909.

W

Warburganal 1428

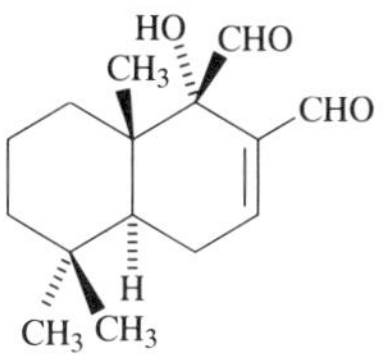

Sesquiterpenoid
(Drimane)

$C_{15}H_{22}O_3$ [62994-47-2] MW 250.34

Bark of *Warburgia salutaris* (Canellaceae).

Potent insect antifeedant, especially against lepidopteran armyworms.

Kubo, I., *J. Chem. Soc., Chem. Commun.*, 1976, 1013.

Wedelolactone 1429

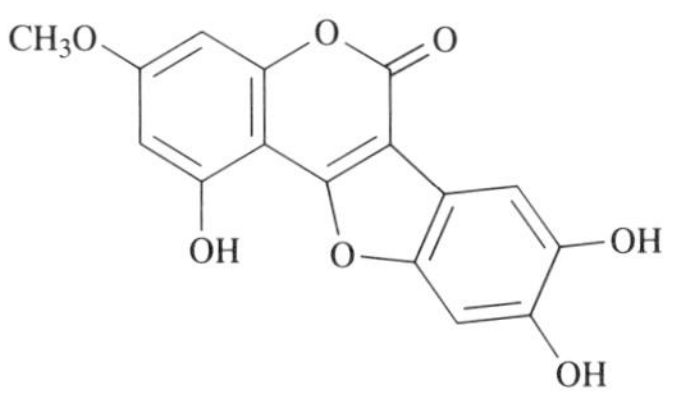

Isoflavonoid
(Coumestan)

$C_{16}H_{10}O_7$ [524-12-9] MW 314.25

Leaves of *Wedelia calendulacea* (Compositae) and heartwood of *Ougeinia dalbergioides* (Leguminosae).

Oestrogenic properties, and has liver protective activity.

Govindachari, T. R., *J. Chem. Soc.*, 1956, 629.

Wedeloside 1430

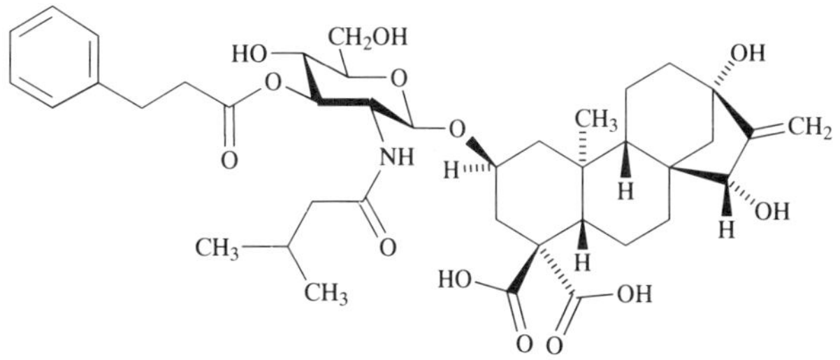

Diterpenoid
(Kaurane)

$C_{40}H_{55}NO_{13}$ [74686-30-9] MW 757.87

Whole plant of *Wedelia asperrima* (Compositae). Co-occurs with the 4′-rhamnoside [74686-30-0].

Major toxin of *Wedelia asperrima*, which when eaten by sheep or other livestock, causes heavy mortality. It is a powerful mitochondrial inhibitor, with a similar action to carboxyatractyloside (q.v.).

Eichholzer, J. V., *Tetrahedron*, 1981, **37**, 1881.

Wighteone 1431

Erythrinin B; Lupinus compound LA-I; 6-Isopentenylgenistein

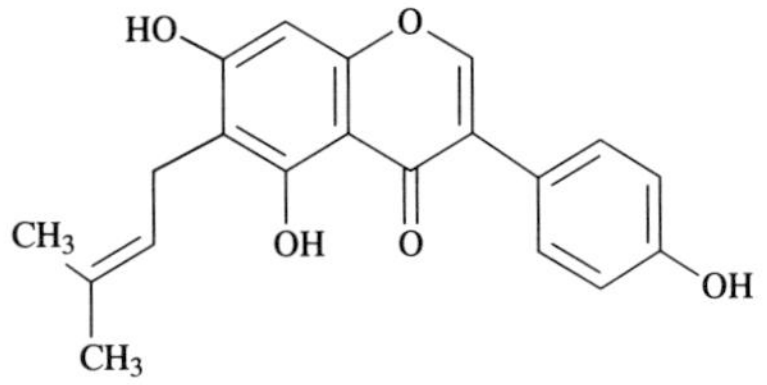

Isoflavone

$C_{20}H_{18}O_5$ [51225-30-0] MW 338.36

Leaves and fruits of *Lupinus albus* and other *Lupinus* spp. (Leguminosae). Also induced as a phytoalexin in leaves of *Laburnum anagyroides* (also Leguminosae).

Antifungal activity.

Ingham, J. L., *Phytochemistry,* 1977, **16**, 1943.

Wilfordine 1432

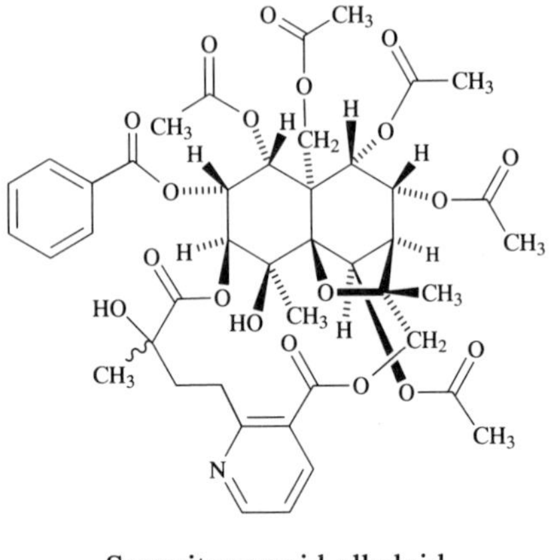

Sesquiterpenoid alkaloid (Eudesmane)

$C_{43}H_{49}NO_{19}$ [37239-51-3] MW 883.86

Seeds of *Trypterygium wilfordii* and *Euonymus alatus* (Calastraceae).

Insecticidal properties.

Yamada, K., *Tetrahedron,* 1978, **34**, 1915.

Withaferin A 1433

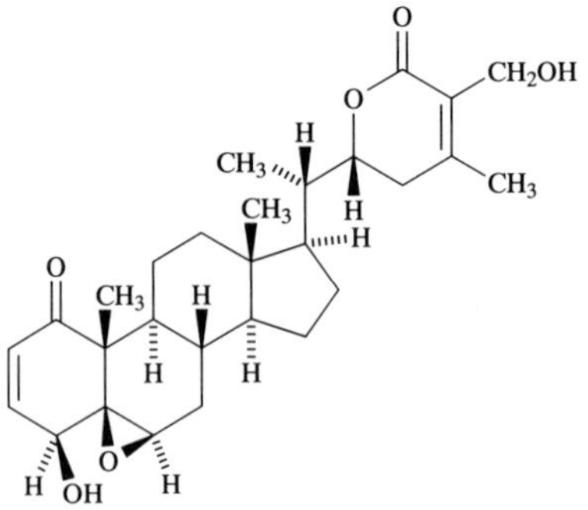

Triterpenoid (Ergostane)

$C_{28}H_{38}O_6$ [5119-48-2] MW 470.61

Leaves of *Acnistus arborescens* and *Withania somnifera* and roots of *W. coagulans* and *W. ashwagandha* (Solanaceae).

Antitumour properties.

Kupchan, S. M., *J. Org. Chem.,* 1969, **34**, 3858.

Withanolide D 1434

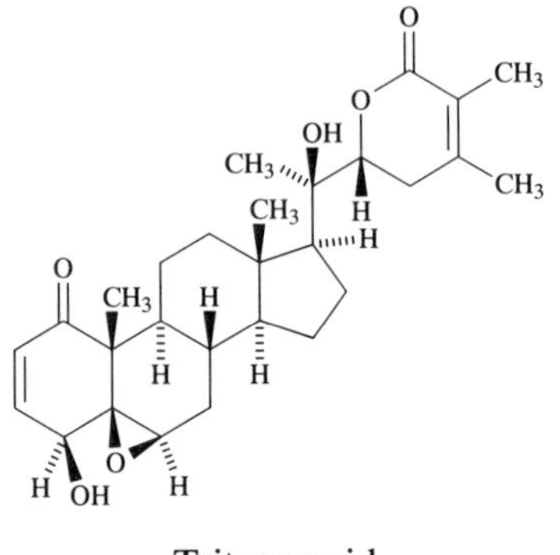

Triterpenoid (Ergostane)

$C_{28}H_{38}O_6$ [30655-48-2] MW 470.61

Leaves of *Withania somnifera* (Solanaceae).

Antitumour properties.

Lavie, D., *Isr. J. Chem.,* 1968, **6**, 671.

Wogonin 1435

Norwogonin 8-methyl ether

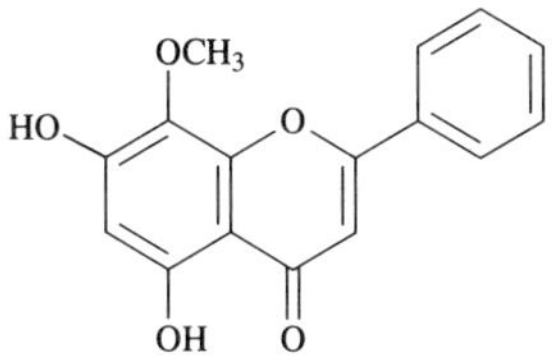

Flavone

$C_{16}H_{12}O_5$ [632-85-9] MW 284.27

Stems of *Anodendron affine* (Apocynaceae) and roots of *Scutellaria baicalensis* (Labiatae).

Oestrogenic and anti-implantation properties in the rat.

Shah, R. C., *J. Chem. Soc.,* 1938, 1555.

Wyerone 1436

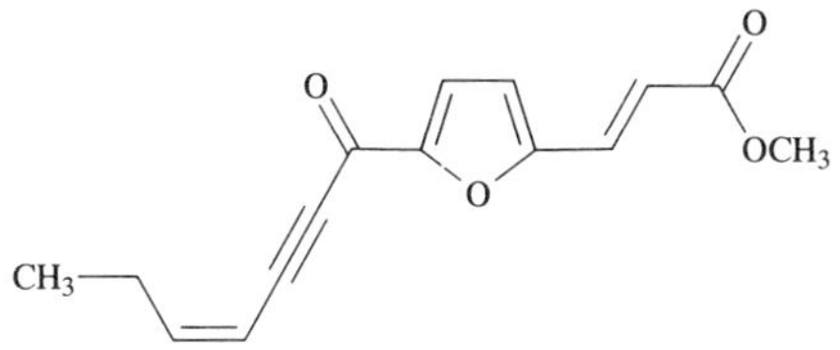

Acetylene

$C_{15}H_{14}O_4$ [20079-30-5] MW 258.27

Broad beans, *Vicia faba* (Leguminosae) infected with *Botrytis fabae*.

Antifungal agent (phytoalexin).

Fawcett, C. H., *J. Chem. Soc., C,* 1968, 2455.

Wyerone acid 1437

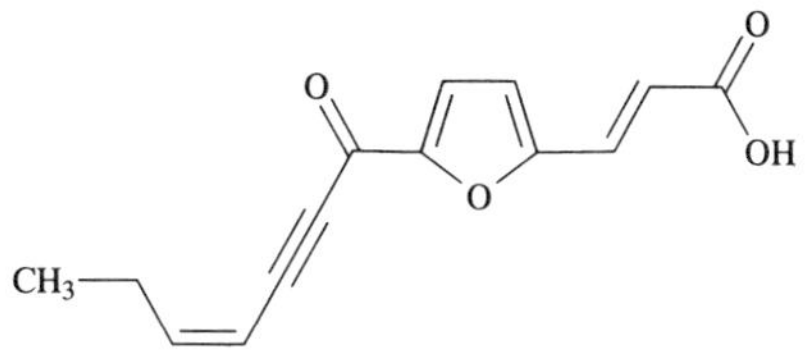

Acetylene

$C_{14}H_{12}O_4$ [54954-14-2] MW 244.25

Broad beans, *Vicia faba* (Leguminosae) infected by *Botrytis fabae*.

Antifungal agent (phytoalexin).

Letcher, R. M., *Phytochemistry,* 1970, **9**, 249.

Wyerone epoxide 1438

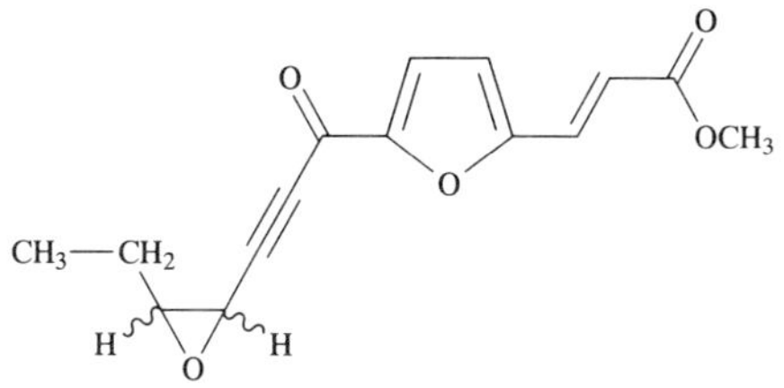

Acetylene

$C_{15}H_{14}O_5$ [60375-16-8] MW 274.27

Broad bean, *Vicia faba* (Leguminosae) and other *Vicia* spp. infected with *Botrytis cinerea*.

Antifungal activity (phytoalexin).

Hargreaves, J. A., *Phytochemistry,* 1976, **15**, 1119.

Xanthatin 1439

Sesquiterpenoid lactone
(Secoguaianolide)

$C_{15}H_{18}O_3$ [26791-73-1] MW 246.31

Xanthium pennsylvanicum, *X. riparium* and *X. sibiricum* (Compositae).

Harmful to insect larvae, inhibiting development.

McMillan, C., *Biochem. Syst. Ecol.*, 1975, **3**, 181.

Xanthohumol 1440

4,2′,4′-Trihydroxy-6′-methoxy-3′-prenylchalcone

Chalcone

$C_{21}H_{22}O_5$ [6754-58-1] MW 354.40

Wood resin of hops, *Humulus lupulus* (Cannabidaceae).

Antifungal activity.

Verzele, M., *Bull. Soc. Chim. Belg.*, 1957, **66**, 452.

Xanthotoxin 1441

Ammoidin; 8-Methoxypsoralen

Furocoumarin

$C_{12}H_8O_4$ [298-81-7] MW 216.19

Seeds of *Ammi majus*, *Pastinaca sativa* and *Angelica archangelica*, roots of *A. officinalis* and *Heracleum sphondylium* (Umbelliferae), and leaves of *Fagara* and *Ruta* spp. (Rutaceae). Formed as a phytoalexin in celery, *Apium graveolens*, infected with *Sclerotina sclerotiorum*.

Toxic to fish, toads and snails. Causes dermatitis in man. It is used as a pigmentation agent in the treatment of vitiligo and psoriasis. Has a bitter taste.

Anderson, T. F., *Annu. Rev. Pharmacol. Toxicol.*, 1980, **20**, 235.

Xanthotoxol 1442

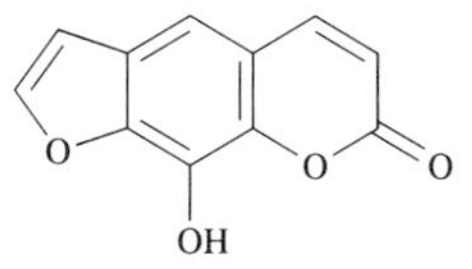

Furocoumarin

$C_{11}H_6O_4$ [2009-24-7] MW 202.17

Seeds of *Angelica archangelica*, parsnip, *Pastinaca sativa* and *Heracleum lanatum* (Umbelliferae) and seeds of *Poncirus trifoliata* (Rutaceae).

Phototoxin. Also inhibits HeLa cell proliferation.

Späth, E., *Ber. Dtsch. Chem. Ges.*, 1937, **70**, 748.

Xanthoxylin 1443

Phloracetophenone 4,6-dimethyl ether; Brevifolin; Hydroxypeonol

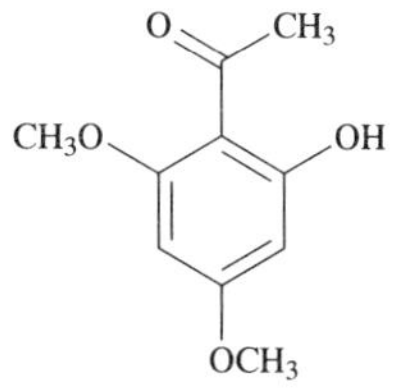

Acetophenone

$C_{10}H_{12}O_4$ [90-24-4] MW 196.20

Bark of *Citrus limon* (Rutaceae) infected with *Phytophthora citrophthora*. Also present constitutively in *Xanthoxylum piperitum* and *X. alatum* (Rutaceae).

Antifungal agent. Also inhibits prostaglandin synthetase and 5-lipoxygenase.

Hartmann, G., *Phytopathol. Z.*, 1974, **81**, 97.

Xanthumin 1444

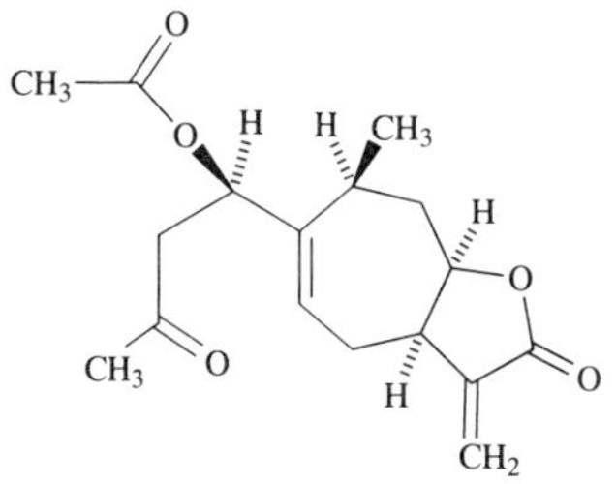

Sesquiterpenoid lactone
(Secoguaianolide)

$C_{17}H_{22}O_5$ [26791-72-0] MW 306.36

Xanthium chasei, *X. chinense*, *X. occidentale* and *X. strumarium* (Compositae).

Insect antifeedant, active against *Drosophila melanogaster*.

Minato, H., *J. Chem. Soc.*, 1965, 7009.

Xanthyletin 1445

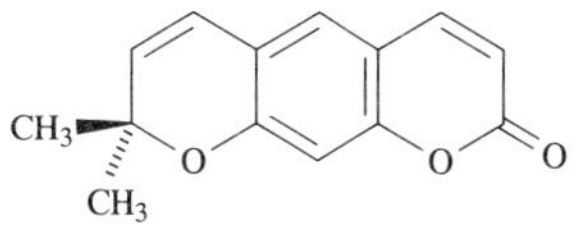

Pyranocoumarin

$C_{14}H_{12}O_3$ [553-19-5] MW 228.25

Bark of *Xanthoxylum americanum*, root of *X. ailanthoides*, fruit of *Luvunga scandens*, wood of *Chloroxylon swietenia* (Rutaceae) and wood of *Brosimum* sp. (Moraceae).

Antitumour properties.

Bell, J. C., *J. Chem. Soc.*, 1937, 1542.

Xerantholide 1446

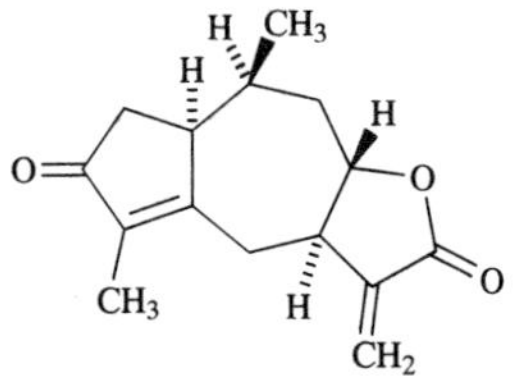

Sesquiterpenoid lactone
(Guaianolide)

$C_{15}H_{18}O_3$ [65017-92-2] MW 246.31

Xeranthemum cylindraceum (Compositae).

Insect antifeedant.

Samek, Z., *Collect. Czech. Chem. Commun.*, 1977, **42**, 2441.

Xylomollin 1447

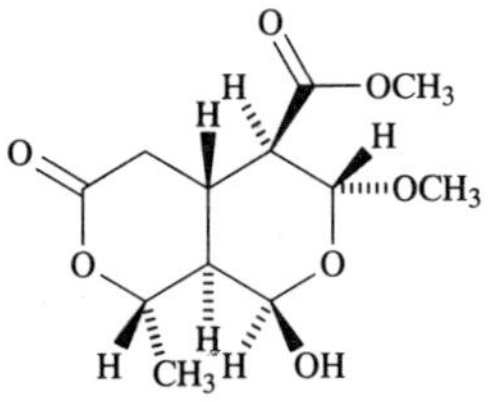

Monoterpenoid
(Seco-iridane)

$C_{12}H_{18}O_7$ [61229-34-3] MW 274.27

Unripe fruit of *Xylocarpus moluccensis* (Meliaceae).

Insect antifeedant.

Kubo, I., *J. Am. Chem. Soc.*, 1976, **98**, 6704.

Y

Yohimbine 1448

Aphrodine; Corynine; Hydroergotocin; Quebrachine

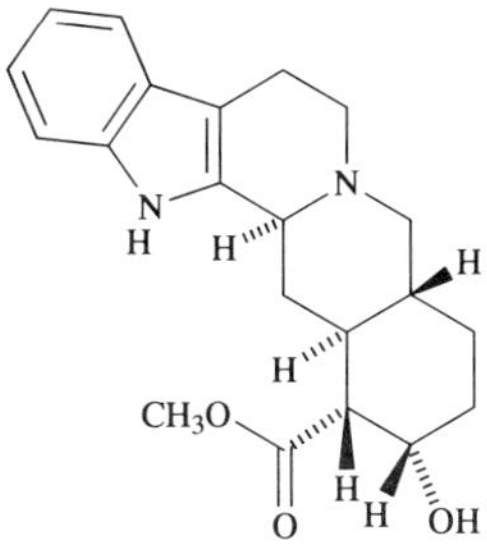

Indole alkaloid
(Yohimbane)

$C_{21}H_{26}N_2O_3$ [146-48-5] MW 354.45
α-form [131-03-3]
β-form [549-84-8]

Yohimbe bark, *Pausinystalia yohimbe* (Rubiaceae) and *Rauwolfia serpentina* (Apocynaceae). Co-occurs as a mixture of stereoisomers, α- and β-yohimbine.

Toxic. It is an α-adrenergic blocking agent, a serotonin antagonist and a mydriatic. Once mistakenly considered to be an aphrodisiac. The main clinical use today is an antidepressant.

Manske, R. H. F., The Alkaloids, 1952, **2**, 369, Academic Press

Yurinelide 1449

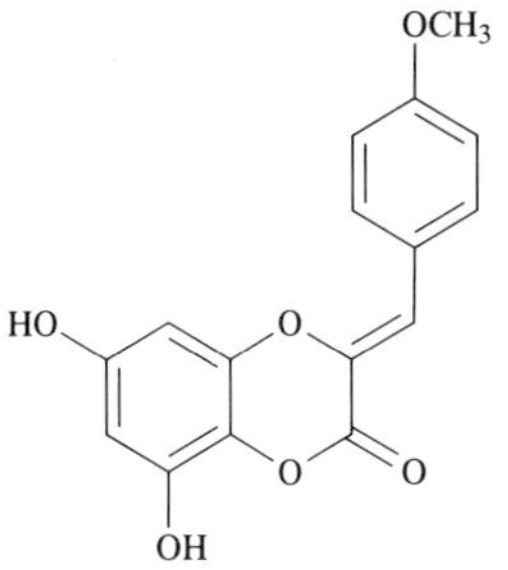

Benzodioxin-2-one

$C_{16}H_{12}O_6$ [144599-00-8] MW 300.27

Infected leaves of *Lilium maximowczii* (Liliaceae).

Antifungal agent (phytoalexin).

Monde, K., *Tetrahedron Lett.*, 1992, **33**, 5395.

Z

Zaluzanin C 1450

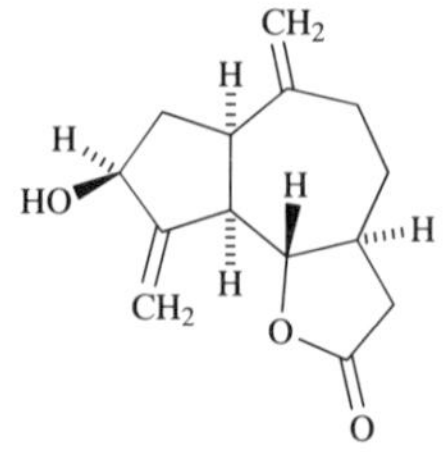

Sesquiterpenoid lactone
(Guaianolide)

$C_{15}H_{18}O_3$ [16838-87-2] MW 246.31

Podachaenium eminens, *Zinnia acerosa* and several *Zaluzania* and *Vernonia* spp. (Compositae).

Cytotoxic and antitumour properties.

Romo de Vivar, A., *Tetrahedron,* 1967, **23**, 3903.

Zexbrevin B 1451

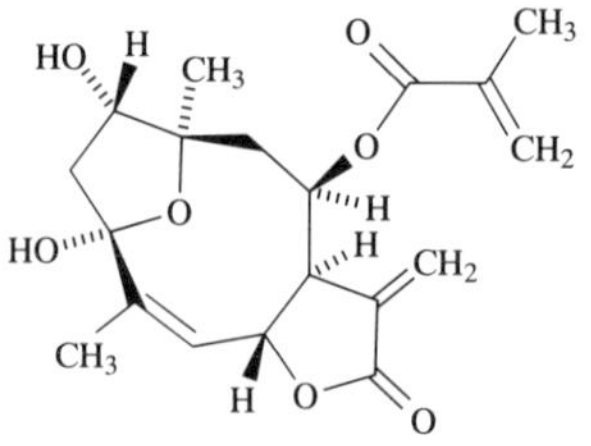

Sesquiterpenoid lactone
(Germacranolide)

$C_{19}H_{24}O_7$ [34302-19-7] MW 364.40

Zexmania brevifolia and *Tithonia tubaeformis* (Compositae).

Cytotoxic and antitumour properties.

Ortega, A., *Rev. Latinoam. Quim.,* 1971, **2**, 38.

Zierin 1452

m-Hydroxysambunigrin

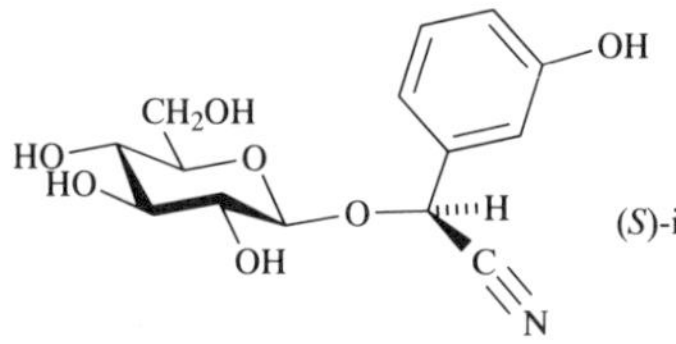

(*S*)-isomer

Cyanogenic glycoside

$C_{14}H_{17}NO_7$ [645-02-3] MW 311.29

Leaves and twigs of *Zieria laevigata* (Rutaceae); Danish populations of *Sambucus nigra* (Caprifoliaceae) and leaves of *Oxytropis campestris* (Leguminosae). A xyloside occurs in achenes of *Xeranthemum cylindricum* (Compositae). The epimer, (*R*)-holocalin [41753-54-2], occurs in seed of *Holocalyx balansae* (Leguminosae).

Bound toxin, releasing cyanide on hydrolysis or tissue damage. Cattle have died after consuming *Zieria* plants.

Finnemore, H., *J. Proc. Roy. Soc. N. S. Wales,* 1936, **70**, 175.

Zingiberene 1453

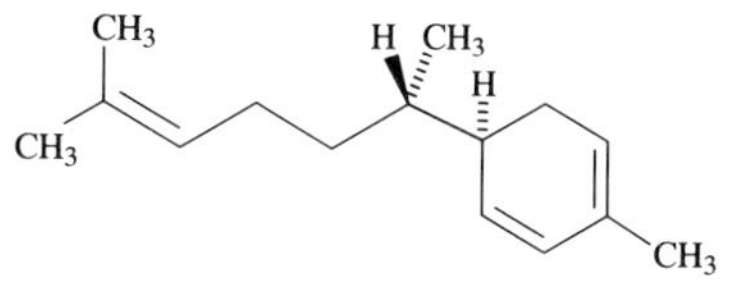

Sesquiterpenoid
(Bisabolane)

$C_{15}H_{24}$ [495-60-3] MW 204.36

Rhizomes of ginger, *Zingiber officinale* and of *Curcuma* spp. (Zingiberaceae), leaves of *Lycopersicon hirsutum* (Solanaceae).

Insecticidal, killing Colorado beetle larvae at a concentration of 12-25 μg per insect.

Arigoni, D., *Helv. Chim. Acta,* 1954, **37**, 881.

Zoapatanol 1454

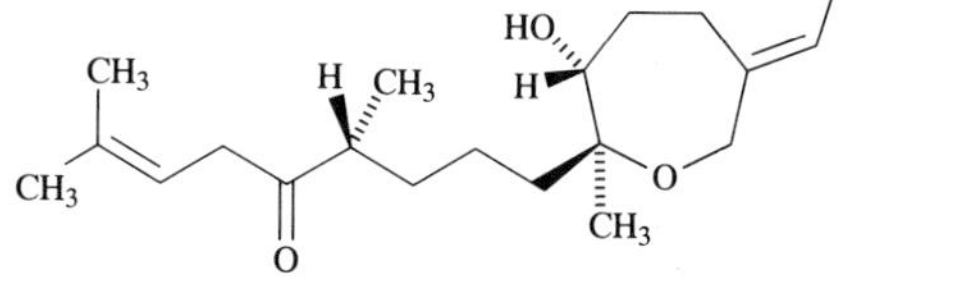

Diterpenoid
(Oxepane)

$C_{20}H_{34}O_4$ [71117-51-6] MW 338.49

Leaves of the Mexican zoapatle tree, *Montanoa tomentosa* (Compositae).

Contragestational activity. The leaves are used by Mexican women to prepare a tea to induce menses and labour.

Kanojia, R. M., *J. Org. Chem.,* 1982, **47**, 1310.

Zygadenine 1455

7-Deoxygermine

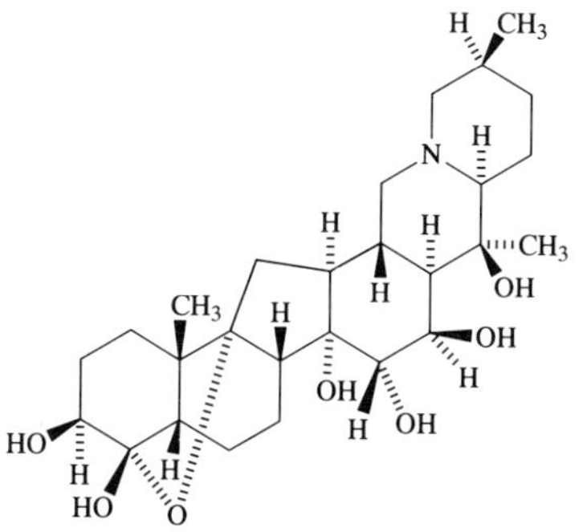

Steroid alkaloid
(Cevane)

$C_{22}H_{43}NO_7$ [545-45-9] MW 493.64

Zigadenus gramineus, Z. venenosus and *Veratrum album* (Liliaceae).

Poisonous principle. Cases of poisoning in humans have been reported from eating *Zigadenus* tubers. The alkaloid also has antitumour and antihypertensive properties.

Kupchan, S. M., *J. Am. Chem. Soc.,* 1959, **81**, 1925.

Subject Index

A

Abrin, 1

Abrus agglutinin *see* Abrin, 1

Absinthiin *see* Absinthin, 2

Absinthin, 2

Absynthin *see* Absinthin, 2

Abyssinone VI, 3

Acacetin, 4

Acacipetalin, 5

Acalyphin, 6

Acamelin, 7

(−)-Acanthocarpan, 8

Acanthoidine, 9

Acerosin, 10

Acetic acid, 11

Acetovanillone, 12

1′-Acetoxychavicol acetate, 13

1′-Acetoxyeugenol acetate, 14

1-Acetoxy-2-hydroxyheneicosa-12,15-dien-4-one, 15

1-Acetoxy-2-hydroxy-4-oxoheneicosa-12,15-diene *see* 1-Acetoxy-2-hydroxyheneicosa-12,15-dien-4-one, 15

Acetylandromedol *see* Grayanotoxin I, 688

Acetylbenzoylaconine *see* Aconitine, 21

Acetylcaranine, 16

O-Acetylcypholophine, 17

7-Acetyl-9-echimidinylretronecine *see* Uplandicine, 1383

3-Acetylhaemanthine *see* 3-Acetylnerbowdine, 18

3-Acetylhemanthine *see* 3-Acetylnerbowdine, 18

5-Acetyl-2-isopropenylbenzofuran *see* Dehydrotremetone, 426

14-Acetylisotalatizidine *see* Condelphine, 341

18-*O*-Acetyllycoctonine *see* Tricornine, 1352

3-Acetylnerbowdine, 18

3-*O*-Acetylnerbowdine *see* 3-Acetylnerbowdine, 18

N-Acetyltetrahydroanabasine *see* Ammodendrine, 77

Achillein *see* (−)-Betonicine, 185

Achillin, 19

Acocantherin *see* Ouabain, 1046

Acomonine *see* Delsoline, 433

Aconifine, 20

Aconitine, 21

Acrifoline, 22

Acronine *see* Acronycine, 24

Acronycidine, 23

Acronycine, 24

Acroptilin, 25

Acrovestone, 26

Acteoside *see* Verbascoside, 1399

Actinidine, 27

Acutumidine, 28

Adiantifoline, 29

Adlumine, 30

Adonitoxin, 31

Adynerigenin 3-diginoside *see* Adynerin, 32

Adynerin, 32

Aeruginosin *see* Microcystin LA, 959

Aescin, 33

Aescusan *see* Aescin, 33

Aethusin, 34

Affinin, 35

Affinine, 36

Affinisine, 37

Agavoside A, 38

Ageratochromene *see* Precocene 2, 1135

Agrimoniin, 39

Agroclavine, 40

Aiapin *see* Ayapin, 163

Ailanthinone, 41

Ajaconine, 42

Ajmalicine, 43

Ajmaline, 44

Akagerine, 45

Akuammicine, 46

Akuammidine, 47

Akuammine, 48

Alantic anhydride *see* Alantolactone, 49

Alantolactone, 49

Alatolide, 50

Albanin F *see* Kuwanone G, 829

Albanol A, 51

Albasapogenin *see* Gypsogenin, 694

Albaspidin AA, 52

Alchorneine, 53

Alchornine, 54

Aldehydoformic acid *see* Glyoxylic acid, 672

Alhanin *see* Alkannin, 56

Alizarin, 55

Alkaloid A *see* Cyclobuxine D, 399

Alkaloid B *see* Dihydrosanguinarine, 478

Alkaloid B *see* Villalstonine, 1416

Alkaloid C *see* Delcosine, 431

Alkaloid F10 *see* Cularidine, 387

Alkaloid F30 *see* Cularimine, 388

Alkaloid III *see* Anagyrine, 85

Alkaloid L *see* Cycloprotobuxine C, 402

Alkaloid L27 *see* Acrifoline, 22

Alkaloid L29 *see* Acrifoline, 22

Alkaloid L32 *see* Cernuine, 290

Alkaloid S-D *see* Integerrimine, 775

Alkaloid V *see* Anisodamine, 96

Alkaloid V *see* Cyclopamine, 401

Alkanna red *see* Alkannin, 56

Alkannin, 56

Alkannin β, β-dimethylacrylate, 57

Allamandin, 58

Alleoside *see* Helveticoside, 717

Allergen Ra5 *see* Ragweed pollen allergen Ra5, 1179

Allicin, 59

Alliin, 60

Allioside A *see* Helveticoside, 717

ent-Alloalantolactone *see* Diplophyllin, 503

Allocryptopine, 61

α-Allocryptopine *see under* Allocryptopine, 61

β-Allocryptopine *see under* Allocryptopine, 61

Alloimperatorin, 62

Allopterin *see* Aloperine, 65

p-Allylanisole *see* Estragole, 564

Allylcatechol methylene ether *see* Safrole, 1215

S-Allyl-L-cysteine *S*-oxide *see* Alliin, 60

Allyl disulfide *see* Diallyl disulfide, 458

β-Allylsulfenylalanine *see* Alliin, 60

Allyl sulfide *see* Diallyl sulfide, 459

Alnusiin, 63

Aloe-emodin, 64

Aloin A *see* Barbaloin, 172

Aloperine, 65

Alstonine, 66

Altholactone *see* Goniothalenol, 679

Amabiline, 67

Amaralin, 68

Amarin *see* Cucurbitacin B, 378

Amarogentin, 69

Amarorine *see* 11-Hydroxycanthin-6-one, 743

Amaryllisine, 70

Ambelline, 71

Ambrosin, 72

Amijitrienol, 73

L-2-Amino-6-amidinohexanoic acid *see* L-Indospicine, 773

(*S*)-2-Amino-4-(aminoxy)butyric acid *see* L-Canaline, 246

4-Aminobutanoic acid *see* γ-Aminobutyric acid, 74

4-Aminobutyric acid *see* γ-Aminobutyric acid, 74

γ-Aminobutyric acid, 74

(*S*)-α-Amino-β-cyanopropionic acid *see* L-β-Cyanoalanine, 394

β-Aminoethylbenzene *see* Phenethylamine, 1089

4-(2-Aminoethyl)benzene-1,2-diol *see* Dopamine, 514

β-Aminoethylglyoxaline *see* Histamine, 729

4-(2-Aminoethyl)imidazole *see* Histamine, 729

4-(2-Aminoethyl)pyrocatechol *see* Dopamine, 514

2-Amino-4-(guanidinoxy)butyric acid *see* L-Canavanine, 247

O-Aminohomoserine *see* L-Canaline, 246

α-Aminoindole-3-propionic acid *see* L-Tryptophan, 1366

L-α-Amino-γ-oxalylaminobutyric acid, 75

L-α-Amino-β-oxalylaminopropionic acid, 76

S-2-Amino-1-oxo-1-phenylpropane *see* D-Cathinone, 281

1-Amino-2-phenylethane *see* Phenethylamine, 1089

(*R*)-2-Amino-3-selenomethylpropanoic acid *see* *Se*-Methyl-L-selenocysteine, 952

(*S*)-Aminosuccinic acid *see* L-Aspartic acid, 139

Ammidin *see* Imperatorin, 770

Ammodendrine, 77

Ammoidin *see* Xanthotoxin, 1441

Ammothamnine, 78

Amurensine, 79

Amurine, 80

Amygdalin, 81

Amygdaloside *see* Amygdalin, 81

Amyl acetate *see under* Acetic acid, 11

(−)-Anabasine, 82

Anacardic acid, 83

Anacardol *see* (15:1)-Cardanol, 259

Anacrotine, 84

Aquilide A *see* Ptaquiloside, 1160

Arborine, 115

Arborinine, 116

Arbutin, 117

Arbutoside *see* Arbutin, 117

Arcelin, 118

Archangelicin, 119

Archin *see* Emodin, 534

(−)-Arctigenin, 120

Arctolide, 121

Arctuvin *see* Hydroquinone, 737

Ardisianone, 122

Arecoline, 123

L-Arginine, 124

Aribine *see* Harman, 700

Aristolindiquinone, 125

Aristolochic acid, 126

Aristolochic acid A *see* Aristolochic acid, 126

Aristolochic acid-I *see* Aristolochic acid, 126

Aristolochine *see* (+)-Bebeerine, 175

Armepavine, 127

Arnebin I *see* Alkannin β,β-dimethylacrylate, 57

Arnebin IV *see* Alkannin, 56

Arnebinol, 128

Arnebinone, 129

Arnicolide A, 130

1-Aromadendranol *see* Palustrol, 1061

Aromaticin, 131

Aromoline, 132

Artabotrine *see* Isocorydine, 789

Arteglasin A, 133

Arvenososide A, 134

Asarabacca camphor *see* β-Asarone, 135

Asarin *see* β-Asarone, 135

β-Asarone, 135

Asarone *see* β-Asarone, 135

α-Asarone *see under* β-Asarone, 135

cis-Asarone *see* β-Asarone, 135

Asarum camphor *see* β-Asarone, 135

Ascapurin *see* Ascaridole, 136

Ascaridol *see* Ascaridole, 136

Ascaridole, 136

Ascarisin *see* Ascaridole, 136

Asclepin, 137

Asebotoxin *see* Grayanotoxin I, 688

Asebotoxin II, 138

L-Asparagic acid *see* L-Aspartic acid, 139

L-Asparaginic acid *see* L-Aspartic acid, 139

Aspartame *see under* L-Aspartic acid, 139

L-Aspartic acid, 139

Aspecioside, 140

Asperulin *see* Asperuloside, 141

Asperuloside, 141

Aspidospermatine, 142

Aspidospermine, 143

Astragaloside III, 144

Astrantiagenin D *see* Gypsogenin, 694

Astrasieversianin XVI, 145

Astringenin *see* Piceatannol, 1105

Astringin, 146

Astrobain *see* Ouabain, 1046

Atanine, 147

Athamantin, 148

Athyriol, 149

Atractylenolide III *see* 8β-Hydroxyasterolide, 739

Atractylin *see* Atractyloside, 150

Atractyloside, 150

Atropamine *see* Apoatropine, 111

Atropine, 151

Atropyltropeine *see* Apoatropine, 111

Aucubigenin *see under* Aucubin, 152

Aucubin, 152

Aucuboside *see* Aucubin, 152

Aurantio-obtusin 6-β-D-glucoside, 153

Auriculasin *see* Cudraisoflavone A, 385

Auriculine, 154

Auriculoside, 155

Austrobailignan 1, 156

Autumnolide, 157

Avadharidine, 158

Avadkharidine *see* Avadharidine, 158

Avenacin A-1, 159

Avenacin B-2, 160

Avenalumin I, 161

Avrainvilleol, 162

Awadcharidine *see* Avadharidine, 158

Axillin *see* Desacetoxymatricarin, 454

Ayapin, 163

Azadirachtin, 164

Azadirachtin H *see under* Azadirachtin, 164

L-Azetidine-2-carboxylic acid, 165

L-2-Azetidinecarboxylic acid *see* L-Azetidine-2-carboxylic acid, 165

Azulon *see* Guaiazulene, 692

B

Baccharinoid B21, 166

Baicalein, 167

Baicalein 6-glucoside *see under* Baicalein, 167

Baicalein 6-glucuronide *see under* Baicalein, 167

Baicalein 7-glucuronide *see* Baicalin, 168

Baicalein 7-*O*-glucuronide *see* Baicalin, 168

Baicalein 7-rhamnoside *see under* Baicalein, 167

Baicalin, 168

Baicaloside *see* Baicalin, 168

Baileyin, 169

Bakkenolide A, 170

Balchanin *see* Santamarin, 1228

Baliospermin, 171

Balm oil *see under* Citronellal, 326

Banisterine *see* Harmine, 701

Baptitoxine *see* Cytisine, 410

Barbaloin, 172

Barbinine, 173

Barosma camphor *see* Diosphenol, 498

Barterin *see* Volkenin, 1426

Base G *see* Europine, 586

Bayogenin 3-glucoside, 174

(+)-Bebeerine, 175

Belamarine *see* Acetylcaranine, 16

Bellamarine *see* Acetylcaranine, 16

Bellidifolin, 176

Bellidifolium *see* Bellidifolin, 176

1, 2-Benzenediol *see* Catechol, 278

1, 4-Benzenediol *see* Hydroquinone, 737

Benzeneethanamine *see* Phenethylamine, 1089

Benzeneethanol *see* 2-Phenylethanol, 1091

2*H*-1-Benzopyran-2-one *see* Coumarin, 364

1, 2-Benzopyrone *see* Coumarin, 364

Benzoylecgonine methyl ester *see* Cocaine, 332
α-Benzoylethylamine *see* D-Cathinone, 281
Benzoylmethylecgonine *see* Cocaine, 332
Benzoylpseudotropeine *see* Tropacocaine, 1364
Benzoyltropein, 177
Benzylglucosinolate *see* Glucotropaeolin, 665
Benzyl isothiocyanate *see under* Glucotropaeolin, 665
Benzyl thiocyanate *see under* Glucotropaeolin, 665
Berbamine, 178
Berbenine *see* Berbamine, 178
Berberastine, 179
Berbericine *see* Berberine, 180
Berbericinine *see* Palmatine, 1058
Berberine, 180
Bergamot oil *see under* Bergapten, 182
Bergamot oil *see under* Dipentene, 501
Bergamottin, 181
Bergapten, 182
Bergaptene *see* Bergapten, 182
Bergaptin *see* Bergamottin, 181
Bergaptol geranyl ether *see* Bergamottin, 181
Betagarin, 183
Betavulgarin, 184
(−)-Betonicine, 185
Betulin, 186
Betulinol *see* Betulin, 186
Betulol *see* Betulin, 186
Betuloside *see* Rhododendrin, 1193
(+)-Bicuculline, 187
Bigitalin *see* Gitoxin, 651
Bikhaconitine, 188
Bilobol, 189
Biochanin A, 190
Biochanin B *see* Formononetin, 610
Bipindogenin 3-digitaloside *see* Bipindoside, 191
Bipindogenin 3-rhamnoside *see* Lokundjoside, 872
Bipindoside, 191
Bisabolol *see* (+)-α-Bisabolol, 192
(+)-α-Bisabolol, 192
2, 4-Bis(prenyl)phenol, 193
Bocconine, 194
Bogoroside *see* Convalloside, 350
Boldine dimethyl ether *see* Glaucine, 655
Borbonol 2 *see* Isoobtusilactone A, 799
Bornan-2-one *see* Camphor, 243
Borneol, 195
endo-Borneol *see* Borneol, 195
Bornyl alcohol *see* Borneol, 195
Borrecapine, 196
Boschnialactone, 197
Bovoside A, 198
Brassilexin, 199
Brassilexine *see* Brassilexin, 199
Brassinin, 200
Brassinine *see* Brassinin, 200
Brevetoxin A, 201
Brevetoxin B, 202
Brevicolline, 203
Brevifolin *see* Xanthoxylin, 1443
Brevilin A, 204
3-Bromo-4,5-dihydroxybenzyl alcohol, 205
Broussonin A, 206
Bruceantin, 207
Brucein B *see* Bruceine B, 208

Bruceine B, 208

Bruceoside A, 209

Brucine, 210

Brunfelsamidine *see* Pyrrole-3-carbamidine, 1172

Bryophyllin A, 211

Bryotoxin A, 212

Bryotoxin B *see under* Bryotoxin A, 212

Bryotoxin C *see under* Bryotoxin A, 212

Buccocamphor *see* Diosphenol, 498

Buchu camphor *see* Diosphenol, 498

Buddledin A, 213

Buddledin B *see under* Buddledin A, 213

Buddledin C *see under* Buddledin A, 213

Buddleoflavonol *see* Acacetin, 4

Budlein A, 214

Bufotenine, 215

Bulbocapnine, 216

Bullatine G *see* Songorine, 1275

Burseran, 217

Butin 7, 3′-diglucoside *see* Butrin, 218

Butrin, 218

3-Butylidene-7-hydroxyphthalide, 219

1-*tert*-Butyl-3-methylbenzene, 220

1-*tert*-Butyl-4-methylbenzene, 221

m-tert-Butyltoluene *see* 1-*tert*-Butyl-3-methylbenzene, 220

p-tert-Butyltoluene *see* 1-*tert*-Butyl-4-methylbenzene, 221

Butyrylmallotochromene, 222

Buxamine *see* Buxamine E, 223

Buxamine A *see under* Buxamine E, 223

Buxamine B *see under* Buxamine E, 223

Buxamine C *see under* Buxamine E, 223

Buxamine E, 223

Buxamine G *see under* Buxamine E, 223

C

(−)-Caaverine, 224

Cactine *see* Hordenine, 732

Cadalene-2, 3-quinone *see* Mansonone C, 910

Cadaverine, 225

Cafesterol *see* Cafestol, 226

Cafestol, 226

Caffeine, 227

Cajanol, 228

Cajeputene *see* Limonene, 861

Cajeputol *see* 1, 8-Cineole, 319

Calabash curare *see under* *C*-Curarine, 390

Calactin, 229

Calanolide A, 230

Calaxin, 231

Calebassine, 232

C-Calebassine *see* Calebassine, 232

Callicarpone, 233

Calligonine, 234

Callinecdysone B *see* Makisterone B, 903

Calophyllin B, 235

Calophyllolide, 236

Calotropin, 237

Calycanthine, 238

Calystegin B_2, 239

Calystegine B_2 *see* Calystegin B_2, 239

Calystigine *see* Palmatine, 1058

Camalexin, 240

Camazulene *see* Chamazulene, 295

Camelliagenin A *see under* Theasaponin, 1330

Chrysanthemum monocarboxylic acid pyrethrolone ester *see* Pyrethrin I, 1168

Chrysanthin *see* Pyrethrosin, 1170

Chrysarobin, 307

Chrysartemin A *see* Canin, 251

Chrysazin, 308

Chrysin, 309

Chrysin 7-glucuronide *see under* Chrysin, 309

Chrysinic acid *see* Chrysin, 309

Chrysophanic acid *see* Chrysophanol, 310

Chrysophanic acid 9-anthrone *see* Chrysarobin, 307

Chrysophanol, 310

Chrysophanol anthrone *see* Chrysarobin, 307

Chrysophanol 8-glucoside, 311

C.I. 75290 *see* Haematoxylin, 696

Cibarian, 312

Cichoralexin, 313

Cichoriin, 314

Cichorioside *see* Cichoriin, 314

Cicutine *see* Coniine, 345

Cicutoxin, 315

Ciguatoxin, 316

Cimarin *see* Cymarin, 405

Cimifugin, 317

C.I. Mordant Red 11 *see* Alizarin, 55

Cinchocatine *see* Cinchonidine, 318

Cinchonan-9-ol *see* Cinchonidine, 318

Cinchonidine, 318

Cinchonine *see under* Cinchonidine, 318

Cinene *see* Limonene, 861

1, 8-Cineole, 319

Cinerin I, 320

Cinerin II, 321

Cinerolone *see under* Chrysanthemic acid, 306

Cinnamic acid, 322

Cinnamoylcocaine, 323

Cinnamoylecgonine methyl ester *see* Cinnamoylcocaine, 323

Cinnamoylmethylecgonine *see* Cinnamoylcocaine, 323

Cinnamylcocaine *see* Cinnamoylcocaine, 323

Cirsilineol, 324

Cirsiliol, 325

Citisine *see* Cytisine, 410

Citrolimonin *see* Limonin, 862

Citronellal, 326

Citronella oil *see under* Dipentene, 501

Citronella oils *see under* Citronellal, 326

Cleistanthin A, 327

Cleomiscosin A, 328

Cleosandrin *see* Cleomiscosin A, 328

Clerodin, 329

Clivorine, 330

Cneorum chromone A *see* Spatheliabischromene, 1278

Cneorum chromone F *see* Ptaeroglycol, 1159

Cnicin, 331

Cocaine, 332

Cocarcinogen A_1 *see* 12-Tetradecanoylphorbol 13-acetate, 1318

Cocculolidine, 333

Codeine, 334

β-Codeine *see* Neopine, 1007

Codonopsine, 335

Coelogin, 336

Coffeine *see* Caffeine, 227

Coffeol *see* Cafestol, 226

Colchiceine methyl ether *see* Colchicine, 337

Croton factor A_1 *see* 12-Tetradecanoylphorbol 13-acetate, 1318

Croton factor F_1 *see* 12-*O*-Palmitoyl-16-hydroxyphorbol 13-acetate, 1059

Cryogenine, 372

Cryptolepine, 373

Cryptopleurine, 374

Cuauchichicine, 375

Cubebin, 376

Cubeb oil *see under* Dipentene, 501

Cucoline *see* Sinomenine, 1267

Cucurbitacin A, 377

Cucurbitacin B, 378

Cucurbitacin D, 379

Cucurbitacin E, 380

Cucurbitacin E 2-*O*-β-D-glucoside *see under* Cucurbitacin E, 380

Cucurbitacin I, 381

Cucurbitacin O, 382

Cucurbitacin P, 383

Cucurbitacin Q, 384

Cudraisoflavone A, 385

Cularicine, 386

Cularidine, 387

Cularimine, 388

Cularine, 389

Cumostrol *see* Coumestrol, 365

Cunaniol *see* Ichthyotherol, 767

C-Curanine II *see* Calebassine, 232

C-Curarine, 390

C-Curarine I *see* *C*-Curarine, 390

Curcin, 391

Curcumin, 392

Curine *see* (+)-Bebeerine, 175

Cusparine, 393

Cuspidoside *see* Lokundjoside, 872

(+)-Cyanidanol *see* (+)-Catechin, 277

(+)-Cyanidan-3-ol *see* (+)-Catechin, 277

3-Cyanoalanine *see* L-β-Cyanoalanine, 394

3-Cyano-L-alanine *see* L-β-Cyanoalanine, 394

β-Cyano-α-alanine *see* L-β-Cyanoalanine, 394

L-β-Cyanoalanine, 394

Cyanoginosin LA *see* Microcystin LA, 959

Cyanoginosin LR *see* Microcystin LR, 960

Cyanoginosin RR *see* Cyanoviridin RR, 395

Cyanoviridin RR, 395

Cycad palms *see under* Cycasin, 396

Cycasin, 396

Cyclamin, 397

Cyclobrassinine sulfoxide *see* Cyclobrassinin sulfoxide, 398

Cyclobrassinin sulfoxide, 398

Cyclobuxine *see* Cyclobuxine D, 399

Cyclobuxine D, 399

(+)-*cis*-β-Cyclocostunolide, 400

Cyclopamine, 401

Cycloprotobuxine A *see under* Cycloprotobuxine C, 402

Cycloprotobuxine C, 402

Cycloprotobuxine D *see under* Cycloprotobuxine C, 402

Cycloprotobuxine F *see under* Cycloprotobuxine C, 402

Cyclovirobuxine C, 403

Cyclovirobuxine D *see under* Cyclovirobuxine C, 403

Cyclovirobuxine F *see under* Cyclovirobuxine C, 403

Cylindrospermopsin, 404

Cymarin, 405

Cymene *see* *p*-Cymene, 406

p-Cymene, 406

D

Digitoxigenin 3-gentiobiosylthevetoside *see* Thevetin B, 1333

Digitoxigenin 3-α-rhamnoside *see* Evomonoside, 587

Digitoxigenin 3-tridigitoxoside *see* Digitoxin, 471

Digitoxin, 471

Digoxigenin 3-tridigitoxoside *see* Digoxin, 472

Digoxin, 472

Dihydroacacipetalin *see* Heterodendrin, 724

2, 3-Dihydroambrosin *see* Damsin, 411

Dihydrocornin aglycone, 473

1, 2-Dihydro-α-elaterin *see* Cucurbitacin B, 378

Dihydrogriesenin, 474

3, 4-Dihydroharmine *see* Harmaline, 699

Dihydrohelenalin *see* Plenolin, 1120

Dihydrohelenalin acetate *see* Arnicolide A, 130

2, 3-Dihydro-3α-hydroxytropidine *see* Tropine, 1365

Dihydromethysticin, 475

Dihydro-oroselol *see* Columbianetin, 339

Dihydropiperlongumine, 476

Dihydroquercetin *see* Taxifolin, 1306

Dihydrosamidin, 477

Dihydrosanguinarine, 478

Dihydrowyerone, 479

1,2-Dihydroxyanthraquinone *see* Alizarin, 55

1,8-Dihydroxyanthraquinone *see* Chrysazin, 308

1,2-Dihydroxybenzene *see* Catechol, 278

6,7-Dihydroxycoumarin 7-glucoside *see* Cichoriin, 314

3,9-Dihydroxycoumestan *see* Coumestrol, 365

6,9-Dihydroxy-7-drimene-11, 12-dial *see* Mukaadial, 984

5,7-Dihydroxy-3-ethylchromone *see* Lathodoratin, 848

7,4′-Dihydroxyflavan, 480

5,7-Dihydroxyflavanone *see* Pinocembrin, 1112

5,7-Dihydroxyflavone *see* Chrysin, 309

5,8-Dihydroxyflavone *see* Primetin, 1140

5,7-Dihydroxyflavonol *see* Galangin, 620

ent-16β, 17-Dihydroxykauran-19-oic acid *see* Diterpenoid SP-II, 506

2′,6′-Dihydroxy-4′-methoxyacetophenone, 481

3′,4′-Dihydroxy-7-methoxyflavan, 482

7,4′-Dihydroxy-5-methoxy-8-prenylflavanone *see* Isoxanthohumol, 805

1,7-Dihydroxy-4-methoxyxanthone, 483

1,8-Dihydroxy-3-methylanthraquinone *see* Chrysophanol, 310

1,8-Dihydroxy-3-methylanthrone *see* Chrysarobin, 307

2,5-Di(hydroxymethyl)-3,4-dihydroxypyrrolidine *see* DMDP, 511

7,4′-Dihydroxy-8-methylflavan, 484

3,4-Dihydroxyphenethylamine *see* Dopamine, 514

3,4-Dihydroxy-L-phenylalanine *see* L-Dopa, 513

α-(3,4-Dihydroxyphenyl)-β-aminoethane *see* Dopamine, 514

3,9-Dihydroxypterocarp-6a-ene *see* Anhydroglycinol, 94

2,7-Dihydroxy-3,4,6-trimethoxydibenzofuran *see* α-Cotonefuran, 360

Dillapiol *see* Dillapiole, 485

Dill apiole *see* Dillapiole, 485

Dillapiole, 485

Dilophic acid, 486

2,6-Dimethoxybenzoquinone, 487

2,6-Dimethoxybenzoquinone *see under* Acamelin, 7

2,6-Dimethoxy-1,4-benzoquinone *see* 2,6-Dimethoxybenzoquinone, 487

3,5-Dimethoxy-4′-biphenylol *see* Isoaucuparin, 784

3,4-Dimethoxydalbergione, 488

6,7-Dimethoxydictamnine *see* Kokusaginine, 828

6,7-Dimethoxy-2,2-dimethylchromene *see* Precocene 2, 1135

2,6-Dimethoxyphenol, 489

4,9-Dimethoxypsoralen *see* Isopimpinellin, 801

2,6-Dimethoxyquinone *see* 2,6-Dimethoxybenzoquinone, 487

10,11-Dimethoxystrychnine *see* Brucine, 210

Dimethulene *see* Chamazulene, 295

β, β-Dimethylacrylalkannin *see* Alkannin β, β-dimethylacrylate, 57

6-(3,3-Dimethylallyl)-1,5-dihydroxyxanthone *see* Calophyllin B, 235

N-(3,3-Dimethylallyl)guanidine *see* Galegine, 622

3-(α, α-Dimethylallyl)psoralen *see* Chalepensin, 293

3-(2-Dimethylaminoethyl)indole *see* *N*, *N*-Dimethyltryptamine, 491

3-(Dimethylaminomethyl)indole *see* Gramine, 685

N-Dimethylconoduramine *see* Gabunamine, 617

Dimethyl disulfide, 490

3,7-Dimethyloct-6-enal *see* Citronellal, 326

N,N-Dimethylseratonin *see* Bufotenine, 215

N,N-Dimethyltryptamine, 491

N,N-Dimethyltyramine *see* Hordenine, 732

1,6-Di(3-nitropropanoyl)glucoside *see* Cibarian, 312

2,6-Di(3-nitropropanoyl)-α-glucoside *see* Coronarian, 356

Dinophysistoxin 1, 492

Dinophysistoxin 2, 493

Dioncopeltine A, 494

Dioscin, 495

Dioscorine, 496

Diosgenin, 497

Diosphenol, 498

Diospyrin, 499

Diosquinone, 500

Dipentene, 501

Diphyllin, 502

Diplophyllin, 503

2,4-Diprenylphenol *see* 2,4-Bis(prenyl)phenol, 193

Di-2-propenyl disulfide *see* Diallyl disulfide, 458

Di-2-propenyl sulfide *see* Diallyl sulfide, 459

Discorea sapogenin *see* Diosgenin, 497

Disenecionyl *cis*-khellactone, 504

Distylin *see* Taxifolin, 1306

Ditaine *see* Echitamine, 523

Diterpenoid EF-D, 505

Diterpenoid SP-II, 506

Dithyreanitrile, 507

Divaricoside, 508

Divostroside, 509

L-Djenkolic acid, 510

DMDP, 511

DMJ *see* Deoxymannojirimycin, 450

DMT *see* *N*, *N*-Dimethyltryptamine, 491

DNJ *see* Deoxynojirimycin, 451

cis-Docosenoic acid *see* Erucic acid, 554

Dolcymene *see* *p*-Cymene, 406

Dolichodial, 512

Donaxine *see* Gramine, 685

Dopa *see* L-Dopa, 513

L-Dopa, 513

Dopamine, 514

Doranine *see* Gramine, 685

Doronine, 515

Dronabinol *see* Δ^1-Tetrahydrocannabinol, 1320

Drummondin A, 516

Drummondin B *see under* Drummondin A, 516

Drummondin C *see under* Drummondin A, 516

Drummondin F *see under* Drummondin A, 516

Dubinidine, 517

Duboisine *see* Hyoscyamine, 759

Dufalone *see* Dicoumarol, 463

4, 8, 13-Duvatriene-1,3-diol *see* α-Cembrenediol, 285

E

Eburnamonine, 518

Ecgonine, 519

Ecgonine cinnamate methyl ester *see* Cinnamoylcocaine, 323

Echimidine, 520

Echinacoside, 521

Echinocystic acid, 522

Echitamine, 523

Echiumine, 524

Elaeagnine *see* Calligonine, 234

Elatericin A *see* Cucurbitacin D, 379

Elatericin B *see* Cucurbitacin I, 381

α-Elaterin *see* Cucurbitacin E, 380

Elaterinide, 525

Elatine, 526

Eldeline *see* Deltaline, 434

Eldoquin *see* Hydroquinone, 737

Eleganin, 527

Elemicin, 528

Elephantin, 529

Elephantopin, 530

Eleutheroside K *see* β-Hederin, 707

Ellipticine, 531

Embeliaquinone *see* Embelin, 532

Embelic acid *see* Embelin, 532

Embelin, 532

Emetine, 533

Emodin, 534

Emodin-3-apioside *see* Frangulin B, 613

Emodin 8-glucoside, 535

Emodin-3-rhamnoside *see* Frangulin A, 612

Enanthotoxin *see* Oenanthotoxin, 1027

Encelin, 536

Enhydrin, 537

L-Ephedrine, 538

7-Epideoxynupharidine *see under* Deoxynupharidine, 452

Epigoitrin *see under* Progoitrin, 1145

(*R*)-Epiheterodendrin *see under* Heterodendrin, 724

(*S*)-Epilotaustralin *see under* Lotaustralin, 875

(*S*)-Epilucumin *see under* Lucumin, 877

(*R*)-Epiproacacipetalin *see under* Proacacipetalin, 1144

13-Episclareol *see under* Sclareol, 1242

Epitetraphyllin B *see* Volkenin, 1426

Epitulipinolide, 539

Epitulipinolide diepoxide, 540

Epivoacorine, 541

Epivolkenin *see under* Volkenin, 1426

8-Epixanthatin, 542

2′,3′-Epoxydiospyrin *see* Diosquinone, 500

1,8-Epoxy-*p*-menthane *see* 1,8-Cineole, 319

Epoxysarmentosin *see* Sarmentosin epoxide, 1235

6,7-Epoxytropine tropate *see* Hyoscine, 758

Equol *see under* Formononetin, 610

Eremanthin *see* Eremanthine, 543

Eremanthine, 543

Eremantholide A, 544

Eremofrullanolide, 545

Eremursine *see* Hordenine, 732

Ergamine *see* Histamine, 729

Ergine, 546

Ergobasine *see* Ergometrine, 547

Ergometrine, 547

F

G

Gitorin, 650

Gitoxigenin 3-glucoside *see* Gitorin, 650

Gitoxigenin 3-glucosyldigitaloside *see* Digitalin, 470

Gitoxigenin 3-*O*-tridigitoxoside *see* Gitoxin, 651

Gitoxin, 651

Gitoxoside *see* Gitoxin, 651

Glabranin, 652

Glaucarubinone, 653

Glaucarubolone, 654

Glaucine, 655

Glaucolide A, 656

α-Gliadin *see under* Gluten, 668

β-Gliadin *see under* Gluten, 668

γ-Gliadin *see under* Gluten, 668

Glucoaurantioobtusin *see* Aurantio-obtusin 6-β-D-glucoside, 153

Glucocapparin, 657

Glucocheirolin, 658

Glucoerysolin, 659

Glucofrangulin A, 660

Glucolepidiin, 661

Gluconasturtiin, 662

Glucoproscillaridin A *see* Scillaren A, 1240

Glucorapiferin *see* Progoitrin, 1145

β-D-Glucosyloxyazoxymethane *see* Cycasin, 396

p-Glucosyloxybenzaldehyde cyanohydrin *see* *p*-Glucosyloxymandelonitrile, 663

p-Glucosyloxymandelonitrile, 663

Glucosyl taraxinate, 664

Glucotropaeolin, 665

L-γ-Glutamyl-L-hypoglycin, 666

Glutaric acid, 667

Gluten, 668

Glutinosone, 669

(−)-Glyceollin I, 670

(−)-Glyceollin II, 671

Glycosine *see* Arborine, 115

Glycyrrhyzin *see under* Abrin, 1

Glyoxalic acid *see* Glyoxylic acid, 672

Glyoxylic acid, 672

Gnidicin, 673

Gnididin, 674

Gnidilatin, 675

Gniditrin, 676

Gofruside, 677

Goitrin *see under* Progoitrin, 1145

Gomezina *see* (−)-Apparicine, 114

Goniodomin A, 678

Goniothalenol, 679

Gonyautoxin I, 680

Gossypol, 681

Goyazensolide, 682

Gracillin, 683

Gradolide, 684

Gramine, 685

Graminiliatrin, 686

Gratibain *see* Ouabain, 1046

Gratiotoxin *see* Elaterinide, 525

Gratus strophanthin *see* Ouabain, 1046

Grayanin, 687

Grayanotoxin I, 688

Greenhartin *see* Lapachol, 842

Grevillol, 689

Grossheimin *see* Grosshemin, 690

Grosshemin, 690

H

I

J

Japaconitine C_1 *see* Hypaconitine, 760

Jateorrhizine *see* Jatrorrhizine, 810

Jatrophatrione, 808

Jatrophone, 809

Jatrorhizine *see* Jatrorrhizine, 810

Jatrorrhizine, 810

Java citronella oil *see under* Citronellal, 326

Jervanin-11-one *see* Jervine, 811

Jervine, 811

Jesaconitine, 812

JH III *see* Juvenile hormone III, 819

Jodrellin B, 813

Judaicin *see* Vulgarin, 1427

Juglone, 814

Justicidin A, 815

Justicidin B, 816

Juvabione, 817

Juvadecene, 818

Juvenile hormone III, 819

Juvocimene 1, 820

Juvocimene 2 *see under* Juvocimene 1, 820

K

Kaempferol 3-(2″, 4″-di-*p*-coumarylrhamnoside), 821

Kalopanaxsaponin A *see* α-Hederin, 706

Kansuinin B *see* Kansuinine B, 822

Kansuinine B, 822

Karacoline *see* Karakoline, 824

Karakin, 823

Karakoline, 824

Karwinskione, 825

Katahdinone *see* Solavetivone, 1274

Katine *see* D-Cathine, 280

Kautschin *see* Limonene, 861

Kessazulen *see* Guaiazulene, 692

7-Ketodehydroabietic acid *see* 7-Oxodehydroabietic acid, 1051

2-Keto-13-epimanool *see* 2-Oxo-13-epimanool, 1052

cis-Khellactone disenecionate *see* Disenecionyl *cis*-khellactone, 504

Kievitone, 826

Kigelinone, 827

α-Kirondrin *see* Glaucarubinone, 653

Kitzuta saponin K_6 *see* α-Hederin, 706

Kokusaginine, 828

Korglykon *see* Convallatoxin, 349

Krimpsiekte *see under* Lanceotoxin A, 837

Kusaginin *see* Verbascoside, 1399

Kuwanone G, 829

Kuwanon G *see* Kuwanone G, 829

L

Labriformidin, 830

Labriformin, 831

Labriformine *see* Labriformin, 831

Lachnophyllum lactone, 832

Lacinilene C 7-methyl ether, 833

Lactucin, 834

Lactucopicrin, 835

Laetrile *see under* Amygdalin, 81

Laevodopa *see* L-Dopa, 513

Lanatoxin *see* Digitoxin, 471

Lanceolatin B, 836

Lanceotoxin A, 837

Lanceotoxin B, 838

Lantadene A, 839

Lantadene B, 840

Lapachenol *see* Lapachenole, 841

Lapachenole, 841

Lapachic acid *see* Lapachol, 842

Lapachol, 842

β-Lapachone, 843

Lapachonone *see* Lapachenole, 841

Lappaconitine, 844

(+)-Lariciresinol, 845

Laserolide, 846

Lasiocarpine, 847

9-Lasiocarpylheliotridine *see* Europine, 586

Lassiocarpine *see* Lasiocarpine, 847

Lathodoratin, 848

Lathyrol, 849

(−)-Laudanidine, 850

Laudanine *see under* (−)-Laudanidine, 850

Laudanine methyl ether *see* Laudanosine, 851

Laudanosine, 851

Lawsone, 852

Lawsone methyl ether *see* 2-Methoxy-1,4-naphthoquinone, 931

LCR *see* Vincristine, 1419

Ledol, 853

Ledum camphor *see* Ledol, 853

Leiokinine A, 854

Lemmatoxin, 855

Lemon oil *see under* Citronellal, 326

Leontin *see* Hederagenin 3-glucoside, 705

Lettucenin A, 856

Leucaenine *see* L-Mimosine, 966

Leucaenol *see* L-Mimosine, 966

Leucenine *see* L-Mimosine, 966

Leucenol *see* L-Mimosine, 966

Leucodin *see* Desacetoxymatricarin, 454

Leucoharmine *see* Harmine, 701

Leucomosin *see* Desacetoxymatricarin, 454

Leukodin *see* Desacetoxymatricarin, 454

Leurocristine *see* Vincristine, 1419

Leurosidine, 857

Leurosine, 858

Levodopa *see* L-Dopa, 513

Liatrin, 859

Ligulatin B, 860

(±)-Limonene *see* Dipentene, 501

Limonene, 861

dl-Limonene *see* Dipentene, 501

Limonin, 862

Linamarin, 863

Linarigenin *see* Acacetin, 4

Linarine *see* Peganine, 1073

Lindazulene *see* Chamazulene, 295

Lindheimerine *see under* Ovatine, 1048

Linifolin A, 864

Linustatin, 865

Lipiferolide, 866

Liriodenine, 867

Lirioresinol B *see* (+)-Syringaresinol, 1300

Lithospermic acid, 868

Lizarinic acid *see* Alizarin, 55

Lobelanidine, 869

Lobelidine *see under* (−)-Lobeline, 870

(−)-Lobeline, 870

Loganin, 871

Loganoside *see* Loganin, 871

Lokundjoside, 872

α-Longilobine *see* Seneciphylline, 1252

β-Longilobine *see* Retrorsine, 1187

Lophophorine, 873

Loroglossol, 874

Lotaustralin, 875

Loturine *see* Harman, 700

LU 1 *see* Deoxymannojirimycin, 450

Lubimin, 876

Lucaconine *see* Delcosine, 431

Luciculine *see* Napelline, 997

Lucumin, 877

Lucuminoside *see* Lucumin, 877

Ludovicin A, 878

Lunacrine, 879

Lunamarine, 880

Lupanine, 881

Lupeol, 882

Lupeol acetate, 883

Lupinidine *see* (−)-Sparteine, 1277

Lupinine, 884

Lupinus compound LA-I *see* Wighteone, 1431

β-Lupulic acid *see* Lupulone, 885

Lupulone, 885

Luteanine *see* Isocorydine, 789

Luteolin 8-*C*-glucoside *see* Orientin, 1037

Luteolinidin, 886

Luteone, 887

Lycaconitine, 888

Lycoctonine, 889

Lycoctonine 18-acetate *see* Tricornine, 1352

Lycopericin *see* Tomatine, 1341

Lycopodine, 890

Lycopsamine, 891

Lycoremine *see* Galanthamine, 621

Lycorenine, 892

Lycoricidine, 893

Lycoricidinol *see* Narciclasine, 1000

Lycorimine *see* Galanthamine, 621

Lycorine, 894

Lyngbyatoxin A, 895

Lysergamide *see* Ergine, 546

Lysergic acid amide *see* Ergine, 546

Lysergic acid propanolamide *see* Ergometrine, 547

Lysergol, 896

M

(−)-Maackiain, 897

Macolidine *see under* Aromoline, 132

Macoline, 898

Macrocarpamine, 899

Macrozamin, 900

Madder *see* Alizarin, 55

Madder purple *see* Purpurin, 1167

Maesanin, 901

Magnolol, 902

Majudin *see* Bergapten, 182

Makisterone B, 903

Mallotochromene, 904

Mallotophenone, 905

Mallotoxin *see* Rottlerin, 1208

Malol *see* Ursolic acid, 1386

Malolic acid *see* Ursolic acid, 1386

Malonic acid, 906

Maltotoxin *see* Candicine, 248

Malvalic acid, 907

Mammea-compound A/AA *see* Mammeisin, 908

Mammeisin, 908

Mancinellin, 909

Mandelonitrile β-gentiobioside *see* Amygdalin, 81

Mandelonitrile glucoside *see* Prunasin, 1152

Mandelonitrile vicianoside *see* Vicianin, 1413

Manihotoxine *see* Linamarin, 863

Mansonone C, 910

Mappine *see* Bufotenine, 215

Margetine *see* Lycoricidine, 893

Marmelide *see* Imperatorin, 770

Marmelosin *see* Imperatorin, 770

Matrine *N*-oxide *see* Ammothamnine, 78

Maytansine, 911

Mecambrine, 912

Medicagenic acid 3-glucoside, 913

Medicagenic acid 3-triglucoside, 914

(−)-Medicarpin, 915

Medicogenic acid 3-triglucoside *see* Medicagenic acid 3-triglucoside, 914

Megaphone, 916

Melampodin A, 917

Melampodinin, 918

Melitoxin *see* Dicoumarol, 463

Mellitoxin, 919

Melochinine, 920

Menthacamphor *see* Menthol, 921

p-Mentha-1, 5-diene *see* α-Phellandrene, 1087

p-Mentha-1(7), 2-diene *see* β-Phellandrene, 1088

Menthol, 921

Mentol *see* Menthol, 921

Mesaconitine, 922

Mescaline, 923

Mesembranol *see* Mesembrinol, 926

Mesembranone *see* Mesembrine, 925

Mesembrenone, 924

Mesembrine, 925

Mesembrinol, 926

Methane carboxylic acid *see* Acetic acid, 11

Methanedicarboxylic acid *see* Malonic acid, 906

Methanethiol *see* Methyl mercaptan, 945

Methanoic acid *see* Formic acid, 609

4′-Methoxyaucuparin, 927

6-Methoxycamalexin *see under* Camalexin, 240

p-Methoxycinnamic acid ethyl ester, 928

8-Methoxycirsilineol *see under* Cirsilineol, 324

7-Methoxy-2, 2-dimethylchromene *see* Precocene 1, 1134

5-Methoxy-*N,N*-dimethyltryptamine, 929

12-Methoxyibogamine *see* Ibogaine, 766

13-Methoxyibogamine *see* Tabernanthine, 1302

6-Methoxymellein, 930

6-Methoxy-2-methyl-4,7-benzofurandione *see* Acamelin, 7

2-Methoxy-1, 4-naphthoquinone, 931

2-Methoxyphenol *see* Guaiacol, 691

8-Methoxypsoralen *see* Xanthotoxin, 1441

5-Methoxypsoralin *see* Bergapten, 182

5-Methoxy-3-stilbenol *see* Pinosylvin methyl ether, 1113

3-Methoxy-1,5,8-trihydroxyxanthone *see* Bellidifolin, 176

3-Methoxy-1,6,7-trihydroxyxanthone *see* Athyriol, 149

N-Methylactinodaphnine *see* (+)-Cassythicine, 270

5-*O*-Methylalloptaeroxylin, 932

3-Methyltheobromine *see* Caffeine, 227

Methyl thioalcohol *see* Methyl mercaptan, 945

Methyl undecyl ketone *see* 2-Tridecanone, 1353

ψ-Methysticin *see* Dihydromethysticin, 475

Mexicanin E, 953

Mexicanin I, 954

Mezcaline *see* Mescaline, 923

Mezerein, 955

Michelenolide, 956

Micheliolide, 957

Michellamine B, 958

Microcystin LA, 959

Microcystin LR, 960

Microcystin RR *see* Cyanoviridin RR, 395

Microhelenin-A, 961

Microhelenin C, 962

Microlenin, 963

Micromélin, 964

Micromelumin *see* Micromelin, 964

Micromerol *see* Ursolic acid, 1386

Milliamine L, 965

L-Mimosine, 966

Minorine *see* Vincamine, 1418

Miotoxin C, 967

Miroestrol, 968

Miserotoxin, 969

Mistletoe lectins, 970

Mitragynine, 971

Molephantin, 972

Molephantinin, 973

Mollugogenol A, 974

Momilactone A, 975

Momilactone B *see under* Momilactone A, 975

Momordin, 976

Monocrotaline, 977

Monofluoroacetic acid, 978

Monogynol B *see* Lupeol, 882

Monolupine *see* Anagyrine, 85

Monotropitoside, 979

Montanine, 980

Moracenin B *see* Kuwanone G, 829

Moracin A, 981

Moranoline *see* Deoxynojirimycin, 451

Moroidin, 982

Morphia *see* Morphine, 983

Morphine, 983

Mucronatine *see* Usaramine, 1390

Mukaadial, 984

Mulberrofuran G *see* Albanol A, 51

Multiflorine, 985

Multigilin, 986

Multiradiatin, 987

Multistatin, 988

Musaroside, 989

Musennin *see under* Echinocystic acid, 522

Muzigadial, 990

Mycoporphyrin *see* Hypericin, 763

Mycosinol, 991

Myomontanone, 992

Myricoside, 993

Myristicin, 994

N

Nagarine *see* Aconifine, 20

Nagilactone C, 995

Nandinin, 996

Nandinin *see under* *p*-Glucosyloxymandelonitrile, 663

Napelline, 997

Napellonine *see* Songorine, 1275

Naphtho[1, 2-*b*]furan-4,5-dione, 998

Narceine, 999

Narciclasine, 1000

Narcissine *see* Lycorine, 894

α-Narcotine, 1001

(−)-β-Narcotine *see under* α-Narcotine, 1001

Narcotoline, 1002

Naringenin 7-methyl ether *see* Sakuranetin, 1219

Naringenin 7-neohesperidoside *see* Naringin, 1003

Naringin, 1003

Narwedine, 1004

Natural Black 1 *see* Haematoxylin, 696

Natural Brown 7 *see* Juglone, 814

Natural Yellow 18 *see* Berberine, 180

NDGA *see* Nordihydroguaiaretic acid, 1018

Neobaicalein *see* Scullcapflavone II, 1244

Neochanin *see* Formononetin, 610

Neoisostegane, 1005

Neolinustatin, 1006

Neopine, 1007

Neoprotoveratrine *see* Protoveratrine B, 1150

Neoquassin, 1008

Neosaxitoxin, 1009

Nepalin 2 *see* α-Hederin, 706

Nepetalactone, 1010

Neprotine *see* Jatrorrhizine, 810

Neriine *see* Conessine, 342

(+)-Ngaione *see* Ipomoeamarone, 779

Nicotimine *see* (−)-Anabasine, 82

Nicotine, 1011

Nicotinimine *see* (−)-Anabasine, 82

Nicouline *see* Rotenone, 1207

Nigakihemiacetal B *see* Neoquassin, 1008

Nigakilactone D *see* Quassin, 1175

Nightshade, black *see under* α-Chaconine, 291

Nilotin, 1012

Nitidine, 1013

Nitogenin *see* Diosgenin, 497

3-Nitro-1-propanol *see under* Miserotoxin, 969

3-Nitro-1-propyl-β-D-glucopyranoside *see* Miserotoxin, 969

Niveusin C, 1014

Nobilin, 1015

Nodakenetin cellobioside *see* Decuroside III, 420

Nodularin, 1016

Nomilin, 1017

Nopinene *see* β-Pinene, 1111

N-Noracutumine *see* Acutumidine, 28

Norathyriol 6-glucoside *see* Tripteroside, 1362

31-Nordebromoaplysiatoxin *see* Oscillatoxin A, 1040

Nordihydroguaiaretic acid, 1018

Nor-ψ-ephedrine *see* D-Cathine, 280

ψ-Norephedrine *see* D-Cathine, 280

Norerythrostachaldine, 1019

Norharman, 1020

Norizalpinin *see* Galangin, 620

Norobtusifolin, 1021

31-Noroscillatoxin B, 1022

Noroxylin *see* Baicalein, 167

Norpseudoephedrine *see* D-Cathine, 280

Nortrachelogenin, 1023

Norwogonin 8-methyl ether *see* Wogonin, 1435

Noscapine *see* α-Narcotine, 1001

NSC 100046 *see* Elephantopin, 530

NSC 33669 *see* Emetine, 533

NSC 403169 *see* Acronycine, 24

NSC 71795 *see* Ellipticine, 531

Nucin *see* Juglone, 814

Nudicauline, 1024

α-Nupharidine *see* Deoxynupharidine, 452

O

Obaculactone *see* Limonin, 862

Ochrolifuanine A, 1025

Octalupine *see* 13-Hydroxylupanine, 747

Odoratol, 1026

Oenanthotoxin, 1027

Oestrone, 1028

Ohchinolide B, 1029

Ohioensin-A, 1030

Oil garlic *see* Diallyl sulfide, 459

Okadaic acid, 1031

Oleandrin monoglucoside *see* Urechitoxin, 1384

Oleanoglycotoxin-A, 1032

Oleanoglycotoxin B *see* Lemmatoxin, 855

Oleum Absinthii *see under* α-Thujone, 1335

Olivacine, 1033

O^2-Methylpancracine *see* Montanine, 980

Onopordopicrin, 1034

Orbicuside A, 1035

Orchinol, 1036

Orientin, 1037

Orizabin, 1038

Oryzalexin A, 1039

Oscillatoxin A, 1040

Oscillatoxin B1, 1041

Oscillatoxin B2, 1042

Oscillatoxin D, 1043

Oscine tropate *see* Hyoscine, 758

Ostruthin, 1044

Otonecine, 1045

Ouabain, 1046

Oubagenin 3-rhamnoside *see* Ouabain, 1046

Ovatifolin, 1047

Ovatine, 1048

Oxalic acid, 1049

Oxaloxanthin *see* Embelin, 532

Oxaloxanthin *see* Rapanone, 1181

Oxoacetic acid *see* Glyoxylic acid, 672

N-(3-Oxobutyl)cytisine, 1050

7-Oxodehydroabietic acid, 1051

S-Oxodiallyl disulfide *see* Allicin, 59

2-Oxo-13-epimanool, 1052

Oxoethanoic acid *see* Glyoxylic acid, 672

19-Oxonorcassaidine *see* Norerythrostachaldine, 1019

2-Oxo-11α-sparteine *see* Lupanine, 881

22-Oxovincaleukoblastine *see* Vincristine, 1419

Oxyayanin A, 1053

Oxyayanin B, 1054

(+)-Oxyfrullanolide, 1055

Oxymatrine *see* Ammothamnine, 78

Oxypeucedanin *see* Prangolarine, 1133

P

Pachycarpidine *see* Ammothamnine, 78

Pachycarpine *see* (−)-Sparteine, 1277

Pachyrrhizone, 1056

Pachysandrine A, 1057

Paclitaxel *see* Taxol, 1311

Palmatine, 1058

12-*O*-Palmitoyl-16-hydroxyphorbol 13-acetate, 1059

Palustrine, 1060

Palustrol, 1061

Panaxynol *see* Falcarinol, 593

Pancratistatin, 1062

Papaverine, 1063

Paramorphine *see* Thebaine, 1331

Parasorbic acid, 1064

Paravallarine, 1065

Parillin, 1066

Parsley camphor *see* Apiole, 108

Parthenicin *see* Parthenin, 1067

Parthenin, 1067

Parthenolide, 1068

Passicapsin, 1069

Passiflorin *see* Harman, 700

Paucin, 1070

Pectenine *see* (±)-Carnegine, 262

Pectenotoxin 1, 1071

Pedunculagin, 1072

Peganine, 1073

Pekilocerin A *see* Calotropin, 237

(−)-Pelletierine, 1074

Pelosine *see* (+)-Bebeerine, 175

α-Peltatin, 1075

β-Peltatin A methyl ether, 1076

3-(8-Pentadecenyl)phenol *see* (15:1)-Cardanol, 259

3,5,7,3′,4′-Pentahydroxyflavanone *see* Taxifolin, 1306

5,6,7,8,4′-Pentamethoxyflavone *see* Tangeretin, 1305

Pentamethylenediamine *see* Cadaverine, 225

Pentane-1,5-diamine *see* Cadaverine, 225

Pentanedioic acid *see* Glutaric acid, 667

Peppermint camphor *see* Menthol, 921

Perforatin A *see* 5-*O*-Methylalloptaeroxylin, 932

Pericalline *see* (−)-Apparicine, 114

Perilla ketone, 1077

1-Peroxyferolide, 1078

Peruvoside, 1079

Petasin, 1080

Petasitenine, 1081

Peyocactine *see* Hordenine, 732

Pfaffic acid, 1082

Pfaffoside A, 1083

PHA *see* Phasin, 1086

Phaeantharine, 1084

Phaeanthine *see under* (+)-Tetrandine, 1322

Phantomolin, 1085

Phaseolunatin *see* Linamarin, 863

Phaseolus substance II *see* Kievitone, 826

Phasin, 1086

α-Phellandrene, 1087

β-Phellandrene, 1088

Phenethyl alcohol *see* 2-Phenylethanol, 1091

Phenethylamine, 1089

Phenol, 1090

3-Phenylacrylic acid *see* Cinnamic acid, 322

2-Phenylethanol, 1091

β-Phenylethylamine *see* Phenethylamine, 1089
2-Phenylethylglucosinolate *see* Gluconasturtiin, 662
2-Phenylethylisothiocyanate *see under* Gluconasturtiin, 662
Phenylheptatriyne *see* 1-Phenylhepta-1,3,5-triyne, 1092
1-Phenylhepta-1,3,5-triyne, 1092
Phenylic acid *see* Phenol, 1090
Phenylmethylglucosinolate *see* Glucotropaeolin, 665
Phloracetophenone 4,6-dimethyl ether *see* Xanthoxylin, 1443
Phloridzin, 1093
Phloroglucinol trimer, 1094
Phoratoxin, 1095
Phorbol 12-tiglate 13-decanoate, 1096
Phoretin 2′-glucoside *see* Phloridzin, 1093
Phrymarolin I, 1097
Phyllanthin, 1098
Phyllanthin *see* Taxiphyllin, 1308
Phyllanthoside *see* Taxiphyllin, 1308
Phyllanthostatin A, 1099
Physoperuvine, 1100
Physosterine *see* Physostigmine, 1101
Physostigmine, 1101
Physostigmine aminoxide *see* Eseridine, 562
Physostigmine oxide *see* Eseridine, 562
Physostol *see* Physostigmine, 1101
Physovenine, 1102
Phytolaccagenin, 1103
Phytolaccatoxin *see under* Phytolaccagenin, 1103
Phytuberin, 1104
Phytuberol *see under* Phytuberin, 1104
Piceatannol, 1105
Picrotin, 1106
Picrotoxinin, 1107
Pilocarpine, 1108
Pilosine, 1109
β-Pinene, 1111
α-Pinene, 1110
Pinene *see* α-Pinene, 1110
2-Pinene *see* α-Pinene, 1110
Pin-2-en-4-one *see* Verbenone, 1401
Pinocembrin, 1112
Pinopalustrin *see* Nortrachelogenin, 1023
Pinoresinol dimethyl ether *see* (−)-Eudesmin, 565
Pinosylvin methyl ether, 1113
Pinosylvin monomethyl ether *see* Pinosylvin methyl ether, 1113
Pipercide, 1114
Piperidic acid *see* γ-Aminobutyric acid, 74
Piperidinic acid *see* γ-Aminobutyric acid, 74
Pisatin, 1115
Pisiferic acid, 1116
Pithecolobine, 1117
Plagiochiline A, 1118
Pleniradin, 1119
Plenolin, 1120
Plicatic acid, 1121
PL Toxin II *see* Okadaic acid, 1031
Plumbagin, 1122
Plumericin, 1123
Podecdysone B, 1124
Podolide, 1125
Podophyllinic acid lactone *see* Podophyllotoxin, 1126
Podophyllotoxin, 1126
Podophyllotoxone, 1127
Polhovolide, 1128
Polycavernoside A, 1129

Polygodial, 1130

Polypodine B, 1131

Ponasterone A, 1132

Ponkanetin *see* Tangeretin, 1305

Potassium atractylate *see* Atractyloside, 150

Prangenidine *see* Alloimperatorin, 62

Prangolarin *see* Prangolarine, 1133

Prangolarine, 1133

Pratensol *see* Biochanin A, 190

Pratol *see* Formononetin, 610

Precocene 1, 1134

Precocene 2, 1135

Prenylbenzoquinone, 1136

Prenyl-1,4-benzoquinone *see* Prenylbenzoquinone, 1136

Prenyl caffeate, 1137

5′-Prenylhomoeriodictyol, 1138

2-Prenyl-1,4-naphthoquinone *see* Deoxylapachol, 448

6-Prenylnaringenin *see* 6-Isopentenylnaringenin, 800

Pretazettine, 1139

Primetin, 1140

Primin, 1141

Pristimerin, 1142

Proacaciberin, 1143

Proacacipetalin, 1144

Proacacipetalin 6′-arabinoside *see* Proacaciberin, 1143

Progoitrin, 1145

1,3-Propanedicarboxylic acid *see* Glutaric acid, 667

Propanedioic acid *see* Malonic acid, 906

Propanethial *S*-oxide, 1146

p-Propenylanisole *see* Anethole, 89

Propolis *see under* Cinnamic acid, 322

2-Propyl-Δ^1-piperidine *see* γ-Coniceine, 343

(*S*)-2-Propylpiperidine *see* Coniine, 345

Prosapogenin CP_2 *see* β-Hederin, 707

Prostratin, 1147

Proteacin, 1148

Protoalba *see* Protoveratrine A, 1149

Protoanemonin *see under* Ranunculin, 1180

Protoveratrine A, 1149

Protoveratrine B, 1150

Provincialin, 1151

Prunasin, 1152

Prunasin xyloside *see* Lucumin, 877

Prunitol *see* Genistein, 633

Prunol *see* Ursolic acid, 1386

Pseudaconitine, 1153

Pseudochelerythrine *see* Sanguinarine, 1227

Pseudoconhydrine, 1154

Pseudomethysticin *see* Dihydromethysticin, 475

Pseudonorephedrine *see* D-Cathine, 280

Pseudopurpurin, 1155

3-Pseudotropanol *see* Pseudotropine, 1156

Pseudotropine, 1156

Pseudotropine benzoate *see* Tropacocaine, 1364

Psoralen, 1157

Psorospermin, 1158

Ptaeroglycol, 1159

Ptaquiloside, 1160

Ptelefolonium *see* O^4-Methylptelefolonium, 948

Pteroglycol *see* Ptaeroglycol, 1159

Pterosterone, 1161

Pterostilbene, 1162

Pulegone, 1163

β-Pulegone *see* Pulegone, 1163

Q

R

Retrofractamide B *see* Pipercide, 1114

Retronecine, 1186

Retrorsine, 1187

Retrorsine *N*-oxide *see* Isatidine, 781

Rhabarberone *see* Aloe-emodin, 64

Rhamnoxanthin *see* Frangulin A, 612

Rhaphiolepsin, 1188

Rhaponticin, 1189

Rhapontin *see* Rhaponticin, 1189

Rhazine *see* Akuammidine, 47

Rheadine *see* Rhoeadine, 1196

Rheic acid *see* Chrysophanol, 310

Rheum-emodin *see* Emodin, 534

Rhinanthin *see* Aucubin, 152

Rhinantin *see* Aucubin, 152

Rhipocephalin, 1190

Rhipocephenal, 1191

Rhodexin A, 1192

Rhodinal *see* Citronellal, 326

Rhododendrin, 1193

Rhodojaponin IV, 1194

ψ-Rhodomyrtoxin, 1195

Rhodotoxin *see* Grayanotoxin I, 688

Rhoeadine, 1196

Rhombinine *see* Anagyrine, 85

Ricin, 1197

Ricin D *see under* Ricin, 1197

Ricinine, 1198

Riddelliine *see* Riddelline, 1199

Riddelline, 1199

Ridentin, 1200

Ridentin A *see* Ridentin, 1200

Rinderine, 1201

Rishitin, 1202

Robin, 1203

Robinia lectin *see* Robin, 1203

Rochessine *see* Conessine, 342

Rodiasine, 1204

Rohdexin A *see* Rhodexin A, 1192

Romanicardic acid *see* Ginkgoic acid, 648

Roquessine *see* Conessine, 342

Roridin A, 1205

Roridin E, 1206

Rotenalone *see* 12a-Hydroxyrotenone, 751

Rotenolon 1 *see* 12a-Hydroxyrotenone, 751

Rotenolone *see* 12a-Hydroxyrotenone, 751

Rotenolone 1 *see* 12a-Hydroxyrotenone, 751

Rotenone, 1207

Rottlerin *see* Aloe-emodin, 64

Rottlerin, 1208

Royline *see* Lycoctonine, 889

Rubichloric acid *see* Asperuloside, 141

Rubrofusarin, 1209

Rugosin D, 1210

Rutamarin, 1211

Ryanex *see* Ryanodine, 1212

Ryanicide *see* Ryanodine, 1212

Ryanodine, 1212

S

Sabinol, 1213

Sacculatal, 1214

Safrole, 1215

Safynol, 1216

Saikosaponin BK1, 1217

Sainfuran, 1218

Sakuranetin, 1219

Sakuraso-saponin, 1220

Salannin, 1221

Salicylic acid, 1222

Salivation factor *see* Anatoxin a(s), 87

Salonitenolide, 1223

(−)-Salsoline, 1224

(+)-Salutaridine, 1225

Samaderin A, 1226

Samaderin B *see under* Samaderin A, 1226

Samaderin C *see under* Samaderin A, 1226

Samaderine A *see* Samaderin A, 1226

(*S*)-Sambunigrin *see under* Prunasin, 1152

Sanguinarine, 1227

Santamarin, 1228

Santamarine *see* Santamarin, 1228

Santolin *see* Achillin, 19

β-Santonin, 1230

Santonin *see* α-Santonin, 1229

α-Santonin, 1229

Sapatoxin A, 1231

Sapelin A, 1232

Sapindoside A *see* α-Hederin, 706

Sapintoxin A, 1233

Sarmentogenin 3-diginoside *see* Divostroside, 509

Sarmentogenin 3-oleandroside *see* Divaricoside, 508

Sarmentogenin 3-rhamnoside *see* Rhodexin A, 1192

Sarmentologenin 3-(6-deoxy-α-L-taloside) *see* Sarmentoloside, 1234

Sarmentoloside, 1234

Sarmentosin epoxide, 1235

Sarmutogenin 3-digitaloside *see* Musaroside, 989

Sarothralin, 1236

Satratoxin D *see* Roridin E, 1206

Saupirin, 1237

Saupirine *see* Saupirin, 1237

Saxitoxin, 1238

Saxitoxin hydrate *see* Saxitoxin, 1238

Scheffleroside *see under* Echinocystic acid, 522

Schizozygine, 1239

Schkuhrin I *see* Hiyodorilactone A, 730

Scillaren A, 1240

Scillarenin 3-glucosylrhamnoside *see* Scillaren A, 1240

Scilliroside, 1241

Scillirosidin 3-glucoside *see* Scilliroside, 1241

Sclareol, 1242

Scopine tropate *see* Hyoscine, 758

Scopoderm *see* Hyoscine, 758

Scopolamine *see* Hyoscine, 758

Scopoline tropate *see* Hyoscine, 758

Scorpioidin, 1243

Scorpioidine *see* Scorpioidin, 1243

Scullcapflavone II, 1244

Scytonemin A, 1245

Secaclavine *see* Chanoclavine-I, 297

Securinine, 1246

L-Selenocystathionine, 1247

Selenomethylselenocysteine *see* *Se*-Methyl-L-selenocysteine, 952

Sempervine *see* Sempervirine, 1248

Sempervirene *see* Sempervirine, 1248

Sempervirine, 1248

Senaetnine, 1249

T

Tripdiolide, 1361
Triphyopeltine *see* Dioncopeltine A, 494
Tripteroside, 1362
Triptolide, 1363
1, 2, 6-Tris(3-nitropropanoyl)-β-D-glucoside *see* Karakin, 823
Trisphaerine *see* Hippeastrine, 727
Trispherine *see* Hippeastrine, 727
Trochol *see* Betulin, 186
Tropacaine *see* Tropacocaine, 1364
Tropacocaine, 1364
1α*H*, 5α*H*-Tropan-3α-ol *see* Tropine, 1365
3β-Tropanol *see* Pseudotropine, 1156
Tropeine *see* Tropine, 1365
Tropine, 1365
Tropine benzoate *see* Benzoyltropein, 177
Tropine tropate *see* Atropine, 151
(*S*)-Tropine tropate *see* Hyoscyamine, 759
ψ-Tropine *see* Pseudotropine, 1156
Tryptophan betaine *see* Hypaphorine, 761
L-Tryptophan, 1366
Tsibulin 1 *see under* Tsibulin 2, 1367
Tsibulin 2, 1367
T34 Toxin *see* Brevetoxin B, 202
T46 Toxin *see* Brevetoxin A, 201
Tubeimoside I, 1368
(+)-Tubocurarine, 1369
Tubotoxin *see* Rotenone, 1207
Tubulosine, 1370
Tulipalin A *see under* Tuliposide A, 1372
Tulipalin B *see under* Tuliposide B, 1373
Tulipinolide, 1371
epi-Tulipinolide diepoxide *see* Epitulipinolide diepoxide, 540
epi-Tulipinolide *see* Epitulipinolide, 539
Tuliposide A, 1372
Tuliposide B, 1373
Tuliposide D, 1374
Tullidinol, 1375
Tullidinol *see under* Karwinskione, 825
Turmeric colour *see* Curcumin, 392
Turmeric yellow *see* Curcumin, 392
Turricolol E, 1376
Tussilagine, 1377
Tutin, 1378
Tyledoside A, 1379
Tyledoside B *see under* Tyledoside A, 1379
Tyledoside D *see under* Tyledoside A, 1379
Tyledoside F *see under* Tyledoside A, 1379
Tyledoside G *see under* Tyledoside A, 1379
(−)-Tylocrebine, 1380
(+)-Tylocrebine *see under* (−)-Tylocrebine, 1380
Tylophorine, 1381

U

Ulexine *see* Cytisine, 410
Umbellatine *see* Berberine, 180
Undulatone, 1382
Uplandicine, 1383
Urechitoxin, 1384
Ursiniolide A, 1385
Ursolic acid, 1386
Urson *see* Ursolic acid, 1386
Urushiol III, 1387
Usambarensine, 1388
Usambarine, 1389

Usaramine, 1390

Uscharidin, 1391

Uscharin, 1392

Uscharine *see* Uscharin, 1392

Usuramine *see* Usaramine, 1390

Uvaretin, 1393

V

Vanillic acid, 1394

Vanillosmin *see* Eremanthine, 543

Vasicine *see* Peganine, 1073

Vasicinol, 1395

Vasicinone, 1396

Vaumigan *see* Guaiazulene, 692

VCR *see* Vincristine, 1419

Veatchine, 1397

(15β)-Veatchine acetate *see* Ovatine, 1048

Veratetrine *see* Protoveratrine B, 1150

Veratramine, 1398

Verbascoside, 1399

Verbenalin, 1400

Verbenaloside *see* Verbenalin, 1400

Verbenone, 1401

Vermeerin, 1402

Vernadigin, 1403

Vernodalin, 1404

Vernodalol, 1405

Vernoflexin, 1406

Vernoflexine *see* Vernoflexin, 1406

Vernoflexuoside, 1407

Vernolepin, 1408

Vernolide, 1409

Vernomenin, 1410

Vernomygdin, 1411

Vertine *see* Cryogenine, 372

Very fast death factor *see* Anatoxin a, 86

Vescalagin *see under* Castalagin, 271

Vestitol, 1412

Viburnine *see* Eburnamonine, 518

Vicianin, 1413

Vignafuran, 1414

Vignatin *see* Kievitone, 826

Viguiestenin, 1415

Villalstonine, 1416

Vinblastine, 1417

Vincaine *see* Ajmalicine, 43

Vincaleukoblastine *see* Vinblastine, 1417

4′α-Vincaleukoblastine *see* Leurosidine, 857

Vincamajoridine *see* Akuammine, 48

Vincamarine *see* Vincamine, 1418

Vincamine, 1418

Vincamone *see* Eburnamonine, 518

Vinceine *see* Ajmalicine, 43

Vincristine, 1419

Vinegar acid *see* Acetic acid, 11

ε-Viniferin, 1420

Vinleurosine *see* Leurosine, 858

Vinrosidine *see* Leurosidine, 857

(*R*)-5-Vinyl-2-oxazolidinethione *see under* Progoitrin, 1145

9-Viridiflorylretronecine *see* Lycopsamine, 891

β-Viscol *see* Lupeol, 882

Viscotoxin, 1421

Viscumin, 1422

Visnacorin *see* Visnagin, 1423

W

X

Y

Z

Species Index

A

Abies spp.
  Car-3-ene, 261
  Juvabione, 817
  β-Phellandrene, 1088

Abies alba
  Limonene, 861

Abies balsamea
  Dehydrojuvabione, 422
  Juvabione, 817

Abrus precatorius
  Abrin, 1

Abuta grisebachii
  Macoline, 898

Acacia spp.
  L-α-Amino-γ-oxalylaminobutyric acid, 75
  L-α, γ-Diaminobutyric acid, 460
  L-Djenkolic acid, 510
  Heterodendrin, 724
  Linamarin, 863
  Proacacipetalin, 1144

Acacia auriculiformis
  Auriculoside, 155

Acacia catechu
  Taxifolin, 1306

Acacia georginea
  Monofluoroacetic acid, 978

Acacia glaucescens
  Prunasin, 1152

Acacia globulifera
  Heterodendrin, 724
  Proacacipetalin, 1144

Acacia hebeclada
  Acacipetalin, 5

Acacia melanoxylon
  Acamelin, 7
  2, 6-Dimethoxybenzoquinone, 487

Acacia sieberiana
  Acacipetalin, 5
  3-Hydroxyheterodendrin, 746
  Proacaciberin, 1143

Acalypha indica
  Acalyphin, 6

Acer spp.
  Corilagin, 354
  Geraniin, 640

Acer pseudoplatanus
  α-(Methylenecyclopropyl)glycine, 938

Acer rubrum
  Gramine, 685

Acer saccharinum
  2, 6-Dimethoxyphenol, 489
  Gramine, 685

Achillea spp.
  Achillin, 19
  Desacetoxymatricarin, 454

Achillea millefolium
  Achillin, 19
  (−)-Betonicine, 185
  Chamazulene, 295

Achillea moschata
  (−)-Betonicine, 185

Acnistus arborescens
  Withaferin A, 1433

Acokanthera ouabaio
  Ouabain, 1046

Aconitum spp.
  Aconitine, 21
  Mesaconitine, 922
  Napelline, 997

Aconitum callianthum
  Hypaconitine, 760

Aconitum carmichaeli
  Hypaconitine, 760
  Karakoline, 824

Aconitum chasmanthum
  Aconitine, 21
  Indaconitine, 771

Aconitum delphinifolium
  Condelphine, 341

Aconitum excelsum
Lappaconitine, 844

Aconitum falconeri
Falaconitine, 591
Indaconitine, 771
Pseudaconitine, 1153

Aconitum ferox
Bikhaconitine, 188
Indaconitine, 771
Pseudaconitine, 1153

Aconitum finetianum
Avadharidine, 158

Aconitum fischeri
Jesaconitine, 812

Aconitum heterophyllum
Heteratisine, 723

Aconitum ibukiense
Delcosine, 431

Aconitum japonicum
(+)-Demethylcoclaurine, 438

Aconitum karakolicum
Aconifine, 20
Karakoline, 824
Songorine, 1275

Aconitum lycoctonum
Lycaconitine, 888
Lycoctonine, 889

Aconitum monticola
Delsoline, 433
Songorine, 1275

Aconitum nagarum
Aconifine, 20

Aconitum napellus
Aconitine, 21
Hypaconitine, 760
Mesaconitine, 922
Napelline, 997

Aconitum orientale
Avadharidine, 158
Lappaconitine, 844

Aconitum ranunculaefolium
Lappaconitine, 844

Aconitum sachalinense
Jesaconitine, 812

Aconitum septentrionale
Lappaconitine, 844

Aconitum soongaricum
Songorine, 1275

Aconitum spicatum
Bikhaconitine, 188

Aconitum spictatum
Pseudaconitine, 1153

Aconitum subcuneatum
Jesaconitine, 812

Aconitum violaceum
Bikhaconitine, 188

Aconitum zeravschanicum
Heteratisine, 723

Acorus calamus
β-Asarone, 135
Methylisoeugenol, 941

Acronychia spp.
Kokusaginine, 828

Acronychia baueri
Acronycidine, 23
Acronycine, 24

Acronychia haplophylla
Acronycine, 24

Acronychia laurifolia
Acrovestone, 26

Acronychia pedunculata
Acrovestone, 26

Acronychia vestita
Acrovestone, 26

Acroptilon repens
Acroptilin, 25
Chlorohyssopifolin A, 304
Repin, 1182

Actinidia arguta
Actinidine, 27

Actinidia polygama
Actinidine, 27
Iridomyrmecin, 780

Adenia fruticosa
Deidaclin, 428

Adenia globosa
Deidaclin, 428

Adenia spinosa
Deidaclin, 428

Adenium honghel
Digitalin, 470

Adhatoda vasica
Peganine, 1073
Vasicinol, 1395
Vasicinone, 1396

Adiantum spp.
Naringin, 1003

Adiscanthus fusciflorus
Dictamnine, 465

Adlumia fungosa
Adlumine, 30
(+)-Bicuculline, 187

Adonis vernalis
Adonitoxin, 31
Convallamaroside, 348
Cymarin, 405
2,6-Dimethoxybenzoquinone, 487
Vernadigin, 1403

Aegiceras corniculata
Rapanone, 1181

Aegiceras corniculatum
Embelin, 532

Aegle marmelos
Alloimperatorin, 62
Dictamnine, 465
Imperatorin, 770

Aesculus hippocastanum
Aescin, 33

Aethusa cynapium
Aethusin, 34

Agastache rugosa
Estragole, 564

Agave americana
Agavoside A, 38

Ageratina spp.
Tremetone, 1347

Ambrosia spp.
Ambrosin, 72
Coronopilin, 358
Damsin, 411
Parthenolide, 1068

Ambrosia abrotanum
1*S*-Hydroxy-α-bisabololoxide A acetate, 740

Ambrosia acanthicarpa
Chamissonin diacetate, 296

Ambrosia artemisiifolia
Ambrosin, 72

Ambrosia camphorata
Tulipinolide, 1371

Ambrosia chamissonis
Epitulipinolide, 539

Ambrosia confertiflora
Santamarin, 1228
Tamaulipin-A, 1304

Ambrosia dumosa
Epitulipinolide, 539
Tamaulipin-A, 1304
Tulipinolide, 1371

Ambrosia eliator
Ragweed pollen allergen Ra5, 1179

Ambrosia maritima
Damsin, 411

Ambrosia psilostachya
Coronopilin, 358
Parthenin, 1067

Ammi spp.
Bergapten, 182

Ammi majus
Alloimperatorin, 62
Xanthotoxin, 1441

Ammi visnaga
Athamantin, 148
Dihydrosamidin, 477
Visnagin, 1423

Ammocharis coranica
Acetylcaranine, 16
Caranine, 257

Ammodendron spp.
Anagyrine, 85

Ammodendron conollyi
Ammodendrine, 77

Ammothamnus lehmannii
Ammothamnine, 78

Ammothamnus songorica
Ammothamnine, 78

Amsinckia hispida
Lycopsamine, 891

Amsinckia intermedia
Echiumine, 524
Intermedine, 776
Lycopsamine, 891

Amsinckia lycopsoides
Intermedine, 776

Amyris pinnata
Austrobailignan 1, 156

Anabaena spp.
Anatoxin a, 86
Microcystin LA, 959

Anabaena circinalis
Saxitoxin, 1238

Anabaena flos-aquae
Anatoxin a(s), 87

Anabasis aphylla
(−)-Anabasine, 82
Lupinine, 884

Anacardium occidentale
Anacardic acid, 83
Bilobol, 189
(15:1)-Cardanol, 259
Ginkgoic acid, 648

Anagallis hirtella
Primin, 1141

Anagyris spp.
Cytisine, 410

Anagyris foetida
Anagyrine, 85

Anamirta paniculata
Picrotin, 1106
Picrotoxinin, 1107

Anaphalis spp.
Helenalin, 708

Anchusa arvensis
Lycopsamine, 891

Anchusa officinalis
Clivorine, 330
Lithospermic acid, 868

Ancistrocladus korupensis
Michellamine B, 958

Andira araroba
Chrysarobin, 307

Andira inermis
(−)-Medicarpin, 915

Androcymbium melanthoides
Androcymbine, 88

Andrographis paniculata
Apigenin 7,4′-dimethyl ether, 107

Anemone spp.
Ranunculin, 1180

Anethum graveolens
Dillapiole, 485

Angelica spp.
Bergapten, 182
Imperatorin, 770
Isopimpinellin, 801

Angelica archangelica
Angelicin, 90
Archangelicin, 119
Prangolarine, 1133
Xanthotoxin, 1441
Xanthotoxol, 1442

Angelica japonica
Cimifugin, 317

Angelica keiskei
Archangelicin, 119

Angelica longeradiata
Archangelicin, 119

Angelica officinalis
Xanthotoxin, 1441

Angelica pubescens
Heptadeca-1,9-diene-4,6-diyne-3,8-diol, 718

Angelica sylvestres
Athamantin, 148

Arnebia euchroma
Arnebinol, 128
Arnebinone, 129

Arnebia nobilis
Alkannin, 56
Alkannin β, β-dimethylacrylate, 57

Arnica longifolia
Arnicolide A, 130

Arnica montana
Arnicolide A, 130

Arrainvillea nigricans
3-Bromo-4, 5-dihydroxybenzyl alcohol, 205

Artemisia spp.
Achillin, 19
Camphor, 243
Canin, 251
Desacetoxymatricarin, 454
Santamarin, 1228
α-Santonin, 1229
β-Santonin, 1230
Vulgarin, 1427

Artemisia absinthium
Absinthin, 2
Chamazulene, 295
Sesartemin, 1258
α-Thujone, 1335

Artemisia apiacea
Methyl caffeate, 935

Artemisia caerulescens
β-Santonin, 1230

Artemisia cana
Canin, 251
Ridentin, 1200

Artemisia cina
β-Santonin, 1230

Artemisia compacta
β-Santonin, 1230

Artemisia douglasiana
Arteglasin A, 133

Artemisia dracunculus
Estragole, 564

Artemisia hispanica
Cirsilineol, 324

Artemisia judaica
Vulgarin, 1427

Artemisia ludoviciana
Ludovicin A, 878

Artemisia maritima
1, 8-Cineole, 319

Artemisia porrecta
Anethole, 89

Artemisia sieversiana
Absinthin, 2

Artemisia taurica
Vulgarin, 1427

Artemisia tridentata
Ridentin, 1200

Artemisia tripartita
Ridentin, 1200

Artemisia vulgaris
Vulgarin, 1427

Arundo donax
Bufotenine, 215
N, *N*-Dimethyltryptamine, 491
Gramine, 685

Asarum canadense
Aristolochic acid, 126

Asarum europaeum
β-Asarone, 135
Methylisoeugenol, 941

Asclepias spp.
Asclepin, 137

Asclepias curassavica
Asclepin, 137
Calactin, 229

Asclepias eriocarpa
Eriocarpin, 550
Labriformidin, 830
Labriformin, 831

Asclepias labriformis
Eriocarpin, 550
Labriformidin, 830
Labriformin, 831

Asclepias speciosa
Aspecioside, 140

Asclepias syriaca
Aspecioside, 140
Nicotine, 1011

Ascophyllum nodosum
Phloroglucinol trimer, 1094

Asparagus officinale
L-Arginine, 124

Asperula spp.
Pseudopurpurin, 1155

Asperula odorata
Alizarin, 55
Asperuloside, 141
Purpurin, 1167

Asphodelus microcarpus
Aloe-emodin, 64

Aspidiosperma marcgravianum
Hydroquinidine, 736

Aspidosperma spp.
Akuammidine, 47
Aspidospermine, 143

Aspidosperma dasycarpon
(−)-Apparicine, 114

Aspidosperma nigricans
Olivacine, 1033

Aspidosperma quebrachoblanco
Aspidospermatine, 142
Aspidospermine, 143

Aspidosperma subincanum
Ellipticine, 531

Aspilia spp.
Thiarubrine A, 1334

Aster tartaricus
Anethole, 89

Aster umbellatus
8β-Hydroxyasterolide, 739

Astragalus atropubescens
Miserotoxin, 969

Astragalus bisulcatus
Se-Methyl-L-selenocysteine, 952

Astragalus canadensis
Cibarian, 312
Karakin, 823

Astragalus cibarius
Cibarian, 312
Coronarian, 356
Karakin, 823

Astragalus falcatus
Cibarian, 312
Coronarian, 356
Karakin, 823

Astragalus flexuosus
Cibarian, 312
Coronarian, 356
Karakin, 823

Astragalus membranaceus
Astragaloside III, 144

Astragalus miser
Miserotoxin, 969

Astragalus pterocarpus
Miserotoxin, 969

Astragalus sieversianus
Astrasieversianin XVI, 145

Astragalus tetrapleurus
Miserotoxin, 969

Astragalus toanus
Miserotoxin, 969

Astralagus spp.
L-Selenocystathionine, 1247

Atalantia roxburghiana
N-Methylflindersine, 939

Athamanta cretensis
Athamantin, 148

Athamanta oreoselinum
Athamantin, 148

Atherosperma moschatum
Berbamine, 178

Athyrium mesosorum
Athyriol, 149

Atractylis gummifera
Atractyloside, 150

Atragene sibirica
Delphinine, 432

Atropa belladonna
Apoatropine, 111
Atropine, 151
Hyoscine, 758
Hyoscyamine, 759
Tropine, 1365

Aucuba japonica
Aucubin, 152

Austrobaileya scandens
Austrobailignan 1, 156

Avena sativa
Avenacin A-1, 159
Avenacin B-2, 160
Avenalumin I, 161

Avicennia marina
Naphtho[1, 2-*b*]furan-4,5-dione, 998

Avrainvillea nigricans
Avrainvilleol, 162

Azadirachta indica
Azadirachtin, 164
Salannin, 1221

B

Baccharis spp.
Tremetone, 1347

Baccharis cordifolia
Miotoxin C, 967
Roridin A, 1205
Roridin E, 1206

Baccharis megapotamica
Baccharinoid B21, 166
Miotoxin C, 967

Baccharis pedunculata
Lachnophyllum lactone, 832

Backhousia anisata
Anethole, 89

Backhousia myrtifolia
Isoelemicin, 795

Baileya multiradiata
Baileyin, 169
Fastigilin B, 597
Fastigilin C, 598
Multigilin, 986
Multiradiatin, 987
Multistatin, 988
Radiatin, 1178

Baileya pauciradiata
Paucin, 1070

Baileya pleniradiata
Baileyin, 169
Paucin, 1070
Pleniradin, 1119
Plenolin, 1120
Radiatin, 1178

Balduina spp.
Helenalin, 708

Balduina angustifolia
Angustibalin, 92

Baliospermum montanum
Baliospermin, 171

Baltimora recta
Encelin, 536

Banisteria caapi
Harmaline, 699
Harmine, 701

Banisteriopsis argentea
Calligonine, 234

Baptisia spp.
Biochanin A, 190
Cytisine, 410
(−)-Sparteine, 1277

Barosma betulina
Diosphenol, 498

Barosma crenulata
Diosphenol, 498

Barosma serratifolia
Diosphenol, 498

Barteria fistulosa
Volkenin, 1426

Bauhinia manca
7,4′-Dihydroxyflavan, 480
3′,4′-Dihydroxy-7-methoxyflavan, 482

Bedfordia solicina
Methyl caffeate, 935

Beilschmiedia podagrica
Glaucine, 655

Berberis spp.
Berberine, 180
Jatrorrhizine, 810
Palmatine, 1058

Berberis laurina
Berberastine, 179

Berberis orthobotrys
Aromoline, 132

Berberis thunbergii
Berbamine, 178

Berberis tinctoria
Tetrahydropalmatine, 1321

Berberis valdiviana
Cularine, 389

Berberis vulgaris
Berbamine, 178
Berberine, 180

Bergenia crassifolia
Arbutin, 117

Berkheya speciosa
Onopordopicrin, 1034
Salonitenolide, 1223

Bersama abyssinica
Hellebrigenin 3-acetate, 714

Beta vulgaris
L-Azetidine-2-carboxylic acid, 165
Betagarin, 183
Betavulgarin, 184
Glutaric acid, 667
Glyoxylic acid, 672
Malonic acid, 906

Betonica officinalis
(−)-Betonicine, 185

Betula spp.
Betulin, 186
Guaiacol, 691
Rhododendrin, 1193
Acacetin, 4

Betula lenta
Salicylic acid, 1222

Betula platyphylla
Betulin, 186

Bidens spp.
1-Phenylhepta-1,3,5-triyne, 1092

Billia hippocastanum
L-Hypoglycin, 764
α-(Methylenecyclopropyl)glycine, 938

Bleekeria vitiensis
Ellipticine, 531

Blighia sapida
L-γ-Glutamyl-L-hypoglycin, 666
L-Hypoglycin, 764

Blumea balsamifera
Borneol, 195

Bocconia spp.
Allocryptopine, 61

Bocconia cordata
Bocconine, 194

Boehmeria cylindrica
Cryptopleurine, 374

Boenninghausenia spp.
Chalepensin, 293

Boenninghausenia japonica
Rutamarin, 1211

Bolbostemma paniculatum
Tubeimoside I, 1368

Boophane distica
3-Acetylnerbowdine, 18

Boophane fischeri
Ambelline, 71

Borago officinalis
Dhurrin, 457
Lycopsamine, 891

Boronia pinnata
Elemicin, 528

Borreria capitata
Borrecapine, 196

Boschniakia rossica
Boschnialactone, 197

Bowiea kilimandscharica
Bovoside A, 198

Bowiea volubilis
Bovoside A, 198

Brachyglottis repanda
Senkirkine, 1254

Brandegea bigelovii
Cucurbitacin O, 382
Cucurbitacin P, 383
Cucurbitacin Q, 384

Brassica spp.
S-Methyl-L-cysteine *S*-oxide, 937

Brassica campestris
Glucocheirolin, 658
Glucoerysolin, 659
Spirobrassinin, 1282

Brassica juncea
Brassilexin, 199
Cyclobrassinin sulfoxide, 398
Progoitrin, 1145

Brassica napus
Glucocheirolin, 658
Glucoerysolin, 659
Progoitrin, 1145

Brassica nigra
Gluconasturtiin, 662

Brassica oleracea
Glucocapparin, 657
Glucocheirolin, 658
Progoitrin, 1145

Brickellia spp.
Tremetone, 1347

Bridelia micrantha
Gallic acid, 624

Brosimum sp.
Xanthyletin, 1445

Broussonetia papyrifera
Broussonin A, 206

Brucea amarissima
Bruceine B, 208
Isobruceine A, 785

Brucea antidysenterica
Bruceantin, 207

Brucea javanica
Bruceoside A, 209

Bruguiera sexangula
Benzoyltropein, 177

Brunfelsia grandiflora
Pyrrole-3-carbamidine, 1172

Brunsvigia rosea
Amaryllisine, 70

Bryophyllum pinnatum
Bryophyllin A, 211

Bryophyllum tubiflorum
Bryotoxin A, 212

Buchnerodendron speciosum
Chaulmoogric acid, 300

Buddleja spp.
Catalpol, 275

Buddleja davidii
Buddledin A, 213

Buddleja globosa
Verbascoside, 1399

Buddleja officinalis
Verbascoside, 1399

Buphane distiche
Acetovanillone, 12

Bupleurum kummingense
Saikosaponin BK1, 1217

Bursera fagaroides
β-Peltatin A methyl ether, 1076

Bursera microphylla
Burseran, 217

Butea monosperma
Butrin, 218

Buxus spp.
Buxamine E, 223

Buxus argentea
Cyclovirobuxine C, 403

Buxus balearica
Cycloprotobuxine C, 402

Buxus harlandi
Cyclobuxine D, 399

Buxus hyrcana
Cyclobuxine D, 399

Buxus malayana
Cycloprotobuxine C, 402
Cyclovirobuxine C, 403

Buxus microphylla
Cyclobuxine D, 399
Cyclovirobuxine C, 403

Buxus sempervirens
(+)-Bebeerine, 175
Buxamine E, 223
Cyclobuxine D, 399
Cycloprotobuxine C, 402
Cyclovirobuxine C, 403

Buxus wallichiana
Cyclobuxine D, 399
Cyclovirobuxine C, 403

C

Cacalia hastata
Bakkenolide A, 170
Integerrimine, 775

Cadia spp.
13-Hydroxylupanine, 747

Cadia ellisiana
Multiflorine, 985

Caesaria graveolens
Micromelin, 964

Cajanus cajan
Cajanol, 228

Calea axillaris
Calaxin, 231

Calendula arvensis
Arvenososide A, 134

Callicarpa candicans
Callicarpone, 233

Calligonum minimum
Calligonine, 234

Callitris drummondii
Podophyllotoxin, 1126

Calocarpum sapota
Lucumin, 877

Calophyllum bracteatum
Calophyllin B, 235

Calophyllum brasiliense
Dehydrocycloguanandin, 421

Calophyllum inophyllum
Calophyllin B, 235
Calophyllolide, 236
Dehydrocycloguanandin, 421

Calophyllum lanigerum
Calanolide A, 230

Calophyllum thalictroides
Hederagenin 3-glucoside, 705

Calophyllum zeylanicum
6-Deoxyjacareubin, 446

Calopogonium fulva
Tecomine, 1312

Calopogonium stans
Tecomine, 1312

Calotropis procera
Calactin, 229
Calotropin, 237
Uscharidin, 1391
Uscharin, 1392

Calycanthus floridus
Calycanthine, 238

Calycanthus glaucus
Calycanthine, 238

Calycanthus occidentalis
Calycanthine, 238

Calycanthus praecox
Calycanthine, 238

Calystegia sepium
Calystegin B_2, 239

Camalium sativa
Camalexin, 240

Camellia japonica
Camellidin I, 241
Camellidin II, 242
Pedunculagin, 1072

Camellia sinensis
Caffeine, 227

Campanula cochleariifolia
Triglochinin, 1356

Campanula medium
(−)-Lobeline, 870

Campanula rotundifolia
Triglochinin, 1356

Camptotheca acuminata
Camptothecin, 244

Cananga odorata
Isosafrole, 802

Canarium commune
Elemicin, 528

Canavalia ensiformis
L-Canaline, 246
L-Canavanine, 247
Concanavalin A, 340

Canella winterana
Mukaadial, 984
Muzigadial, 990

Cannabis sativa
Cannabichromene, 252
Δ^1-Tetrahydrocannabinol, 1320

Capsicum frutescens
Capsidiol, 255

Caragana acanthophylla
(−)-Acanthocarpan, 8

Cardiospermum grandiflorum
Cardiospermin, 260

Cardiospermum hirsutum
Cardiospermin, 260

Carduus spp.
Acanthoidine, 9

Carduus acanthoides
Acanthoidine, 9

Carex brevicollis
Brevicolline, 203

Carica papaya
Carpaine, 265
Glucotropaeolin, 665

Carnegiea gigantea
(±)-Carnegine, 262
Gigantine, 645

Carpesium spp.
Ivalin, 806

Carpotroche brasiliensis
Chaulmoogric acid, 300
Gynocardin, 693

Carthamus tinctorius
Dehydrosafynol, 425
Safynol, 1216
3-*trans*, 11-*trans*-Trideca-1,3,11-triene-5,7,9-triyne, 1354

Carum copticum
p-Cymene, 406
Thymol, 1336

Carya illinoensis
Juglone, 814

Carya ovata
Juglone, 814

Cassia spp.
Chrysarobin, 307

Cassia absus
Chaksine, 292

Cassia angustifolia
Sennoside A, 1255

Cassia glabella
(+)-Cassythicine, 270

Cassia obtusifolia
Aurantio-obtusin 6-β-D-glucoside, 153

Cassia quinquangulata
Quinquangulin, 1177
Rubrofusarin, 1209

Cassia senna
Aloe-emodin, 64
Chrysophanol, 310
Sennoside A, 1255

Cassia siamea
Chrysophanol, 310

Cassia tora
Rubrofusarin, 1209

Cassytha filiformis
Cassyfiline, 269

Cassytha malantha
(+)-Cassythicine, 270

Castanea sativa
Castalagin, 271

Castanopsis indica
Angelicin, 90

Castanospermum australe
Castanospermine, 272

Castela nicholsoni
Chaparrin, 298
Glaucarubolone, 654

Castela tweediei
Castelanone, 273

Castilloa elastica
Cymarin, 405

Casuarina stricta
Casuarictin, 274
Pedunculagin, 1072

Catalpa spp.
Catalpol, 275

Catalpa ovata
Deoxylapachol, 448

Catalpa speciosa
Catalposide, 276

Catha edulis
D-Cathine, 280
D-Cathinone, 281
Pristimerin, 1142

Catharanthus spp.
Leurosine, 858

Catharanthus roseus
Norharman, 1020

Caulerpa taxifolia
Caulerpenyne, 282

Caulophyllum thalictroides
Caulophylline, 283

Cedrelopsis grevei
5-*O*-Methylalloptaeroxylin, 932
Ptaeroglycol, 1159

Cedrus deodara
Centdarol, 286

Celastrus paniculata
Celapanine, 284

Centaurea spp.
Cnicin, 331
Cynaropicrin, 408

Centaurea hyrcanica
Acroptilin, 25
Repin, 1182

Centaurea hyssopifolia
Acroptilin, 25
Chlorohyssopifolin A, 304

Centaurea linifolia
Acroptilin, 25

Centaurea repens
Chlororepdiolide, 305

Centaurea salonitana
Salonitenolide, 1223

Centipeda minima
Brevilin A, 204

Centratherum punctatum
Eremantholide A, 544

Cephaelis acuminata
Emetine, 533

Cephaelis ipecacuanha
Emetine, 533
Tubulosine, 1370

Cephalocereus senilis
Cephalocerone, 287

Cephalotaxus fortunei
Cephalotaxine, 289
Harringtonine, 702

Cephalotaxus hainensis
Harringtonine, 702

Cephalotaxus harringtonia
Cephalotaxine, 289
Harringtonine, 702

Cephalotaxus wilsoniana
Cephalotaxine, 289

Ceratocapnos palaestinus
Cularimine, 388

Ceratonia siliqua
Pyrogallol, 1171

Cerbera odollam
Thevetin B, 1333

Cercidiphyllum japonicum
Magnolol, 902

Cereus pectenaboriginum
(±)-Carnegine, 262

Ceterach officinarum
Naringin, 1003

Chaenactis carphoclinia
Eupatoriopicrin, 583

Chaenactis douglassii
Eupatoriopicrin, 583

Chaerophyllum temulentum
Falcarinone, 594

Chamaecyparis formosensis
Chamaecynone, 294
p-Cresol, 366

Chamaecyparis pisifera
Pisiferic acid, 1116

Chamaemelum nobile
Nobilin, 1015

Chartolepsis intermedia
Grosshemin, 690

Cheilanthes sieberi
Ptaquiloside, 1160

Chelidonium spp.
Allocryptopine, 61
Berberine, 180

Chelidonium majus
Sanguinarine, 1227

Chenopodium ambrosioides
Ascaridole, 136
Cirsilineol, 324
p-Cymene, 406

Chiloscyphus polyanthus
Diplophyllin, 503

Chimaphila corymbosa
Chimaphylin, 301

Chloroxylon swietenia
Rutamarin, 1211
Xanthyletin, 1445

Chondodendron candicans
(+)-Bebeerine, 175

Chondodendron platyphyllum
(+)-Bebeerine, 175

Chondodendron tomentosum
(+)-Bebeerine, 175

Chondria armata
Isodomoic acid A, 793
Isodomoic acid B, 794

Chondrodendron spp.
(+)-Tubocurarine, 1369

Chondrodendron tomentosum
Isochondrodendrine, 787
(+)-Tubocurarine, 1369

Chrysophyllum lacourtianum
Norharman, 1020

Cicer arietinum
Biochanin A, 190
Formononetin, 610
(−)-Maackiain, 897

Cichorium intybus
Cichoralexin, 313
Cichoriin, 314
8-Deoxylactucin, 447
Lactucin, 834
Lactucopicrin, 835

Cicuta douglasii
Cicutoxin, 315

Cicuta maculata
Cicutoxin, 315

Cicuta virosa
Cicutoxin, 315

Cimicifuga simplex
Cimifugin, 317

Cinchona spp.
Quinine, 1176

Cinchona officinalis
Hydroquinidine, 736
Quinine, 1176

Cinchona succirubra
Cinchonidine, 318

Cinchona tucujensis
Cinchonidine, 318

Cinnamomum spp.
Cinnamic acid, 322

Cinnamomum sp.
Carpacin, 264

Crinodendron hookerianum
Cucurbitacin D, 379

Crinum spp.
Crinamine, 367
Galanthamine, 621

Crinum amabile
Hippeastrine, 727

Crinum asiaticum
Crinamine, 367
Crinasiadine, 368
Crinasiatine, 369

Crinum laurentii
Ambelline, 71

Crinum macrantherum
Acetylcaranine, 16

Crithmum maritimum
Apiole, 108
Dillapiole, 485

Crossostylis ebertii
Benzoyltropein, 177

Crotalaria spp.
L-α-Amino-β-oxalylaminopropionic acid, 76
Crotanecine, 370
Retronecine, 1186

Crotalaria anagyroides
Anacrotine, 84

Crotalaria brevifolia
Integerrimine, 775
Usaramine, 1390

Crotalaria crispata
Fulvine, 615
Monocrotaline, 977

Crotalaria dura
Dicrotaline, 464

Crotalaria fulva
Fulvine, 615

Crotalaria globifera
Dicrotaline, 464

Crotalaria incana
Anacrotine, 84
Usaramine, 1390

Crotalaria incarnata
Integerrimine, 775

Crotalaria intermedia
Usaramine, 1390

Crotalaria juncea
Riddelline, 1199
Seneciphylline, 1252

Crotalaria laburnifolia
Anacrotine, 84
Senkirkine, 1254

Crotalaria madurensis
Fulvine, 615

Crotalaria mucronata
Usaramine, 1390

Crotalaria paniculata
Fulvine, 615

Crotalaria paulina
Monocrotaline, 977

Crotalaria quinquefolia
Monocrotaline, 977

Crotalaria spartioides
Retrorsine, 1187

Crotalaria stipularia
Monocrotaline, 977

Crotalaria usaramoensis
Retrorsine, 1187
Usaramine, 1390

Croton balsamifera
(+)-Salutaridine, 1225

Croton flavens
12-*O*-Palmitoyl-16-hydroxyphorbol 13-acetate, 1059

Croton salutaris
(+)-Salutaridine, 1225

Croton tiglium
Crotin, 371
Phorbol 12-tiglate 13-decanoate, 1096
12-Tetradecanoylphorbol 13-acetate, 1318

Cryptocarya pleurosperma
Cryptopleurine, 374

Cryptolepsis sanguinolenta
Cryptolepine, 373

Cryptolepsis triangularis
Cryptolepine, 373

Cucumis spp.
Cucurbitacin D, 379

Cucumis africanus
Cucurbitacin B, 378

Cucumis hookeri
Cucurbitacin A, 377

Cucumis leptodermis
Cucurbitacin A, 377

Cucumis myriocarpus
Cucurbitacin A, 377

Cudrania cochinchinensis
Cudraisoflavone A, 385

Cuminum cyminum
p-Cymene, 406
β-Pinene, 1111

Curcuma spp.
Zingiberene, 1453

Curcuma amada
Curcumin, 392

Curcuma aromatica
Curcumin, 392

Curcuma longa
Curcumin, 392

Curcuma xanthorrhiza
Curcumin, 392

Cuscuta spp.
Agroclavine, 40

Cusparia febrifuga
Galipine, 623

Cycas spp.
Cycasin, 396

Cycas circinalis
Cycasin, 396
3-Methylamino-L-alanine, 933

Cycas media
Macrozamin, 900

Cycas revoluta
Cycasin, 396

Cyclamen europaeum
Cyclamin, 397

Cyclamen purpurascens
Cyclamin, 397

Cyclea peltata
(+)-Tetrandine, 1322

Cylindrospermopsis raciborskii
Cylindrospermopsin, 404

Cymbogon nardus
Citronellal, 326

Cymbopogon spp.
Dipentene, 501

Cymbopogon densiflorus
Diosphenol, 498

Cymbopogon procerus
Elemicin, 528

Cymopholia barbata
Debromoisocymobarbatol, 418

Cymopterus longipes
Prangolarine, 1133

Cynanchum africanum
Cynafoside B, 407

Cynanchum vincetoxicum
Tylophorine, 1381

Cynara cardunculus
Cynaropicrin, 408

Cynara scolymus
Cynaropicrin, 408
Grosshemin, 690

Cynoglossum australe
Amabiline, 67
Heliosupine, 711

Cynoglossum officinale
Heliosupine, 711
Heliotridine, 712

Cynoglossum pictum
Heliosupine, 711

Cyperus iria
Juvenile hormone III, 819

Cypholophus friesianus
O-Acetylcypholophine, 17

Cypripedium calceolus
Cypripedin, 409

Cyrtomium spp.
Farrerol, 596

Cystophora moniliformis
Geranylacetone, 641

Cytisus spp.
Anagyrine, 85
Cytisine, 410
(−)-Sparteine, 1277

Cytisus laburnum
Caulophylline, 283

Cytisus scoparius
Dopamine, 514
13-Hydroxylupanine, 747
Lupanine, 881

D

Dahlia spp.
1-Phenylhepta-1,3,5-triyne, 1092

Dahlia coccinea
Ichthyotherol, 767

Dalbergea spruceata
Elemicin, 528

Dalbergia spp.
Biochanin A, 190

Dalbergia variabilis
(−)-Medicarpin, 915
Vestitol, 1412

Daphnandra aromatica
Aromoline, 132

Daphnandra micrantha
Daphnoline, 414

Daphnandra tennipes
Aromoline, 132

Daphne mezereum
Daphnetoxin, 413
Mezerein, 955

Daphne tangutica
Daphneticin, 412

Daphniphyllum macropodum
Asperuloside, 141

Datura spp.
Anisodamine, 96

Datura innoxia
Hyoscine, 758
Tigloidine, 1337

Datura metel
Hyoscine, 758

Datura pruinosa
Apoatropine, 111

Datura stramonium
Atropine, 151
Hyoscyamine, 759

Daucus carota
Bergamottin, 181
Elemicin, 528
Falcarindiol, 592
Falcarinol, 593
6-Methoxymellein, 930
Methylisoeugenol, 941

Davallia spp.
Vicianin, 1413

Decodon verticillatus
Cryogenine, 372

Deidamia clematoides
Deidaclin, 428

Delonix regia
L-Azetidine-2-carboxylic acid, 165

Delphinium spp.
Anthranoyllycoctonine, 101
Deltaline, 434
Denudatine, 443
Methyl-lycaconitine, 943

Delphinium ajacis
Ajaconine, 42
Delcosine, 431

Delphinium andersonii
14-Deacetylnudicauline, 416
Nudicauline, 1024

E

Elephantopus elatus
Elephantin, 529
Elephantopin, 530

Elephantopus mollis
Molephantin, 972
Molephantinin, 973
Phantomolin, 1085

Elephantopus scaber
Deoxyelephantopin, 445

Elscholtzia nipponica
Phenol, 1090

Embelia ribes
Embelin, 532

Emilia sonchifolia
Senkirkine, 1254

Encelia farinosa
Encelin, 536
Farinosin, 595

Encelia ventuorum
6-Hydroxytremetone, 752

Encelia virginensis
Encelin, 536
Farinosin, 595

Encephalartos spp.
Macrozamin, 900

Encyothalia cliftonii
2,4-Bis(prenyl)phenol, 193

Enhydra fluctuans
Enhydrin, 537

Enicostemma hyssopifolium
Gentianaine, 635

Entandophragma cylindricum
Sapelin A, 1232

Ephedra spp.
D-Cathine, 280
L-Ephedrine, 538

Ephedra equisetina
L-Ephedrine, 538

Ephedra gerardiana
L-Ephedrine, 538

Ephedra sinica
L-Ephedrine, 538

Equisetum arvense
Nicotine, 1011
Palustrine, 1060

Equisetum limosum
Palustrine, 1060

Equisetum palustre
Palustrine, 1060

Equisetum ramossissimum
Palustrine, 1060

Equisetum silvaticum
Palustrine, 1060

Eranthis pinnatifida
Cimifugin, 317

Eremanthus bicolor
Eremantholide A, 544

Eremanthus elaeagnus
Eremanthine, 543
Eremantholide A, 544

Eremanthus goyazensis
Goyazensolide, 682

Eremanthus incanus
Eremanthine, 543
Eremantholide A, 544

Erigeron spp.
Dillapiole, 485

Erigeron philadelphicus
Methyl 2-*trans*, 8-*cis*-matricarate, 944

Eriobotrya japonica
Eriobofuran, 549

Eriophyllum lanatum
Eriolangin, 553

Eriophyllum stachaedifolium
Eupatoriopicrin, 583

Eriostemon tomentellus
Ostruthin, 1044

Erysimum cheiranthoides
Helveticoside, 717

Erysimum crepidifolium
Helveticoside, 717

Erysimum helveticum
Helveticoside, 717

Erythrina spp.
Erysonine, 555
Erysotrine, 556
Erythratidine, 557
α-Erythroidine, 558
β-Erythroidine, 559

Erythrina abyssinica
Abyssinone VI, 3

Erythrina americana
α-Erythroidine, 558
β-Erythroidine, 559

Erythrina berteroana
5′-Prenylhomoeriodictyol, 1138

Erythrina caribea
Erysonine, 555
Erythratidine, 557

Erythrina hypaphorus
Hypaphorine, 761

Erythrina melanacantha
Erysonine, 555
Erythratidine, 557

Erythrina suberosa
Erysotrine, 556

Erythrophleum chlorostachys
Cassaine, 268
Norerythrostachaldine, 1019

Erythrophleum guineense
Cassaidine, 267
Cassaine, 268
Erythrophleguine, 560

Erythroxylum spp.
Cocaine, 332
Ecgonine, 519

Erythroxylum coca
Benzoyltropein, 177
Cinnamoylcocaine, 323
Cocaine, 332
Ecgonine, 519
Geraniin, 640

Erythroxylum truxillense
Cinnamoylcocaine, 323
Tropacocaine, 1364

Escallonia spp.
Asperuloside, 141
Chrysin, 309
Galangin, 620

Escholtzia spp.
Allocryptopine, 61

Eschscholzia californica
Dihydrosanguinarine, 478

Esenbeckia leiocarpa
Leiokinine A, 854

Eucalyptus spp.
Astringin, 146
1,8-Cineole, 319
Citronellal, 326

Eucalyptus citriodora
Citronellal, 326

Eucalyptus cladocalyx
Prunasin, 1152

Eucalyptus globulus
1,8-Cineole, 319

Eucalyptus hemiphloia
(−)-Eudesmin, 565

Eucalyptus viminalis
Casuarictin, 274

Euclea spp.
Diospyrin, 499
Isodiospyrin, 791

Euodia spp.
Kokusaginine, 828

Euodia xanthoxyloides
Arborinine, 116
Evoxine, 588

Euonymus alatus
Wilfordine, 1432

Euonymus europaea
Armepavine, 127
Evomonoside, 587

Eupatorium altissimum
Rinderine, 1201

Eupatorium cannabinum
Eupatolide, 582
Eupatoriopicrin, 583
Rinderine, 1201
Supinine, 1294

Eupatorium compositifolium
Lycopsamine, 891

Eupatorium cuneifolium
Eupacunin, 571
Eupacunolin, 572
Eupacunoxin, 573
Eupaserrin, 579
Eupatocunin, 580
Eupatocunoxin, 581

Eupatorium formosanum
Eupaformonin, 574
Eupaformosanin, 575
Eupatolide, 582

Eupatorium hyssopifolium
Eupahyssopin, 576

Eupatorium lancifolium
Eupacunin, 571
Eupacunolin, 572

Eupatorium mikanioides
Desacetyleupaserrin, 455

Eupatorium riparium
Methylripariochromene A, 951

Eupatorium rotundifolium
Eupachlorin, 568
Eupachlorin acetate, 569
Eupachloroxin, 570
Euparotin, 577
Euparotin acetate, 578
Eupatoroxin, 584
Eupatundin, 585

Eupatorium rugosum
Tremetone, 1347

Eupatorium sachalinense
Hiyodorilactone A, 730

Eupatorium semiserratum
Desacetyleupaserrin, 455
Eupaserrin, 579

Eupatorium serotinum
Supinine, 1294

Eupatorium stoechadosum
Supinine, 1294

Eupatorium tinifolium
Acacetin, 4

Eupatorium urticaefolium
Dehydrotremetone, 426
6-Hydroxytremetone, 752
Tremetone, 1347

Euphorbia esula
Ingenol 3,20-dibenzoate, 774

Euphorbia fortissima
Diterpenoid EF-D, 505

Euphorbia kansui
Kansuinine B, 822

Euphorbia lathyrus
Lathyrol, 849

Euphorbia milii
Milliamine L, 965

Euphorbia poisonii
Candletoxin A, 250
Resiniferatoxin, 1185
Tinyatoxin, 1339

Euphorbia resinifera
Resiniferatoxin, 1185

Euphorbia tirucalli
Sapatoxin A, 1231

Evodia spp.
Berberine, 180

Evodia hupehensis
Atanine, 147

Evodia rutaecarpa
Atanine, 147

F

Fagara spp.
Candicine, 248
Nitidine, 1013
Xanthotoxin, 1441

Fagara chalybaea
N-Methylflindersine, 939

Fagara lepieurii
Arborinine, 116

Fagara macrophylla
Fagaramide, 589

Fagara xanthoxyloides
Atanine, 147
Fagaramide, 589

Fagus sylvatica
(+)-Syringaresinol, 1300

Falcaria vulgaris
Falcarindiol, 592
Falcarinol, 593
Falcarinone, 594

Farfugium japonicum
Senkirkine, 1254

Ferreirea spectabilis
Chrysarobin, 307

Ferula spp.
Isopimpinellin, 801

Ferula communis
Ferprenin, 600
Ferulenol, 601

Ficus carica
Psoralen, 1157

Ficus nitida
Angelicin, 90

Ficus septica
Antofine, 105
Ficuseptine, 603
(−)-Tylocrebine, 1380
Tylophorine, 1381

Filipendula ulmaria
Rugosin D, 1210

Foeniculum vulgare
Anethole, 89
α-Phellandrene, 1087

Forsythia spp.
Verbascoside, 1399

Fragaria spp.
Imperatorin, 770

Fraxinus spp.
Cichoriin, 314

Frullania dilatata
(+)-*cis*-β-Cyclocostunolide, 400
Eremofrullanolide, 545
(−)-Frullanolide, 614
(+)-Oxyfrullanolide, 1055

Frullania nisquallensis
(+)-*cis*-β-Cyclocostunolide, 400
Eremofrullanolide, 545
(+)-Oxyfrullanolide, 1055

Frullania tamarisci
(+)-*cis*-β-Cyclocostunolide, 400
Eremofrullanolide, 545
(−)-Frullanolide, 614
(+)-Oxyfrullanolide, 1055

Fucus vesiculosus
Phloroglucinol trimer, 1094

Fumaria officinalis
Bulbocapnine, 216
Sanguinarine, 1227

Fumaria parviflora
Dihydrosanguinarine, 478

Fumaria vaillantii
Dihydrosanguinarine, 478

Funtumia elastica
Funtumine, 616

Funtumia latifolia
Funtumine, 616

G

Gabunia odoratissima
Gabunine, 618

Gaillardia spp.
Helenalin, 708

Gaillardia fastigiata
Fastigilin B, 597
Fastigilin C, 598

Gaillardia pinnatifida
Mexicanin I, 954

Gaillardia pulchella
Gaillardin, 619
Methyl caffeate, 935

Galanthus spp.
Galanthamine, 621

Galanthus nivalis
Narwedine, 1004

Galega officinalis
Galegine, 622

Galipea officinalis
Cusparine, 393

Galium spp.
Alizarin, 55
Pseudopurpurin, 1155

Galium aparine
Asperuloside, 141

Gambierdiscus toxicus
Ciguatoxin, 316

Gardenia spp.
Acerosin, 10
Geniposide, 632

Gardenia jasminoides
Gardenoside, 625

Garrya laurifolia
Cuauchichicine, 375
Veatchine, 1397

Garrya ovata
Cuauchichicine, 375
Ovatine, 1048
Veatchine, 1397

Garrya veatchii
Garryine, 626
Veatchine, 1397

Gastrolobium spp.
Monofluoroacetic acid, 978

Gaultheria spp.
Catechol, 278

Gaultheria procumbens
Monotropitoside, 979
Salicylic acid, 1222

Geigeria africana
Dihydrogriesenin, 474
Geigerin, 627
Vermeerin, 1402

Geigeria aspera
Dihydrogriesenin, 474
Geigerin, 627
Vermeerin, 1402

Geigeria filifolia
Dihydrogriesenin, 474

Gelsemium sempervirens
Gelsemicine, 628
Gelsemine, 629
12β-Hydroxypregna-4,16-diene-3,20-dione, 749
Sempervirine, 1248

Genipa americana
Genipin, 631

Genista spp.
Anagyrine, 85
Cytisine, 410

Genista cinerea
13-Hydroxylupanine, 747

Genista hystrix
Ammodendrine, 77

Gentiana spp.
Gentianine, 637
Amarogentin, 69

Gentiana cruciata
Gentianaine, 635

Gentiana kaufmanniana
Gentianaine, 635

Gentiana kirilowi
Gentianine, 637

Gentiana lactea
Bellidifolin, 176

Gentiana lutea
Amarogentin, 69
Gentiopicrin, 638
Gentisein, 639

Gentiana olgae
Gentianadine, 634
Gentianaine, 635

Gentiana olivieri
Gentianadine, 634
Gentianaine, 635
Gentianamine, 636

Gentiana turkestanorum
Gentianadine, 634
Gentianaine, 635
Gentianamine, 636

Geranium spp.
Geraniin, 640

Geranium macrorrhizum
Germacrone, 643

Gerbera jamesonii
Amygdalin, 81
Vicianin, 1413

Geum japonicum
Gemin A, 630

Ginkgo biloba
Bilobol, 189
(15:1)-Cardanol, 259
Ginkgoic acid, 648
Ginkgolide A, 649
4′-*O*-Methylpyridoxine, 949

Glaucium spp.
Allocryptopine, 61

Glaucium flavum
Bulbocapnine, 216
Glaucine, 655

Glaucium pulchrum
Bulbocapnine, 216

Glaux maxima
Primin, 1141

Gloriosa superba
Colchicine, 337

Glycine max
Cadaverine, 225
Genistein, 633
(−)-Glyceollin I, 670
(−)-Glyceollin II, 671

Glycosmis arborea
Arborine, 115
Arborinine, 116

Glycosmis cyanocarpa
Sinharine, 1266

Glycosmis mauritania
Methylillukumbin A, 940

Glycyrrhiza glabra
L-Aspartic acid, 139
Formononetin, 610

Glycyrrhiza lepidota
Glabranin, 652

Gnetum gnemon
Malvalic acid, 907
Sterculic acid, 1286

Gnidia spp.
Gnidilatin, 675

Gnidia lamprantha
Gnidicin, 673
Gnididin, 674
Gniditrin, 676

Gochnatra rusbyana
Methyl caffeate, 935

Gomphocarpus fruticosus
Gofruside, 677

Goniodoma pseudogoniaular
Goniodomin A, 678

Goniothalamus giganteus
Goniothalenol, 679

Gonyaulax spp.
Gonyautoxin I, 680

Goodia latifolia
p-Glucosyloxymandelonitrile, 663

Gossypium spp.
Gossypol, 681

Gossypium hirsutum
Lacinilene C 7-methyl ether, 833
Sterculic acid, 1286

Gratiola officinalis
Cucurbitacin E, 380
Cucurbitacin I, 381

Grevillea robusta
Grevillol, 689

Griffonia simplicifolia
5-Hydroxy-L-tryptophan, 753

Grindelia spp.
Tremetone, 1347

Grossheimia macrocephala
Grosshemin, 690

Guaiacum spp.
Guaiacol, 691

Guaiacum officinale
Guaiazulene, 692
Nordihydroguaiaretic acid, 1018

Guaiacum sanctum
Guaiazulene, 692
Nordihydroguaiaretic acid, 1018

Guatteria megalophylla
Isochondrodendrine, 787

Gymnodinium breve
Brevetoxin A, 201
Brevetoxin B, 202

Gynocardia odorata
Gynocardin, 693

Gypsophila spp.
Gypsogenin, 694

Gyrocarpus americanus
Gyrocarpine, 695

H

Haemanthus amarylloides
Montanine, 980

Haemanthus coccineus
Montanine, 980

Haemanthus kalbreyeri
Narciclasine, 1000

Haemanthus montanus
Montanine, 980

Haematoxylon campechianum
Haematoxylin, 696

Haemodorum corymbosum
Haemocorin, 697

Haloragis erecta
Lotaustralin, 875

Haloxylon articulatum
(±)-Carnegine, 262

Haloxylon salicornicum
(±)-Carnegine, 262

Handelia trichophylla
Canin, 251

Hannoa undulata
Undulatone, 1382

Haploclathra paniculata
Gentisein, 639

Haplopappus heterophyllus
Dehydrotremetone, 426
Toxol, 1345
Tremetone, 1347

Haplophragma adenophyllum
Lapachol, 842
β-Lapachone, 843

Haplophyllum spp.
Dictamnine, 465
Evoxine, 588
Kokusaginine, 828

Haplophyllum dubium
Dubinidine, 517

Haplophyllum glabrinum
Haplophyllidine, 698

Haplophyllum hispanicum
Diphyllin, 502

Haplophyllum perforatum
Haplophyllidine, 698

Haplophyllum tuberculatum
Justicidin B, 816

Harrisonia abyssinica
Harrisonin, 703

Harrisonia perforata
5-*O*-Methylalloptaeroxylin, 932

Hedeoma pulegioides
Pulegone, 1163

Hedera helix
Falcarinol, 593
Hederagenin 3-*O*-arabinoside, 704
α-Hederin, 706
β-Hederin, 707

Hedranthera barteri
Voacamine, 1425

Hedychium spicatum
p-Methoxycinnamic acid ethyl ester, 928

Hedysarum polybotris
Sainfuran, 1218

Heimia myrtifolia
Cryogenine, 372

Heimia salicifolia
Cryogenine, 372

Helenium spp.
Helenalin, 708
Mexicanin E, 953
Mexicanin I, 954
Tenulin, 1315

Helenium alternifolium
Brevilin A, 204
Linifolin A, 864

Helenium amarum
Amaralin, 68
Aromaticin, 131
Tenulin, 1315

Helenium arizonicum
Isotenulin, 803

Helenium aromaticum
Aromaticin, 131
Linifolin A, 864

Helenium autumnale
Autumnolide, 157
Florilenalin, 605
Helenalin, 708
Mexicanin I, 954
Plenolin, 1120
Tenulin, 1315

Helenium bigelovii
Isotenulin, 803

Helenium elegans
Tenulin, 1315

Helenium linifolium
Linifolin A, 864

Helenium mexicanum
Mexicanin E, 953

Helenium microcephalum
Isohelenol, 797
Microhelenin-A, 961
Microhelenin C, 962
Microlenin, 963

Helenium plantagineum
Linifolin A, 864

Helenium puberulum
Tenulin, 1315

Helenium quadridentatum
Helenalin, 708

Helenium scorzoneraefolia
Linifolin A, 864

Helenium tenuifolium
Tenulin, 1315

Helianthus spp.
Desacetyleupaserrin, 455

Helianthus agrophyllus
Eupatolide, 582

Helianthus annuus
Ayapin, 163
Helianthoside A, 709
Niveusin C, 1014

Helianthus ciliaris
Calaxin, 231

Helianthus maximiliani
Niveusin C, 1014

Helianthus niveus
Niveusin C, 1014

Helianthus strumosus
Acerosin, 10

Helichrysum spp.
Pinocembrin, 1112

Helietta longifolia
Heliettin, 710

Heliopsis longipes
Affinin, 35

Heliotropium spp.
Heliotridine, 712
Heliotrine, 713

Heliotropium amplexicaule
Indicine, 772

Heliotropium arbainense
Europine, 586
Heliotrine, 713
Lasiocarpine, 847

Heliotropium curassavicum
Heliotrine, 713

Heliotropium europaeum
Europine, 586
Heliotrine, 713
Lasiocarpine, 847

Heliotropium hirsutum
Lasiocarpine, 847

Heliotropium indicum
Heliotrine, 713
Indicine, 772
Supinine, 1294

Heliotropium lasiocarpum
Heliotrine, 713
Lasiocarpine, 847

Heliotropium maris-mortui
Europine, 586

Heliotropium rotundifolium
Europine, 586

Heliotropium steudneri
Lycopsamine, 891

Heliotropium supinum
Heliosupine, 711
Supinine, 1294

Helleborus spp.
Hellebrin, 715

Helleborus niger
Hellebrin, 715

Hemizonia congesta
6-Hydroxytremetone, 752

Heracleum spp.
Angelicin, 90
Bergapten, 182
Imperatorin, 770
Isopimpinellin, 801

Heracleum lanatum
Xanthotoxol, 1442

Heracleum nepalense
Alloimperatorin, 62

Heracleum sphondylium
Xanthotoxin, 1441

Heracleum wallichi
Isochondrodendrine, 787

Heritonia littoralis
Heritonin, 720

Hermidium alipes
Dopamine, 514

Hernandia ovigera
Deoxypodophyllotoxin, 453

Heterodendron oleaefolium
Cardiospermin, 260
Heterodendrin, 724

Hibiscus syriacus
Malvalic acid, 907

Hibiscus tiliaceus
Lapachol, 842

Himantandra baccata
Himbacine, 726

Himantandra belgraveana
Himbacine, 726

Hippeastrum spp.
Ambelline, 71
Galanthamine, 621
Hippeastrine, 727

Hippeastrum candidum
Candimine, 249

Hippomane mancinella
Mancinellin, 909

Histiopteris incisa
Ptaquiloside, 1160

Holarrhena spp.
Conessine, 342

Holarrhena antidysenterica
Conessine, 342

Holarrhena congolensis
Funtumine, 616

Holarrhena febrifuga
Conessine, 342
Funtumine, 616

Holarrhena pubescens
Conessine, 342

Holocalyx balansae
Prunasin, 1152
Zierin, 1452

Holocalyx glaziovii
Prunasin, 1152

Homalanthus nutans
Prostratin, 1147

Homogyne alpina
Bakkenolide A, 170

Hordeum vulgare
Candicine, 248
Gramine, 685
Heterodendrin, 724
Hordenine, 732
Malonic acid, 906

Humbertia madagascariensis
(−)-Eudesmin, 565

Humulus lupulus
6-Isopentenylnaringenin, 800
Isoxanthohumol, 805
Lupulone, 885
Xanthohumol, 1440

Hunteria eburnea
Eburnamonine, 518

Hura crepitans
Huratoxin, 733

Hydnocarpus wightiana
Chaulmoogric acid, 300

Hydrangea spp.
Febrifugine, 599
Loganin, 871

Hydrangea macrophylla
Hydrangenol, 734

Hydrangea umbellata
Isofebrifugine, 796

Hydrastis canadensis
Berberastine, 179

Hymenocallis spp.
Galanthamine, 621

Hymenoclea spp.
Ambrosin, 72

Hymenoclea salsola
Coronopilin, 358
Hymenolin, 756

Hymenodictyon excelsum
Anthrogallol, 102

Hymenoxys spp.
Paucin, 1070

Hymenoxys acaulis
Fastigilin C, 598

Hymenoxys anthemoides
Vermeerin, 1402

Hymenoxys grandiflora
Hymenoflorin, 755

Hymenoxys linearis
Mexicanin I, 954

Hymenoxys odorata
Hymenoxon, 757

Hymenoxys richardsonii
Hymenoxon, 757
Vermeerin, 1402

Hyoscyamus muticus
Hyoscyamine, 759

Hyoscyamus niger
Hyoscine, 758
Hyoscyamine, 759

Hyoscyamus orientalis
Apoatropine, 111

Hypericum spp.
Hypericin, 763

Hypericum brasiliense
Hyperbrasilone, 762

Hypericum chinense
Chinensin I, 302
Chinensin II, 303

Hypericum degenii
Gentisein, 639

Hypericum drummondii
Drummondin A, 516

Hypericum japonicus
Sarothralin, 1236

Hypericum perforatum
Hypericin, 763

Hypolepsis punctata
Ptaquiloside, 1160

Hyptis verticillata
4′-Demethyldeoxypodophyllotoxin, 439

I

Iberis spp.
Cucurbitacin B, 378
Cucurbitacin D, 379
Cucurbitacin E, 380

Iberis amara
Cucurbitacin I, 381

Ichthyothere terminalis
Ichthyotherol, 767

Ilex chinensis
Ilexolide A, 769

Ilex paraguayensis
Caffeine, 227

Ilex pubescens
Ilexolide A, 769

Illicium anisatum
Anethole, 89
Anisatin, 95

Illicium religiosum
Safrole, 1215
Shikimic acid, 1260

Impatiens balsamina
Lawsone, 852

Impatiens glandulifera
2-Methoxy-1,4-naphthoquinone, 931

Indigofera spp.
L-Indospicine, 773

Indigofera linnaei
L-Indospicine, 773

Indigofera spicata
L-Indospicine, 773

Inula spp.
Gaillardin, 619
Isoalantolactone, 783
Ivalin, 806

Inula britannica
Gaillardin, 619

Inula grandis
Alantolactone, 49

Inula helenium
Alantolactone, 49
Isoalantolactone, 783

Inula japonica
Inulicin, 777

Inula magnifica
Alantolactone, 49

Inula royleana
Anthranoyllycoctonine, 101
Lycoctonine, 889

Iphiona aucheri
Carboxyatractyloside, 258

Ipomoea spp.
Agroclavine, 40
Lysergol, 896

Ipomoea sp.
Calystegin B_2, 239

Ipomoea argyrophylla
Chanoclavine-I, 297
Ergine, 546
Ergometrine, 547
Ergosine, 548

Ipomoea batatas
Ipomeamaronol, 778
Ipomoeamarone, 779

Ipomoea cairica
(−)-Arctigenin, 120
Trachelogenin, 1346

Ipomoea parasitica
Lysergol, 896

Ipomoea tricolor
Chanoclavine-I, 297
Ergine, 546

Ipomoea violacea
Chanoclavine-I, 297
Ergine, 546

Iris spp.
Acetovanillone, 12

Isodon shikokianus
Isodomedin, 792
Shikodonin, 1261

Isolona pilosa
(−)-Caaverine, 224

Isotoma longiflora
Lobelanidine, 869

Isotropis forrestii
Iforrestine, 768

Iva spp.
Coronopilin, 358
Ambrosin, 72

Iva acerosa
Acerosin, 10

Iva imbricata
Ivalin, 806

Iva microcephala
Ivalin, 806

Iva nevadensis
Parthenin, 1067

J

Jateorrhiza palmata
Jatrorrhizine, 810
Palmatine, 1058

Jatropha curcas
Curcin, 391

Jatropha gossypiifolia
Jatrophone, 809

Jatropha macrorhiza
Jatrophatrione, 808

Jatropha natalensis
Curcin, 391

Juglans nigra
Juglone, 814

Juglans regia
Juglone, 814

Juniperus spp.
Sabinol, 1213

Juniperus rigida
Anethole, 89

Juniperus sabina
Deoxypodophyllotoxin, 453
β-Peltatin A methyl ether, 1076
Podophyllotoxin, 1126
Sabinol, 1213

Juniperus virginiana
Podophyllotoxin, 1126

Jurinea alata
Alatolide, 50

Jurinea maxima
Salonitenolide, 1223

Justicia hayatai
Diphyllin, 502
Justicidin B, 816

Justicia procumbens
Justicidin A, 815

Justicia prostrata
Carpacin, 264

Justinia hayatai
Justicidin A, 815

K

Kaempferia galangae
p-Methoxycinnamic acid ethyl ester, 928

Kalanchoe lanceolata
Lanceotoxin A, 837
Lanceotoxin B, 838

Kalmia spp.
Grayanotoxin I, 688

Kalmia latifolia
Phloridzin, 1093

Kalopanax septemlobum
α-Hederin, 706

Karwinskia humboldtiana
Karwinskione, 825
Tullidinol, 1375

Karwinskia johnstonii
Tullidinol, 1375

Karwinskia mollis
Tullidinol, 1375

Karwinskia subcordata
Tullidinol, 1375

Karwinskia umbelluta
Tullidinol, 1375

Kielmeyera speciosa
6-Deoxyjacareubin, 446

Kigelia pinnata
Kigelinone, 827
Lapachol, 842

Kochia scoparia
Oxalic acid, 1049

Krameria cystisoides
Conocarpan, 346

L

Lablab niger
Vignafuran, 1414

Laburnum anagyroides
Cytisine, 410
Luteone, 887
Piceatannol, 1105
Wighteone, 1431

Lactuca canadensis
Lactucin, 834
Lactucopicrin, 835

Lactuca sativa
Lettucenin A, 856

Lactuca serriola
8-Deoxylactucin, 447
Lactucin, 834
Lactucopicrin, 835

Lactuca virosa
Lactucin, 834
Lactucopicrin, 835

Lagerstroemia fauriei
Cryogenine, 372

Lantana camara
Lantadene A, 839
Lantadene B, 840

Laportea moroides
Moroidin, 982

Lappa minor
(−)-Arctigenin, 120

Lappa tomentosa
(−)-Arctigenin, 120

Lappula intermedia
Lasiocarpine, 847

Larrea spp.
Nordihydroguaiaretic acid, 1018

Larrea tridentata
Nordihydroguaiaretic acid, 1018

Laser trilobum
Laserolide, 846
Trilobolide, 1358

Laserpitium siler
Gradolide, 684
Isomontanolide, 798
Polhovolide, 1128

Lathyrus spp.
L-α-Amino-β-oxalylaminopropionic acid, 76
L-Homoarginine, 731

Lathyrus cicera
L-Homoarginine, 731

Lathyrus hirsutus
Lathodoratin, 848

Lathyrus latifolius
L-α-Amino-γ-oxalylaminobutyric acid, 75
L-α, γ-Diaminobutyric acid, 460

Lathyrus odoratus
Lathodoratin, 848
Odoratol, 1026

Lathyrus sativus
L-α-Amino-β-oxalylaminopropionic acid, 76
Cadaverine, 225
L-Homoarginine, 731

Lathyrus sylvestris
L-α, γ-Diaminobutyric acid, 460

Laurelia serrata
Dillapiole, 485

Laurencia pacifica
Aplysin, 110

Laurentia longiflora
Lobelanidine, 869

Laurus nobilis
Costunolide, 359

Lavandula spp.
Camphor, 243

Lavandula spica
Borneol, 195

Lawsonia alba
Lawsone, 852

Lawsonia inermis
Lawsone, 852

Lecythis ollaria
L-Selenocystathionine, 1247

Ledebouriella seseloides
Cimifugin, 317

Ledum groenlandicum
Germacrone, 643

Ledum palustre
Ledol, 853
Palustrol, 1061

Lens culinaris
Dihydrowyerone, 479

Lepidium apetalum
Evomonoside, 587

Lepidium sativum
Glucolepidiin, 661
Gluconasturtiin, 662

Leucaena leucophylla
L-Mimosine, 966

Leucojum spp.
Galanthamine, 621

Leucojum aestivum
Pretazettine, 1139

Leucothoe spp.
Grayanotoxin I, 688

Levisticum spp.
Bergapten, 182

Liatris spp.
Tremetone, 1347

Liatris chapmanii
Liatrin, 859

Liatris cylindrica
Isoalantolactone, 783

Liatris elegans
Eleganin, 527

Liatris graminifolia
Graminiliatrin, 686

Liatris provincialis
Provincialin, 1151

Liatris pycnostachya
Spicatin, 1280

Liatris scabra
Eleganin, 527

Liatris spicata
Spicatin, 1280

Liatris tenuifolia
Spicatin, 1280

Libanotis transcaucasica
Athamantin, 148
Isoelemicin, 795

Libertia coerulescens
Alizarin, 55

Libocedrus spp.
Deoxypodophyllotoxin, 453

Ligularia spp.
Tremetone, 1347

Ligularia calthaefolia
Bakkenolide A, 170

Ligularia clivorum
Clivorine, 330

Ligularia dentata
Clivorine, 330

Ligularia elegans
Clivorine, 330

Ligularia intermedia
6-Hydroxytremetone, 752

Ligusticum spp.
Bergapten, 182

Ligusticum acutilobum
Isosafrole, 802

Ligusticum scotinum
Dillapiole, 485

Ligusticum wallichii
3-Butylidene-7-hydroxyphthalide, 219

Ligustrum vulgare
Cinchonidine, 318

Lilium candidum
Cinnamic acid, 322

Lilium maximowczii
Yurinelide, 1449

Linaria vulgaris
Acacetin, 4

Lindackeria dentata
Chaulmoogric acid, 300

Lindelofia spectabilis
Monocrotaline, 977

Lindera triloba
Shiromodiol diacetate, 1263

Linum usitatissimum
Coniferyl alcohol, 344
Linamarin, 863
Linustatin, 865
Lotaustralin, 875
Neolinustatin, 1006

Liparis auriculata
Auriculine, 154

Liparis loeselii
Auriculine, 154

Lippia dulcis
Hernandulcin, 722

Lippia graveolens
Lapachenole, 841

Lippia rehmanni
Lantadene A, 839

Liquidambar orientalis
Cinnamic acid, 322

Liriodendron tulipifera
(−)-Caaverine, 224
Epitulipinolide, 539
Epitulipinolide diepoxide, 540
Lipiferolide, 866
Liriodenine, 867
1-Peroxyferolide, 1078
Taxiphyllin, 1308
Tulipinolide, 1371

Litchi sinensis
α-(Methylenecyclopropyl)glycine, 938

Lithospermum erythrorhizon
Shikonin, 1262

Lithospermum officinale
Lithospermic acid, 868

Lithospermum ruderale
Lithospermic acid, 868

Lobelia berlandieri
N-Methyl-2,6-bis(2-hydroxybutyl)-Δ^3-piperideine, 934

Lobelia hassleri
Lobelanidine, 869
(−)-Lobeline, 870

Lobelia inflata
Lobelanidine, 869
(−)-Lobeline, 870

Lobelia nicotianaefolia
(−)-Lobeline, 870

Lolium perenne
Norharman, 1020

Lomatia spp.
Juglone, 814

Lonchocarpus costaricensis
Deoxymannojirimycin, 450

Lonchocarpus sericeus
Deoxymannojirimycin, 450
DMDP, 511

Lonicera nigra
Hederagenin 3-*O*-arabinoside, 704

Lophophora williamsii
Lophophorine, 873
Mescaline, 923
N-Methylmescaline, 946

Loroglossum hircinum
Hircinol, 728
Loroglossol, 874

Lotus spp.
Vestitol, 1412

Lotus australis
Lotaustralin, 875

Lotus corniculatus
Linamarin, 863
Lotaustralin, 875

Lotus pedunculatus
Cibarian, 312

Lunasia amara
Lunacrine, 879
Lunamarine, 880

M

Mallotus japonicus
Butyrylmallotochromene, 222
Isobutyrylmallotochromene, 786
Mallotochromene, 904
Mallotophenone, 905

Mallotus philippensis
Rottlerin, 1208

Malus spp.
Ursolic acid, 1386

Malus domestica
Phloridzin, 1093

Mammea africana
Mammeisin, 908

Mammea longifolia
Surangin B, 1295

Mammea thwailesii
Mammeisin, 908

Mangifera indica
5-(Heptadec-12-enyl)resorcinol, 719

Manihot esculentum
Linamarin, 863

Mannozia maronii
Dehydrozaluzanin C, 427

Mansonia altissima
Mansonone C, 910

Mappia foetida
Camptothecin, 244

Marrabium vulgare
(−)-Betonicine, 185

Matayba arborescens
Cleomiscosin A, 328

Matricaria chamomilla
(+)-α-Bisabolol, 192
Chamazulene, 295
Guaiazulene, 692

Matricaria suffruticosa
Desacetoxymatricarin, 454

Maytenus spp.
Pristimerin, 1142

Maytenus krukorii
D-Cathine, 280
D-Cathinone, 281

Maytenus ovatus
Maytansine, 911

Maytenus senata
Maytansine, 911

Meconopsis cambrica
Mecambrine, 912

Medicago spp.
Coumestrol, 365

Medicago sativa
L-Arginine, 124
Medicagenic acid 3-glucoside, 913
Medicagenic acid 3-triglucoside, 914
(−)-Medicarpin, 915
Vestitol, 1412

Melaleuca bracteata
Elemicin, 528

Melaleuca leucadendron
1, 8-Cineole, 319

Melampodium americanum
Melampodinin, 918

Melampodium diffusum
Melampodinin, 918

Melampodium heterophyllum
Melampodin A, 917

Melampodium leucanthum
Melampodin A, 917

Melampodium longipes
Melampodinin, 918

Melampodium longipilum
Enhydrin, 537

Melampodium perfoliatum
Enhydrin, 537

Melia azedarach
Ohchinolide B, 1029
Vanillic acid, 1394

Melicope spp.
Kokusaginine, 828

Melicope fareana
Acronycidine, 23

Melicope leptococca
Acronycine, 24

Melilotus spp.
Dicoumarol, 463

Melissa officinalis
Citronellal, 326

Melochia pyramidata
Melochinine, 920

Melodinus balansae
Ajmaline, 44

Melodius spp.
Akuammidine, 47

Menispermum canadense
Dauricine, 415

Menispermum dauricum
Acutumidine, 28
Dauricine, 415

Mentha spp.
Limonene, 861
Menthol, 921
Pulegone, 1163

Mentha piperita
Menthol, 921

Mentha pulegium
Pulegone, 1163

Mentha rotundifolia
Diosphenol, 498

Menyanthes spp.
Deoxyloganin, 449

Menyanthes trifoliata
Loganin, 871

Mesembryanthemum nodiflorum
Oxalic acid, 1049

Mespilus germanica
α-Cotonefuran, 360
6-Hydroxy-α-pyrufuran, 750

Mesua ferrea
Dehydrocycloguanandin, 421

Michelia spp.
Jatrorrhizine, 810

Michelia champaca
Parthenolide, 1068

Michelia compressa
Michelenolide, 956
Micheliolide, 957
Santamarin, 1228

Michelia lanuginosa
Parthenolide, 1068

Miconia spp.
Primin, 1141

Micranda spruceana
Spruceanol, 1284

Microcystis spp.
Microcystin LA, 959

Microcystis aeruginosa
Microcystin LA, 959
Microcystin LR, 960

Microcystis viridis
Cyanoviridin RR, 395
Microcystin LR, 960

Microcystis wesenbergii
Microcystin LR, 960

Micromelum integerrimum
Micromelin, 964

Micromelum minutum
Micromelin, 964

Millettia ovalifolia
Lanceolatin B, 836

Mimosa spp.
L-Djenkolic acid, 510

Mimosa pudica
L-Mimosine, 966

Mitragyna speciosa
Mitragynine, 971

Mollugo hirta
Mollugogenol A, 974

Mollugo pentaphylla
Mollugogenol A, 974

Momordica charantia
Momordin, 976
Trichosanthin, 1350

Monarda punctata
Thymol, 1336

Monnieria trifolia
Evoxine, 588

Montanoa tomentosa
Zoapatanol, 1454

Montezuma speciosissima
Gossypol, 681

Morinda citrifolia
Alizarin, 55

Morinda parvifolia
Digiferrugineol, 468

Morithamnus crassua
Toxol, 1345

Morus spp.
p-Cresol, 366
Deoxynojirimycin, 451

Morus alba
Albanol A, 51
Deoxynojirimycin, 451
Kuwanone G, 829
Moracin A, 981

Morus lhou
Albanol A, 51

Mostuea brunosis
Gelsemicine, 628

Mostuea buchholzii
Sempervirine, 1248

Mucuna spp.
L-Dopa, 513

Mucuna birdwoodiana
2, 6-Dimethoxyphenol, 489

Mucuna pruriens
N,N-Dimethyltryptamine, 491
5-Hydroxy-L-tryptophan, 753

Murraya koenigii
Isosafrole, 802

Musa sapientum
Dopamine, 514
Histamine, 729
Phenethylamine, 1089

Myoporum deserti
Dehydromyodesmone, 423
Dehydrongaione, 424

Myoporum montanum
Myomontanone, 992

Myosotis scorpiodes
Symphytine, 1298

Myristica fragrans
Borneol, 195
Elemicin, 528
Isoelemicin, 795
Myristicin, 994
Safrole, 1215

Myrsine africana
Embelin, 532
Emodin, 534
Norobtusifolin, 1021

N

Nandina spp.
Berberine, 180

Nandina domestica
p-Glucosyloxymandelonitrile, 663
Nandinin, 996

Narcissus spp.
Galanthamine, 621
Lycorine, 894
Narciclasine, 1000

Narcissus pseudonarcissus
7,4′-Dihydroxyflavan, 480
7,4′-Dihydroxy-8-methylflavan, 484
7-Hydroxyflavan, 745

Narcissus tazetta
Pretazettine, 1139

Nasturtium officinale
Gluconasturtiin, 662

Nectandra rodioei
(+)-Bebeerine, 175

Nelumbo nucifera
Anonaine, 97
Armepavine, 127

Nemuaron humboldtii
Safrole, 1215

Neochamaelea pulverulenta
5-*O*-Methylalloptaeroxylin, 932

Neorautanenia amboensis
12a-Hydroxyrotenone, 751

Nepenthes spp.
Histamine, 729

Nepeta cataria
Nepetalactone, 1010

Nepeta leucophylla
1-*tert*-Butyl-4-methylbenzene, 221

Neptunia amplexicaulis
L-Selenocystathionine, 1247

Nerine crispa
3-Acetylnerbowdine, 18

Nerium indicum
Urechitoxin, 1384

Nerium oleander
Adynerin, 32

Nicotiana spp.
Nicotine, 1011
(−)-Anabasine, 82

Nicotiana sp.
Sclareol, 1242

Nicotiana acuminata
(−)-Anabasine, 82

Nicotiana glutinosa
Glutinosone, 669
2-Oxo-13-epimanool, 1052
Sclareol, 1242

Nicotiana sylvestris
α-Cembrenediol, 285

Nicotiana tabacum
Capsidiol, 255
α-Cembrenediol, 285
Nicotine, 1011

Nierembergia hippomanica
Pyrrole-3-carbamidine, 1172

Nodularia spumigena
Nodularin, 1016

Nostoc spp.
Microcystin LA, 959

Notholaena spp.
Acacetin, 4

Notholiron hyacinthinum
α-Chaconine, 291

Nuphar japonicum
Deoxynupharidine, 452

Nuphar luteum
Deoxynupharidine, 452

Nymphaea spp.
Deoxynupharidine, 452

O

Ochrosia confusa
Ochrolifuanine A, 1025

Ochrosia elliptica
Ellipticine, 531

Ochrosia lifuana
Ochrolifuanine A, 1025

Ochrosia miana
Ochrolifuanine A, 1025

Ocimum basilicum
Juvocimene 1, 820
Safrole, 1215

Ocotea glaziovii
Apoglaziovine, 112
(−)-Caaverine, 224

Ocotea variabilis
Apoglaziovine, 112

Ocotea venenosa
Rodiasine, 1204

Oenanthe aquatica
β-Phellandrene, 1088

Oenanthe crocata
Falcarinol, 593
Oenanthotoxin, 1027

Olea europaea
Cinchonidine, 318

Onobrychis viciifolia
Sainfuran, 1218

Onopordum acanthium
Onopordopicrin, 1034

Onosma caucasicum
Shikonin, 1262

Oonopsis condensata
Se-Methyl-L-selenocysteine, 952

Ophiopogon planiscapus
Deltoside, 436

Orchis militaris
Orchinol, 1036

Oreganum vulgare
Naringin, 1003

Origanum majorana
Arbutin, 117

Orixa spp.
Kokusaginine, 828

Orixa japonica
Evoxine, 588

Ornithogalum magnum
Rhodexin A, 1192

Ornithogalum umbellatum
Convallatoxin, 349

Oroxylum indicum
Aloe-emodin, 64
Baicalein, 167

Orthodon spp.
Myristicin, 994

Orthodon formosanus
Dillapiole, 485

Oryza sativa
Momilactone A, 975
Oryzalexin A, 1039
Sakuranetin, 1219
Salicylic acid, 1222

Oscillatoria spp.
Anatoxin a, 86
Microcystin LA, 959

Oscillatoria nigroviridis
Debromoaplysiatoxin, 417
30-Methyloscillatoxin D, 947
31-Noroscillatoxin B, 1022
Oscillatoxin A, 1040
Oscillatoxin B1, 1041
Oscillatoxin B2, 1042
Oscillatoxin D, 1043

Ougeinia dalbergioides
Wedelolactone, 1429

Parthenium incanum
Ligulatin B, 860

Parthenium integrifolium
Tetraneurin-E, 1324

Parthenium ligulatum
Ligulatin B, 860

Parthenium lozanianum
Tetraneurin-E, 1324

Parthenium schottii
Ligulatin B, 860

Parthenium tomentosum
Ligulatin B, 860
Stramonin-B, 1288

Passerina vulgaris
Nortrachelogenin, 1023

Passiflora spp.
Heterodendrin, 724
Linamarin, 863
Linustatin, 865
Lotaustralin, 875
Neolinustatin, 1006

Passiflora capsularis
Passicapsin, 1069

Passiflora coriaceae
Deidaclin, 428

Passiflora incarnata
Harmaline, 699
Harman, 700
Harmine, 701

Pastinaca spp.
Imperatorin, 770
Isopimpinellin, 801

Pastinaca sativa
Sphondin, 1279
Xanthotoxin, 1441
Xanthotoxol, 1442

Paullinia cupana
Caffeine, 227

Pausinystalia yohimbe
Yohimbine, 1448

Peganum harmala
Harmaline, 699
Harmine, 701
Peganine, 1073
Vasicinone, 1396

Peganum nigellastrum
Vasicinone, 1396

Pelea christophersenii
Anethole, 89

Pellia endiviifolia
Sacculatal, 1214

Pennisetum americanum
Orientin, 1037
Vitexin, 1424

Pentaceras australis
Canthin-6-one, 253

Pentachondra pumila
Kaempferol 3-(2″,4″-di-*p*-coumaryl-rhamnoside), 821

Pera ferruginea
Plumbagin, 1122

Pergularia pallida
Tylophorine, 1381

Perilla frutescens
Perilla ketone, 1077

Peripentadenia mearsii
Tropacocaine, 1364

Perovskia angustifolia
Phenol, 1090

Perriera madagascariensis
Glaucarubinone, 653
Glaucarubolone, 654

Persea spp.
Isoobtusilactone A, 799

Persea americana
1-Acetoxy-2-hydroxyheneicosa-12,15-dien-4-one, 15
Catechol, 278

Persea borbonia
Isoobtusilactone A, 799

Persea gratissima
Estragole, 564

Peschiera affinis
Affinine, 36
Affinisine, 37

Peschiera laeta
Affinine, 36

Petalostylis labicheoides
Calligonine, 234

Petasites spp.
Bakkenolide A, 170

Petasites albus
Senkirkine, 1254

Petasites hybridus
Integerrimine, 775
Petasin, 1080
Petasitenine, 1081
Senkirkine, 1254

Petasites japonicus
Petasitenine, 1081

Petroselinum spp.
Bergapten, 182

Petroselinum crispum
Apiole, 108
Myristicin, 994

Peucedanum spp.
Athamantin, 148

Peucedanum austriacum
Aethusin, 34

Peucedanum carvifola
Aethusin, 34

Peucedanum decursivum
Decuroside III, 420

Peucedanum ostruthium
Ostruthin, 1044

Peucedanum palustre
Prangolarine, 1133

Peucedanum rablense
Aethusin, 34

Peucedanum verticillare
Aethusin, 34

Pfaffia iresinoides
Polypodine B, 1131
Pterosterone, 1161

Pfaffia paniculata
Pfaffic acid, 1082
Pfaffoside A, 1083

Phacelia ixodes
Geranylhydroquinone, 642

Phaeanthus ebracteolatus
Phaeantharine, 1084

Phagnalon sordidum
Prenylbenzoquinone, 1136

Phalaris arundinacea
Gramine, 685
Hordenine, 732
5-Methoxy-*N,N*-dimethyltryptamine, 929

Phalaris tuberosa
5-Methoxy-*N,N*-dimethyltryptamine, 929

Phaseolus spp.
γ-Aminobutyric acid, 74

Phaseolus coccineus
Malonic acid, 906

Phaseolus vulgaris
Arcelin, 118
Phasin, 1086

Phebalium argenteum
Psoralen, 1157

Phellodendron amurense
Candicine, 248

Philodendron spp.
Oxalic acid, 1049

Phoebe clemensii
Isocorydine, 789

Phoenix dactylifera
Lupeol acetate, 883
Oestrone, 1028

Phoradendron tomentum
Phoratoxin, 1095

Photinia davidiana
Acetovanillone, 12

Phryma leptostachya
Phrymarolin I, 1097

Phyllanthus acuminatus
Justicidin B, 816
Phyllanthostatin A, 1099

Phyllanthus discoides
Securinine, 1246

Phyllanthus emblica
Lupeol, 882

Phyllanthus gasstroemeri
Taxiphyllin, 1308

Phyllanthus niruri
Phyllanthin, 1098

Phyllanthus reticulatus
Pyrogallol, 1171

Phyllarthron comorense
β-Lapachone, 843

Physalis peruviana
Physoperuvine, 1100

Physoclaina alaica
Apoatropine, 111

Physostigma venenosum
Eseramine, 561
Eseridine, 562
Physostigmine, 1101
Physovenine, 1102

Phytolacca americana
Phytolaccagenin, 1103

Phytolacca dodecandra
Lemmatoxin, 855
Oleanoglycotoxin-A, 1032

Picea spp.
Car-3-ene, 261
β-Phellandrene, 1088
Piceatannol, 1105

Picea sitchensis
Astringin, 146
Rhaponticin, 1189

Picralima klaineana
Akuammicine, 46
Akuammidine, 47
Akuammine, 48

Picrasma spp.
Neoquassin, 1008
Quassin, 1175

Picrasma crenata
Canthin-6-one, 253

Pieris japonica
Asebotoxin II, 138
Phloridzin, 1093

Pierreodendron kerstingii
Ailanthinone, 41

Pilocarpus spp.
Pilocarpine, 1108

Pilocarpus microphyllus
Pilocarpine, 1108
Pilosine, 1109

Pimelea prostrata
Prostratin, 1147

Pimelea simplex
Simplexin, 1265

Pimpinella spp.
Bergapten, 182
Isopimpinellin, 801

Pimpinella anisum
Anethole, 89
p-Cresol, 366

Pinus spp.
1-*tert*-Butyl-3-methylbenzene, 220
Car-3-ene, 261
Chrysin, 309
15-Hydroperoxyabietic acid, 735
β-Phellandrene, 1088
Piceatannol, 1105
α-Pinene, 1110
Pinosylvin methyl ether, 1113

Pinus cembra
Pinocembrin, 1112

Pinus longifolia
Car-3-ene, 261

Pinus palustris
15-Hydroperoxyabietic acid, 735
Nortrachelogenin, 1023
α-Pinene, 1110

Pinus radiata
7-Oxodehydroabietic acid, 1051

Pinus sylvestris
Car-3-ene, 261

Piper aduncum
Dillapiole, 485

Piper angustifolium
Apiole, 108
β-Asarone, 135

Piper cubeba
Cubebin, 376
Dipentene, 501

Piper decurrens
Conocarpan, 346

Piper methysticum
Dihydromethysticin, 475

Piper nigrum
α-Phellandrene, 1087
Pipercide, 1114

Piper novae-hollandiae
Dillapiole, 485
Fagaramide, 589

Piper tuberculatum
Dihydropiperlongumine, 476

Piptadenia macrocarpa
Bufotenine, 215
3,4-Dimethoxydalbergione, 488

Piptadenia peregrina
Bufotenine, 215
N,N-Dimethyltryptamine, 491

Piscidia erythrina
Glabranin, 652
Rotenone, 1207
Sumatrol, 1293

Pisum spp.
γ-Aminobutyric acid, 74

Pisum sativum
Cadaverine, 225
Pisatin, 1115
Salicylic acid, 1222

Pithecolobium bubalinum
L-Djenkolic acid, 510

Pithecolobium lobatum
L-Djenkolic acid, 510

Plagiobothrys arizonicus
Alkannin, 56

Plagiochila hattoriana
Plagiochiline A, 1118

Plagiochila yokogurensis
Plagiochiline A, 1118

Plantago spp.
Catalpol, 275

Plantago asiatica
Hellicoside, 716

Plantago lanceolata
Aucubin, 152

Platanus spp.
Dhurrin, 457

Plocamium telfairiae
Telfairine, 1314

Plumbago europaea
Plumbagin, 1122

Plumeria multiflora
Plumericin, 1123

Plumeria rubra
Plumericin, 1123

Podachaenium eminens
Zaluzanin C, 1450

Podanthus mitiqui
Ovatifolin, 1047

Podanthus ovatifolius
Erioflorin acetate, 551
Erioflorin methacrylate, 552
Ovatifolin, 1047

Podocarpus spp.
Nagilactone C, 995

Podocarpus elatus
Podecdysone B, 1124

Podocarpus gracilior
Podolide, 1125

Podocarpus nagi
Nagilactone C, 995

Podocarpus nakaii
Ponasterone A, 1132

Podophyllum hexandrum
4′-Demethylpodophyllotoxin, 440
Podophyllotoxin, 1126
Podophyllotoxone, 1127

Podophyllum peltatum
α-Peltatin, 1075
β-Peltatin A methyl ether, 1076
Podophyllotoxin, 1126
Podophyllotoxone, 1127

Podophyllum pleianthum
Deoxypodophyllotoxin, 453
Podophyllotoxin, 1126

Pogonopus tubulosus
Tubulosine, 1370

Polyalthia sp.
Goniothalenol, 679

Polyalthia nitidissima
Dauricine, 415

Polycavernosa tsudai
Polycavernoside A, 1129

Polygala macradenia
4′-Demethyldeoxypodophyllotoxin, 439

Polygala nyikensis
1, 7-Dihydroxy-4-methoxyxanthone, 483

Polygala paena
4′-Demethyldeoxypodophyllotoxin, 439

Polygala polygaena
4′-Demethylpodophyllotoxin, 440

Polygonatum multiflorum
L-Azetidine-2-carboxylic acid, 165
L-α, γ-Diaminobutyric acid, 460

Polygonum cuspidatum
Emodin 8-glucoside, 535

Polygonum hydropiper
Polygodial, 1130

Polygonum orientale
Orientin, 1037

Polypodium vulgare
Polypodine B, 1131

Polytrichum ohioense
Ohioensin-A, 1030

Poncirus trifoliata
Alloimperatorin, 62
Xanthotoxol, 1442

Pongamia glabra
Lanceolatin B, 836

Populus spp.
Catechol, 278
Chrysin, 309
Cinnamic acid, 322
Prenyl caffeate, 1137
(+)-Syringaresinol, 1300

Populus balsamifera
(+)-α-Bisabolol, 192

Populus deltoides
Pinocembrin, 1112

Populus tremuloides
Phenol, 1090

Potentilla kleiniana
Agrimoniin, 39

Poupartia fordii
Corilagin, 354

Prangos pabularia
Alloimperatorin, 62
Prangolarine, 1133

Prestonia amazonica
N, *N*-Dimethyltryptamine, 491

Primula elatior
Primin, 1141

Primula mistassinica
Primetin, 1140

Primula modesta
Primetin, 1140

Primula obconica
Primin, 1141

Pristimera indica
Pristimerin, 1142

Prorocentrum lima
Okadaic acid, 1031

Prosopis kuntzei
3, 4-Dimethoxydalbergione, 488

Protea mellifera
Hydroquinone, 737

Protogonyaulax spp.
Gonyautoxin I, 680

Prunella vulgaris
Ursolic acid, 1386

Prunus spp.
Genistein, 633
Pinocembrin, 1112
Prunasin, 1152

Prunus amygdalus
Amygdalin, 81

Prunus cerasus
Glutaric acid, 667

Prunus dulcis
Phenethylamine, 1089

Prunus grayana
Grayanin, 687

Pseudostiffita kingii
Methyl caffeate, 935

Psidium guajava
Casuarictin, 274
Gallic acid, 624

Psilopeganum sinense
Chalepensin, 293

Psilostrophe villosa
Vermeerin, 1402

Psoralea spp.
Psoralen, 1157

Psoralea corylifolia
Angelicin, 90

Psorospermum febrifugum
Psorospermin, 1158

Psorospermum glaberrimum
Emodin, 534

Psychotria granadensis
Tubulosine, 1370

Ptaeroxylon obliquum
5-*O*-Methylalloptaeroxylin, 932
Ptaeroglycol, 1159

Ptelea trifoliata
N-Methylflindersine, 939
O^4-Methylptelefolonium, 948

Pteridium aquilinum
Ptaquiloside, 1160
Shikimic acid, 1260

Pteridium esculentum
Ptaquiloside, 1160

Pteridophyllum spp.
Dihydrosanguinarine, 478
Sanguinarine, 1227

Pteris cretica
Ptaquiloside, 1160

Pterocarpus angolensis
Odoratol, 1026

Pterocarpus officinalis
Hypaphorine, 761

Pterocarpus santalinus
Pterostilbene, 1162

Pueraria mirifica
Miroestrol, 968

Punica granatum
Oestrone, 1028
(−)-Pelletierine, 1074
Punicalagin, 1165

Putterlickia verrucosa
Maytansine, 911

Pycnarrhena manillensis
Berbamine, 178

Pyrola incarnata
Chimaphylin, 301

Pyrus spp.
Ursolic acid, 1386

Pyrus communis
Arbutin, 117
Hydroquinone, 737
α-Pyrufuran, 1173

Q

Quassia africana
Simalikilactone D, 1264

Quassia amara
Neoquassin, 1008
Quassimarin, 1174
Quassin, 1175

R

Rhododendron adamsii
Germacrone, 643

Rhododendron chrysanthum
Rhododendrin, 1193

Rhododendron farrerae
Farrerol, 596
Rhododendrin, 1193

Rhododendron japonicum
Rhodojaponin IV, 1194

Rhododendron ponticum
Grayanotoxin I, 688

Rhodomyrtus macrocarpa
ψ-Rhodomyrtoxin, 1195

Rhus semialata
Corilagin, 354

Rhus undulata
Apigenin 7, 4′-dimethyl ether, 107

Ribes nigrum
Sakuranetin, 1219

Ricinus communis
Ricin, 1197
Ricinine, 1198

Rindera baldschuanica
Rinderine, 1201

Rivea spp.
Agroclavine, 40
Lysergol, 896

Rivea corymbosa
Chanoclavine-I, 297
Ergine, 546

Robinia pseudoacacia
Robin, 1203

Rosa spp.
Eugeniin, 567
Pyrogallol, 1171

Rosa canina
Vanillic acid, 1394

Rosa rugosa
2-Phenylethanol, 1091
Rugosin D, 1210

Rubia tinctorum
Alizarin, 55
Anthrogallol, 102
Pseudopurpurin, 1155
Purpurin, 1167

Rubus spp.
Casuarictin, 274
Pedunculagin, 1072

Rumex spp.
Chrysarobin, 307
Chrysophanol, 310
Emodin, 534

Ruta spp.
Dictamnine, 465
Xanthotoxin, 1441

Ruta chalepensis
Heliettin, 710

Ruta graveolens
Arborine, 115
Arborinine, 116
Bergapten, 182
Chalepensin, 293
Rutamarin, 1211

Ruta montana
Guaiacol, 691

Ryania speciosa
Ryanodine, 1212

S

Saccharum officinarum
L-Aspartic acid, 139
Luteolinidin, 886

Salix caprea
(+)-Catechin, 277
Taxifolin, 1306

Salsola richteri
(−)-Salsoline, 1224

Salvia officinalis
Cirsiliol, 325

Salvia sclarea
Sclareol, 1242

Salvia tomentosa
Cirsilineol, 324

Samanea saman
Pithecolobine, 1117

Sambucus nigra
Prunasin, 1152
Zierin, 1452

Sanguinaria spp.
Allocryptopine, 61

Sanguinaria canadensis
Sanguinarine, 1227

Sanguisorba minor
2′,6′-Dihydroxy-4′-methoxyacetophenone, 481

Santolina oblongifolia
Mycosinol, 991

Sapium indicum
Sapatoxin A, 1231
Sapintoxin A, 1233

Sapium japonicum
Corilagin, 354

Saponaria officinalis
Gypsogenin, 694

Sarothamnus spp.
(−)-Sparteine, 1277

Sarracenia spp.
Histamine, 729

Sarracenia flava
Coniine, 345

Sassafras albidum
Safrole, 1215

Sassafras randaiense
Magnolol, 902

Sauromatum guttatum
Salicylic acid, 1222

Saussurea amara
Cynaropicrin, 408

Saussurea lappa
Costunolide, 359

Saussurea neopulchella
Saupirin, 1237

Saussurea pulchella
Saupirin, 1237

Scabiosa succisa
Methyl caffeate, 935

Sceletium anatomicum
Mesembrenone, 924
Mesembrine, 925
Mesembrinol, 926

Sceletium expansum
Mesembrenone, 924
Mesembrine, 925
Mesembrinol, 926

Sceletium tortuosum
Mesembrenone, 924
Mesembrine, 925
Mesembrinol, 926

Schaefferia cuneifolia
Pristimerin, 1142

Schefflera arboricola
Falcarinol, 593

Schefflera capitata
Echinocystic acid, 522

Schinus terebinthifolius
Bilobol, 189
(15:1)-Cardanol, 259

Schizothrix calcicola
Debromoaplysiatoxin, 417
30-Methyloscillatoxin D, 947
31-Noroscillatoxin B, 1022
Oscillatoxin A, 1040
Oscillatoxin B1, 1041
Oscillatoxin B2, 1042
Oscillatoxin D, 1043

Schizozygia caffaeoides
Schizozygine, 1239

Sciadotenia toxifera
Isochondrodendrine, 787

Scopolia carniolica
Hyoscine, 758
Pseudotropine, 1156
Tropine, 1365

Scutellaria spp.
Baicalein, 167

Scutellaria baicalensis
Baicalin, 168
Scullcapflavone II, 1244
Wogonin, 1435

Scutellaria galericulata
Baicalein, 167
Chrysin, 309

Scutellaria violacea
Clerodin, 329
Jodrellin B, 813

Scutellaria woronowii
Jodrellin B, 813

Scytonema sp.
Scytonemin A, 1245

Securidaca longepedunculata
Securinine, 1246

Securigera securidaca
Hyrcanoside, 765

Securinega suffruticosa
Securinine, 1246

Sedum acre
Cadaverine, 225
Lobelanidine, 869
Nicotine, 1011
(−)-Pelletierine, 1074

Sedum cepaea
Sarmentosin epoxide, 1235

Selaginella willdenowii
Isocryptomerin, 790

Selinum monnieri
Alloimperatorin, 62

Selinum vaginatum
Angelicin, 90

Senecio spp.
Isatidine, 781
Otonecine, 1045
Retronecine, 1186
Senecionine, 1251
Seneciphylline, 1252

Senecio aegypticus
Riddelline, 1199

Senecio aetnensis
Senaetnine, 1249
Senampeline A, 1250

Senecio alpinus
Integerrimine, 775
Jacobine, 807

Senecio ambrosioides
Riddelline, 1199

Senecio angulatus
Angularine, 91

Senecio aucheri
Senampeline A, 1250

Senecio brasiliensis
Integerrimine, 775

Senecio cineraria
Jacobine, 807

Senecio eremophilus
Riddelline, 1199

Senecio filaginoides
Retrorsine, 1187

Senecio grisebachii
Retrorsine, 1187

Senecio illinutus
Senkirkine, 1254

Senecio incarnus
Jacobine, 807

Senecio integerrimus
Integerrimine, 775

Senecio jacobaea
Jacobine, 807
Otonecine, 1045
Senecionine, 1251
Senkirkine, 1254

Senecio kirkii
Otonecine, 1045
Senkirkine, 1254

Senecio longifolius
Precocene 2, 1135

Senecio longilobus
Isatidine, 781
Riddelline, 1199

Senecio othonnae
Doronine, 515

Senecio phillipicus
Retrorsine, 1187
Seneciphylline, 1252

Senecio platyphyllus
Seneciphylline, 1252

Senecio pseudo-orientalis
Retronecine, 1186

Senecio pterophorus
Senampeline A, 1250

Senecio pyramidatus
Bakkenolide A, 170

Senecio renardii
Senkirkine, 1254

Senecio retrorsus
Retrorsine, 1187

Senecio riddellii
Riddelline, 1199

Senecio seratophiloides
Senecivernine, 1253

Senecio squalidus
Integerrimine, 775

Senecio triangularis
Triangularine, 1348

Senecio vernalis
Senecivernine, 1253
Senkirkine, 1254

Senecio vulgaris
Retrorsine, 1187
Riddelline, 1199
Senecionine, 1251

Sesamum indicum
Sesamol, 1257

Sesbania drummondii
Justicidin B, 816
Sesbanimide A, 1259

Sesbania punicea
Sesbanimide A, 1259

Seseli spp.
Bergapten, 182
Disenecionyl *cis*-khellactone, 504
Isopimpinellin, 801

Seseli incanum
Disenecionyl *cis*-khellactone, 504

Seseli libanotis
Athamantin, 148
Disenecionyl *cis*-khellactone, 504

Sida cordifolia
Peganine, 1073
Vasicinol, 1395
Vasicinone, 1396

Sideritis spp.
Cirsilineol, 324

Sideroxylon tomentosum
Gypsogenin, 694

Sigesbeckia pubescens
Diterpenoid SP-II, 506

Simaba multiflora
Cleomiscosin A, 328

Singickia rubra
Harman, 700

Sinomenium acutum
Acutumidine, 28
Sinomenine, 1267

Sisyrynchium spp.
Plumbagin, 1122

Smallanthus fruticosus
Enhydrin, 537

Smallanthus uvedalius
Enhydrin, 537

Smilax aristolochiaefolia
Parillin, 1066

Smyrniopsis armena
Alloimperatorin, 62

Solanum spp.
Demissine, 441
Solanidine, 1269
Solasodine, 1272
Tomatidine, 1340
Tomatine, 1341

Solanum aviculare
Solasonine, 1273

Solanum chacoense
α-Chaconine, 291
Demissine, 441

Solanum demissum
Demissine, 441
Tomatidine, 1340

Solanum dimidiatum
Calystegin B_2, 239

Solanum dulcamara
Solasodine, 1272

Solanum incanum
Solasonine, 1273

Solanum jamesii
Demissine, 441

Solanum kurebense
Calystegin B_2, 239

Solanum laciniatum
Solasodine, 1272

Solanum melongena
Calystegin B_2, 239
Lubimin, 876
Solasonine, 1273

Solanum nigrum
α-Chaconine, 291
Solanidine, 1269
α-Solanine, 1270
Solasodine, 1272

Solanum pseudocapsicum
Solacapine, 1268
Solanocapsine, 1271

Solanum sodomeum
Solasonine, 1273

Solanum torvum
Solasonine, 1273

Solanum tuberosum
Calystegin B_2, 239
α-Chaconine, 291
Glyoxylic acid, 672
Lubimin, 876
Phytuberin, 1104
Rishitin, 1202
Solanidine, 1269
α-Solanine, 1270
Solavetivone, 1274

Solanum viarum
Solasonine, 1273

Solanum xanthocarpum
Solasonine, 1273

Solenanthus circinatus
Rinderine, 1201

Sonchus spp.
Cichoriin, 314

Sophora spp.
Ammodendrine, 77
Anagyrine, 85
Cytisine, 410

Sophora alopecuroides
Aloperine, 65

Sophora flavescens
Ammothamnine, 78

Sophora franchetiana
Ammodendrine, 77

Sorbaria arborea
Cardiospermin, 260
Heterodendrin, 724

Sorbus spp.
Parasorbic acid, 1064

Sorbus aucuparia
Isoaucuparin, 784
4′-Methoxyaucuparin, 927
Parasorbic acid, 1064

Sorghum spp.
Dhurrin, 457

Sorghum bicolor
Luteolinidin, 886

Soulamea soulameoides
Cleomiscosin A, 328

Soulamea tomentosa
Isobruceine A, 785
Soularubinone, 1276

Sparaxis spp.
Plumbagin, 1122

Spartium junceum
Caulophylline, 283

Spathelia glabrescens
Spatheliabischromene, 1278

Spathelia sorbifolia
Spatheliabischromene, 1278

Spinacea oleracea
Coumestrol, 365
Histamine, 729
Oxalic acid, 1049

Sporochnus bolleanus
Sporochnol A, 1283

Stachys sylvatica
(−)-Betonicine, 185

Stachyurus praecox
Casuarictin, 274
Pedunculagin, 1072

Stanleya pinnata
L-Selenocystathionine, 1247

Statice gmelinii
Pyrogallol, 1171

Steganotaenia araliacea
Neoisostegane, 1005
Steganacin, 1285

Stemmadenia spp.
Tabernanthine, 1302

Stemonoporus canaliculatus
Canaliculatol, 245

Stephania discolor
(+)-Tetrandine, 1322

Stephania glabra
Tetrahydropalmatine, 1321

Stephania hernandiifolia
Hernandezine, 721

Stephania sasakii
Berbamine, 178

Stephania tetrandra
(+)-Tetrandine, 1322

Sterculia foetida
Sterculic acid, 1286

Sternbergia lutea
Hippeastrine, 727

Stevia serrata
Methylripariochromene A, 951

Stizophyllum riparium
Stizophyllin, 1287

Streptocarpus dunnii
Digiferrugineol, 468

Strictocardia spp.
Lysergol, 896

Striga asiatica
Apigenin 7, 4′-dimethyl ether, 107

Strophanthus spp.
Lokundjoside, 872

Strophanthus arnotdianus
k-Strophanthoside, 1289

Strophanthus divaricatus
Decoside, 419
Divaricoside, 508
Divostroside, 509
Lokundjoside, 872
Musaroside, 989
Sarmentoloside, 1234

Strophanthus gratus
Ouabain, 1046

Strophanthus kombé
Cymarin, 405
k-Strophanthoside, 1289

Strophanthus sarmentosus
Bipindoside, 191
Lokundjoside, 872
Sarmentoloside, 1234

Strophanthus thollonii
Bipindoside, 191
Lokundjoside, 872

Strychnos spp.
Brucine, 210
Calebassine, 232
Deoxyloganin, 449
Strychnine, 1290

Strychnos aculeata
Brucine, 210

Strychnos angustifolia
Angustine, 93

Strychnos divaricans
Calebassine, 232
C-Curarine, 390

Strychnos dolichothyrsa
Caracurine V, 256

Strychnos froesii
C-Curarine, 390
Toxiferine I, 1344

Strychnos icaja
Strychnine, 1290

T

Tanacetum cinerariifolium
Chrysanthemic acid, 306
Cinerin I, 320
Cinerin II, 321
Pyrethrin I, 1168
Pyrethrin II, 1169
Pyrethrosin, 1170

Tanacetum odessanum
Methyl caffeate, 935

Tanacetum parthenium
Camphor, 243
Canin, 251
Parthenolide, 1068
Santamarin, 1228

Tanacetum vulgare
α-Thujone, 1335

Taraxacum officinale
Glucosyl taraxinate, 664

Taxodium distichum
Taxodione, 1309
Taxodone, 1310

Taxus spp.
Taxol, 1311

Taxus baccata
Cephalomannine, 288
Taxine A, 1307
Taxol, 1311

Taxus brevifolia
Taxol, 1311

Taxus cuspidata
Taxol, 1311

Teclea boiviniana
Evoxine, 588

Teclea natalensis
Arborinine, 116

Tecoma radicans
Actinidine, 27

Tecoma stans
Tecomine, 1312
Tecostanine, 1313

Tectona grandis
Aloe-emodin, 64
Chrysophanol, 310
Deoxylapachol, 448
Lapachenole, 841
β-Lapachone, 843

Telekia speciosa
Isoalantolactone, 783

Tellima grandiflora
Eugeniin, 567

Tephrosia spp.
12a-Hydroxyrotenone, 751

Tephrosia bidwillii
(−)-Acanthocarpan, 8

Tephrosia hildebrandtii
Hildecarpin, 725

Tephrosia lanceolata
Lanceolatin B, 836

Tephrosia purpurea
Lanceolatin B, 836

Tephrosia toxicaria
Toxicarol, 1343

Terminalia chebula
Corilagin, 354

Terminalia oblongata
Punicalagin, 1165
Terminalin, 1316

Tetradymia glabrata
Tetradymol, 1319

Tetragonolobus maritimus
Anhydroglycinol, 94

Tetrapathea tetranda
Deidaclin, 428

Teucrium marum
Dolichodial, 512

Thalictrum spp.
Berberine, 180
Jatrorrhizine, 810
Thalmine, 1327

Thalictrum dasycarpum
(−)-Laudanidine, 850
Thalicarpine, 1325

Thalictrum flavum
Thalicarpine, 1325

Thalictrum foetidum
Fetidine, 602

Thalictrum hernandezii
Hernandezine, 721

Thalictrum kuhistanicum
Thalmine, 1327

Thalictrum lucidum
Aromoline, 132

Thalictrum minus
Adiantifoline, 29
Thalicoside A, 1326
Thalmine, 1327

Thalictrum polygamum
Thalicarpine, 1325

Thalictrum rugosum
Thalsimine, 1328

Thalictrum simplex
Hernandezine, 721
Thalsimine, 1328

Thamnosma montana
Alloimperatorin, 62

Thapsia garganica
Thapsigargin, 1329

Thea sinensis
Theasaponin, 1330

Thermopsis spp.
Cytisine, 410

Thermopsis chinensis
Anagyrine, 85

Thermopsis cinerea
13-Hydroxylupanine, 747

Thermopsis lupinoides
Ammodendrine, 77

Thespesia populnea
Gossypol, 681

Thevetia neriifolia
Thevetin A, 1332
Thevetin B, 1333

U

V

W

X

Xanthium occidentale
Xanthumin, 1444

Xanthium pennsylvanicum
Xanthatin, 1439

Xanthium riparium
Xanthatin, 1439

Xanthium sibiricum
Xanthatin, 1439

Xanthium strumarium
Carboxyatractyloside, 258
Xanthumin, 1444

Xanthorrhoea australis
Aloe-emodin, 64

Xanthoxylum ailanthoides
Xanthyletin, 1445

Xanthoxylum alatum
Xanthoxylin, 1443

Xanthoxylum americanum
Xanthyletin, 1445

Xanthoxylum piperitum
Xanthoxylin, 1443

Xeranthemum cylindraceum
Xerantholide, 1446
Zierin, 1452

Ximenia americana
Prunasin, 1152

Xylocarpus granatum
N-Methylflindersine, 939

Xylocarpus moluccensis
Xylomollin, 1447

Xylopia pancheri
(−)-Laudanidine, 850

Xyris semifuscata
Chrysazin, 308

Z

Zaluzania spp.
Ivalin, 806
Zaluzanin C, 1450

Zaluzania pringlei
Epitulipinolide, 539

Zanthorrhiza simplicissima
Berberastine, 179

Zanthoxylum spp.
Allocryptopine, 61
Berberine, 180
Dictamnine, 465
Nitidine, 1013
Sanguinarine, 1227

Zanthoxylum americanum
Nitidine, 1013

Zanthoxylum budrunga
Arborine, 115

Zanthoxylum clava-herculis
Nitidine, 1013

Zanthoxylum suberosum
Canthin-6-one, 253

Zanthoxylum zanthoxyloides
Fagaronine, 590

Zea mays
Alloimperatorin, 62

Zephyranthes grandiflora
Pancratistatin, 1062

Zephyranthus carinatus
Pretazettine, 1139

Zexmania brevifolia
Zexbrevin B, 1451

Zieria laevigata
Zierin, 1452

Zigadenus gramineus
Zygadenine, 1455

Zigadenus venenosus
Germine, 644
Zygadenine, 1455

Zingiber officinale
Gingerenone A, 646
[6]-Gingerol, 647
Zingiberene, 1453

Zinnia acerosa
Zaluzanin C, 1450

Zinnia elegans
(−)-Anabasine, 82

Zollikoferia eliquiensis
(−)-Anabasine, 82

Zygophyllum fabago
Harman, 700

Molecular Formula Index

$C_7H_6O_3$
Salicylic acid, 1222
Sesamol, 1257

$C_7H_6O_5$
Gallic acid, 624

$C_7H_7BrO_3$
3-Bromo-4,5-dihydroxybenzyl alcohol, 205

C_7H_8O
p-Cresol, 366

$C_7H_8O_2$
Guaiacol, 691

$C_7H_{10}O_5$
Shikimic acid, 1260

$C_7H_{11}NO_2$
L-Hypoglycin, 764

$C_7H_{13}NO_4$
Calystegin B_2, 239

$C_7H_{14}N_2O_4S_2$
L-Djenkolic acid, 510

$C_7H_{14}N_2O_4Se$
L-Selenocystathionine, 1247

$C_7H_{15}N_3O_2$
L-Indospicine, 773

$C_7H_{16}N_4O_2$
L-Homoarginine, 731

$C_7H_{17}N_4O_4P$
Anatoxin a(s), 87

$C_8H_7NO_2$
Gentianadine, 634

$C_8H_8N_2O_2$
Ricinine, 1198

$C_8H_8O_4$
2,6-Dimethoxybenzoquinone, 487
Vanillic acid, 1394

$C_8H_{10}N_2O_4$
L-Mimosine, 966

$C_8H_{10}N_4O_2$
Caffeine, 227

$C_8H_{10}O$
2-Phenylethanol, 1091

$C_8H_{10}O_3$
2,6-Dimethoxyphenol, 489

$C_8H_{11}N$
Phenethylamine, 1089

$C_8H_{11}NO_2$
Dopamine, 514

$C_8H_{13}NO_2$
Arecoline, 123
Heliotridine, 712
Retronecine, 1186

$C_8H_{13}NO_3$
Crotanecine, 370

$[C_8H_{14}NO_9S_2]^-$
Glucocapparin, 657

$C_8H_{15}N$
γ-Coniceine, 343

$C_8H_{15}NO$
(−)-Pelletierine, 1074
Physoperuvine, 1100
Pseudotropine, 1156
Tropine, 1365

$C_8H_{15}NO_3$
Swainsonine, 1296

$C_8H_{15}NO_4$
Castanospermine, 272

$C_8H_{16}N_2O_7$
Cycasin, 396

$C_8H_{17}N$
Coniine, 345

$C_8H_{17}NO$
Pseudoconhydrine, 1154

$C_9H_6N_2S$
Brassilexin, 199

$C_9H_6O_2$
Coumarin, 364

$C_9H_8O_2$
Cinnamic acid, 322

$C_9H_{10}O_3$
Acetovanillone, 12

$C_9H_{10}O_4$
2′,6′-Dihydroxy-4′-methoxyacetophenone, 481

$C_9H_{11}NO$
D-Cathinone, 281

$C_9H_{11}NO_4$
L-Dopa, 513

$C_9H_{13}NO$
D-Cathine, 280

$C_9H_{13}NO_3$
4′-*O*-Methylpyridoxine, 949

$C_9H_{14}O_2$
Boschnialactone, 197

$C_9H_{15}NO_3$
Ecgonine, 519
Otonecine, 1045

$[C_9H_{16}NO_9S_2]^-$
Glucolepidiin, 661

$C_9H_{17}NO_8$
Miserotoxin, 969

$C_9H_{19}N$
(+)-*N*-Methylconiine, 936

$C_{10}H_6O_3$
Juglone, 814
Lawsone, 852

$C_{10}H_6O_4$
Ayapin, 163

$C_{10}H_8O_4$
Acamelin, 7

$C_{10}H_9NO_2$
Gentianine, 637

$C_{10}H_{10}O_2$
Isosafrole, 802
Lachnophyllum lactone, 832
Safrole, 1215

$C_{10}H_{10}O_4$
Methyl caffeate, 935

$C_{10}H_{12}O$
Anethole, 89
Estragole, 564

$C_{10}H_{12}O_3$
Coniferyl alcohol, 344

$C_{10}H_{12}O_4$
Xanthoxylin, 1443

$C_{13}H_{14}O_2$
Tremetone, 1347

$C_{13}H_{14}O_3$
6-Hydroxytremetone, 752
Toxol, 1345

$C_{13}H_{14}O_4$
1′-Acetoxychavicol acetate, 13

$C_{13}H_{15}NO_2$
Securinine, 1246

$C_{13}H_{15}NOS$
Methylillukumbin A, 940

$C_{13}H_{16}O_3$
Flossonol, 606
Precocene 2, 1135

$C_{13}H_{17}NO_3$
Lophophorine, 873

$C_{13}H_{18}N_2O$
5-Methoxy-*N,N*-dimethyltryptamine, 929

$C_{13}H_{19}NO_2$
(±)-Carnegine, 262
Dioscorine, 496

$C_{13}H_{19}NO_3$
Gigantine, 645

$C_{13}H_{21}NO_2$
Tigloidine, 1337

$C_{13}H_{22}O$
Geranylacetone, 641

$C_{13}H_{24}N_2O_{11}$
Macrozamin, 900

$C_{13}H_{24}NO_2$
N-Methyl-2, 6-bis(2-hydroxybutyl)-Δ^3-piperideine, 934

$C_{13}H_{26}O$
2-Tridecanone, 1353

$C_{14}H_8N_2O$
Canthin-6-one, 253

$C_{14}H_8N_2O_2$
11-Hydroxycanthin-6-one, 743

$C_{14}H_8O_4$
Alizarin, 55
Chrysazin, 308

$C_{14}H_8O_5$
Anthrogallol, 102
Purpurin, 1167

$C_{14}H_9NO_3$
Crinasiadine, 368
Dianthalexin, 461

$C_{14}H_{10}O_5$
1,7-Dihydroxy-4-methoxyxanthone, 483
Kigelinone, 827

$C_{14}H_{10}O_6$
Athyriol, 149
Bellidifolin, 176

$C_{14}H_{12}Br_2O_4$
Avrainvilleol, 162

$C_{14}H_{12}N_4O_3$
Iforrestine, 768

$C_{14}H_{12}O_3$
4,5′,8-Trimethylpsoralen, 1360
Xanthyletin, 1445

$C_{14}H_{12}O_4$
Eriobofuran, 549
Piceatannol, 1105
Wyerone acid, 1437

$C_{14}H_{12}O_5$
γ-Cotonefuran, 362

$C_{14}H_{13}NO_4$
Kokusaginine, 828

$C_{14}H_{13}NO_6$
Lycoricidine, 893

$C_{14}H_{13}NO_7$
Narciclasine, 1000

$C_{14}H_{14}O_2$
Ichthyotherol, 767

$C_{14}H_{14}O_3$
Demethylbatatasin IV, 437
Isoaucuparin, 784

$C_{14}H_{14}O_4$
Rhaphiolepsin, 1188

$C_{14}H_{15}NO_8$
Pancratistatin, 1062

$C_{14}H_{16}$
Chamazulene, 295

$C_{14}H_{16}O_3$
Mexicanin E, 953

$C_{14}H_{16}O_4$
Prenyl caffeate, 1137

$C_{14}H_{16}O_5$
1′-Acetoxyeugenol acetate, 14

$C_{14}H_{17}NO_2$
Leiokinine A, 854

$C_{14}H_{17}NO_3$
Fagaramide, 589

$C_{14}H_{17}NO_6$
Prunasin, 1152

$C_{14}H_{17}NO_7$
Dhurrin, 457
p-Glucosyloxymandelonitrile, 663
Taxiphyllin, 1308
Zierin, 1452

$C_{14}H_{17}NO_{10}$
Triglochinin, 1356

$C_{14}H_{18}N_2O_2$
Hypaphorine, 761

$C_{14}H_{18}N_2O_3$
Physovenine, 1102

$[C_{14}H_{18}NO_9S_2]^-$
Glucotropaeolin, 665

$C_{14}H_{18}O$
Chamaecynone, 294

$C_{14}H_{19}NO_5$
Dicrotaline, 464

$C_{14}H_{20}N_2O_9$
Acalyphin, 6

$C_{14}H_{20}O_2$
Glutinosone, 669

$C_{14}H_{21}NO_4$
Codonopsine, 335

$C_{14}H_{22}O_2$
Rishitin, 1202

$C_{14}H_{23}NO$
Affinin, 35

$C_{15}H_8O_5$
Coumestrol, 365

$C_{15}H_8O_7$
Pseudopurpurin, 1155

$C_{15}H_{10}O_2$
Myomontanone, 992

$C_{15}H_{10}O_4$
Anhydroglycinol, 94
Chrysin, 309
Chrysophanol, 310
Digiferrugineol, 468
Primetin, 1140

$C_{15}H_{10}O_5$
Aloe-emodin, 64
Apigenin, 106
Baicalein, 167
Emodin, 534
Galangin, 620
Genistein, 633
Norobtusifolin, 1021

$[C_{15}H_{11}O_5]^+$
Luteolinidin, 886

$C_{15}H_{12}O_3$
Chrysarobin, 307
Lettucenin A, 856

$C_{15}H_{12}O_4$
Hydrangenol, 734
Pinocembrin, 1112

$C_{15}H_{12}O_5$
Rubrofusarin, 1209

$C_{15}H_{12}O_6$
Micromelin, 964

$C_{15}H_{12}O_7$
Taxifolin, 1306

$C_{15}H_{13}NO_5$
Dianthramide A, 462

$C_{15}H_{14}O_2$
Deoxylapachol, 448
7-Hydroxyflavan, 745
Pinosylvin methyl ether, 1113

$C_{15}H_{14}O_3$
7,4′-Dihydroxyflavan, 480
Hircinol, 728
Lapachol, 842
β-Lapachone, 843

$C_{15}H_{14}O_4$
Wyerone, 1436

$C_{15}H_{14}O_5$
α-Pyrufuran, 1173
Wyerone epoxide, 1438

$C_{15}H_{14}O_6$
(+)-Catechin, 277
α-Cotonefuran, 360
6-Hydroxy-α-pyrufuran, 750
Plumericin, 1123
Ptaeroglycol, 1159

$C_{15}H_{15}NO_2$
N-Methylflindersine, 939

$C_{15}H_{15}NO_5$
Acronycidine, 23

$C_{15}H_{16}O_2$
Mansonone C, 910

$C_{15}H_{16}O_3$
Dehydrozaluzanin C, 427
Encelin, 536

$C_{15}H_{16}O_4$
8-Deoxylactucin, 447
Dihydrowyerone, 479
4′-Methoxyaucuparin, 927

$C_{15}H_{16}O_5$
Dihydromethysticin, 475
Lactucin, 834
Vernolepin, 1408
Vernomenin, 1410

$C_{15}H_{16}O_6$
Picrotoxinin, 1107

$C_{15}H_{16}O_7$
Allamandin, 58

$C_{15}H_{16}O_9$
Cichoriin, 314

$C_{15}H_{17}NO_2$
Atanine, 147

$C_{15}H_{17}NO_4$
Dubinidine, 517

$C_{15}H_{18}$
Guaiazulene, 692

$C_{15}H_{18}O_2$
Dehydromyodesmone, 423
Eremanthine, 543
Eremofrullanolide, 545

$C_{15}H_{18}O_3$
Achillin, 19
Ambrosin, 72
Aromaticin, 131
Desacetoxymatricarin, 454
8-Epixanthatin, 542
α-Santonin, 1229
β-Santonin, 1230
Xanthatin, 1439
Xerantholide, 1446
Zaluzanin C, 1450

$C_{15}H_{18}O_4$
Dihydrogriesenin, 474
Farinosin, 595
Grosshemin, 690
Helenalin, 708
Methylripariochromene A, 951
Mexicanin I, 954
Microhelenin-A, 961
Parthenin, 1067
Stramonin-B, 1288

$C_{15}H_{18}O_5$
Canin, 251
Coriamyrtin, 352
Isohelenol, 797

$C_{15}H_{18}O_6$
Tutin, 1378

$C_{15}H_{18}O_7$
Mellitoxin, 919
Picrotin, 1106

$C_{15}H_{19}BrO$
Aplysin, 110

$C_{15}H_{19}NO_2$
Benzoyltropein, 177
Tropacocaine, 1364

$C_{15}H_{19}NO_3$
Cocculolidine, 333

$C_{15}H_{20}N_2O$
Anagyrine, 85

$C_{15}H_{20}N_2O_2$
N-(3-Oxobutyl)cytisine, 1050

$[C_{15}H_{20}NO_9S_2]^-$
Gluconasturtiin, 662

$C_{17}H_{14}O_6$
Pisatin, 1115

$C_{17}H_{14}O_7$
Cirsiliol, 325
Hildecarpin, 725

$C_{17}H_{15}NO_2$
Anonaine, 97

$C_{17}H_{16}Br_2O_4$
Colpol, 338

$C_{17}H_{16}O_4$
3, 4-Dimethoxydalbergione, 488

$C_{17}H_{16}O_5$
Coelogin, 336
Farrerol, 596

$C_{17}H_{16}O_6$
Cajanol, 228

$C_{17}H_{17}NO_2$
(−)-Caaverine, 224

$C_{17}H_{17}NO_5$
Hippeastrine, 727

$C_{17}H_{18}O_5$
Odoratol, 1026

$C_{17}H_{19}N_3$
Brevicolline, 203

$C_{17}H_{19}NO_3$
Erysonine, 555
Morphine, 983
Narwedine, 1004

$C_{17}H_{19}NO_4$
Crinamine, 367
Montanine, 980

$C_{17}H_{20}O_5$
Angustibalin, 92
Arteglasin A, 133
Linifolin A, 864

$C_{17}H_{20}O_6$
Arctolide, 121

$C_{17}H_{21}NO_2$
Apoatropine, 111

$C_{17}H_{21}NO_3$
Galanthamine, 621
Mesembrenone, 924

$C_{17}H_{21}NO_4$
Cocaine, 332
Hyoscine, 758

$C_{17}H_{22}O$
Falcarinone, 594

$C_{17}H_{22}O_2$
Cicutoxin, 315
Oenanthotoxin, 1027

$C_{17}H_{22}O_4$
Epitulipinolide, 539
Tulipinolide, 1371

$C_{17}H_{22}O_5$
Arnicolide A, 130
Eupaformonin, 574
Gaillardin, 619
Isotenulin, 803
Ligulatin B, 860
Lipiferolide, 866
Ovatifolin, 1047
Pyrethrosin, 1170
Tenulin, 1315
Xanthumin, 1444

$C_{17}H_{22}O_6$
Epitulipinolide diepoxide, 540
Tetraneurin-A, 1323

$C_{17}H_{22}O_7$
1-Peroxyferolide, 1078

$C_{17}H_{23}NO_3$
Atropine, 151
Hyoscyamine, 759
Mesembrine, 925

$C_{17}H_{23}NO_4$
Anisodamine, 96

$C_{17}H_{24}O$
Falcarinol, 593

$C_{17}H_{24}O_2$
Falcarindiol, 592
Heptadeca-1, 9-diene-4, 6-diyne-3, 8-diol, 718
Juvadecene, 818

$C_{17}H_{24}O_3$
Buddledin A, 213

$C_{17}H_{24}O_5$
Inulicin, 777

$C_{17}H_{24}O_6$
Tetraneurin-E, 1324

$C_{17}H_{24}O_{10}$
Geniposide, 632
Verbenalin, 1400

$C_{17}H_{24}O_{11}$
Gardenoside, 625

$C_{17}H_{25}NO_3$
Mesembrinol, 926

$C_{17}H_{26}O_4$
Embelin, 532
[6]-Gingerol, 647
Phytuberin, 1104

$C_{17}H_{26}O_9$
Deoxyloganin, 449

$C_{17}H_{26}O_{10}$
Loganin, 871

$C_{17}H_{27}NO_7$
Uplandicine, 1383

$C_{17}H_{28}O_4$
1*S*-Hydroxy-α-bisabololoxide A acetate, 740

$C_{17}H_{29}NO_{11}$
Neolinustatin, 1006

$C_{17}H_{31}N_3O_2$
Palustrine, 1060

$C_{18}H_{14}O_4$
Dehydrocycloguanandin, 421

$C_{18}H_{14}O_5$
6-Deoxyjacareubin, 446

$C_{18}H_{14}O_9$
Phloroglucinol trimer, 1094

$C_{18}H_{15}NO_4$
Lunamarine, 880

$C_{18}H_{16}O_4$
Calophyllin B, 235

$C_{18}H_{16}O_6$
Betagarin, 183

$C_{18}H_{16}O_7$
Cirsilineol, 324

$C_{18}H_{16}O_8$
Acerosin, 10
Oxyayanin A, 1053
Oxyayanin B, 1054

$C_{18}H_{17}NO_3$
Mecambrine, 912

$C_{18}H_{17}NO_4$
Cularicine, 386

$C_{18}H_{18}O_2$
Conocarpan, 346
Magnolol, 902

$C_{18}H_{18}O_6$
Samaderin A, 1226

$C_{18}H_{19}NO_3$
Apoglaziovine, 112

$C_{18}H_{19}NO_4$
Acetylcaranine, 16

$C_{18}H_{19}NO_6$
Candimine, 249

$C_{18}H_{20}N_2$
(−)-Apparicine, 114

$C_{18}H_{21}NO_3$
Codeine, 334
Neopine, 1007

$C_{18}H_{21}NO_4$
Cephalotaxine, 289

$C_{18}H_{21}NO_5$
Ambelline, 71
Pretazettine, 1139

$C_{18}H_{21}NO_6$
Evoxine, 588

$C_{18}H_{22}ClNO_6$
Acutumidine, 28

$[C_{18}H_{22}NO]^+$
N-(*p*-Hydroxyphenethyl)actinidine, 748

$[C_{18}H_{22}NO_4]^+$
O^4-Methylptelefolonium, 948

$C_{18}H_{22}O_2$
Oestrone, 1028

$C_{18}H_{22}O_4$
Arnebinone, 129
Nordihydroguaiaretic acid, 1018

$C_{18}H_{22}O_{11}$
Asperuloside, 141

$C_{18}H_{23}NO_2$
Isococculidine, 788

$C_{18}H_{23}NO_4$
Amaryllisine, 70
Haplophyllidine, 698
Lycorenine, 892

$C_{18}H_{23}NO_5$
Seneciphylline, 1252

$C_{18}H_{23}NO_6$
Riddelline, 1199

$C_{18}H_{25}NO_5$
Integerrimine, 775
Senecionine, 1251
Senecivernine, 1253
Triangularine, 1348

$C_{18}H_{25}NO_6$
Anacrotine, 84
Angularine, 91
Jacobine, 807
Retrorsine, 1187
Usaramine, 1390

$C_{18}H_{25}NO_7$
Isatidine, 781

$C_{18}H_{26}N_2O_2$
Isoajmaline, 782

$C_{18}H_{27}NO_{10}$
Passicapsin, 1069

$C_{18}H_{32}O_2$
Chaulmoogric acid, 300
Malvalic acid, 907

$C_{18}H_{33}FO_2$
ω-Fluorooleic acid, 607

$C_{19}H_{12}O_6$
Dicoumarol, 463

$C_{19}H_{14}O_4$
Columbianetin, 339

$C_{19}H_{16}N_2$
Sempervirine, 1248

$C_{19}H_{16}O_6$
Psorospermin, 1158

$C_{19}H_{17}NO_3$
Cusparine, 393

$C_{19}H_{18}O_8$
Scullcapflavone II, 1244

$C_{19}H_{18}O_{11}$
Tripteroside, 1362

$C_{19}H_{19}NO_4$
Amurensine, 79
Amurine, 80
Bulbocapnine, 216
(+)-Cassythicine, 270

$C_{19}H_{19}NO_5$
Cassyfiline, 269

$C_{19}H_{20}O_6$
Calaxin, 231
Deoxyelephantopin, 445

$C_{19}H_{20}O_7$
Elephantopin, 530
Goyazensolide, 682
Vernodalin, 1404

$C_{19}H_{21}NO_3$
Isothebaine, 804
Thebaine, 1331

$C_{19}H_{21}NO_4$
Cularidine, 387
Cularimine, 388
(+)-Salutaridine, 1225

$C_{19}H_{22}N_2O$
Cinchonidine, 318
Eburnamonine, 518

$C_{19}H_{22}O_3$
Ostruthin, 1044

$C_{19}H_{22}O_4$
Heliettin, 710

$C_{19}H_{22}O_5$
Podolide, 1125

$C_{19}H_{22}O_6$
Çynaropicrin, 408
Molephantin, 972
Saupirin, 1237

$C_{19}H_{22}O_7$
Nagilactone C, 995
Repin, 1182
Vernolide, 1409

$C_{19}H_{22}O_8$
Hydroxyvernolide, 754

$C_{19}H_{23}ClO_7$
Acroptilin, 25
Chlororepdiolide, 305

$C_{19}H_{23}N_3O_2$
Ergometrine, 547

$C_{19}H_{23}NO_3$
Armepavine, 127
Erysotrine, 556

$C_{19}H_{23}NO_4$
Cinnamoylcocaine, 323
Sinomenine, 1267

$C_{19}H_{23}NO_6$
3-Acetylnerbowdine, 18

$C_{19}H_{24}Cl_2O_7$
Chlorohyssopifolin A, 304

$C_{19}H_{24}O_6$
Chamissonin diacetate, 296
Eremantholide A, 544
Onopordopicrin, 1034
Radiatin, 1178
Tagitinin F, 1303

$C_{19}H_{24}O_7$
Vernomygdin, 1411
Zexbrevin B, 1451

$C_{19}H_{25}NO_4$
Erythratidine, 557

$C_{19}H_{25}NO_{10}$
Lucumin, 877
Vicianin, 1413

$C_{19}H_{26}O_6$
Alatolide, 50
Plagiochiline A, 1118

$C_{19}H_{26}O_7$
Orizabin, 1038

$C_{19}H_{26}O_{12}$
Monotropitoside, 979

$C_{19}H_{27}NO_6$
Senkirkine, 1254

$C_{19}H_{27}NO_7$
Petasitenine, 1081

$C_{19}H_{30}O_4$
Rapanone, 1181

$C_{19}H_{30}O_5$
Shiromodiol diacetate, 1263

$C_{19}H_{32}O_2$
Grevillol, 689
Methyl linolenate, 942

$C_{19}H_{32}O_3$
Isoobtusilactone A, 799

$C_{19}H_{33}NO_3$
Melochinine, 920

$C_{19}H_{34}O_2$
Sterculic acid, 1286

$[C_{20}H_{14}NO_4]^+$
Sanguinarine, 1227

$C_{20}H_{14}O_7$
Pachyrrhizone, 1056

$C_{20}H_{15}N_3O$
Angustine, 93

$C_{20}H_{15}NO_4$
Dihydrosanguinarine, 478

$C_{20}H_{16}N_2O_4$
Camptothecin, 244

$C_{20}H_{17}NO_6$
(+)-Bicuculline, 187

$[C_{20}H_{18}NO_4]^+$
Berberine, 180

$[C_{20}H_{18}NO_5]^+$
Berberastine, 179

$C_{20}H_{18}O_5$
(−)-Glyceollin I, 670
(−)-Glyceollin II, 671
Wighteone, 1431

$C_{20}H_{18}O_6$
Luteone, 887

$C_{20}H_{18}O_7$
Styraxin, 1292

$C_{20}H_{18}O_8$
Cleomiscosin A, 328
Daphneticin, 412

$C_{20}H_{18}O_9$
Frangulin B, 613

$C_{20}H_{19}NO_3$
Acronycine, 24

$C_{20}H_{20}N_2O_3$
Schizozygine, 1239

$[C_{20}H_{20}NO_4]^+$
Jatrorrhizine, 810

$C_{20}H_{20}O_4$
Glabranin, 652
Spatheliabischromene, 1278

$C_{20}H_{20}O_5$
6-Isopentenylnaringenin, 800

$C_{20}H_{20}O_6$
Cubebin, 376
Kievitone, 826

$C_{20}H_{20}O_7$
Tangeretin, 1305

$C_{20}H_{21}NO_3$
Galipine, 623

$C_{20}H_{21}NO_4$
Papaverine, 1063

$C_{20}H_{22}N_2O_2$
Akuammicine, 46
Gelsemine, 629

$C_{20}H_{22}O_4$
Pulverochromenol, 1164

$C_{20}H_{22}O_6$
Multiradiatin, 987
Multistatin, 988

$C_{20}H_{22}O_7$
Budlein A, 214
Elephantin, 529
Nortrachelogenin, 1023

$C_{20}H_{22}O_9$
Astringin, 146

$C_{20}H_{22}O_{10}$
Plicatic acid, 1121

$C_{20}H_{23}NO_4$
Cularine, 389
Isocorydine, 789

$C_{20}H_{23}NO_7$
Senaetnine, 1249

$C_{20}H_{24}N_2O$
Affinisine, 37
Borrecapine, 196

$C_{20}H_{24}N_2O_2$
Affinine, 36
Akagerine, 45
Quinine, 1176

$C_{20}H_{24}O_3$
Jatrophone, 809

$C_{20}H_{24}O_4$
Vernoflexin, 1406

$C_{20}H_{24}O_6$
Fastigilin C, 598
(+)-Lariciresinol, 845
Miroestrol, 968
Molephantinin, 973
Multigilin, 986
Triptolide, 1363

$C_{20}H_{24}O_7$
Euparotin, 577
Eupatundin, 585
Tripdiolide, 1361

$C_{20}H_{24}O_8$
Eupatoroxin, 584
Vernodalol, 1405

$C_{20}H_{24}O_9$
Ginkgolide A, 649

$C_{20}H_{25}ClO_7$
Eupachlorin, 568

$C_{20}H_{25}ClO_8$
Eupachloroxin, 570

$C_{20}H_{25}NO_4$
(−)-Laudanidine, 850

$C_{20}H_{26}N_2O$
Ibogaine, 766
Tabernanthine, 1302

$C_{20}H_{26}N_2O_2$
Ajmaline, 44
Hydroquinidine, 736

$C_{20}H_{26}N_2O_4$
Gelsemicine, 628

$C_{20}H_{26}O$
Juvocimene 1, 820

$C_{20}H_{26}O_3$
Jatrophatrione, 808
Momilactone A, 975
7-Oxodehydroabietic acid, 1051
Taxodione, 1309

$C_{20}H_{26}O_5$
Brevilin A, 204
Microhelenin C, 962
Nobilin, 1015

$C_{20}H_{26}O_6$
Desacetyleupaserrin, 455
Eupatoriopicrin, 583
Fastigilin B, 597
Shikodonin, 1261

$C_{20}H_{26}O_7$
Chaparrinone, 299
Cnicin, 331
Eupahyssopin, 576
Niveusin C, 1014

$C_{20}H_{26}O_8$
Glaucarubolone, 654

$C_{20}H_{27}NO_{11}$
Amygdalin, 81

$C_{20}H_{27}NO_{12}$
Proteacin, 1148

$C_{20}H_{28}N_2O_4$
O-Acetylcypholophine, 17

$C_{20}H_{28}O_2$
Spruceanol, 1284

$C_{20}H_{28}O_3$
Cafestol, 226
Cinerin I, 320
Petasin, 1080
Pisiferic acid, 1116
Taxodone, 1310

$C_{20}H_{28}O_4$
Callicarpone, 233

$C_{20}H_{28}O_6$
Eriolangin, 553

$C_{20}H_{28}O_7$
Chaparrin, 298

$C_{20}H_{30}O$
Amijitrienol, 73

$C_{20}H_{30}O_2$
Oryzalexin A, 1039
Sacculatal, 1214

$C_{20}H_{30}O_3$
Hydroxydictyodial, 744

$C_{20}H_{30}O_4$
15-Hydroperoxyabietic acid, 735
Lathyrol, 849

$C_{20}H_{30}O_8$
Ptaquiloside, 1160

$C_{20}H_{31}NO_6$
Echiumine, 524
Symlandine, 1297
Symphytine, 1298

$C_{20}H_{31}NO_7$
Echimidine, 520
Heliosupine, 711

$C_{20}H_{32}O_2$
Dilophic acid, 486
2-Oxo-13-epimanool, 1052

$C_{20}H_{32}O_4$
Diterpenoid SP-II, 506

$C_{20}H_{34}O_2$
α-Cembrenediol, 285

$C_{20}H_{34}O_4$
Zoapatanol, 1454

$C_{20}H_{36}O_2$
Sclareol, 1242

$C_{21}H_{12}O_7$
Dihydrosamidin, 477

$[C_{21}H_{16}NO_5]^+$
Bocconine, 194

$C_{21}H_{16}O_6$
Justicidin B, 816

$C_{21}H_{16}O_7$
Diphyllin, 502

$[C_{21}H_{18}NO_4]^+$
Nitidine, 1013

$C_{21}H_{18}O_7$
Austrobailignan 1, 156

$C_{26}H_{34}O_{11}$
Isobruceine A, 785

$C_{26}H_{34}O_{14}$
Decuroside III, 420

$C_{26}H_{36}O_{8}$
Jodrellin B, 813

$C_{26}H_{38}O_{4}$
Lupulone, 885

$C_{26}H_{38}O_{5}$
Chinensin II, 303

$C_{26}H_{41}NO_{7}$
Delcorine, 430

$C_{26}H_{44}N_{2}$
Buxamine E, 223

$C_{27}H_{22}O_{12}$
Lithospermic acid, 868

$C_{27}H_{22}O_{18}$
Corilagin, 354

$C_{27}H_{30}O_{8}$
Daphnetoxin, 413

$C_{27}H_{30}O_{14}$
Glucofrangulin A, 660

$C_{27}H_{32}O_{10}$
Harrisonin, 703
Spicatin, 1280

$C_{27}H_{32}O_{14}$
Cascaroside A, 266
Naringin, 1003

$C_{27}H_{32}O_{15}$
Butrin, 218

$C_{27}H_{34}O_{10}$
Provincialin, 1151

$C_{27}H_{34}O_{11}$
Undulatone, 1382

$C_{27}H_{36}O_{11}$
Quassimarin, 1174

$C_{27}H_{38}O_{7}$
Diterpenoid EF-D, 505

$C_{27}H_{38}O_{10}$
Trilobolide, 1358

$C_{27}H_{39}N_{3}O_{2}$
Lyngbyatoxin A, 895

$C_{27}H_{39}NO_{2}$
Veratramine, 1398

$C_{27}H_{39}NO_{3}$
Jervine, 811

$C_{27}H_{40}O_{5}$
Chinensin I, 302

$C_{27}H_{41}NO_{2}$
Cyclopamine, 401

$C_{27}H_{41}NO_{8}$
Deltaline, 434

$C_{27}H_{42}O_{3}$
Diosgenin, 497

$C_{27}H_{42}O_{6}$
Podecdysone B, 1124

$C_{27}H_{43}NO$
Solanidine, 1269

$C_{27}H_{43}NO_{2}$
Solasodine, 1272

$C_{27}H_{43}NO_{8}$
Germine, 644
Tricornine, 1352

$C_{27}H_{44}O_{6}$
Ponasterone A, 1132

$C_{27}H_{44}O_{7}$
Pterosterone, 1161

$C_{27}H_{44}O_{8}$
Polypodine B, 1131

$C_{27}H_{45}NO_{2}$
Tomatidine, 1340

$C_{27}H_{46}N_{2}O_{2}$
Solanocapsine, 1271

$C_{27}H_{48}N_{2}$
Cycloprotobuxine C, 402

$C_{27}H_{48}N_{2}O$
Cyclovirobuxine C, 403

$C_{27}H_{48}N_{2}O_{2}$
Solacapine, 1268

$C_{27}H_{50}N_{2}O_{4}$
Carpaine, 265

$C_{28}H_{10}O_{16}$
Terminalin, 1316

$C_{28}H_{22}O_{6}$
ε-Viniferin, 1420

$C_{28}H_{28}O_{11}$
Cleistanthin A, 327

$C_{28}H_{34}O_{9}$
Nomilin, 1017

$C_{28}H_{36}O_{10}$
12α-Hydroxyamoorstatin, 738

$C_{28}H_{36}O_{11}$
Bruceantin, 207

$C_{28}H_{37}NO_{9}$
Harringtonine, 702

$C_{28}H_{38}O_{6}$
Withaferin A, 1433
Withanolide D, 1434

$C_{28}H_{46}O_{7}$
Makisterone B, 903

$C_{29}H_{28}N_{4}$
Usambarensine, 1388

$C_{29}H_{30}O_{13}$
Amarogentin, 69
Phyllanthostatin A, 1099

$C_{29}H_{34}N_{4}$
Ochrolifuanine A, 1025

$C_{29}H_{34}O_{7}$
Microlenin, 963

$C_{29}H_{36}O_{11}$
Labriformidin, 830

$C_{29}H_{36}O_{15}$
Verbascoside, 1399

$C_{29}H_{36}O_{17}$
Hellicoside, 716

$C_{29}H_{37}N_{3}O_{3}$
Tubulosine, 1370

$C_{29}H_{38}O_{7}$
Surangin B, 1295

$C_{29}H_{38}O_{8}$
Roridin E, 1206

$C_{29}H_{38}O_9$
Uscharidin, 1391

$C_{29}H_{38}O_{10}$
Baccharinoid B21, 166

$C_{29}H_{38}O_{11}$
Eriocarpin, 550

$C_{29}H_{40}N_2O_4$
Emetine, 533

$C_{29}H_{40}O_9$
Calactin, 229
Calotropin, 237
Roridin A, 1205

$C_{29}H_{42}O_9$
Corchoroside A, 351
Gofruside, 677
Helveticoside, 717
Peruvoside, 1079

$C_{29}H_{42}O_{10}$
Adonitoxin, 31
Aspecioside, 140
Convallatoxin, 349

$C_{29}H_{42}O_{11}$
α-Antiarin, 103

$C_{29}H_{44}O_3$
Pfaffic acid, 1082

$C_{29}H_{44}O_8$
Evomonoside, 587

$C_{29}H_{44}O_9$
Rhodexin A, 1192

$C_{29}H_{44}O_{10}$
Antioside, 104
Gitorin, 650
Lokundjoside, 872

$C_{29}H_{44}O_{11}$
Sarmentoloside, 1234

$C_{29}H_{44}O_{12}$
Ouabain, 1046

$C_{30}H_{16}O_8$
Hypericin, 763

$C_{30}H_{20}N_4O_6$
Trichotomine, 1351

$C_{30}H_{28}O_8$
Rottlerin, 1208

$C_{30}H_{30}O_8$
Gossypol, 681

$C_{30}H_{34}N_4$
Usambarine, 1389

$C_{30}H_{35}NO_{10}$
Celapanine, 284

$C_{30}H_{36}O_{10}$
Orbicuside A, 1035

$C_{30}H_{37}N_5O_5$
Ergosine, 548

$C_{30}H_{37}NO_7$
Sapintoxin A, 1233

$C_{30}H_{40}O_4$
Pristimerin, 1142

$C_{30}H_{40}O_6$
Absinthin, 2

$C_{30}H_{42}O_7$
Cucurbitacin I, 381

$C_{30}H_{42}O_9$
Decoside, 419

$C_{30}H_{44}K_2O_{16}S_2$
Atractyloside, 150

$C_{30}H_{44}O_6$
11-Deoxocucurbitacin I, 444

$C_{30}H_{44}O_7$
Adynerin, 32
Cucurbitacin D, 379

$C_{30}H_{44}O_8$
Simplexin, 1265

$C_{30}H_{44}O_9$
Cymarin, 405

$C_{30}H_{44}O_{10}$
Musaroside, 989
Vernadigin, 1403

$C_{30}H_{46}O_4$
Gypsogenin, 694

$C_{30}H_{46}O_7$
Cucurbitacin O, 382

$C_{30}H_{46}O_8$
Divaricoside, 508
Divostroside, 509

$C_{30}H_{46}O_{10}$
Bipindoside, 191

$C_{30}H_{48}O_3$
Ursolic acid, 1386

$C_{30}H_{48}O_7$
Cucurbitacin P, 383

$C_{30}H_{50}O$
Lupeol, 882

$C_{30}H_{50}O_2$
Betulin, 186

$C_{30}H_{50}O_4$
Sapelin A, 1232

$C_{30}H_{52}O_4$
Mollugogenol A, 974

$C_{31}H_{21}O_{10}$
Isocryptomerin, 790

$C_{31}H_{34}O_8$
Sarothralin, 1236

$C_{31}H_{38}O_9$
Toonacilin, 1342

$C_{31}H_{38}O_{11}$
Tyledoside A, 1379

$C_{31}H_{39}NO_{10}S$
Labriformin, 831

$C_{31}H_{41}NO_8S$
Uscharin, 1392

$C_{31}H_{42}O_8$
Oscillatoxin D, 1043

$C_{31}H_{42}O_{10}$
Asclepin, 137
Cotyledoside, 363

$C_{31}H_{42}O_{11}$
Miotoxin C, 967

$C_{31}H_{43}NO_7$
Anopterine, 98

$C_{31}H_{44}O_9$
Bovoside A, 198

$C_{31}H_{44}O_{10}$
31-Noroscillatoxin B, 1022

$C_{31}H_{45}NO_8$
Auriculine, 154

$C_{31}H_{46}O_{10}$
Oscillatoxin A, 1040

$[C_{31}H_{46}O_{18}S_2]^{2-}$
Carboxyatractyloside, 258

$C_{31}H_{48}O_7$
Phytolaccagenin, 1103

$C_{32}H_{32}O_7$
Karwinskione, 825

$C_{32}H_{32}O_8$
Tullidinol, 1375

$C_{32}H_{34}O_{14}$
Haemocorin, 697

$C_{32}H_{36}O_{12}$
Filixic acid ABA, 604

$C_{32}H_{38}N_2O_8$
Deserpidine, 456

$C_{32}H_{42}O_8$
Acrovestone, 26

$C_{32}H_{42}O_{12}$
Bryotoxin A, 212

$C_{32}H_{42}O_{16}$
Bruceoside A, 209

$C_{32}H_{44}N_2O_8$
Lappaconitine, 844

$C_{32}H_{44}O_7$
Sapatoxin A, 1231

$C_{32}H_{44}O_8$
Cucurbitacin E, 380
30-Methyloscillatoxin D, 947

$C_{32}H_{44}O_{11}$
Lanceotoxin B, 838

$C_{32}H_{44}O_{12}$
Lanceotoxin A, 837
Scilliroside, 1241

$C_{32}H_{46}N_2O_8$
Anthranoyllycoctonine, 101

$C_{32}H_{46}O_8$
Cucurbitacin B, 378

$C_{32}H_{46}O_9$
Cucurbitacin A, 377

$C_{32}H_{46}O_{10}$
Oscillatoxin B1, 1041
Oscillatoxin B2, 1042

$C_{32}H_{47}BrO_{10}$
Aplysiatoxin, 109

$C_{32}H_{48}O_8$
Cucurbitacin Q, 384

$C_{32}H_{48}O_{10}$
Debromoaplysiatoxin, 417

$C_{32}H_{50}O_8$
Baliospermin, 171

$C_{32}H_{52}O_2$
Lupeol acetate, 883

$C_{33}H_{40}N_2O_9$
Reserpine, 1184

$C_{33}H_{43}O_{19}$
Myricoside, 993

$C_{33}H_{44}O_4$
Eudesobovatol A, 566

$C_{33}H_{45}NO_9$
Delphinine, 432

$C_{33}H_{45}NO_{10}$
Hypaconitine, 760

$C_{33}H_{45}NO_{11}$
Mesaconitine, 922

$C_{33}H_{50}N_2O_3$
Pachysandrine A, 1057

$C_{33}H_{52}O_9$
Agavoside A, 38

$C_{34}H_{24}O_{22}$
Pedunculagin, 1072

$C_{34}H_{26}O_8$
Albanol A, 51

$C_{34}H_{36}O_7$
Ingenol 3,20-dibenzoate, 774

$C_{34}H_{41}NO_{17}$
Anthemis glycoside B, 100

$C_{34}H_{44}O_9$
Salannin, 1221

$C_{34}H_{46}ClN_3O_{10}$
Maytansine, 911

$C_{34}H_{47}NO_{10}$
Falaconitine, 591
Indaconitine, 771

$C_{34}H_{47}NO_{11}$
Aconitine, 21

$C_{34}H_{47}NO_{12}$
Aconifine, 20

$C_{34}H_{48}O_8$
Huratoxin, 733

$C_{34}H_{48}O_{14}$
Hyrcanoside, 765

$C_{34}H_{50}O_{12}$
Thapsigargin, 1329

$C_{35}H_{34}N_2O_5$
Trilobine, 1357

$C_{35}H_{36}O_6$
Daphnoline, 414

$C_{35}H_{39}N_2O_8$
Milliamine L, 965

$C_{35}H_{42}N_2O_9$
Rescinnamine, 1183

$C_{35}H_{44}O_9$
Candletoxin A, 250

$C_{35}H_{44}O_{10}$
Ohchinolide B, 1029

$C_{35}H_{44}O_{16}$
Azadirachtin, 164

$C_{35}H_{46}O_{13}$
Trichilin A, 1349

$C_{35}H_{46}O_{20}$
Echinacoside, 521

$C_{35}H_{47}NO_{10}$
Taxine A, 1307

$C_{35}H_{49}NO_{12}$
Jesaconitine, 812

$C_{35}H_{52}O_5$
Lantadene A, 839
Lantadene B, 840

$C_{35}H_{52}O_8$
Phorbol 12-tiglate 13-decanoate, 1096

$C_{35}H_{52}O_{15}$
Convalloside, 350

$C_{35}H_{54}O_{7}$
Ilexolide A, 769

$C_{35}H_{54}O_{14}$
Thevetin B, 1333

$C_{35}H_{56}O_{8}$
Hederagenin 3-*O*-arabinoside, 704

$C_{35}H_{62}O_{4}$
Diepomuricanin A, 467

$C_{36}H_{36}N_{2}O_{5}$
Tiliacorine, 1338

$C_{36}H_{36}O_{10}$
Gnidicin, 673

$C_{36}H_{38}N_{2}O_{6}$
Aromoline, 132
(+)-Bebeerine, 175
Isochondrodendrine, 787

$C_{36}H_{38}O_{8}$
Tinyatoxin, 1339

$C_{36}H_{46}N_{2}O_{10}$
Barbinine, 173

$C_{36}H_{48}N_{2}O_{10}$
14-Deacetylnudicauline, 416
Lycaconitine, 888

$C_{36}H_{51}N_{3}O_{10}$
Avadharidine, 158

$C_{36}H_{51}NO_{11}$
Bikhaconitine, 188

$C_{36}H_{51}NO_{12}$
Pseudaconitine, 1153

$C_{36}H_{52}O_{8}$
Mancinellin, 909

$C_{36}H_{52}O_{13}$
Scillaren A, 1240

$C_{36}H_{52}O_{15}$
Hellebrin, 715

$C_{36}H_{54}O_{8}$
Synaptolepsis factor K_1, 1299

$C_{36}H_{56}O_{8}$
12-Tetradecanoylphorbol 13-acetate, 1318

$C_{36}H_{56}O_{11}$
Medicagenic acid 3-glucoside, 913

$C_{36}H_{56}O_{14}$
Digitalin, 470

$C_{36}H_{58}O_{9}$
Hederagenin 3-glucoside, 705

$C_{36}H_{58}O_{10}$
Bayogenin 3-glucoside, 174

$C_{37}H_{40}N_{2}O_{6}$
Berbamine, 178
Gyrocarpine, 695
Thalmine, 1327

$C_{37}H_{40}O_{9}$
Resiniferatoxin, 1185

$[C_{37}H_{41}N_{2}O_{6}]^{+}$
Macoline, 898

$[C_{37}H_{42}N_{2}O_{6}]^{+}$
(+)-Tubocurarine, 1369

$C_{37}H_{42}O_{10}$
Gniditrin, 676

$C_{37}H_{44}O_{10}$
Gnididin, 674

$C_{37}H_{48}O_{10}$
Gnidilatin, 675

$C_{37}H_{50}N_{2}O_{10}$
Methyl-lycaconitine, 943

$C_{38}H_{38}O_{10}$
Mezerein, 955

$C_{38}H_{40}N_{2}O_{7}$
Thalsimine, 1328

$C_{38}H_{40}N_{4}O_{2}$
Caracurine V, 256

$C_{38}H_{42}N_{2}O_{6}$
Rodiasine, 1204
(+)-Tetrandine, 1322

$C_{38}H_{42}O_{14}$
Kansuinine B, 822

$C_{38}H_{44}N_{2}O_{6}$
Dauricine, 415

$C_{38}H_{50}N_{2}O_{10}$
Elatine, 526

$C_{38}H_{50}N_{2}O_{11}$
Nudicauline, 1024

$C_{38}H_{54}N_{2}O_{11}$
Delavaine A, 429

$C_{38}H_{54}O_{13}$
Elaterinide, 525

$C_{38}H_{58}O_{14}$
Urechitoxin, 1384

$C_{38}H_{60}O_{9}$
12-*O*-Palmitoyl-16-hydroxyphorbol 13-acetate, 1059

$C_{39}H_{32}O_{14}$
Kaempferol 3-(2″, 4″-di-*p*-coumaryl-rhamnoside), 821

$[C_{39}H_{40}N_{2}O_{6}]^{2+}$
Phaeantharine, 1084

$C_{39}H_{44}N_{2}O_{7}$
Hernandezine, 721

$C_{39}H_{49}NO_{21}$
Anthemis glycoside A, 99

$C_{39}H_{56}O_{12}$
Spinoside A, 1281

$C_{40}H_{36}O_{11}$
Kuwanone G, 829

$[C_{40}H_{44}N_{4}O]^{2+}$
C-Curarine, 390

$C_{40}H_{46}N_{2}O_{8}$
Fetidine, 602

$[C_{40}H_{46}N_{4}O_{2}]^{2+}$
Toxiferine I, 1344

$C_{40}H_{48}N_{4}O_{2}$
Tabernamine, 1301

$[C_{40}H_{48}N_{4}O_{2}]^{2+}$
Calebassine, 232

$C_{40}H_{52}O_{14}$
Nilotin, 1012

$C_{40}H_{55}NO_{13}$
Wedeloside, 1430

$C_{40}H_{60}O_{13}$
Pfaffoside A, 1083

$C_{41}H_{26}O_{26}$
Alnusiin, 63
Castalagin, 271

$C_{41}H_{28}O_{26}$
Casuarictin, 274
Eugeniin, 567

$C_{41}H_{28}O_{27}$
Geraniin, 640

$C_{41}H_{46}N_4O_3$
Macrocarpamine, 899

$C_{41}H_{48}N_2O_8$
Thalicarpine, 1325

$C_{41}H_{48}N_4O_4$
Villalstonine, 1416

$C_{41}H_{60}N_8O_{10}$
Nodularin, 1016

$C_{41}H_{63}NO_{14}$
Protoveratrine A, 1149

$C_{41}H_{63}NO_{15}$
Protoveratrine B, 1150

$C_{41}H_{64}O_{13}$
Digitoxin, 471

$C_{41}H_{64}O_{14}$
Digoxin, 472
Gitoxin, 651

$C_{41}H_{64}O_{15}$
Diginatin, 469

$C_{41}H_{64}O_{19}$
k-Strophanthoside, 1289

$C_{41}H_{66}O_{11}$
β-Hederin, 707

$C_{41}H_{66}O_{12}$
α-Hederin, 706

$C_{41}H_{68}O_{14}$
Astragaloside III, 144

$C_{42}H_{32}O_9$
Canaliculatol, 245

$C_{42}H_{38}O_{20}$
Sennoside A, 1255

$C_{42}H_{50}N_2O_9$
Adiantifoline, 29

$C_{42}H_{50}N_4O_5$
Gabunamine, 617
Gabunine, 618

$C_{42}H_{64}O_{19}$
Thevetin A, 1332

$C_{42}H_{70}O_{14}$
Thalicoside A, 1326

$C_{43}H_{49}NO_{19}$
Wilfordine, 1432

$C_{43}H_{52}N_4O_5$
Voacamine, 1425

$C_{43}H_{52}N_4O_6$
Epivoacorine, 541

$C_{43}H_{60}O_{12}$
Goniodomin A, 678

$C_{43}H_{68}O_{15}$
Polycavernoside A, 1129

$C_{44}H_{68}O_{13}$
Dinophysistoxin 2, 493
Okadaic acid, 1031

$C_{45}H_{53}NO_{14}$
Cephalomannine, 288

$C_{45}H_{70}O_{13}$
Dinophysistoxin 1, 492

$C_{45}H_{72}O_{16}$
Dioscin, 495

$C_{45}H_{72}O_{17}$
Deltonin, 435
Gracillin, 683

$C_{45}H_{73}NO_{14}$
α-Chaconine, 291

$C_{45}H_{73}NO_{15}$
α-Solanine, 1270

$C_{45}H_{73}NO_{16}$
Solasonine, 1273

$C_{46}H_{48}N_2O_8$
Michellamine B, 958

$C_{46}H_{56}N_4O_9$
Leurosine, 858

$C_{46}H_{56}N_4O_{10}$
Vincristine, 1419

$C_{46}H_{58}N_4O_9$
Leurosidine, 857
Vinblastine, 1417

$C_{46}H_{67}N_7O_{12}$
Microcystin LA, 959

$C_{47}H_{51}NO_{14}$
Taxol, 1311

$C_{47}H_{66}N_{14}O_{10}$
Moroidin, 982

$C_{47}H_{70}O_{15}$
Pectenotoxin 1, 1071

$C_{47}H_{78}O_{18}$
Astrasieversianin XVI, 145

$C_{48}H_{28}O_{30}$
Punicalagin, 1165

$C_{48}H_{76}O_{19}$
Echinocystic acid, 522

$C_{48}H_{76}O_{21}$
Medicagenic acid 3-triglucoside, 914

$C_{48}H_{78}O_{17}$
Saikosaponin BK1, 1217

$C_{48}H_{78}O_{18}$
Arvenososide A, 134
Lemmatoxin, 855
Oleanoglycotoxin-A, 1032

$C_{49}H_{70}O_{13}$
Brevetoxin A, 201

$C_{49}H_{74}N_{10}O_{12}$
Microcystin LR, 960

$C_{49}H_{75}N_{13}O_{12}$
Cyanoviridin RR, 395

$C_{50}H_{70}O_{14}$
Brevetoxin B, 202

$C_{50}H_{83}NO_{20}$
Demissine, 441

$C_{50}H_{83}NO_{21}$
Tomatine, 1341

$C_{51}H_{84}O_{22}$
Parillin, 1066

$C_{51}H_{84}O_{23}$
Deltoside, 436

$C_{53}H_{84}O_{24}$
Camellidin II, 242

$C_{53}H_{86}O_{21}$
Helianthoside A, 709

$C_{54}H_{80}O_{20}$
Avenacin B-2, 160

$C_{55}H_{83}NO_{21}$
Avenacin A-1, 159

$C_{55}H_{86}O_{24}$
Aescin, 33

$C_{55}H_{86}O_{25}$
Camellidin I, 241

$C_{56}H_{84}O_{21}$
Cynafoside B, 407

$C_{57}H_{94}O_{27}$
Convallamaroside, 348

$C_{58}H_{94}O_{27}$
Cyclamin, 397

$C_{59}H_{92}O_{27}$
Theasaponin, 1330

$C_{60}H_{56}O_{11}$
Conocurvone, 347

$C_{60}H_{86}O_{19}$
Ciguatoxin, 316

$C_{60}H_{98}O_{27}$
Sakuraso-saponin, 1220

$C_{63}H_{98}O_{29}$
Tubeimoside I, 1368

$C_{71}H_{106}N_{12}O_{21}$
Scytonemin A, 1245

$C_{82}H_{54}O_{52}$
Agrimoniin, 39

$C_{82}H_{56}O_{52}$
Gemin A, 630

$C_{82}H_{58}O_{52}$
Coriariin A, 353
Rugosin D, 1210

Common Names Index

A

Ackee
L-γ-Glutamyl-L-hypoglycin, 666
L-Hypoglycin, 764

Aconite
Hypaconitine, 760

African daisy
Arctolide, 121

Alder buckthorn
Glucofrangulin A, 660

Alfalfa
L-Arginine, 124
Medicagenic acid 3-triglucoside, 914

American wormseed
Ascaridole, 136
p-Cymene, 406

Angostura bark
Cusparine, 393

Angostura
Galipine, 623

Apple
Phloridzin, 1093
Ursolic acid, 1386

Arrow grass
Triglochinin, 1356

Artichoke
Grosshemin, 690

Asparagus
L-Arginine, 124

Aspirin
Salicylic acid, 1222

Aubergine
Calystegin B_2, 239

Australian blackwood
Acamelin, 7

Avocado
1-Acetoxy-2-hydroxyheneicosa-12,15-dien-4-one, 15
Catechol, 278

Ayan satinwood
Oxyayanin A, 1053

Ayan wood
Oxyayanin B, 1054

B

Balsam fir
Dehydrojuvabione, 422
Juvabione, 817

Banana
Dopamine, 514
Histamine, 729
Phenethylamine, 1089

Barley
Gluten, 668
Gramine, 685
Heterodendrin, 724
Hordenine, 732
Malonic acid, 906

Bay laurel
Costunolide, 359

Bean
Arcelin, 118
Malonic acid, 906
Phasin, 1086

Bearberry
Ursolic acid, 1386

Beech tree
(+)-Syringaresinol, 1300

Beechwood tar
Guaiacol, 691

Beet
Malonic acid, 906

Bergamot
Bergapten, 182

Bergamot oil
Bergamottin, 181

Betel nut
Arecoline, 123

Bhat tree
Clerodin, 329

Birch
Acacetin, 4
Betulin, 186
Salicylic acid, 1222

Birdsfoot trefoil
Linamarin, 863

Birthwort
Aristolochic acid, 126

Bitter almonds
Amygdalin, 81

Bitter gourd
Momordin, 976

Bitter leaf
Vernolide, 1409

Bitter Macadamia nuts
Proteacin, 1148

Bittersweet
Solasodine, 1272

Bitterweed
Hymenoxon, 757

Black mustard
Gluconasturtiin, 662

Black nightshade
α-Chaconine, 291
Solanidine, 1269

Black pepper oil
α-Phellandrene, 1087

Black walnuts
Juglone, 814

Blackcurrant
Sakuranetin, 1219

Blazing star
Graminiliatrin, 686
Liatrin, 859
Provincialin, 1151

Blessed thistle
Cnicin, 331
Salonitenolide, 1223

Blindgrass
Stypandrol, 1291

Bloodroot
Sanguinarine, 1227

Borage
Lycopsamine, 891

Borneo camphor
Borneol, 195

Borneo camphor oil
Borneol, 195

Box
Cyclobuxine D, 399
Cycloprotobuxine C, 402

Bracken fern
Ptaquiloside, 1160

Bracken
Shikimic acid, 1260

Brazil gingseng
Pfaffoside A, 1083

Brazil ginseng
Pfaffic acid, 1082

Broad bean
Dihydrowyerone, 479
Wyerone epoxide, 1438

Broad beans
L-Dopa, 513
Wyerone, 1436
Wyerone acid, 1437

Broom
Caulophylline, 283
Dopamine, 514
13-Hydroxylupanine, 747
Lupanine, 881

Brown mustard
Progoitrin, 1145

Buchu
Diosphenol, 498

Bur buttercup
Ranunculin, 1180

Butterbur
Petasin, 1080

Buttercups
Ranunculin, 1180

C

Cabbage
Glucocapparin, 657
Progoitrin, 1145

Cabbage leaves
Spirobrassinin, 1282

Cabbages
S-Methyl-L-cysteine *S*-oxide, 937

Cactus
Cephalocerone, 287

Calabar bean
Eseramine, 561
Eseridine, 562
Physostigmine, 1101
Physovenine, 1102

Calabash curare
Toxiferine I, 1344

Calamus oil
β-Asarone, 135

Californian mistletoe
Phoratoxin, 1095

Calumba root
Jatrorrhizine, 810
Palmatine, 1058

Camphor wood
Apiole, 108

Canary grass
Gramine, 685
Hordenine, 732

Cannabis
Cannabichromene, 252

Cape gooseberry
Physoperuvine, 1100

Caper
Glucocapparin, 657

Carnation
Dianthalexin, 461
Dianthramide A, 462

Carolina jessamine
Gelsemicine, 628
Gelsemine, 629
Sempervirine, 1248

Carpano tree
Carpacin, 264

Carrot
Elemicin, 528
Falcarindiol, 592
Falcarinol, 593
6-Methoxymellein, 930

Carum
p-Cymene, 406

Cashew nut
Anacardic acid, 83
(15:1)-Cardanol, 259
Ginkgoic acid, 648

Cassava
Linamarin, 863

Castor bean
Ricin, 1197

Castor oil
Ricinine, 1198

Catalpa tree
Deoxylapachol, 448

Cauliflower
Glucocapparin, 657
Glucocheirolin, 658

Celery
Columbianetin, 339
Psoralen, 1157
4,5′,8-Trimethylpsoralen, 1360
Xanthotoxin, 1441

Celery seed
Guaiacol, 691

Chamomile
(+)-α-Bisabolol, 192
Chamazulene, 295
Guaiazulene, 692
Nobilin, 1015

Chaparro amargosa
Chaparrin, 298

Chaulmoogra oil
Chaulmoogric acid, 300

Checkerberry
Monotropitoside, 979

Chenopodium oil
Ascaridole, 136

Cherry
Prunasin, 1152

Cherry trees
Genistein, 633

Chickling vetch
L-α-Amino-β-oxalylaminopropionic acid, 76

Chickpea
Biochanin A, 190
Formononetin, 610
(−)-Maackiain, 897

Chicory
Cichoralexin, 313
8-Deoxylactucin, 447
Lactucin, 834
Lactucopicrin, 835

Clover
Biochanin A, 190
Coumestrol, 365
Linamarin, 863

Coca
Cinnamoylcocaine, 323
Cocaine, 332
Ecgonine, 519

Cocklebur
Carboxyatractyloside, 258

Coffee
L-Aspartic acid, 139
Cafestol, 226
Caffeine, 227

Coffee bean oil
Cafestol, 226

Cola
Caffeine, 227

Colophony
15-Hydroperoxyabietic acid, 735

Coltsfoot
Senkirkine, 1254
Tussilagine, 1377

Comfrey
Echimidine, 520
Heliosupine, 711
Lasiocarpine, 847
Symphytine, 1298

Common barberry
Berberine, 180

Common bean
Phasin, 1086

Common groundsel
Senecionine, 1251

Corn poppy
Rhoeadine, 1196

Costus root oil
Costunolide, 359

Cotton
Gossypol, 681
Lacinilene C 7-methyl ether, 833

Cottonseed oil
Sterculic acid, 1286

Cowbane
Cicutoxin, 315

Cowhage
N, *N*-Dimethyltryptamine, 491

Cowpea
Kievitone, 826
Vignafuran, 1414

Cranberry
Ursolic acid, 1386

Creosote bush
Nordihydroguaiaretic acid, 1018

Crown vetch
Hyrcanoside, 765

Cumin
p-Cymene, 406

Cumin oil
β-Pinene, 1111

Curare
Isochondrodendrine, 787

Cyclamen
Cyclamin, 397

Grasshopper's cyperus
Juvenile hormone III, 819

Greater periwinkle
Vincamine, 1418

Grey mangrove
Naphtho[1, 2-*b*]furan-4,5-dione, 998

Guaiac resin
Guaiacol, 691

Guaiac wood
Guaiazulene, 692

Guarana
Caffeine, 227

Gum rosin
15-Hydroperoxyabietic acid, 735

Gymnosperm tree
Plicatic acid, 1121

H

Hashish
Cannabichromene, 252

Heavenly bamboo
Nandinin, 996

Hedge hyssop
Cucurbitacin I, 381

Heliotrope
Heliotrine, 713
Indicine, 772

Hemlock
γ-Coniceine, 343
Coniine, 345
(+)-*N*-Methylconiine, 936
Pseudoconhydrine, 1154

Hemp agrimony
Eupatolide, 582
Eupatoriopicrin, 583

Henbane
Hyoscine, 758
Hyoscyamine, 759

Hop
Lupulone, 885

Hops
6-Isopentenylnaringenin, 800
Isoxanthohumol, 805
Xanthohumol, 1440

Horse beans
L-Dopa, 513

Horse-chestnut
Aescin, 33

Horseradish
Glucocapparin, 657
Glucocheirolin, 658
Glucolepidiin, 661
Glucotropaeolin, 665

Horsetail
Palustrine, 1060

Hound's tongue
Heliosupine, 711
Heliotridine, 712

Hura tree
Huratoxin, 733

Hyssop-leaved thoroughwort
Eupahyssopin, 576

I

Ignatius beans
Strychnine, 1290

Indian birthwort
Aristolochic acid, 126

Indian gentian
Amarogentin, 69

Indian pea
Cadaverine, 225

Indian podophyllin
4′-Demethylpodophyllotoxin, 440

Ipecacuanha
Emetine, 533

Ironweed
Glaucolide A, 656

Ironweeds
Vernodalin, 1404
Vernodalol, 1405

Ivy
Falcarinol, 593
Hederagenin 3-*O*-arabinoside, 704
α-Hederin, 706
β-Hederin, 707

J

Jack bean
Concanavalin A, 340

Jackbean
L-Canaline, 246
L-Canavanine, 247

Japanese oil of camphor
Camphor, 243

Japanese spurge
Pachysandrine A, 1057

Japanese star anise
Anisatin, 95

Japanese yew
Taxol, 1311

Javanese coca
Cinnamoylcocaine, 323
Tropacocaine, 1364

Jequirity
Abrin, 1

Jerusalem cherry
Solacapine, 1268
Solanocapsine, 1271

Jumbie bean
L-Mimosine, 966

Jute
Corchoroside A, 351

K

Kat
D-Cathine, 280
D-Cathinone, 281

Khat
D-Cathine, 280
D-Cathinone, 281

Kino gum
(−)-Eudesmin, 565

N

Nasturtium
Erucic acid, 554
Glucotropaeolin, 665

Neem tree
Azadirachtin, 164
Ohchinolide B, 1029

Nettle
Histamine, 729

Ngai camphor
Borneol, 195

Nigerian satinwood
Oxyayanin A, 1053

Nutmeg
Borneol, 195
Elemicin, 528
Myristicin, 994

Nux-vomica
Strychnine, 1290

O

***O*-Acetylsalicylic acid**
Salicylic acid, 1222

Oak tannins
Casuarictin, 274

Oat
Avenalumin I, 161

Oats
Avenacin A-1, 159
Avenacin B-2, 160
Gluten, 668

Oil of balm
Citronellal, 326

Oil of bergamot
Bergamottin, 181
Bergapten, 182
Dipentene, 501

Oil of bitter almonds
Phenethylamine, 1089

Oil of Borneo camphor
Borneol, 195

Oil of black pepper
α-Phellandrene, 1087

Oil of caraway
Limonene, 861

Oil of cashew nuts
(15:1)-Cardanol, 259

Oil of catnip
Nepetalactone, 1010

Oil of chamomile
(+)-α-Bisabolol, 192

Oil of chenopodium
Ascaridole, 136

Oil of citronella
Dipentene, 501

Oil of citrus
Limonene, 861

Oil of coffee beans
Cafestol, 226

Oil of costus root
Costunolide, 359

Oil of cubeb
Dipentene, 501

Oil of cumin
β-Pinene, 1111

Oil of dill
Limonene, 861

Oil of elecampane
Alantolactone, 49

Oil of eucalyptus
1,8-Cineole, 319

Oil of fennel
α-Phellandrene, 1087

Oil of hedeoma
Pulegone, 1163

Oil of lemon
Citronellal, 326

Oil of Levant wormseed
1,8-Cineole, 319

Oil of neroli
Limonene, 861

Oil of nutmeg
Isoelemicin, 795

Oil of orange flowers
Limonene, 861

Oil of parsley seed
Apiole, 108

Oil of pine resin
1-*tert*-Butyl-3-methylbenzene, 220

Oil of pulegium
Pulegone, 1163

Oil of savin
Sabinol, 1213

Oil of spike
Borneol, 195

Oil of tansy
α-Thujone, 1335

Oil of turpentine
α-Pinene, 1110
β-Pinene, 1111

Oil of wintergreen
Monotropitoside, 979

Oil of wormwood
α-Thujone, 1335

Oils of citronella
Citronellal, 326

Oleander
Adynerin, 32

Olive
Cinchonidine, 318

Ololiuqui
Ergine, 546
Chanoclavine-I, 297

Onion
Allicin, 59
Alliin, 60
Diallyl disulfide, 458
Dimethyl disulfide, 490
S-Methyl-L-cysteine *S*-oxide, 937
Propanethial *S*-oxide, 1146

Onion bulbs
Tsibulin 2, 1367

Rhubarb
Oxalic acid, 1049

Rice
Momilactone A, 975
Oryzalexin A, 1039
Sakuranetin, 1219
Salicylic acid, 1222

Rose oil
2-Phenylethanol, 1091

Rosemary
Ledol, 853
Palustrol, 1061

Rough chervil
Falcarinone, 594

Round-leaved thoroughwort
Eupachlorin, 568
Eupachlorin acetate, 569
Eupachloroxin, 570
Euparotin, 577
Euparotin acetate, 578
Eupatoroxin, 584
Eupatundin, 585

Rowan
Isoaucuparin, 784
4′-Methoxyaucuparin, 927

Rowan tree
Parasorbic acid, 1064

Rue
Arborine, 115
Bergapten, 182
Chalepensin, 293

Russian comfrey
Intermedine, 776
Lycopsamine, 891
Symlandine, 1297
Symphytine, 1298
Uplandicine, 1383

Russian knapweed
Chlorohyssopifolin A, 304
Chlororepdiolide, 305

Rutgers tomato
Tomatidine, 1340

Rye
Gluten, 668

S

Safflower
Dehydrosafynol, 425
Safynol, 1216

Sage
Cirsiliol, 325

Sago palm
Cycasin, 396

Saguaro cactus
Gigantine, 645

Salicylic acid, acetyl derivative
Salicylic acid, 1222

Sandalwood
Pterostilbene, 1162

Sarsaparilla
Parillin, 1066

Scotch thistle
Onopordopicrin, 1034

Sea hare
Aplysiatoxin, 109

Sea onion
Bovoside A, 198
Scillaren A, 1240
Scilliroside, 1241

Sedge
Brevicolline, 203

Senna
Chrysophanol, 310

Siamese ginger
1′-Acetoxyeugenol acetate, 14

Silver maple
Gramine, 685

Sitka spruce
Astringin, 146
Rhaponticin, 1189

Sneezeweed
Amaralin, 68
Autumnolide, 157
Florilenalin, 605
Linifolin A, 864
Mexicanin I, 954
Plenolin, 1120

Snowdrop
Narwedine, 1004

Soapwort
Gypsogenin, 694

Solomon's seal
L-Azetidine-2-carboxylic acid, 165
L-α, γ-Diaminobutyric acid, 460

Soya bean
Genistein, 633
(−)-Glyceollin I, 670
(−)-Glyceollin II, 671

Soybean
Cadaverine, 225

Spike oil
Borneol, 195

Spinach
Coumestrol, 365
Histamine, 729
Oxalic acid, 1049

Squill
L-Azetidine-2-carboxylic acid, 165
Scillaren A, 1240
Scilliroside, 1241

St. John's wort
Hypericin, 763

Star-of-Bethlehem
Convallatoxin, 349

Stinging nettles
Formic acid, 609

Stonecrop
Cadaverine, 225

Strawberries
Imperatorin, 770

Styrax
Cinnamic acid, 322

Subterranean clover
Cadaverine, 225

Sugar beet
L-Azetidine-2-carboxylic acid, 165
Betagarin, 183
Betavulgarin, 184
Glutaric acid, 667
Glyoxylic acid, 672

Sugar cane
L-Aspartic acid, 139
Luteolinidin, 886

Sundews
Histamine, 729

Sunflower
Ayapin, 163
Helianthoside A, 709
Niveusin C, 1014

Surinam quassia wood
Quassimarin, 1174
Quassin, 1175

Swamp cypress
Taxodone, 1310

Swede
Glucocheirolin, 658
Glucoerysolin, 659

Sweet basil
Juvocimene 1, 820

Sweet herb
Hernandulcin, 722

Sweet pea
Odoratol, 1026

Sweet potato
Ipomeamaronol, 778
Ipomoeamarone, 779

Sycamore
α-(Methylenecyclopropyl)glycine, 938

T

Tangerine
Limonene, 861

Tannic acid
Corilagin, 354

Tansy oil
α-Thujone, 1335

Tansy ragwort
Jacobine, 807
Riddelline, 1199
Senecionine, 1251

Tea
Caffeine, 227
Theasaponin, 1330

Teak
Aloe-emodin, 64
Chrysophanol, 310
Deoxylapachol, 448

Thistle
Cirsiliol, 325

Thornapple
Atropine, 151
Hyoscyamine, 759

Thyme
Cirsilineol, 324
p-Cymene, 406

Tinya
Tinyatoxin, 1339

Toadflax
Acacetin, 4

Tobacco
Capsidiol, 255
Nicotine, 1011

Tomato
Bergapten, 182
α-Solanine, 1270
Tomatine, 1341
2-Tridecanone, 1353

Tuberous comfrey
Echimidine, 520

Tulip bulbs
Tuliposide A, 1372
Tuliposide B, 1373

Tulip tree
Epitulipinolide, 539
Epitulipinolide diepoxide, 540
Lipiferolide, 866
1-Peroxyferolide, 1078
Taxiphyllin, 1308
Tulipinolide, 1371

Turnip
Glucocheirolin, 658
Glucoerysolin, 659

Turpentine oil
α-Pinene, 1110
β-Pinene, 1111

Turpentine oils
Car-3-ene, 261

U

Upas tree
α-Antiarin, 103
Antioside, 104
Convallatoxin, 349
Convalloside, 350

V

Verbena oil
Verbenone, 1401

Vetch
L-β-Cyanoalanine, 394

Vine leaves
Pterostilbene, 1162

Viper's bugloss
Heliosupine, 711

Vomiting bush
Dihydrogriesenin, 474

Voodoo lily
Salicylic acid, 1222

W

Water dropwort
β-Phellandrene, 1088

Water forget-me-not
Symphytine, 1298

Water hemlock
Falcarinol, 593
Oenanthotoxin, 1027

Water pepper
Polygodial, 1130

Watercress
Gluconasturtiin, 662

Water-lily
Deoxynupharidine, 452

Weir vine
Calystegin B_2, 239

Welensali
12-*O*-Palmitoyl-16-hydroxyphorbol 13-acetate, 1059

Welsh poppy
Mecambrine, 912

Wheat
2, 6-Dimethoxybenzoquinone, 487
Gluten, 668

White hellebore
Cyclopamine, 401
Protoveratrine A, 1149
Protoveratrine B, 1150

White lupin
Multiflorine, 985

White snakeroot
6-Hydroxytremetone, 752

White squill
Scillaren A, 1240

Wild fig tree
(−)-Tylocrebine, 1380

Wild ginger
Aristolochic acid, 126

Wild parsnip
Sphondin, 1279

Wild potato
α-Chaconine, 291

Willow catkin
(+)-Catechin, 277

Winged pea
Anhydroglycinol, 94

Wintergreen
Salicylic acid, 1222

Woody nightshade
α-Solanine, 1270

Wormwood
Absinthin, 2
Chamazulene, 295

Wormwood oil
α-Thujone, 1335

Y

Yarrow
Achillin, 19
Chamazulene, 295

Yellow gentian
Gentiopicrin, 638

Yellow jessamine
Gelsemicine, 628
Gelsemine, 629
Sempervirine, 1248

Yellow lupin
Lupeol, 882

Yellow nightshade
Urechitoxin, 1384

Yellow oleander
Peruvoside, 1079

Yellow pheasant's eye
Adonitoxin, 31
Cymarin, 405
Vernadigin, 1403

Yellow wood
Punicalagin, 1165
Terminalin, 1316

Yew
Cephalomannine, 288
Taxine A, 1307
Taxol, 1311

Ylang-ylang
Isosafrole, 802

Yohimbe
Yohimbine, 1448

Z

Zoapatle tree
Zoapatanol, 1454